Advances in Intelligent Systems and Computing

Volume 1282

The series "Advances in Intelligent Systems and Computing" contains publications on theory, applications, and design methods of Intelligent Systems and Intelligent Computing. Virtually all disciplines such as engineering, natural sciences, computer and information science, ICT, economics, business, e-commerce, environment, healthcare, life science are covered. The list of topics spans all the areas of modern intelligent systems and computing such as: computational intelligence, soft computing including neural networks, fuzzy systems, evolutionary computing and the fusion of these paradigms, social intelligence, ambient intelligence, computational neuroscience, artificial life, virtual worlds and society, cognitive science and systems, Perception and Vision, DNA and immune based systems, self-organizing and adaptive systems, e-Learning and teaching, human-centered and human-centric computing, recommender systems, intelligent control, robotics and mechatronics including human-machine teaming, knowledge-based paradigms, learning paradigms, machine ethics, intelligent data analysis, knowledge management, intelligent agents, intelligent decision making and support, intelligent network security, trust management, interactive entertainment, Web intelligence and multimedia.

The publications within "Advances in Intelligent Systems and Computing" are primarily proceedings of important conferences, symposia and congresses. They cover significant recent developments in the field, both of a foundational and applicable character. An important characteristic feature of the series is the short publication time and world-wide distribution. This permits a rapid and broad dissemination of research results.

**** Indexing: The books of this series are submitted to ISI Proceedings, EI-Compendex, DBLP, SCOPUS, Google Scholar and Springerlink ****

More information about this series at http://www.springer.com/series/11156

John MacIntyre · Jinghua Zhao ·
Xiaomeng Ma
Editors

The 2020 International Conference on Machine Learning and Big Data Analytics for IoT Security and Privacy

SPIoT-2020, Volume 1

Springer

Editors
John MacIntyre
David Goldman Informatics Centre
University of Sunderland
Sunderland, UK

Jinghua Zhao
University of Shanghai for Science and Technology
Shanghai, China

Xiaomeng Ma
Shenzhen University
Shenzen, Guangdong, China

ISSN 2194-5357 ISSN 2194-5365 (electronic)
Advances in Intelligent Systems and Computing
ISBN 978-3-030-62742-3 ISBN 978-3-030-62743-0 (eBook)
https://doi.org/10.1007/978-3-030-62743-0

This Springer imprint is published by the registered company Springer Nature Switzerland AG
The registered company address is: Gewerbestrasse 11, 6330 Cham, Switzerland

Foreword

SPIoT 2020 is an international conference dedicated to promoting novel theoretical and applied research advances in the interdisciplinary agenda of Internet of Things. The "Internet of Things" heralds the connections of a nearly countless number of devices to the Internet thus promising accessibility, boundless scalability, amplified productivity, and a surplus of additional paybacks. The hype surrounding the IoT and its applications is already forcing companies to quickly upgrade their current processes, tools, and technology to accommodate massive data volumes and take advantage of insights. Since there is a vast amount of data generated by the IoT, a well-analyzed data is extremely valuable. However, the large-scale deployment of IoT will bring new challenges and IoT security is one of them.

The philosophy behind machine learning is to automate the creation of analytical models in order to enable algorithms to learn continuously with the help of available data. Continuously evolving models produce increasingly positive results, reducing the need for human interaction. These evolved models can be used to automatically produce reliable and repeatable decisions. Today's machine learning algorithms comb through data sets that no human could feasibly get through in a year or even a lifetime's worth of work. As the IoT continues to grow, more algorithms will be needed to keep up with the rising sums of data that accompany this growth.

One of the main challenges of the IoT security is the integration with communication, computing, control, and physical environment parameters to analyze, detect, and defend cyber-attacks in the distributed IoT systems. The IoT security includes: (i) the information security of the cyber space and (ii) the device and environmental security of the physical space. These challenges call for novel approaches to consider the parameters and elements from both spaces and get enough knowledge for ensuring the IoT's security. As the data has been collecting in the IoT and the data analytics has been becoming mature, it is possible to conquer this challenge with novel machine learning or deep learning methods to analyze the data which synthesize the information from both spaces.

We would like to express our thanks to Professor John Macintyre, University of Sunderland, Professor Junchi Yan, Shanghai Jiaotong University, for being the keynote speakers at the conference. We thank the General Chairs, Program Committee Chairs, Organizing Chairs, and Workshop Chairs for their hard work. The local organizers' and the students' help are also highly appreciated.

Our special thanks are due also to editors Dr. Thomas Ditzinger for publishing the proceedings in Advances in Intelligent Systems and Computing of Springer.

Organization

General Chairs

Bo Fei (President)	Shanghai University of Medicine & Health Sciences, China

Program Chairs

John Macintyre (Pro Vice Chancellor)	University of Sunderland, UK
Jinghua Zhao	University of Shanghai for Science and Technology, China
Xiaomeng Ma	Shenzhen University, China

Publication Chairs

Jun Ye	Hainan University, China
Ranran Liu	The University of Manchester

Publicity Chairs

Shunxiang Zhang	Anhui University Science and Technology, China
Qingyuan Zhou	Changzhou Institute of Mechatronic Technology, China

Local Organizing Chairs

Xiao Wei	Shanghai University, China
Shaorong Sun	University of Shanghai for Science and Technology, China

Program Committee Members

Paramjit Sehdev	Coppin State University, USA
Khusboo Pachauri	Dayanand Sagar University, Hyderabad, India
Khusboo Jain	Oriental University of Engineering and Technology, Indore, India
Akshi Kumar	Delhi Technological University, New Delhi, India
Sumit Kumar	Indian Institute of Technology (IIT), India
Anand Jee	Indian Institute of Technology (IIT), New Delhi, India
Arum Kumar Nachiappan	Sastra Deemed University, Chennai, India
Afshar Alam	Jamia Hamdard University, New Delhi, India
Adil Khan	Institute of Technology and Management, Gwalior, India
Amrita Srivastava	Amity University, Gwalior, India
Abhisekh Awasthi	Tshingua University, Beijing, China
Dhiraj Sangwan	CSIR-CEERI, Rajasthan, India
Jitendra Kumar Chaabra	National Institute of Technology, Kurkshetra, India
Muhammad Zain	University of Louisville, USA
Amrit Mukherjee	Jiangsu University, China
Nidhi Gupta	Institute of Automation, Chinese Academy of Sciences, Beijing, China
Neil Yen	University of Aizu, Japan
Guangli Zhu	Anhui Univ. of Sci. & Tech., China
Xiaobo Yin	Anhui Univ. of Sci. & Tech., China
Xiao Wei	Shanghai Univ., China
Huan Du	Shanghai Univ., China
Zhiguo Yan	Fudan University, China
Jianhui Li	Computer Network Information Center, Chinese Academy of Sciences, China
Yi Liu	Tsinghua University, China
Kuien Liu	Pivotal Inc, USA
Feng Lu	Institute of Geographic Science and Natural Resources Research, Chinese Academy of Sciences, China
Wei Xu	Renmin University of China, China
Ming Hu	Shanghai University, China

2020 International Conference on Machine Learning and Big Data Analytics for IoT Security and Privacy (SPIoT-2020)

Conference Program

November 6, 2020, Shanghai, China

Due to the COVID-19 outbreak problem, SPIoT-2020 conference will be held online by Tencent Meeting (https://meeting.tencent.com/).

Greeting Message

SPIoT 2020 is an international conference dedicated to promoting novel theoretical and applied research advances in the interdisciplinary agenda of Internet of Things. The "Internet of Things" heralds the connections of a nearly countless number of devices to the Internet thus promising accessibility, boundless scalability, amplified productivity, and a surplus of additional paybacks. The hype surrounding the IoT and its applications is already forcing companies to quickly upgrade their current processes, tools, and technology to accommodate massive data volumes and take advantage of insights. Since there is a vast amount of data generated by the IoT, a well-analyzed data is extremely valuable. However, the large-scale deployment of IoT will bring new challenges and IoT security is one of them.

The philosophy behind machine learning is to automate the creation of analytical models in order to enable algorithms to learn continuously with the help of available data. Continuously evolving models produce increasingly positive results, reducing the need for human interaction. These evolved models can be used to automatically produce reliable and repeatable decisions. Today's machine learning algorithms comb through data sets that no human could feasibly get through in a year or even a lifetime's worth of work. As the IoT continues to grow, more algorithms will be needed to keep up with the rising sums of data that accompany this growth.

One of the main challenges of the IoT security is the integration with communication, computing, control, and physical environment parameters to analyze, detect, and defend cyber-attacks in the distributed IoT systems. The IoT security includes: (i) the information security of the cyber space and (ii) the device and environmental security of the physical space. These challenges call for novel approaches to consider the parameters and elements from both spaces and get enough knowledge for ensuring the IoT's security. As the data has been collecting in the IoT, and the data analytics has been becoming mature, it is possible to conquer this challenge with novel machine learning or deep learning methods to analyze the data which synthesize the information from both spaces.

We would like to express our thanks to Professor John Macintyre, University of Sunderland and Professor Junchi Yan, Shanghai Jiaotong University, for being the keynote speakers at the conference. We thank the General Chairs, Program Committee Chairs, Organizing Chairs, and Workshop Chairs for their hard work. The local organizers' and the students' help are also highly appreciated.

Our special thanks are due also to editors Dr. Thomas Ditzinger for publishing the proceedings in Advances in Intelligent Systems and Computing of Springer.

Conference Program at a Glance

Friday, Nov. 6, 2020, Tencent Meeting		
9:50–10:00	Opening ceremony	Jinghua Zhao
10:00–10:40	Keynote 1: John Macintyre	
10:40–11:20	Keynote 2: Junchi Yan	
11:20–11:40	Best Paper Awards	Xiaomeng Ma
14:00–18:00	Session 1	Shunxiang Zhang
	Session 2	Xianchao Wang
	Session 3	Xiao Wei
	Session 4	Shaorong Sun
	Session 5	Ranran Liu
	Session 6	Jun Ye
	Session 7	Qingyuan Zhou
	Short papers poster	

SPIoT 2020 Keynotes

The Next Industrial Revolution: Industry 4.0 and the Role of Artificial Intelligence

John MacIntyre

University of Sunderland, UK

Abstract. The fourth industrial revolution is already approaching—often referred to as "Industry 4.0"—which is a paradigm shift which will challenge our current systems, thinking, and overall approach. Manufacturing industry will be transformed, with fundamental changes possible in how supply chains work, and how manufacturing becomes agile, sustainable, and mobile. At the heart of this paradigm shift is communication, connectivity, and intelligent manufacturing technologies. Artificial intelligence is one of the key technologies that will make Industry 4.0 a reality—and yet very few people really understand what AI is, and how important it will be in this new industrial revolution. AI is all around us already—even if we don't realize it!

Professor John MacIntyre has been working in AI for more than 25 years and is Editor-in-Chief of Neural Computing and Applications, a peer-reviewed scientific journal publishing academic work from around the world on applied AI. Professor MacIntyre will give a picture of how Industry 4.0 will present both challenges and opportunities, how artificial intelligence will play a fundamental role in the new industrial revolution, and provide insights into where AI may take us in future.

Unsupervised Learning of Optical Flow with Patch Consistency and Occlusion Estimation

Junchi Yan

Shanghai University, China

Junchi Yan is currently an Associate Professor in the Department of Computer Science and Engineering, Shanghai Jiao Tong University. Before that, he was a Senior Research Staff Member and Principal Scientist with IBM Research, China, where he started his career in April 2011. He obtained the Ph.D. in the Department of Electrical Engineering of Shanghai Jiao Tong University, China, in 2015. He received the ACM China Doctoral Dissertation Nomination Award and China Computer Federation Doctoral Dissertation Award. His research interests are mainly machine learning. He serves as an Associate Editor for IEEE ACCESS and a Managing Guest Editor for IEEE Transactions on Neural Networks and Learning Systems. He once served as Senior PC for CIKM 2019. He also serves as an Area Chair for ICPR 2020 and CVPR 2021. He is a Member of IEEE.

Oral Presentation Instruction

1. Timing: a maximum of 10 minutes total, including speaking time and discussion. Please make sure your presentation is well timed. Please keep in mind that the program is full and that the speaker after you would like their allocated time available to them.
2. You can use CD or USB flash drive (memory stick), make sure you scanned viruses in your own computer. Each speaker is required to meet her/his session chair in the corresponding session rooms 10 minutes before the session starts and copy the slide file(PPT or PDF) to the computer.
3. It is suggested that you email a copy of your presentation to your personal inbox as a backup. If for some reason the files can't be accessed from your flash drive, you will be able to download them to the computer from your email.
4. Please note that each session room will be equipped with a LCD projector, screen, point device, microphone, and a laptop with general presentation software such as Microsoft PowerPoint and Adobe Reader. Please make sure that your files are compatible and readable with our operation system by using commonly used fronts and symbols. If you plan to use your own computer, please try the connection and make sure it works before your presentation.
5. Movies: If your PowerPoint files contain movies please make sure that they are well formatted and connected to the main files.

Short Paper Presentation Instruction

1. Maximum poster size is 0.8 meter wide by 1 meter high.
2. Posters are required to be condensed and attractive. The characters should be large enough so that they are visible from 1 meter apart.
3. Please note that during your short paper session, the author should stay by your short paper to explain and discuss your paper with visiting delegates.

Registration

Since we use online meeting way, no registration fee is needed for SPIoT 2020.

Contents

Novel Machine Learning Methods for IoT Security

The Innovation of UI Design Courses in Higher Vocational Colleges Based on the Internet Perspective

Jing Lei(✉)

Wuhan City Polytechnic, Wuhan, Hubei, China
wcplei126@eiwhy.com

Abstract. With the development of the information industry in the information age, the total amount of information in various fields has shown a spurt of growth. The rapid expansion of the total amount of social information has greatly accelerated the flow of information, and the amount of information people receive per unit of time has increased dramatically. How to find effective information in redundant information has become an urgent problem to be solved. To a certain extent, flat UI design is the first screening for users, presenting information in the most concise and easiest way to convey. However, with the popularization of flat UI design, new problems have also emerged. The number of uniform flat designs increases, and new information groups will be formed. In this case, we need to make the UI design more characteristic and easier to be recognized, so we began to pay attention to Internet tools. Analyze user needs and preferences from the perspective of the Internet, and then conduct research on the innovation of UI design courses in higher vocational colleges based on Internet technology.

Keywords: Internet · Perspective · Higher vocational colleges · UI design · UI courses

1 Introduction

With the development of the Internet, full-process media, holographic media, full-employment media, and full-effect media have emerged. Information is ubiquitous, omnipotent, and no one doesn't need it. Faced with profound changes in the ecology of public opinion, media patterns, and communication methods. The expression of media is also becoming more and more abundant. Design objects have changed from traditional two-dimensional planes to multi-dimensional networks, mobile communications, LEDs, etc. Interface design and interaction design are widely used in smart phones, tablets and other touch screen mobile products [1]. UI design teaching content is also changing with the rapid development of the times. The half-lives of new theories and new technologies applied in all walks of life are getting shorter and shorter. This will bring about continuous changes in professional jobs and changes in market demand. The content of professional courses in colleges every academic year and even every semester poses challenges. As professional teachers in vocational schools, we must respond to this challenge with a spirit of reform [1].

J. MacIntyre et al. (Eds.): SPIoT 2020, AISC 1282, pp. 3–8, 2021.
https://doi.org/10.1007/978-3-030-62743-0_1

Higher vocational colleges take the training of technical talents as the main task to meet the most basic needs of the educated for learning skills, forming professional abilities, and seeking survival and development. This requires teachers not only to cultivate students' professional abilities and professional qualities, but more importantly, let students master the know-how of learning, start with the methods and rules of teaching learning knowledge, and effectively improve students' adaptability and competitiveness to the requirements of future social work positions [2].

2 UI Design Development History

2.1 User Interface Design

User interface design has gradually attracted the attention of the public with the popularization of digital terminals in the information age. It is also called UI (User Interface), which is the interface presented to users on the display of digital terminals. The design of the user interface directly affects the user's perception and use, and is the way of presenting information. Humans follow certain physiological laws and habits when viewing external things [3]. The final presentation form of the user interface design should fit human visual psychological logic, meet the needs of the user group as much as possible, and be targeted, including the 7 major elements described in Fig. 1. The development process of UI design follows the sequence from early flattening to pseudo-materialization and then to new flattening. This process is the result of technological advancement and to a certain extent reflects the progress of design methods, from functionalism to humanism [3].

Fig. 1. Seven elements that the UI interface needs to include

2.2 Functional User Interface Design

Microsoft's MS-DOS system is a landmark case of early user interface design. In the mid-1980s, the technology was not fully developed. The DOS system was a command-line, function-oriented operation mode, that is, the user entered a command and the system made Response. During this period, the terminal display had low resolution and monotonous colors, and the interface could only display the commands entered by the user into the system, with almost no graphic information and no good human-computer interaction effects. Such a situation creates a relatively high access baseline for users to use computers, and command input interaction requires the audience to have a certain degree of relevant professionalism [4].

2.3 Humanistic User Interface Design

Due to the difficulty of operating the DOS system, manufacturers began to develop interface designs that can be understood by ordinary people. During this period, Microsoft and Apple have successively introduced quasi-materialized interface and icon designs, and UI design has become "function-centric". To the revolutionary change in the concept of "user-centered", good human-computer interaction is regarded as the requirement standard of product design, and more attention is paid to the user experience [5]. In the WINDOWS XP system and the MAC OS X system of the same period, the interface design has begun to be materialized. The picture shows the icon design in the OS system and the XP system, which pursues realistic three-dimensional, light and shadow effects. UI with decorative real effects the interface began to show the problem of difficult to highlight effective information in the "information overload era" [4].

2.4 Humanistic User Interface Design

In 2006, Microsoft launched the famous Zune player, and began to use a similar flat interface design style [5]. UI design has been explored in the pseudo-material style for more than ten years. It suddenly ushered in a subversive change and began to streamline information. Research and exploration of flat style interface. In this trend, Apple also began to abandon the materialized style of previous generations of IOS's fine simulation, canceling the pseudo-materialized details such as shadows, gradients, and highlights, and replacing it with a flat visual design [6]. The purpose of the flat design is to streamline the overly cumbersome decorative information in the picture, leaving only accurate and accurate information, and to quickly convey effective messages in the information ocean.

3 The Current State of UI Design Courses

3.1 The Teaching Status of UI Design Courses in Vocational Colleges

Higher vocational colleges pay more attention to and pay attention to the training of vocational skill-based talents. UI design, as an emerging industry at this stage, has good employment opportunities and platforms to a certain extent [7]. Therefore, most vocational colleges have opened UI design exhibition industry courses. The main courses of the UI design course mainly include web UI design, interactive UI interface design, mobile APP application development, and image and graphics processing. According to the analysis of professional courses: First, UI design technology and digital media technology need to be fully integrated. Only in this way can students not only learn the operating technology of the software, but also improve their own design capabilities; secondly, students can creative design performance is studied systematically to cultivate logical thinking ability. Although most vocational colleges offer UI design-related courses, the teaching system has not yet reached the mature stage [7]. Most people think that UI design is interface art, and UI design is actually a profession that combines technology and art.

3.2 Deficiencies in UI Design Course Teaching

First of all, professional integration is relatively lacking. UI design courses integrate UI design theory, UI design creative ideas, layout design, color composition, and related software operations, etc., so as to improve students' practical operation ability, but for the direction of professional development A complete teaching model and concept has not yet been formed. Secondly, the update of professional knowledge has certain limitations. With the development of science and technology, the iOS system and the Android system are the two main system platforms of the current mobile terminal. The update speed is relatively fast, and the relevant design specifications are gradually improved, but professional Knowledge system update has a certain lag [8]. In the process of UI design professional course teaching, the concepts and knowledge content imparted to students will be replaced by new technology and design rules after a period of time. In the process of UI design learning, students will systematically study the overall production methods of design schemes, frameworks and effects, but the concepts of interaction design and user experience are relatively vague, and their logical thinking ability has not yet reached the mature stage [9].

4 Innovative Suggestions for UI Design Courses from the Perspective of the Internet

4.1 Offer Relevant Courses in Combination with Employment Orientation

UI design courses need to be set up in accordance with the current employment direction, and APP application development requires related positions such as product managers, user experience designers, and interface designers. First of all, a qualified product must be comprehensively analyzed by the product manager, the actual needs of customers and product selling points must be reasonably planned, and the actual needs must be sorted, so as to scientifically design the logical function of the product [9]. Therefore, it is necessary to set up corresponding Structural design and visual design scheme analysis and other related courses; secondly, user experience designers mainly design the interaction between products and users, so as to clearly stipulate functions and communication methods between users [10].

4.2 The Improvement of Teachers' Professional Skills

Under the influence of the continuous development and progress of science and technology, UI design teachers must have a comprehensive understanding and proficiency in UI design-related content, and they are required to have the spirit of keeping pace with the times to a certain extent [11]. First of all, teachers need to constantly broaden their horizons and study hard to improve their professional knowledge and provide corresponding support for teaching. Secondly, cooperation between schools and enterprises. Actively communicate with enterprises in related industries, establish a school-enterprise cooperation mechanism, invite enterprises to conduct academic lectures on UI design-related knowledge and content in schools, and participate in the training of UI design talents. At the same time, the school mainly adopts the forms of knowledge forums, seminars, and on-the-job training to encourage teachers to understand the actual situation of the UI design industry [11].

4.3 Using the Internet to Adopt Mathematical Design

The mathematical design method believes that pixelation, sense of sequence, and proportion system constitute the wholeness of the expression method or composition of design works of art, architecture, products, logos, etc. Although it does not take into account the influence of concepts, culture and media, it does clarify some design rules and methods, helping us to discover how mathematics plays a role in the research process of how to create and realize beauty, and use mathematical methods And the law is related to the creation [12]. Including pixelation, sense of sequence, symmetry, proportion coordination, and friendly experience, as shown in Fig. 2.

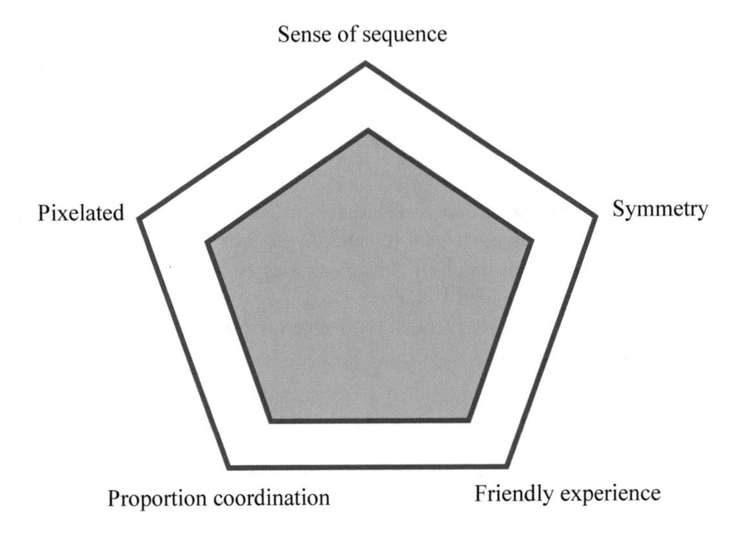

Fig. 2. Five major requirements for the digital design of the UI interface internet

5 Conclusion

In summary, with the rapid development of social economy and science and technology, UI design has become more popular and diversified in Internet design, and it has become richer and more vivid in form and content. The rational use of UI design in Internet design is an inevitable trend of social development. It can not only promote the rapid development of the Internet design industry, but also provide users with a more convenient Internet space.

Acknowledgments. Fund Project: This paper is the outcome of the study, Research on Micro-Course Resource Construction for *UI Interface Design* Based on Campus Network, which is supported by the Foundation of Wuhan Education Bureau for Key Teaching Research Projects of Municipal Universities in 2017. The project number is 2017046.

References

1. Lei, J.: Exploration of the teaching practice of UI interface design course in higher vocational education. Mod. Inf. Technol. **3**(04), 96–97(2019). (in Chinese)
2. Wu, J.: On the teaching strategy of UI interface design course. Chin. Foreign Entrep. **8**(32), 158–161 (2018). (in Chinese)
3. Shen, H.: Exploration of teaching reform in UI interface design course. Educ. Teach. Forum **12**(43), 154–155 (2018). (in Chinese)
4. Zhou, T.: The conception and creation of UI interface design curriculum in higher vocational colleges. Guangdong Seric. **51**(4), 58–60 (2017). (in Chinese)
5. Lu, P.: Analysis of the teaching reform of UI interface design courses in independent colleges. Fine Arts Educ. Res. **9**(19), 123–124 (2017). (in Chinese)
6. Lin, T.: Exploration and thinking on the teaching reform of UI interface design course. Popul. Lit. **9**(15), 248–250 (2016). (in Chinese)
7. He, D.: Analysis and research of icon design elements based on user experience. Shandong Ind. Technol. **7**(13), 230–233 (2019). (in Chinese)
8. Zheng, Y.: The design and implementation of the library Internet visualization analysis system. Technol. Wind. **11**(30), 17–20 (2018). (in Chinese)
9. Zhu, Z., Tian, J., Lin, J.: Design of a personalized music recommendation system based on user location in the Internet environment. Wirel. Internet Technol. **16**(02), 79–80 (2019). (in Chinese)
10. Li, P.: Research on the training model of visual communication design professionals under the background of mobile Internet. Art Technol. **6**, 32–34 (2015). (in Chinese)
11. Qi, M.: Exploration and analysis of UI design teaching content for digital media art major. Popul. Lit. **9**, 186–189 (2017). (in Chinese)
12. Qi, M.: Exploration of UI design teaching content for Internet majors. Popul. Lit. **10**(9), 187–189 (2017). (in Chinese)

Predicting the Getting-on and Getting-off Points Based on the Traffic Big Data

JinLei Wang[1], Feng Chen[1], Keyu Yan[2], Lu Wang[2], FangChen Zhang[2], Li Xia[2], and Zhen Zhao[2](✉)

[1] The North China Sea Data and Information Service, SOA, Qingdao, China
[2] Information Science and Technology School, Qingdao University of Science and Technology, Qingdao, China
zzxm2000@126.com

Abstract. Taxi is an important part of urban passenger transportation. Taxi traffic big data has become an important research topic in smart transportation cities. In order to dispatch taxi resources more reasonably, this paper based on the GPS big data of Baota District of Yan'an City, predicting the getting-on point and getting-off point. Firstly, the historical data is cleaned, and then the DBSCAN clustering algorithm is used to divide the area with sufficient high density into several clusters. Finally, the high-definition map API is used to visualize the prediction results, which solves the problem of forecasting the taxi loading point. In real life, it has a practical reference value for the travel of local residents.

Keywords: Boarding point prediction · Traffic big data · Clustering algorithm

1 Introduction

With the increase of urban population and the increase of the number of cars, the problem of urban traffic [1, 2] has become increasingly prominent. In the big cities such as Beijing, due to the excessive number of cars, traffic congestion, traffic accidents, and air pollution have emerged. According to the survey, 80% of the roads and 90% of the roads in China have reached the limit, and the riding environment has deteriorated severely. Traffic problems have become increasingly serious affecting people's work efficiency and physical health, threatening the urban traffic situation and social and economic development.

At present, taxi drivers cannot accurately predict the passenger's pick-up location. They can only wait for pick-up in places with a large number of people, and they cannot guarantee that they can receive passengers in a short period of time, which may cause traffic congestion and waste of taxi resources. The problem of wasted time has exacerbated the seriousness of urban traffic problems [3]. Based on this phenomenon, a study to predict the boarding point is proposed. We will use the GPS and other driving information generated by the taxis in Yan'an City as the data to study the problem of predicting the getting on the train. By analyzing the existing data, predicting possible pick-up points, adding this to the taxi system, allowing the taxi driver to more accurately find the passenger's pick-up point to receive the guest, to alleviate the traffic

J. MacIntyre et al. (Eds.): SPIoT 2020, AISC 1282, pp. 9–16, 2021.
https://doi.org/10.1007/978-3-030-62743-0_2

congestion problem. Maximize the use of taxi resources and reduce the time wastage of passengers.

In the research process of this subject, the following six parts are included: data acquisition, data cleaning, algorithm selection, model establishment based on DBscan algorithm, data visualization and implementation process and experimental results.

2 Data Cleaning

2.1 First Round of Data Cleaning

1. Cleaning purpose:
   Remove inaccuracies in the data to improve the accuracy of the predictions.
2. Cleaning process:
   (1) Remove unreasonable data. For example, data that does not belong to the Yan'an area, the altitude does not meet the data of the elevation in Yan'an area, and the speed is not within the reasonable range.
   (2) Remove duplicate data.

2.2 Second Round of Data Cleaning

1. Cleaning purpose:
   According to the research in this paper, valuable data needs to be obtained.
2. Cleaning process:

By observing the data file, it is known that there are several important data types of addressHistory, driverWork, evaluate 1802, meter, and state in the entire data. In the file of the meter, there are two 'ISUNUM' 'UPTIME'. The data tag, then there are several data tags such as 'isuNUM' 'speed' 'time' in the address type file, combined with the research topic of this article - predict the boarding point.

The first step: cleaning all the data of the boarding point, the data of the boarding point can be obtained from the 'Uptime' in the meter.

Step 2: Search for the corresponding file of the address type for a time range of 10 s before the cleaning of the boarding point. The address type file is known to store every taxi in the form of every second, so a more accurate pick-up point can be found based on 'isunum' and the time period described above. Find the more accurate boarding point by these two methods, and store the points in the address class file into the new table. This new table is the data set representing the boarding point.

One of the issues to be aware of is the data type, where the data type of the time is a datetime type, and normal addition and subtraction operations cannot operate on it.

With this operation of datetime.timedelta(), you can get a time interval and use the regular expression in pymongo for range search.

Use query as an address type file search condition and use regular expressions to indicate its scope. Then go to the address type file search on the condition of query, if found. Then output '1' and save it in the new table. At this point, the data cleaning has ended.

3 Algorithm Selection

According to the research of this paper, the clustering algorithm is selected to analyze the data. The goal of clustering is to make the similarity of the same kind of objects as large as possible, and to make the similarity between different kinds of objects as small as possible. At present, there are many methods for clustering. Since the research of this paper is to predict the getting on the point, firstly, the goal of this study is to predict, so we need an algorithm that can represent the range of the boarding area. By comparison, it is found that K-means and DBscan are compared. Two suitable algorithms. Combining the results of clustering the cleaned data with the two algorithms and the advantages and disadvantages of both, choose the algorithm that is suitable for the research of this topic.

3.1 K-Means

According to the flow of K-means [4], the data is processed:

Four points are randomly selected for clustering, as four points that have not been iterated at the beginning, as shown in Fig. 1.

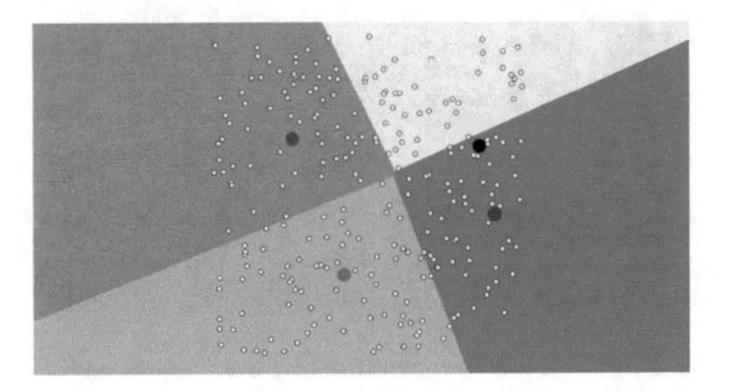

Fig. 1. Four points that have not been iterated

Start the first four clusters, as shown in Fig. 2.

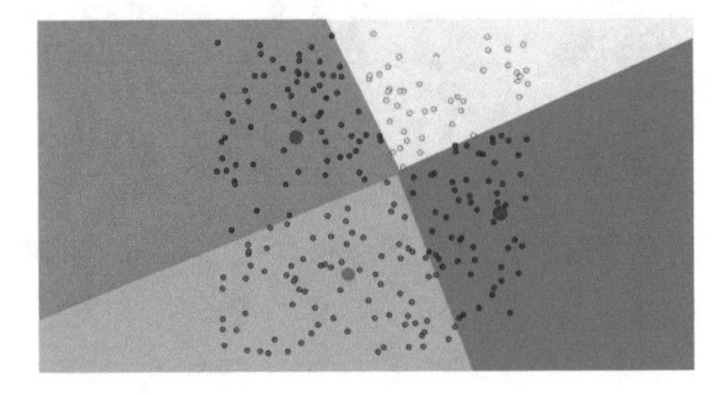

Fig. 2. Four clusters created for the first time

Then the cluster center point is continuously updated. As the cluster center changes, the range of the cluster also changes, and the data set in the cluster also changes, as shown in Fig. 3.

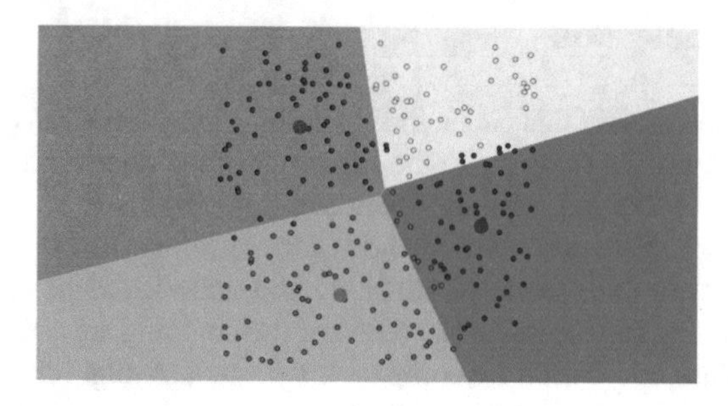

Fig. 3. Four clusters after updating the cluster center point

After continuous iteration, no matter how iterative, the cluster center will not change, then these four points are the clustering center of K-means, indicating that the algorithm has converged, as shown in Fig. 4.

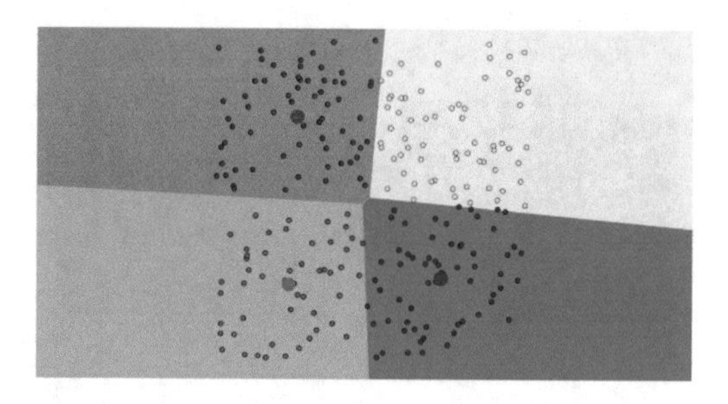

Fig. 4. Four center points of the final cluster

3.2 DBscan Algorithm

According to the flow of DBscan [5], the data is processed:

The neighborhood radius is determined to be 0.8 and the minimum number of MinPts is 4, as shown in Fig. 5.

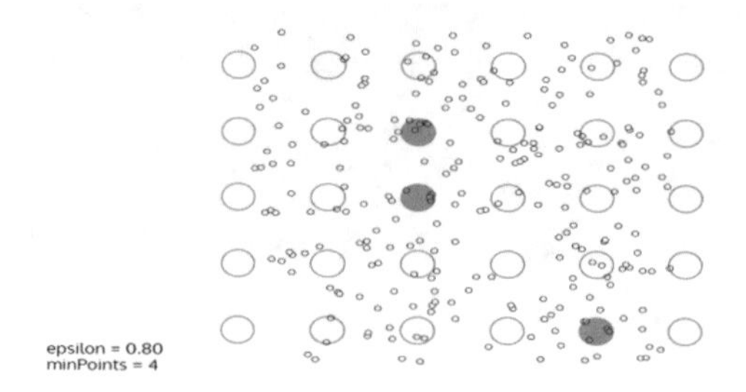

Fig. 5. Field radius is 0.8 and MinPts is 4

Since the selection of the DBscan algorithm is random, we are unable to operate on its selected points. Let it continue to iterate, and the final result is shown in Fig. 6.

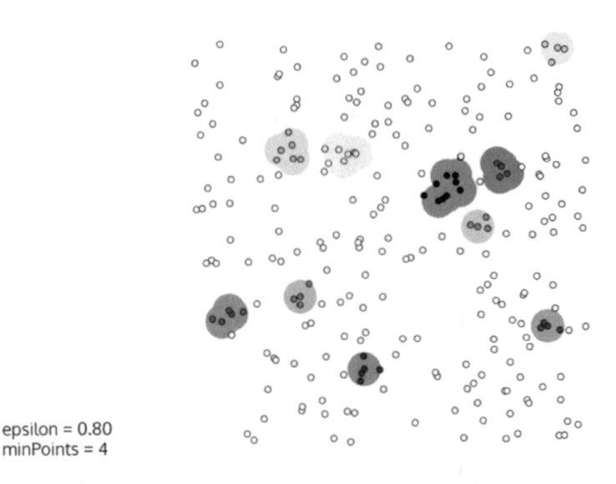

Fig. 6. Results after continuous iteration

The results of clustering show clustering in complex situations, unlike global clustering of the K-means algorithm, in which case the DBscan algorithm The clustering is similar to the clustering cluster of the boarding point, which corresponds to the part of the vehicle.

3.3 Algorithm Selection

By using these two algorithms to visualize the data after clustering the data, it can be seen that different algorithms will produce different results. Therefore, according to the advantages and disadvantages of the algorithm and the final result of the algorithm, it is determined that the DBscan algorithm can effectively avoid the trouble caused by the outliers and cluster the required boarding points. Therefore, the DBscan algorithm is finally selected. As an algorithm used in this research.

4 Model Establishment Based on Dbscan Algorithm

Before using the algorithm, you must first understand the principle of the DBscan algorithm, and know how to use the corresponding sklearn library to build a model for data prediction. To build a model, you need to use it. The fit method, whose return value needs to be used. The lables_ method is used to get.

4.1 Modeling

Create a model and check its label values. The following plt.scatter is the result of the model, and the results are shown in Fig. 7.

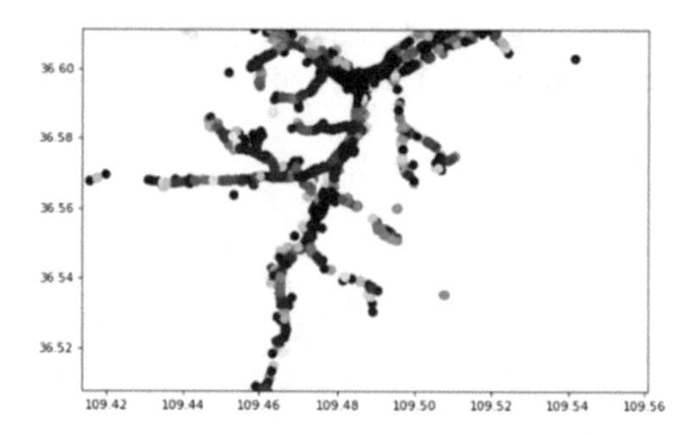

Fig. 7. Showing model results

4.2 Clustering

After obtaining the corresponding flag set, we cluster it once, generally only need to cluster once, but in order to get the average value of the corresponding cluster as the mark point, we carry out the second cluster and corresponding The clusters are averaged.

Finally, we gather and recently express the surrounding car points, and finally write the data to the Excel table using the pandas operation. At this point, the model is established.

5 Visualization Section

5.1 Implementation Process

We use the Gaode Map API to visualize the data. When importing the initial data into the map, we found that most of the points are actually on the mountain or in the lake. Therefore, the geographic data derived by MongoDB has trajectory deviation, so it also needs trajectory correction, on various web-end platforms, or The coordinates obtained by Gao De, Tencent, and Baidu are not GPS coordinates, but GCJ-02 coordinates, or their own offset coordinate system. For example, the coordinates of the Google Maps API and the Gold Map API are GCJ-02 coordinates. They are all generic and are also suitable for most map API products and their map products. Therefore, we took 50 points for high-order coordinate picking work, and then calculated the two data. The longitude of the original data plus 0.005409454, the latitude of the original data plus 0.000088, so that most of the data can be track corrected, but the problem is that there is an error. However, if the latitude and longitude coordinates of the boarding point calculated by using the DBscan algorithm using these coordinates are not the coordinates of the high German common, then the trajectory correction needs to be performed after the calculation.

5.2 Track Correction

The data showing no track correction is shown in Fig. 8.

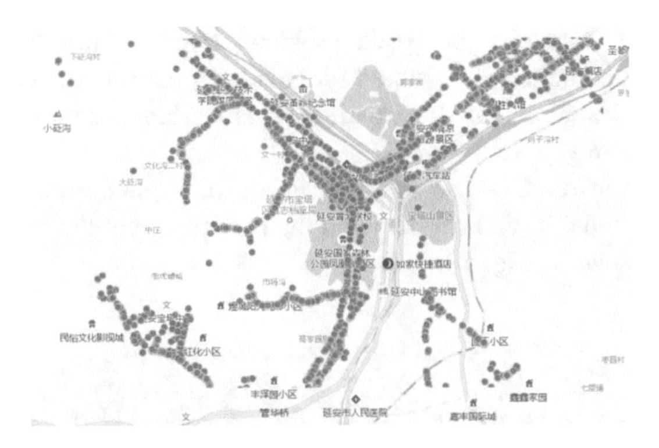

Fig. 8. Data display without track correction

The data after the track correction is performed is shown in Fig. 9.

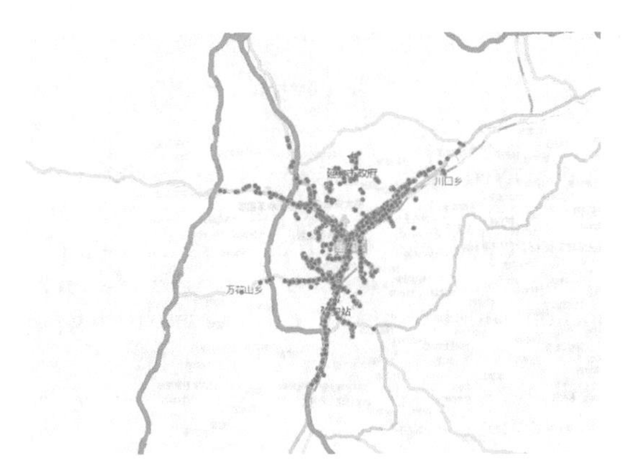

Fig. 9. Data display after track correction

Another problem is that the format of the known data is different from the format of the heat map data, so we modified it with the CONCATENATE function of Excel. For example, if the data given is 109.490486, 36.553871, then change it to {"lng": 109.490486, "lat": 36.553871, "count": 25}. This puts the data in heatmapData=[].

6 Implementation Process and Experimental Results

If there is only path planning, there is no obvious indication of the boarding point predicted by the display of the heat map [6]. It is not clear where the other boarding points are. We take a starting point as an example, the nearest to him. The pick-up point can be represented by the path plan, and the distance from the pick-up point is displayed, and how to go will also be displayed on the map, but the problem is how to plan the heat map and path on a map. Shown in it.

In response to the above proposed problem, after learning to select the Loca container, the Loca container is a method for creating a map container or associating an existing map. For example, in the existing AMap. Map instance, you can directly specify the instance creation.

The var map can directly use the map in the path plan, and then use the Loca container to connect the basic heat map and the path plan, but the heat map data needs to be added, as shown in Fig. 10.

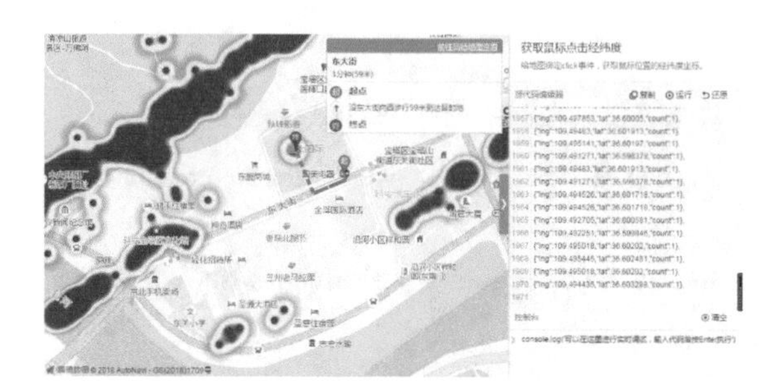

Fig. 10. Heat map to add data

The dense point displayed by the heat map is the possible boarding point, and the end point is the boarding point closest to the starting point. If you need to set multiple starting points and predict multiple pick-up points to use path planning to represent, you only need to define several functions such as walking search ([start point data], [end point data], function (status, result)). The data of the starting point can be repeated, but the planned starting point is the closest to the starting point, so if the data is not much different, it will be difficult to find the nearest boarding point. Therefore, the starting point of the selection is best.

References

1. Mcafee, A., Brynjolfsson, E.: Big data: the management revolution. Harv. Bus. Rev. **90**(10), 60–66 (2012)
2. Qian, W., Jiang, X. Liao, Y., et al.: Taxi passenger point recommendation algorithm based on spatiotemporal context collaborative filtering. Comput. Appl. **35**(6), 1659–1662 (2015)
3. Li, W.: Applied Statistics, pp. 55–56. Tsinghua University Press, Beijing (2014)
4. Hartigan, J.A., Wong, M.A.: Algorithm AS 136: a k-means clustering algorithm. J. R. Stat. Soc. **28**(1), 100–108 (1979)
5. Birant, D., Kut, A.: ST-DBSCAN: an algorithm for clustering spatial-temporal data. Data Knowl. Eng. **60**(1), 208–221 (2007)
6. Deng, W., Wang, Y., Liu, Z., et al.: Heml: a toolkit for illustrating heatmaps. PLOS ONE, **9**(11), e111988 (2014)

The Prediction Analysis of Passengers Boarding and Alighting Points Based on the Big Data of Urban Traffic

JinLei Wang[1], Li Zhang[1], Lu Wang[2], Keyu Yan[2], Yuqi Diao[2], and Zhen Zhao[2(✉)]

[1] The North China Sea Data and Information Service, SOA, Qingdao, China
[2] Information Science and Technology School, Qingdao University of Science and Technology, Qingdao, China
zzxm2000@126.com

Abstract. Real-time road condition prediction based on GPS big data plays a very important role in traffic management. Based on a large number of taxi GPS data and accurately predict the passengers point of problem, using historical taxi GPS data to predict the fact that the speed and comparing with the real speed, so as to realize the analysis of urban traffic in real time, and discusses its implementation in K - means algorithm, in data cleaning, data clustering, visual display. The research work for urban traffic points, data mining and forecasting analysis for the passenger transportation and distribution of public resources, is to reduce the taxi empty loading rate and improve the traffic efficiency.

Keywords: Prediction of boarding point · Traffic big data · K-means algorithm

1 The Introduction

Data is the foundation of the intelligent transportation and lifeblood, urban traffic control positioner can accurately provide passengers guest the location, direction, up and down, when the induced traffic are limited by the lack of traffic data speed and taxi passenger taxi operation workshop condition data, such as between the data contains the long, wide roads, collection of low cost, convenient for centralized management, the dynamic changes of the urban traffic information, caused the wide the academia become the domestic and foreign cities floating car system typical ultimate concern, and has spawned many research on intelligent transportation service, carrying vehicles. The GPS device installed on the taxi has recorded the urban research. It shows that the travel time of the road section can be obtained after calculating the dynamic changes of traffic and crowd movement in these taxi track data and mining the research theory and knowledge of urban traffic. Section average speed has a great deal of data and important information, can provide than traditional questionnaire and parameters such as road congestion degree, and learn that the driver of the selected road, the meter ICGPS routes and measures for guest information, which can provide real-time traffic flow rate for another city for the operation management and operation characteristics of

J. MacIntyre et al. (Eds.): SPIoT 2020, AISC 1282, pp. 17–24, 2021.
https://doi.org/10.1007/978-3-030-62743-0_3

urban traffic analysis provides important decision basis, such as card richer and more accurate information.

Promote the reform of urban traffic intelligent, in a data-driven intelligent transportation development needs and background, at home and abroad were reviewed in this paper, a taxi GPS trajectory data in intelligent application research progress of urban transportation, and points out that the facing problems and progress in this field of research direction is equipped with a GPS taxi data can not only dynamic continuous trajectory tracking a moving vehicle, is also able to situational awareness of urban road network running state, the condition of congestion. For example, the INRIX system in the United States, through the analysis of GPS track data, determines the location of traffic congestion, and provides travelers with a complete map of road traffic conditions. These data are of high volume and high dimension. Change fast, is a typical big data, behavior geography has become a geographic information system, intelligent transportation systems, urban planning research hotspot in fields such as new technology under the background of big data development, starting from the data-driven intelligent transportation, this article mainly reviews the taxi GPS trajectory data in intelligent traffic fields of research and application status, the future research was forecasted [1, 2].

2 Data Cleaning

Data cleaning can clean a lot of dirty data to make the remaining data more accurate and provide a more reliable basis for big data analysis. As an important part of data analysis, this paper sorted out the following cleaning ideas according to the characteristics of GPS traffic data in baota district of yan'an. According to the data uploaded by the taxi, data cleaning and non-standard data processing are carried out to achieve the purpose of accurate data. The taxi data is sorted twice and the longitude and latitude coordinates of passengers are extracted to complete the data cleaning.

2.1 Data Cleaning in Non-Yan'an Area

Due to inaccurate positioning or packet loss in the process of data return, many of the obtained data are not within the geographic area studied. In this paper, the cleaning was carried out according to the latitude and longitude of yan'an city. After consulting relevant information, the selected longitude and latitude were determined to be [107.41–110.31] and [35.21–87.31].

2.2 Cleaning up Wrong Data and Missing Key Data Items

In this paper, data with normal driving speed between [0, 80] are selected for analysis and research, and data with speed greater than 80 are deleted. Data missing items and incomplete data are cleaned up, such as the incomplete data caused by the absence of information such as longitude and latitude, time and height, or the absence of data is empty, which shows the lack of rigor of data and is biased in terms of data accuracy [1, 2].

2.3 Extract the Boarding Point and Move the Boarding Point to a Collection

Why extract all collection entry points into a collection? This is because after moving all the boarding points to a set, it is easier to classify the data later, and it is more intuitive to move the data to the following boarding points in a set.

So how do you find the entry point? State is the State of the car, that is, 03 is the State of the car, and 02 is the State of the car. At this time, our idea is like this: we can sort all the data in time, and then use the data of State from 02 to 03 as the boarding point. However, some problems were found during this operation.

After looking at the number of data in the collection, the result shows that the amount of data in each collection is too large. Through trial and error, we found that when sorting time, the time consumption is relatively large. In view of the low timeliness of this method, another method is adopted in this paper, that is, to find the data with speed of 0 and state of 3 as the boarding point. This data processing can eliminate some abnormal boarding point data. For example, if only the speed is 0, the data that the driver stops at a certain place for a certain period of time may be treated as boarding point.

But it still can not rule out some abnormal on data, such as when the traffic lights, the speed of 0, the state is 3, but not get in the car, it also has a solution, that is in the original data in the screen data to the driver when the traffic lights, so that we can guarantee the accuracy and correctness of the vast majority of data [3, 4].

3 Clustering

3.1 Combination of DBSCAN Clustering and Traffic Big Data

(1) partition based k-means algorithm

A typical partition clustering algorithm, it USES the center of a cluster to represent a cluster, that is, the cluster point selected in the iteration process is not necessarily a point in the cluster. Its purpose is to minimize the Sum of Squared Error between the data points in each cluster (k in total) and the center of mass of the cluster, which is also the evaluation standard for evaluating the final clustering effect of k-means algorithm.

(2) based on the density of DBSCAN algorithm

A typical density-based clustering algorithm, which USES spatial indexing technology to search the neighborhood of objects, introduces the concepts of "core objects" and "density-reachable", and forms a cluster of all density-reachable objects from the core objects. Simply put, it is based on a process of scaling up according to the density of the object.

Through multiple considerations and verification, DBSCAN algorithm was finally selected for clustering, and the processing process combining DBSCAN algorithm with traffic big data [3, 4].

In DBSCAN algorithm, data points are divided into three categories: (1) Core points: points containing more than MinPts within the radius Eps; (2) Boundary points: the number of points within the radius Eps is less than MinPts, but falls within the neighborhood of the core point; (3) Noise points: points that are neither core nor boundary points. There are two quantities, one is the radius Eps, the other is the specified number. Both methods can accurately cluster the data and integrate the big data of taxi traffic operation so that the integrated data can be displayed on the map. The clustering effect is shown in Fig. 1.

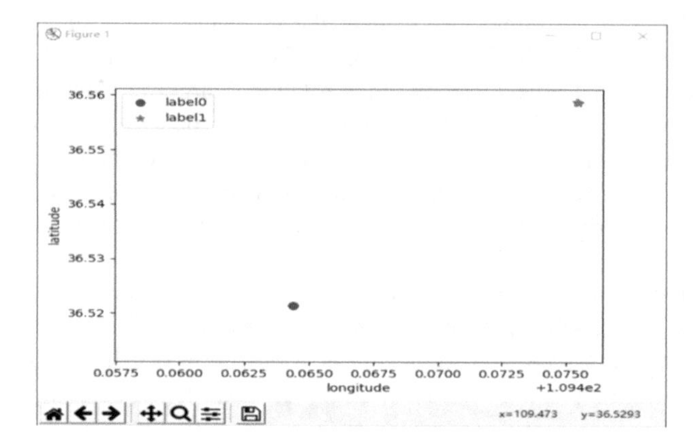

Fig. 1. Cluster point display

3.2 Classification of Data

Mongodb operating instructions are used to select the time set. Here, the built-in operating instructions of mongodb are used to classify the data. The data is classified according to time, from Monday to Sunday, respectively, and one hour is taken as a time period for classification [5].

After sorting all the data, we found that there is a function in pandas that can directly classify the data by time. Using this function, we can classify the data by time more easily and efficiently.

4 Visual Display

The MAPV tool is used to import the data and display the coordinates. The position effect is shown in Fig. 2.

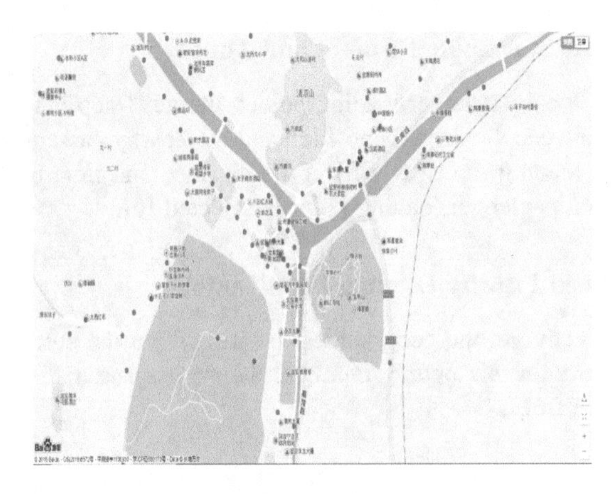

Fig. 2. MAPV shows the boarding point

4.1 Integrate the Code of Important Steps

This step is to consolidate all the code into a single file, so we can directly use this file to cluster the data from the CSV file of the specified date, and convert the coordinates after the clustering, and import it into the new CSV file after the conversion.

4.2 Passenger Load Point Displayed

The current state is to show the global load point, so if we need local information, we need to hide some points. The effect is shown in Fig. 3.

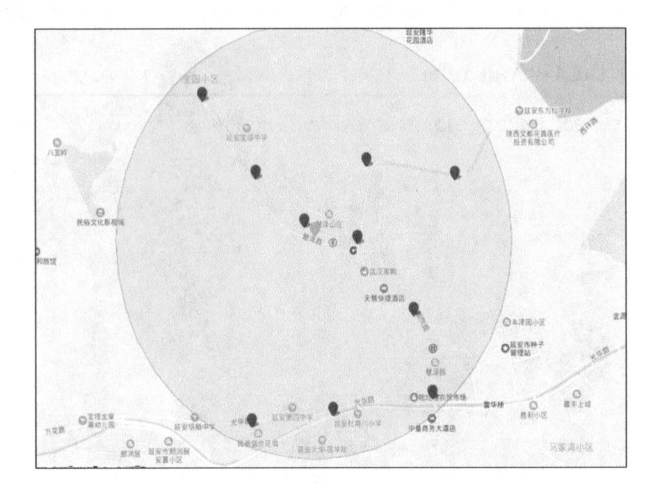

Fig. 3. Shows local passenger loading points

4.3 Obtain Route and Route Distance and Time

After marking the local passenger loading points, the next step is to obtain the route, that is, the route between the driver's position and the nearby passenger loading points, and the passenger loading route, as well as the distance and time of the route will be displayed after each passenger loading point is selected [6].

4.4 Recommended Priority Loading Point Button

The driver can directly get the geographical position from the current position to the passenger point through this button, and can choose the automatic route generation. The effect is shown in Fig. 4.

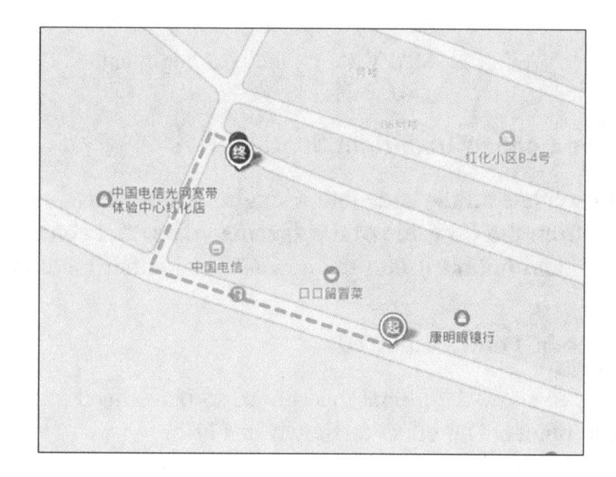

Fig. 4. Effect diagram of the recommended priority loading point button

4.5 Visualize the Overall Effect

The partial passenger loading points are shown in Fig. 5.

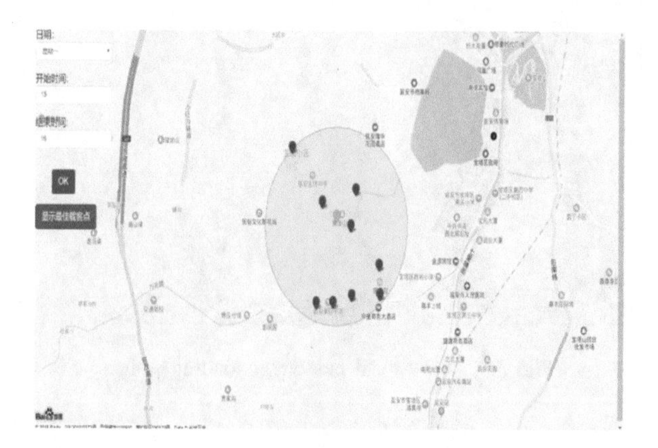

Fig. 5. Shows local passenger loading points

Judge the correctness of user input time, to determine whether a difference of 1 between start time and end time, then cannot use num + 1, num is 10, for example, num + 1 after the command, the result showed 101, this is because the js is a weakly typed language, it argues that digital is also a string, so in operation, add 1 to use num − 0 + 1 operation, the cases of (1) num = 20; Num + = 1; Num = 201; (2) num = 20; Num = num − 0 + 1; Num = 21.

The total route length is shown in Fig. 6.

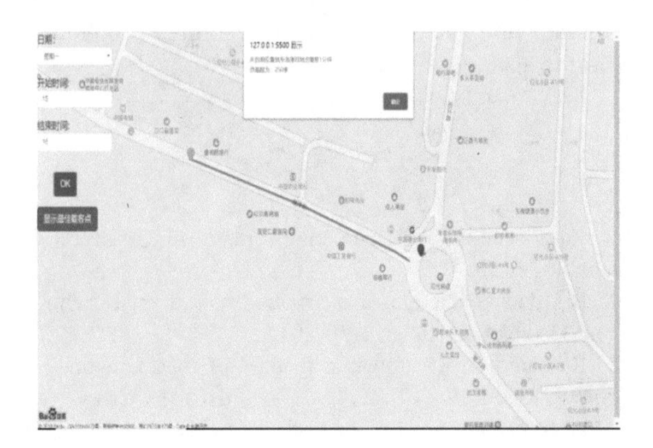

Fig. 6. Indicate the length and time of the route

When adding various functions of the webpage, multiple prompt boxes often appear. This is because the function is not performed only once, so we can add a mark after the corresponding function operation command to make the function perform only once. At the same time, the more functions we add, the more mark there will be. The recommended passenger loading function is shown in Fig. 7.

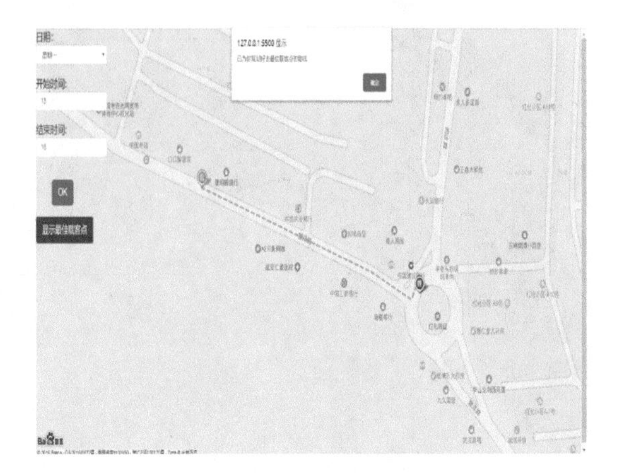

Fig. 7. Function of recommended pick-up points

5 Summary

With the continuous development of science and technology, big data technology is more mature, and it stands out from numerous identification technologies and occupies the mainstream position. Compared with the actual taxi data, big data technology has absolute advantages and is relatively simple to realize. With the development of big data, the prediction technology will become more mature, and the more mature prediction technology will make full use of the existing data to conduct a higher degree of analysis on data mining and prediction analysis points, which will make an indispensable contribution to passenger travel.

References

1. Han, H.: Research on intelligent transportation platform based on big data. Chengdu University of Technology (2014)
2. Li, X., Meng, D.: Big data prediction of distributed incremental traffic flow based on road network correlation. Geogr. Sci. (02), 52–59 2017
3. Yang, W., Ai, T.: Research on the updating method of road network based on big data of vehicle track. Comput. Res. Dev. (12), 2681–2693 (2016). 13 pages
4. Tang, S., Li, Q.: Real-time analysis of urban traffic conditions based on GPS big data of taxi. Shang PS Big Data R.-Time Anal. Urban Traffic Cond. (30), 219–219 (2016)
5. Ji, Q., Wen, H.: Research on big data platform architecture of public transport. Electron. Sci. Technol. (02), 127–130 (2015)
6. Liang, J., Lin, J., Du, Y., etc.: Research on identification of urban land type under the condition of big data based on dynamic perception of taxi GPS data. Shanghai Land Resour. **37**(1), 28–32 (2016)

Verifiable Random Number Based on B-Spline Curve

Jiaqi Qiu[1] and Weiqing Wang[2(✉)]

[1] School of International, Guangxi University, Nanning, Guangxi, China
[2] School of Business, Southwest University, Chongqing, China
2500680163@qq.com

Abstract. In this paper, we focus on the randomness of random numbers and pseudorandom numbers, which cannot be verified without disclosing the seed key. Based on the B-spline curve formula, a widely validated random number production scheme is constructed. The verifiable random number based on B-spline curve has the characteristics of low computational complexity and high efficiency. When the number of participants increases exponentially, the computation increases linearly. The scheme is still efficient when there are many participants in the protocol and suitable for mobile uses.

Keywords: Random number generation · Verification · B-spline curve

1 Introduction

Random numbers play an important role in cryptography, which requires random numbers in authentication, digital signature, password protocol, etc. [1, 2]. With the need of information security protocol construction, new requirements are put forward for the nature of random numbers. For example, in the electronic lottery scheme, lottery buyers are allowed to verify whether they participate in the generation of random numbers. In the mobile meeting system, n users share the meeting password, and the meeting participants verify whether they participate in the generation of the meeting password. This requires verifiable random Numbers.

The ideal verifiable random number should be: participable, unpredictable, reproducible, verifiable [3]. N users $U_1, U_2, \ldots U_n$ select $r_1, r_2, \ldots, r_n$, respectively to participate in the generation of random number r. Moreover, U_i can verify that r_i is involved in the generation of r by its own r_i without requiring the $r_j (j \neq I)$ of other participants, so as to ensure that r is not controlled by people and that the generation process of r is fair and just.

At present, the generation of verifiable random numbers mainly includes: a verifiable random number based on the Diffie-Hellman difficulty, and a verifiable random number based on the dual-sex difficulty, both of which require complex exponential operation and have low operating efficiency. Literature [4] take advantage of Lagrange interpolation polynomials to generate verifiable random numbers. A method of generating verifiable random Numbers by constructing interpolation polynomials is as follows Formula (1): The user first uses n + 1 points to construct:

J. MacIntyre et al. (Eds.): SPIoT 2020, AISC 1282, pp. 25–30, 2021.
https://doi.org/10.1007/978-3-030-62743-0_4

$$A(x) = a_0 + a_1x + \cdots + a_nx^n \tag{1}$$

Make it through the given n + 1 points, then (where, || represents the connector), r is the random number generated by the interpolation coefficient. This random number can be verified. The user substituted x_i into formula (1) through. If $A(x) = y_i$, the user participated in the generation of random number r. However, when the random number is long, n + 1 point $(x_i, y_i), i = 0, 1, \cdots n$, when solving $r = a_0||a_1||\cdots||a_n$, the running time is long and the efficiency is not high.

Therefore, a method of fast generation of verifiable random numbers with low computational complexity is needed. B-spline is useful in many ways [5, 6]. In this paper, a verifiable random number generation method based on B-spline is invented. The calculation speed is fast, and the generation time increases linearly with the increase of n.

2 B-Spline Curve Value Point and Control Point Formula

2.1 B-Spline Curve Calculation

Formula (2) is the formula for calculating B-spline curve of n times. Where,

$$P(t) = \sum_{k=0}^{n}(P_k \frac{1}{n!}\sum_{j=0}^{n-k}((-1)^j C_{n+1}^j (t+n-k-j)^n)) \quad 0 \le t \le 1 \tag{2}$$

$P_0, P_1 \ldots P_k$ is the control point of the curve, n is the number of B-spline curves, the n of Third order B-spline is 3; t is a decimals value between 0 and 1; k is the kth point, and the value range of k is $0, 1, 2 \ldots n$.

Third order B-spline curve is expanded as follows:

$$P(t) = \frac{1}{6}[P_0 \quad P_1 \quad P_2 \quad P_3]\begin{bmatrix} 1 & -3 & 3 & -1 \\ 4 & 0 & -6 & 3 \\ 1 & 3 & 3 & -3 \\ 0 & 0 & 0 & 1 \end{bmatrix}\begin{bmatrix} 1 \\ t \\ t^2 \\ t^3 \end{bmatrix} \tag{3}$$

From the above formula, it is easy to get the following formula:

$$\begin{aligned} &P(0) = (P_0 + 4P_1 + P_2)/6 \quad P(1) = (P_1 + 4P_2 + P_3)/6 \\ &\ldots \qquad P(m-3) = (P_{m-3} + 4P_{m-2} + P_{m-1})/6 \\ &P'(0) = (P_2 - P_0)/2 \quad P'(1) = (P_3 - P_1)/2 \\ &\ldots \qquad P'(m-2) = (P_{m-1} - P_{m-3})/2 \end{aligned} \tag{4}$$

If $Q_i = p(i)$, then Q_i is third order B-spline value point of through type, $k = 1, 2 \ldots, m-1$,

There are:

$$\begin{aligned} & P_2 - P_0 = 2Q_1' \quad P_3 - P_1 = 2Q_2' \\ & \cdots \qquad\qquad P_{i-1} + 4P_i + P_{i+1} = 6Q_i \\ & P_{i-1} + 4P_i + P_{i+1} = 6Q_i \\ & P_{m-1} - P_{m-3} = 2Q_{m-2}' \end{aligned} \tag{5}$$

According to formula (5), the B-spline data point control and point formula can be obtained:

$$\begin{bmatrix} -1 & 0 & 1 & & & \\ 1 & 4 & 1 & & & \\ & 1 & 4 & 1 & & \\ & & \cdots & \cdots & & \\ & & & 1 & 4 & 1 \\ & & & -1 & 0 & 1 \end{bmatrix} \begin{bmatrix} P_0 \\ P_1 \\ P_2 \\ \cdots \\ P_{m-2} \\ P_{m-1} \end{bmatrix} = \begin{bmatrix} 2Q_1' \\ 6Q_1 \\ 6Q_2 \\ \cdots \\ 6Q_{m-2} \\ 2Q_{m-2}' \end{bmatrix} \tag{6}$$

2.2 Formula Calculation About Data Point and Control Point of B-Spline Curve

Given that $Q_1', Q_1, Q_2, \cdots Q_{m-2}, Q_{m-2}'$ can be used to solve $P_0, P_1, P_2, \cdots P_{m-2}, P_{m-1}$ by the chasing and driving method. The chasing and driving method is a very common method, in the baidu network there are specific solutions and source program, here is not to say more.

3 Verifiable Random Number Construction Based on B-Spline Curve

$U_1, U_2, \cdots U_n$ needs to share a random number r in the communication activity. The number of r is m, and m is greater than n. The generation of r requires the participation of the $U_i (1 \le i \le n)$, and the $U_i (1 \le i \le n)$ can verify whether it participates in the generation of random number r. The construction steps are as follows:

Step 1: the computer center randomly generates m random Numbers and assigns m random Numbers to $Q_1', Q_1, Q_2, \cdots Q_{m-2}, Q_{m-2}'$. Use the variable k to record the number of unused $Q_1', Q_1, Q_2, \cdots Q_{m-2}, Q_{m-2}'$. K is initialized to m.

Step 2: Given $Q_1', Q_1, Q_2, \cdots Q_{m-2}, Q_{m-2}'$, the formula of B-spline curve data point and control point is as formula (6).

We can solve for $P_0, P_1, P_2, \cdots P_{m-2}, P_{m-1}$ using the chasing and driving method. Then $r = P_0 || P_1 || \cdots || P_n$. (||is the initial random number).

Step 3: To participate in the communication, the user U_i should apply to the computer center. After the audit by the computer center, it randomly finds an unmarked location j from $Q_1, Q_2, \cdots Q_{m-2}$ and sends the j to the user U_i.

Step 4: After receiving j, the user enters a random X_i and sends (j, X_i) to the computer center.

Step 5: After the computer center receives (j, X_i), mark Q_j as guaranteed and k minus 1.

Step 6: The computer center generates k random numbers at random and assigns them to numbers not marked for use in $Q_1^{'}, Q_1, Q_2, \cdots Q_{m-2}, Q_{m-2}^{'}$.

Step 7: According to $Q_1^{'}, Q_1, Q_2, \cdots Q_{m-2}, Q_{m-2}^{'}$, the following B-spline curve data points and control points are obtained formula (6):

If $P_0, P_1, P_2, \cdots P_{m-2}, P_{m-1}$ is solved by the chasing and driving method, $r = P_0||P_1||\cdots||P_n$ (|| represents the connector) is a verifiable random number. The computer center consists of a verifiable random number r.

Step 8: Repeat steps 3 through 7 if a new user wants to participate in the communication

Step 9: if a user U_i wants to exit the communication, the computer center marks the U_i's Q_j as unused, adds k to 1, and repeats steps 6 to 7.

4 Verifiable Random Number Verification Based on B-Spline Curve

If the user U_i doubts the randomness of r, it can be verified by the following process:

Step 1: User U_i extracts P_j, P_{j+1}, P_{j+2} according to j and $r = P_0||P_1||\cdots||P_n$, and calculates:

$$A(j) = (P_j + 4P_{j+1} + P_{j+2})/6 \tag{7}$$

Step 2: U_i verifies whether A(j) = X_i is valid. If so, the user participates in the generation of random number r. Otherwise, random number r is not reliable.

5 Performance Analysis of Verifiable Random Numbers Based on B-Spline Curve

It can be seen from the pseudo-code solved by the catch-up method that the computational complexity of the random number can be verified in this paper to be O (n). The running time on the Intel i7-3770 CPU, 3.40 GHz, 8 GB of memory, and windows 7 operating system is as follows (Table 1):

Table 1. Multiplexer truth

Random number length	The elapsed time
50	0.009 ms
100	0.0187 ms
200	0.0325 ms
300	0.0485 ms
400	0.0639 ms
500	0.0833 ms

Comparison literature [2] take advantage of Lagrange interpolation polynomials to generate verifiable random numbers, and it takes 1 200 s to calculate a 400-times interpolation polynomial. so the verifiable random numbers generated by the invention can be generated quickly.

6 Example of Verifiable Random Number Generation Based on B-Spline Curve

Take 20 random numbers for example.

Perform the First Step: The computer center randomly generates 20 random numbers, with each random number ranging from 0 to 100. Assign 20 random numbers to $Q_1^{'}, Q_1, Q_2, \cdots Q_{18}, Q_{18}^{'}$, with 20 random numbers as follows:

$$78, 15, 44, 52, 02, 65, 89, 28, 20, 14, 73, 87, 27, 10, 96, 04, 38, 64, 39, 87.$$

Perform the Second Step: The solved $P_0, P_1, P_2, \cdots P_{18}, P_{19}$, $r = P_0||P_1||\cdots||P_n$, and the initial random password is:

$$-156.0, 164.1, -32.5, 56.0, 72.7, -34.7, 78.1, 112.4, 6.3, 30.4,$$
$$-8.0, 85.6, 103.5, 22.5, -31.5, 163.6, -46.8, 47.7, 84.1, 0.0.$$

Perform Steps 3 to 5: If the new user confirms (5, 47), send it to the computer center.

Perform the sixth step: the computer center randomly generates k random Numbers and assigns the random numbers to the numbers not marked for use in $Q_1^{'}, Q_1, Q_2, \cdots Q_{18}, Q_{18}^{'}$. The 20-bit random numbers are updated as:

$$01, 61, 19, 46, 47, 86, 12, 06, 33, 14, 35, 28, 74, 25, 78, 55, 43, 81, 11, 86.$$

Perform the seventh step: solve $P_0, P_1, P_2, \cdots P_{18}, P_{19}$ with the chasing and driving method, and update the random password as follows:

$$-2.0, -23.2, 100.6, -13.4, 66.8, 22.1, 126.9, -13.6, -0.3, 50.9, -5.2, 53.9,$$
$$-0.4, 115.8, -18.8, 109.4, 49.3, 23.2, 115.7, 0.0.$$

When the password is published, each digit is unified into five digits. We agree that the first digit of each number is the sign bit, when the number is positive, the first digit is 0, and when the number is negative, the first digit is 1. The published password is:

$$10020||10232||01006||10134||00668||00221||01269||10136||10003||00509||$$
$$10052||00539||10004||01158||10188||01094||00493||00232||01157||00000$$

After new user confirmed (5, 47), then the published random code can be verified Firstly, we find out P5, P6, P7 is 66.8, 22.1, 126.9, and calculate

$$A(5) = (P_5 + 4P_6 + P_7)/6 = (66.8 + 4 * 22.1 + 126.9)/6 = 47.01$$

A(5) = 47 is established, so the user participates in the generation of random number r.

7 Conclusion

Refer To sum up, based on the random number generation method of B-spline curve, the paper can calculate the verifiable random number in a very short time with a low-complexity operation method, and the calculation time increases linearly with the digit of random numbers increasing. The number of B-spline curves and the number of bits of random numbers can be adjusted flexibly without using scenarios, and the calculation accuracy and calculation amount are considered. Each user verifies the random number based on the random value of the operation, which has strong privacy and is not easy to be deceived. The verification method is simple in flow and small in computation. It is suitable for use on mobile terminals.

References

1. Ye, J., Li, L.: Group signature scheme based on verifiable random number. J. Discrete Math. Sci. Cryptogr. **20**(2), 525–533 (2017)
2. Christodoulou, K., et al.: RandomBlocks: a transparent, verifiable blockchain-based system for random numbers. J. Cell. Autom. **14**(5–6), 335–349 (2019)
3. Liu, J., Liu, Y.: Random number generation and verification scheme for two entities. Comput. Eng. Appl. Chin. **54**(18), 121–124 (2018)
4. Zhou, J., Zou, R.: Hierarchical group key distribution scheme based on Lagrange interpolation polynomials. J. Xiamen Univ.: Nat. Sci. **46** (Suppl.), 75–78 (2007)
5. Mirzaee, F.: An efficient cubic B-spline and bicubic B-spline collocation method for numerical solutions of multidimensional nonlinear stochastic quadratic integral equations. Math. Methods Appl. Sci. **43**(1), 384–397 (2020)
6. Liu, J., Xie, J., Li, B., Hu, B.: Regularized cubic b-spline collocation method with modified L-curve criterion for impact force identification. IEEE Access **8**, 36337–36349 (2020)

An Improved Particle Swarm Optimization Algorithm Based on DFC&HRS

Wu Zhou[1(✉)], Taizhi Lv[2], and Yijun Guo[1]

[1] College of Mechanical and Electrical Engineering, Huangshan University, Huangshan, Anhui, China
allen_zhou5@163.com

[2] School of Information Technology, Jiangsu Maritime Institute, Nanjing, Jiangsu, China

Abstract. Particle swarm optimization (PSO) has been used in various fields due to its excellent performance. However, two weaknesses, slow convergence and getting easily trapped into local optimum, limit the standard PSO algorithm's application. The PSO research in this paper is done to accelerate the convergence speed and to achieve the global optimum instead of the local optimum. We design herein an improved PSO algorithm, dynamic fraction calculus and hybrid resample PSO (DFC&HRS-PSO), to achieve the two goals. Two approaches are introduced in this algorithm. One is using dynamic fractional calculus to update the particle velocity. This approach combines the time-varying controlling strategy and fractional calculus. By reflecting the history information about particle movement and adapting to different computational periods, the approach improves the convergence speed and searching efficiency. The other is a hybrid resample approach. The hybrid resample approach consists of partial resample and total resample. Both accelerating convergence speed and avoiding the local optimum are considered in this approach. Experimental results show that dynamic fractional calculus and hybrid resample dramatically improve the PSO algorithm's performance. DFC&HRS-PSO performs well on both convergence speed and global searching capability under test functions.

Keywords: Particle swarm optimization · Dynamic fractional calculus · Hybrid resample

1 Introduction

Kennedy and Eberhart put forward the Particle swarm optimization (PSO) algorithm in 1995. Due to its efficiency in solving optimization problems, the PSO algorithm has been used in various fields, including finance, social biomedicine, engineering design, robot, automation, and pattern recognition [1–3]. There are two key weaknesses restricting wide applications of the standard PSO algorithm. Firstly, the convergence speed of the standard PSO algorithm is slow. On the other hand, it could easily slip into the local optimum. For the aim of accelerating the convergence speed and avoiding the local optimum, various improved PSO algorithms have been proposed [4–6]. However, it seems to be difficult to solve the two weaknesses simultaneously. To achieve both goals, an improved particle swarm optimization algorithm based on dynamic fractional

J. MacIntyre et al. (Eds.): SPIoT 2020, AISC 1282, pp. 31–38, 2021.
https://doi.org/10.1007/978-3-030-62743-0_5

calculus and hybrid resample (DFC&HRS-PSO) is proposed. Two approaches, dynamic fraction calculus and hybrid resample, are introduced in this algorithm.

1.1 Dynamic Fraction Calculus

The time-varying controlling strategy is one of the most salient approaches to perfect the PSO algorithm [7]. And the fractional PSO [4] is effective to improve the convergence speed. The fractional calculus makes use of the fractional order's memory property. With the memory property, fractional order velocity can improve the convergence speed. Hence, in this paper, an adaptive fractional order strategy, dynamic fraction calculus, is developed. The fraction calculus is adjusted dynamically to perfect the fractional PSO.

1.2 Hybrid Resample

The hybrid resample approach consists of partial resample and total resample. By using different resample methods in the searching process, this approach not only increases the convergence speed, but also improves the global searching capability. Aiming at evaluating the performance of DFC&HRS-PSO, 3 classical functions are used for numerical analysis.

2 Particle Swarm Optimization

Each particle of the Particle Swarm Optimization (PSO) algorithm represents a potential solution. Each particle stores the speed and location of its solution. In the evolving process, each particle's speed and location are computed iteratively with Eq. (1) and Eq. (2).

$$\mathbf{V}_{t+1,i} = \omega \mathbf{V}_{t,i} + c_1 r_1 \left(\mathbf{P}_{t,i} - \mathbf{X}_{t,i}\right) + c_2 r_2 \left(\mathbf{P}_{t,g} - \mathbf{X}_{t,i}\right) \tag{1}$$

$$\mathbf{X}_{t+1,i} = \mathbf{X}_{t,i} + \mathbf{V}_{t+1,i} \tag{2}$$

Where $\mathbf{X}_{t,i}$ is the i-th particle location at the t-th iteration. The optimization process of PSO is similar with the flying process of a bird. The particle moves to the next location $\mathbf{X}_{t+1,i}$ with the velocity $\mathbf{V}_{t+1,i}$, which depends on three components, the momentum component $\omega \mathbf{V}_{t,i}$, the cognitive component $c_1 r_1 \left(\mathbf{P}_{t,i} - \mathbf{X}_{t,i}\right)$ and the social component $c_2 r_2 \left(\mathbf{P}_{t,g} - \mathbf{X}_{t,i}\right)$. The momentum component models a flying bird's inertial behavior. ω represents the inertia weight of the velocity at the last time instant. The cognitive component represents the best choice of the bird's visited positions. The social component plays the role of attracting the bird to the best position visited in its neighborhood. c_1 is the cognitive coefficient, and c_2 is the social acceleration coefficient. r_1 and r_2 are random numbers less than 1. $\mathbf{P}_{t,i}$ represents the i-th particle's best position at time t, and $\mathbf{P}_{t,g}$ represents the particle swarm's best global position at time t.

3 DFC&HRS-PSO Algorithm

3.1 Dynamic Fractional Calculus

Fractional calculus [4, 8] is used to compute the particle velocity iteratively, which is formulated as the following equation.

$$\mathbf{V}_{t+1,i} = \alpha\mathbf{V}_{t,i} + \frac{1}{2}\alpha\mathbf{V}_{t-1,i} + \frac{1}{6}(1-\alpha)\mathbf{V}_{t-2,i} + \frac{1}{24}\alpha(1-\alpha)(2-\alpha)\mathbf{V}_{t-3,i} + c_1 r_1(\mathbf{P}_{t,i} - \mathbf{X}_{t,i}) + c_2 r_2(\mathbf{P}_{t,g} - \mathbf{X}_{t,i}) \tag{3}$$

In this algorithm, the fractional order varies at different computational periods. At the onset period of a PSO search, particles may be scattered in various areas, thus low fractional order can obtain fast convergence. At the latter period, the population becomes more easily trapped into local optimum with low factional order, so the order value should be higher. The initial value of the order is 0.5 and the region is [0.4, 0.8]. The adjustment process of the fractional order is as follows:

1) Calculate the sum of the distances between each particle and the other particles.

$$d_i = \frac{1}{M-1}\sum_{j=1,j\neq i}^{M}\sqrt{\sum_{k=1}^{D}\left(x_{i,k} - x_{j,k}\right)^2} \tag{4}$$

Where M is the particle number, and D is the solution dimension.

2) Calculate the sum of the distance $d_{t,g}$ between the best global position and the positions of the other particles.
3) Recalculate the fractional order every 50 iterations.

$$\beta = \beta + 0.005 \tag{5}$$

$$\alpha = \frac{0.55}{1 + 1.5e^{-2.6\times\left|\frac{d_{t,g}-d_{\min}}{d_{\max}-d_{\min}}\right|}} + \beta \tag{6}$$

Where $d_{\max}$ and $d_{\min}$ are max distance and min distance between these particles.

3.2 Hybrid Resample

3.2.1 Partial Resample

By adjusting the worst particle position, partial resample can improve convergence speed and global searching capacity. Four methods are used to adapt different functions and stages. Accelerating convergence speed and avoiding the local optimum are both considered in these methods.

The first method adjusts the worst particle to the neighbor of the best global position. The second method adjusts the worst particle to the other side of the best particle. The third method adjusts the worst particle to the best global position with a different value in a random dimension. The last method adjusts the worst particle to a random position.

The partial resample process is shown in Fig. 1.

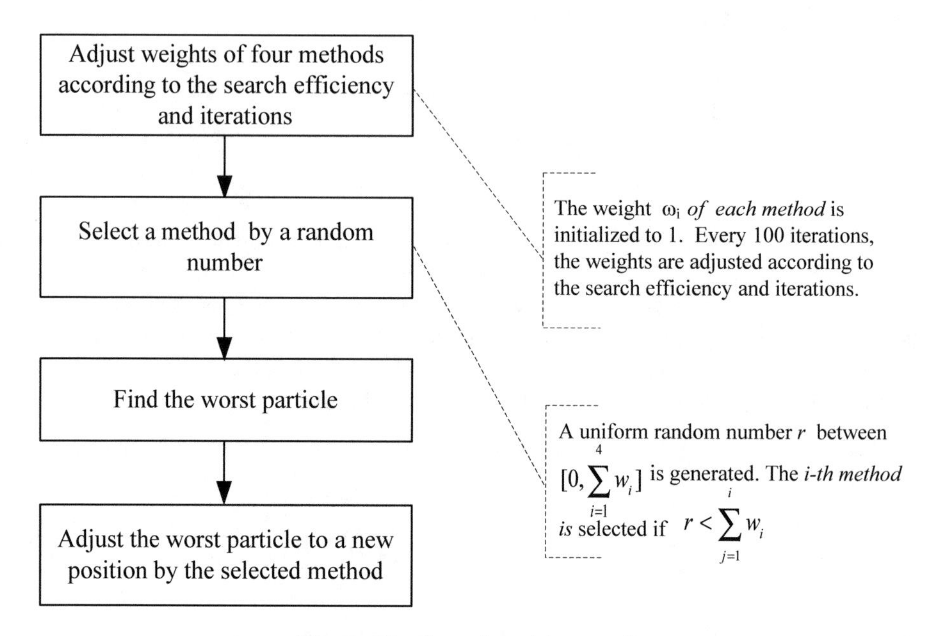

Fig. 1. The flow of partial resample

3.2.2 Total Resample

The PSO search often stays in a relatively stable stage, and the convergence is very slow or even stagnant. This stage may take a long time, and the best global position changes very little or not at all. A stable stage generally occurs at the onset and latter periods of the searching process. Resample methods are the external force which may cause particles to escape from this stable stage. Different resampling methods are used in different searching periods.

(1) Onset Period

At the onset period of the searching process, fast convergence causes the particles to vibrate and makes no detailed search, resulting in the process being easily to be stuck in a stable stage [9, 10]. With an external force, the search may escape from the stable stage quickly. The total resample flow at the onset period is as follows:

Step 1: Check the best global position

If there is no change on the best global position every 200 iterations, the resample process is called.

Step 2: Resample all particles

The same method as the first method in partial resample is used to adjust half of the particles. The third method in partial resample is used to adjust the other half of the particles.

Step 3: Check the resample effect

If there is still no change on the best global position after 100 iterations, the resample method is called again and the coefficient c_4 is expanded.

$$c_4 = c_4 \times 100 \tag{7}$$

(2) Latter Period

At the latter period of the searching process, particles tend to slip into the local optimum. It is difficult for the particles to avoid slipping into the local optimum, so the search is often stuck into the stable stage before achieving the global optimum. Without an external force, no better solution can be found. By using the total resample, particles can escape from the local optimum. For this stage, all particles are reset to

$$X_{t,1:M} = floor + r(upper - floor) \tag{8}$$

Where $X_{t,1:M}$ represents all the particles at time t, r is a random number during [0, 1], *upper* and *floor* represent the search boundaries. If there is still no change on the best global position through 400 iterations after the resample, the resample method is called again.

3.3 DFC&HRS-PSO Algorithm

The whole flow of DFC&HRS-PSO algorithm is shown in Fig. 2.

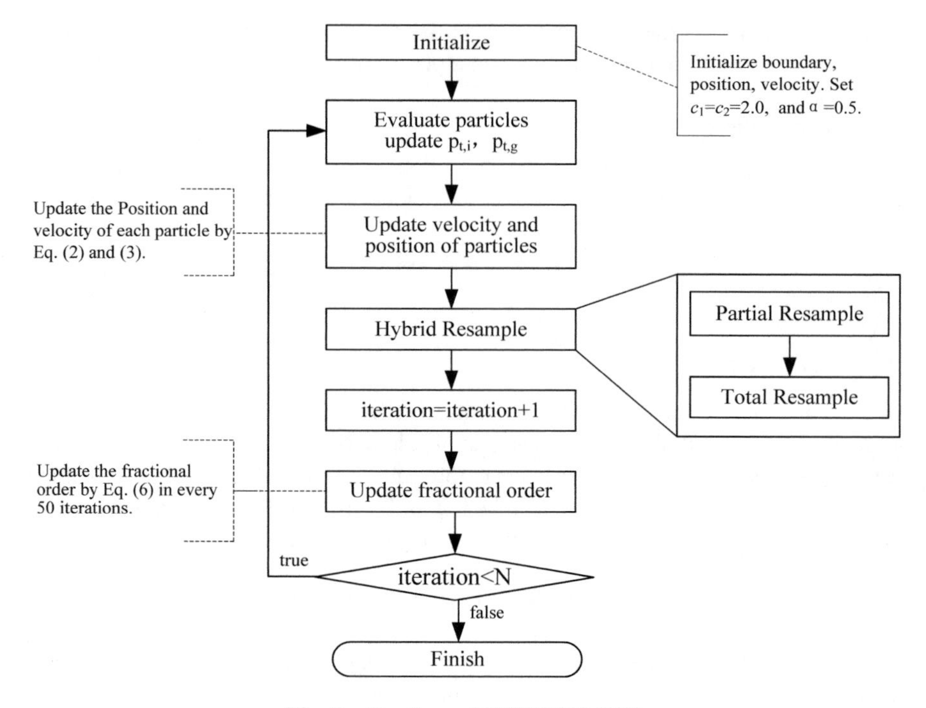

Fig. 2. The flow of DFC&HRS-PSO

4 Results and Discussion

We carry out experiments to validate the performance of DFC&HRS-PSO, which is compared with existing PSOs. Three test functions, listed in Table 1, are adopted for the experiments here. PSO and its improved algorithm [4–6] are compared with the DFC&HRS-PSO.

Table 1. Three test functions

Test function	Search space	Name
$f_1(x) = \sum_{i=1}^{D} x_i^2$	$[-100, 100]^D$	Sphere
$f_2(x) = \sum_{i=1}^{D} ix_i^2$	$[-5.12, 5.12]^D$	Axis parallel hyper-ellipsoid
$f_3(x) = \sum_{i=1}^{D} (\sum_{j=1}^{i} x_j)^2$	$[-100, 100]^D$	Quadric

To carry a fair comparison among all PSO algorithms, the number of particles is set to 20 for all PSO algorithms. In the experiments, the inertia weight is in [0.4, 0.95], acceleration constant $c_1 = c_2 = 2$, and the iteration number is set to 10,000. All the experiments are carried out in a computer with an Intel Core i5 2.60 GHz CPU, 8 GB memory, and Windows 7 operating system. In order to improve the reliability, every function is tested 50 times. Mean value of the 50 results is adopted for comparison.

The performance of the comparative tests on f_1–f_3 functions is listed in Table 2. The results show that DFC&HRS-PSO offers the highest accuracy among these functions. Dynamic fractional order and partial resample can make the DFC&HRS-PSO converge faster than other PSOs, and the higher convergence speed to reach the global optimum region results in higher accuracy. The execution time of DFC&HRS-PSO is less than that of GAPSO and CLPSO, and more than that of standard PSO and fractional PSO.

Table 2. Comparison between DFC&HRS-PSO and other PSOs on test functions

Function		PSO	Fractional PSO	GAPSO	CLPSO	DFC&HRS-PSO
f_1	Mean	2.0710e−03	1.0879e−80	3.9655e−09	1.5627e−60	1.1273e−246
	Std. Dev.	5.6943e−03	3.4398e−80	1.2352e−08	2.3319e−60	<2.2251e−308
	Best	5.8408e−09	1.0669e−98	3.5243e−15	6.2390e−62	8.3917e−257
	Time(s)	11.4617	15.7592	18.0776	16.7256	16.5137
f_2	Mean	3.5640e−03	1.5153e−78	2.0399e−09	3.3875e−62	2.8168e−246
	Std. Dev.	1.1212e−02	4.5950e−78	6.4305e−09	3.3875e−62	<2.2251e−308
	Best	6.1874e−11	1.5453e−100	1.2620e−17	1.4775e−63	9.8041e−256
	Time(s)	12.6219	15.7730	18.3831	16.5729	16.4953
f_3	Mean	83.5012	1.0805e−06	4.8435	6.2641e−01	1.7642e−11
	Std. Dev.	962.0634	2.3518e−07	4.7427	3.5379e−01	9.0590e−1
	Best	67.3757	3.1767e−08	5.1676e−02	2.2297e−01	1.7001e−15
	Time(s)	13.1247	16.4193	18.9697	17.9167	17.7313

5 Conclusion

An improved PSO, DFC&HRS-PSO, is put forward in this paper. Two approaches are introduced here to improve the performance of PSO. Firstly, the dynamic fractional calculus makes the algorithm extremely efficient, presenting a substantially fast convergence speed. Secondly, the hybrid resample method enables particles to escape from the local optimum. The hybrid resample method consists of partial resample and total resample. Experimental results show that DFC&HRS-PSO performs well in both convergence speed and global searching capability.

Acknowledgments. This research is financially supported by the National Natural Science Foundation of China (No. 51405450), the Start-up Research Project of Huangshan University (No. 2018xkjq009), and the Scientific Research Foundation of the Education Department of Anhui Province (KJ2019A0615).

References

1. Li, X.D., Engelbrecht, A.P.: Particle swarm optimization: an introduction and its recent developments. In: Proceedings of the 9th Annual Conference Companion on Genetic and Evolutionary Computation, New York, pp. 3391–3414 (2007)
2. Hamta, N., Ghomi, S.F., Jolai, F., Shirazi, M.A.: A hybrid PSO algorithm for a multi-objective assembly line balancing problem with flexible operation times, sequence-dependent setup times and learning effect. Int. J. Prod. Econ. **141**(1), 99–111 (2013)
3. Selvakumar, A.I., Thanushkodi, K.: A new particle swarm optimization solution to nonconvex economic dispatch problems. IEEE Trans. Power Syst. **22**(1), 42–51 (2007)
4. Solteiro Pires, E.J., Tenreiro Machado, J.A., de Moura Oliveira, P.B.: Particle swarm optimization with fractional-order velocity. Nonlinear Dyn. **61**(2), 295–301 (2010)
5. Liang, J.J., Qin, A.K., Suganthan, P.N., Baskar, S.: Comprehensive learning particle swarm optimizer for global optimization of multimodal functions. IEEE Trans. Evol. Comput. **10** (3), 281–295 (2006)
6. Xie, B., Chen, S., Liu, F.: Biclustering of gene expression data using PSO-GA hybrid. In: Proceedings of the 1st International Conference on Bioinformatics and Biomedical Engineering, Wuhan, China, pp. 302–305 (2007)
7. Zhan, Z.H., Zhang, J., Li, Y., Chung, H.S.: Adaptive particle swarm optimization. IEEE Trans. Syst. Man Cybern. Part B (Cybern.) **39**(6), 1362–1381 (2009)
8. Machado, J.T., Kiryakova, V., Mainardi, F.: Recent history of fractional calculus. Commun. Nonlinear Sci. Numer. Simul. **6**(3), 1140–1153(2011)
9. Lv, T.Z., Zhou, W., Xia, P.P.: Novel particle swarm optimization algorithm. Comput. Appl. Res. **31**(8), 2303–2306 (2014). (in Chinese)
10. Yao, X., Liu, Y., Lin, G.M.: Evolutionary programming made faster. IEEE Trans. Evol. Comput **3**(2), 82–102 (1999)

Application and Practice of ID3 Algorithms in College Students' Education

Liandong Wei(✉)

Changchun University of Finance and Economics, Changchun, Jilin, China
liping85856@163.com

Abstract. In order to understand the application of ID3 algorithm in college students' education, relevant analysis will be carried out in practical testing. Firstly, this paper summarizes the ID3 algorithm and its relationship with college students' education, and then, the application mode and practical effect of ID3 algorithm will be understood through theory and case study. The results show that ID3 algorithm has good performance in college students' education, can improve the quality of education, optimize the details of work, and has good application value. In the traditional university student education, many problems can not be solved because of the unavoidable influence of human resources. ID3 algorithm can break through the limitations of human resources and achieve good results.

Keywords: University student education · ID3 algorithm · Application in practice

1 Introduction

In recent years, with the wide application of database technology and computer network, the amount of data people have increased dramatically. The abundance of data brings about the need for powerful data analysis tools [1]. A large amount of data is described as "data-rich, but information-poor". Massive data is stored in databases, and it is difficult to understand them without powerful tools. At present, the data analysis tools are difficult to deal with the data in depth, so people can only look at the "number" sigh. Data mining is to solve the shortcomings of traditional analysis methods, and for the analysis and processing of large-scale databases. Data mining extracts useful information hidden behind data from a large number of data, and achieves good results, which provides great help for people to make correct decisions. This phenomenon also appears in the education of College students, so data mining is needed to deal with it. ID3 algorithm (decision tree algorithm) is a typical data mining method.

In order to realize the application of ID3 algorithm, this paper firstly analyses the function of ID3 algorithm and summarizes it, then analyses its theoretical application mode, and finally confirms the effectiveness of ID3 algorithm in University Students' education according to the case performance.

J. MacIntyre et al. (Eds.): SPIoT 2020, AISC 1282, pp. 39–46, 2021.
https://doi.org/10.1007/978-3-030-62743-0_6

2 Analytical Methods

The principle of ID3 algorithm is that when selecting attributes at all levels of decision tree nodes, information gain is used as the criterion for selecting attributes, so that when testing each non-leaf node, the maximum category information about the tested record can be obtained. The specific method is to detect all attributes, select the attributes with the greatest information gain to generate decision tree nodes, establish branches from different values of the attributes, and then recursively call the method to establish branches of decision tree nodes for each subset until all subsets contain only the same category of data. Finally, a decision tree is obtained, which can be used to classify new samples. The application steps can be divided into two parts, namely, data preprocessing and classification rules mining. The details are listed below.

(1) Data preprocessing

The function of data preprocessing of ID3 algorithm is that the original business data is often complex, repetitive and incomplete, so a complete data mining system must include data preprocessing part. It aims at discovering tasks, guided by domain knowledge, organizes original business data with brand-new business model, discards attributes unrelated to objectives, and provides clean, accurate and more targeted data for internal accounting method of data mining, thus reducing the amount of data processing of the algorithm, improving the efficiency of mining, and improving knowledge dissemination. The starting point and the accuracy of knowledge. Processing steps can be divided into three parts: data cleaning, correlation analysis and data transformation. Data cleaning is aimed at eliminating or reducing data noise (e.g. using smoothing technology) and processing vacancies (e.g. using the average value of the attribute or the most common value of the attribute). Relevance analysis is as follows: Sex analysis is to select data suitable for data mining applications from all data related to business objects and discard data unrelated to data mining. For example, in four years of university, there are many examinations in many subjects, and their contributions to mining results are different. If all the courses are considered, there will be too many attributes involved in data mining in the future, which will be time-consuming and unnecessary. By choosing attributes that customers are really interested in to mine, we can meet the actual needs. Data transformation is to find the characteristic representation of data, reduce the number of valid variables or find the invariants of data, and convert data into a form suitable for mining. The main work of data transformation in this paper includes: generalizing the attributes of continuous values such as student's course performance into discrete intervals, such as excellent, good, medium and poor, and generalizing the classification attributes to high-level [2, 3] with only two values.

(2) Mining Classification Rules

Data classification is a two-step process. In the first step, a model is established to describe a predetermined set of data classes or concepts. The model is constructed by analyzing the described database tuples. Assuming that each tuple belongs to a predefined class, it is determined by an attribute called a class label. In order to build the model, a part of data tuples to be analyzed is selected as training set and a part of test set. Generally, learning models are provided in the form of decision trees or mathematical formulas using classification rules. For example, given a database of customer credit information, customers can be identified according to their reputation by learning the classification rules. These rules can be used to classify future data samples and provide a better understanding of the content of the database. The second step is to use the model to classify. Firstly, the prediction accuracy of the model (classification method) is evaluated. Keeping and K-fold cross-validation are two commonly used techniques for evaluating the accuracy of classification methods based on random sampling of given data [4]. If the accuracy of the model is acceptable, it can be used to classify data tuples or objects with unknown class labels [5, 6].

ID3 algorithm has a variety of application modes in college students' education, such as college students' psychological counseling, comprehensive assessment of learning level, and the application processes of different ways are quite similar. In view of this, this paper will take college students' psychological counseling as an example to analyze. In college students' psychological counseling, the application of ID3 algorithm can be divided into seven detailed steps, the specific content is shown below.

(1) Determining the mining objects and targets

The object of data mining is college students [7]. The results of UPI and some basic information of students are obtained by psychological counseling. Then the decision tree model of whether students may have mental illness is constructed by ID3 algorithm.

(2) Data acquisition

The questionnaire was used to collect the data of psychological indicators, basic data, family situation, personality characteristics, health status and learning situation of College students. The data collected above can be input to the computer. The data generated can be xls file, DBF file or MDB file.

(3) Data preprocessing

This step can check incomplete, noisy and inconsistent data. Data cleaning can remove noise from data and correct inconsistencies. Data selection can extract the most relevant fields in the data set to simplify the training sample set. Data integration combines data from multiple sources into a consistent data store. Data transformation can discretize continuous data, code numerical data and reduce attribute values. After pretreatment, a training sample set of data mining composed of several records is obtained [8–10]. Table 1 shows the training sample set cases.

(4) Construction of Decision Tree Model

Using ID3 algorithm for classification mining, the steps of building a decision tree model of whether students may have mental illness are as follows: 1. Calculating the information gain rate of the attribute for each test attribute in Table 1; 2. Selecting the attribute with the greatest information gain rate as the root node, and dividing the data set according to its value, if the attribute has only one. One value stops partitioning; 3.

Table 1. Training sample set cases

Introversion	Family harmony	Only child	Poor students	Psychological diseases
No	No	Yes	Yes	No
No	No	No	Yes	No
No	Yes	No	No	Yes
No	Yes	Yes	No	No
Yes	Yes	No	No	Yes
Yes	Yes	Yes	Yes	Yes
Yes	Yes	Yes	No	Yes

Recursive execution of 1–2 for each partitioned sub-data set. After pruning, the decision tree model is finally constructed as shown in Fig. 1.

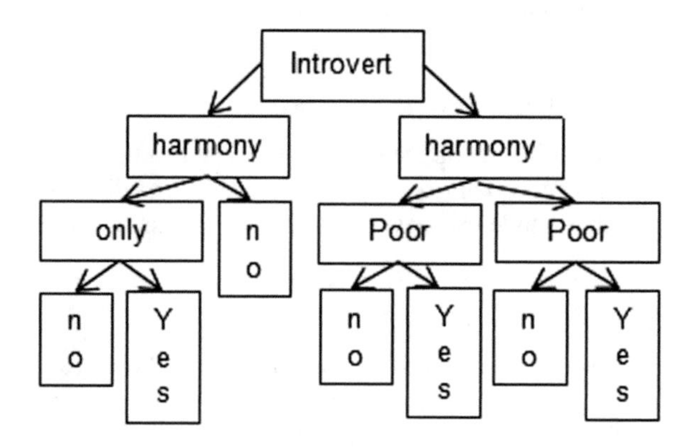

Fig. 1. A decision tree model of students' possibility of psychological diseases

(5) Generation of classification rules

IF can be obtained directly from the decision tree model shown in Fig. 1. The classification rules of students' psychology in THEN form are as follows:

IF Introvert = No AND Harmony = No AND Only = No THEN Asymptomatic
IF introversion = whether AND is harmonious = whether AND is only child = THEN has symptoms
IF introversion = whether AND harmony = THEN asymptomatic
IF introversion = AND harmony = AND poverty = THEN asymptomatic
IF introversion = AND harmony = AND poverty = THEN symptoms
IF introversion = AND harmony = AND poverty = THEN asymptomatic

(6) Analysis of experimental results

From the above rules, we can see that the most relevant attribute is introverted personality, followed by family harmony, poverty and only child. Therefore, in psychological counseling, special attention should be paid to introverted and family disharmony students, and timely psychological counseling should be given to them. Through the prediction of psychological problems, the pertinence and effectiveness of psychological counseling can be really improved [11].

(7) Model Accuracy Assessment

After the decision tree model is generated, it is usually evaluated by error rate or accuracy rate. A test data set containing 160 records is prepared for testing and calculating, and the classification accuracy is close to 80% [12, 13].

3 Case Verification

The data source is the survey data of the practical course of a university in the semester of 2017-2018-02. Because the data of the whole university is very large, including various colleges and courses, the representative compulsory experimental course of the College of Electronic Engineering: the survey data of the digital logic circuit experiment course is chosen as the training sample. The data structure of the Practice Course Questionnaire is shown in Table 2.

Table 2. Questionnaire of learning situation

Serial number	Factor (Attribute)	Attribute value
1	Is the purpose of this experiment clear (G1)	A: To make clear B: Basic clarity C: Vague
2	Do you preview this experiment (G2)	A: Preview experiment B: No Preview
3	Is there any difficulty in completing this experiment (G3)	A: No difficulty, can be completed independently B: Basically achievable C: Difficulties are more difficult to accomplish
4	Are you interested in this course (G4)	A: Have interest in B: commonly C: No interest

Then the information entropy is obtained by data preprocessing, and the decision tree model is generated, as shown in Fig. 2.

Then the steps of extracting classification rules are carried out, and the following classification rules are obtained:

Rule 1: IF G4 = "A" and G7 = "A" THEN class = "A".

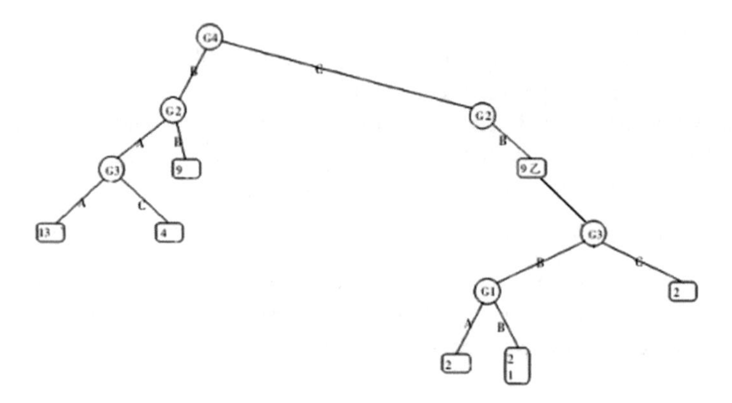

Fig. 2. Decision tree model

Rule 2: IF G4 = “A” and G7 = “B” THEN class = “A”.
Rule 3: IF G4 = “A” and G7 = “C” THEN class = “B”.
Rule 4: IF G4 = “B” and G2 = “A” and G3 = “A” THEN class = “A”.

4 Result Analysis

Data Mining Results Analysis: The more the decision tree model graph nodes are generated, the more important their role is. The top root node is G4. It can be seen that the attribute of “interest in the experimental course” is the key factor for students to evaluate the teaching effect of practical courses. Combining Figs. 2 and 2, we can draw the following conclusions: 1) For the branch of G4 = “A”, we can see from Rules 1 to 3 that “satisfaction with the teacher’s guidance” occupies the work. Secondly, the two functions can directly get the students’ overall evaluation of the curriculum. Therefore, it is important for teachers to stimulate students’ interest in learning. Secondly, in the process of teaching, teachers collect more students’ suggestions on the curriculum and improve the teaching content in time, so as to achieve a more satisfactory teaching effect. 2) For the branch of G4 = “B”, in Fig. 1, the order from top to bottom nodes is G2 and G3, so the importance of these two factors is “preview experiment”, “whether it is difficult to complete the experiment”, “can submit the experiment report on time”. Combining with Rules 1–4, it can be seen that teachers can take measures to supervise the completion of students [14]. Preview by experiment, such as writing a preview report, and then setting up homework with moderate difficulty to ensure that students submit experimental reports on time, can achieve good teaching results. 3) For the branch of G4 = “C”, that is, students who are not interested in experimental courses but do pre-class preparations, we can see from Fig. 2 and Rule 3 that “whether they can submit experimental reports on time” and “whether they feel that the experiment has been harvested” also have some influence on curriculum evaluation. From Rule 1 to 2, we can see that the factors are important. The order of sex is “whether to submit the experiment report in time”, “whether it is difficult to complete the experiment”, "whether to satisfy the teacher's guidance" and "whether to clarify the purpose of the experiment".

5 Conclusion

From the point of view of practical application, this paper analyses and processes the school achievement database by using data mining technology, and obtains some rules and information, such as student achievement prediction, curriculum relevance, etc., to provide reference for school teaching management. In the future, we will study the factors affecting students' quality and the qualifications of innovative talents in a more comprehensive way. We will introduce data mining technology into the field of mental health education in Colleges and universities, use data mining classification technology to classify psychological data, and use ID3 algorithm to achieve classification mining. The experimental results achieved the expected goal, and excavated a series of psychological classification rules, which provided a scientific reference for our school's mental health education and contributed to the improvement of the level of mental health education. ID3 classification algorithm was used to build decision tree model for the survey data of digital logic circuit practice course. The accuracy of the model is 80%. The teaching suggestions have great reference value and reliable theoretical basis. Combining the decision tree algorithm with our specific survey data, we can not only classify and predict, but also find new problems.

Acknowledgements. Fund project: 2019 annual Social Science Foundation Project of Jilin province, "The Research on the Network Communication of Historical Nihilism".

References

1. Elhoseny, M., Shankar, K.: Reliable data transmission model for mobile adhoc network using signcryption technique. IEEE Trans. Reliab. (2019, in Press)
2. Zhai, G., Liu, C.: Research and improvement on ID3 algorithm in intrusion detection system. Comput. Secur. **6**, 3217–3220 (2010)
3. Li, J.F., Lei, J.H., Zhao, X.X., et al.: An improved ID3 algorithm. Appl. Mech. Mater. **444–445**, 723–727 (2013)
4. Puri, V., Jha, S., Kumar, R., Priyadarshini, I., Son, L.H., Abdel-Basset, M., Elhoseny, M., Long, H.V.: A hybrid artificial intelligence and internet of things model for generation of renewable resource of energy. IEEE Access **7**(1), 111181–111191 (2019)
5. Zhu, H.D., Zhong, Y.: Optimization of ID3 algorithm. J. Huazhong Univ. Sci. Technol. **38** (5), 9–12 (2010)
6. Kumar, A.B.R.: Threshold extended ID3 algorithm. Proc. SPIE Int. Soc. Opt. Eng. **8334**, 153 (2012)
7. El-Hasnony, I.M., Barakat, S., Elhoseny, M., Mostafa, R.R.: Improved feature selection model for big data analytics. IEEE Access **8**(1), 66989–67004 (2020)
8. Zhang, Q., You, K., Gang, M.: Application of ID3 algorithm in exercise prescription. In: Lecture Notes in Electrical Engineering, p. 99 (2011)

9. Yang, S., Guo. J.Z, Jin, J.W.: An improved ID3 algorithm for medical data classification. Comput. Electr. Eng. **65**, S004579061732517X (2017)
10. Rani, P., Mishra, N., Diwedi, S.: An efficient multi-set HPID3 algorithm based on RFM model. Int. J. Comput. Appl. **69**(1), 44–47 (2014)
11. Rutkowska, B.D.: A new version of the fuzzy-ID3 algorithm. In: Lecture Notes in Computer Science, pp. 1060–1070 (2006)
12. Jiang, M.H., Luo, X.S.: Classification of student achievement using ID3 algorithm. Appl. Mech. Mater. **220–223**(8), 2540–2545 (2012)
13. Phu, V.N., Tran, V.T.N., Chau, V.T.N., et al.: A decision tree using ID3 algorithm for English semantic analysis. Int. J. Speech Technol. **20**(4), 1–21 (2017)
14. Zaher, M., Shehab, A., Elhoseny, M., Farahat, F.F.: Unsupervised model for detecting plagiarism in internet-based handwritten Arabic documents. J. Organ. End User Comput. (JOEUC) **32**(2), 42–66 (2020)

Construction of Human Knee Joint Mechanics Model and Study on Mechanical Characteristics of Flexion Movement Based on Neural Network Algorithm

Huige Li(✉)

Department of Physical Education, Huanghuai University, Zhumadian 463000, Henan, China
hgli816916@163.com

Abstract. The analysis of the mechanical characteristics of the knee flexion movement can effectively improve the athlete's competitiveness and provide an important theoretical basis for the study of human mechanics. The human knee is an important analysis goal. In the force analysis of the flexion movement of the athlete, the flexion movement characteristics of the knee joint of the athlete are mainly analyzed. The traditional biological method only analyzes the force of all the human joints in the athlete, but ignored the athlete's knee test. In this paper, we propose a method to analyze the mechanical characteristics of knee flexion movement in the human body. Taking the throwing movement as an example, the professional far-infrared system is used to test the whole process of the flexion movement of the athletes and the mechanics model of multi-rigid body is established to carry out the mechanical characteristics analysis of the knee flexion movement, and analyze the knee force situation of the athletes under different movements in detail. In order to verify the effectiveness of the proposed method, the experiment is carried out. From the experimental results, it can be seen that the method for researching the mechanical characteristics of the knee flexion movement can be used to effectively analyze the knee force.

Keywords: Human knee joint · Mechanical model · Flexion movement · Mechanical characteristics

1 Introduction

With the continuous strengthening of national strength, China has changed from a big sports country into a sports power [1]. All kinds of sports have been rapid development [2]. Throwing movement in the technical level has a higher demand [3]. Throwing sports are generally the project of short-term, high-intensity, and continuous operation, only through the continuous flexion action can fully demonstrate the strength and speed of the project requirements [4–6]. As it has far-reaching significance of that the modeling method for the knee joint injury of the throwing athletes caused by over-training, it has become the focus of research in the industry [7–9], has been widespread concern, but also appeared a lot of good methods.

J. MacIntyre et al. (Eds.): SPIoT 2020, AISC 1282, pp. 47–55, 2021.
https://doi.org/10.1007/978-3-030-62743-0_7

In reference [10], the modeling method for the knee joint injury caused by overtraining based on neural network algorithm is studied. Through the topological relationship between the athletes' over - training and the influence factors on knee injury constructed by this method, the model of the knee joint injury caused by overtraining is established, and analyzes the mechanical characteristics under the flexion movement. The method is more adaptable, but there are too cumbersome problems of calculation. In reference, base on the parallel optimization principle of particle swarm optimization to establish the model of knee joint injury caused by overtraining is proposed. The method is simple, but the current algorithm cannot analyze the factors that affect the joint injury of the athletes in the establishment of the model, and there is a problem of large modeling error. In reference, based on ant colony algorithm, the method for modeling the knee joint injury caused by overtraining is proposed.

This paper proposes a method to analyze the mechanical characteristics of knee flexion movement in the human body. Taking the throwing movement as an example, the professional far-infrared system is used to test the whole process of the flexion movement of the athletes and the mechanics model of multi-rigid body is established to carry out the mechanical characteristics analysis of the knee flexion movement, and analyze the knee force situation of the athletes under different movements in detail. In order to verify the effectiveness of the proposed method, the experiment is carried out. From the experimental results, it can be seen that the method for researching the mechanical characteristics of the knee flexion movement can be used to effectively analyze the knee force, with high accuracy and strong robustness.

2 Research Object of Knee Flexion Movement

A total of 34 throwing athletes of different projects are selected from a sports school in Liaoning Province. These throwing sports included shots, hammers, javelins, discus, frisbees and so on. The grade of the athletes participating in the test are all two or above. The basic information of the athletes participating in the test are shown in Table 1.

Table 1. The basic information of the test athletes

Item	Number	Age (years)	Height (cm)	Weight (kg)	Grade
Shot-put	6	20 ± 1	1.80 ± 0.05	92 ± 3	2nd grade
Hammer-throwing	7	19 ± 2	1.82 ± 0.04	73 ± 4	2nd grade
Item	Number	Age (years)	Height (cm)	Weight (kg)	Grade
Javelin-throwing	6	20 ± 1	1.83 ± 0.03	68 ± 4	1st grade
Frisbee-throwing	8	19 ± 2	1.82 ± 0.05	72 ± 3	1st grade
Discus-throwing	7	19 ± 1	1.81 ± 0.04	78 ± 4	1st grade

The whole process of the flexion movement of athletes is tested by Swedish QUALI SYS-M CU500 far infrared system, made in Sweden, to get three-dimensional kinematics data of athletes' joints, and the data acquisition frequency is 240 fps. The number of data acquisition points is 11, and the acquisition parts mainly include shoulder, elbow, wrist, hand, chest, back, hip, etc.

Each participant is subjected to three flexion tests, and a set with the minimum data interval is used as the final data. The interval part of data is automatically padded by the QTrc software. The 3D coordinate data of the test points are obtained by using QUALI SYS motion analysis system. The experimental data are analyzed and processed by using Excel, Origin and other software, and the kinematic data which can be directly used are obtained.

3 Construction of Human Knee Joint Mechanics Model and Study on Mechanical Characteristics of Flexion Movement

3.1 The Mechanical Model Construction of Athletes Knee Joint

In order to analyze the force of knee joint accurately, an accurate mechanical model of knee joint is needed. In the course of throwing, the key influence is the athlete's limbs, and therefore the emphasis is on the analysis of the limbs. It includes the lower limbs, forearms, and hands, and the limb modeling needs to be simplified as follows:

(1) In the analysis of motion biomechanics, the shoulder joint is regarded as a ball hinge with fixed center of rotation, and the rotation of shoulder joint during throwing is neglected;
(2) The motion of the radioulnar joint is regarded as the rotational motion of the knee joint. The knee joint and wrist joint are taken as the steering axis and the rotation is ignored.
(3) The effects caused by finger movements are negligible, because in the actual process of flexion motion, the effect of finger movement on the arm is so small and it can neglect the finger freedom in the modeling process.
(4) The inertial coefficient of each part of the athlete's body is referenced to the adult standard value published by the state.

3.2 Mechanical Characteristics Analysis of Flexion Motion of Human Knee Joint

In order to accurately describe the force condition of knee flexion movement, it is necessary to study the mechanical characteristics and establish an accurate coordinate system. The specific method is as follows:

(1) Establish a fixed standard coordinate system OXYZ on the ground;
(2) Establish the coordinate system $1-C_0xyz$ of athletes body, in which C_0 is the center of mass of the athletes' body, x is the axis pointing to the back of the body, y is the vertical axis, and z is the horizontal axis;
(3) Establish the coordinate system $2-O_0x_0y_0z_0$ of the athletes in the center of athlete's shoulder joint, in which O_0 coincides with and O_1, as the center of the

shoulder joint, O_0x_0 is the rear axis pointed to the athletes body, O_0y_0 is the bottom axis pointed to the athletes body and O_0z_0 is the outside axis pointed to the athletes body;

(4) Establish the coordinate system $O_1x_1y_1z_1$ of the lower limbs, where O_1 is the center of the shoulder joint, O_1x_1 is the axis pointing to the rear of the athlete's body, O_1y_1 is the axis of the knee joint pointing to the shoulder joint, O_1z_1 is the axis pointing to the right side of the athlete's body;

(5) Establish the coordinate system $O_2x_2y_2z_2$ of forearm, and the coordinate system $O_3x_3y_3z_3$ of wrist, in accordance with the establishment of the lower limb coordinate system. The coordinate system $O_jx_jy_jz_j(j = 0, 1, 2, 3)$ is parallel to each other when the arms are naturally drooping.

By establishing the coordinate system of different joints, we can establish the physical model of multi-rigid body of the athlete's body. According to the established model, we can know that the throwing process of the athlete can be transformed into the dynamic problem of the multi-rigid body. Therefore, it can get the following kinetic equations:

$$F^{(r)} + F^{*(r)} = 0, (r = 1, 2, \ldots, 7) \tag{1}$$

$$q\&_r = q\&(u_r, q_r) = 0, (r = 1, 2, \ldots, 7) \tag{2}$$

If the q_r, u_r is known in the throwing process, it can be calculated the muscle torque of every joints. Assuming that the force is concentrated in the center of the articular surface, the Newton's second law can be used to calculate the stress of each joint in the R_0 coordinate system.

The $\vec{H}_i$ is the movement torque of any one of the joints in the body of the athlete relative to the body coordinate system $2(O_0x_0y_0z_0)$. It can be seen from the principle of rigid body mechanics that the movement of the rigid body can be decomposed into two parts with respect to O_1, which are the movement with the $O_1x_1y_1z_1$ level and the rotation of with the center of the body, namely:

$$\vec{H}_i = \vec{r}_i \times m_i\vec{v}_i + \vec{H}_{ci} = \vec{r}_i \times m_i\vec{v}_i + \tilde{I}_i\vec{k}_i \tag{3}$$

Where, m_i is the mass of the i-th joint, $\vec{r}$ is the position vector of the focus of joint i relative to O_1, $\vec{v}_i$ is the motion vector of the joint i, $\tilde{I}_i$ is the rotational vector of the joint i, and $\vec{k}$ is the angular velocity of the joint i.

4 Test Results and Analysis

4.1 Analysis of the Action Curve of Each Joint in Athletes' Throwing Movement

In the process of the throwing movement, due to different movements have different operating forms, in order to facilitate a more accurate analysis, hand forward acceleration in the flexion movement is as the movement to divide the different stages of movement. The key of the flexion movement is to get the maximum outward velocity of the end of the athlete's body. Therefore, the athletes in the throwing movement, the joints of the body will make the corresponding action characteristics, so that the action has continuity. Athletes in the flexion movement, the action of each joint can be described in Fig. 1:

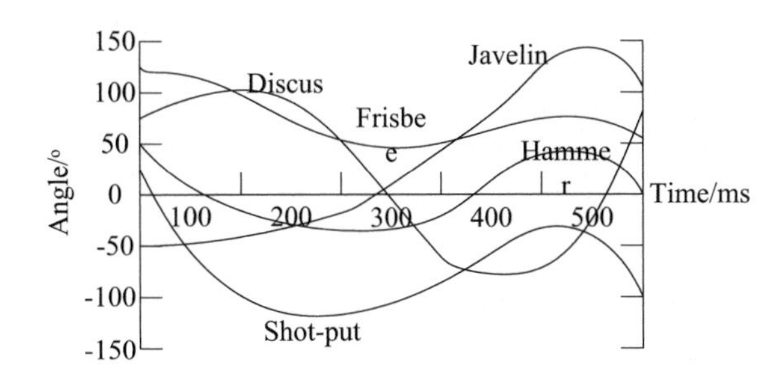

Fig. 1. The movement curve of each joint in the throwing movement

Figure 1 shows that when the athletes in the throwing exercise, the lower limbs extend outward 90° in the horizontal, the knee is forward movement, then, the knees begin to stretch, then the wrist begins to move forward. Finally, the wrist is subjected to rapid flexion movement, where the hand gets the largest forward movement.

The graphs of the joints obtained from the athletes' throwing movement of each throwing item can get the movement timing characteristics of the joints in the five kinds of throwing sports items, as shown in Table 2:

Analysis of Table 2 shows that the actions of different kinds of throwing movements are basically the same, they all follow the movement timing of "forward twisting of the body - lower limb adduction - knee joint extension-lower limb pronation - wrist movement". This movement timing has a greater difference to the motion order of large joints to drive the small joints that many people think.

The movement of the athlete's body can be equivalent to the process of rotation around the joint. The power is from the torque generated by the muscle stretching and contraction of the parts of body in the process of throwing. But the relationship between muscle strength and torque is not the proportional. Such as supporting the muscles of the lower limbs, including the chest muscles, deltoid muscles, back muscles, etc., but the role of these muscles for the internal rotation of lower limbs is smaller, so the generated torque of muscle group is also small, which is difficult to support the throwing movement to reach the highest speed in a short time. Athletes in the process of movement, the limbs is in the posture of horizontal extension, outward rotation and elbow bending, in such an action state, the body twisted forward, driving lower limbs adduction, and elbow acceleration movement in the direction of forward.

In short, the performance form of the coordination and cooperation of the athletes' joints is in line with the movement which muscle torque is larger, so that the athlete's posture can be more reasonable, and thus it can more effectively take the inward rotated

Table 2. Joint characteristics of different throwing items

Sequence	Shot-put	Hammer-throwing	Javelin-throwing	Frisbee-throwing	Discus-throwing
1	Torso twist 0. 224	Torso twist 0. 139	Lower limb adduction 0. 164	Torso twist 0. 180	Torso twist 0. 121
2	Lower limb adduction 0. 119	Lower limb adduction 0. 066	Torso twist 0. 154	Lower limb adduction 0. 156	Lower limb adduction 0. 124
3	Elbow extension 0. 077	Elbow extension 0. 058	Elbow extension 0. 088	Elbow extension 0. 075	Elbow extension 0. 068
4	Internal rotation of lower limb 0. 035	Internal rotation of lower limb 0. 037	Internal rotation of lower limb 0. 046	Internal rotation of lower limb 0. 064	Internal rotation of lower limb 0. 073
5	Wrist flexion 0. 018	Wrist adduction 0. 023	Wrist flexion 0. 033	Wrist movement 0. 041	Wrist movement 0. 037

muscles in the shoulder joint, so that these muscles have greater flexion posture and store greater elastic potential energy, improving the muscle torque, which is conducive to the body to do further acceleration action.

4.2 The Joint Changes in the Index During Concentric Isokinetic Exercise

Table 3. The changes of peak torque and its ratio of flexor and extensor muscle

Angular velocity (%)	Peak torque (Nm)		Peak torque ratio of flexor and extensor
	Flexor muscle	Extensor muscle	
60	90.00 ± 19.26	189.53 ± 32.21	0.47 ± 0.06
120	84.34 ± 19.55	159.02 ± 25.41	0.53 ± 0.08
180	69.28 ± 14.92	139.55 ± 22.13	0.54 ± 0.10
P value	0.044	0.000	0.306

It can be seen from Table 3, The peak torque changes of flexor and extensor muscle are consistent, and all of them showed a significant decrease with the increase of movement speed, when the subjects are in eccentric isokinetic exercise. The peak torque of flexor muscle at 180°/s and 240°/s were significantly lower than those at 60°/s ($P < 0.05$). The peak torque changes of the extensor muscle are significance ($P < 0.05$) between the angular velocities except 120°/s and 180°/s. The peak torque ratio changes of flexor and extensor muscle increase slowly with the increase of movement speed, but there is no significant difference.

Table 4. The changes of the relative peak torque and the mean power of the flexor and extensor muscle with movement speed

Angular velocity (%)	Relative peak torque (Nm)		Average power (W)	
	Flexor	Extensor	Flexor	Extensor
60	1.40 ± 0.25	2.96 ± 0.42	44.12 ± 12.94	93.15 ± 15.74
120	1.32 ± 0.27	2.48 ± 0.30	73.21 ± 16.70	140.84 ± 21.56
180	1.16 ± 0.19	2.18 ± 0.24	87.14 ± 17.71	161.09 ± 22.00
240	1.08 ± 0.20	2.02 ± 0.28	104.47 ± 17.72	180.78 ± 25.56
P value	0.014	0.001	0.000	0.000

It can be seen from Table 4 that the relative peak torque of the flexor and extensor muscle decreases with the increase of the angular velocity at different angular velocities during eccentric isokinetic exercise, and the change of the extensor muscle is very significant ($P < 0.01$); at the same angular velocity, the relative peak torque of flexor muscle is greater than that of extensor muscle obviously ($P < 0.05$). The mean power of flexor and extensor muscle is consistent with the angular velocity, and all of them showed a significant increase with the increase of angular velocity. The statistical test was very significant ($P < 0.01$). At the same angular velocity, the average power of extensor muscle is significantly greater than that of flexor muscle ($P < 0.01$).

4.3 The Joint Changes in the Index During Eccentric Isokinetic Exercise

It can be seen from Table 5 that the peak torque of the flexor and extensor muscle decreases slightly with the increase of the angular velocity during eccentric exercise, but the change does not show significant; However, at the same angular velocity, the peak torque of the flexor muscle is significantly larger than that of the extensor muscle ($P < 0.05$), so does the peak torque ratio of flexor muscle and extensor muscle.

Table 5. The changes of peak torque and its ratio of flexor and extensor muscle

Angular velocity (%)	Peak torque (Nm)		Peak torque ratio of flexors and extensors
	Flexor	Extensor	
60	208.29 ± 39.65	101.44 ± 24.64	108.12 ± 27.52
120	203.04 ± 47.77	100.62 ± 23.74	149.33 ± 44.12
180	196.39 ± 58.27	98.55 ± 25.53	171.55 ± 51.02
P value	0.815	0.976	0.000

It can be seen from Table 6 that the relative peak torque of the flexor and extensor muscle decreases slightly with the increase of the angular velocity during eccentric isokinetic exercise, but the change does not show significant; at the same angular velocity, the relative peak torque of flexor muscle is greater than that of extensor muscle obviously ($P < 0.05$). The mean power increases significantly with the increase of the angular velocity significantly, and so do the mean power of flexor and extensor muscle, and the statistical test has significant difference ($P < 0.01$). At the same angular velocity, the average power of flexor muscle is significantly greater than that of extensor muscle ($P < 0.01$).

Table 6. The changes of the relative peak torque and the mean power of the flexor and extensor muscle with movement speed

Angular velocity (%)	Relative peak torque (Nm)		Average power (W)	
	Flexor	Extensor	Flexor	Extensor
60	3.16 ± 0.50	1.57 ± 0.33	108.12 ± 27.52	51.31 ± 11.78
120	3.08 ± 0.69	1.55 ± 0.34	149.33 ± 44.12	83.81 ± 22.29
180	2.97 ± 0.83	1.51 ± 0.36	171.55 ± 51.02	99.43 ± 23.65
P value	0.815	0.97	0.000	0.000

5 Conclusions

The traditional biological method only analyzes the force of all the human joints in the athlete, but ignored the athlete's knee test. To this end, we propose a method to analyze the mechanical characteristics of knee flexion movement in the human body. Taking the throwing movement as an example, the professional far-infrared system is used to test the whole process of the flexion movement of the athletes and the mechanics model of multi-rigid body is established to carry out the mechanical characteristics analysis of the knee flexion movement, and analyze the knee force situation of the athletes under different movements in detail. In order to verify the effectiveness of the proposed method, the experiment is carried out. From the experimental results, it can be seen that the method for researching the mechanical characteristics of the knee flexion movement can be used to effectively analyze the knee force.

References

1. Arun, S., Kanagaraj, Ṡ.: Mechanical characterization and validation of poly (methyl methacrylate)/multi walled carbon nanotube composite for the polycentric knee joint. J. Mech. Behav. Biomed. Mater. **50**, 33–42 (2015)
2. Zhang, J., Si, Y., Zhang, Y., et al.: The effects of restricting the flexion-extension motion of the first metatarsophalangeal joint on human walking gait. Bio-Med. Mater. Eng. **24**(6), 2577–2584 (2014)
3. Tarnita, D., Catana, M., Tarnita, D.N.: Modeling and finite element analysis of the human knee joint affected by osteoarthritis. Key Eng. Mater. **601**(3), 147–150 (2014)
4. Choi, J., Hong, K.: 3D skin length deformation of lower body during knee joint flexion for the practical application of functional sportswear. Appl. Ergon. **48**, 186–201 (2015)

5. Pierrat, B., Millot, C., Molimard, J., et al.: Characterisation of knee brace migration and associated skin deformation during flexion by full-field measurements. Exp. Mech. **55**(2), 1–12 (2015)
6. Bilston, L.E., Tan, K.: Measurement of passive skeletal muscle mechanical properties in vivo: recent progress, clinical applications, and remaining challenges. Ann. Biomed. Eng. **43**(2), 261–73 (2015)
7. Bobrowitsch, E., Lorenz, A., Wülker, N., et al.: Simulation of in vivo dynamics during robot assisted joint movement. BioMedical Eng. Line **13**(1), 167 (2014)
8. Cyr, A.J., Shalhoub, S.S., Fitzwater, F.G., et al.: Mapping of contributions from collateral ligaments to overall knee joint constraint: an experimental cadaveric study. J. Biomech. Eng. **137**(6), 397–397 (2015)
9. Danso, E.K., Mäkelä, J.T., Tanska, P., et al.: Characterization of site-specific biomechanical properties of human meniscus-importance of collagen and fluid on mechanical nonlinearities. J. Biomech. **48**(8), 1499–507 (2015)
10. Liu, H., Zhao, Y., Hu, Y., et al.: Microstructural characteristics and mechanical properties of friction stir lap welding joint of Alclad 7B04-T74 aluminum alloy. Int. J. Adv. Manuf. Technol. **78**(9), 1415–1425 (2015)

Application of Artificial Intelligence Technology in Physical Fitness Test of College Students

Xin Wang(✉)

Jilin Engineering Normal University, Changchun, Jilin, China
2254688755@qq.com

Abstract. A comprehensive evaluation of college students is important for schools and employment units. This has to do with students' own development. School education reform and employment units can obtain high-quality human resources to study the factors affecting the overall quality of college students, establish a comprehensive quality evaluation index system for college students, quantitatively analyze each index, and use the analytic hierarchy process to determine the weight of each index. In order to solve the ambiguity and uncertainty of student quality, this paper proposes a fuzzy comprehensive evaluation method, and calculates the overall quality of each student and the sub-project quality module score. However, when applying ambiguous mathematics to the overall quality assessment of students, the calculation process is cumbersome and cannot be automatically adjusted according to functions and ambiguous rules. The network convergence speed has not been fundamentally improved. In order to solve this problem, the number of hidden nodes in the model is improved by using the similarity measure method in fuzzy mathematics. Experiments prove that the model increases the network convergence speed and optimizes the network structure. This paper expounds the advantages of career prediction combined with wavelet analysis and neural network, builds a career prediction model based on wavelet neural network, classifies occupations into four categories, and trains the network separately. The model has good simulation and prediction accuracy. High, training time is short, the prediction results provide a strong basis for students to understand their career direction in all directions.

Keywords: Comprehensive quality of college students · Comprehensive quality evaluation · Neural network · Physical fitness test

1 Introduction

Good physical fitness is the foundation of healthy learning and life for modern college students, and it is also one of the essential elements for achieving comprehensive development [1]. The purpose of physical examination for college students is to encourage college students to give up exercise and improve their physical health. Health is vital to their life and development [2]. Having a healthy body is an important foundation for success. Participating in physical activities and developing good habits to continue exercising can strengthen your physique and improve your health. In the

J. MacIntyre et al. (Eds.): SPIoT 2020, AISC 1282, pp. 56–62, 2021.
https://doi.org/10.1007/978-3-030-62743-0_8

past few years, with the continuous reform of our education system, middle school sports has attracted worldwide attention. High school sports management is further standardized [3, 4]. The construction speed of gyms and other facilities has been greatly increased. Comprehensively promote the sports reform curriculum. Weightlifting is becoming more active. The level of sports training has been greatly improved. The number of physical education teachers is increasing. If the scientific research level is not good, it will increase sharply [5]. With the development of computer technology, artificial intelligence technology has been greatly improved, and its application in sports games and testing has become more and more widespread. Objectively and impartially evaluate sports competitions and tests to reduce referee disputes between referees and athletes. This is also a coach. Athletes and referees have strong expectations for artificial intelligence technology. This makes future sports development more convenient and fair, and allows more follow-up in the Internet age. It also points out the direction of sports researchers [6].

Sports practical computer science has developed rapidly and promoted sports in just ten years, but we must calmly see that sports practical computer technology lags far behind the development of computer science. When the Department of Artificial Intelligence is very successful based on the "knowledge process" The actual calculation of motion machine technology still follows the traditional "data structure + algorithm = program" programming method [7, 8]. How to be organized on the basis of system analysis, how to apply the latest developments in computer science to the field of sports, and to tap the potential of computer science to a greater extent, which has taken sports science to a new level and has become sports research Important issues for personnel [9].

This article analyzes the current status of sports artificial intelligence and looks forward to the future of sports artificial intelligence [10]. The goal is to increase the interest of sports workers in sports artificial intelligence and make more people engaged in this work. This paper designs a new mathematical model for comprehensive evaluation. Combining the advantages of fuzzy analysis and neural network, a comprehensive algorithm and fuzzy neural network algorithm are proposed [11]. We created a fuzzy neural network model for college students. In order to improve the convergence speed of the network, a cluster analysis method is used to cluster each index [12]. This method can realize the intelligence of overall quality and high calculation accuracy, solve the problem that fuzzy evaluation cannot automatically adjust the degree of slavery, and better evaluate the quality of students [13].

2 Method

2.1 Conceptual Analysis of Artificial Intelligence Technology

A broad and in-depth understanding of the concept of artificial intelligence can give us a better understanding of its applications. Artificial intelligence refers to a new way of using computers. It uses computers to imitate the way the human brain thinks, fires the way people behave, and achieves better technical development requirements. The scope of use of this technology It is very extensive and is well known to us. For example, the use of artificial intelligence technology in some factories for family planning labor, the use of artificial intelligence technology in the field of psychology, etc., the use of such technology can help us better save labor Cost to improve our standard of living. The use of artificial intelligence technology is very demanding, and this is not just a simple imitation. The use of artificial intelligence technology is first based on the analysis and research of human big data. Artificial intelligence belongs to the pioneers of intelligent technology. It is based on the analysis of human behavior and thinking methods, and inputs the results of the analysis. In computer systems, human life is imitated.

2.2 Artificial Neural Network Test Diagnosis

The neural network builds a mathematical model based on the simulated biological nervous system. The mathematical model simulates the evolution of the living nervous system and is connected through all the ring nodes of the model to form a nonlinear system. Compared with traditional diagnosis methods, neural networks have higher fault tolerance, learning ability and operability, and are one of the most effective methods in intelligent diagnosis and analysis. Neural networks can analyze different signals and performance, put rules together, and predict uncertainty and future events. The basic principle is to collect a lot of template information, create a complete training sample, and output what will happen as input. At present, the algorithm and radial basis function neural network technology have been widely used in diagnosis. Several studies have introduced ambiguous cellular fault diagnosis methods based on radial basis function neural networks. In addition, the validity and superiority of the algorithm are verified through several simulation conditions. Although neural networks have strong fault tolerance and learning capabilities, there are still some problems. First, we need to enter the information and then predict the uncertainty. Create a complete example that requires a complete and large number of database resources. Third, the learning time is long and the learning effect cannot be guaranteed. There are currently many neural network-derived technologies, such as quantum neural network algorithms, genetic wavelet neural networks, and fuzzy neural networks.

2.3 Single Hidden Layer Neural Network Classifier

Discriminant model based on neural network. A neural network consists of non-linear basis functions, each of which is itself a non-linear function of the input variables. Now suppose $\{(x_1, y_1), \ldots (x_n, y_n)\}$ is the data sample of the input layer, and the input data dimension is d, the number of hidden neuron nodes is M, and the number of output layer nodes is K, then a single hidden layer fully connected neural network structure can be constructed.

$$u_k^{(2)}(x; w) = \sum_{j=1}^{M} w_{kj}^{(2)} h(\sum_{i=1}^{n} w_{ij}^{(1)} x_i + b_j^{(1)} + b_k^{(2)}) \tag{1}$$

$$o_k^{(2)}(x; w) = \sigma(u_k(x;w)) \tag{2}$$

Among them, w_{ij} indicates that the weight of the i-th neuron in the input layer to the j-th neuron in the hidden layer is $b_j^{(1)}$, the offset of the j-th neuron, and the superscript (1) indicates that the parameter belongs to the nerve Meta's first layer network. h(.), σ(.) is a differential non-linear activation function for the hidden layer and the output layer, respectively. The activation function of hidden layer h(.) usually uses sigmoid function, such as hyperbolic tangent function. σ(.) In binary classification problems, usually use Logistic Sigmoid activation function. The Sigmoid function maps the input variable to the (0, 1) interval, which can be used to represent the conditional probability $p(c_k|\mathrm{x})$. In the case of two classifications:

$$p(c_1|\mathrm{x}) = \mathrm{sigmoid}(\mathrm{u_k}) = \frac{1}{1 + \exp(-\mathrm{u_k})} \tag{3}$$

The normalization function can be extended to the multi-class (K > 2) situation:

$$p(c_k|\mathrm{x}) = \frac{\exp(u_k)}{\sum_j^k \exp(u_j)} \tag{4}$$

Equation (4) is also called Soft-max activation function.

The universal approximation theory shows that with a sufficient number of neurons in the hidden layer, at least one layer of feed-forward neural network with "squeeze" property activation function and linear output layer can approach any finite space with any accuracy The measurement function maps to another finite space. The "squeeze" attribute refers to squeezing a range of input values into a fixed output range.

3 Experiment

Step1: Acquisition of fitness test data. Obtain and record the traditional fitness test data, and analyze the shortcomings of fitness test. Through the analysis of the shortcomings, we should find out the problems that artificial intelligence technology needs to pay attention to during the construction of college students' physical fitness test.

Step2: Model construction and detection. Apply mathematical algorithm based on artificial intelligence technology to college students' physical fitness test model to calculate and classify various levels and types of different physical fitness. In the calculation process, we must ensure the accuracy of the data, and then build artificial intelligence technology test models for college students' physical fitness according to mathematical algorithms. After the model construction is completed, the model is tested, and the overall effect, error, and accuracy are tested. The experiments are used to test whether the artificial intelligence-based physical fitness test model based on artificial intelligence technology makes up for the shortcomings of traditional testing.

Step3: Design and application of the model. According to the above-mentioned experimental result data and the characteristics of artificial intelligence technology itself, the physical fitness test object model is designed, and the relevant design is tested for practical application. Confirm the consistency between the design and application of the model. Record the difficulties encountered in the actual design and the shortcomings in the real application, and further modify the model according to the actual situation.

4 Discussion

4.1 Basic Indicators and Methods of Physical Fitness Test

One of the positive effects of performing a fitness test is that it can teach students about all aspects of physical fitness and health and prompt them to think about what needs to be done to improve these areas. The physical fitness test will involve the shape, function and quality of the body. Considering that the physical development of students in higher vocational colleges has been completed, the physical fitness test should focus on the function and quality. From a practical perspective, the physical test A large number of professional instruments are required, so the quality testing and evaluation is more operable and feasible.

Table 1. Basic indicators and methods of physical fitness test

Evaluation index	Index connotation	Common test methods
Body composition	Body fat percentage	Fat caliper measurement, weighing
Flexibility	Range of motion around a given joint	Seated forward bend, prone supine
Power	Ability to exert force	Push-ups, horizontal bar pull-ups
Strength endurance	The ability to exert force over a period of time	Resistance squat, flat support
Explosive force	Ability to exert force in the shortest time	Vertical jump, standing triple jump
Speed	Ability to pass a certain distance in the shortest time	Sprint test
Sensitivity	Ability to change direction quickly	Hexagonal reaction ball test
Cardiovascular endurance	Ability of the body to supply cardiovascular endurance energy for exercise	800-m, 1000-m, Cooper 12-min test run

It can be seen from Table 1 that the basic indicators and methods of physical fitness tests are very common. These indicators and methods are also often involved in the teaching of physical education. It can be said that physical fitness tests are both an evaluation indicator and a teaching content. In higher vocational physical education courses, most of the evaluation indicators can be completed through normal lectures, such as forward flexion for flexible sitting, pull-ups for horizontal bars for strength, sit-ups for strength and endurance, and standing for explosive power. Triple jump, 100-m sprint run for speed, T test for sensitivity, 12-min run for cardiovascular endurance, and special skills test for coordination (Fig. 1).

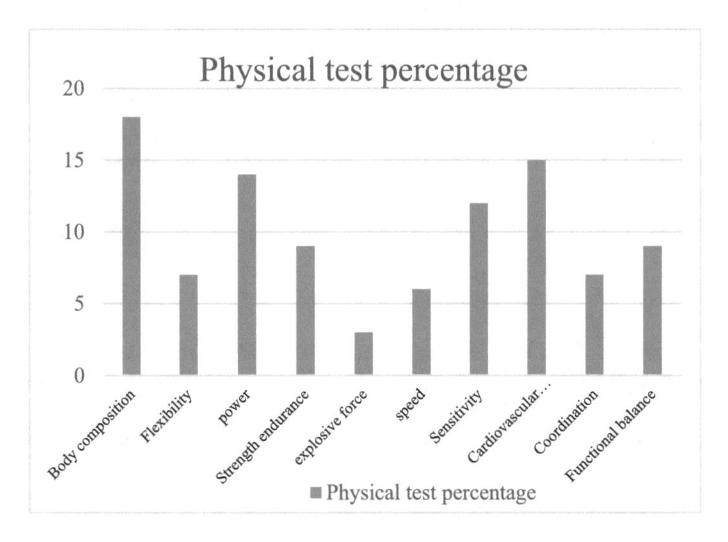

Fig. 1. Percentage of physical indicators in physical examination

4.2 Physical Fitness Test Under Artificial Intelligence Technology

Physical education courses in higher vocational education often ignore the ability test of functional balance. Functional balance includes static balance and dynamic balance. Most teachers in higher vocational physical education courses think that static balance is not important and tend to think that dynamic balance can pass. Other quality tests reflect this. However, the functional balance of the body is very important. It can not only help the body maintain a certain posture and promote the body to complete a certain action, but also reflect the defects and deficiencies of the body. It is an important guarantee for physical exercise, improving exercise efficiency and reducing sports injuries. Therefore, this research actively advocates functionally balanced testing and evaluation of vocational PE courses in order to improve the effectiveness of vocational PE courses. The artificial intelligence technology can test physical fitness differently from the previous test methods. In the past tests, certain items were often used to test the comprehensive quality of college students. Among them, there are often many shortcomings. Like the 50-m run, many bodybuilders often check the stopwatch. It's not until after that started, as well as many generations of testing.

5 Conclusion

This article uses neural network technology to test athletes 'physical testing needs to create a database of athletes' physical tests. We have laid the foundation for athletes' physical information testing. By introducing the B/S mode, many teachers can enter the professional test results of athletes through the webpage, thereby avoiding the differences in computer skills among PE teachers of different levels. Increased usability. System; MD 5 algorithm is introduced into the system to enhance system confidentiality, and users can only enter the system through authentication. In the design of test indicators, the project is set according to the test needs of the athletes, which greatly improves the start-up ability of the project and meets the training needs of the athletes. However, when combined with the current mainstream technology, system design needs to be expanded, such as the development of system-based operating interfaces so that teachers can enter scores on the phone during the test. Therefore, the next research needs further optimization and design.

References

1. Cai, S., Aimukdes, Y., Li, R., et al. Application of artificial intelligence-assisted diagnosis based on deep learning in early esophageal cancer. Chin. J. Dig. Endosc. **36**(4), 246–250 (2019)
2. Cai, Z., Gai, Z.: Application of artificial intelligence in picking up the first to the middle of seismic P-waves: taking the aftershock sequence of the Wenchuan earthquake as an example. J. Peking Univ. (Nat. Sci.) **55**(3), 451–460 (2019)
3. Lin, Y., Dong, S., Zhu, W., et al.: Application of machine learning algorithms in the prediction of coronary heart disease and myocardial infarction. Int. Med. Health Guid. News **24**(17), 2580 (2018)
4. Yin, M., Cui, G.: Opportunities and challenges of artificial intelligence to the employment of college students. Cult. Mater. (16), 162–163 (2017)
5. Yue, M., Wang, W., Zhang, Y.: Application and prospect of artificial intelligence in China aerospace. Comput. Meas. Control. (6). 1–4 (2019)
6. Xu, M.: Research on security face recognition technology and test method. Inf. Commun. Technol. Policy (5), 75–82 (2019)
7. Lin, Y., Dong, S., Zhu, W., et al.: Application of machine learning algorithms in the prediction of coronary heart disease and myocardial infarction. Int. Med. Health Guid. News **24**(17), 2580–2585 (2018)
8. Feng, C., Zhao, R.: Current situation and solutions of college students' employment under artificial intelligence. Mod. Commun. (16), 130–130 (2017)
9. Gu, X., Liu, X., Tian, Y.: Application of online English translation in English learning and entrepreneurship. Mod. Commun. (9), 55–56 (2019)
10. Gu, H.: On the employment situation and countermeasures of college students under artificial intelligence. Yangtze River Ser. (17), 206–206 (2018)
11. Zhong, X.: Tongue elephant physical examination: modern artificial intelligence and traditional wisdom. Coll. Stud. (6), 15–16 (2019)
12. Qi, B.: Application of financial robot process automation based on artificial intelligence perspective. Finance Account. (17), 58–59 (2018)
13. Chen, Y., Yu, Y.: Research on library intelligent management strategy under the background of artificial intelligence——taking ZQ library as an example. China Manag. Inf. (11), 156–157 (2019)

Theoretical Research on College Students' Professional Literacy Design Based on Deep Learning

Longquan Huang(✉)

Guangdong Polytechnic of Science and Trade,
Guangzhou 510430, Guangdong, China
2858093479@qq.com

Abstract. University is a critical period for cultivating college students' professional qualities, and counselors' guidance to college students during this period is very important. This article mainly studies the theory of professional literacy design for college students based on deep learning. Based on the status quo of college students' professional literacy in China, this article analyzes the necessity of cultivating college students' professional literacy, and proposes some strategies on how counselors can cultivate college students' professional literacy, with a view to helping college students' professional literacy improve. The research results in this article show that the average score of "influence" competency is only 2.48, and there is a huge gap between competences with a score of 4 and above, and only 29.41% of students achieve a competency of 4 or more. The average score of "interpersonal insight" competency is 3.79, which is closer to 4 competences, and the proportion of students who reach 4 or above is 79.59%. The "interpersonal competence" professionalism is in the four competence features. The large gap indicates that there is a serious imbalance in the interpersonal communication ability of local college students, that is, the "influence" competence of local college students is seriously insufficient, and the proportion of college students with organizational leadership in professional positions is low.

Keywords: Deep learning · Professional literacy of college students · Professional literacy · College students

1 Introduction

With the rapid development of Internet technology, many innovative technologies such as big data and cloud computing have begun to develop rapidly at home and abroad. Artificial intelligence technology has entered a substantial application phase with each passing day. The grand information technology revolution has pushed human society to a sudden. Deep learning has and will continue to replace humans to complete many tasks, like the manufacturing, construction Industry, mining, textile and other industries, as the industrial revolution corresponds to a large number of people laid off and re-employed, the increasingly advanced artificial intelligence technology will trigger a new round of layoffs and re-employment in many industries. Deep learning is

J. MacIntyre et al. (Eds.): SPIoT 2020, AISC 1282, pp. 63–68, 2021.
https://doi.org/10.1007/978-3-030-62743-0_9

unstoppable. How can college students improve their employment competitiveness and innovation and entrepreneurship?

Professional literacy is the internal norms and requirements of a profession. It is a comprehensive quality exhibited by a person in the course of employment. This quality is relatively stable and has a decisive effect on work [1, 2]. Therefore, professional literacy is an important indicator of a person's professional maturity. There are many factors that affect and restrict professional literacy, such as education level, practical experience, social environment, work experience, and some basic conditions of themselves [3]. College students are recognized as a high-quality group, and society and families have high hopes for them. College students' professional quality determines their professional ability and affects the speed of social development [4, 5]. As a university counselor, we must shoulder the heavy responsibility of improving college students' professional literacy, make full use of various opportunities to strengthen the professional literacy education of college students, and maximize the development of college students' professional literacy [6, 7].

Colleges and universities are in an important position to transport innovative talents for the new era to society [8], facing new educational challenges, and cultivating college students' professional literacy and employment and entrepreneurship capabilities have become as important as disseminating cutting-edge knowledge and professional skills [9, 10]. Only by cultivating high-quality talents required by the industry can we enhance their professional skills, job market competitiveness, and innovation and entrepreneurship capabilities, and truly become high-quality talents needed for socialist modernization.

2 Method

2.1 Deep Learning

Deep learning is an important concept proposed by contemporary learning science. From the perspective of Bloom's educational goal taxonomy, deep learning is the acquisition of advanced cognitive skills, such as applying, analyzing, synthesizing, and evaluating information rather than simply remembering, copying, and memorizing information. It is not satisfied with mechanically and passively receiving knowledge, storing information in isolation, paying more attention to reflection after acceptance, emphasizing and paying attention to learners' active and critical learning, requiring learners to understand the meaning of learning content, Integration integrates them into the original cognitive structure, establishes the connection between existing knowledge and new knowledge, can apply the learned knowledge to new situations, make decisions and solve problems, which is conducive to training learners Critical thinking, innovative ideas, and the ability to find and solve problems. For learners, deep learning needs to focus on cultivating innovative thinking; pay more attention to key content; learn to integrate knowledge across time and disciplines; learn in practice; constantly evaluate and reflect on the learning process.

2.2 Professional Literacy of College Students

Professional literacy is also called professional competence or professional competence in the field of business management practice. It has been pointed out that competence is a potential characteristic of the person necessary to produce high performance in a work situation, and this characteristic can bring practical measurable results, and based on this, an iceberg model of competence is proposed and considered to exist in Knowledge and skills, social roles, self-image, personality and motivation, of which knowledge and skills on the surface can be observed and evaluated, but other characteristics under the water are difficult to observe and can only be achieved through specific actions Only speculative evaluation can be made. As a professional reserve group that is about to enter the post, college students should have the basic literacy requirements before graduation to meet the basic needs of professional development. The composition of college students' professional literacy elements is a systematic layered system. From the perspective of corporate human resources management, for all employees, their professional literacy includes professionalism, teamwork awareness, integrity, proactive attitude, communication and coordination. Five elements of ability. For different professional positions, professional literacy also includes specific professional literacy, such as sales, finance, professional technology and other professional literacy. For people at different organizational levels, professional literacy also includes professional literacy in key positions, such as the professional literacy of senior, middle and grass-roots managers.

3 Experiment

The status quo of professional literacy of local college students not only directly reflects the quality of talent training in local colleges, but also directly affects the employment competitiveness of local college graduates. It is also an important realistic basis for further deepening education and teaching reform in local colleges. The survey of professional literacy of local college students based on the competency model can provide more direct and effective theoretical analysis tools and evaluation methods for evaluating the status of professional literacy of local college students.

The competency model questionnaire for the survey of occupational literacy of local college students has a total of 11 questions, and all questions are measured using the Likert scale 5 level measurement. Based on the representativeness of the survey sample and the availability of survey data, the survey subjects mainly selected senior students from a local university and conducted the survey through a network questionnaire. After excluding invalid questionnaires, a total of 153 valid questionnaires were recovered.

4 Discuss

4.1 Analysis of College Students' Professional Quality

(1) Analysis of the status of "personal motivation" professional literacy

For the status quo of "personal motivation" professional literacy, the survey is mainly based on the "achievement" competence characteristics of college students for evaluation. The survey results show that the average score of "achievement" among college students is 3.92. Although they have not reached a competency of 4 or above, they are relatively close, and the proportion of students with 4 or above is 81.7%. This shows that local college students have a better current status on the "personal motivation" element of professional literacy. Most students can meet the requirements of future professional positions, but there is still room for improvement.

Table 1. Evaluation of professional competence characteristics of local college students

Core professional quality of college students	Competence characteristics	The average score	Match ratio/%
Personality motivation	Desire for achievement	3.93	81.72
Self-knowledge	Confidence	3.33	45.78
Problem solving ability	Analytical thinking	3.84	73.24
	Initiative	3.71	67.33
	Information seeking	3.95	77.81

(2) Analysis of the status of "self-awareness" professional literacy

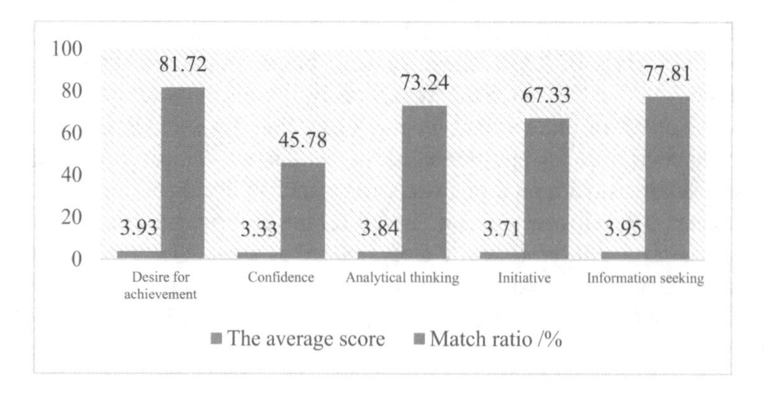

Fig. 1. Competency characteristics of professional literacy of local college students

"Self-knowledge" is an important core element of college students' professional literacy. The status quo of this professional literacy is mainly evaluated by the "self-confidence" of college students. As shown in Table 1 and Fig. 1, the average score of "self-confidence" competence for college students is only 3.31, and there is a significant gap from the competence status of 4 points. If the proportion of students who achieve competence of 4 points or more is only 45.75% Of students meet the competence status, and more than half of the students fail to meet the competence requirements for professional positions in the important professional literacy of "self-awareness". This is an important content and goal of the next step for local colleges and universities to improve the professional quality of college students.

4.2 Suggestions on the Way to Cultivate College Students' Professional Quality

The university stage is a crucial period for exercising various abilities, mastering professional knowledge, and cultivating good professional qualities. As a college counselor, we must recognize the importance of improving college students' professional literacy. We must use various opportunities and methods to improve college students' professional literacy.

(1) Cultivate the professional awareness of college students

The first step in cultivating college students' professional awareness is to guide them to plan for their future. College students lack practical experience and lack of cognition of occupation. As a counselor, you must gradually instill relevant knowledge about careers in college students, so that they have a basic understanding of the workplace and occupations, recognize the nature of their majors, and understand the main content of future work so that they can be fully ideologically. Preparation. Counselors should guide and help students make full use of the holiday time to feel the workplace. On the one hand, they should find opportunities for students through various methods and channels to enter workplace internships; on the other hand, they should encourage students to exert their own initiative and find out what to do. opportunity. In a real workplace, students are more likely to enter roles, understand the importance of professional skills, and understand the significance of solidarity and collaboration. When you return to school with such an understanding, you will have greater motivation and a clearer direction in your studies, and you will have an inherent motivation to improve your college students' professionalism.

(2) Guide students to improve their hidden ethics such as professional ethics and professional style

Professional ethics and professional style are the fundamentals for improving college students' professional quality. Occupational implicit literacy is reflected in many aspects, such as independence, responsibility, professionalism, teamwork and so on. A person has a good professional hidden literacy, can work harder and bear hardships at work, has a strong sense of social responsibility, can consciously delve into professional skills, and make a great contribution at work; conversely, if a person does not have a good professional hidden literacy, It's easy to get lost in work, unable to work down-to-earth, and difficult to do their own job. As a facilitator, take advantage of various opportunities to make students aware of the significant impact of professional ethics and style on a person's professional development.

5 Conclusion

Deep learning plays a very important role for learners to learn. Whether or not deep learning is a decisive factor affecting the quality of learning and the development of academic ability. In this paper, deep learning methods are used in the process of college students 'professional literacy education. According to the characteristics and needs of college students' professional literacy education, combined with deep learning methods, this paper proposes a problem-solving-based vocational education strategy for college students.

References

1. Han, J., Zhang, D., Cheng, G.: Advanced deep-learning techniques for salient and category-specific object detection: a survey. IEEE Signal Process. Mag. **35**(1), 84–100 (2018)
2. Ye, Y., Chen, L., Hou, S.: DeepAM: a heterogeneous deep learning framework for intelligent malware detection. Knowl. Inf. Syst. **54**(2), 265–285 (2018)
3. Bernard, O., Lalande, A., Zotti, C.: Deep learning techniques for automatic MRI cardiac multi-structures segmentation and diagnosis: is the problem solved? IEEE Trans. Med. Imaging **37**(11), 2514–2525 (2018)
4. Yang, S.J., Berndl, M., Ando, D.M.: Assessing microscope image focus quality with deep learning. BMC Bioinform. **19**(1), 77 (2018)
5. Ranjan, R., Sankaranarayanan, S., Bansal, A.: Deep learning for understanding faces: machines may be just as good, or better, than humans. IEEE Signal Process. Mag. **35**(1), 66–83 (2018)
6. Khan, S., Yairi, T.: A review on the application of deep learning in system health management. Mech. Syst. Signal Process. **107**(1), 241–265 (2018)
7. Jing, Y., Bian, Y., Ziheng, H.: Deep learning for drug design: an artificial intelligence paradigm for drug discovery in the big data era. AAPS J **20**(3), 58 (2018)
8. Ryu, J.Y., Kim, H.U., Lee, S.Y.: Deep learning improves prediction of drug–drug and drug–food interactions. Proc. Natl. Acad. Sci. U. S. A. **115**(18), 201803294 (2018)
9. Veltri, D., Kamath, U., Shehu, A.: Deep learning improves antimicrobial peptide recognition. Bioinformatics **34**(16), 2740–2747 (2018)
10. Elliott, A., Byrne, M.: Professional values competency evaluation: comparing student written assignments to literature findings. Nurs. Educ. Perspect. **39**(3), 1 (2018)

Artificial Intelligence Is the Technical Guarantee of Network Security

Ying Wu(✉)

Shanghai University of Political Science and Law, Shanghai, China
wuying@shupl.edu.cn

Abstract. Artificial intelligence can rely on its powerful learning capabilities and data analysis capabilities to detect unknown threats and prevent malicious software applications, transforming previously passive defenses into active preventions. Its technical advantages and distinct application characteristics greatly improve the safety detection efficiency, accuracy and automation.

Artificial intelligence can use its powerful self-learning memory and data analysis and computing capabilities to detect unknown threats, prevent malware and file execution, and turn passive defense into active prevention, which will greatly enhance Network security capabilities, especially their technical advantages, have distinct application characteristics, greatly improving the efficiency, accuracy and automation of security detection.

Keywords: Artificial · Intelligence · Network security

1 Introduction

With the continuous development of the Internet industry, the network boundary is gradually blurred, and all kinds of intelligent devices access to the network, which will lead to the growing network threats. We will meet new challenges in new ways, which will be a tough battle.

Due to the continuous growth of various black industries, there are endless means of attack. In the defense process, we often encounter the previously unknown type of malware, but artificial intelligence can quickly find out tens of millions of events by virtue of its powerful learning ability and data analysis ability, so as to quickly find anomalies, risks and future threats. This is the advantage of artificial intelligence in the field of defense, which makes the former passive defense become active prevention, greatly improves the network security ability and protection efficiency [1–4].

2 Application Characteristics of Artificial Intelligence in Network Security

The explosive growth of data volume, the optimization and improvement of deep learning algorithm, and the dramatic improvement of computing power have promoted the leapfrog development trend of artificial intelligence technology. The application of

J. MacIntyre et al. (Eds.): SPIoT 2020, AISC 1282, pp. 69–74, 2021.
https://doi.org/10.1007/978-3-030-62743-0_10

artificial intelligence technology in the field of network security has great technical advantages.

2.1 Big Data Analysis to Identify Threats

Big security can be achieved based on big data. The AI security analysis engine based on machine learning and deep learning algorithm can better deal with fuzzy, non-linear and massive data. By aggregating, classifying and serializing a large number of data of different data types, it can effectively detect and identify various network security threats, and greatly improve the efficiency and accuracy of security detection. And the degree of automation.

2.2 Relevant Security Situation Analysis

Internal and external security threats can be fully perceived. Network security situational awareness (NSSA) is a means of quantitative analysis of network security and a fine measurement of network security by acquiring, understanding, evaluating and predicting the future development trend of many factors affecting network security. Artificial intelligence technology can merge, correlate and fuse all kinds of data of network security elements. Through a large number of security risk data, it can analyze the correlative security situation, comprehensively analyze the elements of network security, evaluate the status of network security, predict its development trend, and then construct a situation awareness system of network security threats [5–8].

2.3 Self-learning Emergency Response Defense

Active security defense system can be constructed. Network security defense is evolving towards faster (machine learning, artificial intelligence, automation) and more accurate (behavior recognition, visualization). With the learning and evolutionary ability of AI, we can cope with unknown and changing attacks, and combine current security strategies and Threat Intelligence to form security wisdom, and actively adjust existing security protection strategies. This is also the key to form an active security defense system of comprehensive perception, intelligent collaboration and dynamic protection.

3 The Applicable Contents of Artificial Intelligence in the Field of Network Security

In recent years, multi-agent systems, neural networks, consulting systems, machine learning and other artificial intelligence technologies have emerged in the network security defense. In general, artificial intelligence is currently used in areas such as network security intrusion detection, malware detection, and situation analysis.

3.1 In Terms of Network Intrusion Detection

Intrusion detection technology uses various means to collect, filter, and process abnormal network traffic and other data, and automatically generate security reports for users, such as DDoS detection and botnet detection. At present, neural networks, distributed agent systems, and advisory systems are all important artificial intelligence intrusion detection technologies.

3.2 In Terms of Predictive Malware Defense

Predictive malware defense techniques use machine learning and statistical models to find malware family characteristics, predict evolutionary directions, and defend against advance. At present, with the continuous increase of virus malware and the emergence of ransomware, enterprises have a very urgent need for protection against malware. A number of related product systems using artificial intelligence technology have emerged on the market [9, 10].

3.3 In Terms of Network Security Dynamic Perception

Network security situational awareness technology uses data fusion, data mining, intelligent analysis and visualization technologies to visually display and predict network security postures, and provide protection for network security early warning protection, which can improve the system's defense level in the continuous self-learning process. Invincea Inc. of the United States has developed X by Invincea flagship products for detecting unknown threats based on artificial intelligence technology, and the British company Darktrace has developed a corporate security immune system. Domestic Weida Security demonstrated its "Intelligent Dynamic Defense" technology independently developed by the industry, as well as six "Fantasy" series products that incorporate "Artificial Intelligence" and "Dynamic Defense" technologies.

4 Application Case of Artificial Intelligence in the Security Field

4.1 Deep Learning Based Phishing URL Detection

Phishing is a social engineering technique that captures sensitive user information such as usernames, passwords, and credit card details by attempting to pretend to be a trusted entity in electronic communications. Phishing typically induces users to enter detailed information on URLs and fake websites that look and feel very similar to legitimate websites, thereby achieving illegal purposes.

4.2 Password Cracking

Passwords Application Trend of Artificial Intelligence in Network Security

CB Insights, a strength data company, used Trends tools to analyze millions of media articles, track the development trend of hotspot technology, and found that the frequency of network security and artificial intelligence coexisted. "Artificial intelligence + network security" has become one of the most important hot technologies. The application of artificial intelligence technology such as machine learning and deep learning in the field of network security is triggering the growth of new technology research and development and the growth of new security industry.

4.3 Academic Research on the Application of Artificial Intelligence in the Field of Network Security Is in Full Swing

In the academic world, Several provincial and municipal laboratories and university laboratories have conducted in-depth research in this field, mainly including MIT Computer Science and Artificial Intelligence Laboratory, Massachusetts. The Lincoln Laboratory of Science and Technology, the laboratories of the provinces and cities in the Pacific Northwest of the United States, and the laboratories of the provinces and cities of Sandia, USA.

4.4 A Group of Companies Committed to "Artificial Intelligence + Network Security" Is Developing Well in the Industry

CB Insights has found that a considerable number of companies have integrated artificial intelligence into security technologies, with innovative capabilities and long-term network security enterprises due to the outstanding performance of technology research and development has won the favor of a large number of investors. For example, Cylance used artificial intelligence algorithms to predict, identify, and block malware, mitigate the damage caused by 0Day attacks, and acquired a total of $177 million in investment from KOSLA Ventures, Phil Haven Capital, and Citi Ventures; LogRhythm will labor Intelligent mechanisms applied to threat intelligence analysis, enabling compliance automation to quickly detect, respond to and neutralize threats, and received $126 million in investment from Access Venture Partners, Siemens Ventures, and Exclusive Ventures; Darktrace leverages behavior analysis and advanced mathematics automation Detecting abnormal network security behaviors in the enterprise, and obtaining a total investment of $107 million from Softbank Group, Samsung Ventures, and Ten Eleven Ventures.

4.5 Traditional Large-Scale IT Enterprises Turned to "Artificial Intelligence + Network Security" Strategy

In the 2017 strength quarter, three network security startups focused on artificial intelligence technology were acquired by technology giants. Among them, Sophos acquired Invincea for $100 million in cash, Hewlett-Packard acquired Niara (undisclosed amount), and Amazon acquired Harvest.ai for $19 million. In May 2017, Microsoft acquired Israeli online security startup Hexadite for $100 million in transactions. The company uses artificial intelligence to automatically analyze threats and helps internal network security teams manage and prioritize potential risks by accepting multiple source alerts. The above acquisition highlights the obvious shift from traditional large-scale IT companies to the "artificial intelligence + network security" strategy, which is becoming more and more important for future network security.

4.6 "Artificial Intelligence + Network Security" Gradually Rises to the Network Security Level of Each (State) Province and City

Western major developed state government agencies have begun to attach importance to "artificial intelligence + network security" and actively transform traditional military network defense ideas and technology. Occupy the commanding heights of military and network security strategies at home and abroad. In October 2016, the US State Science and Technology Commission issued a report entitled "Preparing for the Future of Artificial Intelligence", delineating the development route and strategy of artificial intelligence in the United States, using special chapters to elaborate on the "artificial intelligence and cybersecurity". Blue print. In February 2017, at the RSA2017 conference of the annual and influential security industry at home and abroad, scientists from the US Department of Homeland Security demonstrated new cybersecurity technologies, including artificial intelligence-supported technologies, ready to enter the market. In March 2017, the Pentagon plans to introduce more options into the field of cybersecurity. As an innovator in the field of military cyber security, the US Department of Defense has carefully created the Research and Planning Bureau to actively promote the research of "artificial intelligence + network security". The projects promoted include: research and development of artificial intelligence attack and defense software, weapon system network security. Research on artificial intelligence projects, research on electronic warfare artificial intelligence projects against adaptive wireless communication threats.

5 Conclusion

Although artificial intelligence is very hot, we also see that in the field of network security, there is no new effective solution based on AI technology. In the field of network security, artificial intelligence technology has been used very early. For example, the correlation analysis engine in SOC products is an expert system based on rule reasoning, but the new artificial intelligence technology has not been widely used. For the deep learning in recent years, we have also seen positive attempts in the industry, such as malicious file detection, application identification, abnormal behavior analysis, etc., but the real effect is far from the requirements of the industry. In addition, deep learning is inherently less prone to interpretability and lack of robustness, and no promising solution has yet been seen.

References

1. Guo, H.: Network security technology based on artificial intelligence. Electron. Technol. Softw. Eng. **23**, 181–182 (2017)

2. Huang, Z., Qiu, B.: On the application of artificial intelligence technology in the defense of cyberspace security. China's Sci. Technol. Invest. (8), 288 (2017)
3. Wang, Z.: A point of view on realizing network security with artificial intelligence. Comput. Knowl. Technol. **15**(02), 27–28 + 30 (2019)
4. Shao, J.: Artificial intelligence to help network security detection and response. Inf. Secur. Commun. Secur. (295(07)), 29–30 (2018)
5. Li, Z.: Application of artificial intelligence technology in network security defense. Inf. Commun. (1), 196–197 (2018)
6. Jin, J., Zou, J: The development of artificial intelligence in the field of cyberspace security. Def. Technol. **39**(4), 43–46, 51 (2018)
7. Wang, H.: Research on information security situation awareness system based on big data and artificial intelligence technology. Netw. Secur. Technol. Appl. **3**, 60–63 (2018)
8. Zheng, Y.: Research and practice of artificial intelligence application and analysis technology in information security situational awareness system. Digit. Commun. World (160(4)), 229 (2018)
9. Fang, Z.: Information security in the age of artificial intelligence. Inf. Secur. Res. **3**(11), 966–967 (2017)
10. Wenbing, D.: Challenges and countermeasures for information security supervision in the age of artificial intelligence. China Inf. Secur. **106**(10), 106–108 (2018)

Intelligent Question Answering System of Medical Knowledge Map Based on Deep Learning

Yafei Xue(✉)

Department of Information Science and Technology, Nanjing Normal University Zhongbei College, Nanjing, Jiangsu, China
3116851297@qq.com

Abstract. The combination of knowledge map and knowledge card enables users to not only read the text to obtain detailed explanation, but also perceive the relevance between knowledge ontology through visual graphics. The text and graphics complement each other and cooperate properly. This intelligent question answering system is to sort out the disordered user corpus information scientifically and orderly, process and extract the natural language keyword information through CRF word segmentation technology, and obtain the final answer to the user based on the basic principle of knowledge map. As the auxiliary recommendation information of drug use, knowledge map and attribute list are presented at the same time.

Keywords: Deep learning algorithm · Knowledge map · Medical field · Intelligent Q&A system

1 Introduction

On the basis of the existing natural language processing technology and knowledge map construction related research, combined with the advantages of various ways, at the same time, using as rich data sources as possible, this paper puts forward more suitable sentence analysis rules for medical field query; at the same time, it provides a combination of knowledge map and knowledge card to show a Chinese medicine more clearly Body specific information. By combining knowledge map and knowledge card, users can not only read text to obtain detailed explanation, but also perceive the relevance between knowledge ontology through visual graphics. Text and graphics complement each other, enriching the connotation of the system, making the resources provided by the system for users no longer monotonous, and improving the perception of the system.

J. MacIntyre et al. (Eds.): SPIoT 2020, AISC 1282, pp. 75–80, 2021.
https://doi.org/10.1007/978-3-030-62743-0_11

2 Overall Design

2.1 Design Emphasis

Based on the method of feature template, we use large-scale corpus to learn the annotation model, and then annotate the sentences. The feature template is usually some binary feature functions defined manually, mining the characteristics of the composition of the named entity and context [1–3]. The Chinese entity recognition method based on the bidirectional lstm-crf model is used to identify the medical domain and other entities in the professional medical classic documents. Different from the traditional LSTM, the bidirectional LSTM considers both the past and future features, and the results are more accurate.

2.2 Design Difficulties

(1) The biggest difference between Chinese and other languages is that there is no space between Chinese words, which makes it more difficult for language processing algorithms to segment sentences and understand semantics.
(2) Because this system is based on the medical field, we must pay more attention to the authority of source data. At the same time, in the process of building medical domain knowledge base, there will be special problems of dictionary requirements, which need special part of speech tagging.
(3) In sentences with ambiguous words, unexpected results often appear. When processing the natural language input by users, we need to call the corresponding method of hanlp to integrate the segmentation. However, in sentences with ambiguous words, unexpected results will inevitably appear.
(4) Visual knowledge map layout balance problem. It needs to use the corresponding algorithm to realize the beauty and readability of knowledge map layout. Otherwise, the nodes of the atlas will cross each other and reduce the readability of the atlas.

3 Function Realization

3.1 Bkg-dl Framework Based on Deep Learning Algorithm

In the construction of knowledge map, the most important three links are the extraction of knowledge units, the recognition of the relationship between knowledge units, and the drawing of knowledge map. Among them, knowledge unit extraction and knowledge unit relationship recognition are the most important. In the field of discipline, researchers can directly use key information such as keywords, abstracts, subject words, authors and so on to construct knowledge map, while the information sources in the field of medicine are much more complex than in the field of discipline, although there are structured data, but more semi-structured and unstructured text information. How to achieve knowledge extraction and relationship recognition with as little human intervention as possible has always been the focus and difficulty of knowledge map construction in the medical field. By using machine learning method, we can learn a small number of knowledge unit patterns in medical field information, and then automatically distinguish a large number of related field information, which can significantly improve the efficiency of knowledge recognition and greatly reduce the human intervention [4–8].

3.2 BKG-DL Framework Process

After the extraction of knowledge units and the recognition of knowledge relationships are mapped to named entity recognition and entity relationship recognition, the bkg-dl process based on deep learning algorithm in Fig. 1 can be obtained.

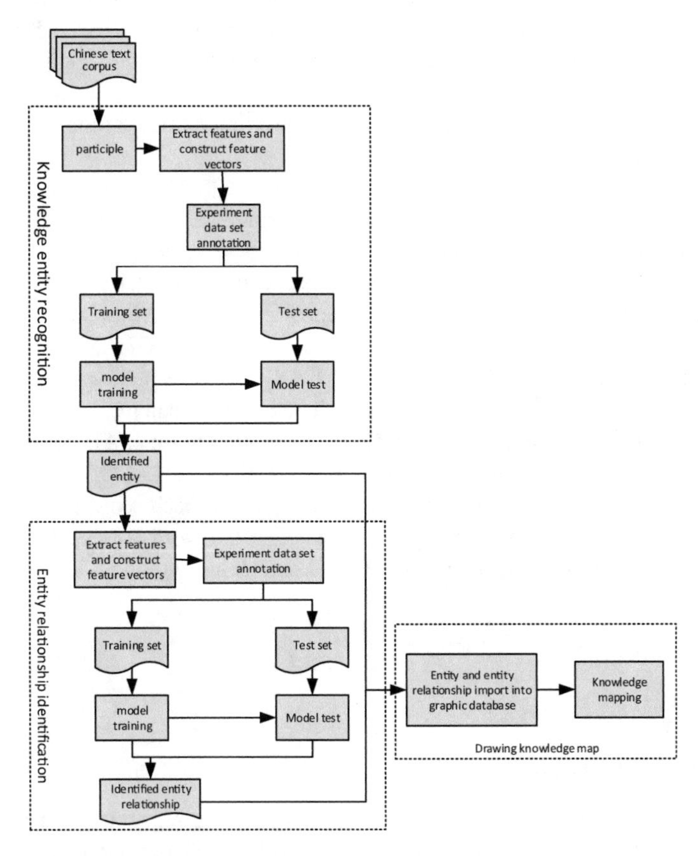

Fig. 1. Construction process of medical domain knowledge map based on deep learning algorithm

This process includes three parts: knowledge entity recognition, entity relationship recognition and knowledge map drawing. In the stage of knowledge entity recognition, because there is no clear word segmentation boundary in Chinese text, we need to segment the text first, then select the appropriate features for the specific scene and build the feature vector for the subsequent model learning and testing, then the model training and testing. In the stage of entity relationship recognition, it is still necessary to select appropriate features and construct feature vectors, then train and test the model. The result of entity relationship recognition is still saved, together with the result of knowledge entity recognition, to the graphic database. After that, the query language of the graphic database can be used to automatically generate the knowledge map, or the rest API of the graphic database can be used for further visual development.

3.3 Chinese Knowledge Entity Recognition Based on Depth Confidence Network

In the recognition of knowledge entities, it is necessary to extract the features of knowledge entities to describe them. The features that can be selected include word features, word features, part-of-speech features, context window features, etc. This paper chooses word features, part of speech features, context window features, dictionary features and other statistical features to describe named entities.

1) Word Features

Because the Chinese text lacks natural word segmentation markers, the text needs to be segmented first. The ICTCLAS system of Chinese Academy of Sciences can be used for Chinese word segmentation. The system can support Chinese word segmentation, part-of-speech tagging and other functions. ICTCLAS also supports user-defined dictionaries and has wide applicability. At present, ICTCLAS3.0 has a speed of 996 KB/s and a precision of 98.45%. ICTCLAS has corresponding modules under Python and R that can be called directly, which is very convenient, and the results after word segmentation can be put into storage and processed in time. All words after word segmentation are formed into a character table d, $d = \{D1, D2, \ldots, dn\}$ where di represents a word, $i \in [1, n]$. The word feature vector of each word e is expressed as $v = \{v1, v2, \ldots, vn\}$, where vi represents whether the word corresponds to di, vi in the character table d. the calculation method is as follows 31:

$$v_i = \begin{cases} 1 & d_i = E \\ 0 & d_i \neq E \end{cases} \quad (1)$$

2) Part of Speech Features

Part-of-speech features are also very important for the recognition of named entities. For example, the part of speech of naming entities such as person names and place names are generally nouns. Chinese POS tagging can also be done by ICTCLAS algorithm of Chinese Academy of Sciences. The construction of part-of-speech features is consistent with the construction of word features. Firstly, the part-of-speech table d, $d = \{D1, D2, \ldots, DM\}$ is constructed, assuming that the part-of-speech of the word is p, then the part-of-speech feature vector of the word is $v = \{v1, v2, \ldots, VM\}$, where vi represents whether the part-of-speech of the word corresponds to di, vi in the part-of-speech table d in the following calculation method:

$$v_i = \begin{cases} 1 & d_i = p \\ 0 & d_i \neq p \end{cases} \tag{2}$$

3) Contextual window features

In a text, the context window composed of consecutive words sometimes has certain rules. For example, when a person's point of view is expounded in the news, a similar expression of "momo CFO Zhang Xiaosong" is usually used. At this time, analyzing the context window of the word can provide basis for the recognition of the word. The word "express" after "Zhang Xiaosong" can initially determine "Zhang Xiaosong" as a company name while "momo" can initially determine "momo" as a company name through "CFO" after "momo". The context window can be set according to specific scenarios. If it is set to 3, it means that the previous word and the latter word of the word are selected for analysis. If it is set to 5, it means that the first two words and the latter two words of the word are selected for analysis.

4) Dictionary features

You can choose words related to entities to form a dictionary, such as the appellation of a person and the suffix of an organization to form a dictionary. The above-mentioned feature vector construction method is also adopted to construct the feature vector of each word to assume that the dictionary is d $= \{D1, D2, \ldots, dn\}$, and the word feature vector of each word e is expressed as v $= \{v1, v2, \ldots, VN\}$, wherein vi represents whether the word corresponds to di, vi in the dictionary d. the calculation method is shown in formula (3-1).

5) Other statistical characteristics

When identifying named entities, some statistical features can also be added for calculation, such as TF-IDF, which is often used in natural language processing. TF-IDF is usually used to evaluate the importance of a word to a corpus. Generally speaking, in a text, named entities are more important to the text than some common words. TF in TF-IDF represents word frequency, which mainly measures the frequency of words appearing in text. IDF is the frequency of reverse files, which measures the universal importance of words in all document sets. TF-IDF has the following formula:

$$W_{\mathrm{fj}} = tf_{\mathrm{fj}} \times idf_{\mathrm{j}} = tf_{\mathrm{ij}} \times \log(\frac{N}{n_j}) \tag{3}$$

Where tfij represents the frequency of occurrence of the word tj in the df mountain of the document, n represents the total number of documents in the corpus, and nj represents the document tree in which the word tj appears.

In addition to the above features, some other features can also be used to identify named entities, such as word features, word formation pattern features and so on [], which will not be repeated here.

4 Conclusion

The system realizes Chinese language annotation according to the written or spoken language input by the user; sentence pattern template matching according to the signal words, generating the corresponding query sentences according to certain matching principles, retrieving the information that needs to be provided to the user in the existing medical knowledge base, and presenting it through knowledge card, visual atlas and drug recommendation [9, 10].

References

1. Hu, F.: Chinese knowledge based on multiple data sources a study on the construction method of atlas. East China University of Science and Technology Learn (2015)
2. Waltz, Zhu, Y., Zhao, X.: Natural language answer system. Xinjiang Soc. Sci. Inf. (3), 10-Fourteen (2013)
3. Chen, F., Liu, Y., Wei, C., et al.: Conditional random the discovery of new words in open field by field method. Softw. Sci. Rep. **5**, 1051–1060 (2013)
4. Xie, T.: Natural language question answering system based on Web a study of. Southeast University (2001)
5. Bo, W.: Research on intelligent question answering system based on ontology. Jiangxi Normal University (2009)
6. Zhao, Y., Hu, Y.H., Liu, J.: Random triggering-based sub-nyquist sampling system for sparse multiband signal. IEEE Trans. Instrum. Meas. **66**(7), 1789–1797 (2017)
7. Zhou, D., Nguyen, T., Breaz, E., Zhao, D., Clénet, S., Gao, F.: Global parameters sensitivity analysis and development of a two-dimensional real-time model of proton-exchange-membrane fuel cells. Energy Convers. Manag. **162**, 276–292 (2018)
8. Lv, Z., Li, X., Li, W.: Virtual reality geographical interactive scene semantics research for immersive geography learning. Neurocomputing **254**, 71–78 (2017). https://doi.org/10.1016/j.neucom.2016.07.078
9. Zhou, D., Ravey, A., Al-Durra, A., Gao, F.: A comparative study of extremum seeking methods applied to energy management strategy of fuel cell hybrid electric vehicles. Energy Convers. Manag. **151**, 778–790 (2017)
10. Yang, L., Chen, H.: Fault diagnosis of gearbox based on RBF-PF and particle swarm optimization wavelet neural network. Neural Comput. Appl. https://doi.org/10.1007/s00521-018-3525-y

Construction and Practice of Red Teaching Resources Based on Machine Learning

Meihua Shuai(✉) and Xia Tang

Jiangxi V&T College of Communications, Nanchang, China
284308963@qq.com, 629065138@qq.com

Abstract. With the rapid development of the socialist market economy, computer technology has been very rapid development and promotion, the application of computer technology in education and teaching, is a very big challenge but also can get excellent results. The computer has now entered the society in each profession, the effective use of the computer, not only can simplify the work, but also can improve the work level. The purpose of this paper is to effectively improve the shortcomings of traditional teaching methods, give full play to the advantages of computer technology, enrich the teaching content and improve the teaching quality through the use of computers in red education and teaching resources. In this paper, by combining the wiki technology to explore the significance of education, the shortcomings of the traditional teaching way, red the advantages of the application of computer in education teaching, the role of computer in the education teaching and the teaching reform of computer application in the education teaching and teaching reform are studied, and discusses the red building and practice of teaching resources.

Keywords: Red education · Computer technology · Conservative constructivism · Application measures

1 Introduction

The application of computer provides diversified teaching forms and interests for education and teaching, provides convenience for both teachers and students, and provides development space for the development of the education industry. How to use computer in education and teaching and the reform of computer application teaching is a topic that needs to be discussed.

Taking the joint teaching resources of undergraduates, postgraduates and doctoral students in a certain university as the research object, Zhang Dezhen studied the problem of course arrangement in colleges and universities in combination with practical problems [1]. With the development of computer application technology, corpus based on linguistics and computer science has been paid more and more attention. Yan, Kong took hainan province as an example to analyze the construction and application of English classroom corpus system in primary schools [2]. Traditional medical education teaching concept does not make full use of modern information

J. MacIntyre et al. (Eds.): SPIoT 2020, AISC 1282, pp. 81–86, 2021.
https://doi.org/10.1007/978-3-030-62743-0_12

technology. Daniel c. Baumgart's goal was to objectively determine the impact of tablet intensive training on the learning experience and MKSAP ® test scores. In this study, the methods were prospective, controlled studies in which medical students and residents in the final year of the study were assigned alternately to active testing or traditional education (control) groups [3].

Information technology has become the main dependence of today's social development, and the computer as the main implementation tool of information technology, its technology has been applied in all walks of life. As the forefront of academic research, the application of computer technology should be further strengthened. Based on machine learning, this paper discusses the advantages of computer technology in the construction of red teaching resources, explains its significance for red education and teaching management, and analyzes the problems faced by computer in teaching management and the solutions.

2 Proposed Method

(1) Advantages of computer in teaching management
  1) In the era of multimedia teaching, the application of computer aided teaching of various subjects in colleges and universities can help to improve the image and intuitiveness of teaching, which is a new teaching auxiliary means and can effectively improve the teaching effect and quality.
  2) With the help of computer technology, students can get rid of the boring feeling under the traditional teaching mode when learning knowledge of various subjects, accept fresh learning materials and completely different teaching hardware environment, and truly experience learning
     Diversity and strong interest will help students improve their ability of independent innovation.
  3) The application of computer aided teaching in colleges and universities can mobilize the interactive atmosphere in the classroom, stimulate the communication between students and teachers, and help teachers to timely discover various problems in the course teaching, so as to facilitate teaching and research improvement [4].
  4) Due to the high transmission efficiency of computer for all kinds of teaching information, there are many information transmission channels, teachers use the computer for teaching, can help save teaching funds, can be regarded as a relatively low cost, good technical content of teaching means.
  5) The computer can realize scenario simulation and experiential teaching, help students to establish a teaching environment suitable for the characteristics of the subject, change the traditional single theory teaching into a context-rich experiential teaching of the subject, change the learning method of rote memorization, and improve the method of learning and mastering knowledge points [5, 6].

6) Poor software adaptation. Currently, colleges and universities have a variety of choices of teaching software, but in the specific choice, colleges and universities can not combine with their actual needs of teaching management, which leads to the wrong choice. This is not only bad for the development of daily teaching management, but also bad for the normal play of the role of computer technology in teaching. Although it is extremely difficult to choose reasonable and scientific teaching management software, colleges and universities should take measures according to local conditions in the face of their own education management practice, and match their own needs with software functions, so as to improve the adaptability of software [7, 8].
7) Change teaching methods. Under the background of quality education, the reform of college education system is speeding up day by day. For teachers, it is necessary to change the original single and boring teaching method, combine the teaching practice, apply the computer technology, and combine the subject characteristics to help transform the teaching environment, improve the classroom teaching atmosphere, enhance students' initiative and interest in learning and creation, and create high-quality teaching results [9, 10].

3 Experiments

In this survey, a self-compiled questionnaire was used as a survey tool, and the questionnaire star platform was used to conduct a survey on the status quo of teacher professional development of all teachers of a school named Marx college in March 2019, in order to provide data guidance for the research on the platform construction and application of teacher education resources of Marx college under Wiki technology. The contents of the questionnaire include basic information and core information. There are 12 questions about basic information, including gender, age, teaching age, professional title, educational background and other basic information, which are used to compare and analyze different individuals in each core information. There were a total of 22 core information questions, which were scored on a likert 5-point scale. The scoring scale ranged from 1 to 5, with 1 point for highly consistent description, 2 points for relatively consistent description, 3 points for generally consistent description, 4 points for insufficiently consistent description, and 5 points for completely inconsistent description. In this questionnaire survey, a total of 29 questionnaires were issued, and 29 effective questionnaires were recovered, with the effective recovery rate of 100%. In this study, the questionnaire data were reorganized and imported into SPSS22.0 for analysis. The results are shown in Table 1.

Table 1. Statistics of questionnaire results of teachers of Marx college

The dimension	Variable	The number of people	The percentage	Effective percentage
Gender	Male	9	31	31
	Female	20	69	69
Age	Under the age of 25	2	6.9	6.9
	26 to 30 years old	9	31	31
	31 to 35 years old	10	34.5	34.5
	More than 35 years old	8	27.6	27.6
Professional life	1–3 years	12	41.4	41.4
	4–6 years	6	20.7	20.7
	7–9 years	3	10.3	10.3
	More than 9 years	8	27.6	27.6

4 Discussion

4.1 Analysis on the Construction of Educational Resources Based on Machine Learning

Through the above survey and the interview analysis shows that the teachers were very interested in the repository and the extent of very be fond of more than 50%, interest and the degree of 30%, that is to say, the extent of teachers' interest in the repository, and more than 80%, as shown in Fig. 1, it shows that they are interested in learning through the repository, like through the repository for autonomous learning, have the realistic requirement of construction repository. There is a lack of systematic design in the content construction of the resource library, which does not conform to the corresponding professional teaching logic of the red education, and the construction of "quality" is not enough to meet the requirements of students' learning and teachers' teaching. In the teaching use of the resource library, the utilization rate of both students and teachers is very low, which is the direct result of the lack of "quality" of the resource library. In addition, the low utilization rate of teachers also indirectly affects the enthusiasm of students to use the resource library. The design and construction of the resource library does not start from fully understanding the actual needs of students' learning and teachers' teaching, but more from the realization of the acceptance standards and technical paths of the national demonstration school, lacking of "connotation" characteristic design.

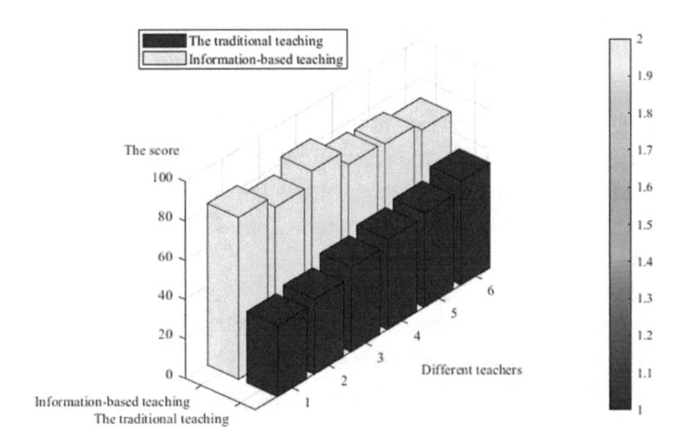

Fig. 1. Evaluation data of teachers on traditional teaching and teaching using resource bank

4.2 Construction and Practical Suggestions of Red Teaching Resources Based on Computer Technology

Using computer technology to improve teaching quality. Examination is an important way to check the quality of students' study and teachers' teaching. At the same time, teachers can also improve the quality of teaching by reasonably designing exams. In the traditional teaching management, teachers often design a large number of papers to implement the term exam, for the students' score correction and statistics are a huge task. In the computer environment, teachers can arrange the test time, design the test content scientifically, and apply the computer intelligent system to realize the statistics of students' test scores and rankings, and greatly reduce the pressure of teachers' test preparation and management.

5 Conclusions

Efficient teaching management is the focus and difficulty of its normal operation, and the computer, as a tool of information, has been popularized. As a manager, it is necessary to make good use of the management function of the computer through scientific and effective means, so as to make the teaching management of colleges and universities more effective and promote colleges and universities to create greater value for students and the society. Based on machine learning, this paper discusses and analyzes the construction and practice of red teaching resources. At last, it concludes that information technology can meet the needs of red teaching resources construction through questionnaire survey.

Acknowledgement. This work was supported by key research on the teaching reform of higher education in Jiangxi Province in 2019, Project title: Construction and Practice of Red Teaching Resources Based on Machine Learning (NO. JXJG-19-53-3).

References

1. Dezhen, Z., Gang, C., Ying, W., et al.: University course timetabling problem using a heuristic approach based on uniform teaching resources. J. Syst. Eng. **12**(1), 22 (2015)
2. Yan, K.: The construction and application of the corpus system in primary school English classroom: taking Hainan Province as an example. J. Comput. Theor. Nanosci. **13**(12), 10385–10390 (2016)
3. Baumgart, D.C., Wende, I., Grittner, U.: Tablet computer enhanced training improves internal medicine exam performance. PLoS ONE **12**(4), 211–214 (2017)
4. Lin, M., Preston, A., Kharrufa, A.: Making L2 learners' reasoning skills visible: the potential of computer supported collaborative learning environments. Think. Ski. Creat. **22**(3), 15 (2016)
5. Qian, M., Levis, J., Chukharev-Hudilainen, E.: A system for adaptive high-variability segmental perceptual training: implementation, effectiveness, transfer. Lang. Learn. Technol. **22**(1), 25 (2018)
6. Ke, Z., Ding, J., Wang, X.: The investigation on the present situations of the implementation of the whole course tutorial system on the engineering technology undergraduates in NUDT. J. High. Educ. Res. **12**(2), 412 (2015)
7. Baporikar, N.: Technology integration and innovation during reflective teaching. Int. J. Inf. Commun. Technol. Educ. **12**(2), 14–22 (2016)
8. Liu, P., An, B., Zhong, Y.: Virtual cluster based resource sharing approach for private cloud environment. J. Front. Comput. Sci. Technol. **32**(4), 279 (2017)
9. Masic, I., Mujanovic, O.B., Racic, M.: Comparative analysis of family medicine education and exams at cathedras of family medicine of Universities in Southeastern Europe–"Splitska inicijativa", Sarajevo, 2017. Acta Inform. Med. **255**(1), 61–72 (2017)
10. Wang, P., Ma, Y., Geng, M.: New teaching-learning-based optimization with neighborhood structure based on small world. J. Front. Comput. Sci. Technol. **23**(5), 67–72 (2016)

Reasonable Approach and Development Trend of Artificial Intelligence Sports Development

Longqiang Chen(✉)

Ministry of Public Sports, Fujian Agriculture and Forestry University, Fuzhou 350002, China
lql5060682716@163.com

Abstract. Based on the literature, logical analysis, and progressive reasoning are used to sort out the development ideas of artificial intelligence sports, explore the law of development, and propose a path for future development. Artificial intelligence sports is an inevitable way to integrate sports informatization and sports industrialization. The core point is based on sports big data and the use of deep learning algorithms to mine valuable information. At present, it has been continuously promoted in the field of competitive sports and is also being used for national fitness. The field is gradually spreading, which has the effect of improving the level of professional sports competition and facilitating public fitness. At the industrial level, the artificial intelligence sports industry is dominated by small and micro enterprises, and the industrial development is uneven. To promote the further development of artificial intelligence sports, from the policy level, it is necessary to introduce macro-encouraging policies and establish supporting regulatory systems; from the technical level, it is necessary to build an interconnected sports big data platform to solve negative problems through technical solutions; It is necessary to optimize the industrial structure and promote sports consumption.

Keywords: Artificial intelligence · Sports · Big data · Competitive sports · National fitness

1 Introduction

In September 2019, the General Office of the State Council's "Opinions on Promoting National Fitness and Sports Consumption to Promote the High-Quality Development of the Sports Industry" further pointed out that the promotion of smart manufacturing, big data, artificial intelligence and other emerging technologies in the field of sports manufacturing, and support for ice and snow, football Development of smart sports events with sports, basketball, racing and other sports as the main content [1]. There is no doubt that the future of artificial intelligence sports has come. However, the development of artificial intelligence sports is still in its infancy, and the concepts of artificial intelligence sports, the specific application direction of artificial intelligence in sports, and the development laws of artificial intelligence sports have not been clarified. AI sports already exist in China. The functional expectation of "Science and Technology Assistance" also faces the realistic challenge of discomfort between new

J. MacIntyre et al. (Eds.): SPIoT 2020, AISC 1282, pp. 87–93, 2021.
https://doi.org/10.1007/978-3-030-62743-0_13

technologies and old industries. If the above problems are not resolved at the institutional and technical level, the development of artificial intelligence sports in China is likely to have twists and turns. Therefore, exploring the complementary advantages and risk regulation of artificial intelligence technology and sports superposition, and thinking about the development path of intelligent sports from a macro level, so that the best coupling of the sports industry to "sports + artificial intelligence" has theoretical foresight and practical application.

2 Overview of Artificial Intelligence Sports

2.1 Definition of AI-Related Sports Concepts

To understand the connotation of intelligent sports, we must first understand the meaning of artificial intelligence. It is generally believed that artificial intelligence is a new scientific technology that researches and develops theories, methods, technologies and application systems for simulation, extension and expansion of human intelligence. Artificial intelligence is not only a scientific consensus, but also a popular trend And commercial development, if artificial intelligence is structured, infrastructure layer, algorithm layer, technology layer, and application layer are in order from bottom to top. According to the intelligence level of the machine, the mainstream view envisages that artificial intelligence will experience weak artificial intelligence and strong artificial intelligence. Weak artificial intelligence and super artificial intelligence. From the current point of view, the boom of artificial intelligence in recent years has been accompanied by breakthrough developments in big data. It is based on machine intelligence based on data, algorithms and computing power, and also belongs to the weak artificial intelligence stage.

Therefore, artificial intelligence sports uses the new generation of information technology such as artificial intelligence as a means to comprehensively perceive and deeply analyze sports big data to gain insight into the patterns, relationships, changes, abnormal features and distribution structures behind it, so as to form warning, The knowledge system of prediction, decision-making and analysis provides theoretical methods and supporting technologies for sports decision-making. The core lies in the first need to comprehensively and accurately sense various sports behaviors and sports scenes in order to obtain a large amount of sports big data, such as athlete statistics, tactical information, and even the weather information on the competition day; It is necessary to adopt various intelligent algorithms to mine knowledge from big data and explore the hidden laws and logics, such as automatically digging fans' consumption preferences from the fans' watching behavior; thirdly, to provide decision theory for the optimization of sports behavior and the transformation of sports scene And key technologies, and apply knowledge to all aspects of economic activities such as sports production, distribution, exchange, and consumption.

It should be noted that in the current academic research, the concept of "AI sports" is often accompanied by "Internet + sports", "digital sports", and "smart sports". These concepts are related but not identical. The concept of "Internet + sports" was first proposed, and to a certain extent has the rudiment of intelligent sports thought, which refers to the use of the Internet to dissolve the essence of time and space to transform the sports industry [2]. The concept of digital sports originates from the "digital earth" and is an important part of the digital earth. It involves information technology, sports, management sports, sports development, and sports experience. It is the product of a comprehensive combination of information technology and sports. Smart sports is to realize a comprehensive perception of various sports behaviors through various ubiquitous sensors. It uses intelligent processing technologies such as cloud computing to process and analyze massive amounts of perceptual information. This kind of demand makes intelligent response and intelligent decision support. The concept of smart sports comes from The "smart city" proposed by the IMB in 2008 aims to apply specific information technology systems to urban management to achieve comprehensive perception, ubiquitous interconnection, pervasive computing, and integration applications. The integration of industrialization and urbanization needs to lead the development of urbanization through the development of industries [3]. Therefore, Chinese artificial intelligence 2.0 Experts from the development strategy research project team suggested that "smart cities" are more suitable to describe China's development status than "smart cities". China's smart cities will follow The process is constantly evolving. It can also be inferred that artificial intelligence sports are more suitable for the current development status of our country than smart sports.

2.2 Development Characteristics of Artificial Intelligence Sports

From the perspective of sports industry ecology, artificial intelligence sports include both the traditional development in the sports industry and the new developments after the application of artificial intelligence technology, as well as the emergence of new formats that are driven by artificial intelligence technology. Starting from the core content of artificial intelligence sports and taking traditional sports as a reference, three characteristics of intelligent sports can be summarized: one is digital. The core point of artificial intelligence sports is to generate massive data through video recognition, image recognition, gesture recognition, etc., to quantify the qualitative information in traditional sports, thereby changing the value judgment based on "experience" in traditional sports, which is derived from The result is professionalism and reliability compared to traditional sports. Digitalization can provide more reliable evidence for the scientific planning and decision-making of event organizers and participants; provide more accurate and personalized exercise and learning suggestions for bodybuilders and sports skill learners; and provide more accurate and flexible for sports media practitioners The means of information distribution and more lively and lively game relay technology can also provide sports bettors with a more reliable basis for decision-making. The second is entertainment. Compared with the traditional sports industry, the entertainment of artificial intelligence sports has increased significantly. Intelligent sports equipment increases the fun of sports activities through more powerful functions and more effective interaction with users; on the other hand, artificial intelligence sports provide the possibility for a wider range and more forms of interpersonal interaction, which enhances The sociality of sports meets the social attributes of sports [4]. In fact, the integration of sports and entertainment industry driven by artificial intelligence sports has become a new economic growth point for the sports industry. The most important thing is convenience. Demand for modern sports consumption is more and more focused on fast, flexible, personalized, and inclusive. Artificial intelligence sports breaks the time and space constraints of traditional sports activities, making people more flexible in scheduling exercise time and participating in rich sports and exercise programs in more diverse locations. Artificial intelligence sports also use the application of intelligent equipment and algorithms, So that people can more easily and quickly get the sports information they want, and better meet people's fast and flexible consumption needs. Artificial intelligence sports can maximize the effect of tailor-made and personal customization. According to the user's specific conditions (including physical conditions, exercise habits, exercise intensity, exercise posture, etc.), formulate exercise guidance methods suitable for the user's own, to meet individualization To achieve the best user experience [5].

3 Progress of Artificial Intelligence Sports

3.1 Technical Perspective

According to the general law of the development of the artificial intelligence industry in China, three integration paths of "technology-business-market" are generally required. "Artificial Intelligence + Sports" is also a multi-layered, multi-content development process. The integration of artificial intelligence technology and sports industry is the prerequisite and foundation for the realization of artificial intelligence sports. Business integration is an important content. Market integration is a necessary condition. Only by creating sufficient market demand can the value of technology integration and business integration be realized.

From a technical perspective, sports, like other disciplines, require technology. People hope to use artificial intelligence research to simulate and expand human intelligence, and assist or even replace people to achieve multiple functions in sports competitions, sports training, and mass fitness. To understand the vision of artificial intelligence in sports, it is necessary to understand the current developments in artificial intelligence and the problems that can be solved. Different from the flash of the previous two waves of artificial intelligence, this wave of artificial intelligence is booming along with the rapid development of information technology and the popularity of the Internet. It is based on the formation of big data resources, breakthroughs in computing power technology, and algorithm enhancement. On the one hand, the development of "Internet +" integrates online and offline, and promotes the interconnection of various devices. People use social networks based on the Internet and the Internet of Things to generate and retain a large amount of information for analysis and Quantitative data; on the other hand, the computing power that supports the storage and operation of these data has also made breakthrough progress [6].

3.2 Business Perspective

According to economic principles, supply is driven by demand. The reason why artificial intelligence has a wide spread effect in the sports field is related to the business needs of sports. First, the blessing of technology is urgently needed in competitive sports. The improvement of the competitive level is inseparable from efficient scientific training and accurate match analysis. The integration and application of intelligent technology in competitive sports can bring huge improvements in these two aspects. For example, smart practice venues and wearable training monitoring equipment can analyze athletes' heart rate, blood pressure, brain activity, joint pressure and other indicators in real time, so as to improve training quality and avoid fatigue and sports injuries caused by over training The intelligent game analysis system can sort out massive game data, analyze the gains and losses of tactics through machine learning, and find the most effective way to complete the game. The application of these intelligent technologies can help China's competitive sports to continuously optimize the structure of competitive events, and on the basis of maintaining and consolidating the advantages and achievements of existing sports, the pursuit of new levels of development and beyond. In addition to the need for artificial intelligence to improve competitive performance, the innovation and efficiency of competitive sports management models also depend on artificial intelligence. Whether at the level of individual competitive development management, or in the management of sports teams or leagues, artificial intelligence can collect individual data through big data means, thereby providing managers with assistance in related decisions. Secondly, based on project intelligent selection, expert cloud guidance and other methods, choose to meet your own health status, and formulate fitness programs and fitness plans based on seasonal, environmental, equipment and other objective conditions. Not only does public fitness require the intervention of artificial intelligence, the government better manages national fitness, but it can also collect personal data through big data means, connect with social security, education, medical and other systems, use in venues, facility maintenance, event organization, public welfare training, exercise supervision And emergency assistance, assist government departments to provide scientific decision-making and precise services at the macro level. Finally, expanding the development space within the sports industry also requires the help of artificial intelligence. It is necessary to promote the integration of the sports industry with industries such as medical care, pensions, tourism, and education. And how to integrate, this intermediary is actually artificial intelligence. For example, analyzing the consumption characteristics of event viewers through machine learning can promote the integration of the event viewing industry and the advertising industry; through big data analysis of the sports habits and health status of middle-aged and elderly people, it can promote the fitness and leisure industry and medical, pension and other industries Integration. These industrial integrations can not only drive the traditional sports business lines, but also continuously expand and tap new sports businesses.

3.3 Market Perspective

The "artificial intelligence boom" in sports is closely related to the prosperity of the sports market. The nature of marketization is that as long as a technology is profitable, it can promote operations. The development of supercomputing power from computers and the development of big data have brought new value and development resources. Data show that at the beginning of smart sports, as of the first half of 2019, the number of new entrepreneurial projects in this field has reached 916, and 366 projects have received capital market investment. In addition, the prosperity of the market also brings official promotion, because artificial intelligence can solve many problems in the reform of the traditional sports industry, but from the current situation, the ability to use capital investment and labor to promote the development of the sports industry has significantly decreased. These two levers are the traditional driving force for production. They can no longer use their own characteristics to promote the growth and prosperity of the total scale of the sports industry. The application of new sports technology and the growth of new business formats have become important growth points in the sports industry. Therefore, the official has given artificial intelligence sports a functional expectation that can solve problems and promote the development of the sports industry. For example, at the provincial and municipal levels, the Guangdong Sports Expo has created an artificial intelligence sports zone; Zhejiang Province has hosted the first artificial intelligence sports competition; Jiangsu Province has arranged hundreds of millions of sports industry guidance funds each year to build a "smart sports" service covering the province Networks and platforms, promote the combination of the sports industry and e-commerce, encourage the use of artificial intelligence to expand sports consumption, and strengthen the intelligent construction of stadiums; official assistance has further promoted the entry of private capital, and this cycle has made the market for smart sports increasingly rolling bigger.

4 Conclusion

The sports industry has always attached great importance to the role of science and technology. Under the new normal of the economy, the scientific and technological revolution is even more needed to promote the transformation and upgrading of the sports industry. The development of new sports events helps maintain the high-level development of China's competitive

Sports and the high-speed growth of the sports industry. It is also an effective weapon to promote sports consumption. It can effectively promote the change of people's sports consumption concepts and obtain smooth sports consumption Experience and strengthen continuous sports consumption behavior. In the wave of science and technology, we must vigorously develop artificial intelligence sports and promote the transformation and upgrading of the sports industry. At the same time, we must also be aware of the current weakness of China's sports big data technology and the reality of the initial stage of the development of the intelligent sports industry. Under the fast variant, focus on the slow variant of sports. On the one hand, it further promotes its development at the policy level, and on the other hand, it effectively controls its social risks from the technical and legal level. All in all, in the age of science and technology, artificial intelligence sports must be embraced and treated cautiously, and the coordinated progress of competitive sports and mass sports must be promoted in order to realize the benefits of artificial intelligence sports for the progress of human civilization.

References

1. Zhang, X.: Exclusive rights of sports events under the background of artificial intelligence technology. J. Shanghai Inst. Phys. Educ. **44**(2), 64–73 (2020)
2. Li, H.: Internet restructuring sports industry and its future trend. J. Shanghai Inst. Phys. Educ. **16**(6), 8–15 (2016)
3. Zhang, F.: Research on the prospects of the age of artificial intelligence. Soc. Sci. Yunnan **1**, 53–58 (2020)
4. Li, Y.: A comparative study of tourism business models in the internet era. Int. J. Front. Sociol. **1**(1), 99–108 (2019)
5. Feng, Y., Luo, X.: Performance evaluation of sewage treatment plant under sustainable development. Int. J. Front. Sociol. **1**(1), 1–11 (2019)
6. Sun, Z.: A study on the educational use of statistical package for the social sciences. Int. J. Front. Eng. Technol. **1**(1), 20–29 (2019)

Returnee Migrant Workers' Entrepreneurship Based on Artificial Intelligence

Dejun Zhou(✉)

School of Law, Jiangsu University, Zhenjiang, Jiangsu, China
1469463202@qq.com

Abstract. With the development of modern science and technology, the Internet, as an important scientific and technological product, has penetrated into the vast rural areas of China. Various artificial intelligence tools, methods and products which based on the Internet have become the impact on agricultural production and development, rural social progress, farmers' important factors for income generation and income growth. Returning rural migrant workers to start a business is an important part of China's strategy to promote rural rejuvenation. Based on the perspective of artificial intelligence, we will analyze the practical dilemma of artificial intelligence in the process of entrepreneurship of returning rural migrant workers in China, and find solutions to the problem. For the development of agricultural industry production, the promotion of rural social progress, the promotion of farmers' income and employment, and finally the realization of the rural revitalization strategy. The goal has important theoretical value and practical significance.

Keywords: Artificial intelligence · Returning rural migrant workers to start a business · Intelligent identification APP · Increasing farmers' income

1 Introduction

As China's urban-rural integration progresses, more and more farmers choose to return to their hometowns to start businesses. The agricultural entrepreneurial activities of returning migrant workers not only increase their own income, but also promote the common development of the local economy and society. Returning rural migrant workers to start a business is an important part of the implementation of China's rural revitalization strategy. The advancement of modern science and technology and the development of agricultural artificial intelligence have provided technical possibilities and material guarantees for migrant workers to return home to start their own businesses. Agricultural artificial intelligence has not only enriched the marketing channels for returning peasant workers to start their own agricultural products, but also has expanded the return path for returning peasant workers to start their own businesses. Aiming at the current plight of the development of agricultural artificial intelligence in the current rural society in China, combined with the urgent needs of returning rural migrant workers to start a business, this paper proposes specific countermeasures for the application of artificial intelligence to returning rural migrant workers to start a

J. MacIntyre et al. (Eds.): SPIoT 2020, AISC 1282, pp. 94–101, 2021.
https://doi.org/10.1007/978-3-030-62743-0_14

business, which is a new era of increasing farmers' income, rural progress and agriculture Useful exploration for development.

2 Background of Artificial Intelligence Development for Migrant Workers Returning Home

The development and improvement of human society's science and technology is prompting the transformation of agriculture to modernization [1]. With the advancement of social science and technology, the Internet has penetrated the lives of ordinary people. Even ordinary farmers in rural areas have felt the impact of the Internet on their lives. Due to the gradual fading of the old generation of migrant workers, the cultural level of the new generation of migrant workers has relatively improved, and the basic text communication has become a basic requirement for work and life, and mobile phones have become a basic living tool. As far as ticket purchase is concerned, online ticket purchase has become the first choice for the new generation of migrant workers. Online shopping has also become the norm in rural society. The courier company's site has reached the village. As an important communication tool, WeChat has become an indispensable communication method for ordinary farmers, providing a convenient communication tool for farmers to start their own businesses. The development of artificial intelligence has penetrated all aspects of modern rural development in China, and has had an important impact on farmers' production and life. Modern rural families are pursuing the diversified information and safe, comfortable and convenient living environment brought about by the intelligence of the family [2]. Whether it is ordinary employment of farmers or innovation and entrepreneurship, artificial intelligence is inseparable. Returning rural migrant workers who have mastered artificial intelligence tools first have already tasted the sweetness in the process of starting a business. However, we should also realize that although the overall cultural level of the new generation of returning rural migrant workers has improved compared with the older generation of rural migrant workers. But if it is compared with the requirements of high-tech development of artificial intelligence, there are still large gap. The development of artificial intelligence is both a severe challenge and an important opportunity for the entrepreneurship of returning migrant workers.

3 The Intelligent Dilemma Faced by Returning Rural Migrant Workers

3.1 The Intelligence Network Platform Has Fewer Resources for Peasant Entrepreneurship, Which Weakens the Possibility of Returning Migrant Workers to Start a Business

At present, the network training resources on the intelligent network platform are huge and comprehensive, and there are many types, but the resources on knowledge and skills training mainly focus on high-level knowledge transfer, and less specifically involve rural areas. Recreation resources and youth learning resources are relatively abundant, but training resources for returning rural migrant workers are significantly less, and intelligent resources for employment and entrepreneurship guidance in specific areas and villages are even more scarce. Especially under the realistic national conditions of vast rural space and obvious geographical differences in China, returnee peasant workers can only rely on self-exploration to start their own businesses. They do not know much about market conditions and product production, and the probability of successful entrepreneurship is not optimistic. Because specialized vocational skills training is often expensive and not necessarily practical, ordinary return farmers are reluctant to participate in such skills training. Unless local government departments specifically organize returnee farmers to start their own businesses, returnee farmers do not have the opportunity to receive special guidance. Even sometimes, because the training activities are more focused on form, even if returning farmers participate in learning, they will not learn practical and useful entrepreneurial knowledge.

3.2 The Application of Smart Product Methods to Agricultural Development Has Less Hindered the Enthusiasm of Returning Rural Workers to Start Their Own Businesses

Although scientific research institutions have developed a lot of intelligent products for agricultural production, such as intelligent identification systems for pest detection, soil detection APP, and agricultural market demand analysis APP, etc., these intelligent products are more at the product development stage. It is rarely used in agricultural production. Due to the uneven water evaluation of the current agricultural modernization development in China, the farmers 'cultural level is generally low, and high-tech intelligent products and methods have not yet been universally adopted. The returnees' migrant workers rely on individual experience and traditional methods for entrepreneurship. High-tech smart products are often expensive due to their high development costs, and the application of agricultural smart products has not yet become widespread. Especially for some rural places in China that have just completed their poverty alleviation, it is obviously impractical to emphasize the application of modern smart products. The survey shows that, although China has made considerable progress in the development of agricultural development, the era of "Internet + agriculture" has arrived. In rural areas of China, the development of agricultural intelligent products is still a long way from practical application, and the overall promotion is still slower, hindering the enthusiasm and creative realization of returning rural migrant workers.

3.3 The Achievements of Smart Tools and Facilities for Agricultural Development Have Limited the Creativity of Returning Rural Migrant Workers

Since the creation of smart tools and facilities, it has had an important impact on the production and life of human society. However, compared with the change of smart tools to urban life, the results of smart facility tools that have important effects on agricultural production are relatively small. The investment and use of these smart tools and facilities often require a large amount of capital. For returning rural migrant workers who are still in the low-cost initial stage, the lack of entrepreneurial funds is a common practical dilemma. Without funding guarantees, advanced agricultural intelligent tools and facilities cannot be obtained; without a consumer market, the development, development and promotion of agricultural intelligent facilities will be slower, and facility innovation will lag behind. Compared with the development of the entire information society, the development of agricultural intelligent tools and facilities is obviously insufficient. Even if returnee migrant workers have innovative ideas in the process of entrepreneurship, they may not be able to achieve it without corresponding intelligent tools and facilities.

3.4 Insufficient Coverage of Intelligent Platform Development for Agricultural Products Restricts the Continuity of Entrepreneurship for Returning Rural Migrant Workers

Although with the overall development of our society and economy, the road traffic in rural areas in China has improved significantly, compared with cities, there are still many shortcomings. With the extensive application of logistics information technology in the field of logistics, information technology, intelligent technology and corresponding new transportation technologies have been integrated [3]. The development and use of artificial intelligence platforms in the process of entrepreneurship for returning farmers must rely on the general development of supporting industries such as logistics. Otherwise, the transactions completed through the online sales platform cannot be realized, and the entrepreneurial results cannot be effectively converted into farmers' income. Therefore, in terms of the coverage of the current development of artificial intelligence, cities and towns have achieved full coverage, but for the vast rural areas, the network has not yet achieved full coverage, especially for remote areas with relatively backward economic development. Constrained by the uneven level of rural economic development, the use of agricultural intelligent information management platforms is greatly restricted, coupled with the lack of supporting systems such as agricultural product logistics, the realization of the value of returnee migrant workers' entrepreneurial results is still greatly affected.

4 The Countermeasures and Ideas of Returning Rural Migrant Workers in the Perspective of Intelligence

4.1 Access to More Entrepreneurial Information Resources and Entrepreneurship Training with the Help of Intelligent Network Platforms

Artificial intelligence platforms provide a wider space for knowledge transfer. Returning rural migrant workers have survived in the city for many years and accumulated some work experience. They have an urgent need and desire for entrepreneurial knowledge. In the age of information networks, the transfer and expression of knowledge can be achieved through a variety of swift channels. Such as small video of trembling sound, various live websites, etc., can realize the rapid transfer and popularization of entrepreneurial knowledge. As long as returning migrant workers understand basic network usage knowledge and can perform general operations on smartphones, they can click to play videos to obtain the latest entrepreneurial information, learn entrepreneurial knowledge, and improve entrepreneurial capabilities. Therefore, local government departments at all levels should strengthen the construction of rural-oriented intelligent network platforms, strengthen the training of farmers' Internet platforms, smart tools such as mobile phones, and basic network practical skills [4]. On the one hand, increase the information network resources of the intelligent platform, so that returnee entrepreneurs can learn something; on the other hand, increase capital investment to ensure the supply of public service resources with good information network material conditions. The timely and rapid information and abundant resources are the outstanding advantages of the agricultural intelligent network platform. Local governments should actively develop production skills training with local agricultural characteristics for the local population, and provide rich vocational training guarantees for returning rural migrant workers to improve their entrepreneurial capabilities. Government departments at all levels should increase investment and actively build intelligent platforms. Through the intelligent network, provide more information channels for returnee farmers to start their own businesses, stimulate innovation consciousness, provide entrepreneurial references, and exchange entrepreneurial experiences. Rural government departments should provide a full range of artificial intelligence support for returnee migrant workers to start their businesses by accurately positioning the information exchange function, regulating the information release channels and content [5].

4.2 Strengthening Agricultural Entrepreneurial Risk Response and Entrepreneurial System Planning Through Smart Tool Means

The advent of the "Internet + Agriculture" era has provided more options for the development of modern agriculture. Judging from the status quo of the development of modern agriculture in various countries around the world, artificial intelligence has possessed a full range of service functions from the production environment of agricultural products, production processes, and marketing of finished products. At present, the results of existing agricultural artificial intelligence have also fully demonstrated that intelligent tools can achieve various functional functions such as soil detection, pest and disease protection, yield prediction, and livestock and poultry disease early warning in the agricultural production field, which can effectively eliminate traditional agricultural development. Various problems occurred in the process. Through the collection of information from artificial intelligence tools, and the use of data processing and information feedback from artificial intelligence platforms, returnee entrepreneurs in the planting industry can find potential pest and disease crisis in time, and can get immediate feedback on the types, cycles and countermeasures of the crisis. In order to carry out targeted timely treatment, the loss of natural disaster risks such as pests and diseases of traditional agriculture is reduced to a minimum. The application of agricultural artificial intelligence tools to the planning of entrepreneurial systems can achieve the full-scale control of agricultural product production, ensure the orderly progress of the production process of the product, ensure the quality of the product, and thus achieve the effective supply of consumer demand for agricultural products. For example, with the help of artificial intelligence, returnee entrepreneurs can analyze and organize product information data through intelligent information data processing equipment, scientifically predict the market conditions of agricultural products, adjust entrepreneurial projects and entrepreneurial goals in time, and ultimately achieve the optimal results of entrepreneurship.

4.3 Using Intelligent Facility Management to Improve the Development of Agricultural Entrepreneurship Industry and Production Technology Management

Agricultural informatization and intelligence are the objective requirements of China's agricultural modernization construction laws [6]. The application of artificial intelligence facilities to product design and process has become one of the important characteristics of modern industrial enterprises. The application of artificial intelligence facilities to the production and control of agricultural products is an important measure to promote the success of farmers' entrepreneurship and improve the quality of agricultural products. On the basis of information network conditions in the vast rural areas of China, modern entrepreneurial ideas and business ideas must be combined with modern science and technology in order to play a more active role in promoting China's agricultural development and increasing farmers' income. In the process of agricultural product production and operation and technical management, the full use of artificial intelligence facilities for management and control will realize the return of ordinary returning rural migrant workers to start a business from traditional manual labor to technical brain activities. The comprehensive use of modern business management concepts and high-tech artificial intelligence facilities can effectively control various production and operation activities of agricultural entrepreneurial enterprises. In the application process of intelligent facilities, not only can it increase the productivity of agricultural entrepreneurial activities, improve the quality and quantity of products, and increase the income of returning entrepreneurial farmers; it can also form the basis of scientific raw data. Through systematic analysis, it can timely discover agricultural product production Problems in the process facilitate entrepreneurs to take immediate and effective response measures. For example, government agencies can arrange specialized agencies to use a large amount of satellite image data related to agriculture to analyze the relationship between climate change and crop growth in the region, so as to make accurate predictions on the output of crops involved in entrepreneurial products. The choice of entrepreneurial goals for returnee entrepreneurs provides positive guidance and response.

4.4 Using Intelligent Platforms to Strengthen Sales Management and Channel Expansion of Agricultural Entrepreneurship Products

Due to the prominent advantages of big data and artificial intelligence in information processing, the application of intelligent platforms to economic operations and management has become an important feature of modern economic society and has had a positive impact. The most important thing for returning rural migrant workers to carry out entrepreneurial activities is the realization of the economic value of entrepreneurial products. The intelligent and shared management system for agricultural and sideline product logistics turnover boxes is a positive performance of the development of artificial intelligence in the process of realizing the value of agricultural entrepreneurial products [7]. We should use intelligent information management platform to provide technical support for the sales management and channel expansion of agricultural entrepreneurial products. For example, through intelligent management methods, intelligent wearable products are used in animal husbandry production, and individual information of livestock and poultry is collected in real time, and the growth of livestock is analyzed through artificial intelligence platforms, which can analyze the health of livestock and detect estrus. In order to take active and effective response measures. Another example is to build an intelligent sales platform to show the production process of entrepreneurial agricultural products. For example, the control of artificial intelligence facilities is applied to livestock breeding, so that consumers can understand information about livestock feeding status, location services, and breeding environment. On one hand, it provides consumers with an opportunity to learn about goods or services and attracts consumer interest. On the other hand, it can effectively expand the social influence of agricultural products, increase its brand value and social effects, and successfully realize the economic value of farmers' entrepreneurship. By embedding the farmer's platform into artificial intelligence and collecting user consumption data, intelligent planning is provided for returning rural migrant workers to start businesses. In addition, in order to make up for the shortcomings of traditional logistics, it is also necessary to speed up the construction of an intelligent platform for rural logistics and increase the sales of agricultural products. Based on the intelligent platform, integrating GIS into agricultural logistics can effectively improve the quality and quality of agricultural logistics [8].

5 Conclusion

With the continuous advancement of human society in science and technology, information technology and artificial intelligence facilities have been widely used in people's lives, which have also had a profound impact on industrial and agricultural production. The analysis of the actual dilemma faced by returning rural migrant workers based on the perspective of artificial intelligence is an important content to improve the level of artificial intelligence of farmers in China, an important measure to realize the modernization of rural society in China, and an important tool to improve farmers 'professional skills and increase farmers' entrepreneurial income. Artificial intelligence is widely used in returning rural migrant workers to start businesses, and will definitely promote the development of rural society and effectively promote the realization of the strategic goals of rural revitalization.

Acknowledgements. This work is supported by the key project of the 2016 Jiangsu Social Science Fund Project "Research on the Employment Capacity Enhancement Path of Jiangsu New Generation Migrant Workers Based on Supply-Side Structural Reform" (Project Number: 16SHA003).

References

1. Weili, T.: Internet-based agricultural logistics development strategy. Agric. Eng. **9**, 156–158 (2019)
2. Futing, W., Li, R.: Application of intelligent building concept in temperature control system of rural houses. J. Liaoning Inst. Sci. Technol. **20**, 10–12 (2018)
3. Genlong, C.: Research on the application of logistics information technology in the development of agricultural logistics. Natl. Circ. Econ. (028), 12–14 (2019)
4. Ministry of Agriculture held farmer mobile phone application skills training week. Agricultural Broadcasting School released training materials and smart mobile phone app. Sci. Technol. Train. Farmers, 55 (2017)
5. Yueyong, D.: Discussion on ways to enhance the information exchange function of the new rural website of Wancun Networking Project. Zhejiang Agric. Sci. 943–945 (2014)
6. Yueyong, D.: Discussion on the development of agricultural informationization in Zhejiang under the synchronization of the "four modernizations". J. Zhejiang Agric. Sci. **25**, 1435–1443 (2013)
7. Jiachen, Y., Xiao, Y.: Common system of intelligent turnover box for agricultural and sideline products. Sci. Technol. Vis. (014), 27–28, 78 (2018)
8. Xuemei, C.: The application of GIS in agricultural logistics. Hebei Enterp. (1), 89–90 (2020)

Improvement of College Teachers' Teaching Ability Under the Background of the Development of Artificial Intelligence Platform

Jing Su(✉)

Department of Human Resources, Jiangsu University, Zhenjiang, Jiangsu, China
cheer_su@126.com

Abstract. With the advent of the "Internet + Education" era, educational artificial intelligence platforms are playing an increasingly important role in higher education. It provides rich learning resources and convenient learning methods for college students, and is generally welcomed by college students. As an important element of higher education talent training, college teachers are facing a severe impact from educational artificial intelligence platforms. Facing the development of artificial intelligence platforms, college teachers must take proactive measures to continuously enhance their informationization level, improve the teaching ability of using artificial intelligence platforms, and cultivate outstanding talents for society. Based on the realistic background of the development of educational artificial intelligence platform, this article will analyze the problems existing in the course teaching of the era of artificial intelligence in China's college teachers from the perspective of supply side, and seek solutions to the problems.

Keywords: Artificial intelligence platform · College teachers · Improvement of teaching ability · Career development

1 Introduction

The key to competition between countries is the competition for talents, and higher education is the most important link in training excellent and high-quality talents. In the context of the "Internet +" era of the development of artificial intelligence, how teachers in universities use the educational artificial intelligence platform to effectively carry out teaching and improve the quality of higher education talent training is a key link for the implementation of national artificial intelligence strategies [1]. How to make new technology better serve education and teaching so that students have greater gains, the teaching ability of teachers is particularly important. The rapid development of technologies such as artificial intelligence, big data, the internet of things, and blockchains have created a great opportunity for the unprecedented development of artificial intelligence education, and has trained more and more innovative teachers with high professional quality and technical capabilities [2]. However, we should also see that there are still serious shortcomings in terms of the actual demand for

J. MacIntyre et al. (Eds.): SPIoT 2020, AISC 1282, pp. 102–109, 2021.
https://doi.org/10.1007/978-3-030-62743-0_15

high-quality talents in China's social development. Therefore, based on the development of artificial intelligence, it is of positive theoretical value and practical significance to deeply analyze the problem of improving the teaching ability of college teachers.

2 The Background of the Development of Higher Education Artificial Intelligence Platform

At present, human society is in the era of the new industrial revolution, and the development of educational artificial intelligence is one of the important topics in the field of educational technology in various countries. With the advent of the "Internet + Education" era, the role of artificial intelligence platforms in the training of higher education talents has drawn increasing attention from all walks of life. The development of artificial intelligence platforms in the field of higher education has greatly promoted teaching reform. It has broken the disadvantages of traditional teaching methods and effectively alleviated the pressure on teaching due to individual differences among college students [3]. Every student can receive a good education, effectively alleviating various disadvantages brought by traditional teaching methods [4], and improving students' satisfaction with teaching work. Artificial intelligence has become one of the most important currents in the development of science and technology in the world by virtue of its strong learning ability, rational judgment and super strong work force [5]. For the development of higher education, the development of educational artificial intelligence has become the important constituent elements of the teaching ecological structure of colleges and universities are restructuring the traditional teaching ecological structure with its unique operating mechanism. On the one hand, it provides new possibilities for college teaching, and on the other hand, it brings teachers' teaching ability. Serious challenges. Continuously improving the level of artificial intelligence teaching is an inevitable choice for the professional development of college teachers in the new era.

3 The Practical Dilemma of Artificial Intelligence Platform Teaching for College Teachers

3.1 Incomplete Hardware Facilities of the Information Network Platform Has Weakened the Enthusiasm of the Artificial Intelligence Platform for College Teachers

The development of modern computer technology has promoted the emergence and development of artificial intelligence to a certain extent. The progress of society and the development of science and technology have made artificial intelligence applied in more and more fields [6]. However, at present, there is still a lot of room for improvement in the use of artificial intelligence platforms in universities in China. The basic conditions of the information network of different schools are significantly different. The multimedia facilities in many classrooms are not configured uniformly, and the stability of the campus network is not sufficient to guarantee the basic operation of teaching. Teachers cannot use the Internet effectively in the classroom teaching process, and still retain the traditional teaching mode. In addition, the slow update of classroom computer software facilities is also an important reason affecting the use of artificial intelligence platforms for teachers.

3.2 The Content of the Artificial Intelligence Platform Is Not Systematically Affected the Use Planning of the Artificial Intelligence Platform for College Teachers

The teaching resources of the existing artificial intelligence platform are complicated, the system is unclear, and the teachers are dazzled when choosing, which seriously affects the use planning of college teachers. Because there is no systematic guidance, teachers lack the ability and level to screen resources, and they have a fear of massive resources. In order to avoid trouble, teachers are reluctant to use artificial intelligence platforms in the courses they teach, and prefer the traditional teaching model. In fact, the traditional indoctrination theory teaching model can not meet the teaching needs of artificial intelligence courses [7], which affects the practical advancement of the use of educational artificial intelligence platforms.

3.3 Inadequate Training on the Use of Artificial Intelligence Platforms Hinders the Use of Artificial Intelligence Platforms by College Teachers

The use of artificial intelligence platforms requires systematic training. Although existing educational artificial intelligence platforms are not complicated to use, not every teacher has a sensitive ability to accept innovative technologies. Because the use of artificial intelligence platform organized by the school is often a unified centralized training, there is no mandatory requirement for participation, and there is no corresponding assessment test at the end of the training. In fact, after only one or two trainings, it is difficult for teachers to master the system usage of artificial intelligence platform. After the introduction of the artificial intelligence platform, the actual use effect also lacks corresponding supervision and management, which is not optimistic.

3.4 Insufficient Assessment and Evaluation of Artificial Intelligence Platform Reduces the Use Effect of Artificial Intelligence Platform for College Teachers

The practical effect of the use of artificial intelligence platforms by college teachers in teaching is the decisive factor that influences teachers' decision on whether to use the platform for a long time. Whether the artificial intelligence platform can effectively achieve the assessment of students is the basic function of the platform. Although the existing artificial intelligence platforms take the assessment function of learners as one of the important functions of the platform, due to the existence of many system loopholes and defects, relying on artificial intelligence platforms alone cannot actually reflect the learning situation of students comprehensively and objectively.

3.5 Incomplete Use of Artificial Intelligence Platforms Deprives College Teachers of Their Willingness to Use Artificial Intelligence Platforms

At present, the assessment of college teachers is mainly concentrated on the two aspects of scientific research and teaching. In fact, most colleges and universities only focus on the assessment of scientific research work. Teachers who teach wholeheartedly do not have an advantage in assessment. The use of educational artificial intelligence platforms for teaching has greatly increased the invisible workload, but it is difficult to have an accurate calculation tool or performance evaluation mechanism. The imperfect performance system for the use of artificial intelligence platforms in college education has seriously reduced teachers' willingness to use artificial intelligence platforms.

4 Countermeasures and Ideas to Improve the Teaching Ability of Artificial Intelligence Platform for College Teachers

4.1 Focus on the Construction of Information Network Platform Hardware Facilities to Stimulate the Enthusiasm of Artificial Intelligence Platform for College Teachers

Strengthening education investment and focusing on the construction of hardware facilities of university information network platforms are the premise and basis for motivating teachers to use artificial intelligence platforms, and also an important guarantee for promoting education equity [8]. The biggest advantage of artificial intelligence education platform is data processing and resource sharing. The local universities in China are very different. They develop personalized artificial intelligence platforms and intelligent teaching products suitable for different colleges and universities. They can use the advanced data processing functions of the artificial intelligence platform to detect teaching effects and stimulate the use of artificial intelligence platforms for teachers. And it will enthuse to improve the teaching level of teachers. Creating a "professional + artificial intelligence" teaching teacher team is an important measure to improve the teaching ability of college teachers in the new era [9].

4.2 Enriching the Content of the Artificial Intelligence Platform System to Build and Improve the Use Plan of the Artificial Intelligence Platform for College Teachers

The construction of a rich artificial intelligence platform's resource system is an important factor in rationally formulating a platform usage plan. Whether there is a complete curriculum resource system is an important basis for teachers to choose an artificial intelligence platform. Education authorities at all levels should increase the systematic guidance and construction of higher education network resources. Teachers can take advantage of the perfect content of the artificial intelligence platform to continuously enrich the teaching content, and finally form a high-quality curriculum system, and realize inter-school sharing through the Internet, and ultimately achieve the accumulation and improvement of national high-quality higher education resources. After long-term development, it is inevitable for college teachers to formulate a systematic and comprehensive artificial intelligence platform use plan.

4.3 Strengthen the Use of Artificial Intelligence Platforms in Education and Training Increase the Use of Artificial Intelligence Platforms by College Teachers

Institutions of higher learning should conduct systematic teaching and training on the introduction of artificial intelligence teaching platforms, and increase the assessment of training, including not only the assessment of the training process, but also the assessment of practical applications. The training centers used in artificial intelligence platforms can take a variety of channels and methods. For example, special training instructors for training, school teacher demonstration training, etc. are all better training methods. The introduction of artificial intelligence education platforms in schools should maximize the role of the platform and apply it to teaching practice. By training teachers on the use of artificial intelligence platform skills, they can not only ensure that teachers are proficient in the process of actually applying artificial intelligence teaching platforms, effectively improve the actual use of teaching, but also effectively improve the teachers' artificial intelligence platform teaching capabilities.

4.4 Improve the Evaluation and Evaluation System of Artificial Intelligence Platform Pay Attention to the Use of Artificial Intelligence Platform by College Teachers

The assessment of students is an important criterion for achieving teaching goals. The key to the successful application of artificial intelligence platform in teaching lies in the platform's evaluation and assessment of student learning effects. As users of the platform, college teachers must systematically master the platform's assessment and evaluation functions, actively and effectively evaluate and evaluate course students, motivate outstanding students, and urge late students. Through the assessment of the artificial intelligence platform, teachers can also test the effectiveness of the course teaching in a timely manner, which is convenient for teachers to adjust teaching strategies in a timely manner and adjust the use of the platform. To improve the assessment and evaluation system of artificial intelligence platforms for students, we must fully consider the actual situation of the use of information networks. We must not only urge students to make full use of artificial intelligence platforms for learning, but also give full play to the assessment and evaluation functions of artificial intelligence platforms.

4.5 Improve the Use of Artificial Intelligence Platform Performance Methods to Enhance the Willingness of College Teachers to Use the Artificial Intelligence Platform

The use of artificial intelligence platforms poses serious challenges to teachers' professional knowledge and business capabilities, which is beyond doubt. Teachers using artificial intelligence platforms for teaching have increased the workload of class preparation, and are no longer limited to 45 min of classroom teaching. Therefore, colleges and universities should systematically consider when measuring and evaluating teachers' teaching workload. Under the realistic background that artificial intelligence platforms are widely used in teaching, colleges and universities should build a comprehensive evaluation mechanism for teaching workload, especially improve the performance evaluation of teachers 'use of artificial intelligence platforms for teaching, affirm teachers' labor and dedication, and increase teachers. A sense of honor and happiness, enhance the willingness to use artificial intelligence platforms for teaching, and effectively improve the teaching ability of teachers' artificial intelligence platforms.

5 Misunderstandings that Universities Should Avoid Using Artificial Intelligence Platforms for Teaching

5.1 Avoid the "Universal Theory" of Artificial Intelligence Platform for College Education

The advent of the "Internet +" era has indeed brought about drastic changes in our lives and brought about a revolution in higher education. But we should clearly realize that the role of artificial intelligence tools is not everything. Talent training in higher education is a comprehensive education for students to master theoretical knowledge and improve their practical ability. Teachers in colleges and universities must systematically construct teaching processes from the aspects of curriculum system construction, case teaching, independent learning ability training, and strengthening school-enterprise cooperation. Artificial intelligence platforms can do it. In fact, even in the era of artificial intelligence, the goal of China's higher education talent training is still to cultivate students 'imagination, stimulate students' creativity, and guide students to establish a correct outlook on the world and life. The widespread use of artificial intelligence platforms in higher education has brought opportunities to college teachers and severe challenges at the same time. College teachers need to improve their ability to use artificial intelligence platforms and artificial intelligence tools, and then improve their teaching capabilities on this basis, instead of completely deviating from the original intention of higher education talent training. In the long run, the status of teachers in higher education cannot be completely replaced by artificial intelligence, and the class teaching system will continue to exist for a long period of time [10].

5.2 Ignore Restructuring of Teacher-Student Relationship in Higher Education Talent Training

The era of artificial intelligence not only requires and provides higher education, but the basic process is to use the rich teaching resources provided by the artificial intelligence platform to achieve personalized and customized teaching. The deep integration of artificial intelligence and education has become an important trend of future education reform, and the traditional teaching ecology will be restructured under the influence of artificial intelligence [11]. However, it cannot be ignored that in the

"Internet + education" era, the teacher-student relationship in the training of higher education talents has not been eliminated, but has become more important on the contrary. Teaching sensitivity is the ability and quality of teachers to perceive, perceive and respond to the factors and changes involved in teaching in the classroom teaching process. It is an important part of teachers' teaching ability. In particular, the teaching objects faced by college teachers are adult citizens, and the reconstruction of teacher-student relationships in the application of artificial intelligence platforms is particularly important. We must attach importance to the recognition of the role of teachers and students in the application of artificial intelligence tools, the interaction of professional abilities, the endogenous ability of reflection and the protection of the practical environment [12], and cultivate healthy, orderly, and positive university-teacher relationship.

6 Conclusion

With the advent of the "Internet + Education" era, the role of artificial intelligence platforms in higher education talent training will become increasingly important. The development of artificial intelligence provides unprecedented opportunities for higher education reform, and has positive significance in terms of enriching teaching content, improving teaching methods, and improving teaching efficiency [13]. The application of artificial intelligence platforms to education has changed the learning environment, enriched learning content, and initiated the educational revolution. For college students born in the new century, they pay more attention to the application of teachers' intelligent teaching methods. College teachers must adapt to the advancement of science and technology and the requirements of the times, and constantly improve their own artificial intelligence teaching level. At the same time, they should also pay attention to avoid misunderstandings in the use of artificial intelligence education platforms and cultivate more excellent high-quality talents for society.

Acknowledgments. This work was financially supported by the fund of the 2019 Jiangsu Higher Education Teaching Reform Research Project "Theoretical and Practical Research on the Construction of a Hybrid "Gold Course" for Law Majors Based on the Outstanding Rule of Law Education Training Plan 2.0" (Project Number: 2019JSJG324) and Jiangsu University's Research Project on Higher Education Teaching Reform in 2019 "Research on the theory and practice of the construction of a hybrid" gold course "online and offline for law majors based on the Outstanding Rule of Law Education Training Plan 2.0" (Project Number: 2019JGZZ009).

References

1. Xiaoming, W., Zengxi, H.: Exploration of artificial intelligence teaching in local universities. Fujian Comput. **35**, 150–153 (2019)
2. Xiaoting, L., Xu, F.: Investigation and research on the status quo of the application of artificial intelligence teaching in college teachers. China Educ. Inf. **5**, 78–81 (2019)
3. Ge, S., Liu, Y.: The impact of artificial intelligence on teaching methods. Mod. Educ. **6**, 195–196 (2019)

4. Shasha, Z.: Paths to improve the teaching ability of college teachers under the background of artificial intelligence. Contemp. Educ. Pract. Teach. Res. **6**, 39–40 (2020)
5. Zeyuan, Y., Jinghua, Z.: Teaching reconstruction from the perspective of artificial intelligence. Res. Mod. Distance Educ. **31**, 37–46 (2019)
6. Yanfang, L.: Research on college teaching strategies under the background of artificial intelligence. Comput. Prod. Distrib., 166 (2019)
7. Yunliang, L.: Thoughts on teaching reform of artificial intelligence course based on innovative teaching concept. World Digit. Commun. **6**, 247 (2019)
8. Zimin, W., Huang, Z.: Research on promoting artificial intelligence to promote education and teaching reform. Prod. Res. **3**, 128–131 (2019)
9. Tingting, L., Zhiguo, S., Yanli, J.: Analysis on the hotspots and trends of domestic artificial intelligence teaching research. High. Educ. Sci. **06**, 98–107 (2019)
10. Haiyan, Z.: Classroom teaching in the age of artificial intelligence. Sci. Educ. Wenhui (Late Issue) **2**, 58–59 (2019)
11. Zelin, L., Juan, Y.: Reconstruction of school teaching ecology in the age of artificial intelligence. Curric. Teach. Mater. Teach. Method **39**, 34–41 (2019)
12. Yue, L., Zhang, M.: Teachers' teaching sensitivity and generation path in the age of artificial intelligence. J. Bingtuan Educ. Inst. **29,** 45–48 (2019)
13. Xiaoxu, H.: Analysis on the limits and paths of teaching reform in the age of artificial intelligence. Educ. Watch. **8**, 97–98 (2019)

Influence Factors of Using Modern Teaching Technology in the Classroom of Junior Middle School Teachers Under the Background of Artificial Intelligence—Analysis Based on HLM

Yemei Wei(✉)

School of Teacher Education, Nanjing Xiaozhuang University, Nanjing 211171, China
wym0721@163.com

Abstract. The application of AI in classroom teaching not only creates a vivid simulation situation, but also improves students' direct "on-site" experience through human-computer interaction, which is helpful for students' thinking development and deep understanding. But at present, the ability of teachers to use information technology is not satisfactory. Based on the data of CEPS survey, we make a study that uses an analysis of the HLM in two levels to explore factors which influence junior teachers' use of modern education technology. The research based on the result of a questionnaire of 797 teachers. The result firstly shows that there is a strong relationship between a week's workload, the pressure of school management, job title, class performance and teachers' use of modern education technology. Secondly, we find at the school level the better school hardware condition is, the more likely they are to use modern education technology. There is no significant relationship between learning atmosphere and teachers' use of modern education technology. Thirdly, at the perspective of the regulation of cross level, school hardware condition promotes the relationship between the teachers' a week's workload and teachers' use of modern education technology.

Keywords: Junior school teachers · Modern education technology · HLM

1 Introduction

The application of AI in teaching changes and innovates the traditional teaching mode. It's no longer just learning knowledge in the classroom, but also learning knowledge online, such as one to one online teaching. Although AI can comprehensively analyze learners' knowledge accumulation and intelligence development, it challenges teachers' educational technology literacy. Since the new century, educational informatization has been one of the key points in China's education work. The government has made great efforts to promote the construction of educational informatization through the standardization construction of compulsory education schools, the improvement of weak schools and other projects. From "School-school Network Project" to "flipped

J. MacIntyre et al. (Eds.): SPIoT 2020, AISC 1282, pp. 110–118, 2021.
https://doi.org/10.1007/978-3-030-62743-0_16

classroom", a new teaching mode based on mobile Internet, and "E-Book Package" which provides the functions of students' growth history management, home school communication, digital education resources, etc., all these promote the development of education information in China in depth. However, some scholars point out that not all primary and secondary school teachers are willing to use modern teaching media, many teachers reject or even refuse to use modern teaching media in the classroom, and the frequency of using modern teaching media in the classroom is not high [1]. What factors need to be considered to affect the frequency of teachers using modern teaching media in the classroom? This paper attempts to use the data of China Education tracking survey (CEPs) and hierarchical linear regression model to explore the influencing factors of the use of modern teaching media in the classroom of junior middle school teachers in China.

2 Literature Review

The Office of Technology Assessment (OTA 1995) published a study, which showed that American teachers encountered certain obstacles in using educational technology, including time, cost of acquiring skills, external support and training [2]. Blankenship (1998) studied teachers in different grades in rural areas of the United States, and found that teachers' own age, gender, attitude towards the use of new technology and external school support and other factors significantly affected teachers' use of new technology in the classroom [3]. Jaber (1999) focuses on the correlation between the frequency of teachers using computers for teaching and teachers' time, external training and access to equipment [4]. Although different foreign students find that the influencing factors of the frequency of using modern teaching media in classroom are not the same, they are basically considered from the two aspects of teachers themselves and the external environment, and the research methods are mainly quantitative statistical analysis. In China, there are also studies on related topics, such as research has found that the nature of subjects, position factors, educational ideas, expectations and personality traits all affect the frequency of primary and secondary school teachers using new media [5]. Through a questionnaire survey, Zhao Yu (2002) found that the factors that affect the use of modern education media by middle school teachers are: lack of software, bad media environment, no hardware, poor hardware quality, poor software quality and poor management [6]. Liu Yan (2002) found that college teachers have realized the importance of modern teaching media to college teaching, but the overall level of using modern teaching media is not very high, and there are differences in the frequency of using modern teaching media with different professional backgrounds and different abilities [7].

Compared with foreign countries, most of the related researches in China are about teachers in a certain region, a certain province, and lack of empirical research on the use of teachers' classroom teaching media nationwide. In addition, when analyzing the influence of various influencing factors on dependent variables, the existing researches usually put the influencing factors of different levels into one layer for consideration, which is easy to produce problems such as overall deviation, error of parameter estimation accuracy and heterogeneity of regression [8]. Based on this, this study attempts to use the data of CEPs, using software HLM to carry out multi-level linear regression to explore the influencing factors of the frequency of using modern technology for junior middle school teachers.

3 Methodology

3.1 Data Source

The data of this study comes from the junior high school stage of CEPS (2013–2014). The sampling method of the survey is stratified, multi-stage, probability and scale proportional sampling. Based on schools, 112 schools and 448 classes are selected nationwide, with a total of 1752 teachers. The author deleted any sample with missing research variables, the number of teachers adopted in this study is 797.

3.2 Variable Manipulation

Dependent Variable: Frequency of modern teaching media used by junior middle school teachers. The question title is "do you use the following teaching media when teaching in the investigated class?" Teaching media can be divided into four aspects. The teacher's alternative answers are: never, occasionally, sometimes, often and always.

Independent Variables at the Teachers' Individual Level: (1) Assessment pressure: including administrative measures, enrollment rate and class performance, which are divided into five levels from none to very large, with 1–5 points assigned successively, as a continuous variable; (2) Teaching age as a continuous variable into regression analysis; (3) Class performance as a continuous variable; (4) Education background as a binary variable virtualization processing. The code of junior college and below is 0, and the code of undergraduate and above is 1; (5) Professional Title: senior professional title and non senior professional title. The code of non senior professional title is 0, and the code of senior professional title is 1. As a binary variable virtualization processing.

Independent Variables at School Level: (1) School hardware conditions: From very dissatisfied to very satisfied, it is divided into five levels, with 1–5 points assigned successively; (2) School learning atmosphere: From very bad to very good, it can be divided into five levels, with 1 to 5 points assigned in turn. The results show that the VIF values of all the explanatory variables are less than 10, indicating that the independent variables of the sample data do not have multicollinearity.

3.3 Research Model

Because the frequency of teachers' using modern teaching media is influenced by two levels of factors, the individual level variables are nested in the school level variables. HLM model establishes the regression equation by layering the independent variables, sets the intercept and slope of the individual level regression equation as the function of the group level variables, so as to associate the data of different levels, and is suitable for processing the layering nested data.

4 Analysis Results

4.1 Zero Model

We use the zero model of HLM to confirm whether the frequency of modern teaching media used by junior middle school teachers will be different due to different schools. Establish the following model:

Layer one equation: Yij = β 0j + Rij; Layer two equation: β 0j = ϒ 00 + u0j.

In the above two equations, Yij is the frequency of modern teaching media used by the i teacher of J school, β0j is the frequency of modern teaching media used by the teachers of J school, Rij represents the difference between the average frequency of modern teaching media used by the I teacher of J school and the school, which is the random error at the individual level; ϒ00 is the total average frequency of modern teaching media used by the middle school teachers of each school, and u0j is the total frequency of modern teaching media used by the j school and the school The difference between the mean value and the mean value is the random error at the school level. According to the results of the empty model, we can see the proportion of the total variance of the dependent variable is caused by the difference of the second level units through calculating the cross-level correlation coefficient $\rho = \tau_{00}/(\sigma^2 + \tau_{00})$. According to the two levels of random effects, the correlation coefficient ICC (1) = 1.87/(1.87 + 6.05) = 23.61% was calculated, which indicated that 23.61% of the total variation of the frequency of modern teaching media used by junior high school teachers came from the difference between schools. According to the standard proposed by Cohen, ICC (1) >0.138 is highly correlated, which shows that the analysis of teachers' turnover intention is very suitable for HLM analysis.

4.2 Random Coefficient Regression Model

The random coefficient regression model is based on the zero model, and the first level equation includes the variables at the teacher level, including the number of hours per week, school assessment pressure, education background, professional title and class performance, to investigate the impact of individual factors on the use frequency of modern teaching media.

$$Y_{ij} = \beta_{0j} + \beta_{1j}X_{i1j} + \beta_{2j}X_{i2j} + \ldots\ldots \beta_{Qj}X_{iQj} + r_{ij}$$

In the above formula, β1J, β2J……β QJ are independent variables X1, X2 Partial regression coefficient of the influence of XQ on Yij. According to the results of Table 1, the number of hours per week, the pressure of school assessment, professional title and class performance have significant predictive effect on the dependent variables, while the academic degree has no significant effect on the frequency of using modern teaching media.

4.3 Intercept Model

In the second level equation of the zero model, the school level variables are included, including school hardware conditions and learning styles. The influence of school level factors on Teachers' frequency of using modern teaching media is investigated.

$$\text{Yij} = \text{r00} + \beta\text{01Xi} + \beta\text{02X2} + \text{u0j} + \text{rij}$$

In the above formula, β01 and β02 are the partial regression coefficients of school hardware conditions and learning styles on Yij. It can be seen from Table 1 that under the standard of significance level 0.1, the school hardware conditions have a significant prediction effect on the frequency of modern teaching media, and the school learning atmosphere has no significant effect on the frequency of teachers' use of modern teaching media. The β coefficient of school hardware condition is positive, indicating that the better the school hardware condition is, the higher the frequency of teachers using modern media in class.

4.4 Slope Prediction Model

We need continue to study whether school level independent variables can regulate the relationship between individual level independent variables and dependent variables, so we get the slope prediction model, also known as the whole model. Through the analysis of the results of the previous model, we have found that the number of hours per week, the pressure of school assessment and the performance of the class have a significant effect on the use frequency of modern teaching media. In the whole model, we explore the moderating effect of school level variables on the relationship between these three independent variables and dependent variables. In order to prevent the collinearity problem of independent variables, we should focus on the number of hours per week, the pressure of school assessment and the class performance.

4.5 Data Analysis

Table 1. Analysis of factors influencing the frequency of classroom use of modern teaching media

Variable	Zero model	Random coefficient regression model	Intercept model	Slope prediction model
Teacher level	Fixed effect			
Hours per wee		−0.042*		−0.028
School assessment pressure		0.171**		0.151**
Education background		0.358		0.399
Title		−0.452**		−0.442**
Class grade		0.175***		0.176***
Hours per week *learning atmosphere				−0.145 **
Hours per week *school hardware condition				0.140**
School assessment pressure * learning atmosphere				−0.097
School assessment pressure * school hardware condition				0.103
Class grade * learning atmosphere				−0.072
Class grade * school hardware condition				−0.011
School level				
School hardware condition			1.130***	1.058***
Learning atmosphere			0.063	0.136
Random effects				
$\sigma 2$	6.05	5.04	6.06	4.98

(notes: *P < 0.1; **P < 0.05; ***P < 0.01)

It can be seen from Table 1, school assessment pressure, professional title and class performance still have a significant predictive effect on the frequency of teachers' use of modern teaching media. At the same time, in the cross-level interaction, the interaction items of weekly hours and school hardware conditions, weekly hours and school atmosphere have significant influence on the dependent variables. The coefficient of cross level interaction between weekly hours and school hardware conditions is 0.140 (SIG. = 0.01 < 0.05), indicating that school hardware conditions play a significant positive role in regulating the number of weekly hours and the frequency of teachers' use of modern teaching media. The coefficient of the cross-level interaction between the number of hours per week and the school learning atmosphere is −0.145 (SIG. = 0.013 < 0.05), which shows that the school learning atmosphere plays a significant negative role in regulating the number of hours per week and the frequency of using modern teaching media. There is no significant interaction between school assessment pressure, class performance and school level variables.

5 Conclusion and Discussion

5.1 Research Conclusion

On the individual level of teachers, the number of hours per week, the pressure of school assessment, professional title and class performance have significant predictive effect on the dependent variables. The heavier the workload of teachers every week, the lower the frequency of using modern teaching media in the classroom; the greater the assessment pressure teachers feel, the higher the frequency of using modern media teaching; the frequency of using modern teaching media for teachers with senior titles is significantly lower than that of teachers with non-senior titles; the class performance and the frequency of using modern media teaching are significantly positive correlation.

On the school level, the better the school hardware conditions are, the higher the frequency of teachers using modern media in the classroom; school learning atmosphere has no significant effect on the frequency of teachers using modern teaching media.

In terms of cross-level adjustment, school hardware conditions play a significant positive role in regulating the number of hours per week and the frequency of using modern teaching media; school learning atmosphere plays a significant negative role in regulating the number of hours per week and the frequency of using modern teaching media.

5.2 Discussion and Suggestions

The research of Oxford University in 2013 shows that in the future society, teachers are one of the most unlikely occupations to be replaced by AI, but the information technology literacy of teachers is facing a huge challenge. Teachers should improve their ability to use technology, especially for their own use of intelligent technology.

Pay attention to the influence of teachers' own factors on the use of modern intelligent technology, and change passive use to rational use. As some scholars have pointed out, "it's really worrying to examine the current situation of the selection of teaching media. One of the most prominent problems exposed is the lack of attention to the factors of teachers' own choice". The research shows that professional titles, class hours, assessment pressure and class performance have a significant impact on Teachers' use of modern teaching media in the classroom. These factors provide more focus for us to improve the frequency of teachers' use of modern teaching media. However, it is worth noting that many junior middle school teachers are still passive in using new media and lack the motivation to actively use modern teaching media to improve their teaching level. Education administrative departments and schools should set up correct ideas through national training, showing the effect of modern teaching media, so as to stimulate teachers' internal drive to use modern teaching media in the classroom.

Strengthen the construction of hardware and software of modern education technology and create an atmosphere to encourage the use of modern teaching technology. Although the country has accelerated the construction of educational information facilities in basic education schools in recent years, many schools, especially in rural areas in backward areas, still lack of modern educational technology and equipment, and the whole school has only one or two multimedia classrooms, which is obviously not conducive to teachers' use of modern teaching media in the classroom. Schools should set up special funds to purchase modern educational technology and equipment, increase the number of modern teaching media to improve the teaching environment, and strive to achieve complete supporting facilities. We should also maintain and manage the existing hardware equipment, and regularly maintain and moderately repair the easily damaged conventional teaching equipment. In addition to hardware, we also need to pay attention to the development of school educational technology software resources, such as various application platforms, network curriculum resources, information management system, etc. The school can organize more modern educational technology teaching competitions, give certain rewards to teachers who have outstanding performance in using modern teaching media, and create an atmosphere to encourage the use of modern teaching media.

The school actively carries out school-based training to improve the transfer ability of teachers' Educational Technology Application. In recent years, various units have organized various forms of teacher information technology training, which has achieved some positive results, but the disconnection between learning and using has always been one of the outstanding problems. It is difficult for teachers to use the information technology knowledge and ability in the real classroom teaching situation, and the transfer ability of teachers' application of technology needs to be improved. Due to the large number of teachers in basic education and the uneven knowledge and ability of teachers' educational technology, school-based trainer is one of the most economical and feasible ways to improve the application and transfer ability of teachers' educational technology. The school can invite experts of modern educational technology to enter the school. According to the difference of each teacher's educational technology knowledge level, the school can formulate different levels of training contents, such as basic knowledge literacy, improving the use of teachers' teaching software, educational technology and subject knowledge integration.

Acknowledgments. The Project of Philosophy and Social Science Research in Colleges and Universities in Jiangsu Province "Research on the performance mode and influence mechanism of primary and secondary school teachers' participation in school governance"(2019SJA0425).

References

1. Yu, Z.: Research on the current situation and countermeasures of Gansu normal school teachers using modern education media. e-Educ. Res. (3), 73–80 (2002)
2. Novaes, B.S.: Teachers and technology: making the connection. Off. Technol. Assess. (6), 20–25 (1995)
3. Blankenship, S.E.: Factors related to computer use by teachers in classroom instruction. Virginia Polytechnic Institute and State University (1998)
4. Jaber, W.E.: A Survey of factors which influence teachers' use of computer-based technology. Int. J. Instr. Media (26), 253–266 (1999)
5. Tao, L., Li, Z.: Research on the use of new media by primary and secondary school teachers: the perspective of teacher-student interaction. e-Educ. Res. (2), 102–107 (2016)
6. Liu, Y., Jiang, B., et al.: The current situation and analysis of the use of modern teaching media by university teachers in Hubei Province. China Educ. Technol. (7), 16–18 (2002)
7. Davidian, M.: Hierarchical linear models: applications and data analysis methods. In: Raudenbush, S.W., Bryk, A.S. (eds.) Economics of Education Review, 2nd edn., vol. 98, no. 1, pp. 767–768 (2003)

Analysis and Design of Personalized Learning System Based on Decision Tree Technology

Qiaoying Ming[1(✉)] and Ran Li[2]

[1] Department of Electronic Science, School of Engineering and Technology, Xi'an Fanyi University, Xi'an, China
184072575@qq.com

[2] Department of English Language, School of English Language and Literature, Xi'an Fanyi University, Xi'an, China

Abstract. Advancements in Science and Social Progress present opportunities for the development of education, as well as colossal challenges. At present, on-line learning penetrates the whole world, however, most on-line learning systems are traditional learning mechanisms lacking consideration of students' personalized needs, which leads to problems such as poor learning effects and interactivity. Targeted at these issues, the article presents a thorough analysis and detailed design of personalized learning system on the basis of the construction of student model and the implementation of data mining techniques in on-line learning system, analyzing and addressing the problem of personalizing on-line learning from the perspective of data mining, to better serve the students with personalized learning and guarantee the quality and effects of on-line teaching.

Keywords: Personalized learning · Decision tree technique · System analysis and design

1 Introduction

At present, on-line learning has already become a prevalent way of learning for its informativity and ubiquity. However, as time progresses, apart from free allocation of time and sharing of learning resources, learners' demands focus more on personalized learning tailored to individual situation. Currently, a sea of web-based learning systems fails to address students' personal learning needs because learner's needs and habits have not been given due consideration, besides, the real-time learning assistance and effective supervision is lacking and the teaching resources have not been fully exploited, which caused the deviation in student's learning objective and loss of learning motivation. To address the above-mentioned issues, the article adopts the decision tree technique in the analysis of learner's learning style and personality mining, informing the decision-making process of personalized learning system, to ensure its particularity in learning style, multi-dimensionality in learning resources and longitudinality in learning process.

J. MacIntyre et al. (Eds.): SPIoT 2020, AISC 1282, pp. 119–126, 2021.
https://doi.org/10.1007/978-3-030-62743-0_17

2 Theory and Technology of Personalized On-Line Learning

2.1 The Notion of Personalized Learning

Personalized learning is a learning style catering to students' learning aptitude, it emphasizes the recognition of students' individual characteristics and the corresponding influence on the entire process of teaching.

2.2 The Construction of Personalized Student Model

IEEE1484.2 PAPI, is a comparatively speaking mature student model, however, its indexation of the information of students' learning process is only limited and it fails to reflect the influence of non-intellectual factors in the learning process, such as learning motivation, learning style and learning preferences [1]. Directed at these issues, the IEEE1484.2PAPI model has been refined to specify students' learning process, learning style, as well as the individual personality traits [2]. It categorizes and delineates student model information with a five-tuple set, as is illustrated below in Fig. 1:

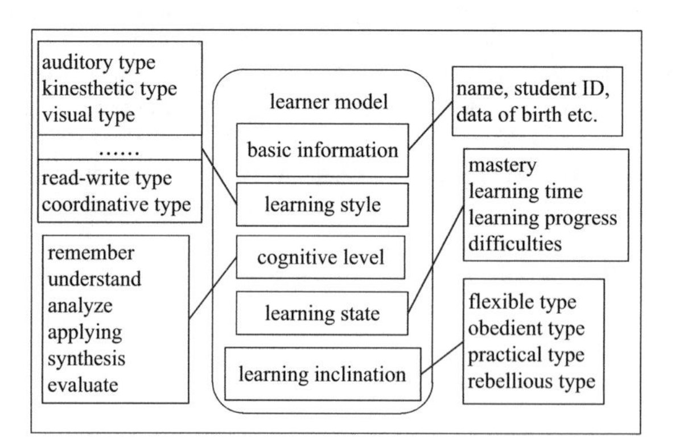

Fig. 1. The design of student model

2.3 Decision Tree Technology

Data mining is the process by which valuable hidden information is extracted from the gigantic, incomplete, noise-corrupted, random fuzzy data sets. Data mining methods can be classified into two categories, namely the statistical type and machine learning type. The decision tree method derived from machine learning constructs classification model from the training sample set, which is represented in the form of a tree. Decision tree categorizes examples from the root node to the leaf node, generating classification rules. Each node on the decision tree corresponds to one trait, each branch represents the possible trait values of the upper node to which it is connected. Decision tree reaches a conclusion on the leaf node, therefore, one route from the root node to the leaf node corresponds to one rule, the whole decision tree corresponds to a disjunctive expression set.

3 The Application Process of Decision Tree in Personalized Learning

1) Data Collection
   The primary data source of the present research falls into two general categories, namely, the underlying database of learning system and student's non-intellectual factors.
2) Data Pre-processing
   Data pre-processing filters the data obtained in the first step meanwhile masks out the invalid data or data of low reliability. Finally, based on the analysis of the correlative coefficient and covariance of groups of data, element property may be concluded.
3) Data Quantization
   The present research aims to conduct respective quantification of different categories of data.
4) The Construction of Decision Tree
   Decision tree may be constructed after the data which has been split and processed is inputted and configurated in the open source platform in the form of training sample set.
5) Decision Tree Pruning and Evaluation
   In the generation of decision tree, its depth and the number of node elements are limited. The test results will undergo random sequence distribution analysis and cross validation. Then, the verification results may be used in rule-screening. Afterwards, the reliability of rules and the supported attributes of variables may be analyzed and screened out to select the more optimal decision tree as prediction model.
6) Generating rules and conclusion

After the above-mentioned five procedures are completed, a well-pruned decision tree model of intuitive expression is generated. Disjunction rules will be derived via the sequential decision made from the root node of the decision tree to the leaf nodes.

Figure 2 presents the analysis of students' learning style based on the utilization of learning resources. Regulations of time-length and frequency have been set regarding retention time, click ratio, number of participations [3].

According to the decision tree in Fig. 2, each time the course from root node to leaf node has been completed, a rule may be generated, which can be described as follows:

(1) If (learning resources = media and audio/video = long retention time, high click ratio) then (students' learning style = auditory and visual type)
(2) If (learning resources = media and audio/video = short retention time, low click ratio and teaching plan/teaching notes = long retention time, high click ratio) then (students' learning style = read-write type).

The rest learning style is deducible from the same procedure, which will not be detailed to avoid repetition. Teachers can recommend suitable learning resources to the students in accordance with the classification rules of learning style.

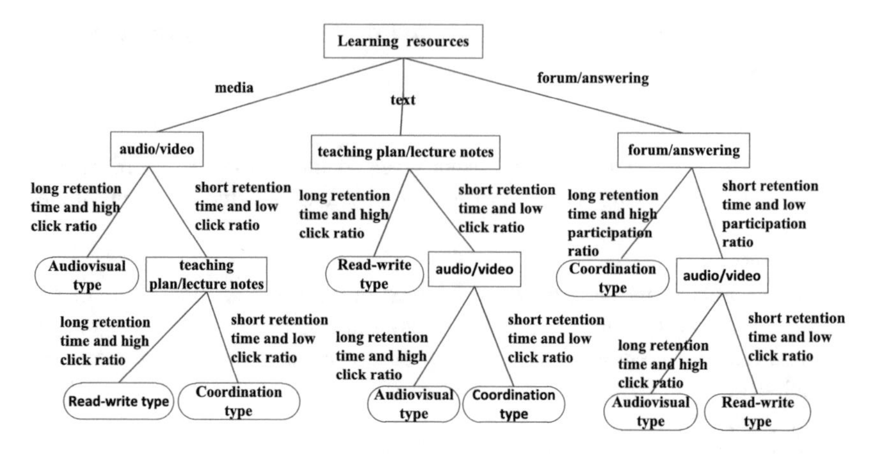

Fig. 2. Decision tree of learning style classification

4 The Design of Personalized Learning System

4.1 The Overall Architectural Design

The overall architectural design of the system is illustrated in Fig. 3 [4].

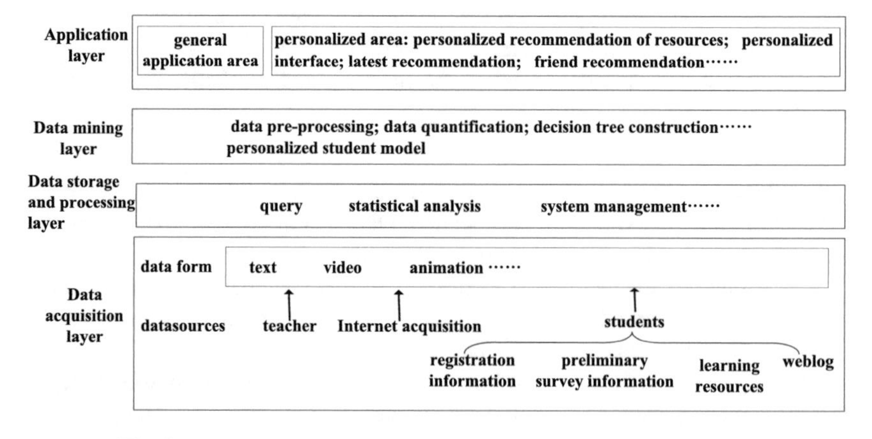

Fig. 3. The Overall architectural design of on-line learning system

1) Data Acquisition Layer
   Principal data sources for this layer include: the learning materials uploaded by teachers, relevant learning materials collected from the internet via information acquisition tools, and the information acquired from student data.
2) Data Storage and Processing Layer
   Major functions of this layer include the storage, analysis, calculation and scheduling of multiple data.

3) Data Mining Layer
   Major functions of this layer include the pre-processing, quantification of relevant data, utilizing supporting tools in the construction and pruning of decision tree, generating corresponding rules and conclusions of personalized recommendation.
4) Data Application Layer
   This layer guarantees the attainment of students' learning objectives through operation interface.

4.2 The Design of Personalized Learning Model

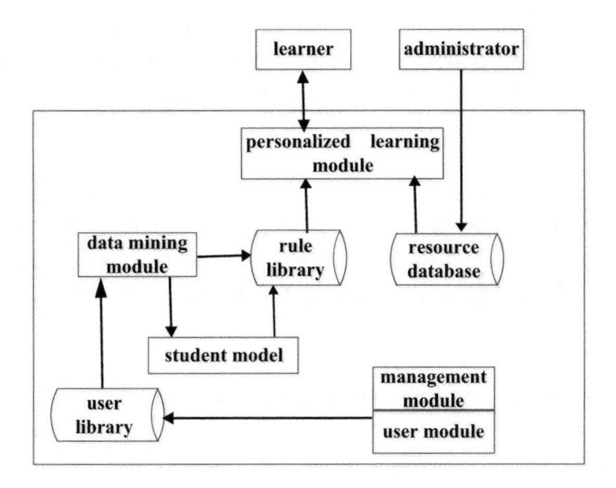

Fig. 4. The structure of personalized learning model

As is illustrated in Fig. 4 [5], the principal function modules of personalized learning system are detailed as follows:

1) Database

The databases in the system encompasses: resource database, user information database, rule library. The resource database has a vast storage of learning materials. The primary information stored in user information database includes the basic information of registration, predictive information of learning style, learning log etc. Rule library is the crucial functioning factor in learning system, which encompasses the following rules: generally acknowledged rules in the education circle preloaded in the database, the rules mined via datamining, the basis for organizing learning materials, such as the learning materials like charts, Figures and videos provided for the visual learners. The rules in the library should be updated constantly and authorized by experts.

2) Management Module

Management module embodies administrator's operation administration and maintenance of the system and has higher authority. Teachers enjoy certain administration authority.

3) User Module

User module embodies the subscribed leaners of the system. Learners can choose the intended learning content in the system or use the learning resources recommended by the system.

4) Learner Model Module

Learner model presents the analysis, classification and representation of student's registration information in the system, test result, learning process, and learning style, as is illustrated in Fig. 1.

5) Data Mining Module

Data mining module extracts from the massive data stored in the system the information of great value to students' personalized learning.

6) Personalized Learning Module

Based on the integration of student model with data mining, personalized learning module generates learning content in accordance with the dynamic data, which will be provided to different students in different manifestation. The objective-oriented personalized service of learning system is realized based on students' individual characteristics [6].

4.3 The Design of Data Mining Module Based on Decision Tree

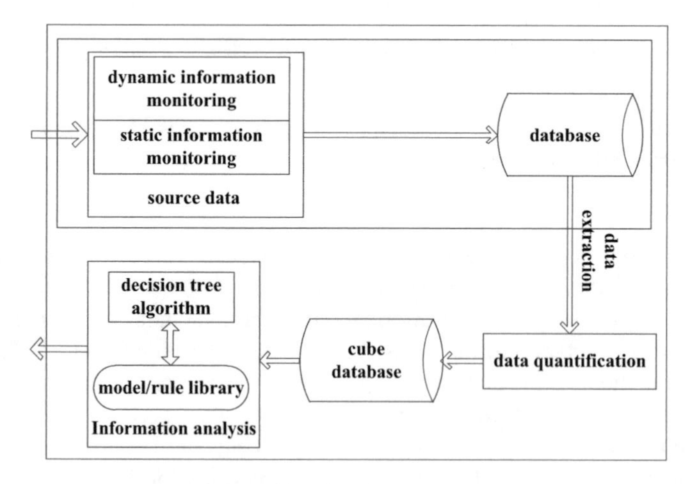

Fig. 5. Data mining module

The Module acquires information student information from the server, then, after making path supplement of students' access log, the module extracts and preprocesses data in the database, clearing the information item irrelevant to the mining, finally getting the cube database of students' original data. Meanwhile, the decision tree method of the data mining algorithm will be employed to process the cube database. The corresponding mining results may supplement and refine students' personality model, eventually presenting learning contents to the students in an explicit and acceptable way, conforming to existing rules, also giving students server feedbacks as to the appropriate learning strategies and suggestions [7] (Fig. 5).

4.4 Personalized Recommendation Module Design

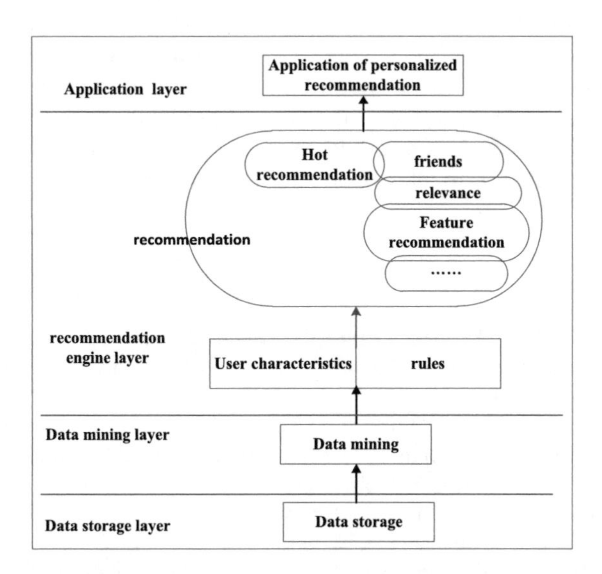

Fig. 6. The hierarchical structure of personalized recommendation module

The hierarchical structure of personalized recommendation module of the learning system consists is illustrated in Fig. 6: Data storage layer stores all kinds of relevant information of the students. On account of its large memory capacity, it is based on HDFS (Hadoop Distributed File System) distributed data storage. Data mining layer performs mining analysis of the massive data stored in the system, digging out potential valuable information. The recommendations generated by the recommendation engine layer are grounded in students' individual characteristics and rules, recommending the resource and information type favored by the students, meanwhile exploring the correlation between resources or information itself to effect the recommendation in accordance with object relativity [8, 9]. For students with the same or similar characteristics and preference, the engine may provide friend recommendations to form interest groups. The resources with high click ratio may become hot recommendations, the newly uploaded resources may become latest recommendations. There may also be the optimal path recommendations etc. [10].

5 Conclusion

The application of data mining technology in learning system not only elevates the level of personalized service but also provides supplementary means for teaching decision analysis. The analysis and design of personalized learning system based on decision tree technology take full account of students' personality traits, cognitive level etc., in the research, personalized student model has been constructed and the analysis and design of the learning system has been accomplished, which in a sense is of high reference value to the study, promotion and personalized application of on-line learning system. A full-scale and in-depth research on the subject will be conducted in the future, with the prospect of the deep integration of data mining technology with personalized learning system, to realize its mature application in every step of on-line learning, enhancing students' internal motivation, so as to promote their learning initiative, learning efficiency, to effect individualized teaching.

References

1. Mu, S., Cui, M., Wang, X., Qiao, J., Tang, D.: Learners' attention preferences of information in online learning: an empirical study based on eye-tracking. Interact. Technol. Smart Educ. **16**(3), 186–203 (2019)
2. Romero, L., Gutierrez, M., Caliusco, M.L.: Conceptual modeling of learning paths based on portfolios: strategies for selecting educational resources. In: 2018 13th Iberian Conference on Information Systems and Technologies (CISTI), pp. 1–6 (2018)
3. Wang, R., Qin, X., Wang, B.: Design of mobile user behavior analysis system based on big data. China Comput. Commun. (11), 88–90 (2019). (in Chinese)
4. Xin, G., Wei, S.: Mobile learning analysis system design based on big data technology. China Comput. Commun. (06), 98–99 (2019). (in Chinese)
5. Wang, X.: Research and implementation of test-based personalized learning guiding system. Comput. Educ. (14), 56–60, 65 (2015). (in Chinese)
6. Wu, H.: Research on personalized learning system under the perspective of smart learning. China Educ. Technol. (06), 127–131 (2015). (in Chinese)
7. Maldonado-Mahauad, J., Pérez-Sanagustín, M., Kizilcec, R.F., Morales, N., Munoz-Gama, J.: Mining theory-based patterns from big data: identifying self-regulated learning strategies in massive open online courses. Comput. Hum. Behav. **80**, 179–196 (2018)
8. Zhou, Y., Huang, C., Hu, Q., Zhu, J., Tang, Y.: Personalized learning full-path recommendation model based on LSTM neural networks. Inf. Sci. **444**, 135–152 (2018)
9. Wang, L.: Research on personalized recommendation technology of online learning based on collaborative filtering. Microcomput. Appl. **33**(05), 49–51 (2017). (in Chinese)
10. Liu, H., Li, X.: Learning path combination recommendation based on the learning networks. Soft. Comput. **24**, 4427–4439 (2019)

Deep Learning Classification and Recognition Model Construction of Face Living Image Based on Multi-feature Fusion

Chunyan Li(✉) and Rui Li

Engineering College, Honghe University, Mengzi 661199, Yunnan, China
dale225@126.com

Abstract. Convolutional neural network, as a common method of deep learning, is excellent in image recognition and classification performance. As one of the key technologies of biometrics recognition, face recognition is a research hotspot in the field of pattern recognition. Based on the basic model, a multi-layer feature fusion face recognition model structure is proposed. This model performs feature extraction on each convolution pooled feature map in the basic model and fuses the features obtained from each layer as the final face representation features. PCA (Principle Component Analysis), LDA (Linear Discriminate Analysis) and LPP (Locality Preserving Projection) are effective feature extraction methods. We fused them into the basic model to build a multi-layer feature fusion face recognition based on the basic model. Experimental results show that the feature fusion model method is significantly higher in recognition rate than the basic model method. Compared with the basic model, the feature fusion model can extract face features more effectively, thereby improving the face recognition rate.

Keywords: Face recognition · Deep learning · Convolutional neural network · Multi-feature fusion

1 Introduction

The human face is the most important human feature in the entire human structure [1]. Its structure is very complicated, and its details also vary. At the same time, it also contains a lot of different information. Using face information for identity verification is very natural, and it is direct, friendly, and convenient [2]. Compared with the feature information of other parts of the human body, it is easier to be accepted by the user [3]. Therefore, face recognition has been widely researched and applied in the course of social development [4]. Face recognition technology is a very advanced and comprehensive technical field, which involves many interdisciplinary disciplines, including image processing, pattern recognition, biological statistics, etc. [5]. The two most important steps of face recognition are face detection and face recognition. Face recognition is basically divided into two categories: one is face verification, which is a one-to-one matching problem; the other is face identification, which is a one-to-many matching process.

J. MacIntyre et al. (Eds.): SPIoT 2020, AISC 1282, pp. 127–133, 2021.
https://doi.org/10.1007/978-3-030-62743-0_18

The methods for extracting global features of human faces are mainly PCA (Principle Component Analysis) and LDA (Linear Discriminate Analysis) [6]. The principal component analysis method is an unsupervised learning method. Linear discriminant analysis can effectively use the category identification information [7]. It is a supervised learning method. The advantage of PCA is that it can greatly increase the calculation speed and reduce the interference of noise, but the shortcomings of PCA are also obvious. The method of PCA is insensitive to background changes on the face image, changes in light intensity, and different facial expressions of people. The LDA method can maximize the difference between classes of different faces while minimizing the difference within the class of the same face. Relevant scholars conducted a recognition experiment on 10 face images of a total of 16 people [8]. The recognition rate of the PCA method was 81%, while the recognition rate of the Fisher face method reached 99.4%. The advantages of LDA make it widely used in pattern recognition, but the method of LDA may overfit the data and LDA dimensionality reduction is more suitable for sample data of Gaussian distribution [9]. With the advent of the era of big data and the substantial improvement of computer computing power, deep learning has achieved the best results in the field of face recognition [10]. On the one hand, it is driven by data, because the deep learning method itself is a set of features. With the integration of extraction and classification, the network can better fit the data. However, the shortcomings of current deep learning recognition methods are that as the network level continues to increase, the characteristics of input data are more highly represented and there is a disadvantage of information loss, and how to effectively combine with other traditional recognition methods is gradually being used by people.

Combined with traditional feature extraction methods, a multi-layer feature fusion face recognition network model based on a convolutional neural network is constructed. The network model first extracts and fuses each layer of convolution pooled feature maps of the basic model to obtain the final face feature. By inputting the classifier, the effectiveness of the features extracted by the feature fusion model and the basic model is compared and analyzed. Experiments prove that the feature fusion model has better performance in facial feature extraction.

2 Deep Learning Face Recognition Model

2.1 Layering of Deep Learning Network

The learning process of deep learning models is the model network trying multi-level nonlinear information processing process to complete unsupervised feature extraction, conversion, classification, recognition and other tasks.

The reason why deep learning is "deep" is because of its "layered" idea, that is, the network structure usually contains multiple hidden layers in addition to the input layer and the output layer, and the layers are interconnected by a large number of neuron networks. The feature extraction and conversion of each layer from low to high are also deepened. The retention of the original information between layers is also reduced layer by layer, similar to the human brain. When we want to remember a person, usually it is the entire picture including a person. According to your requirement to remember the "look", first you look for the individual "person" in the picture seen by the eyes, then find the "head" position, find the "face", and finally remember each face feature. Even so, the brain is not completely able to portray the "face" of a person, but it can be distinguished among many people, which is a further abstraction of the brain. The layered abstraction principle of deep learning is similar to this.

2.2 Convolutional Neural Network

CNN (Convolutional Neural Network) as a classic neural network has significant advantages in image processing. The response of a single neuron to stimuli in its received field can be approximated by mathematical convolution operations. CNN uses a supervised training method to adjust the parameters of the network, input the original data into the network to obtain its actual output, compare the actual output with its real label, and obtain its loss error, and adjust the whole by obtaining the minimum value of the error.

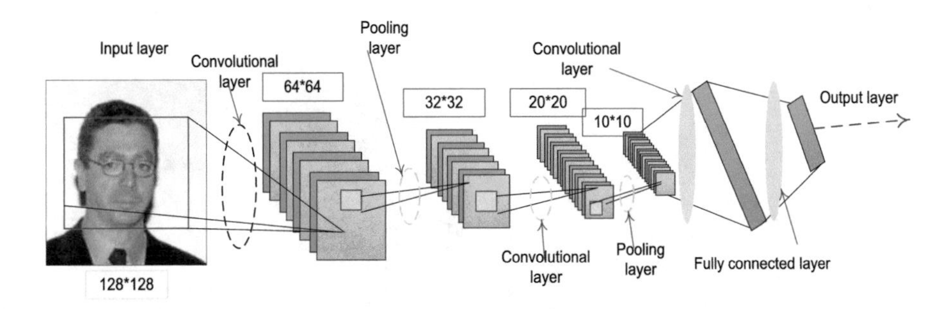

Fig. 1. Convolutional neural network structure diagram

As shown in Fig. 1, the basic structure of CNN is formed by the stacking of different layers, the most important of which are the convolution layer and the sub-sampling layer. The convolutional layer is the core structure of CNN. The parameters of the convolutional layer are composed of several learnable filters. Each filter is convolved with the corresponding size of the image block in the input picture, and the filter is obtained by the activation function. The number of convolutional layer filters determines the number of output feature maps. Different feature maps can represent different features of the original image, and CNN-specific weight sharing solves the problem of excessive network parameters. After obtaining the feature map through the convolutional layer, CNN reduces the dimensionality of these features through the sub-sampling layer, reducing the number of parameters and calculations in the network, thus preventing overfitting. CNN finally vectorizes the obtained series of feature maps through a fully connected layer, so that we can classify and predict the input data under the action of a multi-layer perceptron or classifier. Like other deep models, CNN can autonomously learn the intrinsic features of input data, avoiding the time-consuming and laborious extraction of artificial features. As an effective feature extraction method, CNN has achieved great success in image classification and recognition.

The face image defined on the original image is mapped to the feature map of the shared layer after the operation of the convolution layer and the pooling layer. At this time, the feature map is divided through the sliding window. The movement of the sliding window here is similar to the movement of the convolution kernel on the feature map, and the size and step size of the sliding window need to be set to appropriate values. If the sliding window is set too small, most of the segmented feature maps retain only local information, and there is no more global information, and some structural information of the face will be lost; if the sliding window is too large, you will lost some local specific information of the face. For the sliding step size of the sliding window, if it is too small, it will cause too many overlapping regions of the divided feature blocks, and the amount of calculation will increase significantly; if the step size is too large, the information on the edge of the feature block will be lost due to cutting. Therefore, the size and step size of the sliding window will give the best results in multiple experiments. This article introduces how to use the sliding window to segment on the feature map of the shared layer. The segmentation process is shown in Fig. 2.

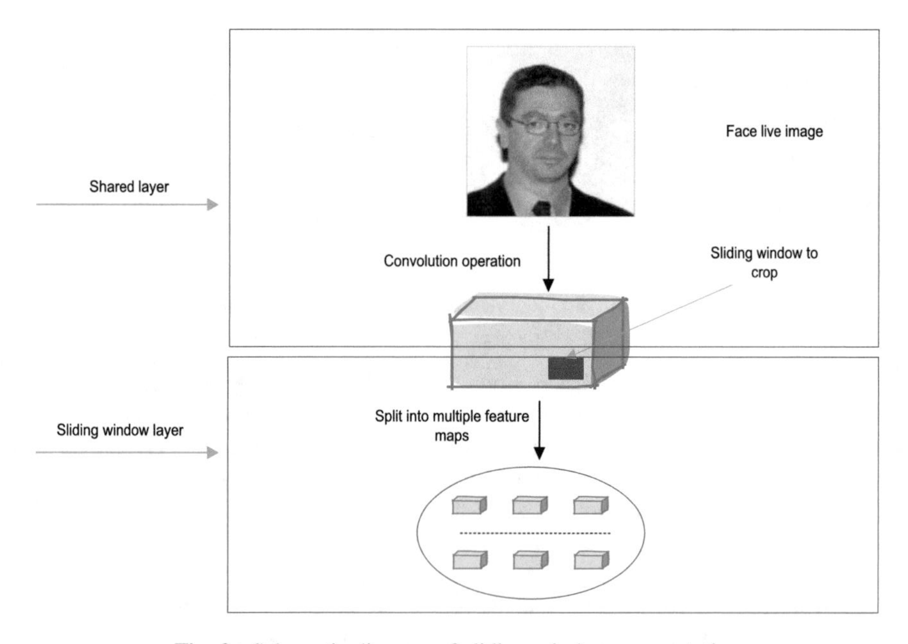

Fig. 2. Schematic diagram of sliding window segmentation

3 Experimental Results and Analysis

3.1 Experimental Results

In the experiment, we randomly selected 6 pictures of each person and the remaining 4 pictures on the ORL dataset for training and testing. We repeat this procedure ten times to get the average to get the final recognition rate and the time taken for each training test. Figure 3 shows the test comparison results of different feature fusion models under the ORL data set. The feature fusion model is based on SVM classification and the input features of the classifier are 120-dimensional. The CNN (Re LU) + PCA + LDA method in the figure represents the feature fusion model composed of the second feature extraction using the PCA + LDA method on different layers of the basic model. ReLU means that CNN network selects ReLU activation function, and Sigmoid means that CNN network activation function is selected as Sigmoid function.

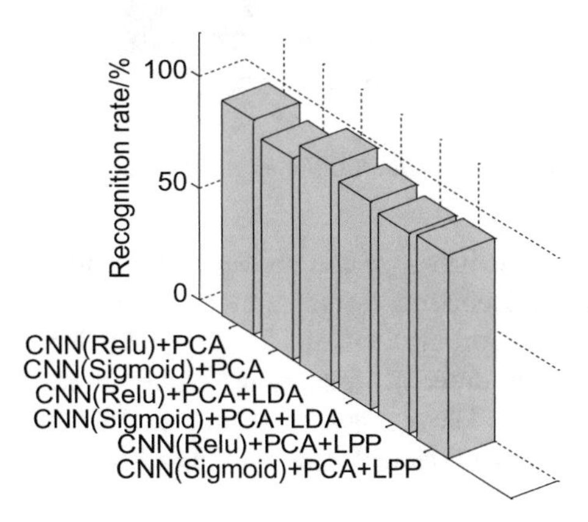

Fig. 3. Recognition rate of different algorithm models under SVM classifier

3.2 Results Analysis

The effect of the activation function on the CNN network can be seen in Fig. 3. The recognition rate of the four methods using the ReLU function is better than the Sigmoid function. This is because the Sigmoid function has a problem of gradient disappearance. The neuron may be in saturation state and gradient dispersion occurs, which affects the network learning, so that the features extracted on the convolution pooling layer of each part of the convolutional neural network are not ideal. In the recognition result based on KNN classification shown in Fig. 4, the activation function used by the basic model is ReLU.

The recognition rate of other feature fusion methods is higher than the recognition rate of the method based on the basic model. The recognition rate of the CNN (Re LU) method is 87.4%, and the worst effect of the feature fusion method CNN (Re LU) + PCA + LPP (Locality Preserving Projection) also reached 90.3%. This shows that the feature fusion method can extract face features more effectively.

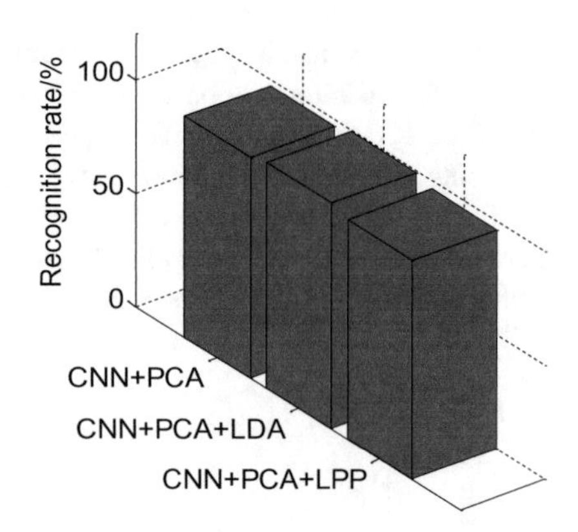

Fig. 4. Recognition rate of different models under KNN classifier

The input feature dimensions of the above experimental classifiers are all 120-dimensional, and the influence of feature dimension changes on the feature fusion model is discussed below. Among them, the basic model (CNN) method (the activation function uses ReLU) sets different dimensions in the output feature dimension and classifies the recognition. The extracted features get different feature dimensions through dimensionality reduction, and input into the SVM classifier. The experimental results are shown in Fig. 5.

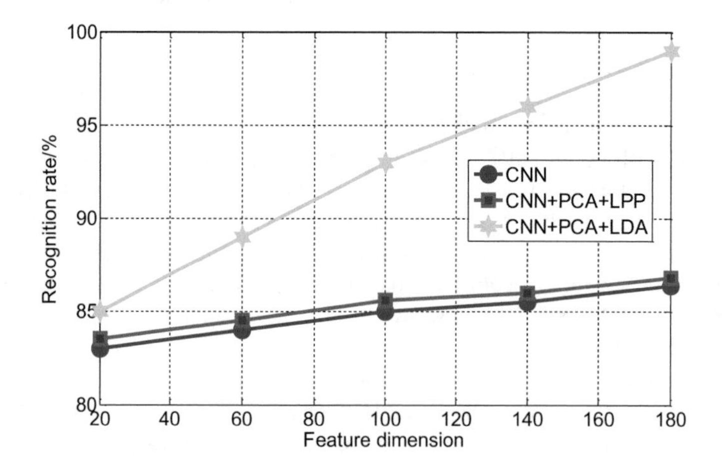

Fig. 5. Comparison of recognition rates under different feature dimensions

As can be seen from Fig. 5, in the experiments on the ORL dataset, the recognition rates of the CNN + PCA + LDA and CNN + PCA + LPP methods under different feature dimensions are higher than the basic model (CNN) method. The figure shows that the recognition rate of the CNN + PCA + LDA method fluctuates greatly, and the recognition rate differs by 14.1% in different feature dimensions, indicating that the feature fusion model is more sensitive to changes in feature dimensions.

4 Conclusion

This paper introduces a multi-layer feature fusion face recognition model based on convolutional neural network. This method is based on the basic model to extract features from each convolution pooled feature map and fuse to obtain the final face features. Face features are input into SVM and KNN classifiers for face recognition testing. This model uses principal component analysis, linear discriminant analysis, and local preserving projection to extract features from convolution pooled feature maps. The result of the analysis of the experimental results shows that the feature fusion model has a recognition rate of 98.6% on the ORL data set. It proves that the feature fusion model has better performance in facial feature extraction. The CNN + PCA + LDA method has different feature dimensions. The recognition rate of it is generally higher than that of the basic model (CNN), which further proves the superiority of the feature fusion algorithm model in feature extraction.

References

1. Deng, Z.: Mutual component convolutional neural networks for heterogeneous face recognition. IEEE Trans. Image Process. **28**(6), 3102–3114 (2019). (in Chinese)
2. Zhou, L.: Separability and compactness network for image recognition and superresolution. IEEE Trans. Neural Netw. Learn. Syst. **30**(11), 3275–3286 (2019). (in Chinese)
3. Kumar, P.M.: Intelligent face recognition and navigation system using neural learning for smart security in internet of things. Clust. Comput. **22**(4), 7733–7744 (2019)
4. Li, J.: Micro-expression recognition based on 3D flow convolutional neural network. Pattern Anal. Appl. **22**(4), 1331–1339 (2019). (in Chinese)
5. Ma, Z.: Lightweight privacy-preserving ensemble classification for face recognition. IEEE Internet Things J. **6**(3), 5778–5790 (2019). (in Chinese)
6. Zhao, Z.-Q.: Object detection with deep learning: a review. IEEE Trans. Neural Netw. Learn. Syst. **30**(11), 3212–3232 (2019). (in Chinese)
7. Haider, K.Z.: Deepgender: real-time gender classification using deep learning for smartphones. J. R.-Time Image Process. **16**(1), 15–29 (2019)
8. Zhang, H.: Sitcom-star-based clothing retrieval for video advertising: a deep learning framework. Neural Comput. Appl. **31**(11), 7361–7380 (2019). (in Chinese)
9. Litvin, A.: A novel deep network architecture for reconstructing RGB facial images from thermal for face recognition. Multimed. Tools Appl. **78**(18), 25259–25271 (2019)
10. Sonkusare, S.: Detecting changes in facial temperature induced by a sudden auditory stimulus based on deep learning-assisted face tracking. Sci. Rep. **9**(1), 1–11 (2019)

A Prediction Method of Blood Glucose Concentration Based on Nonlinear Auto-Regressive Model

Xiong-bo Huang[1], Ling Qiu[2(✉)], and Ye-quan Liao[2]

[1] School of Electronic and Information Engineering, Foshan Professional Technical College, Foshan 528137, Guangdong, China

[2] Department of Pharmacy, Foshan Hospital of Traditional Chinese Medicine, Foshan 528000, Guangdong, China

fshxb@126.com

Abstract. In view of the highly nonlinear and complex non-stationary characteristics of blood glucose sequence, a blood glucose concentration prediction algorithm based on nonlinear auto-regressive model is proposed. First, the nonlinear auto-regressive model is used to describe the existing blood glucose sequence, and the polynomial expansion function is used to approximate the description model; then, the relaxation iterative search algorithm is used to estimate the nonlinear least square parameters of the polynomial approximation model in the hierarchical calculation mode. The experimental results show that the proposed algorithm has excellent computational efficiency while ensuring more accurate prediction accuracy.

Keywords: Nonlinear auto-regressive model · Least squares estimation · Prediction of blood glucose concentration · Relaxed iterative search algorithm

1 Introduction

Diabetes is a chronic and progressive disease of the whole body. It is the β cell of the islet of Langerhans can not normally secrete insulin, which leads to the absolute or relative deficiency of insulin, and causes the metabolic disorder of the three major nutrients of sugar, fat and protein in the body. Insulin injection is an important means to maintain the normal blood glucose level in diabetic patients. In order to improve the treatment effect, it is necessary to accurately predict the blood glucose concentration of patients in the future (generally 30 min), so as to effectively determine the injection dose and injection time of insulin [1]. R. Hovorka simulated the short acting insulin injected subcutaneously and the absorption process of intestinal tract by nonlinear dynamic system, and determined the parameters of time-varying model by Bayesian parameter estimation, so as to obtain an effective nonlinear blood glucose prediction model [2]. H. Efendic based on the transition probability of physiology and blood glucose change, designed and implemented a short-term prediction algorithm of blood glucose concentration with interval probability using Gaussian function, which has a very good prediction effect on blood glucose concentration in the time range of 10–

J. MacIntyre et al. (Eds.): SPIoT 2020, AISC 1282, pp. 134–140, 2021.
https://doi.org/10.1007/978-3-030-62743-0_19

30 min [3]. It is noted that the blood glucose sequence is highly non-linear and complex non-stationary. A blood glucose concentration prediction algorithm based on non-linear auto-regressive model is proposed, which effectively improves the calculation efficiency under the condition of ensuring the accuracy of prediction.

2 The Nonlinear Auto-regressive Description of Blood Glucose Sequence

Record the original sequence of blood glucose is $y_t(t = 1, 2, \cdots, n)$, describe it with nonlinear auto-regressive model, and get the expression of input-output relationship as shown in formula (1) [4]

$$y_t = f\left(y_{t-1}, y_{t-2}, \cdots, y_{t-p}, e_{t-1}, e_{t-2}, \cdots, e_{t-q}\right) \tag{1}$$

In Formula (1), y_{t-i} represents the value of blood glucose before time i, e_{t-j} represents the state of white noise before time j, f is a nonlinear function, p and q represent the order of y_t and e_t in the model respectively.

According to Weierstrass approximation theorem, any continuous function f defined in the bounded closed interval $[a, b]$ can always be approximated by polynomial $\{p_n(t)\}$, and its approximation error approaches to 0 at $n \to \infty$. Therefore, Eq. (1) is approximated by polynomial $\{p_n(t)\}$, which is [5, 6].

$$\begin{aligned}\hat{y}_t \approx\ & \alpha_{10}y_{t-1} + \alpha_{20}y_{t-2} + \cdots + a_{p0}y_{t-p} + \alpha_{11}y_{t-1}^2 + \alpha_{22}y_{t-2}^2 \\ & + \cdots + \alpha_{pp}y_{t-p}^2 + \beta_{10}e_{t-1} + \beta_{20}e_{t-2} + \cdots + \beta_{q0}e_{t-q} + \beta_{11}e_{t-1}^2 \\ & + \beta_{22}e_{t-2}^2 + \cdots + \beta_{qq}e_{t-q}^2 + \theta_{11}y_{t-1}e_{t-1} + \theta_{12}y_{t-1}e_{t-2}\, \theta_{21}y_{t-2}e_{t-1} + \cdots + \theta_{pq}y_{t-p}e_{t-q} + e_t.\end{aligned} \tag{2}$$

Without losing generality, if the system described in formula (2) has the characteristics of zero input response and zero state response, then the system has no input before the time $t = 0$, and accordingly, the system has no zero input response output. It is easy to know that at the time $t = 1$, there are

$$\hat{y}_1 = e_1 \tag{3}$$

Starting from Eq. (2), calculate y_2 corresponding to time $t = 2$, and then get

$$\hat{y}_2 = \alpha_{10}y_{t-1} + \alpha_{11}y_{t-1}^2 + \beta_{10}e_{t-1} + \beta_{11}e_{t-1}^2 + \theta_{11}y_{t-1}e_{t-1} + e_2 \tag{4}$$

Substituting Eq. (3) into Eq. (4) to get $\hat{y}_2$ represented by $\hat{y}_1$

$$\hat{y}_2 = e_2 - (\alpha_{10} + \beta_{10})y_1 - (\alpha_{11} + \beta_{11} + \theta_{11})y_1^2 \tag{5}$$

The operation of formula (2) - formula (5) is repeated continuously, as shown in formula (6), can obtain the polynomial approximation expression of the nonlinear auto-regressive description of blood glucose sequence y_t

$$\hat{y}_t = \sum_{i=1}^{z} \phi_{i0} y_{t-i} + \sum_{i=1}^{z} \sum_{j=1}^{z} \phi_{ij} y_{t-i} y_{t-j} + \cdots + \sum_{i=1}^{z} \cdots \sum_{k=1}^{z} \phi_{i,\cdots,k} \prod_{w=1}^{z} y_{t-w} + e_t \quad (6)$$

In formula (6), $z = \max(p, q)$, it is not difficult to find that the identification and modeling of blood glucose sequence y_t can be completed by solving the parameter vector ϕ of formula (6).

3 Relaxation Iterative Search Solution of Nonlinear Auto-regressive Model

3.1 Relaxation Iterative Search Solution Principle

The solution of parameter vector ϕ in Eq. (6) is a nonlinear least square solution problem. Generally speaking, there are Gauss Newton method and Levenberg Marquardt method to solve nonlinear least squares. However, for the solution of some practical nonlinear auto-regressive models, due to the singularity or serious ill condition of the coefficient matrix, the Gauss Newton method or Levenberg Marquardt method have strict requirements for the selection of initial value in the iterative solution process. In addition, in the process of each iteration, the linear equations need to be solved, which to a certain extent leads to the calculation Calculate the cost increase.

In order to solve the heavy calculation and difficulties of the existing algorithm, the relaxation iterative search algorithm is introduced to solve the nonlinear least squares problem. By transforming the high-dimensional problem into a successive one-dimensional search operation, the relaxation iterative search algorithm can avoid a large number of matrix operations and the solution of linear equations, so it has more excellent computing performance.

Set the objective function as

$$Q = \sum_{t=1}^{n} \frac{1}{y_t} [y_t - \hat{y}_t(\phi)] \quad (7)$$

Its minimum point $\phi^* = \left(\phi^*_{1,\cdots,1}, \cdots, \phi^*_{i,\cdots,k}\right)^T$ must satisfy

$$\frac{\partial Q}{\partial \phi_{i,\cdots,k}} = -2 \sum_{t=1}^{n} \frac{1}{y_t} [y_t - \hat{y}_t(\phi)] \frac{\partial \hat{y}_t}{\partial \phi_{i,\cdots,k}} = 0 \, (i, k = 1, 2, \cdots, z) \quad (8)$$

Therefore, the parameter vector ϕ problem of solving the nonlinear auto-regression of blood glucose sequence y_t can be transformed into the linear equation set shown in Eq. (9)

$$\psi_{i,\cdots,k}(\phi) = \sum_{t=1}^{n} \frac{1}{y_t}[y_t - \hat{y}_t(\phi)]\frac{\partial \hat{y}_t}{\partial \phi_{i,\cdots,k}} = 0\,(i,k = 1,2,\cdots,z) \tag{9}$$

For Eq. (9), given the initial value $\phi^{(0)} = \left(\phi_{1,\cdots,1}^{(0)}, \cdots, \phi_{i,\cdots,k}^{(0)}\right)^T$ of the optimal solution ϕ^*, the corresponding relaxation iterative search method is [7].

$$\phi^{(k)} = \phi^{(k-1)} + \sigma_k \delta_g^{(k-1)} \tag{10}$$

$$\delta_g^{(k-1)} = -\frac{\psi_g}{\frac{\partial \psi_g}{\partial \varphi_g}\Big|_{\varphi^{(k-1)}}}\xi_g \; k = 1,2,\cdots,\; g = (k \bmod n) \tag{11}$$

In formula (11), $\xi_p = (0,\cdots,0,1,0,\cdots,0)^T$, σ_k is relaxation factor. The selection principle of σ_k is to make Eq. (12) hold

$$Q\left(\phi^{(k)}\right) < Q\left(\phi^{(k-1)}\right) \tag{12}$$

Given the accuracy $\varepsilon > 0$, if $\left|Q\left(\phi^{(k+1)}\right) - Q\left(\phi^{(k)}\right)\right| < \varepsilon$ is true, the iteration is over.

3.2 Algorithm Design

In conclusion, the following non-linear auto-regressive model can be used to predict blood glucose concentration.

Algorithm name: blood glucose concentration prediction algorithm based on nonlinear auto-regressive model

Input: the original sequence of blood glucose is $y_t (t = 1,2,\cdots,n)$, and the fitting accuracy is e

Output: m predictive values of blood glucose sequence $y_t (t = n+1, n+2,\cdots,m)$

Step (1): describe y_t with the non-linear auto-regressive model shown in Eq. (6), and make the preset $z = 1$;

Step (2): use the relaxation iterative search method shown in Eq. (10) to Eq. (11) to solve the polynomial approximation expression of Eq. (6), and obtain the optimal parameter vector solution $\phi_{(Z)}^*$ of the current order;

Step (3): calculate the corresponding $\hat{y}_t$ with the parameter vector solution $\phi_{(Z)}^*$ obtained in step (2), if $\frac{1}{n}\sum_{t=1}^{n}\left|\frac{\hat{y}_t - y_t}{y_t}\right| < e$ holds, skip to step (4); otherwise, $z = z+1$, skip to step (2);

Step (4): print out the model parameters of the nonlinear auto-regressive model, and calculate the predicted values of the subsequent m blood glucose sequences based on the model, and the algorithm ends.

4 Experiment and Result Analysis

4.1 Experimental Process and Method

This paper selects the clinical data set of diabetes (diabetes research in children network) published by the National Institutes of health. The data set has continuous blood glucose values of more than 100 children with diabetes, and the sampling interval is 5 min. The blood glucose sequence of 10 patients in a day (24 h) was randomly selected as the experimental sample, and the dynamic neural network (NAR) algorithm, differential auto-regressive moving average (ARIMA) algorithm and the algorithm in this paper were used to model and analyze. In the experimental process, we focus on the comparison of the existing algorithm and the algorithm in the identification accuracy and calculation costs and other technical indicators, and analyze and discuss the relevant experimental results.

In this paper, the mean square error (MSE) and mean absolute percentage error (MAPE) are introduced to evaluate the modeling accuracy of various algorithms

$$MSE = \frac{1}{n}\sum_{t=1}^{n}(\hat{y}_t - y_t)^2 \tag{13}$$

$$MAPE = \frac{1}{n}\sum_{t=1}^{n}\left|\frac{\hat{y}_t - y_t}{y_t}\right| \tag{14}$$

In formula (13)–(14), y_t is the input sequence, $\hat{y}_t$ is the fitting sequence, and n is the length of blood glucose sequence.

4.2 Experimental Results and Analysis

Table 1. MSE and MAPE values of various algorithms

	NAR algorithm	ARIMA algorithm	Algorithm in this paper
MSE	17.63	676.17	59.24
MAPE	99.78%	91.55%	99.26%

The blood glucose samples of 10 patients were modeled and analyzed by three algorithms, and their average MSE and MAPE were calculated. The results are shown in Table 1. Taking No. 5 patients as an example, the fitting curves of various algorithms are shown in Fig. 1.

From the data in Table 1 and the fitting curve in Fig. 1, it can be seen that the NAR algorithm has the highest fitting accuracy. The algorithm in this paper is slightly lower than the former, while the ARIMA algorithm is the lowest. In fact, the NAR algorithm has powerful nonlinear mapping function, so in the process of network learning

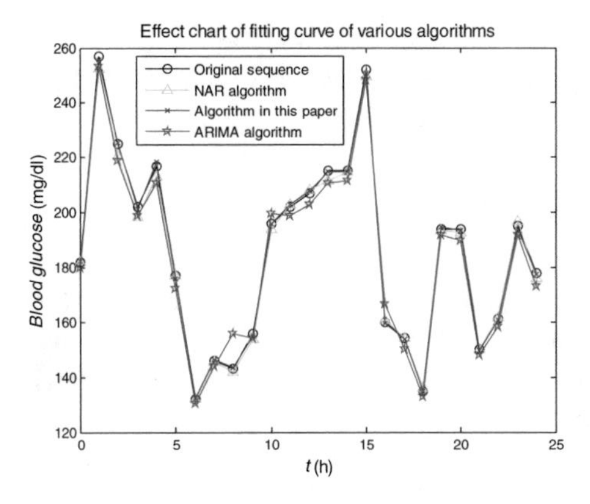

Fig. 1. Effect diagram of fitting curve of various algorithms

training, excellent fitting effect can be obtained by adjusting the network size and structure. In theory, the algorithm in this paper can approximate the original blood glucose sequence with infinitesimal error, but because of the inherent accumulated error in the iterative calculation process of parameters, its fitting accuracy is slightly lower than that of the NAR algorithm. As for the ARIMA algorithm, after several difference operations, the residual sequence still has some non-linear and non-stationary characteristics, so the fitting error is more obvious under the treatment of linear stationary modeling.

From the calculation time data in Table 2, it can be found that the algorithm in this paper needs the least calculation time, the ARIMA algorithm takes the second place, and the NAR algorithm needs the most calculation time.

Table 2. Average calculation time of various algorithms (unit: s)

	NAR algorithm	ARIMA algorithm	Algorithm in this paper
Calculation time	67.39	8.55	6.42

Compared with other algorithms, the dynamic neural network (NAR) algorithm needs a lot of computing time. The reason is that the network size and structure (input node, hidden layer number and hidden layer number) need to be adjusted repeatedly to determine the optimal network structure, and each adjustment leads to the relearning of network parameters. This blind adaptation test leads to the low computing efficiency Effective. Different from that, in the process of determining the optimal parameter vector of the nonlinear auto-regressive model, the algorithm in this paper introduces the relaxation iterative search method, which effectively avoids a large number of matrix operations and the solution of linear equations, so it only needs a small amount of calculation time to complete the solution of the parameter vector. In addition, the ARIMA algorithm takes less time because of its mature iterative algorithm.

From the above analysis, we can see that the algorithm in this paper has excellent calculation efficiency, and its fitting accuracy is only slightly lower than the NAR algorithm. Therefore, the algorithm in this paper is correct and effective.

5 Conclusion

Based on the existing algorithm of predicting blood glucose concentration, an algorithm of predicting blood glucose concentration based on nonlinear auto-regressive model is proposed. By introducing the relaxation iterative search algorithm to estimate the parameter vector of the nonlinear least squares of the polynomial approximation model, the calculation efficiency is effectively improved under the condition of ensuring the more accurate prediction accuracy, so it has certain practical application value.

Acknowledgements. This work was supported by natural science characteristic innovation project of Guangdong Provincial Department of Education (2018gktscx048), Foshan medical research project (20200351).

References

1. Rudenko, P.A., Pozhar, K.V., Litinskaia, E.L., et al.: Development of the short-term blood glucose prediction algorithm for using in closed-loop insulin therapy device. In: 2018 IEEE Conference of Russian Young Researchers in Electrical and Electronic Engineering (EIConRus). IEEE (2018)
2. Hovorka, R., Canonico, V., Chassin, L.J., et al.: Nonlinear model predictive control of glucose concentration in subjects with type 1 diabetes. Physiol. Meas. **25**(4), 905–920 (2004)
3. Efendic, H., Kirchsteiger, H., Freckmann, G., et al.: Short-term prediction of blood glucose concentration using interval probabilistic models. In: Mediterranean Conference on Control & Automation. IEEE (2014)
4. Wessel, N., Malberg, H., Bauernschmitt, R., et al.: Nonlinear additive autoregressive model-based analysis of short-term heart rate variability. Med. Biol. Eng. Comput. **44**(4), 321–330 (2006)
5. Ren, H., Xu, F., Ruwen, C., et al.: Linear/nonlinear modeling based on approximation theory. J. Southeast Univ. (Nat. Sci. Ed.) **048** (001), 30–37 (2018)
6. Ruwen, C., Ren, H.: Research on nonlinear autoregressive time series model and its prediction application. Theory Pract. Syst. Eng. (09), 196–205 (2015)
7. Muldowney, P.: Numerical Calculation. A Modern Theory of Random Variation: With Applications in Stochastic Calculus, Financial Mathematics, and Feynman Integration. Wiley (2012)

Signal Processing Based on Machine Learning Optical Communication

Yibo Guo(✉)

Sichuan Vocational and Technical College, Suining 629000, Sichuan, China
501180482@qq.com

Abstract. With the rapid development of machine learning, machine learning provides powerful tools to deal with problems in many fields. Its typical characteristic is the ability of self-learning and evolution. Signal equalization and optical spectral measurement are important problems in signal processing in optical communication, but there are few researches on the application of machine learning to solve this problem. This paper mainly focuses on the signal quality improvement and performance parameter monitoring in optical communication and applies machine learning to visible light communication equalization and spectrum analysis. The combination of machine learning algorithm and equalization technology enhances the ability of tracking channel characteristics, achieves intelligent learning and updating of equalizer, and introduces machine learning into spectrum analysis. The analysis of the spectrum was designed according to the intelligent mechanism of different machine learning algorithms. The original material of the analysis was the input data, which was adjusted according to the gap between the labels and output results corresponding to each set of data. All the test time was less than 0.8 s. SVM test time on wavelength, OSNR and bandwidth estimation is minimum (less than 0.34 s). Experiments show that the defined plane generated by support vector machine is more suitable for spectrum analysis, and the ability to summarize features is more suitable for spectrum analysis.

Keywords: Machine learning · Visible light communication · Unsupervised learning · Deep learning

1 Introduction

Over the past decade, machine learning has been successfully used in the fields of prediction, classification, pattern recognition, data mining, feature extraction, and behavior recognition [1]. With the development of 5G communication and the challenge brought by mass communication data processing, the combination of communication system and machine learning has become an irresistible trend [2]. Previous studies have shown that many algorithms in the field of machine learning can be used to solve nonlinear problems in communication systems, such as estimating parameters from noise, determining complex mapping relationship between input and output, inferring probability distribution of received signals, and estimating output values based on input samples, etc. [3]. For example, artificial neural network (ANN), deep

J. MacIntyre et al. (Eds.): SPIoT 2020, AISC 1282, pp. 141–147, 2021.
https://doi.org/10.1007/978-3-030-62743-0_20

neural network (DNN), support vector machine (SVM) and principal component analysis (PCA) can be used to detect and realize damage detection and performance monitoring in optical communication systems [4]. At the same time, traditional machine learning algorithms such as K-means, ANN, PCA, variable Db Bayesian expectation maximization, etc. have also been proved to have good effects in the face of signal demodulation, channel equalization and bit rate recognition [5]. However, recent research results show that DNN can obtain better system performance than traditional channel estimation algorithm when it comes to channel estimation of orthogonal frequency division multiplexing (OFDM) wireless communication systems and multi-input and multi-output (MIMO) systems under bad channels [6].

Machine learning is widely used in optical communication, which greatly promotes the development of intelligent systems [7]. Machine learning has the ability of self-learning and evolution. As long as there is new data, new mapping network can be established by adjusting the structure and parameters to further create new capabilities, which is of great significance for solving the problem of timely tracking channel characteristics and intelligent analysis of signal performance parameters in signal equalization. Machine learning in the future more technology will be constantly optical communication field of application of machine learning self-improvement will be take full advantage of the ability of self learning, real time to adapt to the characteristics of the different scenarios will be fully digging, different output optical communication system [8], continuously optimizing the resources distribution of communication system, improve the quality of communication system, testing and monitoring the level of communication, performance parameters of the output system unceasingly, in turn as the optimization of system data to support [9, 10].

This paper mainly focuses on the signal quality improvement and performance parameter monitoring in optical communication and applies machine learning to visible light communication equalization and spectrum analysis. The combination of machine learning algorithm and equalization technology enhances the ability of tracking channel characteristics, achieves the intelligent learning and updating of equalizer, and introduces machine learning into the spectrum analysis to make more accurate qualitative analysis and more accurate quantitative analysis of performance parameters from the spectrum.

2 Blind Equalization Algorithm for High-Speed Visible Light Communication

2.1 WDM-DCO-OFDM Offline VLC System

Traditional OFDM is complex form and bipolar signal, so it is not suitable for visible light communication with intensity modulation. In the case of visible light system with intensity modulation, the traditional complex form and bipolar signal are transformed into real form through conjugate stacking and adding DC bias for special transformation. The modulation adopted here will transform the signal S into a complex signal Sk, and then into a real signal in the following ways. First, it becomes a conjugate symmetric sequence:

$$X = \left[0, S_{k(k=1)}^{(N/2-1)}, 0, S_{k(k=N/2-1)}^{*1}\right]$$
$$X = \left[0, \{S_k\}_{k=1}^{\frac{N}{2}-1}, 0, \{S_k^*\}_{k=\frac{N}{2}-1}^{1}\right] \quad (1)$$

The Discrete Fourier Transform is expressed as:

$$X(k) = \sum\nolimits_{n=0}^{N-1} x(n) W_N^{kn} = \sum\nolimits_{n=0}^{N/2} x(n) W_N^{kn} + \sum\nolimits_{n=\frac{N}{2}+1}^{N-1} x(n) W_N^{kn} \quad (2)$$

2.2 Blind Equalization Method Based on Unsupervised Learning for Visible Light Communication System Combining WDM and DCO-OFDM

WDM and DCO-OFDM technologies are applied to the visible light communication system to create a high-speed communication environment. The blind equalization method of unsupervised learning is designed and tested to see whether it has a good equalization effect when applied to high-speed communication. A visible light communication system with high transmission rate is adopted to improve the spectrum utilization rate. The adoption of DCO-OFDM mode can increase the spectrum utilization rate of visible light communication system by three times, and the transmission rate of the system is greatly improved.

Light passes through the focusing lens together to form along the beam through the red, green and blue three color filter, filter separation is the main purpose of light signals of different frequency band filter out the light, light signals of different frequency band to the photodetector main function is to transfer to the light signal into electrical signal, signal more easily than optical signal processing, conversion, further into the oscilloscope in parallel three way signal samples into digital signal processing.

3 Basic Principles of Machine Learning Algorithms

In this paper, three machine learning algorithms are mainly used: artificial neural network, decision tree and K-nearest neighbor algorithm to analyze the spectrum diagram and establish a mapping model. The principle of the algorithm is as follows.

3.1 Machine Learning

Machine learning based on a large number of neurons in the model is similar to the neurons of the node link to each other, the hierarchy has a corresponding special corresponding relationship between neighboring neurons, connection between neurons represents the weight of each other and influence each other, each neuron as a excitation function, weights and the adjustment of the excitation function to force the output of the network structure is close to label values, in turn, the label value and the difference between the output value will force the entire network constantly adjust the weight of internal continued self-improvement.

3.2 Decision Tree

The principle of decision tree is feature extraction, which takes the feature extraction as the current classification standard and evaluation standard, recursively generates child nodes from top to bottom, divides data according to features continuously from the root node, distributes to the child nodes, each child node represents a feature, and finally reaches the leaf node. Special attention should be paid to feature selection, and the optimal feature should be selected recursively according to information gain and information gain ratio. The principle diagram of decision tree is shown in Fig. 1. Because the decision tree is easy to overfit, it needs to be pruned to reduce the size of the structure.

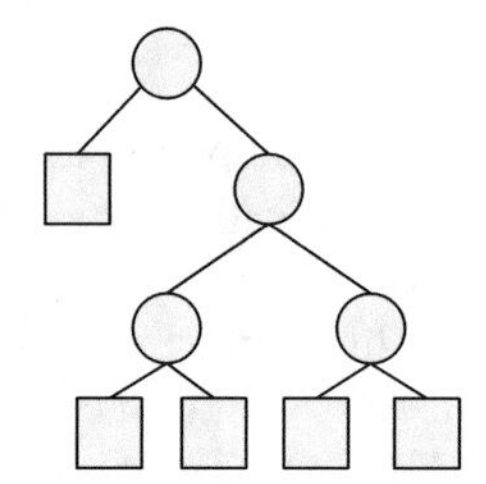

Fig. 1. Schematic diagram of decision tree

3.3 K Nearest Neighbor Algorithm

K neighbor algorithm is the basic principle of the training data and the condition of known labels, by measuring the distance between different characteristic value, in the space of the adjacent sample large probability of belonging to the same category, the characteristics of the test data and the features of training data comparison, find the most similar multiple data in the training data for the classification is the class of multiple data, the use of the test number of many characteristics in the statistical sample will mark it on the ownership of the category.

4 Optical Spectrum Analysis Based on Machine Learning

4.1 Simulation Results and Analysis

In order to achieve intelligent processing and analysis of signals, spectrum is selected as the basic analysis tool, and an intelligent spectrum graph analysis method based on machine learning is proposed. Next, train the mapping network according to the data and label; the third step is to input the spectrum diagram into the trained intelligent spectrum analysis module for feature extraction and performance analysis, and finally output the analysis results.

In the training mapping network, based on the simulation system, the control variable method is selected to set the signals to be processed, and the machine learning algorithm is applied to comprehensively analyze them. The wavelength in the training data set changes from 0.l nm to 1551.7 nm to 1553.2 nm under the same bandwidth and SNR. The bandwidth of the second training data set varies from LNM to 8 nm at the same wavelength and SNR. The SNR of the third training data set is extended from 15 dB to 30 dB at the same wavelength and bandwidth as they are defined by the ITU-T G.964 standard as WDM or the channel bandwidth of the finer WDM. At different bandwidth, the central wavelength was tuned at 0.1 nm (1551.7–1553.2 nm) and OSNR at a step size of L.0 dB (15–30 dB). Each set of training data has three tags, representing 9 bandwidth categories, 15 OSNR and 15 wavelengths respectively. In the training data set, each group of training data represents a spectrum, and each group has 20 spectral data collected, thus the total training data collected is 40500 (20 $\times$ 9 15 $\times$ 15). The same data size is used to collect test data sets.

The structure of the mapping network is determined by the machine learning algorithm adopted, and support vector machine (SVM) has the best test effect in the present invention. Therefore, we take support vector machine (SVM) as an example, SVM as a binary classifier can generate a boundary (called hyperplane) to classify two groups of data. Based on statistical theory learning, SVM tries to find the most appropriate hyperplane to ensure that the data closest to the hyperplane of each category is as far away from the hyperplane as possible, that is, to maximize the two categories and achieve the best classification effect of data.

4.2 Intelligent Spectrum Analysis Results

Referred to stay in the test of spectrum data input into the trained mapping network, the output of each group of test data label, and the results of the validation of different machine learning algorithms to prove in the advantages and disadvantages of the present invention, we also use the other machine learning algorithms to build training network mapping, applied in the control variable method, one by one, change the wavelength, OSNR and bandwidth, three kinds of performance parameters respectively using decision tree, artificial neural network (ANN), K neighbor (KNN) algorithm and support vector machine (SVM) four machine learning algorithm to measure the performance parameters of the three performance parameters is obtained respectively under the four kinds of machines such as identification precision See Fig. 2.

As shown in Fig. 2, SVM implements the optimal result, SVM implements the wavelength, the signal-to-noise ratio, the identification precision of the optimal bandwidth. Due to the availability of spectral data, the K-nearest neighbor algorithm (KNN) also achieves satisfactory accuracy. Decision trees are easy to implement, but only with low precision. Surprisingly, poor performance of artificial neural network, especially for the evaluation accuracy of OSNR is poorer, according to the theoretical analysis, mainly because the spectral data has a very high dimensions, a large number of neurons leads to extremely complex network structure, therefore, in a given period of training, artificial neural network is not completely adapt to the intelligent analysis of the spectrum of this new application scenarios.

In addition to accuracy, the spectrum analyzer has strict requirements on processing speed, and the test time is shown in Fig. 3. Therefore, we calculated the test time for each algorithm, based on a normal desktop computer. As long as the training is completed, the model does not need to be trained again. Therefore, compared with the

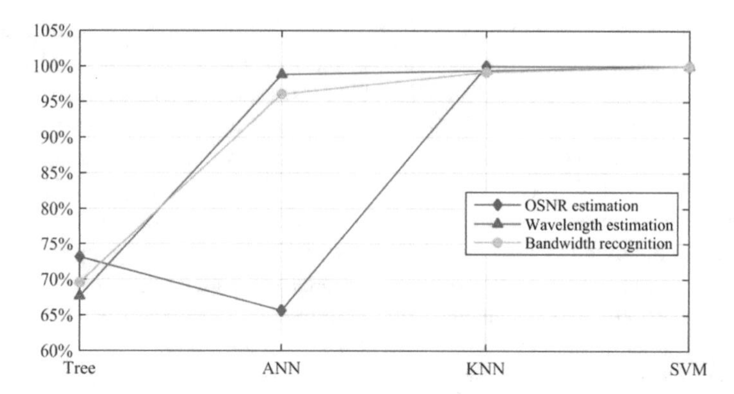

Fig. 2. Identification accuracy of spectrum parameters of the model constructed by four machine learning algorithms

training time, the test time is a key factor to measure whether the model can be analyzed in real time. The test results showed that all the test times were less than 0.8 s. SVM has the least test time on wavelength, 0SNR, and bandwidth estimation (less than 0.34 s), which is an acceptable speed for real-time processing and can be further improved with the use of high-performance computers. KNN is the most time-consuming algorithm, which is characterized by traversal calculation, so it requires a long test time. In summary, considering recognition performance and processing speed, SVM is obviously the best choice for wavelength, SNR and bandwidth analysis.

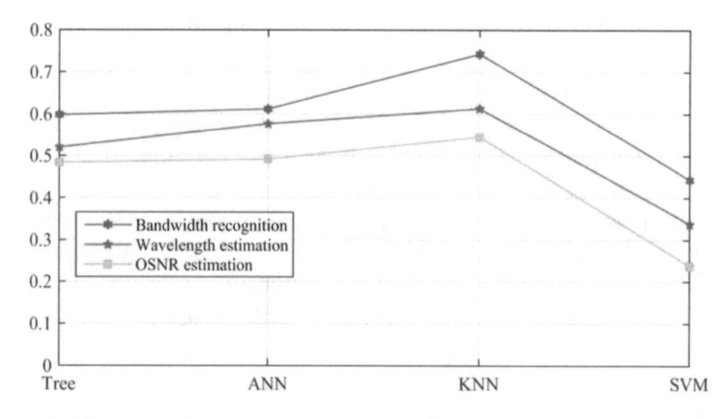

Fig. 3. Test time of spectrum parameters for models constructed by four machine learning algorithms

We believe that the proposed invention has the potential to be embedded in test instruments for intelligent spectrum analysis or applied to OPM modules to ensure the robustness of network operations. As a spectrometer, the proposed scheme is feasible for intelligent signal analysis and optical performance monitoring by embedding the spectrum analysis module of equipment such as OPM, which can convert signals from time domain to frequency domain to obtain spectral data.

5 Conclusion

SVM has the lowest test time on wavelength, OSNR, and bandwidth estimation (less than 0.34 s), which is an acceptable speed for real-time processing and can be further improved by using high-performance computers. Experiments show that the defined plane generated by support vector machine is more suitable for spectrum analysis, and the ability to summarize features is more suitable for spectrum analysis. In this paper, the application of machine learning to blind equalization and spectrum analysis in visible light communication is only a small step. In the future, machine learning will definitely create new applications, and even create new systems in optical communication.

References

1. Sidiropoulos, N.D., De Lathauwer, L., Fu, X., et al.: Tensor decomposition for signal processing and machine learning. IEEE Trans. Signal Process. **PP**(13), 3551–3582 (2017)
2. Tomislav, H., Jorge, M.D.J., Heuvelink, G.B.M., et al.: SoilGrids250m: global gridded soil information based on machine learning. PLoS ONE **12**(2), e0169748 (2017)
3. Jiang, C., Zhang, H., Ren, Y., et al.: Machine learning paradigms for next-generation wireless networks. IEEE Wirel. Commun. **24**(2), 98–105 (2017)
4. Wang, J.X., Wu, J.L., Xiao, H.: Physics-informed machine learning approach for reconstructing Reynolds stress modeling discrepancies based on DNS data. Phys. Rev. Fluids **2**(3), 1–22 (2017)
5. Baltrusaitis, T., Ahuja, C., Morency, L.P.: multimodal machine learning: a survey and taxonomy. IEEE Trans. Pattern Anal. Mach. Intell. **PP**(99), 1 (2017)
6. Voyant, C., Notton, G., Kalogirou, S., et al.: Machine learning methods for solar radiation forecasting: a review. Renew. Energy **105**, 569–582 (2017)
7. Chen, J.H., Asch, S.M.: Machine learning and prediction in medicine - beyond the peak of inflated expectations. N. Engl. J. Med. **376**(26), 2507 (2017)
8. Chernozhukov, V., Chetverikov, D., Demirer, M., et al.: Double/Debiased/Neyman machine learning of treatment effects. Am. Econ. Rev. **107**(5), 261–265 (2017)
9. Zhou, L., Pan, S., Wang, J., et al.: Machine learning on big data: opportunities and challenges. Neurocomputing **237**, 350–361 (2017)
10. Liu, S., Wang, X., Liu, M., et al.: Towards better analysis of machine learning models: a visual analytics perspective. Vis. Inform. **1**(1), 48–56 (2017)

Refined Management of Installation Engineering Cost Based on Artificial Intelligence Technology

Shu Zong[1(✉)], Bin Chen[1], and Wei Yan[2]

[1] Yunnan Institute of Technology and Information Career College, Kunming, Yunnan, China
928288652@qq.com

[2] Department of Engineering, Yunnan Institute of Technology and Information Career College, Kunming, Yunnan, China

Abstract. This paper studies the fine management of installation project cost based on artificial intelligence technology. Through investigation and research, this paper finds out the problems in the current project cost management, and puts forward the fine management scheme of installation project cost based on artificial intelligence technology. This scheme is mainly divided into five stages, the first stage is decision-making stage, accounting for 18.6% of the total project task The second stage is the design stage, accounting for 15.8% of the project task; the third stage is the bidding stage, accounting for 7.8% of the project task; the fourth stage is the construction stage, which is the most important stage of the whole project task, accounting for 35.1%; the last stage is the completion settlement stage, and the proportion of the project task is only in the construction stage, accounting for 22.7%.

Keywords: Artificial intelligence technology · Engineering cost · Refined management · Installation engineering

1 Introduction

In the field of construction and installation engineering, with the continuous expansion of the project scale, the data involved in cost management of relevant units increases rapidly, making cost management more difficult [1, 2]. Fine management of installation project cost can ensure reasonable and sufficient funds for project construction and control the cost of each stage of project construction within a reasonable range [3, 4].

The wide application of artificial intelligence technology has an inevitable impact on the construction and installation project cost [5, 6]. Through the integration of data and information model of construction and installation engineering, sharing and transmission are carried out in the whole life cycle process of project planning, construction, operation and maintenance, so that engineering and technical personnel can correctly understand and effectively respond to various building information [7, 8]. In this way, it can provide the basis of collaborative work for the design team and all parties including the construction and operation units, and play an important role in improving production efficiency, saving costs and shortening the construction period [9, 10].

J. MacIntyre et al. (Eds.): SPIoT 2020, AISC 1282, pp. 148–154, 2021.
https://doi.org/10.1007/978-3-030-62743-0_21

This paper first introduces the artificial intelligence technology used in the fine management of the installation project cost, then expounds the related concepts of the project cost management, summarizes and analyzes the existing problems in the current project cost management through investigation and research, and finally puts forward the fine management scheme of the installation project cost based on the artificial intelligence technology, which is divided into five stages: decision-making stage Section, design stage, bidding stage, construction stage and completion settlement stage, and details the main tasks of each stage.

2 Artificial Intelligence Technology and Project Cost Management

2.1 Artificial Intelligence Technology

Assuming that the number of neurons i n the input layer of a neural network is n, the input values are X_1, X_2, X_3... X_n, and the connection weight values between the input layer neurons and a neuron I in the next layer are W_1, W_2, W_3... Wn, the signal transmission process is as follows:

$$\sum\nolimits_{i=1}^{n} x_i * w_i \tag{1}$$

In general, when information is transmitted to a neuron, a corresponding bias data B will be added, that is:

$$\sum\nolimits_{i=1}^{n} x_i * w_i + b \tag{2}$$

In the artificial neural network, the calculation formula of a single neuron is:

$$y = f\left(\sum_{i=1}^{N} x_i * w_i + b\right) \tag{3}$$

Where f is the activation function. The activation function is an exponential non-linear function or a piecewise linear function.

2.2 Project Cost Management

On the surface, project cost management is the cost management of construction projects. But the meaning is different from different perspectives. The first kind: for the investor, project cost management refers to the management of all costs in project investment decision-making, design, bidding and construction stages. The second kind: for the contractor, it is only recognized as the management of the project contract price, which is jointly recognized by the owner and the Contractor through the bidding process.

3 Research Methods Used in This Paper

3.1 Qualitative and Quantitative Analysis

In this paper, the qualitative analysis method is used to study, analyze and judge the current situation and existing problems of construction project cost management in China, and the importance of fine management of installation project cost is obtained. On this basis, the paper introduces the combination of artificial intelligence technology and fine management of installation project cost, and analyzes the significance and specific strategies of introducing artificial intelligence technology into fine management of installation project cost by using quantitative analysis method and operation and data support.

3.2 Literature Research Method

In this paper, by means of computer network and other methods, relevant books, papers, literature and other materials at home and abroad are collected for extensive reading, so as to understand the application of artificial intelligence technology at home and abroad and the current situation of cost management of installation projects at home and abroad.

3.3 Survey Statistics

In this paper, through a survey of several construction companies in a certain area, to understand the current methods of installation project cost management, combined with the current artificial intelligence technology, to carry out comparative analysis, summed up the current problems of installation project cost management and make statistics.

4 Refined Management of Installation Engineering Cost Based on Artificial Intelligence Technology

4.1 Analysis of Current Situation of Cost Management of Installation Project

In this paper, the methods of cost management of installation engineering in China are understood through visit investigation, and the current artificial intelligence technology is combined to carry out comparative analysis, and the current problems of cost management of installation engineering are summarized. The specific statistics are shown in Fig. 1.

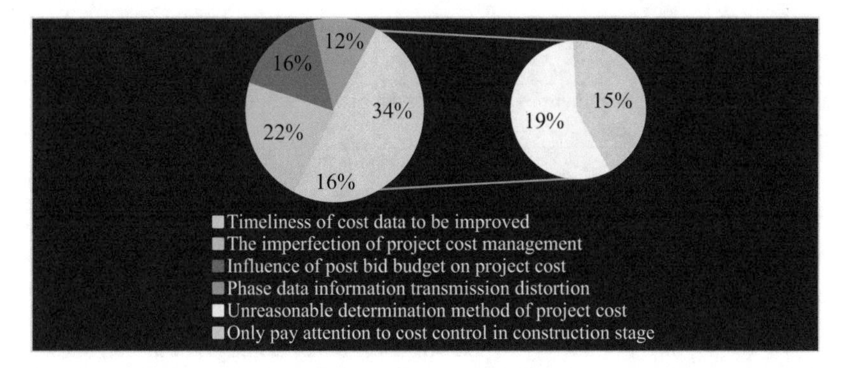

Fig. 1. Summary of installation project cost management

It can be seen from Fig. 1 that at present, the cost management of domestic installation project mainly has the following problems:

(1) The timeliness of cost data needs to be improved, accounting for 15.7%
Confirmation of completion settlement, as an important link reflecting the actual cost of the construction project, is the last critical stage of the project cost control. It is the main task of the whole stage to check the effectiveness of the actual completed quantities, visas, hidden data, design changes and other data of the construction project. However, in the construction stage, all kinds of data are not confirmed on site in time, and the validity cannot be confirmed until the final completion of the project.

(2) The project cost management is not perfect, accounting for 22.4%
The lack of a sound cost management system is the main factor leading to the lack of effective control of cost management, especially when the construction unit is lack of awareness of the whole process control, there will be serious cost management confusion, resulting in the project can not continue, and then affect the construction progress.

(3) Impact of post bid budget on project cost, accounting for 16.3%
It is necessary to strengthen the communication between various departments and ensure the orderly development of follow-up work on the basis of active cooperation. The workload of construction engineering is large. Due to the long construction period, there will be various unstable factors, which will seriously affect the final project quality.

(4) Stage data information transmission distortion, accounting for 11.7%
The management of each stage presents different characteristics, and there are certain deviations in the information transmission of the adjacent stages.

4.2 Application of Artificial Intelligence Technology in Fine Management of Installation Project Cost

According to the problems existing in the cost management of domestic installation projects, this paper summarizes the application of artificial intelligence technology in the refined cost management of installation projects. The specific steps are shown in Table 1 and Fig. 2.

Table 1. Artificial intelligence technology in the stage of refined management of installation project cost

Each stage of refined management of installation engineering cost	Main tasks of each stage
Decision making stage	Through the design and construction of the proposed project artificial intelligence model, and the evaluation of the project
Design phase	Find the basic data to segment the material calculation to control the cost
Bidding stage	Extract engineering quantity from engineering information, so that bidding information and design information can be related to each other
Construction stage	Adjust plan, improve work efficiency and reduce construction cost
Completion settlement stage	Improve the speed of project settlement and shorten the time of project completion settlement

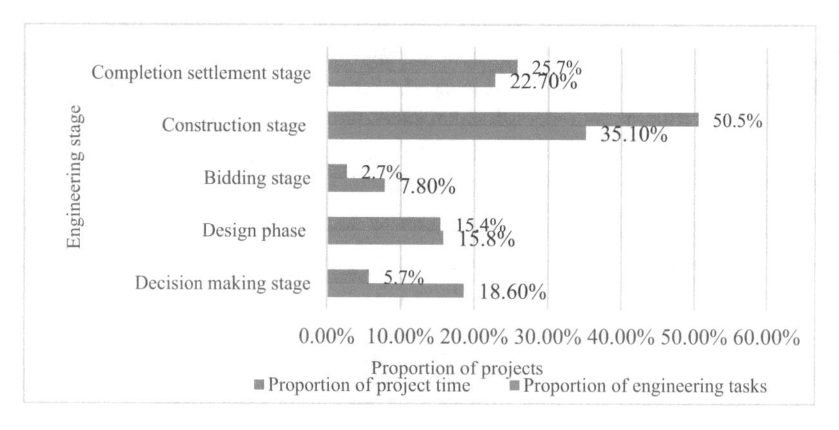

Fig. 2. Proportion of artificial intelligence technology in each stage of refined management of installation project cost

It can be seen from Table 1 and Fig. 2 that the application of artificial intelligence technology in the refined management of installation project cost is mainly divided into five stages: The first stage is the decision-making stage, accounting for 18.6% of the total project volume task, accounting for about 5.7% of the time; the second stage is the design stage, accounting for 15.8% of the project task, accounting for about 15.4% of the time; the third stage is the bidding stage, accounting for 7.8% of the project task, The time accounts for about 2.7%, the fourth stage is the construction stage, which is the most important stage of the project task, accounting for 35.1%, and the time accounts for about 50.5%. The last stage is the completion settlement stage, which is only the construction stage, accounting for 22.7%, and the time accounts for about 25.7%.

(1) Decision making stage
Using the visual characteristics of artificial intelligence technology, through the design and construction of the proposed project artificial intelligence model, and the evaluation of the project, combined with the artificial intelligence model to collect relevant cost software, carry out statistical calculation, and finally complete the project investment budget. In the project decision-making stage, it is necessary to carry out the refined cost management of its project to select the optimal investment scheme from a variety of construction schemes, and the decision-making scheme is also the premise of high-quality cost management.

(2) Design stage
In the design stage, according to the preliminary design scheme, through artificial intelligence technology, find out the basic data segmentation material calculation to control the cost. In the stage of construction drawing design, artificial intelligence technology is used to build a model for virtual construction, which provides basis for subsequent progress payment, material procurement plan formulation, labor planning and other work.

(3) Bidding stage
Combined with the development and control of artificial intelligence technology, the work efficiency of bidding management department can be greatly improved. Supervise the whole bidding process and promote the open, fair and equitable development of the construction industry. In the project cost, investment and bidding should stand on the overall point of view, in line with the principle of preciseness, strengthen the decision-making power and influence of subsequent projects in investment, ensure the actual income of project investment, and realize the maximization of economic benefits of project construction.

(4) Construction stage
Through the plan management of artificial intelligence, progress measurement and payment, project change management, timely adjust the plan, improve work efficiency, reduce construction cost, and timely pay the project schedule according to the project quantity. The reasonable application of artificial intelligence technology to the construction progress can improve the management of the construction link of the project cost, real-time control and feedback of the construction progress, thus shortening the expected construction time, achieving the goal of reducing the construction cost and improving the management level.

5 Conclusions

In this paper, the research on the refined management of installation engineering cost based on artificial intelligence technology is carried out, and it is concluded that with the support of artificial intelligence technology, the effectiveness of the application of artificial intelligence technology can be maximized in the refined management of engineering cost, so as to promote the engineering benefit to a new height and lay a solid foundation for strengthening the core competitiveness of the construction unit.

The combination of the two can realize the efficient management of the project, mostly for the establishment of the model, providing help for the pre construction verification, shortening the adjustment time of the pre construction scheme, and effectively improving the quality of the project construction.

References

1. Sardehaei, E.A., Mehrjardi, G.T., Dawson, A.: Full-scale investigations into installation damage of nonwoven geotextiles. Geomech. Eng. **17**(1), 81–95 (2019)
2. Koh, K.X., Wang, D., Hossain, M.S.: Numerical simulation of caisson installation and dissipation in kaolin clay and calcareous silt. Bull. Eng. Geol. Env. **77**(3), 953–962 (2018)
3. Koerner, R.M., Koerner, M.R., Koerner, G.R.: Utilizing PVDs to provide shear strength to saturated fine-grained foundation soils. Int. J. Geosynth. Ground Eng. **3**(4), 31 (2017)
4. Passini, L.D.B., Schnaid, F., Salgado, R.: Experimental study of shaft resistance of model piles in fluidized and nonfluidized fine sand. J. Offshore Mech. Arct. Eng. **139**(5), 0520011–05200112 (2017)
5. Howard, A.: Samuel Howard and the music for the installation of the duke of grafton as chancellor of Cambridge university, 1769. Eighteenth Century Music **14**(02), 215–234 (2017)
6. Esmaili, D., Hatami, K.: Comparative study of measured suction in fine-grained soil using different in-situ and laboratory techniques. Int. J. Geosynth. Ground Eng. **3**(3), 27 (2017)
7. Benassai, G., Luccio, D.D., Corcione, V., et al.: Marine spatial planning using high-resolution synthetic aperture radar measurements. IEEE J. Oceanic Eng. **43**, 586–594 (2018)
8. Weaver, A.M., Parveen, S., Goswami, D., et al.: Pilot intervention study of household ventilation and fine particulate matter concentrations in a low-income urban area, Dhaka, Bangladesh. Am. J. Trop. Med. Hyg. **97**(2), 615–623 (2017)
9. So, A., Chan, W.L.: A study of linear PMSM driven ropeless elevators. Building Serv. Eng. Res. Tech. **40**(1), 93–108 (2019)
10. Mahawish, A., Bouazza, A., Gates, W.P.: Effect of particle size distribution on the bio-cementation of coarse aggregates. Acta Geotech. **13**(4), 1019–1025 (2018)

Application of Artificial Intelligence Technology in International Trade Finance

Shuangmei Guo(✉)

Department of Economics and Management,
Yunnan Technology and Business University, Kunming, Yunnan, China
13741168@qq.com

Abstract. Artificial intelligence technology is a new method proposed by people inspired by the biological evolution mechanism and some natural phenomena. Because it can well solve the modeling and optimization problems of complex systems, it has received extensive attention and applications in various fields. As the scale, quantity, and scope of international trade continue to expand, the complexity and uncertainty of trade increase, so artificial intelligence is more widely used in international trade. The purpose of this article is the application of artificial intelligence technology in international trade finance. Aiming at the deficiencies of Artificial Bee Colony (ABC), this paper proposes two types of improvement strategies: In order to improve the convergence accuracy and operational stability of ABC, this paper proposes an improved artificial bee colony algorithm, and also proposed a hybrid artificial bee colony algorithm with predictive selection ability, this paper tested it through 23 benchmark optimization problems The simulation results show that the improved algorithm has a good global search capability and convergence speed, so that it can play an important role in the healthy development of China's international trade and has practical significance.

Keywords: Artificial intelligence technology · Artificial bee colony algorithm · International trade · Global search

1 Introduction

Compared with traditional methods, artificial intelligence technology has better adaptive capabilities. It usually does not need to understand the relevant knowledge of complex systems to solve many complex problems, which mainly include artificial neural networks, fuzzy logic, intelligent optimization algorithms and other methods. This article focuses on the in-depth research and application of artificial bee colony algorithm in intelligent optimization algorithm [1, 2]. As a key area of global technology and industrial transformation, artificial intelligence has become a strong driving force for sustainable economic development. Compared with developed countries such as Europe and the United States, the international trade of artificial intelligence in my country is mostly in the early stage of growth [3]. Therefore, in-depth research on the financing efficiency of the artificial intelligence industry is of great significance to the sustainable and healthy development of the industry [4].

J. MacIntyre et al. (Eds.): SPIoT 2020, AISC 1282, pp. 155–162, 2021.
https://doi.org/10.1007/978-3-030-62743-0_22

A series of heuristic intelligent optimization algorithms, such as genetic algorithms, particle swarms, biogeographic optimization algorithms, and leapfrog algorithm [5]. Traditional optimization algorithms are mainly proposed for differentiable convex optimization problems. For non-convex functions, the It is easy to fall into a local minimum [6, 7]. In addition, with the continuous improvement of the development level of international trade, it is difficult to obtain all relevant information about the optimization problem, resulting in the limitation of traditional optimization algorithms in practical trade applications [8]. In recent years, inspired by natural phenomena, proposed artificial bee colony algorithms, etc., these algorithms can effectively and quickly find the optimal solution, and overcome the shortcomings of traditional optimization algorithms, have a wider range of applications, and have been successfully applied Various fields. However, they still have some shortcomings, so this chapter focuses on one of the intelligent optimization algorithms to further improve its performance [9, 10].

Artificial Bee Colony (ABC) is an intelligent optimization algorithm inspired by bee colony foraging behavior. Many literatures have demonstrated that the ABC algorithm has good optimization accuracy and convergence speed through comparison with optimization algorithms such as genetic algorithm, particle swarm and ant colony. Due to the advantages of simple concept, easy implementation and fast convergence of ABC algorithm, ABC has attracted more and more attention and has been successfully applied to many fields. However, ABC still has some shortcomings, such as being easy to fall into a local optimal solution sometimes. Therefore, in view of the shortcomings of the ABC algorithm, this chapter proposes two types of improved artificial bee colony algorithms and verifies the effectiveness of these algorithms through classic optimization problems. At the same time, it shows that it has strong global search capabilities and fast convergence speed, thus It laid the foundation for the optimization of international trade financing in the future.

2 Method

2.1 Principle of Artificial Bee Colony Algorithm

In the ABC algorithm, the entire bee colony contains three types of bees: hire bees, wait-and-see bees and scout bees. In the algorithm, hired bees and wait-and-see bees account for half of the entire bee colony. Only after the food source is abandoned, the corresponding hired bees become scout bees, and the number of hired bees is equal to the number of food sources. Each food source represents a feasible solution to an optimization problem, and the amount of pollen contained in the food source represents the quality of the feasible solution (adaptability). First, the ABC algorithm randomly generates an initial population $P(C = 0)$ of SN food sources, where each food source x_i (i = 1, 2…, SN) is a D-dimensional vector, and D represents the parameter of the optimization problem. number. The probability that the food source shared by the wait-and-see bees selected by the hired bees is completely still in the pollen amount of a related food source, and the probability value can be calculated by formula (2–1).

$$P_i = \frac{fit_i}{\sum_{j=1}^{SN} fit_j} \quad (1)$$

$$fit_i = \begin{cases} \frac{1}{1+fi} & fi \geq 0 \\ 1 + abs(fi) & fi < 0 \end{cases} \quad (2)$$

fit_i——the pollen value of the i-th food source (fitness value);
fi——The i-th solution of the optimization problem;

2.2 Improved Artificial Bee Colony Algorithm (I-ABC)

With the deepening of research, a large number of experimental results show that the artificial bee colony algorithm, like other intelligent optimization algorithms, also has its own limitations when solving multi-objective optimization problems, such as weak local search ability and slow later convergence speed. In addition, in scientific research and engineering practice, most multi-objective optimization problems are subject to certain conditions. When solving these multi-objective optimization problems, these conditions are required to be met while making multiple objectives reach the Pareto optimal. For the performance of ABC algorithm, this section proposes an improved artificial bee colony algorithm (I-ABC). In I-ABC, three brand-new variables (current optimal solution, inertia weight and acceleration coefficient) are introduced into the original ABC algorithm. In addition, in order to better balance the exploration and mining capabilities, the modified formulas for hired bees and wait-and-see bees are different, which is mainly reflected in the second acceleration factor. The modified formulas for hire bees and wait-and-see bees are:

$$V_{ij} = x_{ij}w_{ij} + 2(\Phi_{ij} - 0.5)(x_{ij} - x_{kj})\Phi_1 + \varphi_{ij}(x_j - x_{kj})\Phi_2 \quad (3)$$

Where: w_{ij}——inertia weight, which can control the influence of the original food source x_i on the new food source V_i;

Φ_1, Φ_2—— positive parameters, which can control the maximum step length of the change;

Φ_{ij}, φ_{ij}—— is a random number between [0,1];

2.3 Predictive Selection Artificial Bee Colony Algorithm (PS-ABC)

Although I-ABC can find the global optimal solution very quickly for most of the high-dimensional single-peak and multi-peak benchmark functions, for a few benchmark functions I-ABC sometimes easily fall into local optimal solutions, and even some It is not as good as the solutions found by ABC and GABC. In order to achieve the goals involved in the above in a variety of fields, while having the advantages of the three algorithms of ABC, GABC and I-ABC, this section proposes one Efficient hybrid artificial bee colony algorithm (Artificial Bee Colony with the abilities of Prediction and Selection, PS-ABC) with predictive selection ability. The operators for hire bees and wait-and-see bees to obtain candidate food sources are as follows:

$$V_{ij}^{1} = x_{ij}w_{ij} + 2\left(\Phi_{ij}^{1} - 0.5\right)\left(x_{ij} - x_{kj}\right)\Phi_1 + \varphi_{ij}^{1}\left(y_j - x_{ij}\right)\Phi_2 \tag{4}$$

$$V_{ij}^{2} = x_{ij}w_{ij} + 2\left(\Phi_{ij}^{2} - 0.5\right)\left(x_{ij} - x_{kj}\right) + \varphi_{ij}^{2}\left(y_j - x_{ij}\right) \tag{5}$$

$$V_{ij}^{3} = x_{ij} + \Phi_{ij}^{3}\left(x_{ij} - x_{kj}\right) \tag{6}$$

$$V_{ij} = \begin{cases} V_{ij}^{1} \text{ if } \mathrm{fit}\left(V_i^1\right) \geq \max\left\{\mathrm{fit}\left(V_i^2\right), \mathrm{fit}\left(V_i^3\right)\right\} \\ V_{ij}^{1} \quad \text{else if } \mathrm{fit}\left(V_i^2\right) \geq \mathrm{fit}\left(V_i^3\right) \\ V_{ij}^{3} \quad \text{others} \end{cases} \tag{7}$$

3 Experiment

3.1 Experimental Subjects

The research object of the thesis is undoubtedly an example of international trade investment portfolio. The research area is limited to the use of artificial intelligence technology in international trade financing, and the fundamental foothold is to be implemented on the algorithm after all. First of all, select the optimal algorithm in artificial intelligence, and use the artificial bee colony algorithm to calculate the optimal combination in the international trade investment portfolio, so as to maximize the profit. In addition, the artificial bee colony algorithm has certain defects. The ABC algorithm has been effectively promoted and improved, and a reasonable test has been carried out based on China's investment in international trade, and the test is based on artificial intelligence.

3.2 Experimental Method

a. Graphic and text complementary method: Research on this topic, through example operations and a combination of images and text, to illustrate the advantages and methods of artificial bee colony algorithm in interpreting the combination in international trade finance through example operations and the standardization of the combination. The proportional component in trade is more scientific.
b. Comparative induction: compare the difference between the traditional artificial bee colony algorithm and the improved artificial colony algorithm, thereby strengthening the advantages of artificial intelligence technology in international trade finance and increasing the rate of return on risky assets.

c. Design practice method: Practice summarizes theoretical experience under artificial intelligence technology, theoretical experience further guides the practice of international trade finance, and then returns to practice to guide practice, which further deepens the role of the research theme.

4 Discuss

4.1 Performance Comparison of I-ABC, ABC, GABC

In order to verify the performance of the I-ABC and PS-ABC algorithms, this paper applies them to 23 classic benchmark optimization problems. Under the set parameters, the I-ABC algorithm can find the theoretical optimal value of two high-dimensional test functions (f1, f2) in different dimensions, which does not include the test function f3 when the dimension is 20. The optimal solution found by I-ABC on the other three test functions is very close to the theoretical optimal value, which can be approximately regarded as the theoretical optimal value. The optimal solution found by the I-ABC algorithm on 13 test functions (f1, f2, f3) is much better than the optimal solution found by the basic ABC algorithm, while the basic ABC algorithm only compares the results of the optimization of the test function f3 Okay, and the test function f2 with dimension 30 is not included. For most test functions, the number of iterations for I-ABC to reach convergence is significantly smaller than that of the ABC algorithm. From the number of convergence iterations, it can be seen that I-ABC requires fewer iterations to reach the optimal solution. In general: Most optimization problems I-ABC have good optimization effects and extremely fast convergence speed, but I-ABC still has some shortcomings, such as sometimes falling into a local optimal solution, and even the optimization effect of individual optimization problems is not as good as ABC or GABC. The optimization effect of I-ABC is good. In addition, there is a shortcoming for the fixed-dimensional test function I-ABC, that is, there is a small descending gradient in the initial stage of optimization, but I-ABC can reach the optimal solution faster. Based on this, it can be said that I-ABC is an efficient optimization algorithm for most test functions (Table 1).

Table 1. The running results of algorithms ABC and I-ABC for high-dimensional test functions

Function	Dimension	ABC I-ABC					
		C.I.	Mean	S.D.	C.I.	Mean	S.D.
f_1	20	998	6.1873×10^{-16}	2.1148×10^{-16}	208	0	0
f_2	20	1000	1.3569×10^{-10}	7.1565×10^{-11}	726	0	0
f_3	20	1000	3.1314×10^{3}	1.1867×10^{3}	1000	4.5408×10^{3}	2.6919×10^{3}

4.2 Test the Effect of the Improved I-ABC Algorithm in Solving the Multi-objective International Trade Portfolio Optimization Problem

Table 2 gives the probability distribution of risky asset yield and liquidity. In addition, assuming that the transaction fee rate of all risky assets is 0.007, the minimum requirement of investors' liquidity expectations for risky securities is 0.005, and investors expect that the upper limit of the investment ratio of a single risky security does not exceed 0.8. At the same time, in order to verify the effectiveness and advancement of the improved multi-objective artificial bee colony algorithm proposed in this paper, a series of simulation experiments are carried out here. In the simulation experiments, in view of the fairness principle, all algorithms set the same parameters. The initial population Pop size of each multi-objective optimization algorithm is set to 50, and the maximum number of iterations is 100. In order to reduce the adverse effect of randomness on algorithm evaluation, each multi-objective optimization algorithm runs independently 20 times.

Table 2. I-ABC algorithm to solve multi-objective international trade portfolio

Asset	Return	Liquidity
A1	$(-0.38, -0.175, 0.456, 0.79)$	$(0.0065, 0.0116, 0.0206, 0.024)$
A2	$(-0.56, -0.204, 0.548, 0.76)$	$(0.0002, 0.0008, 0.0027, 0.0034)$
A3	$(-.0384, -0.003, 0.51, 0.63)$	$(0.0003, 0.00126, 0.00515, 0.01146)$
A4	$(-0.456, -0.096, 0.624, 0.76)$	$(0.0013, 0.0017, 0.0048, 0.008)$

4.3 With or Without the Investment Return Ratio Under the Algorithm

It can be seen intuitively that the ratio of the return on investment under artificial intelligence technology and the traditional environment can be seen, the use of high-quality algorithms can significantly improve the return on investment, the use of artificial intelligence can quickly and accurately search out The most investment portfolio to maximize returns. Comparison of investment returns with or without the use of algorithms as show in Fig. 1.

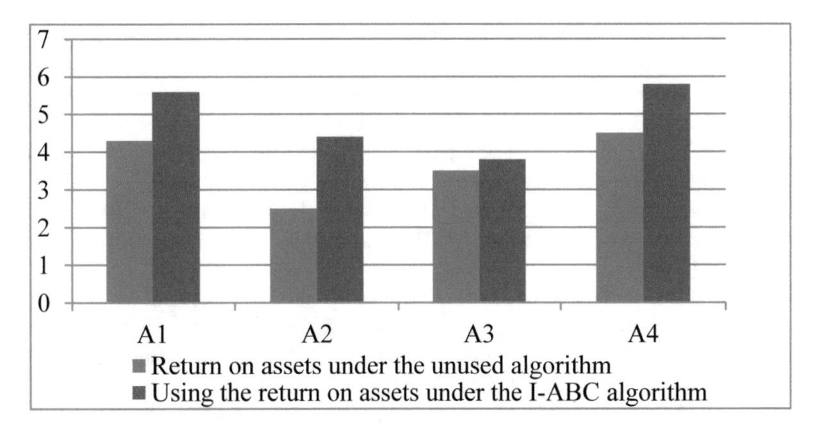

Fig. 1. Comparison of investment returns with or without the use of algorithms

5 Conclusion

The problem of portfolio optimization in international trade finance has always been one of the current research hotspots of financial theory. Its main research content is to effectively allocate financial capital in a complex and full of uncertain financial markets, thereby Achieve a balance between the returns and risks of financial assets. In many cases, the direction of national investment is often not able to accurately predict the probability distribution of asset returns, and its description of returns and risks can usually only be summarized by some vague concepts, which makes asset returns to a large extent Performance is fuzzy uncertainty. Therefore, in order to solve the problem of fuzzy uncertainty in international finance, we discussed the investment portfolio optimization model based on the possibility theory. Because the proposed multi-objective portfolio optimization model is a multi-objective optimization problem with complex constraints, its calculations are relatively complicated, and the problem has also been proved to be an NP-Hard problem in combinatorial optimization problems, which is difficult to solve by conventional methods. Therefore, for the proposed multi-objective portfolio optimization model with complex constraints, this paper designs an improved constrained multi-objective artificial bee colony algorithm (Improved Constrained Multi-objective Artificial Bee Colony, ICMOABC), which uses artificial intelligence technology. The improved artificial bee colony algorithm can reasonably solve the optimal combination, so as to maximize the return on investment in international trade, which is of great help to China's international trade financing.

References

1. Guo, T., Eckert, R., Li, M.: Application of big data and artificial intelligence technology in industrial design. Int. J. Adv. Trends Comput. Sci. Eng. **5**(1), 10–14 (2020)
2. Gonzalez-Cancelas, N., Serrano, B.M., Soler-Flores, F.: Seaport sustainable: use of artificial intelligence to evaluate liquid natural gas utilization in short sea shipping. Transp. J. **58**(3), 197–221 (2019)

3. Weber, F.D., Schutte, R.: State-of-the-art and adoption of artificial intelligence in retailing. Digit. Policy Regul. Gov. **21**(3), 264–279 (2019)
4. Wu, T.: AI industry in China and the United States: "convergence" should exceed "competition". China's Forgn Trade **567**(03), 44–45 (2018)
5. Li, Y., Peng, Y., Luo, J., et al.: Spatial-temporal variation characteristics and evolution of the global industrial robot trade: a complex network analysis. PLoS One **14**(9), e0222785 (2019)
6. Ben Hajkacem, M.A., Ben N'Cir, C.E., Essoussi, N.: STiMR k-Means: an efficient clustering method for big data. Int. J. Pattern Recogn. Artif. Intell. **33**(8), 1950013.1–1950013.23 (2019)
7. Dulebenets, M.A., Golias, M.M., Mishra, S.: A collaborative agreement for berth allocation under excessive demand. Eng. Appl. Artif. Intell. **69**, 76–92 (2018)
8. Bonnar, T.: The first steps in mitigating contemporary risks to our strategic sea lines of communication. Martime Report. Eng. News **81**(8), 26 (2019)
9. Velichko, A.: Multi-commodity flows model for the pacific Russia inter-regional trade. Int. J. Artif. Intell. **16**(1), 158–166 (2018)
10. Liu, Z., Zhang, J., Zhang, M., et al.: Hash-based block matching for non-square blocks in screen content coding. Int. J. Pattern Recogn. Artif. Intell. **32**(8), 1850027.1–18500270.13 (2018)

An Improved Genetic Algorithm for Vehicle Routing Problem

Jiashan Zhang(✉)

Chongqing Vocational Institute of Engineering, Chongqing 402260, China
zh_jiashan@163.com

Abstract. Evolutionary algorithms, including genetic algorithm, usually appear premature convergence. In this paper, new crossover and mutation operators are introduced. This paper addresses an application of improved genetic algorithms (IGA) for solving the Vehicle Routing Problem (VRP). After the introduction of new genetic operators, diversity of population becomes abundant. The ability of global search is obviously improved in new algorithm. Simulations indicate that new genetic algorithm is competitive with other modern heuristics.

Keywords: Premature convergence · Genetic operator · Global search · Simulations

1 Introduction

In the optimization of distribution networks, the vehicle routing problem (VRP) [1, 2] remains a challenging issue. Traditional exact algorithms [3] play a very important role as well as its limitation. They can only solve VRPs less than 50 nodes, not solve large-scale VRPs. Approximate algorithms, which aim at finding approximate solutions in the finite time, are applied to large-scale instances widely. For example, [4] solves VRP based on tabu search. [5] applies simulated annealing to VRP. An improved particle swarm optimization (PSO) is proposed to solve the vehicle routing problem in[6]. Jun Zhang [7] proposed an Ant Colony Optimization for VRP with Time Windows. An adaptive memory strategy for VRP is proposed in [8]. Since genetic algorithm (GA) [9, 10] was put forward, it has gained extensive attention of researchers. It is robust and flexible. It has been applied to many combinatorial problems. For example, genetic algorithm is used to optimize the capacitated clustering problem [11]. However, the general Genetic Algorithm usually appear premature convergence. Numerous successful applications intensively favor improved algorithm. In this paper, a competitive improved genetic algorithm (IGA), in which new crossover and mutation operators are designed, is proposed.

2 Formulation for VRP

There are m vehicles in the distribution center, which are used to delivery cargo for n customers $(v_1, v_2, \ldots, v_n)$. Demand level of each customer q_i is known. All vehicles are required to start from distribution center. All vehicles are required to return back to

J. MacIntyre et al. (Eds.): SPIoT 2020, AISC 1282, pp. 163–169, 2021.
https://doi.org/10.1007/978-3-030-62743-0_23

distribution center after finishing distribution. Every customer's demands are satisfied by just one vehicle' service. Here, let the total distance traveled represent distribution cost. To find an path with the minimum cost is this paper objective.

Set 0–1variables as follows

$$x_{ijk} = \begin{cases} 1, & \textit{if} \text{ vehicle k visits customer j after i} \\ 0, & \textit{else} \end{cases} \qquad y_{jk} = \begin{cases} 1, & \textit{if} \text{ vehicle k visits client j} \\ 0, & \textit{else} \end{cases}$$

The objective function is:

$$\min z = \sum_{i} \sum_{j} \sum_{k} c_{ij} x_{ijk} \tag{1}$$

S.t.

$$\sum_{i=1}^{n} q_i y_{ik} \leq Q, \qquad k = 1, 2, \ldots, m \tag{2}$$

$$\sum_{i} \sum_{j} d_{ij} \cdot x_{ijk} \leq \mathrm{L}, \quad k = 1, 2, \ldots, m \tag{3}$$

$$\sum_{k=1}^{m} y_{ik} = \begin{cases} 1, & i = 1, 2, \ldots, n \\ m, & i = 0 \end{cases} \tag{4}$$

$$\sum_{i=0}^{n} x_{ijk} = y_{jk}, \qquad j = 1, 2, \ldots, n; \quad k = 1, 2, \ldots, m \tag{5}$$

$$\sum_{j=0}^{n} x_{ijk} = y_{ik}, \qquad i = 1, 2, \ldots, n; \quad k = 1, 2, \ldots, m \tag{6}$$

3 The Proposed IGA for VRP

3.1 Framework

In this paper, calculate an initial population at first, i.e. the first generation. The paper suppose that the initial population is made up of n individuals. Population size (popsize) should be appropriate. For each individuals, there is a fitness value. We can compute them, and rank individuals according to fitness value. Then, we choose a pair of individuals with high fitness value(parents), and apply cross operation on them. Two new individuals are get, called children. Subsequently, we apply mutation operation to the children produced. The algorithm stops until termination condition is satisfied.

The improved algorithm framework is shown below:

```
Initialization
Randomly generate the initial chromosome population
Repeat
k=1
 For p =l:mdo
Choose two chromosomes Si,Sjfrom previous generation
Produce the chromosome  Sn based on the proposed crossover;
  If fit(Sn)> fit(Si), and fit(Sn)>fit(Sj)
         Add Sn to population
  Else
         Abandon Sn
       Excute mutation operation;
End for
  k=k+ 1
 Until (termination conditions)
```

3.2 Coding and Fitness Function

Coding plays an important role. For each VRP solution, good coding is helpful to identify the number of vehicles [12]. Actually, a chromosome I(n) simply is corresponding to a sequence S of nodes. For example, there are 9 customers, i.e. 9 nodes. Randomly generated a chromosome: 5-3-8-2-7-1-6-9-4. We can divide the chromosome 5-3-8-2-7-1-6-9-4 into three parts: 0-5-3-8-2-0, 0-7-1-6-0, and 0-9-4-0. Each part stands for an feasible route. For each vehicle, the total demand of customer nodes must less than the capacity of vehicle. Total number of vehicles required does not exceedm. Then the solution is legal; otherwise, it is illegal.

To measure quality of solution, each chromosome has a fitness value assigned. Here, This paper choose the total distance of travelling for all vehicles as fitness value.

$$fit(Si) = \frac{1}{total\_distance(Si)}$$

3.3 Crossover

Crossover is performed with two chromosomes, called parent chromosomes. The crossover aims to generate off springs, child chromosomes. It is a probabilistic process. In most cases, it is possible to get one child or two children from two parents. It leads to take information from both parents and transfer it to the children, which accelerates search process. To prevent premature convergence, the crossover probability p_c is not constant. When the iteration is lower than ten percent of the prespecified number of generation N_{cmax}, the crossover probability remains fixed, or shown below.

$$p_c = \begin{cases} k_1, & f_c \le f_{avg} \\ \frac{k_1(f_{\max} - f_c)}{f_{\max} - f_{avg}}, & f_c > f_{avg} \end{cases} \quad (7)$$

Where k_1 is a constant, and $0 < k_1 < 1$.

Let the two parent solutions be $P_1 = (123456789)$, $P_2 = (452187693)$. The crossover procedures are shown below.

(1) Choose two positions randomly in the parents, see Fig. 1(a).
(2) Scramble the order of genes randomly. Repeat the operation, see Fig. 1(b).
(3) Exchange two substrings, see Fig. 1(c) and (d).

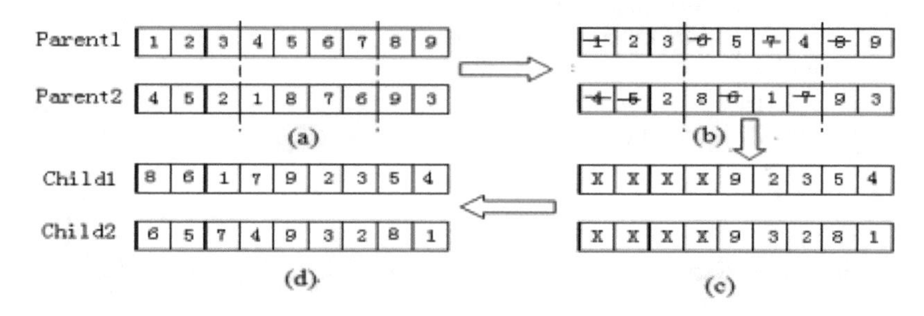

Fig. 1. Crossover operators

After crossover, calculate fitness values of Parent1, Parent2, Child1 and Child2 and evaluated them. Choose two solutions with better fitness value, labeled C1, C2. They are preserved into next iteration.

3.4 Mutation

Mutation happens with a small probability. In this paper, when the iteration is lower than ten percent of the prespecified number of generation N_{cmax}, the mutation probability remains fixed, or shown below.

$$p_m = \begin{cases} k_2, & f_m \le f_{avg} \\ \frac{k_2(f_{\max} - f_m)}{f_{\max} - f_{avg}}, & f_m > f_{avg} \end{cases} \quad (8)$$

The mutation operator *Inversion* [12] is a widely used. See procedures as follows.

1. Randomly choose mutation points: $P_1 = 2$, $P_2 = 6$.
2. Reverses the segment between these two mutation points.
3. Repeated reverse operation for n/10 times.

Inversion operation is shown below (Fig. 2).

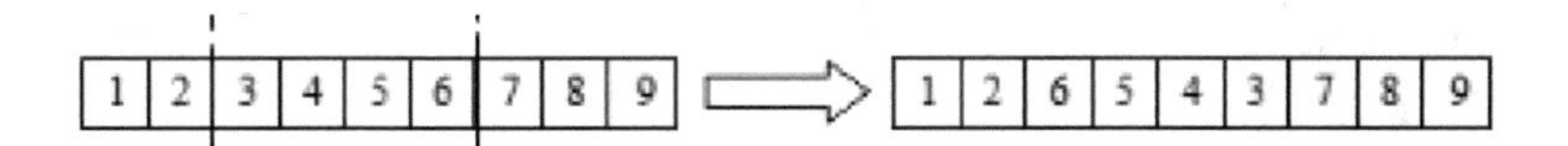

Fig. 2. Inversionmutationoperators

4 Experiments

In this section, the improved genetic algorithm (IGA) proposed has been executed on an Intel Pentium 5 computer, which is equipped with 4 GB memory, running 10 times. Our simulation is based on the instance China Traveling Salesman Problem (CTSP).

Parameters are set as below:

Popsize = 50, N_{cmax} = 1000, probability of crossover operation p_c = 0.90, and mutation probability: p_m = 0.005. The results of computation are shown below (Table 1).

Table 1. Comparison of general genetic algorithm and improved genetic algorithm

Number	General genetic algorithm	Improved genetic algorithm
1	15460	15415
2	15745	15480
3	15871	15736
4	15457	15381
5	15743	15590
6	15621	15471
7	15634	15482
8	15449	15381
9	15478	15454
10	15452	15381

The best solution obtained is showed in Fig. 3.

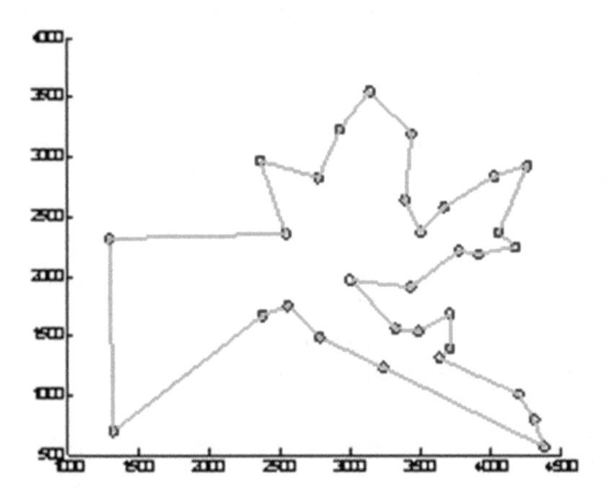

Fig. 3. Best solution found for problem

The also compared the general GA, IGA and other methodes. The results of simulation are presented in Table 2.

Table 2. Comparison of different algorithms

Method	Best solution	Number of iterations
General ACA	15512	234
Or-opt method [10]	15437	–
General GA	15449	192
Improved genetic algorithm	15381	627

5 Conclusion

To solve Vehicle Routing Problem, an improved genetic algorithm (IGA) is proposed in this article. An effective crossover operator is introduced, which helps to improve the global searching ability. Compared with general GA, the crossover probability varies adaptively, which prevents premature convergence of GA. Inversion mutation gradually add some new characteristics to the population. The improved genetic algorithm can effectively solve a variety of vehicle routing problems and retain optimum.

Acknowledgements. Foundation item: Supported by Chongqing Education Department Science and Technology Project (No. KJ1603201), and Chongqing Education Department project (No. 173245).

References

1. Christofides, N.: The vehicle routing problem. Oper. Res. **59**(2), 55–70 (2010)
2. Paolo, T., Daniele, V.: Models, relaxations and exact approaches for the capacitated vehicle routing problem. Discrete Appl. Math. **123**, 487–512 (2002)
3. Laporte, G.: Thevehicle routing problem: an overview of exact and approximate algorithms. Eur. J. Oper. Res. **59**, 345–348 (1992)
4. Gloyer, F.: Future paths for integer programming and links to artificial intelligence. Comput. Oper. Res. **13**, 533–549 (1986)
5. Metropolis, N., Rosenbluth, A., Rosenbluth, M., et al.: Equation of state calculations by fast computing machines. J. Chem. Phys. **21**, 1087–1092 (1953)
6. Ai, T.J., Kachitvichyanukul, V.: A particle swarm optimization for the vehicle routing problem with simultaneous pickup and delivery. Comput. Oper. Res. **36**, 1693–1702 (2009)
7. Gajpal, Y., Abad, P.: An ant colony system (ACS) for vehicle routing problem with simultaneous delivery and pickup. Comput. Oper. Res. **36**, 3215–3223 (2009)
8. Zachariadis, E.E., Tarantilis, C.D., Kiranoudis, C.T.: An adaptive memory methodology for the vehicle routing problem with simultaneous pick-ups anddeliveries. Eur. J. Oper. Res. **202**, 401–411 (2010)
9. Holland, J.H.: Adaptation in Natural and Artificial System. University of Michigan Press, Ann Arbor (1975)

10. Jong De, K.A.: An Analysis of the Behavior of a Class of Genetic Adaptive Systems, Ph.D. Dissertation, University of Michigan, U.S.A. (1975)
11. Shieh, H.M., May, M.D.: Solving the capacitated clustering problem with genetic algorithms. J. Chin. Inst. Ind. Eng. **18**, 1–12 (2001)
12. Liu, G.: Improved ant colony algorithm for solving VRP problem under capacity and distance constrains. J. Guangxi Univ. Nationalities (Nat. Sci. Edn). **02**, 51–53 (2010)

A Machine-Learning Based Store Layout Strategy in Shopping Mall

Yufei Miao(✉)

Shanghai Lixin University of Accounting and Finance, Shanghai, China
myf@myf.cloud

Abstract. In nowadays society, shopping mall is a great part of peoples' life. For the operator of a shopping mall, store layout is paramount due to the fact that it enhances the satisfaction level of the customers and increases the profit of the entrepreneurs as well. However, finding a reasonable layout for the shopping mall is still a complex and long-term issue. Motivated by this observation, a machine learning based strategy to estimate the preferable layout of shopping mall is proposed. In this strategy, transaction information from banks is collected as input data. Then K-Means algorithm is provided to classify stores according to three derived indicators. In addition, relationships between stores using Apriori algorithm is also provided. The experiments results demonstrate that our proposed strategy can be used to make decisions to deploy the stores in the shopping mall with a reasonable mode.

Keywords: Machine learning · Store layout · Shopping mall · Transaction

1 Introduction

According to research data from China Commerce Association for General Merchandise in 2018, the total retail sales of social consumer goods was 38 trillion RMB, an increase of 9.0% over 2017 [1]. The increasing importance of shopping mall in people's life motivates the operator of a shopping mall to consider and provide the service with higher and better quality to the customers. Among all possible factors that impact the operation of shopping mall, the determination of various store locations in a mall is paramount, as the operators are capable of effectively managing the stores from the viewpoints of location. In the other hand, well-organized store layout of mall is able to enhance the experiments of customers, due to the fact that they can quickly find the appropriate items with high probability. Profits for business owners can increase because customers need less time to find products, and the number of satisfied customers rises.

Actually, finding the rule of determining the ideal stores location is still a complicated and long-term issue. Fortunately, machine learning (ML) provides the possible solution to this problem. As a principle approach to artificial intelligence (AI), ML is the scientific study of algorithms and statistical models that computer systems use to

J. MacIntyre et al. (Eds.): SPIoT 2020, AISC 1282, pp. 170–176, 2021.
https://doi.org/10.1007/978-3-030-62743-0_24

perform a specific task without using explicit instructions but relying on patterns and inference instead. By now, many ML-based methods have applied to the various fields, such as Medicare [2], Industry [3], Transportation [4], etc. However, to the best of our knowledge, less literatures have proposed effective solutions to store layout in shopping mall.

In this paper, by introducing the appropriate data analytical models, the approach of store location layout in malls according to customers' transaction data is proposed.

The rest of the paper is organized as follows. Section 2 the notations, dataset and the problem formulation are presented. Given the transaction dataset, the solution is proposed in Sect. 3. In Sect. 4, the extensive experimental results are provided. Section 5 concludes the paper and gives the future directions.

2 Preliminaries

2.1 Notations

The notations throughout the paper are listed in Table 1.

Table 1. Notations

Symbol/Abbreviation	Description
TDS	Transaction data set of customers
$\|S\|$	Cardinality of set *S*
$TDS(i)$	A subset of *TDS*, which includes all transaction records related to store i
K	Number of clusters for store classification

2.2 Dataset Description

Transaction data collected from UnionPay of China collected within six months (from 2019/1/1 to 2019/6/30) at a certain shopping mall was used for analysis, which is shown in Fig. 1. In this figure, each row includes the detailed information of a transaction. For a transaction, *store_id* means the identification of a store, *transaction_date* and *transaction_time* mean the date and time that the transaction happened, *amount* means the transaction amount, *acq_bank* means the bank which the transaction data was acquired from, *last_4_numbers* and *ref_number* are used to identify a specific credit card.

store_id	transaction_date	transaction_time	amount	acq_bank	last_4_numbers	ref_number
15	2019/5/17	12:53	¥393.61	BCM	9138	474799116705
32	2019/5/6	14:12	¥1,642.00	HSBC	8623	302299873543
40	2019/4/28	17:29	¥3,518.00	ABC	0143	552487880489
34	2019/4/28	14:01	¥2,170.08	ABC	5401	105194385951
4	2019/6/13	11:50	¥208.05	CCB	1286	836626260747
37	2019/4/1	15:56	¥984.00	CIB	9510	828278403354
24	2019/4/13	11:30	¥1,215.06	CIB	4435	188847802133
7	2019/4/27	17:16	¥478.30	SPDB	1394	919582932596
32	2019/6/21	16:38	¥177.90	BCM	4089	804964346642
7	2019/6/5	17:06	¥93.50	CIB	5602	318933403214
50	2019/5/18	18:14	¥981.00	BCM	9790	742534016662
24	2019/5/1	12:56	¥83.00	ICBC	6967	732866029953
37	2019/6/10	12:33	¥149.90	HSBC	4196	285144743578
15	2019/6/16	15:40	¥461.80	ABC	5898	010123805737
13	2019/4/15	18:28	¥822.90	ICBC	9524	058916419559
48	2019/4/21	17:07	¥112.70	ICBC	1383	488419573234
49	2019/4/12	19:40	¥1,111.70	BCM	2159	531929841808
20	2019/6/19	13:19	¥46.60	CIB	7965	878120885730
57	2019/4/7	18:35	¥872.10	ABC	1399	979206496215
23	2019/4/7	14:00	¥184.00	ABC	1568	296753190651

Fig. 1. Transaction dataset

2.3 Problem Formulation

Based on the transaction records, the aim is to provide the reasonable store layout to improve the operation efficiency and the customers' experience. To achieve this, the following two problems will be investigated:

- The stores should be divided into multiple groups, and the stores in the same group should show the similar indicators of performance, such that the operator of a mall could provide customized suggestions to the different groups.
- The relationship among stores should be found, such that the stores with higher association should have the nearer distance.

3 Solution

3.1 Framework

As seen in Fig. 2, the whole framework is composed of four main steps. Firstly, data is collected as input. Then, ETL (Extraction-Transformation-Load) operation is performed. Thirdly, proper data models for analysis are used. At last, the result data is visualized to aid the operator making decisions.

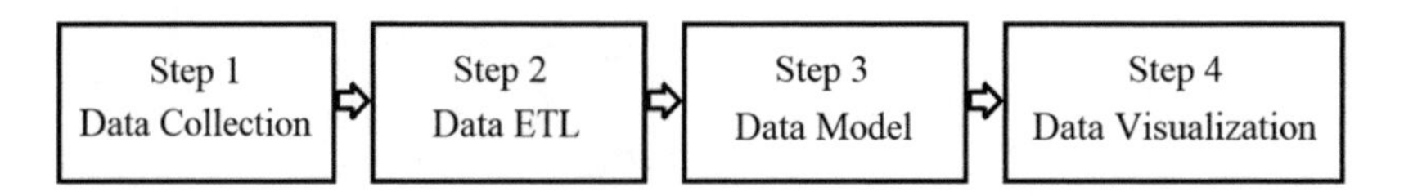

Fig. 2. Overall steps of the whole framework

3.2 Store Classification

(1) ETL

ETL is the general procedure of copying data from various sources into a destination system which represents the data differently from the source or in a different context than the source [5]. To obtain the features of store i for store classification, three indicators are provided as follow.

Average amount Per Transaction (APT): $\mathrm{APT(i)} = (\sum \mathrm{TDS(i).amount})/|\mathrm{TDS(i)}|$

Number of WeekDays (NWD): NWD(i) is calculated as the number of weekdays in TDS(i).

Number of WeekEnds (NWE): NWE(i) is calculated as the number of weekends in TDS(i).

(2) Dimension reduction

Principle Component Analysis (PCA) is an algorithm of finding the principal components of given data so that the dimension of the data can be reduced. In this paper, PCA is introduced for the purpose of data visualization.

(3) Store Clustering

K-Means is a popular distance-based partitional clustering algorithm that is simple yet efficient [6]. It is an unsupervised learning model which divides n points into K clusters, so that each point belongs to the cluster corresponding to its nearest mean (cluster center). As *TDS* is not labeled, K-Means is an appropriate solution for classification. In addition, SSE (Sum of Squared of Errors) is derived as follows, which can be used to obtain the optimal value of K.

$$SSE = \sum Distance(A_i, Center)^2$$

A cluster center is the representative of the cluster which it belongs to. The squared distance between each point and its cluster center is the required variation. The aim of K-Means is to find these K clusters and their centers while reducing the total error. One feasible way is to calculate the sink rate of SSE, which can be presented by inertia of K-Means.

3.3 Store Association

Besides store clustering, store association is also the important issue in store layout. Here, store association means the connection between each pair of stores and the order that the customers prefer to visit.

Apriori algorithm is a mining algorithm based on Boolean association rules. It is used to mine the associations between data items [7]. Generally, for a given set of items (for example, a set of retail transactions, where each set lists the purchase information for a single product), the algorithm is designed to mine frequent itemsets based on the support and confidence.

4 Experiment

4.1 Experiment Description

In this paper, Python is programmed to analyze the transaction data. Firstly, three indicators (NWD, NWE and APT) are calculated based on the transaction data. In order to visualize the data, PCA is then introduced to reduce the indicators from three to two. Thirdly, K-Means model is performed based on the result of PCA. At last, Apriori algorithm is used to explore association rules between stores.

4.2 K-Means Result

In order to show the change under the different number of clusters K, given the different values of K, the results of SSE is plotted in Fig. 3.

Figure 3 denotes that it has the largest sink rate when $K = 3$, so $K = 3$ is the optimal value of K. This can be interpreted that it is better to classify stores into three types. Based on this information, the shopping mall operator should put stores of the same cluster together so that customers can find specific store easier.

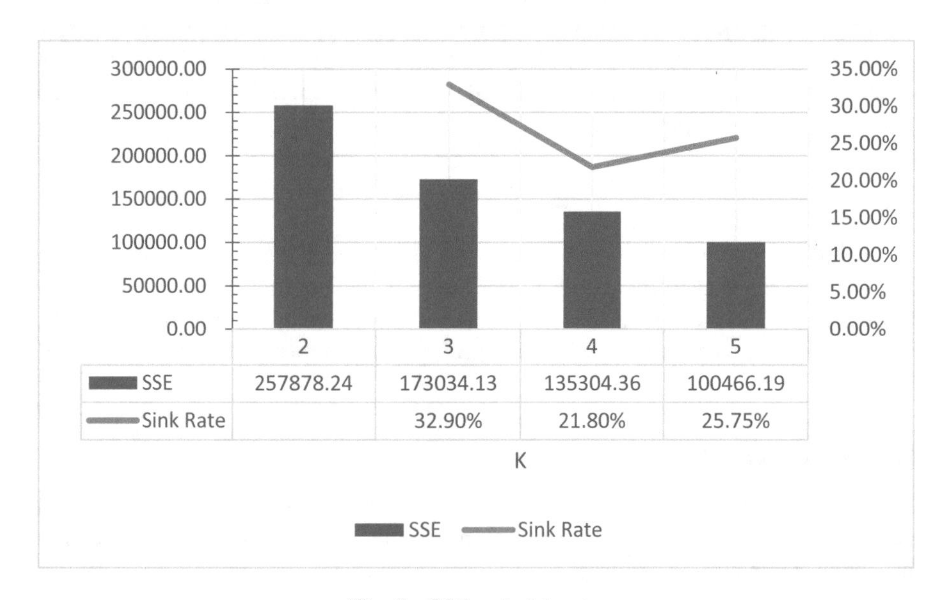

	2	3	4	5
SSE	257878.24	173034.13	135304.36	100466.19
Sink Rate		32.90%	21.80%	25.75%

Fig. 3. SSE and sink rate

When the value of K is 3, the result of K-Means algorithm is shown in Fig. 4, where star symbol represents the cluster center and cross symbol represents every piece of data after dimensionality reduction.

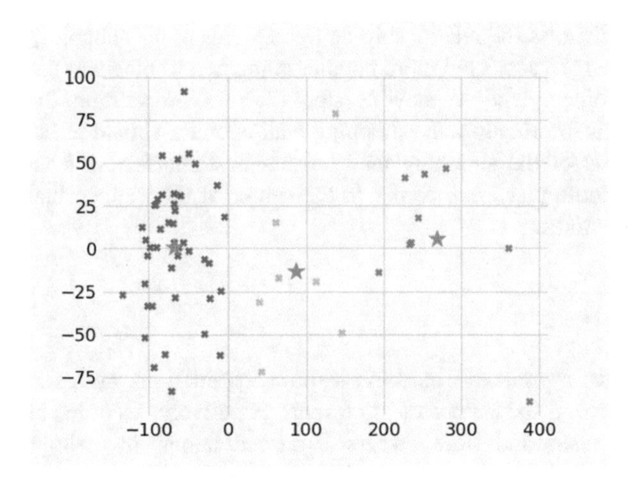

Fig. 4. Result of K-Means algorithm

4.3 Store Association Result

In the next step, each customer is identified by a unique number generated from credit card number. Each line of data represents a specific customer and contains numbers of store id which corresponds to the store where the customer had shopped. Frequent itemset is retrieved and association rules are calculated. In order to simplify the analysis, only relationships between two stores are considered. The result of Apriori algorithm can be seen as Fig. 5.

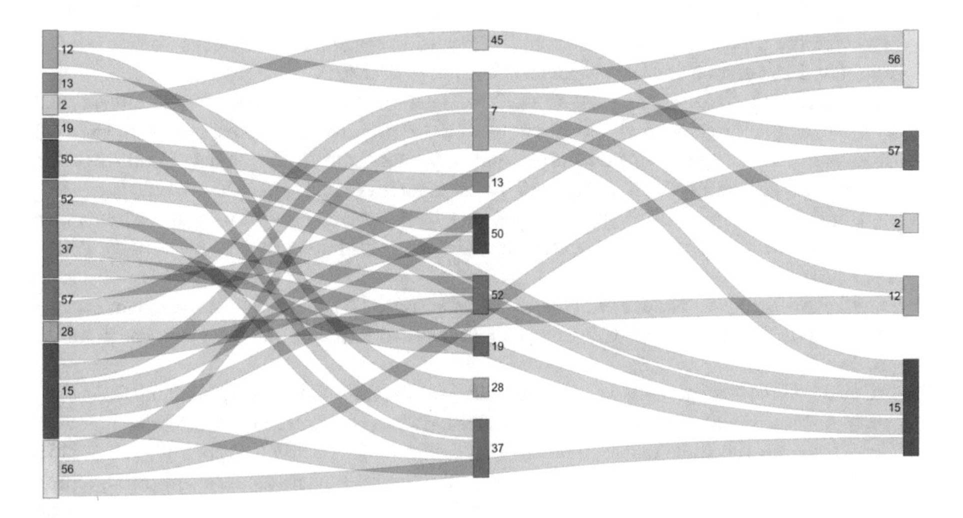

Fig. 5. Result of association algorithm

This figure denotes the relationship between stores at the shopping mall. Firstly, to some extent, store 15 and 7 are more popular compared to others in the shopping mall. Secondly, customers tend to stay at store 7 between visiting two other stores. According to this information, the shopping mall operator should place stores 15 and 7 in different areas of the shopping mall so that customers can be more evenly distributed, and should place the store 7 in the center of the mall so that customers can patronize more stores.

5 Conclusion

In this paper, we introduce a machine learning based framework for shopping mall operators to manage the layout of store more effectively, in order to provide better service to customers and store owners. Transaction data at a shopping mall from UnionPay are used as input and extract three features from the data set. In the proposed model, each store in the shopping mall is categorized into clusters by K-Means algorithm. Then, the underlying relationship between stores is calculated by Apriori algorithm. According to visualization of the model output, operators can arrange the layout of shopping mall in a way that balances customers in the mall, making customers easier to find the desired store and maximizing the profit for stores.

References

1. 2018–2019 China Department Store Retail Development Report. http://www.ccagm.org.cn/bg-yj/4319.html. December 2019
2. Bauder, R.A., Herland, M., Khoshgoftaar, T.M.: Evaluating model predictive performance: a medicare fraud detection case study. In: IEEE 20th International Conference on Information Reuse and Integration for Data Science (IRI), pp. 9–14 (2019)
3. Caesarendra, W., Wijaya, T., Pappachan, B.K., Tjahjowidodo, T.: Adaptation to industry 4.0 using machine learning and cloud computing to improve the conventional method of deburring in aerospace manufacturing industry. In: 12th International Conference on Information & Communication Technology and System (ICTS), pp. 120–124 (2019)
4. Jahangiri, A., Rakha, H.A.: Applying machine learning techniques to transportation mode recognition using mobile phone sensor data. IEEE Trans. Intell. Transp. Syst. **16**(5), 2406–2417 (2015)
5. Diouf, P.S., Boly, A., Ndiaye, S.: Variety of data in the ETL processes in the cloud: state of the art. In: IEEE International Conference on Innovative Research and Development (ICIRD), pp. 1–5 (2018)
6. Banerjee, S., Choudhary, A., Pal, S.: Empirical evaluation of K-Means, bisecting K-Means, fuzzy C-Means and genetic K-Means clustering algorithms. In: IEEE International WIE Conference on Electrical and Computer Engineering (WIECON-ECE), pp. 168–172 (2015)
7. Ya, L., Lei, Y., Li, W., Chun, M., Guiming, Y.: Application research of apriori algorithm based on matrix multiplication in children's drug interaction. In: 12th International Conference on Measuring Technology and Mechatronics Automation (ICMTMA), pp. 507–512 (2020)

Risk Analysis in Online Peer-to-Peer Loaning Based on Machine Learning: A Decision Tree Implementation on PPDai.com

Chengxun Wu(✉)

Honors Mathematics, New York University Shanghai, Shanghai 200122, China
cw2961@nyu.edu

Abstract. With the developing network technologies and the formation of complete mobile payment systems, the world is witnessing a growing trend of P2P lending. This research paper analyzes the data from a Chinese P2P lending platform PPDai. Also, the research utilizes the Decision Tree algorithm based on attributes related to customers' loan features, verification statuses and loan history to predict and determine whether a customer could fulfill his/her repayment on time. Hence, P2P lending platforms could have predictions for future customers concerning whether they are potentially failing to pay back money timely. In this way, P2P lending platforms can prevent some potential losses and discouragements from the less trusted customers. Experimental results from the Decision Tree algorithm demonstrate that: the verification processes play a significant role in identifying customers as trusted or not. Moreover, the results propose some broader implications for the healthy development of the online P2P industry for both borrowers and lenders.

Keywords: Machine learning · Classification · Decision tree · Risk analysis

1 Introduction

Peer-to-peer (P2P) lending, also called crowdlending, is a popular model for capital circulation between investors and borrowers. This model features direct communications and Transactions between lenders and borrowers without the intervention from traditional financial institutions. With the development of internet, most of such transactions have moved to an online mode. The processes of online P2P lending contain applying for borrow requests, submitting personal information, platform's announcement of borrowers' situations and lenders' making decisions in whether to invest or not. The convenience compared to traditional loans has attracted many customers. Customers' crowd psychology and group effect always increase the probability of getting a load from P2P, while these effects decrease risk control [1]. On a global scale, its popularity has been growing fast since its inception and has now taken comparatively stable steps in expanding its popularity.

Compared to the traditional modes, online P2P lending is more flexible since it matches lenders with potential borrowers. However, such flexibility also features high risks: the asymmetry in the market, where the lack of veracity in the information

J. MacIntyre et al. (Eds.): SPIoT 2020, AISC 1282, pp. 177–186, 2021.
https://doi.org/10.1007/978-3-030-62743-0_25

provided by the borrowers might perplex the lenders [2]. The lenders may make bidding decisions based on the inaccurate transaction information and, in turn, suffer from the losses from the unexpected defaults by the borrowers. Also, as Jiang et al. point out, online P2P lending differs from the traditional P2P lending (where transactions usually occur among acquaintances), as it generally involves transactions with strangers. To reduce potential risks, many scholars have investigated the risk analysis in online P2P loaning. Such researches cover various fields such as studying the strategies of lenders in making P2P loans, different influences on default probability based on borrowers' personal information, and various other factors. This part will be further discussed in detail in the section of the literature review in this paper [3].

Different from the smooth growth of the service from a global scale, peer-to-peer loaning service in China fluctuates more: it has witnessed dramatic rises and falls during the 2010s [4]. Due to the drastic growth of the sophisticated network infrastructure in China and people's growing interest in investing, the promotion of P2P lending service surged. According to statistics in China P2P industry report from wdsj.com, in the first several years of the 2010s, the growth in the annual turnover of online peer-to-peer loaning platforms was almost exponential [5].

However, the fantastic growth and the expanding trend of P2P lending among citizens foreshadow the underlying risks. Deng points out primary risks that preclude the continuing skyrocketing of the P2P lending industry in China: credit risk, technological risk and policy risk. The credit risk is the most influential one. Sun also mentions the significant loss that the lenders undergo caused by false information of the borrowers [6]. Such cumulative frauds lead to the final bankruptcy of many Chinese P2P lending platforms and more strict supervision from governmental departments. More strict policies have regulated the market heterogeneity and expelled unqualified lenders, borrowers and platforms, leading to an overall decreasing in China's P2P lending service. The figure below reveals the sharp decrease in the amount of China P2P platforms.

The shrinking P2P lending market in China serves as an alarm for the existing P2P lending platforms (not only in China but also in the world): all the platforms should reflect on the failure of peer platforms to avoid bankruptcy in the future. Moreover, such challenges in the development of P2P lending are the same in different areas around the world. Hence, it is indispensable to implement risk analysis beforehand to evaluate the risks for defaults and frauds from potential customers. Among the online P2P lending platforms in China, PPDai.com is one of the most characteristic ones. It is the first online P2P lending platform in China. Established in 2007, PPDai.com is one of the most successful and long-lasting P2P platforms in China. Comparatively, PPDai.com has a stable growth during the dramatic changes in China's P2P lending service [7].

2 Literature Review

The topic of risk analysis in online P2P lending has already received much attention. The primary researches conducted on investigating the pattern between online P2P lending transactions can be divided into two categories: traditional modeling approaches and machine learning approaches. The first category includes the construction of empirical models: the researches build models based on data features (e.g., via correlation analysis) and test accuracy (i.e., the matching degree of data to the models) of actual data. The second category applies machine learning algorithms (e.g., KNN, K-Means, Random Forests). Models come from results of training sets and the accuracy of test data indicates the effectiveness of analysis based on that machine learning algorithm. Though the two categories differ in their methodologies, they commonly aim at getting a consistent result from the widely dispersed data [8].

2.1 Traditional Modeling Approaches

The traditional modeling approaches focus on synthesizing multiple features of data to obtain a comprehensive model. According to E. Lee and B. Lee, a multinomial logit market fits well in the construction of the empirical model concerning P2P customers' behaviors based on data from Popfunding, a South Korean online P2P platform. By proposing and verifying experimental hypotheses, E. Lee and B. Lee obtain a conclusion that both soft information (e.g., influences from age) and hard information (e.g., borrowers' personal histories) mutually contribute to how much a lender trusts his/her borrowers. Another research also applies the empirical modeling approach to a data set from PPDai.com. Based on correlation analysis and the statistics identification, another model categorizes the borrowers into three levels of borrowers to investigate borrowers' strategies. Their model reveals that the P2P lending transaction studies should consider the borrowers' strategies as a significant factor in analyzing lenders' actions. Greiner & Wang also build their empirical model with multifarious personal information from the borrowers and conclude that online P2P lenders should "pay special attention to the economic status of borrowers." All these studies use the approach of building empirical models and have obtained different results, which commonly play an indispensable role in investigating the pattern behind the online P2P lending mechanisms.

Differently, Zhao et al. apply a sequential approach based on stochastic processes to obtain the study's experimental model. By computing and comparing the modeling accuracy of listing-Bayesian hidden Markov model and listing and marketing-Bayesian hidden Markov model, their study reveals the relationship between the frequency of online P2P transactions and borrowers' corresponding marketing interests. Another modeling method applied in the analysis of online P2P is the Fuzzy Analytic Hierarchy Process. Liao et al. implement the Fuzzy Analytic Hierarchy Process to compare the weights of risk analysis among various data features, provide concrete conclusions: the administrative regulations of online P2P platforms play a more significant role in the transactions than borrowers' personal information. These studies do not construct the empirical models; following a similar modeling strategy, their discoveries are also decent.

2.2 Machine Learning Approaches

Researches in the second category implement machine learning methodologies. Compared with the modeling processes, machine learning does better in dealing with large quantities of data [9]. Concerning the accuracy of machine learning training results in analyzing the data from LendingClub, the algorithms of Decision Tree,

Random Forest, and Bagging (Bootstrap Aggregating) show excellent performances in reaching high accuracy based on massive amounts of input data. Implementing deep learning algorithms in the analyses and comparisons among data sets from Australia, Germany, and LendingClub, the LDA classification algorithm stands out as the one with the best predictive performance among several methods (including KNN, KMeans, linear SVM). Ha, et al. endorse the effectiveness of deep learning algorithms in risk analysis of online P2P lending. These studies have shown the feasibility and efficiency of machine learning algorithms in risk analysis, especially when there is a large database of various features. Moreover, Wu uses the decision tree algorithm in the risk analysis in medical cosmetic surgeries and discovers the relativity between district differences and the P2P lending default rate. Also, as Guo et al. point out, in their instance-based credit risk assessment, machine learning algorithms perform well in the portfolio optimization problem with boundary constraints with the advantages of dynamic stability: the analyses do not depend much on statistic predictions. Hence, the machine learning approaches feature excellent accuracy and comprehensiveness of data selection.

2.3 Inspirations on the Research Methodology

Two types of approaches in online P2P lending risk analysis both give inspirations to the approach selection in this study. Results from empirical modeling show that generally, personal information is less likely to influence the behaviors of borrowers and lenders in P2P lending. It is essential in deciding the targeting features during the cleaning of data to decrease the negative influences from outside unnecessary data [10]. Also, since this research works on a database with a large size, machine learning algorithms become potential choices to facilitate the analysis of the data. The existing results from machine learning approaches have demonstrated the feasibility of them in analyzing the data. Hence, this study will mainly take the approach of a machine learning algorithm: the Decision Tree algorithm. According to the studies above, the Decision Tree algorithm has an excellent performance in classification problems.

3 Theoretical Methodology

3.1 Decision Tree: Overview

The main algorithm that the research implements is the decision tree model. It is a widely applied, supervised machine learning strategy. Based on the different attributes of the data, the decision tree model generates a tree-like structure to present the relationships between the target attribute and the other attributes.

3.2 Decision Tree Principles

In this research, several essential concepts facilitate the successful implementation of the decision tree algorithm.

Information Entropy: the measurement for the uncertainty in a random variable's possible outcomes. The calculation formula for the information entropy of a certain data set D with n different groups of outcomes is: $Entropy(D) = -\sum_{i=1}^{n} p_i \log p_i$, where p_i is defined as the ratio of the number of the *i*-th value (outcome) in the target attribute to the total amount of values in the data set D with respect to the target attribute. The goal of the Decision Tree algorithm is try to reduce the information entropy of a data set.

Information Gain: the Information Gain of an attribute A is defined as the following: InfoGain(D,A) = Entropy(D) - Entropy(D|A). In this formula, Entropy D) is the information gain on for the whole data set with respect to the target attribute. Entropy(D|A) is the information entropy of D given A. The data set is divided into separate subsets based on the different values of attribute A, then the conditional entropy is the weighted average of the information entropy on the subsets.

Principles for branching in the Decision Tree Algorithm: the Decision Tree algorithm operates the branching process after choosing the attribute with the most significant information gain. On a recursive basis, when making the subsequent decisions (branching processes), the remaining attributes will be calculated again based on the subsets after classification to form sub decision trees as outcomes of the branching processes.

When to stop branching: the algorithm stops the branching processes if (1) the values in the subset concerning the target attribute are identical or (2) all the attributes except for the target attributes have undergone some branching processes. On that basis, the algorithm forms a complete decision tree by connecting all the subtrees generated with respect to each operation of the branching processes.

3.3 Advantages of the Decision Tree Algorithm

The Decision Tree algorithm is good at categorizing data sets with large capacity with different attributes. It does not require data normalization or data scaling, making it comparatively efficient and straightforward to implement. Also, to the general audience, outcomes of the Decision Tree algorithm is easily understandable. In this research, the target attribute is whether the loan borrower could be trusted, and the other attributes are potential factors affecting a borrower's punctuality in repayment. Hence, the loan platforms can better predict whether a new borrower could be trusted: whether he/she satisfies the criteria for paying back on time. The judgment process is simple, given a complete decision tree: the platform only needs to check the "if then" structures when receiving the related information of a new borrower.

There are several versions of the Decision Tree algorithm: the initial algorithm of ID3 and its optimized versions C4.5, C5.0 and CART. Since the CART decision tree algorithm does not handle multiple branches within a node (i.e., the CART decision tree is always binary), considering the variety of attributes in the P2P lending field, this research implements the C4.5 decision tree for the data analysis.

4 Data Features and Data Processing

4.1 Data Features

The data source in this research paper comes from 158,976 customers' information from PPDai.com. The original data set has around 241,091 customers' information. Based on the principle of supervised learning, this research has selected customers with personal borrowing records for the implementation of the Decision Tree algorithm. Primarily, the data is categorized into 20 features. They are further divided into four different categories by the homogeneity among some different attributes.

The first category (4 attributes): basic personal information. This category contains the customer's ID number, customer's sexes, ages, and their initial credit evaluation.

The second category (5 attributes): loan information. This category stores the duration of loaning, the date of loaning, the loan amount, the interest rate of current loaning, and the types of borrowing. Attributes in this category reflect the features of a customer's loan.

The third category (6 attributes): verification status. The data from PPDai.com is very representative because it features a multi-dimensional verification mechanism for the borrowers. Also, this online P2P lending platform values the status of verification a lot in deciding whom to trust. Such verification covers mobile phone verification, registered residence verification, video verification, diploma verification, credit investigation verification, and Taobao verification.

The fourth category (5 attributes): personal historical borrowing data. Since the research focuses on the customers with previous borrowing records, data in this category is essential to have credit predictions for future customers. This category contains the total amount of times and money of successful former borrowings, the total amount of unpaid repayment, the total times of repayment on time, and the overdue repayment records.

4.2 Data Processing

The original data contains miscellaneous features where some of them might not contribute to the desired classification result. Also, some data forms were hard for the Decision Tree algorithm to identify.

4.2.1 Data Cleaning

This step selects and removes the features unrelated to the research concentrations: the research focuses on the relationship between a customer's loan behaviors and the degree of being trusted. In the first category, the study removes information about the necessary personal information (i.e., customers' ID numbers, sexes, and ages). Only the customers' initial credit evaluation is maintained since the P2P lending platforms highly emphasize a borrower's credit evaluation as an essential indicator of trust. In the second category, the date, amount, and duration of current loaning are removed. The date of loaning is an extraneous variable, as it does not reflect the behavior of loaning. Compared with the duration and rate of a customer's current loaning, the correlation coefficient between loaning interest rate and the ratio of successful repayment is significantly larger (see the figure below). Hence, we keep the rate of current loaning as a feature for further algorithm analysis.

The research keeps all the data from the third category since they jointly reflect a status of multifarious verification for a borrower. For features in the last category, the research keeps the data of the total times of repayment on time and the overdue repayment records. This process aims at considering a borrower's historical defaults and on-time repayment as an indicator of whether a customer can be trusted.

4.2.2 Data Transformation

Table 1. Correlation analysis between success ratio and loan features

	Ratio of success	Loan duration	Interest rate	Loan amount
Ratio of success	1			
Loan duration	−0.0484	1		
Interest rate	−0.2579	0.3781	1	
Loan amount	0.0151	−0.0013	−0.1469	1

After the selection of the key features to investigate in the previous step, this step aims at changing the feature values into easily identifiable ones for the algorithm (Table 1).

Loan interest rate. The distribution of the interest rates in the original data is as the following: 42% of the data has interest rates of higher than or equal to 22%, 28% of the data shows interest rates of less than or equal to 18%, and the rest 30% data has interest rates in the range of 18% to 22%. Hence, in order to avoid too many discrete values in classification, the research classifies the interest rate into three categories: high (with an interest rate $\geq 22\%$), low (with an interest rate $\leq 18\%$), and medium.

Initial evaluation. The original classification of customers' initial evaluation contains 6 different values. In the actual implementation of the Decision Tree algorithm, too many different values within one attribute often result in the low-efficiency and unnecessary complexity. Hence, we further group the attributes based on the distribution of initial evaluation. In proportion, 49% of data has an initial evaluation level of C, followed by D, with a proportion of 34%. In terms of the remaining evaluation levels, 10% of the total data has a level of B. Level A and Level E have a similar proportion of 3%, and level F has a proportion of 1%. The research takes the following rule: grouping level A and B as "excellent initial evaluation level", combining level D, E, F as "average evaluation level" and C as "good initial evaluation level" (Table 2).

Table 2. Sample data after initial processing

Interest rate (%)	Initial evaluation	Loan type	Phone verification	Residential verification	Video verification	Diploma verification	Credit verification	Taobao verification	Historical successes	Historical defaults
18	C	Else	Yes	No	Yes	No	No	No	57	16
20	D	Else	No	Yes	No	No	No	No	16	1
20	E	Normal	Yes	No	No	No	No	No	25	3
18	C	Else	Yes	Yes	Yes	No	No	No	41	1
16	C	EC	Yes	Yes	Yes	No	No	No	118	14

Verification status. The verification comes from separate fields; then, it is vital to integrate the different verification types to obtain an overall view of a customer's verification status. The research transforms the "Yes" or "No" to integer indicator values 0 and 1. The new attribute, overall verification, is determined by the sum of all the six indicator values. Hence, the values of the new attribute efficiently combine six different attributes into a new one. It also reduces the complexity and time cost for running the Decision Tree algorithm with a smaller number of total attributes. For the values in the attribute "overall verification," a sum in the range 2 to 6 yields an "excellent," a sum of 1 yields an "good," and a sum of 0 yields an "average".

The target attribute: trust status. The rate of trust in this research is defined as: The formula is valid for all the data since the data only contains customers with historical repayment record. In order to avoid the cases with insufficient experience of P2P borrowing (i.e., an insufficient amount of successful historical borrowings that tolerate many peculiarities), the threshold based on the rate of trust of whether a customer is trusted or not will be determined on the data sets of customers who have successful historical borrowings for more than or equal to 3 times. The average rate of trust $\text{Rate}OfTrust = \frac{HistoricalSuccesses}{HistoricalSuccesses + HistoricalDefaults}$ in the selected subset of data is 95.04%, with variance 0.01. This project chooses the first quartile in the selected subset as the threshold for a customer being trusted or not, which yields the threshold rate of trust: 94.03%. Equivalently, for each data, the new attribute "Trust Status" shows the value of "Yes" when that data has a rate of trust larger or equal to 94.03% and the value of "No" otherwise. Hence, for each customer in the data, the attributes historical successes and historical defaults will be combined as a criteria for whether the customer could be trusted. The Table 3 below is a fully processed sample table as presented above:

Table 3. Sample data after data transformation

Interest rate	Initial evaluation	Loan type	Overall verification	Trust status
Low	Good	Else	Good	No
Medium	Average	Else	Average	Yes
Medium	Average	Normal	Average	No
Low	Good	Else	Good	Yes
Low	Good	EC	Good	No

4.3 Algorithm Implementation

Based on such a processed data set, the research implements the Decision Tree algorithm with fewer difficulties. After the data cleaning and transforming processes, the research starts to train the data using the Decision Tree Algorithm. Based on the different percentages of data being selected as training and testing sets, the accuracy ratio (after decision tree pruning) also changes accordingly (see Fig. 1 below).

Fig. 1. Accuracy of decision tree classification

The overall verification is of the highest importance in deciding whether a customer is worthy of being trusted. A status of being "excellent" in overall verification, i.e., have accomplished more than 2 out of 6 terms of verification, results in a customer's being categorized as a trusted customer. The completeness of various verification reflects a customer's credibility. For the rest, the comparative incompleteness might also be an outcome of failing to complete on time, so the other attributes are considered to make further decisions.

Another important indicator is the loaning type. Among the three types of loaning in the data set, E-Commercial loaning is more likely to be trusted. The result is because of higher risks compared with other types (more general ones, normal and else). Hence, such customers tend to take more prudent strategies in making loans: they tend to consider their future capabilities to avoid defaults. For customers with other loan types, further determination is necessary.

The rest of the indicators are loan rates and initial credit evaluation level. The decision-making processes of the decision tree show that an initial credit evaluation with level "good" and "excellent" (with the credit level in the original data A, B, C) is more likely to results in a status of "trusted". Also, more credit is likely to be given to the customers with a higher loan interest rate: a high loan interest rate is scarcely negligible by the borrowers. In order to minimize their loss of money after defaulting, such customers will put better efforts into avoiding defaults, thus being more "trusted".

5 Conclusion

Based on the results from the algorithm implementation on the data from PPDai.com, one can see that the Decision Tree algorithm is accurate and efficient in classifying data about P2P lending. With the particular focus of PPDai.com on various verification processes (verification in 6 aspects and the detailed consideration of the initial evaluation), the decision tree reveals that the verification of personal credit in P2P lending is crucial. Hence, a broader implication for all P2P platforms is that completing a sophisticated credit verification system is highly effective in filtrating the potential defaults away. Although some extraneous factors might affect the accuracy of the verification system in evaluating one's credits, generally, through the verification processes, online P2P platforms can reduce the risk of possible defaults when the platform receives more customers. Implementing such a Decision Tree model, the online P2P platforms could reduce their losses caused by potential defaults. Apart from a more comprehensive verification system, online P2P platforms could also take action to collect authentic information from both lenders and borrowers. An excellent example from Zhao and Liu shows that in verification of personal information, borrowers that upload personal photos generally have a lower default rate and in turn more likely to win trust from the lenders to make successful loans.

Besides, online P2P customers should be more aware of their personal credit evaluations before attempting to make transactions on online platforms. As more people are starting to increase their personal credit evaluation levels, a better environment of online P2P learning will be maintained together by the platforms and the customers. The borrowers should take the initiative to make sure the veracity in their personal information to reduce the potential risks of fraud. Conclusively, the implementation of the Decision Tree model is helpful not only to the online P2P platforms but also the potential P2P customers. The results from data of PPDai.com are not limited to the online P2P platforms alone: they have pointed out a healthier and more sustainable way for further development of the P2P lending service, which benefits both platforms and customers.

References

1. Deng, C.: A comparative research on online peer-to-peer lending risk analysis models: an analysis of prosper and PPDai.com (title translated by the author of this paper). China J. Commer. **24**, 63–66 (2019)
2. Feng, Y., Fan, X., Yoon, Y.: Lenders and borrowers' strategies in online peer-to-peer lending market: an empirical analysis of PPDai.com. J. Electron. Commer. Res. **16**, 242–260 (2015)
3. Greiner, M.E., Wang, H.: Building consumer-to-consumer trust in e-finance marketplaces: an empirical analysis. Int. J. Electron. Commer. **15**(2), 105–136 (2010)
4. Guo, Y., Zhou, W., Luo, C., Liu, C., Xiong, H.: Instance-based credit risk assessment for investment decisions in P2P lending. Eur. J. Oper. Res. **249**, 417–426 (2015)
5. Ha, S., Ld, N., Choi, G., Nguyen, H.-N., Yoon, B.: Improving credit risk prediction in online peer-to-peer (P2P) lending using feature selection with deep learning. In: 21st International Conference on Advanced Communication Technology (ICACT), pp. 511–515 (2019)
6. Jiang, X., Zhang, Q., Cheng, J.: Research on credit risk identification of peer-to-peer lending market. China Bus. Market. **34**(04), 67–75 (2020)
7. Vinod, L., Subramanyam, N., Keerthana, S., Chinmayi, M., Lakshmi, N.: Credit risk analysis in peer-to-peer lending system. In: 2016 IEEE International Conference on Knowledge Engineering and Applications (ICKEA), pp. 193–194 (2016)
8. Sun, J.: The analysis of profit model of online loaning platforms. Modern Bus. **34**, 117–118 (2019)
9. Zhu, J., Ghosh, S., Wu, W.: Group influence maximization problem in social networks. IEEE Trans. Comput. Soc. Syst. **6**(6), 1156–1164 (2019)
10. Liu, L., Zhao, J.: Research on the effects of borrowers' active uploading photos in online peer-to-peer lending market. China Price **3**, 108–113 (2020)

The Application of BP Neural Network in the Opioid Crisis

Dongbing Liu[1] and Wenbin Liu[2,3](✉)

[1] College of Mathematics and Computer, Panzhihua University, Panzhihua, China
[2] Department of Public Course, Wuhan Technology and Business University, Wuhan, China
812926096@qq.com
[3] College of Information Engineering, Wuchang Institute of Technology, Wuhan, China

Abstract. In order to solve the opioid crisis which broke out in the United States, the diffusion model suitable for the problem is established, inspired by the error inverse propagation model. By observing the diffusion characteristics of opioids between the five states and their counties, the locations where each state began to use specific opioids are identified and the diffusion characteristics of the corresponding programmes are summarized. Ultimately, the opportunity for the government to focus on such cases and the threshold of the corresponding drug is found, and the location and time of the case are predicted.

Keywords: BP neural network · The opioid crisis · The diffusion model

1 Introduction

Today, the United States is experiencing a national crisis of illicit use of opioids, with the increasing negative impact of overdose or drug extraction. It is understood that from 1986 to date, the number of deaths from opioid overdose has exceeded 350,000. The United States Government and its corresponding organizations were actively addressing the negative effects of opioids, but the implementation of existing laws was a complex challenge for organizations and the United States Government [1, 2].

In the annual report issued by UNDCP, we can see information on cases of toxic substances in a number of counties in the five States of the United States, from which we have developed a mathematical model aimed at focusing on the characteristics of the spread of synthetic opioids and heroin incidents between the five States and their counties, as well as identifying locations where each State began to use characteristic opioids [3, 4]. The appeal model is refined to determine which drugs will occur under the threshold of the corresponding cases, thus determining the timing of changes of concern to Governments and organizations [5–7].

At the same time, we pay attention to the socio-economic data provided by the United States census, analyze and construct the corresponding model in a large amount of data, we find the relationship between the information provided and the trend of opioid crisis, and give the strategy to deal with this crisis [8–10].

J. MacIntyre et al. (Eds.): SPIoT 2020, AISC 1282, pp. 187–193, 2021.
https://doi.org/10.1007/978-3-030-62743-0_26

2 The Foundation of the Error Inverse Propagation Model

The error inverse propagation model is used to conduct a large number of training analysis of the data given in the problem. Training columns extracted from the given data (x_k, y_k), the diffusion model is obtained as follows:

$$Y_j^k = f(\beta_j - \theta_j) \tag{1}$$

Y_j^k means output value of the j neuron in the model output layer. β_j means input value of the j neuron in the model output layer. θ_j means a threshold for the j neuron in the output layer of the model, which determines how much input is activated.

In the diffusion model, the specific method is:

$$\beta_j = \sum_{h=1}^{a} W_{hj} b_h \tag{2}$$

W_{hj} is the connection weight in the hidden layer. It is adjusted for the error of the neurons in the hidden layer, and it is iteratively repeated until we reach the desired result. The updated formula for any weight is:

$$W \leftarrow W + \Delta W \tag{3}$$

The solution of ΔW is:

$$\Delta W = -\eta \frac{\partial E_k}{\partial W_{hj}} \tag{4}$$

E_k is the error of mean variance about (x_k, y_k), and we want it to be close to 0 as possible. Based on the gradient descent method, we find the optimal solution, that is, we adjust the parameters in the negative gradient direction of the target through multiple iterations, the new weight parameters will gradually approach the optimal solution. It was easy to see that the expressions of E_k is as follow.

$$E_k = \frac{1}{2} \sum_{i=1}^{l} \left(Y_j^k - y_j^k \right)^2 \tag{5}$$

In formula (5), l is the number of input layer weights. And the coefficient $\frac{1}{2}$ is the constant coefficient that can be offset, when we find the derivative formula transformation.

By using MATLAB software to bring into the model and observe the results, we can see that the changes of W_{hj} can affect the input value β_j of the output layer neurons, then affect the output value Y_j^k, and then affect the error E_k. Then we can get as follow:

$$\frac{\partial E_k}{\partial W_{hj}} = \frac{\partial E_k}{\partial Y_j^k} \frac{\partial Y_j^k}{\partial \beta_j} \frac{\partial \beta_j}{\partial W_{hj}} \tag{6}$$

Combined with $\beta_j = \sum_{h=1}^{a} W_{hj} b_h$, we can finally get as follow:

$$\frac{\partial \beta_j}{\partial W_{hj}} = b_h \tag{7}$$

The above is the explanation of β_j in the model. Next we'll go on with the function $f(x)$ in the model.

$f(x)$ is the activation function of the output layer. For $f(x)$, first of all, we think of adopting step function, but the disadvantage of step function is very obvious, that is, discontinuous, undirectable, not smooth. Therefore, we combine the particularity of the problem and the flexibility of the model, and decide to use sigmoid function as activation instead of step function. The functional graph of the two functions respectively is in Fig. 1.

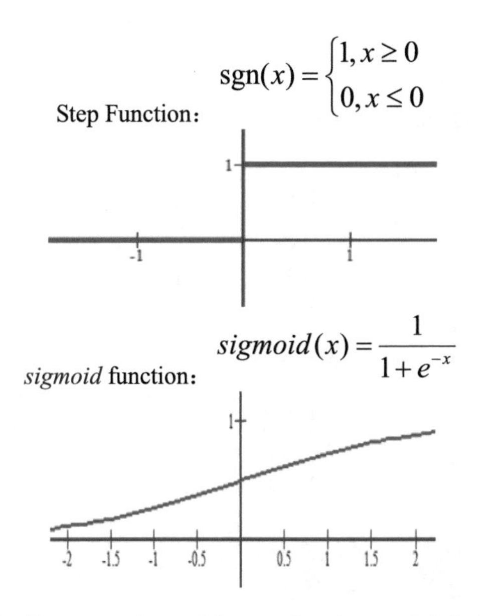

Fig. 1. The expressions of the two functions and their images

Second, for the activation function $sigmoid(x) = \frac{1}{1+e^{-x}}$, it's easy for us to derive the following properties.

$$f'(x) = f(x)[1 - f(x)] \tag{8}$$

Assumptions:

$$g_j = -\frac{\partial E_k}{\partial Y_j^k}\frac{\partial Y_j^k}{\partial W_{hj}} \tag{9}$$

According to $E_k = \frac{1}{2}\sum_{i=1}^{l}\left(Y_j^k - y_j^k\right)^2$ in (5)

Then, we can conclude as follows.

$$\begin{aligned} g_j &= -\frac{\partial E_k}{\partial Y_j^k}\frac{\partial Y_j^k}{\partial W_{hj}} = -\left(Y_j^k - y_j^k\right)f'(\beta_j - \theta_j) \\ &= \left(y_j^k - Y_j^k\right)f(\beta_j - \theta_j)\left[1 - f(\beta_j - \theta_j)\right] \end{aligned} \tag{10}$$

According to $Y_j^k = f(\beta_j - \theta_j)$ in (1), then

$$g_j = \left(y_j^k - Y_j^k\right)Y_j^k\left(1 - Y_j^k\right) \tag{11}$$

Combining (6), (7) and (9), then

$$\frac{\partial E_k}{\partial W_{hj}} = -\left(y_j^k - Y_j^k\right)Y_j^k\left(1 - Y_j^k\right)b_h \tag{12}$$

$$\Delta W_{hj} = \eta\left(y_j^k - Y_j^k\right)Y_j^k\left(1 - Y_j^k\right)b_h \tag{13}$$

In the same way, we can get as follow.

$$\begin{cases} \Delta\theta_j = -\eta g_j \\ \Delta V_{ih} = \eta e_h x_i \\ \Delta\gamma_h = -\eta e_h \end{cases} \tag{14}$$

$$e_h = -\frac{\partial E_k}{\partial B_h}\frac{\partial B_h}{\partial \alpha_h} = -\sum_{j=1}^{l}\frac{\partial E_k}{\partial B_h}\frac{\partial B_h}{\partial B_h}f'(\alpha_h - \gamma_h) = b_h(1 - b_h)\sum_{j=1}^{l}W_{hj}g_j \tag{15}$$

3 Solution to the Model

We decided to solve the practical problem in two steps.

3.1 Processing Existing Data and Observing Patterns

After processing and analysis of the given data using python programming, the number of synthetic opioids and heroin identified in the counties of the five states 2010 in Fig. 2 was obtained. We visually found that for the states, the five counties with the largest number of substances identified in 2010 remained largely unchanged and far more than the other counties. For the changing trend, the number of identified corresponding substances changed significantly in the five most counties and stabilized after the tenth county. This intuitive data is very helpful for us to determine the two threshold values θ_j and γ_h in the model and can greatly reduce the range predicted by the model, so that the final results Y_j^k obtained by our model are as accurate as possible.

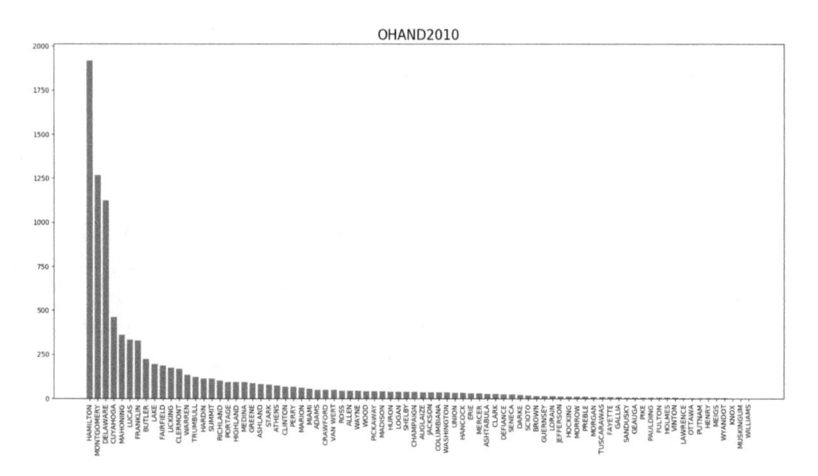

Fig. 2. The values of the corresponding substances identified in all counties of Ohio in 2010

3.2 Determine the Parameters and Bring Them into the Model to Get the Predicted Results

At the first step, by using python programming, we analyze a lot of data and get the model parameters θ_j and γ_h and the number of input neurons in the model is 17. θ_j and γ_h is a random value obtained after repeated training of the model.

We can then speculate on specific locations in each state where specific opioids are being used. When $0 \leftarrow Y_j^k = f\left(\beta_j - \theta_j\right)$, We can find the corresponding coordinates of latitude and longitude to find the source. Taking Ohio as an example, the coordinates of longitude and latitude obtained by our model training are as follows in Fig. 3.

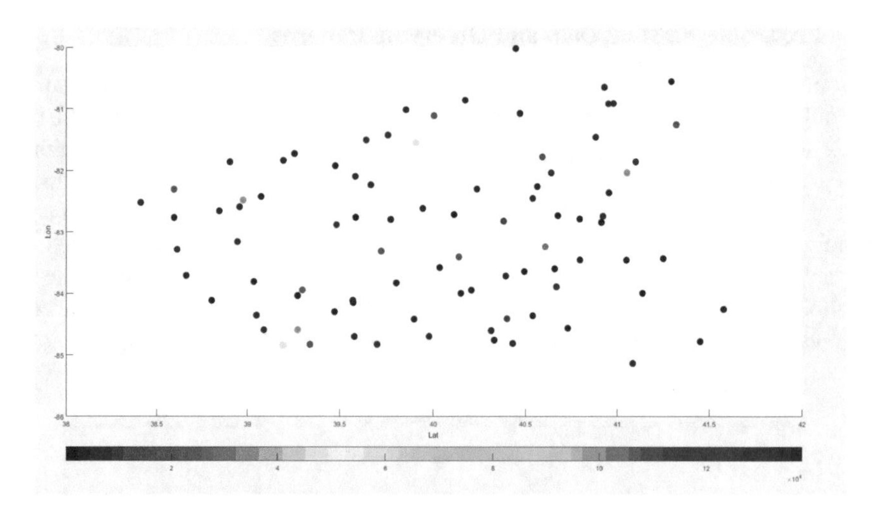

Fig. 3. The model predicted the opioid origin map

From Fig. 3, we can roughly determine where opioids are starting to be used in Ohio at coordinates (39.22, 85.1) and (39.57, 81.4). In the same way, we found the origin of the other four states. The coordinates of Virginia is (37.68, 77.52), the coordinates of Pennsylvania is (39.86, 75.26), the coordinates of Kentucky is (38.08, 85.72), and the coordinates of West Virginia is (39.17, 78.26).

4 Conclusions

Based on the analysis of relevant data, a diffusion model is established, which can automatically repair and deal with a large amount of data. After analyzing the nflis data, we get the following conclusion.

The reported spread and characteristics of synthetic opioids and heroin incidents between the five states and their counties are as follows. Its Character diffuse from the origin to the periphery, but the diffusion range is small and increases with time. The number of counties in the model data that are below the threshold is small and the growth is extremely slow.

When heroin, morphine and methadone reach the threshold, all drugs experience an opioid crisis.

Acknowledgements. This work was supported by Sichuan Province Science and Technology Program (2019YJ0683).

References

1. Deming, W., Li, W., Zhang, G.: Short-term wind speed prediction model based on genetic BP neural network. J. Zhejiang Univ. (Eng. Sci. Edn) **46**(05) 837–841,904. (2012)

2. Sun, Y.: Research on recommendation algorithm based on big data. Xiamen University (2014)
3. https://baike.baidu.com/item/%E7%A5%9E%E7%BB%8F%E7%BD%91%E7%BB%9C%E6%A8%A1%E5%9E%8B
4. Li, H., He, G., Guo, Q.: Similarity retrieval method of organic mass spectra based on Pearson correlation coefficient. Chemical analysis and metrology (2015)
5. Alexander, M.J., Mathew V.K., Barbieri, M.: Trends in black and white opioid mortality in the United States, 1979–2015. Epidemiology **29**(5) 707–715 (2018). www.com/epidem/Fulltext/2018/09000/Trends_in_Black_and_White_Opioid_Mortality_in_the.16.aspx
6. Anselin, L.: An introduction to spatial regression analysis (2003). http://R.labs.bio.unc.edu/buckley/documents/anselinintrospatregres.pdf
7. BAART Programs: Vermont's opioid addiction: A family crisis (2018). https://baartprograms.com/vermonts-opioid-addiction-a-family-crisis/
8. Berezow, A.: White overdose deaths 50% higher than blacks, 167% higher than (2018) https://www.hispanics.acsh.org/news/2018/04/05/white-overdose-deaths-50-higher-blacks-167-higher-hispanics-12804
9. Bivand, R.: predict.sarlm: Prediction for spatial simultaneous autoregressive linear model objects. Documentation reproduced from package spdep version 0.8-1 (n.d.). https://www.rdocumentation.org/packages/spdep/versions/0.8-1/topics/predict.sarlm
10. Blau, M.: Stat forecast: Opioids could kill nearly 500,000 Americans in the next decade (2017). https://www.statnews.com/2017/06/27/opioid-deaths-forecast/

Design of Personalized Intelligent Learning Assistant System Under Artificial Intelligence Background

Ping Xia(✉)

Guangzhou College of Technology and Business, Guangdong, China
710398795@qq.com

Abstract. In the context of the era of artificial intelligence, integration and innovation of big data, cloud computing, Internet of Things, and 5G and other cutting-edge technologies in the field of higher education have the development of smart education. It provides a valuable opportunity for fully implementing adaptive and personalized learning based on teaching according to aptitude. This article is driven by AI and student-centered, starting from the personalized needs of learning. Analyze human-computer interaction technology founded on natural language processing, and design anthropomorphic learning assistance and communication functions. Collect and process educational big data, mining quantitative data on learning behavior characteristics, learning status and learning effects, and build a staged student portrait model. Intelligently generate personalized learning programs and plans, recommend high-quality learning resources, and implement adaptive personalized learning program updates based on learning effect feedback. Through the research and design of the framework and hierarchical functions of the intelligent learning assistant system, we are committed to providing students with anthropomorphic learning assistance functions and an ecological learning environment, promoting students' personalized learning and sustainable development, and realizing the function of intelligent tutoring.

Keywords: Artificial intelligence · Intelligent learning assistant · Natural language processing · Personalized learning · Student portrait model

1 Introduction

In 2017, the State Council of my country issued the "New Generation Artificial Intelligence Development Plan", which proposed the development of intelligent education and the construction of a new education system for intelligent learning and interactive learning [1]. In 2018, the Ministry of Education released the "Innovative Action Plan for Artificial Intelligence in Higher Education" to promote the implementation of artificial intelligence to the field of higher education [2]. In May 2019, at the International Conference on Artificial Intelligence and Education held in Beijing, in-depth discussion on the use of AI to reshape education and learning, assist intelligent learning, build a personalized learning environment and other content, and proposed "AI+ education" global cooperation development framework [3]. Under the strong

J. MacIntyre et al. (Eds.): SPIoT 2020, AISC 1282, pp. 194–200, 2021.
https://doi.org/10.1007/978-3-030-62743-0_27

support of national policies and the influence of the international research atmosphere, the field of higher education will be driven by the core technologies of AI, and great changes and innovations will occur in teaching modes, learning methods, management and evaluation methods.

In recent years, the student-centered teaching concept has gradually been accepted by colleges and universities. In the process of education and teaching, the enthusiasm, concentration, and participation of students' subjective learning, personalized development, and individualized teaching has become important factors that affect students' learning effects. It determines the effectiveness of personnel training in colleges and universities, and profoundly affects the comprehensive development of students' professional skills and comprehensive quality. This article is mainly about students, and deeply explores the design of the intelligent learning assistant system in the learning field. Exploring the use of the progressive nature of AI to fully arouse students' interest in learning and guide students to explore independently through all links in and out of class. Adopt big data technology to analyze students' learning status, monitor process quality, scientifically formulate and adaptively update learning programs, solve practical problems in the learning process, improve students' learning efficiency, and thus make some contributions to the development of wisdom education.

2 Research Status of the Integration of Artificial Intelligence and Intelligent Learning

My country's research on the application of artificial intelligence to the field of education is at an initial stage. Lack of education big data and the low value density has led to a more complicated process of big data integration and analysis. In addition, the difficulty of the integration of artificial intelligence technology and education, human-computer collaboration and human-computer interaction lack of sufficient trust and other problems, hinder the application and realization of artificial intelligence technology in the field of education [4]. Research hotspots of domestic scholars in the field of intelligent learning are mainly reflected in the topics of deep learning, machine learning, adaptive learning, personalized learning, hybrid learning and lifelong learning. Committed to building sustainable learning programs and a strong intelligent learning platform to achieve a fair distribution of educational resources [5].

At present, relevant research abroad is mainly reflected in the topics of intelligent tutor system, educational robot, deep learning, intelligent learning and computational thinking. The research trend is toward the development of emotional cognition and learning cognition based on neuroscience [6]. For example, Samarakou and others have developed a digital learning system to extract text understanding and learning style features to help students recognize their own learning style [7]. Oshima et al. explored how robots can be used as intelligent tutors to assist and support groups in collaborative learning [8]. The analysis of the development status and trend of intelligent learning shows that there are few domestic researches on artificial intelligence-based intelligent learning systems, foreign research is relatively frontier, and the results are more prominent, which provides better experience and reference for the research of this topic. Based on in-depth exploration and application of cutting-edge technologies at this stage, this paper studies the intelligent learning assistant system that meets the need for personalized learning based on the reality of Chinese students.

3 Framework Design of Intelligent Learning Assistant System

Driven by artificial intelligence, using big data analysis and mining, cloud computing and cloud storage technologies, research and design an adaptive personalized learning assistant system. The overall framework of the system is shown in Fig. 1. It consists of four core functional layers from bottom to top: human-computer interaction layer; data analysis layer; student portrait layer; system application layer. The lower layer provides interface services and system data for the upper layer, and the upper layer uses the services offered by the lower layer. All levels collaborate with each other to jointly achieve structured system functions. With students as the main body, intelligent learning assistants are used as learning guidance and auxiliary tools, emphasizing the important role of learning initiative and internalization of knowledge on students' learning ability and personalized development. Through intelligent human-computer interaction, stimulate students' interest in learning, fully mobilize learning enthusiasm and participation, and cultivate students' habit of self-study and self-inquiry in and out of class. Based on the quantitative data of students' daily learning as the basis of big data analysis, scientifically formulate personalized learning plans based on learning characteristics, status and effects, intelligently recommend high-quality learning resources, provide anthropomorphic learning assistants and an ecological learning environment. The assistant system aims to promote students' personalized learning and sustainable development, and realize the function of intelligent tutoring.

4 The Main Function Level Design of the Intelligent Learning Assistant System

4.1 Human-Computer Interaction Layer

The human-computer interaction layer is the main module for virtual communication between the user and the system, and its processing flow is shown in Fig. 2. Understand and generate human language through natural language processing and sentiment analysis technology, and provide students with interface functions such as knowledge quiz, topic discussion, question feedback, and classroom assistance based on knowledge graph.

First, pre-process the voice or text input by the students, clean up the noise and interference signals in the voice data set, and filter the redundant information. Use speech recognition and machine translation to convert speech into text, and perform a round of information filtering based on error detection and correction. For the special content where ambiguity cannot be judged, the characteristic parameters of speech content presentation are extracted through sentiment analysis technology. In terms of emotion recognition and understanding, we analyze from three levels to eliminate ambiguity to the greatest extent: speech (speech speed, intonation, stress, emotional changes, etc.), morphology (facial expressions, gestures, posture, etc.), physiological signals (Heart rate, blood pressure, EEG, etc.). Normalize and classify texts, and based on contextual content understanding, query the language model library to generate inference and association feedback results, eliminating the diversification, ambiguity and variability of Chinese text in different contexts and situations.

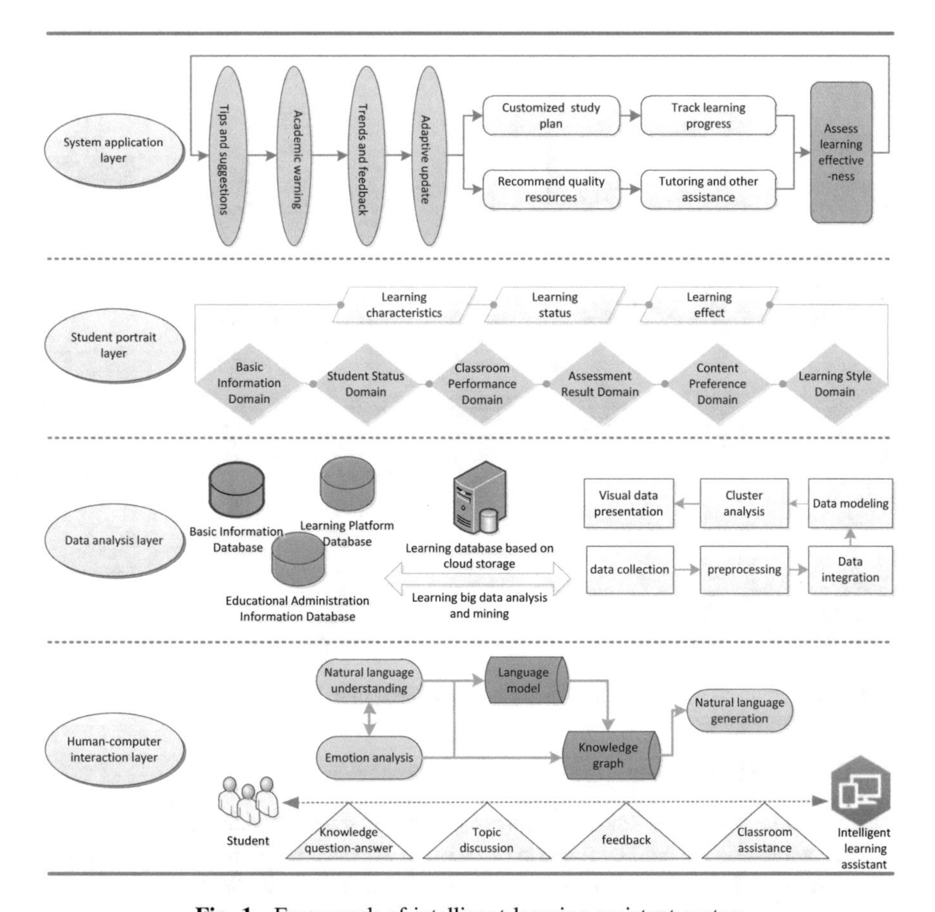

Fig. 1. Framework of intelligent learning assistant system

Secondly, using automatic summarization technology to understand the original meaning of the text, extract key information and reorganize the semantic structure to produce high-quality and fluent summary sentences. By extracting keywords from the summary information, the knowledge system matching the keywords is deeply searched the knowledge graph database, and the query results are visualized and presented. In the interaction with users, the use of natural language generation technology to achieve anthropomorphic knowledgeable answers and discussions. Through three-stage planning sentence, implementation syntax, planning text flow processing language generation, and then utilizing deep learning technology for training and generation, to enhance the rigor and precision of sentences. Finally, the system outputs the result of the interaction and feeds it back to the user.

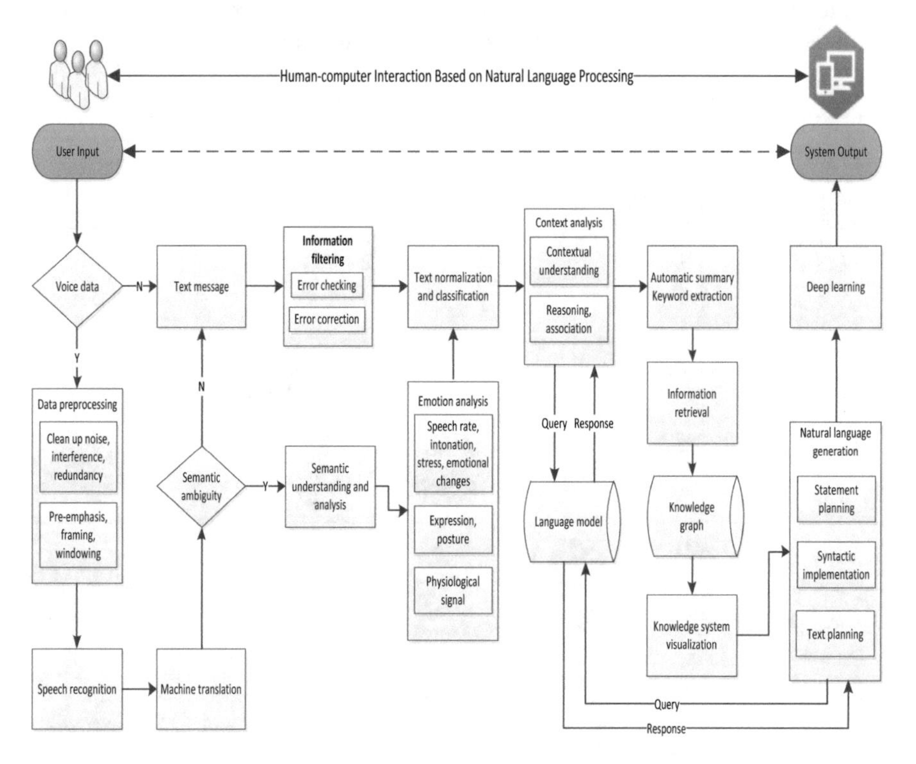

Fig. 2. Flow chart of human-computer interaction design based on natural language processing

4.2 Data Analysis Layer

Education data is the key source of information that reveals students' behavior laws, learning characteristics, and evaluation of learning effects, and is the objective basis for the realization of the function of the entire system. Connect the basic information database to obtain detailed records of the learner's basic situation, such as name, gender, age, household registration, school registration, etc. [9]. Call the interface of the educational administration system information database to extract process data reflecting the learning status, such as course selection, attendance, social practice, internship, book borrowing, digital resource utilization, etc. Collect result data that reflects the learning effect, such as subject scores, comprehensive evaluation, papers, experimental reports, scientific research results, etc. Obtain data from various online learning platforms, such as learning records, progress, collections, information retrieval, and download content. In addition, supplement other types of education data through manual collection to build a related education database with students as the main body. In the process of data flow processing, technologies such as preprocessing, integration, modeling, cluster analysis, and visual presentation are used to provide a data support system for upper-level student portraits.

4.3 Student Portrait Layer

The portrait layer is based on the integration and mining of lower-level information sources, using students as independent models to build the main body, and generates six major label information fields based on learning features, learning status and learning effects as the core elements, including basic information domain, student information domain, classroom performance domain, assessment result domain, content preference domain, learning style domain [10]. Based on the deep knowledge discovery of the essential content and internal connection of association rules, the label information domain is aggregated to reveal the user's periodic learning path and growth law. The establishment of a student data portrait model dominated by multi-level label domains and implicit relationship interpretation provides an explicit basis for the realization of personalized development and precise management of students.

4.4 System Application Layer

The system application layer is the long-term management and scientific utilization of student portraits, and is the realization of the system's scene and intelligent functions. Through multi-dimensional analysis and application based on portraits, provide dynamic learning tips and suggestions, conduct appropriate interventions in learning behaviors and timely academic warnings, so as to achieve an organic combination of primary prevention and full control. Depending on students' learning characteristic points and process status, intelligently customize personalized learning programs and implementation plans. And track the learning progress in real time, give long-term dynamic process monitoring and quality assessment. According to the learning plan, recommend high-quality learning resources, provide specialized learning guidance and other life-type auxiliary functions. According to the feedback and visual analysis of learning progress, periodic learning effect, dynamically adjusts the learning plan and the recommendation of learning resources to realize the adaptive personalized learning plan update to meet the periodic fluctuation and change need of students.

5 Conclusion

This research is mainly based on the actual needs of students, using artificial intelligence and big data analysis as the main means of implementation, designing an intelligent learning assistant system based on four core functional levels, dedicated to providing students with personalized guidance services and an ecological learning environment. Due to the limitation of frontier technology research, the systematic research work carried out at this stage still has a certain gap from the actual application, and needs further in-depth and detailed exploration. In the future work, we will continue to improve and optimize the system functional design to ensure the feasibility of the actual application of the system.

References

1. The State Council of the People's Republic of China. A new generation of artificial intelligence development plan. http://www.gov.cn/zhengce/content/2017-07/20/content_5211996.htm. (in Chinese)
2. Ministry of Education of the People's Republic of China. Artificial intelligence innovation action plan for colleges and universities. http://www.cac.gov.cn/2018-04/11/c_1122663790.htm. (in Chinese)
3. International Conference on Artificial Intelligence and Education. https://www.edu.cn/c_html/xxh/2019rgzn/. (in Chinese)
4. Wenyan, Z., Anqi, M.: Application status and development path analysis of domestic artificial intelligence education. Elem. Mid. Sch. Audio-Visual Educ. **Z2**, 99–102 (2019). (in Chinese)
5. Xu, F.: Analysis of research hotspots and trends of "artificial intelligence + education" in the past decade. Curr. Teach. Res. (06) 47–54, 67 (2020). (in Chinese)
6. Rongfei, Z., Leilei, Z., Yuehong, L., Keyun, Z.: Topics and trend analysis of educational artificial intelligence research abroad——visual analysis based on web of science literature keywords. Modern Educ. Tech. **29**(12), 5–12 (2019). (in Chinese)
7. Samarakou, M., Tsaganou, G., Papadakis, A.: An e-learning system for extracting text comprehension and learning style characteristics. Educ. Tech. Soc. **1**, 126–136 (2018)
8. Oshima, J.: Collaborative reading comprehension with communication robots as learning partners. In: Proceedings of ICLS 2012, no. 2, pp. 182–186 (2012)
9. Shi, M., G, Y., Min, Z.J., Qiu, H., Chen, Z.: Probe into the construction of student portrait system based on big data analysis of campus behavior. J. Chin. Multimedia Network Educ. (Early Issue) **04**, 70–71 (2020). (in Chinese)
10. Xueying, X., Yuhui, M.: Research on personalized learning guidance strategy based on student portraits. J. Heilongjiang Vocat. Coll. Ecol. Eng. **33**(03), 125–128 (2020). (in Chinese)

Application of Artificial Intelligence in Intelligent Decision-Making of Human Resource Allocation

He Ma[1](✉) and Jun Wang[2]

[1] A7 Dalian Neusoft University of Information, 8 Ruanjianyuan Road, 116023 Dalian, China
mahe@neusoft.edu.cn
[2] Department of Human Resource Management, Dalian Neusoft University of Information, Dalian, China

Abstract. In the process of social development, artificial intelligence has exerted varying degrees of influence on the production mode of modern society or the lifestyle of modern people. Artificial intelligence has been extensively used in many fields, and it will have a certain impact on human resource management. The object of this paper is to study intelligent decision-making of human resource allocation in the era of artificial intelligence. This research mainly use case study and data analysis to explore the artificial intelligence in human resources deployment of the application of intelligent decision-making, analysis of the artificial intelligence and the concepts related to the human resource allocation, artificial intelligence are analyzed based on the background now, in the current human resource allocation in application advantages, explore how artificial intelligence and human resource allocation, to better implement intelligent decision of human resource allocation. The experimental results show that the combination of artificial intelligence and human resource allocation together can make managers real-time monitoring human resource configuration, as much as possible to achieve information-based regulation, a real-time dynamic, and it can fully tap human resources potential, as well as improve management work efficiency and management efficiency, realize the enterprise or the high quality and rapid development of colleges and universities.

Keywords: Artificial intelligence · Human resources · Resource allocation · Intelligent decision

1 Introduction

The era of artificial intelligence emerges with continuous growth of computer science and information technology, which fully reflects highly developed era of information technology. In this context, we should closely follow the pace of The Times, the traditional allocation of human resources into the allocation of human resource information, in this process, the combination of artificial intelligence and human resources is essential. Artificial intelligence could obviously improve the efficiency of human resource management, as well as is beneficial to promote effectiveness of work, strengthening research

J. MacIntyre et al. (Eds.): SPIoT 2020, AISC 1282, pp. 201–207, 2021.
https://doi.org/10.1007/978-3-030-62743-0_28

on artificial intelligence application in human resource management, can play its important role, not only can improve efficiency of enterprise human resource allocation, but also improve efficiency of their job, and the management efficiency, effectively allocate and manage cost saving, new management solutions for human resource allocation in China, in order to promote the evolution of human resource management industry in China.

In China, many scholars have analyzed human resource allocation in era of artificial intelligence. Some researchers believe that human resource management under the Internet background should be reformed and innovated to adapt to the evolution trend of The Times. Four new thoughts of human resource management are listed [1]. They believe that in the Internet era, more attention should be paid to the management of human resources so as to effectively deal with and solve the problem of brain drain [2]. Some scholars also believe that we are in the era of artificial intelligence and new normal of economy, so it is essential to improve management and deployment of human resources under the background of multiple factors [3]. However, the existing research is only theoretical, and there is no relevant empirical research to prove the feasibility of theoretical research. This kind of problem also appeared in the study of foreign scholars. In the journal Science, some scholars conducted a systematic mapping study on human resource allocation, which showed that the combination of human resource allocation and artificial intelligence was feasible, but this study alone was not enough to prove that the combination of large-scale human resource allocation and artificial intelligence was of practical significance [4, 5].

Of this study is to make up for deficiencies in the research of current scholars, not only from the theoretical knowledge, combined with the related concepts of artificial intelligence and human resource allocation, analyze feasibility and necessity of combining of artificial intelligence and human resource allocation, validated at the same time, combined with specific case analysis, further evidence that artificial intelligence merge with human resource allocation is imperative as well as the huge advantage; Moreover, data analysis is conducted according to relevant data [6, 7]. Here is mainly to use the human resources deployment of intelligent decision-making and human resource deployment of traditional company for data collection, using comparative analysis of these data according to the analysis of the results of the analysis of artificial intelligence application advantages in human resources deployment of intelligent decision making, for the previous theoretical knowledge data provide strong theoretical support, fully demonstrates the research process and the research content of this article has the feasibility and practical significance, will provide new solutions for our country human resource allocation management [8, 9].

2 Method

2.1 Key Points and Difficulties

This study focuses on analyzing the existing theoretical knowledge, systematically sorting out the combination of artificial intelligence and human resource allocation, and enumerating possibility as well as necessity of association of artificial intelligence and human resource allocation. At the same time, the case of ant financial using intelligent dispatching center to optimize resource allocation of customer service center was used to verify and analyze the possibility and necessity [10]. The difficulty of this study lies in how to combine theoretical knowledge with data knowledge according to the results of data survey [11]. Data knowledge provides data basis for theoretical knowledge, which can better verify the correctness of association of artificial intelligence and human resource allocation sorted out in this paper, and add new research ideas to the theoretical research of this paper [12].

2.2 Research Ideas and Methods

This study mainly USES case analysis method and the data analysis method to conduct the research. Among them, case molecular method refers to combining theoretical knowledge with specific analysis of a case. Data analysis method can provide a more solid data basis for theoretical knowledge, thus making the research of this paper more complete. First of all, the case study mainly focuses on the case of ant financial's use of intelligent dispatching center to optimize resource allocation of customer service center. The case of ant financial is chosen as the research case of this paper because the intelligent decision-making development of human resource allocation of ant financial has achieved good results, which is typical of cases, and also has reference significance for enterprises that want to develop intelligent decision-making of human resource allocation. Combined with the related concepts of artificial intelligence and human resource allocation to analyze case, more enriched the theoretical research of this paper.

Secondly, the collected data are used for comparative analysis, and charts are used for visual analysis. Data object is to use the company human resources deployment of intelligent decision-making and put to a traditional human resource allocation decisions, acquisition of data, including to promote the efficiency of work, the human resources allocation efficiency, corporate annual performance and efficiency of emergency response, an independent analysis and comparative analysis of these data, exploring the human resources allocation efficiency and adopted the intelligent decision of company related human resource allocation efficiency of traditional decision-making company related, if there are obvious differences in both prove that association of artificial intelligence and human resource allocation can effectively improve the company's operating efficiency, If there is no obvious difference between the two, it proves that association of artificial intelligence and human resource allocation cannot effectively improve the operating efficiency of the company in all aspects.

Finally, based on the molecular results of data and theoretical knowledge, it provides feasible application Suggestions for intelligent decision-making of human resource allocation, and actively solves various problems existing in human resource management of enterprises. Enterprises are encouraged to establish advanced human resource management concepts as well as formulate a set of human resource management systems and systems that adapt to the changing times based on their own actual conditions.

3 Experiment

According to the case analysis of ant financial, the usage of artificial intelligence in human resource allocation can be divided into three aspects: perception, decision making and response. Among them, decision-making refers to that artificial intelligence can reasonably assign the connection situation between customer service personnel and users, and assist operation personnel to make reasonable judgment and optimal choice according to real-time status of the site. In the process of case analysis, we can learn that the decision-making module directly or indirectly leads to the efficiency of human resource allocation and emergency response.

Data analysis is mainly for the improvement of work efficiency, human resources allocation efficiency, enterprise annual performance improvement and emergency response efficiency four kinds of data for independent and comparative analysis. Independent analysis is used to find out the possibility of adopting intelligent decision making. By comparing and analyzing the data of the two companies, we can see whether intelligent decision making is beneficial to the improvement of efficiency in all aspects of the company. Combined with the result of data analysis and case analysis shows that human resource allocation of the intelligent decision, greatly increasing the efficiency of the company's various aspects, association of artificial intelligence and human resources deployment of intelligent decision-making and development is that we must solve the problem now, the growth of artificial intelligence is a kind of new form of interaction, can gradually take the place of human to solve the problem and reduce the cost of the corresponding.

4 Discuss

4.1 Experimental Results Show

The data analysis results are shown in Table 1 and Table 2. Table 1 is the data of companies (C1) that have adopted the intelligent decision-making of human resource allocation, and Table 2 is the data of companies (C2) that have adopted the traditional decision-making of human resource allocation. The header Iwe in the table represents the improvement of work efficiency, Hrde represents the allocation efficiency of human resources, Copi represents the improvement of annual corporate performance, and Ere represents the efficiency of emergency response. We can see from Table 1 and Table 2 that the data of human resource allocation decisions made in different ways are quite different.

Table 1. Intelligent decision

	Iwe	Hrde	Copi	Ere
C1	68%	93%	75%	84%

Table 2. Traditional decision

	Iwe	Hrde	Copi	Ere
C2	60%	74%	45%	39%

Combined with the relevant results of the collected data, we can obtain the data representation shown in Fig. 1. We can see from Fig. 1 USES human resources deployment of intelligent decision-making of the company related data were higher than the traditional decision of company data, especially the human resources allocation efficiency and the efficiency of emergency response on the two data, the addition of artificial intelligence can greatly improve efficiency of human resource allocation efficiency and emergency response.

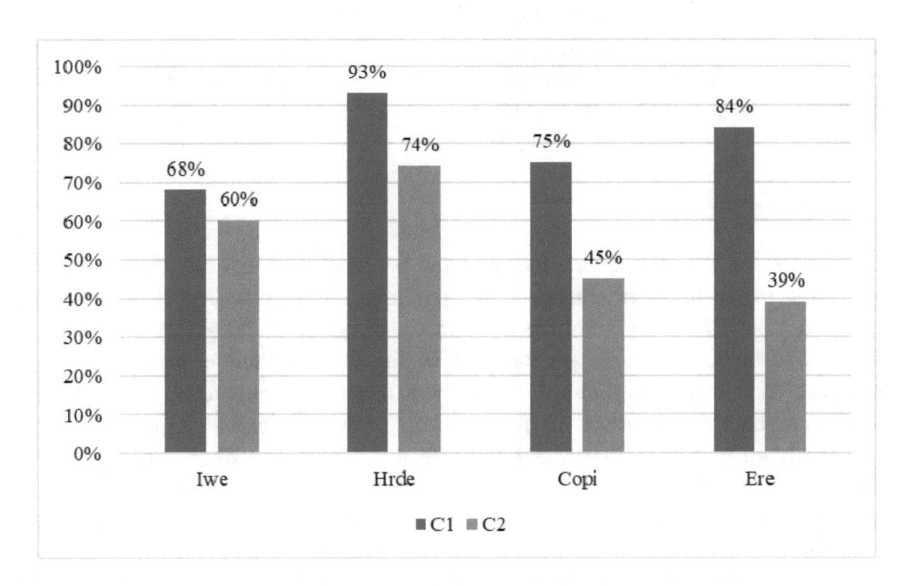

Fig. 1. Comparative analysis

4.2 Countermeasure Analysis

(1) Establish the cognitive concept with a far-sighted vision

We should realize that in the foreseeable future, artificial intelligence technology would not fully substitute all the labor force in human resource management departments. Therefore, for popularization of artificial intelligence technology, we should not only comply with the development trend of technology, but also actively introduce and improve the advanced technology and equipment in terms of human resource management, so as to enhance effectiveness and quality of human resource management. At the same time, it is required to see advantages of human in thinking, logic, judgment and other aspects as a dynamic production factor, and realize that in a predictable long time, human and artificial intelligence machines will present a complex cooperative relationship of competition and cooperation.

(2) Deepening the application of artificial intelligence in human resources

In process of practical application, artificial intelligence can handle problems through various aspects and measures. For example, the construction of scenarios can be used to simulate various methods, which can provide good processing effect for complex problems and some support for decision-making.

(3) Enterprise is based on optimizing the human resource allocation management system

It is not a blind process for enterprises to construct the human resource allocation management mechanism. It is necessary to have sufficient theoretical basis, practical experience, operation practice and development prospect support, so as to avoid "blind decision-making", "patting head decision-making" and other problems. First, the establishment of talent incentive and restraint mechanism consistent with the management mechanism. Second, enterprises should not only provide the incentives with the nature of "stock", but also construct the incentives with the nature of "flow" – talent growth and career development system.

5 Conclusions

According to the research in this paper, it can be concluded that combining with the application of artificial intelligence, strengthening human resource management, improving level of human resource allocation management and giving play to core functions of talents act as a pivotal part in the development of enterprises. The emergence of artificial intelligence, for human resource management, could act as a good part in the improvement of traditional management model, but at the same time it is necessary to correctly understand the role and effect of artificial intelligence in human resource management, which can not only bring artificial intelligence into full play, but also improve level of human resource management.

Acknowledgements. This work was supported by Science and Technology Innovation Think Tank Project of Liaoning Science and Technology Association (project number: LNKX2018-2019C37), and Social Science Planning Fund Project of Liaoning Province (project number: L19BRK002).

References

1. Asma Enayet, Md., Razzaque, A., Hassan, M.M.: A mobility-aware optimal resource allocation architecture for big data task execution on mobile cloud in smart cities. IEEE Commun. Mag. **56**(2), 110–117 (2018)
2. Leatherdale, S.T., Lee, J.: Artificial intelligence (AI) and cancer prevention: the potential application of AI in cancer control programming needs to be explored in population laboratories such as COMPASS. Cancer Causes Control **30**(7), 671–675 (2019)
3. Powell, M.E., Cancio, M.R., Young, D.: Decoding phonation with artificial intelligence (DeP AI): Proof of concept. Laryngoscope Invest. Otolaryngol. **4**(3), 9–14 (2019)

4. Sellak, H., Ouhbi, B., Frikh, B.: Towards next-generation energy planning decision-making: An expert-based framework for intelligent decision support. Renew. Sustain. Energy Rev. **80**, 1544–1577 (2017)
5. Yao, M., Sohul, M., Marojevic, V.: Artificial Intelligence Defined 5G Radio Access Networks. IEEE Commun. Mag. **57**(3), 14–20 (2019)
6. Wang, J., Xu, S.X., Xu, G.: Intelligent decision making for service and manufacturing industries. J. Intell. Manufact. **5**(7), 1–2 (2019)
7. Zuo, J., Yang, R., Zhang, Y.: Intelligent decision-making in air combat maneuvering based on heuristic reinforcement learning. Acta Aeronautica Et Astronautica Sinica **38**(10), 401–416 (2017)
8. Mohajer, A., Barari, M., Zarrabi, H.: Big data based self-optimization networking: a novel approach beyond cognition. Intell. Autom. Soft Comput. **24**(12), 1–7 (2017)
9. Grant, K., White, J., Martin, J.: The costs of risk and fear: a qualitative study of risk conceptualisations in allied health resource allocation decision-making. Health Risk Soc. **53** (10), 1–17 (2019)
10. Lane, H., Sturgess, T., Philip, K.: What factors do allied health take into account when making resource allocation decisions? Int. J. Health Policy Manage. **7**(5), 412–420 (2017)
11. Stamos, A., Bruyneel, S., De Rock, B.: A dual-process model of decision-making: the symmetric effect of intuitive and cognitive judgments on optimal budget allocation. J. Neurosci. Psychol. Econ. **11**(1), 1–27 (2018)
12. Aggarwal, M.: Soft information set for multicriteria decision making. Int. J. Intell. Syst. **5**(3), 21–26 (2019)

Research on Early Warning of Security Risk of Hazardous Chemicals Storage Based on BP-PSO

Conghan Yue, Jun Ye, and Zhen Guo(✉)

School of Computer Science and Cyberspace Security, Hainan University, Haikou, China
guozhen@hainanu.edu.cn

Abstract. The storage and transportation of hazardous chemicals has always been a serious problem. In this paper a scheme for the safety monitoring and management of hazardous chemicals storage is proposed. Multiple sensors are used for real-time detection of external and internal environmental, and the data is transferred to the cloud. Further more the data processing in the cloud. Even if the corresponding early warning is made, it will help the storage security management of hazardous chemicals. For data in the cloud, BP neural network, and multi-particle swarm algorithm (PSO) are used to optimize the weights and thresholds of the network, and to avoid the shortcomings of using BP network alone, converge faster, and obtain the global optimal solution. Through the simulation study of hydrogen storage tank data, the results show that the results predicted by the BP-PSO neural network algorithm are more consistent with the actual situation, and are significantly better than the BP network results. This model can accurately make early warnings of hazardous chemicals and promptly help operators analyze existing security risks, which has a high reference value.

Keywords: Hazardous chemicals storage · BP neural network · PSO

1 Introduction

The storage and transportation of hazardous chemicals has always been a very important issue. How to make a security risk assessment and management of hazardous chemical storage is necessary. Li, Huiqin et al. (2017), use RFID technology to establish the storage management system and traceability system of hazardous chemicals based on Internet of Things [1]. Pang, Yajing et al. (2018), developed and proposed a real-time intelligent monitoring system for hazardous chemicals based on Internet of Things (IoT) technology [2]. Feng, Lixiao et al. (2018), proposed to put the hazardous chemicals data into the existing industrial inspection network for preprocessing, store it in a database, and then export the data to a real-time entity model, which is processed by big data and automatically reasoned, so that it can be hazardous at any time The real-time status of product equipment and the prediction of safety risks [3]. Zhou, Ning et al. (2020), according to the different storage modes, used the fuzzy comprehensive evaluation method to scientifically and effectively predict the storage of hazardous chemicals [4]. Bo, Dai et al. (2018), uses GA (genetic algorithm) associated

J. MacIntyre et al. (Eds.): SPIoT 2020, AISC 1282, pp. 208–213, 2021.
https://doi.org/10.1007/978-3-030-62743-0_29

with the residual rectangle algorithm to reasonably manage the storage space of hazardous chemicals under the condition of safety [5]. Li, Haisheng et al. (2018), designed and implemented the chemical logistics supervision and early warning system based on cloud computing of the Internet of Things, and the real-time monitoring of the whole hazardous chemical logistics chain is well applied to the monitoring and early warning of hazardous chemical logistics [6]. Liu, Xuanya et al. (2017), put forward the framework of risk management technology system for inflammable and explosive hazardous chemicals and a dynamic safety monitoring system based on the bow-tie model, the risk analysis of hazardous chemicals storage area is carried out [7]. Zhengbing Li et al. (2019), used the Markov Chain Monte Carlo (MCMC) algorithm to establish a transient hydrothermal and leakage risk assessment model to determine the leakage volume and risk level of hazardous chemicals [8].

Artificial neural network (ANN) has been a research hotspot in the field of artificial intelligence since the 1980s. In the last decade, the research of ANN has been deepening and great progress has been made. In 1943, the mathematical model of formal neurons was proposed and the era of theoretical research in neuroscience was created [9]. In 1997, a new learning method of multilayer neural network was proposed [10]. In 1999, the stability study of neural network was proposed in [11]. In 2008, the neural network structure was directly determined before the transformation. The algorithm has the advantages of simple operation, less parameters to be adjusted, strong global search ability and fast convergence speed [12]. In 2016, data extraction based on BP neural network methods was proposed [13].

These studies put forward different safety early warning methods in the process of chemical transportation and storage, which have certain reference significance for related personnel. This paper uses the data transmitted from multiple sensors to the cloud to learn based on the BP-PSO model, which has a faster neural network convergence speed, thus avoiding the situation where only a local optimal solution is found without the global optimal solution. Make more accurate warnings for the safety risks of hazardous chemicals.

2 Materials and Methods

2.1 BP Neural Network

BP neural network is a multi-layer feedforward neural network trained according to error inverse propagation algorithm, and its model topology consists of input layer, hidden layer and output layer. The basic idea is the gradient descent method, which uses the gradient search technology to minimize the squared error sum between the actual output value and the expected output value, and finally obtains the mapping relationship between the input and output. But it tends to get stuck in a locally optimal solution.

Figure 1 shows a three-layer neural network structure, including an input layer, a hidden layer and an output layer. Where, x_1, x_2, x_3, ..., x_M is the input value, y_1, y_2, ..., y_L is the output value; w_{ij} *and* w_{kj} are the connection weights between neurons, reflecting the connection strength between neurons and revealing the mapping relationship between input and output. BP neural network has been widely used, but there are some problems in the training process, such as easy to fall into the local minimum and slow convergence rate.

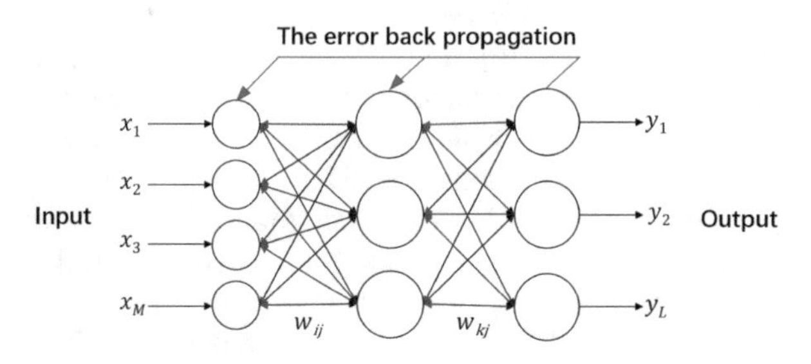

Fig. 1. BP network topology.

2.2 Multiparticle Swarm Optimization

The basic PSO algorithm is to randomly initialize a group of particles without volume and mass, and treat each particle as a feasible solution to the optimization problem. Each particle has a fitness value determined by an optimized function and passes a velocity variable. To determine the direction and position of each particle. The PSO algorithm finds the optimal solution through repeated iterations. In each iteration, the particles updated by tracking two extreme values. One of these two extreme values is the individual optimal solution pbest, and the other is the global optimal solution gbest. During the optimization process, the velocity and position of the particles are updated according to the following formula:

$$v_{i,j}(t+1) = wv_{i,j}(t) + c_1r_1\left[p_{i,j} - x_{i,j}(t)\right] + c_2r_2\left[p_{g,j} - x_{i,j}(t)\right] \tag{1}$$

$$x_{i,j}(t) = x_{i,j}(t) + v_{i,j}(t+1) \tag{2}$$

If $v_{i,j} > v_{max}$, then $v_{i,j} = v_{max}$; if $v_{i,j} < -v_{max}$, then $v_{i,j} = -v_{max}$. Assuming there are m particles and d-dimensional space, in the above formula, i represents the current particle; j represents the current dimensional space; c_1 and c_2 are learning factors; r_1 and r_2 are two random values in [0,1] Number; w is the weight of inertia.

2.3 BP-PSO

Because the BP neural network algorithm has the disadvantages of slow training speed and local convergence, a hybrid algorithm combining PSO algorithm and BP neural network is adopted. In order to improve the shortcomings of the BP neural network algorithm, the PSO algorithm is used to optimize the relevant parameters of the neural network. The weights and thresholds in the BP neural network are mapped to the particles in the PSO, and the speed and position of the particles are updated through iteration to optimize the parameters. Proceed as follows:

(1) Initialize the neural network and determine its topology. Determine the topology according to the input and output parameters of the system. Map the weights and thresholds of the neural network to particles, and determine the dimension d of the particles

$$d = n_h + n_o + n_i \times n_h + n_h \times n_o \tag{3}$$

In the formula, n_i, n_h, n_o are the number of neurons in the input layer, hidden layer and output layer, respectively.

(2) Initialize the particle swarm size M, learning factors c_1 and c_2, weighting factor w, and the position and velocity of each particle, and calculate the fitness value of each particle accordingly. The fitness function is as follows

$$E = \frac{1}{M}\left(\sum_{k=1}^{M}\sum_{j=1}^{n_O}\left(y_{kj} - \overline{y_{kj}}\right)^2\right) \tag{4}$$

In the formula, M is the number of samples; $k = 1, 2,\ldots, M$; $j = 1, 2,\ldots, n_o$; y_{kj} is the theoretical output value of the neural network, and $\overline{y_{kj}}$ is the actual output value of the neural network.

(3) Calculate the fitness value of each particle, and determine the individual extreme value pbest and the global optimal value gbest by comparing their sizes.
(4) Update the particle based on the particle speed and position formula, and judge whether it exceeds the limit.
(5) Update the fitness value of each particle, and use pbest to record the best position experienced.
(6) Search for the global optimal solution from all the individual optimal pbest values, and update the global optimal solution gbest.
(7) Determine whether the preset accuracy or the maximum number of iterations is met. If the conditions are met, the iterative calculation is terminated and the result is output; otherwise, it returns to step (3) to continue searching.

3 Results and Analysis

Collect the data of a representative hydrogen tank storing hazardous chemical hydrogen, with storage quality, temperature, internal pressure, humidity and surrounding concentration indicators as input variables, and safety (a value of 1 represents safety) as output variables. After determining the relevant parameters, use MATLAB to establish a BP neural network for learning related data, and then use the network model after multi-particle swarm optimization for comparison and verification.

Set the error accuracy to 0.0001, and use the BP neural network algorithm and the PSO-BP neural network algorithm to train the hydrogen tank security risk assessment model. and the training error curve is shown in Fig. 2 and Fig. 3.

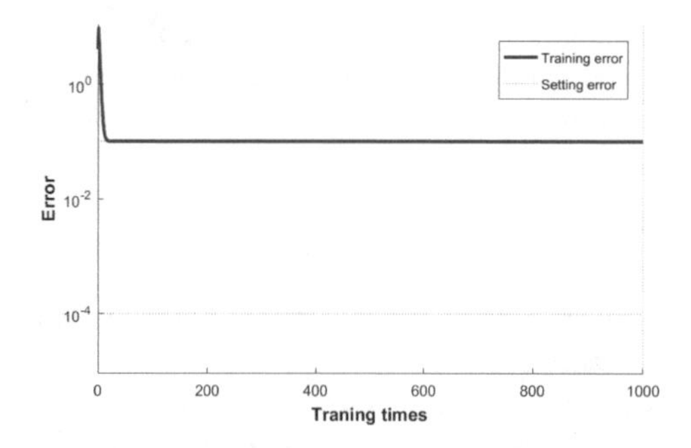

Fig. 2. BP neural network model training error curve.

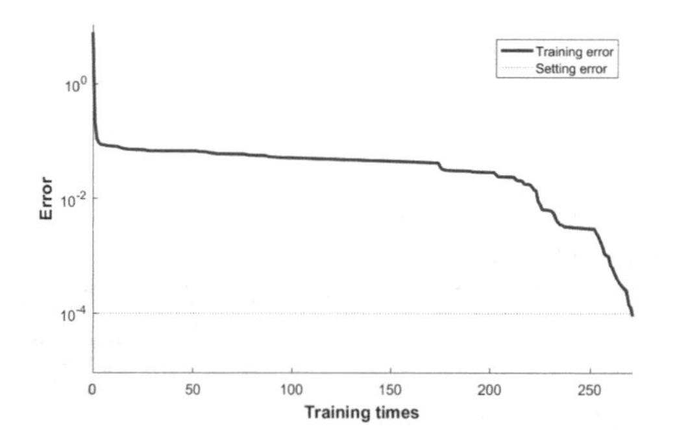

Fig. 3. BP-PSO neural network model training error curve.

It can be shown from the simulation test results that the number of iterations of the PSO-BP neural network algorithm in Fig. 3 is relatively small, the convergence speed is faster, and the accuracy is higher than the traditional BP algorithm in Fig. 2.

4 Conclusion

(1) Based on the analysis of the security risk related parameters of hazardous chemicals, the PSO-BP neural network algorithm was used to establish the relationship model of storage quality, temperature, internal pressure, humidity, ambient concentration and safety, and construct the hazardous chemicals Security risk assessment model.
(2) Comparative analysis with the risk assessment of the BP neural network model. The test results show that the PSO-BP neural network hybrid algorithm overcomes the shortcomings of being easy to fall into the local optimum, and has significantly improved both the convergence speed and the calculation accuracy.
(3) This method can use the latest data to update the model at any time, so it has a good guiding significance for the storage and management of hazardous chemicals.

Acknowledgments. This work was partially supported by the Science Project of Hainan Province (No.619QN193), the Science Project of Hainan University (KYQD(ZR)20021).

References

1. Li, H., Zhang, Y.: Research on storage management of hazardous chemicals based on internet of things. Chem. Eng. Trans. **62**, 1405–1410 (2017)
2. Pang, Y., Jia, S.: Design of intelligent monitoring system for hazardous chemicals based on internet of things technology. Chem. Eng. Trans. **71**, 199–204 (2018)
3. Feng, L., Chen, G., et al.: Ontology faults diagnosis model for the hazardous chemical storage device. In: Proceedings of 2018 IEEE 17th International Conference on Cognitive Informatics and Cognitive Computing, ICCI*CC 2018, pp. 269–274 (2018)
4. Zhou, N., Xu, B., et al.: An assessment model of fire resources demand for storage of hazardous chemicals. Process Safety Progress (2020)
5. Bo, D., Yanfei, L., et al.: Research on optimization for safe layout of hazardous chemicals warehouse based on genetic algorithm. IFAC-PapersOnLine **51**(18), 245–250 (2018)
6. Li, H.: Implementation of chemicals logistics supervision forewarning platform based on IoT cloud computing. Chem. Eng. Trans. **71**, 727–732 (2018)
7. Liu, X., Li, J., Li, X.: Study of dynamic risk management system for flammable and explosive dangerous chemicals storage area. J. Loss Prevent. Process Ind. **49**, 983–988 (2017)
8. Li, Z., Feng, H., et al.: A leakage risk assessment method for hazardous liquid pipeline based on Markov chain Monte Carlo. Int. J. Crit. Infrastruct. Prot. **27**, 100325 (2019). ISSN 1874-5482
9. Tong-Tong, W., Jian, Z., Chuan, T.U., et al.: Application of IPSO-BP neural network in water quality evaluation for Tianshui section of Wei river. Environ. Sci. Technol. **36**(8), 175–181 (2013)
10. Monjezi, M., Mehrdanesh, A., Malek, A., et al.: Evaluation of effect of blast design parameters on flyrock using artificial neural networks. Neural Comput. Appl. **23**(2), 349–356 (2013)
11. Hirota, M., Fukui, S., Okamoto, K., et al.: Evaluation of combinations of in vitro sensitization test descriptors for the artificial neural network-based risk assessment model of skin sensitization: evaluation of the descriptor for ANN risk assessment of skin sensitization. J. Appl. Toxicol. **35**(11), 1333–1347 (2015)
12. Shafabakhsh, G.A., Talebsafa, M., Motamedi, M., Badroodi, S.K.: Analytical evaluation of load movement on flexible pavement and selection of optimum neural network algorithm. KSCE J. Civil Eng. **19**(6), 1738–1746 (2015)
13. Mansouri, M., Golsefid, M.T., Nematbakhsh, N.: A hybrid intrusion detection system based on multilayer artificial neural network and intelligent feature selection. Arch. Med. Res. **44**(4), 266–272 (2015)

The Application of Artificial Intelligence and Machine Learning in Financial Stability

Chongyan Li(✉)

Guangzhou University Sontan College, Guangzhou 511370, China
lichongyan123@163.com

Abstract. In recent years, with the rapid development of artificial intelligence (AI) and machine learning, they have been widely used in various aspects of the financial field, and have a significant impact on the financial market, financial institutions, financial supervision and so on. The financial stability Council (FSB) has issued the development of artificial intelligence and machine learning in the financial service market and its impact on financial stability. The people's Bank of China has also put forward the regulatory requirements for "intelligent investment advisers" in the "guidance on regulating the asset management business of financial institutions (Draft)". The application in the financial field is becoming more and more common. With the deepening of the application of artificial intelligence and machine learning, it will bring a series of far-reaching impacts on the financial field at the micro and macro levels, and will also bring challenges to financial stability. The report suggests that at this stage, based on the practical assessment of the application risks of artificial intelligence and machine learning in data privacy, operational risk, network security and other fields, continuously improve the interpretability of the application models related to artificial intelligence and machine learning, and strengthen the supporting supervision of the application of artificial intelligence and machine learning in the financial industry. This paper summarizes the development of AI and machine learning in the field of financial services, analyzes the impact on micro, macro-economy and financial stability, and puts forward some suggestions on strengthening financial supervision by AI and machine learning.

Keywords: Artificial intelligence · Machine learning · Financial stability

1 Introduction

Artificial intelligence is the research, simulation, extension and expansion of human intelligence [1]. It is a subject that uses computer to study data of different types, sources and quality, and extract valuable information from it. In recent years, with the development of computer technology and the improvement of data availability, quality and quantity, all parties have regained their enthusiasm for artificial intelligence, which has been widely used in various fields including financial industry. Artificial intelligence is closely related to big data (analysis) and machine learning. Big data (analysis) refers to the behavior of processing and analyzing large or complex unstructured or semi-structured data. Machine learning is a branch of artificial intelligence, which studies how computers simulate or realize human learning behavior, so as to acquire

J. MacIntyre et al. (Eds.): SPIoT 2020, AISC 1282, pp. 214–219, 2021.
https://doi.org/10.1007/978-3-030-62743-0_30

new knowledge and skills, and reorganize the existing knowledge structure to improve its performance. According to the ability of automatic optimization, machine learning can be divided into four categories: supervised learning, unsupervised learning, intensive learning and deep learning. Among them, deep learning is usually called deep structure learning or hierarchical learning, which comes from the research of human brain neural network. It is an algorithm that attempts to use multiple processing layers including complex structures or constructed by multiple nonlinear transformations to abstract data. In recent years, deep learning has made significant achievements in many fields, such as image recognition, natural language processing, etc., thus laying the foundation for financial institutions to provide self-service financial services (Table 1).

2 Research on Machine Learning

For machine learning researchers, the algorithm SVM derived from statistical machine learning theory seems to be more attractive. However, if researchers forget the statistical basis based on SVM, it is contrary to the original intention of Vapnik. In fact, the theory of finite samples is the essence of statistical machine learning theory. Generally speaking, the statistical basis of machine learning is based on the assumption of minimum empirical risk. In other words, the estimation of generalization ability (empirical risk) of the model established by machine learning algorithm is based on the statistical distribution of hypothesis:

$$R(f) = \int \left| f(x) - y \right| P(x,y)dxdy, R_{emp}(f) = \sum_{i=1}^{l} \frac{1}{l} \left| f(x_i) - f_i \right| \tag{1}$$

Where x_i is independently and identically distributed in the probability density function $P(x, y)$ [2].

According to the law of large numbers in statistics, for learning model f, when the number of sample points l tends to infinity, the empirical risk $R_{emp}(f)$ converges to the expected risk $R(f)$ in probability. Generally, machine learning algorithms take the minimum empirical risk $R_{emp}(f)$ as the objective function. However, in 1971, Vapnik pointed out that the lower bound of $R_{emp}(f)$ may not converge to the lower bound of $R(f)$ in probability, This means that it is unreasonable to use the lower bound of $R_{emp}(f)$ as the objective function of machine learning, The lower bound of $R_{emp}(f)$ converges to the lower bound of $R(f)$ in probability if and only if $R_{emp}(f)$ uniformly converges to $R(f)$ (i.e. the law of large numbers in functional spaces) in probability. This is the statistical theory of finite samples. Since this statistical theory can be described by the VC dimension of the structure of the set of samples, then, The objective function of machine learning needs to be based on the structure of sample set, rather than on the minimum empirical risk such as mean square deviation, which is the essence of statistical machine learning theory.

3 The Development of AI and Machine Learning in Finance

(1) Basic situation of development

AI can be defined as the ability of a computer system to undertake tasks traditionally performed by humans relying on their intelligence. AI covers a wide range of fields, and machine learning is one of the branches. Machine learning can be defined as designing a series of "algorithms" to automatically make the optimal choice through empirical learning without or limited human intervention. In recent years, AI technology has developed rapidly and has been widely used in the financial field. According to the report of the World Economic Forum (WEF), many factors promoting the development of financial technology have promoted the development of AI and machine learning in the field of financial services. Financial institutions also have the motivation to use AI and machine learning to reduce costs, manage risks, improve business levels and increase revenue.

(2) Four applications in financial field [3]

The application of AI and machine learning in financial field can be divided into four aspects. First, customer-oriented (front desk) applications, including credit scoring, insurance and customer-oriented service robots; second, management level (back office) applications, including capital optimization, risk management model and market impact analysis; third, trading and portfolio management in financial markets; fourth, AI and machine learning are used by financial institutions for regulatory compliance management ("regte") The use of "psuch" in financial regulation.

Table 1. Application cases of AI and machine learning

Application area	Application content
Application for financial customers	Credit rating For pricing, marketing and managing insurance business Customer oriented chat robot
Application of financial institutions management	Risk management model (backtesting and model validation) and stress testing Market impact analysis (block trading model, etc.)
Trading and portfolio management	Application in transaction execution Application in portfolio management Compliance TECHNOLOGY (Application in financial regulatory compliance)
Application of financial compliance and supervision	Macro prudential supervision and data quality assurance Regulatory technology used by financial regulators

4 Application of Artificial Intelligence and Machine Learning in Finance

One is the credit score. Due to the adoption of machine learning and complex data analysis model, data sources are no longer limited to traditional credit data (structured data), but a large number of new data (unstructured or semi-structured) such as social media and communication tools are added. This not only benefits more people who don't enjoy the traditional financial services, but also classifies the borrowers quickly and accurately through qualitative analysis of the borrower's consumption behavior and willingness to pay, so as to improve the efficiency of loan decision-making. However, there is no evidence that the new model based on machine learning is superior to the traditional model in evaluating reputation. Moreover, the wide application of the new model faces two major constraints: one is the privacy of personal data and data protection; the other is that the method is based on many variables and models, and the calculation process is difficult to explain, which is not conducive to the communication between departments and customers [4, 5].

Second, insurance pricing, marketing and internal management of insurance companies. Big data analysis based on machine learning can identify high-risk cases by integrating real-time and highly detailed data, and provide more competitive pricing and improve marketing. At the same time, artificial intelligence and machine learning can also improve underwriting and claims efficiency, reduce operating costs and improve profitability. It is estimated that the total investment in insurance technology in 2016 is US $1.7 billion. 26% of insurance companies provide financial or non financial support to digital start-ups.

Third, intelligent customer service. Chat robot is a virtual human assistant, helping customers to handle business or solve problems. These automated programs use natural language processing methods (via text or voice) to communicate with customers and constantly improve themselves with machine learning algorithms. Some financial services companies have introduced chat robots into mobile applications or social media. At present, the service provided by chat robot is still relatively elementary, such as providing balance information for customers. In the future, chat robots will provide more consulting services to help customers make financial decisions.

5 The Influence of Artificial Intelligence and Machine Learning on Financial Field at Macro Level

From the perspective of economic growth, artificial intelligence and machine learning will promote economic growth through the following mechanisms: first, improve the efficiency of financial services. Artificial intelligence and machine learning can improve the efficiency of asset risk management and capital allocation of financial institutions, accelerate the speed of payment and settlement, save costs, and promote the development of the real economy. The second is to increase the interaction between finance and other industries to form a new "scale economy". The extensive application of artificial intelligence and machine learning in the financial industry will enhance the cooperation between financial services and e-commerce, sharing economy and other industries to form economies of scale and promote economic growth. The third is to attract investment in artificial intelligence and machine learning. Many industries outside the financial sector also show high enthusiasm for artificial intelligence and machine learning, which has led to related R & D investment and economic growth.

From the perspective of macro finance, the medium and short-term impact of artificial intelligence and machine learning on financial structure and financial market may be more complex, as follows: first, the impact on market concentration and systemically important institutions. Artificial intelligence and machine learning may change financial market concentration and enhance the importance of some financial businesses. The ability to obtain big data determines the importance of institutions in the system. If we can use big data to achieve scale effect, it will further consolidate the importance of institutions. In addition, due to the large amount of financial support needed to develop new technologies, new technologies are increasingly concentrated in a few large institutions that can afford high R & D costs. It is difficult to judge the impact of AI and machine learning on systemically important banks. The new technology may break the traditional banking business, attract new enterprises into the financial service industry, and reduce the systematic importance of large all-round banks.

(1) Suggestions on strengthening financial supervision in China

First, financial institutions should strengthen internal audit and risk management. Financial institutions should establish a professional internal and external audit system and an effective risk management mechanism for the application of artificial intelligence and machine learning. Second, the financial regulatory authorities should establish a special reporting and monitoring mechanism for the application and development of financial technology. Relevant innovative products or applications should be reported to the regulatory authorities in a timely manner, and effective statistical and evaluation systems should be established to master the application and impact of relevant technologies. The third is to monitor the external third-party technology providers, fully study and judge the relevance of the external institutions with financial institutions and financial markets. If necessary, relevant external institutions should be directly or indirectly included in the scope of financial supervision.

(2) Development direction of relevant financial supervision

In the "impact of the development of artificial intelligence and machine learning in the financial service market on financial stability" released by FSB, it puts forward three directions for financial supervision of AI and machine learning. One is to pay attention to the risk problems in related fields. At present, the application of AI and machine learning will bring data privacy problems, operational risks, network security issues and other risks. The second is to improve the interpretability of AI and machine learning technology applications. The increasingly complex model will limit the ability of developers and users to fully control and interpret the model, and even lead to their inability to understand the operation mode of the model. In order to better carry out risk management, the results of algorithm and decision-making process should be "explainable" while developing AI and machine learning technology. Thirdly, the financial regulatory authorities should continuously monitor the application of AI and machine learning, and it is necessary to strengthen the monitoring and evaluation of relevant innovation.

6 Conclusions

In conclusion, the application of artificial intelligence and machine learning should be paid close attention to and monitored continuously. At this stage, based on the practical assessment of the application risks of artificial intelligence and machine learning in data privacy, operational risk, network security and other fields, we should constantly improve the interpretability of the application models related to artificial intelligence and machine learning, and strengthen the supporting supervision of the application of artificial intelligence and machine learning in the financial industry.

Acknowledgements. Characteristic specialty construction project of teaching quality and teaching reform project of Guangdong Undergraduate University in 2019 ("Finance"); Teaching quality and teaching reform engineering course construction project of Guangzhou University Sontan College in 2019 ("**International Finance**").

References

1. Rissanen, J.: Modeling by shortest data description. Automatica **14**, 465–471 (1978)
2. Quinlan, J.: Induction of decision trees. Mach. Learn. **1**(1), 81–106 (1986)
3. Quinlan, J.: Improved use of continuous attributes in C4.5. Journal of Artificial Intelligence Research, 4: 77–90 (1996)
4. Min, Y: Machine learning and its development direction
5. The 31st International Conference on machine learning: J. Intell. Syst. no. 01 (2014)

Application of Alternative Routing Configuration Mechanism Based on Genetic Algorithm in Power Communication Network

Hong Tao(✉)

Shenzhen Power Supply Bureau Co., Ltd., Shenzhen 518001, China
cdan1268@163.com

Abstract. Aiming at a large number of key business scenarios in power communication network, considering the business performance requirements and network risks, an alternative routing configuration mechanism based on genetic algorithm is proposed. Firstly, the routing configuration problem of critical services is modeled, the channel pressure index is defined, and the routing configuration mathematical model of minimizing channel pressure is constructed. Then, combined with the characteristics of the model, genetic algorithm is used to solve the problem. Finally, the simulation experiment is carried out based on the existing network topology. The experimental results show that, compared with the traditional method, this mechanism can obtain the alternative route configuration scheme with lower global channel pressure value for the scenario with large network scale.

Keywords: Power communication network · Alternative routing configuration · Dispatch algorithm

1 Introduction

With the continuous development of power communication, the networking mode and structure of power communication network are more and more complex, especially in the transmission network side, the continuous application of SDH, PTN and OTN technology makes the capacity of the network continue to improve. At the same time, the key business, such as relay protection business, dispatching automation business, security control business, is also more and more, so the network planning and optimization It has brought great challenges. At present, there are strict levels of power communication security risk, which specifies the business type, quantity and affected degree of each risk level. When planning alternative routes for multiple critical services, it is necessary to balance the traffic distribution to avoid the load pressure and disruption impact caused by the excessive concentration of key services [1, 2]. In order to meet the above conditions, it is required that the alternative routing configuration algorithm can efficiently realize the simultaneous configuration of multiple services in the complex initial service environment, and effectively balance the network risk. At present, the traditional routing algorithm or intelligent optimization algorithm is usually used to configure the business routing of power communication network. In this paper, we use the heuristic algorithm

J. MacIntyre et al. (Eds.): SPIoT 2020, AISC 1282, pp. 220–225, 2021.
https://doi.org/10.1007/978-3-030-62743-0_31

[3, 4] to achieve load balancing based on the shortest available path in respectively. However, these algorithms only aim at the service balanced distribution of bandwidth, and do not consider the factor of service importance, which makes them not suitable for solving the problem of business risk distribution based on service importance in power communication network. At the same time, most of the routing configuration strategies are sequential configuration services, which can not handle the problem of multiple services configuring routes at the same time.

In order to minimize the overall risk of deploying multiple services at the same time, this paper proposes a key service routing configuration algorithm based on genetic algorithm. Considering the channel pressure, the algorithm models the risk minimization of the whole network, and proposes a service routing configuration algorithm based on genetic algorithm. Through the topology simulation close to the existing network, the reliability and optimization ability of the algorithm are verified; the algorithm can effectively optimize the channel pressure value of the service path in large-scale network topology, so that the total value of network channel pressure is the lowest and the risk is reduced.

2 Key Business Routing Configuration Modeling

2.1 Network Topology Model

In order to abstract the routing configuration problem of key services in power communication network, the network topology graph G (V, E) is defined, where $V = \{v_1, v_2, \cdots, v_n\}$ represents the set of nodes and $E = \{e_{12}, e_{13}, \cdots, e_{n-1n}\}$ represents the set of links between nodes, and its subscript represents the node numbers at both ends of the link. At the same time, the service set $S = \{s_1, s_2, \cdots, s_n\}$. Among them, the service Sn has the importance parameter d_n, which is the quantitative index of the impact of service interruption, and the weight is set according to the service classification. For any combination of nodes in the network, there is $p(i,j) = \{v_i, v_k, \cdots, v_j\}$, whose starting point and ending point are v_i and v_j; and respectively. If $\forall v_m, v_n \in p(i,j), v_m \neq v_n$. In other words, if there is no ring on the path and there is at least one edge between all the adjacent nodes on the path, then p (i, j) is called node y; for a path between and, the set of paths between nodes v_i and v_j is p (i, j).

When the service paths of all services have been determined, the one-to-one corresponding service paths can be obtained:$P_n(v_n^{start}, v_n^{end}) \in P(v_n^{start}, v_n^{end})$, where the n subscript represents the corresponding service, and the start and end subscripts respectively represent the start and end nodes of the service. For the link e_{ij} between any two nodes v_i and v_j we can know which services are carried by the service path. At the same time, the service path has time delay T_n, and the delay threshold T_{th}. At this time, the service path delay is calculated as follows:

$$T_n = \frac{L_n \gamma}{c} + \sum_{v \in p_n} t(v) \tag{1}$$

Where T_n is the delay of service path p_n, L_n is the total length of the path fiber, γ is the refractive index of the fiber core, c is the speed of light in vacuum. t (v) represents node delay.

2.2 Definition of Channel Pressure

In order to analyze the risk of channels carrying key services in the network, the concept of channel pressure value is defined here. For different services carried on the link, different quantitative importance can be given according to the types of services, such as line protection service and security and stability control service, which are the two most important services, i.e., the highest importance; After calculating the sum of the importance of all services on the link, we can get a quantitative indicator of the traffic load on the link, that is, the channel pressure value pr (E). The value of channel pressure is the sum of the importance of all services carried by the link. It can be used to effectively distinguish high-voltage and low-voltage links:

$$\Pr(e_{ij}) = \delta_{e_{ij}} \sum_{s_i} n_{s_i} \times d(s_i) \tag{2}$$

Where $\delta_{e_{ij}}$ is the inherent weight of edge e_{ij}, n_{s_i} And d (s_i) are the number and weight of traffic s_i on edge e_{ij}.The channel pressure values on the link are classified according to the sum of importance, so as to distinguish different levels of channel pressure.

2.3 Problem Analysis

In the service configuration of electric power communication network, one service is required to be equipped with three routes: primary route, alternative route and migration route. According to the operation of power communication network and routing configuration mode, this paper puts forward the configuration parameter requirements of three routes. Among them, the main route requires the minimum delay, the alternative route requires the optimization of channel pressure value, and the detour route requires the optimal delay in the emergency scenario. At the same time, the three routes can not have duplicate nodes and links. In the outage scenario, the interruption of a link will lead to the problem of rerouting for multiple services carried by it. For the main route and circuitous route, the shortest path algorithm can be used to optimize the delay. However, for alternative routes, channel pressure is the sum of the importance of link carrying services, and it is related to service deployment dynamically. Therefore, intelligent optimization algorithm is needed to unify and plan multiple services. Therefore, this paper mainly takes the alternative route planning scheme as the optimization model. Based on the above analysis, it can be seen that the optimization object of the algorithm is all service routes:

$$P = \{P_n(v_n^{start}, v_n^{end}), s_n \in S\} \tag{3}$$

The optimization objective is the global channel pressure value after service deployment, and the constraint condition is that the path delay of any service path is less than or equal to the delay threshold value:

$$\min_{P} \sum_{e_q \in E} \Pr(e_{ij}), s.t. \forall p_n \in P, T_n \leq T_{th} \quad (4)$$

In the traditional intelligent optimization algorithm, the real coded genetic algorithm is easy to be applied to the path planning problem, and the core crossover and mutation of the algorithm can effectively global search. Therefore, this paper proposes a genetic optimization algorithm, which is based on the traditional intelligent optimization algorithm genetic algorithm. This algorithm unifies the original single gene genome, realizes the simultaneous distribution of multiple services, and optimizes the global channel pressure value.

3 Genetic Algorithm

Genetic algorithm starts from a population representing the potential solution set of the problem, and a population is composed of a certain number of chromosomes encoded by genes. In the process of implementing the algorithm, chromosome coding is needed first. After the generation of the first generation population, according to the principle of survival of the fittest and survival of the fittest, a better and better approximate solution is generated. In each iteration process, chromosomes are selected according to the fitness of chromosomes in the problem domain, and the genetic operators of natural genetics are used for combination crossover and mutation, the population representing the new solution set is generated. This process will lead to the species group like the natural evolution of the posterity population more adapt to the environment than the previous generation, the optimal individual in the last generation population can be decoded as the approximate optimal solution.

3.1 Real Genome Coding

When genetic algorithm is used to solve the problem of optical transmission network line planning, each chromosome represents a service path. In this paper, real number coding is used. Each gene bit represents the node on a business path, and the sequence of the business path passing through the nodes is represented by the sequence of gene bits. One chromosome of multiple operations is bound into a group of genomes, which keeps synchronization in cross selection variation. When selecting, two groups of genomes were selected according to the total fitness value of the genome as the standard: when crossing, the two groups of genomes were crossed in pairs; when 8 | changed the sign, the whole genome was randomly mutated and downloaded with high definition and no water.

3.2 Fitness Function

The fitness function is determined by reliability and time delay. The reliability is expressed by the channel pressure value PR, which indicates that the higher the value, the lower the risk of business path and the higher the network reliability. The delay is the total delay TM of the service path. Since the optimization direction of the objective function corresponds to the direction of increasing antibody affinity, the fitness function of chromosome s is as follows.

$$\mathrm{f}(s_v) = \begin{cases} Z - \sum\limits_{e_{ij} \in S_v} \Pr(e_{ij}) \\ 0, \quad other \end{cases} \tag{5}$$

Where Z is a large number and the value of f (s_v) is guaranteed to be positive.

3.3 Selection Operator

In the selection of chromosomes, we should not only ensure that the excellent chromosomes can be selected with a large probability, but also ensure the diversity of the offspring population, so we use the selection operator which is proportional to the chromosome usage and inversely proportional to the chromosome concentration.

$$\mathrm{Q}(s_v) = \frac{f(s_v)}{\sum\limits_{v=1}^{N_s} f(s_v)} \tag{6}$$

Among them, $\mathrm{Q}(s_v)$ is the selection probability of chromosome s_v and N_s is the population size.

3.4 Crossover Operator

In this paper, the crossover used in this algorithm is contraposition crossover. First, we judge whether the chromosomes are crossed according to the set probability. If it should, we select the contraposition chromosomes in two groups of genomes in turn, and judge whether there are the same nodes except the starting point in the two paths. If there is the same node, the node is taken as the boundary and the nodes after the node are exchanged to mix the two paths.

3.5 Mutation Operator

In this algorithm, insertion point mutation is used to determine whether the chromosome satisfies the ring density constraint 1. If not, two path nodes are randomly selected and inserted into all nodes in turn to calculate the genome fitness; if there is a node to increase the fitness of the new path after the insertion point, the path after the insertion point is saved.

4 Conclusions

In order to solve the problem of high voltage caused by unbalanced pressure of multi service alternative routes in power communication network, this paper proposes an alternative routing configuration mechanism based on genetic algorithm. Firstly, after integrating the channel pressure and service characteristics of equipment and lines, setting real coding gene and fitness function, an improved genetic algorithm based on genome is proposed to determine the service path. Through the topology simulation close to the existing network, it is verified that the mechanism can effectively optimize the channel pressure index of the network when the network scale is large, and reduce the possibility of high voltage state of equipment and lines. The experimental results show that the proposed algorithm can well handle the problem of multi service alternative routing configuration in large-scale networks. However, the improvement of optimization degree is not obvious compared with the shortest path algorithm, and it needs to continue optimization. The design of genome interaction rules of genetic algorithm is also worth further mining. The next step is to improve the optimization algorithm based on the actual power communication network topology to enhance the application value and persuasiveness.

References

1. Power dispatching communication center of Guangdong Power Grid Corporation. Quantitative assessment method for security risk of Guangdong electric power communication network. Guangdong Power Grid Corporation, Guangzhou (2010)
2. Liu, N., Xiao, C.: Overlay multicasting at a path-level granularity for multi homed service nodes. In: International Conference on Information Science and Technology (ICIST2011), Nanjing, pp. 939– 946 (2011)
3. Chang, H.: A multipath routing algorithm for degraded-bandwidth services under availability constraint in WDM networks. In: 26th International Conference on Advanced Information Networking and Applications Workshops (WAINA2012), Fukuoka, pp. 881–884 (2012)
4. Santos, J., Pedro, J., Monteiro, P., et al.: Impact of collocated regeneration and differential delay compensation in optical transport nctworks. International Conference on Computer as a Tool (EUROCON 2011), vol. 11, Lisbon, pp. 1–4 (2011)

Safety Situation Assessment of Underwater Nodes Based on BP Neural Network

Haijie Huang, Kun Liang, Xiangdang Huang(✉), and Qiuling Yang

School of Computer and Cyberspace Security,
Hainan University, Haikou 570228, China
990623@hainanu.edu.cn

Abstract. With the wide application of underwater wireless sensor network, underwater node positioning technology also plays an important role in underwater wireless sensor network. However, nodes are very easy to be captured as malicious nodes. In order to deal with the problem of node security in a more comprehensive way, a node security situation assessment technology based on BP neural network is proposed to better identify malicious nodes. Through the training of historical interaction data between nodes, a prediction model is obtained, and the reliability of nodes is determined according to the calculated situation value in the assessment system. The algorithm proposed in this paper can evaluate the security situation of nodes more comprehensively, identify malicious nodes more accurately, and maintain the security of nodes.

Keywords: Underwater node security · BP neural network · Situation assessment

1 Introduction

With the application of underwater sensor networks in the ocean, most of the underwater nodes are in a complex and open network environment. Beacon nodes are easy to be captured as malicious nodes, resulting in great losses. Therefore, the underwater node positioning technology which can accurately identify malicious nodes has become a very important research. Nowadays, many researches have applied trust management to wireless sensor networks. By identifying malicious nodes and selfish nodes, and screening error information and data, the traditional security mechanism can be supplemented and the security and reliability of wireless sensor networks can be improved comprehensively [1]. Beth et al. introduced the concept of experience in the trust model based on experience and probability statistics [2] to measure trust relationship and used Bayes to conduct trust assessment. However, this method cannot effectively detect the behavior of bad nodes to attack by changing policies. Tang Wen et al. [3] applied the fuzzy set theory to the trust evaluation, and provided the general evaluation mechanism and the derivation rules of trust relationship. However, it does not provide the evaluation method of recommended trust and the meaning of trust evaluation scale. Wang Jingpei et al. [4] put forward the evaluation method of trust model integrating fuzzy theory and Bayesian network, and put forward the security application scheme of Internet of things integrating trust from both qualitative and

J. MacIntyre et al. (Eds.): SPIoT 2020, AISC 1282, pp. 226–232, 2021.
https://doi.org/10.1007/978-3-030-62743-0_32

quantitative perspectives, but did not put forward a more suitable algorithm in the evaluation system to verify and modify the proposed algorithm.

The introduction of trust management can eliminate the impact of bad nodes, but there are also shortcomings [5]: high resource overhead, bad nodes attack by changing the strategy, malicious nodes recommend false experience to slander other nodes. For the lack of trust evaluation, in order to deal with the security problems of underwater nodes more comprehensively and perfectly, in this paper, on the basis of trust management is introduced into Cyberspace Situation Awareness (CAS) technology, an underwater node security situation assessment model based on BP neural network is proposed. By analyzing the historical interaction information of neighbor nodes, the trust evidence which can reflect the trust degree of neighbor nodes is calculated.

BP neural network belongs to feed-forward neural network, and adopts error back propagation algorithm. Data is input from the input layer, processed by the hidden layer, and finally output from the output layer [6]. The calculated trust evidence is taken as the input data of BP neural network, and the situation prediction model is obtained through training. The pending location node calculates the situation value of neighbor node according to the situation prediction model, and judges the trust degree of neighbor node according to the established situation assessment system, so as to more accurately identify the malicious node.

2 Research on the Status of Underwater Node Location Security

Underwater positioning is one of the most basic tasks in the underwater acoustic sensor network. Generally, acoustic signals are used as information carrier to realize the positioning of underwater nodes. Positioning methods can be divided into range-free and range-based methods [7]. At present, the main research purpose of node positioning technology is to improve the positioning accuracy and energy efficiency, ignoring the possible attacks in the positioning process. Many scholars have done research on situation awareness, such as applying trust to network situation awareness technology, which can reflect the network security situation more comprehensively and real-time and predict the development trend of network security situation.

At present, network security situation awareness technology is one of the important research methods of network security mechanism. Many situation awareness models have been proposed abroad, such as JDL model and Endsley model. JDL model [8] is mainly divided into three parts: data acquisition, data fusion, human-machine interface. Endsley model [9] divides situation awareness into three steps: situation element acquisition, situation understanding and situation prediction. It is also the most widely used model at present.

The mechanism of situation awareness based on trust can be divided into three parts: acquisition of trust situation elements, analysis and evaluation of trust situation, and prediction of trust situation. In reference [10], Bayesian model is introduced into dynamic trust evaluation, and Kalman filter is used to filter information. Sample information is learned many times under Bayesian theorem, and the model is updated according to system context factors, which can well cope with the dynamic characteristics of trust ambiguity in large-scale networks. The situation awareness mechanism based on comprehensive trust proposed in reference [11] is a network security scheme with trust situation as the main reference element. Through the interaction and cooperation between nodes, the trust situation of each node is perceived, and the interaction objects are selected according to the trust between nodes. However, in response to the extreme dynamic of large-scale network environment, it is not accurate and effective to obtain the trust situation of nodes, and more comprehensive methods are needed to obtain the situation element information. The decision-making trust mechanism proposed in reference [12] divides the trust between nodes into four types, namely, real, opposite, overestimate and underestimate, and makes different decisions according to different types, but does not consider the time decay when calculating the trust. In reference [13], a decision-making trust evaluation model is proposed, which uses fuzzy theory to calculate the uncertainty and fuzziness, and solves the problem that nodes only obtain information that is not shared, but does not consider load balancing.

At present, the mechanism of situation awareness based on trust mainly considers how to build a trust model. There is no decision-making on malicious behavior, nor threat assessment. The main work of this paper is to use BP neural network to train evidence data on the basis of calculating the trust evidence, to get the situation prediction model. In the established situation evaluation system, according to the calculated situation value, we can judge which level the credibility of the node belongs to, so as to distinguish the malicious node.

3 Evaluation of Malicious Nodes Based on BP Neural Network

The communication behavior of underwater nodes mainly includes the sending and receiving of data. Attackers will interfere with the normal communication behavior between nodes by various means, take the interaction information between nodes as the input of BP neural network, judge the credibility of nodes through the training results of the new input data, identify malicious nodes, and predict the future behavior of nodes.

3.1 BP Neural Network

The historical interaction information between nodes becomes the key factor to determine whether a node is trustworthy, and the successful and failed interactions between nodes i and j in the past are taken as the input element of the neural network.

The basic BP neuron (input n) node passes the weighted value ωi $(i = 1, 2, \ldots, n)$ input parameters χi $(i = 1, 2, \ldots, n)$, neuron threshold value θ, excitation function f, output parameter y is expressed as:

$$y = f(\sum_{i=1}^{n} \omega_i x_i - \theta) \tag{1}$$

The hidden layer of BP neural network requires that the excitation function be continuous. In this case, the excitation function is Sigmoid function:

$$f(x) = \frac{1}{1+e^{-x}} \tag{2}$$

BP neural network model includes information propagation and error correction.

3.2 Information Dissemination

As shown in Fig. 1. the input neuron N in the first layer, namely the n-dimensional vector $X \in R^n$, where $X = (x_1, x_2, \ldots, x_n)^T$; the output of l neurons in the hidden layer is $X' \in R^l$, $X' = (x_1', x_2', \ldots, x_n')^T$; The threshold value θ_i; the output of m neurons in the output layer is $Y \in R^m$, $Y = (y_1, y_2, \ldots, y_m)^T$, threshold θ_i', $i \in (1, m)$. The weight from the input layer to the hidden layer can be an $n \times 1$ matrix, that is $W_{ij}\{i \in (1,n), j \in (1,l)\}$, similarly, the weight matrix from hidden layer to output layer is $W_{ij}'\{i \in (1,n), j \in (1,l)\}$, each element of neural network output is:

$$x_i' = f(\sum_{j=1}^{n} \omega_{ji} x_j - \theta_i), i = 1, \ldots, l; y_i = f(\sum_{j=1}^{k} \omega_{ji} x_i' - \theta_i'), i = 1, \ldots, m \tag{3}$$

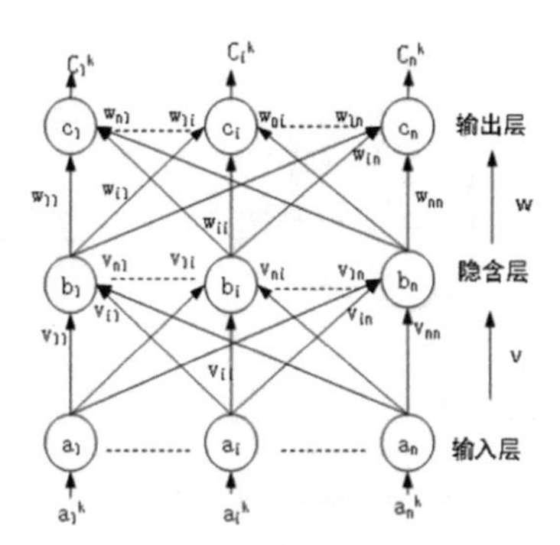

Fig. 1. Three-layer structure diagram of BP neural network algorithm

When the training sample s is sent to the output layer, it is compared with the expected one:

$$E^{(i)} = \frac{1}{2} \sum_{i=0}^{m-1} (d_i^{(s)} - y_i^{(s)})^2 \tag{4}$$

The sum of training sample error is the total error:

$$E = \sum_{s=1}^{count} E^{(s)} = \frac{1}{2}\sum_{s=1}^{count}\sum_{i=0}^{m-1}\left(d_i^{(s)} - y_i^{(s)}\right)^2 \tag{5}$$

Where count is the total number of samples.

3.3 Reverse Correction of Errors

According to the formula in Sect. 2, if $E_i(l) \leq \varepsilon(i = 1, 2, \ldots, m)$($\varepsilon$ is the specified minimum error), then the training is completed and the threshold value of neurons and corresponding weight are determined at the same time. Adjust the weight along the direction of negative gradient, namely:

$$\Delta W_{ij} = -\eta \frac{\partial E}{\partial W_{ij}} \tag{6}$$

Where η is called the learning coefficient. Equation (5) is converted to:

$$\Delta W_m^{l,l-1} = -\eta \delta_m^l \gamma_j^{l-1} \tag{7}$$

Where, l is the number of layers, δ_m^l is the m-th neuron of the layer, γ_j^{l-1} is the output of the j-th neuron of the layer:

$$\delta_m^l = \begin{cases} (\gamma_m^l - d_m) f'(I_m^l), & l = level \\ (\sum_m \delta_{mj}^{l+1,k}) f'(I_m^l), & l < level \end{cases} \tag{8}$$

3.4 Establish an Evaluation System

As shown in Table 1, the node trust assessment system is divided into five levels.

Table 1. Evaluation system table

Situation value	Assessment level
(0,0.2]	1
(0.2,0.4]	2
(0.4,0.6]	3
(0.6,0.8]	4
(0.8,1]	5

According to the situation value obtained to judge the credibility of the node, this paper selects the interaction data among 30 underwater nodes, uses BP neural network to predict the network security situation, obtains the state value as shown in Table 2, judges the node credibility level according to the situation value, so as to more accurately identify malicious nodes.

Table 2. Safety situation value

Node number 1 2… 30
Situation value 0.7 0.3… 0.5

4 Summary and Future Work

For the problem that underwater nodes are easy to be captured as malicious nodes, the situation awareness technology based on BP neural network is used to evaluate the credibility of nodes. For the two nodes that have never interacted, it is impossible to determine whether the nodes are reliable due to the lack of historical information. BP neural network can deal with this uncertain problem and make a correct judgment for the nodes. However, BP neural network is easy to fall into the local minimum. In the future, it is hoped that an optimized neural network can be established to evaluate node security.

Acknowledgment. This work was supported by the following projects: the National Natural Science Foundation of China (61862020); the key research and development project of Hainan Province (ZDYF2018006); Hainan University-Tianjin University Collaborative Innovation Foundation Project (HDTDU202005).

References

1. Zhang,Y.: Research on safe location algorithm of Underwater Sensor Network based on Malicious Anchor Node detection. Tianjin University (2013)
2. Beth, T., Borcherding, M., Klein, B.: Valuation of trust in open network. In: Proceedings of the European Symposium on Research in Security, pp. 3–18. Springer, Brighton (1994)
3. Tang, W., Chen, Z.: Research on subjective trust management model based on fuzzy set theory. J. Software **14**(8), 1401–1408 (2003)
4. Wng, J.P.: Research on key technologies of trust management and model evaluation in distributed scenarios. Beijing University of Posts and Telecommunications (2013)
5. Ye, A.Y.: Node localization security in wireless sensor networks. Xi'an University of Electronic Science and Technology (2009)
6. Xiao, P., Xian, M.: Network security situation prediction method based on MEA-BP. In: IEEE International Conference on Computational Intelligence and Communication Technology (2017)
7. Zhu, G.M.: Research on target tracking algorithm of asynchronous underwater sensor network. Zhejiang University, pp. 2–10 (2015)
8. Dasarathy, B.V.: Revisions to the JDL data fusion model. Proc. SPIE Int. Soc. Opt. Eng. **3719**(12), 430–441 (1999)

9. Endsley, M.R.: Design and evaluation for situation awareness enhancement. In: Human Factors & Ergonomics Society Annual Meeting Proceedings, vol. 32, no. 1, pp. 97–101 (1988)
10. Melaye, D., Demazeau, Y.: Bayesian dynamic trust model. In: Pěchouček, M., Petta, P., Varga, L.Z. (eds.) Multi-Agent Systems and Applications IV, pp. 480–489. Springer, Heidelberg (2005)
11. Li, J.Y.: Research on distributed cooperative situational awareness based on trust in wireless networks. Chongqing University of Posts and Telecommunications (2017)
12. Jelenc, D., Hermoso, R., Sabater-Mir, J., et al.: Decision making matters: a better way to evaluate trust models. Knowl.-Based Syst. **52**, 147–164 (2013)
13. Yin, G., Wang, Y., Dong, Y., et al.: Wright–fisher multi-strategy trust evolution model with white noise for Internetware. Expert Syst. Appl. **40**, 7367–7380 (2013)

Big Data Analytics for IoT Security

The Innovation of College Counsellor's Work Based on Big Data Analysis

Yunshan Liu(✉)

School of Marxism, Wuxi Vocational Institute of Commerce, Wuxi, Jiangsu, China
lubianinfo@foxmail.com

Abstract. The Internet is an main position for counsellors to realize ideological and political education. This research studies the current situation of using of campus network (CN) and network public opinion, discusses the relationship between counsellors' daily work and network public opinion, and analyze methods of using the Internet to innovate the theory, method, content and position construction. The aim is to guide the practice of network public opinion in campus.

Keywords: Network public opinion · Ideological and political education · Innovation

1 Introduction

Ideological and political education through the Internet is a crucial front that college counsellors must stick to. Network ideological and political education is an effective way to publish correct public opinions, realize the online ideological and political education in higher education institutions, and guide students to reasonably screen information on the Internet and establish correct outlooks on life, values and the world.

2 Analysis on the Current Situation of Network Public Opinion in Colleges and Universities

According to statistics, by the end of March 2020, China has 904 million Internet users, and 897 million mobile Internet users. 99.3% middle school students use mobile phones; middle school students also account for the largest proportion of Internet users, which is 26.9%. By March 2020, China has 423 million online education users, accounting for 46.8% of total Internet users. At the beginning of 2020, primary and secondary schools in China did not reopen the school on time; 265 million students turned to online courses; applications of online education boomed [1]. Therefore, network has been extended to every corner of life, completely subverting people's way of production and life. Colleges and universities are places with high degree of information; the diversified information and network technology applications provide students with various convenient conditions for learning and living. Taking Jiangnan

J. MacIntyre et al. (Eds.): SPIoT 2020, AISC 1282, pp. 235–241, 2021.
https://doi.org/10.1007/978-3-030-62743-0_33

University as an example, the network is a necessary item for every students' dormitory. According to the survey, 15% students surf the Internet for about one hour every day; 46% surf the Internet for 2–3 h every day; 20% surf the Internet for 3–4 h every day; 19% surf the Internet for more than 4 h every day. The survey shows that 48.9% of college students' attention to online information focuses on reports different from traditional media, 32.4% on positive reports and 18.7% on negative reports.

The content of campus network public opinion can be mainly divided into two parts: inside the school and outside the school. The part outside the campus mainly include important news at home and abroad, information about cultural and sports activities, social focus issues and so on; the part inside the campus mainly include matters related to college students, such as evaluation and awards, examination information, grant of awards, canteen prices, water and electricity network in campus life and so on. From the survey of students' public opinion participation in hot events on the campus forum [2], 16.8% of them surprise at the occurrence of the event; 45.2% of them rationally analyze the event and form their own views; 17.5% of the students post and forward the information; 20.5% of them are simply spectators. The attitudes of students towards unexpected events and negative news on the Internet are as follows. 16.8% of the students just browse the information with low attention; 45.3% of the students do not pay much attention; only 17.4% of the students are very concerned and would actively participate in the discussion of online interaction; 20.5% of the students would discuss with the people around them. From the perspective of students' acceptance of network information, the active cases account for 25.3%; the passive cases account for 15.3%; 59.5% cases depend on the situation. In terms of whether students will reprint or recommend the Internet news to others, 6.4% of the students often do so; 58.8% of the students occasionally do so; 31.1% of the students seldom do so; 3.7% of the students never do so.

In 2004, the CPC Central Committee and the State Council published the *Opinions on Further Strengthening and Improving the Ideological and Political Education in Colleges*. The opinion points out, "we need to take the initiative to take the ground of network ideological and political education. We need to keep an eye on online trends, understand the ideological status of students, form a online education system, and take the initiative of the network education on politics and ideology." College students are a group with strong plasticity. The characteristics of higher education determine that college students' ideology has the characteristics of obvious transformation from campus culture to social culture. Due to the openness of the network, college students have more channels to obtain information, and the amount of information they get is large, but the good and bad are intermingled, which increases the difficulty of distinguishing right from wrong. All kinds of possible influences on college Students' ideology and behaviour cannot be ignored. The information and speech on the Internet are hard to distinguish between the true and the false; they are easy to cause the confusion in student's thoughts, even affect the political stability of the university and society [3].

3 The Basic Relationship Between Campus Network Public Opinion and Counsellor's Work

3.1 Campus Network Public Opinion is an Effective Way to Carry Out Counsellor's Work

As we all know, the most important work of counsellors is the ideological and political education; the Internet is a new position for that education [4]. Campus network culture is becoming the main media and an important channel for the education on campus. The network brings new opportunities and challenges for the education on politics and ideology. Therefore, in the network environment, how to grasp the main position of the Internet, promote the innovate development of education, and effectively play the role of education, has become urgent issues for college counsellors.

College students can get different of information and publish information at will through mobile phones, social networking sites, and various instant chat tools. College students have strong ability to receive new things, and pay attention to inclusive information. From current affairs to people's livelihood, from school policies to daily life, they are willing to express their views and have a high enthusiasm for participation. Most hot spots on the Internet that college students participate reflect the current situation of campus learning and life, and also reflect the general ideology and collective dynamics of college students.

The network public opinion in higher education institutions can directly and objectively reflect the essence of some social and campus phenomena, and can truly reflect the value demands and psychological states of different groups of teachers and students. These groups have similar age and experience, so they tend to have similar ideas on issues of social problems, which makes it easy for them to reach agreement, and promote the emergence of a specific public opinion. Therefore, online public opinion has the characteristics of group controllability, which brings convenience to the education, as well as management and guidance. As students' instructors in colleges, counsellors are good at grasping the important position of network public opinion education [5].

3.2 Counsellor's Work Provides an Important Guarantee for the Effective Guidance of Campus Network Public Opinion

Public opinion online can directly reflect social public opinion. Due to the weak restriction of law and morality on the Internet, improper speeches are easy to appear. In interactive websites, the information can be manipulated artificially, which makes the information develop to a bad trend. Therefore, the response to the network public opinion means to use modern information technology to analyze the network public opinions, so as to control and guide the information reasonably [6].

Public opinion online itself has the deviation, and is easy to cause the deviation cognition of netizen. However, after the formation of biased cognition, college students are often blinded by the surface phenomena of things, which lead to the emotional trend of negative dissatisfaction. They may express some vulgar, negative and extreme remarks or behave improperly without careful consideration and irrationality, which will bring certain challenges to the ideological and political education. Once the opinion becomes difficult to control, it can affects the effective development and correct guidance of counsellors on the education.

From the point of campus stability and harmony, the ideological and political education carried out by counsellors at the level of network public opinion can guide students' negative emotions, and can promote students to establish sense of values in line with the mainstream of society. Through counsellor's daily work on network ideological and political education, student netizens which account for 26.9% of Internet users of our country become more rational and objective when they express their opinions on some hot events, which provides important support and guarantee for resolving conflicts, and promoting the network and campus harmony.

4 Innovation in the Ideological and Political Education Based on Campus Network Public Opinion

4.1 Innovation of the Concept of Counsellor's Work

"People oriented" is the core element of the scientific outlook on development. In colleges and universities, we should do a good job in the ideological and political education, and put "student-oriented" at the top of our work. We should keep an eye on students' ideological trends, understand their thoughts in time, listen to their voices and solve their practical difficulties. To understand the status of students in time and effectively, we must change the concept of work and interact with them through the way that students are willing to participate and accept [7]. In daily work, counsellors should timely capture the ideological changes of students through public opinions on various platforms of the Internet, and effectively integrate them into the education; they should also analyze the value orientation of college students from the content of network public opinions on campus and effectively carry out in-depth dialogue with students.

Any single public opinion on the Internet is a battlefield; a network public opinion that we did not concern may produce butterfly effect. If a college student's individual public opinion is not valued, it may have unpredictable consequences. The consequences of bad network public opinion may affect the reputation of the school and even its safety and stability. Therefore, as a counsellor dealing with students all day long, we need to change the working concept, changing from offline to online and from passive communication to active listening. In the daily education, we need to focus on students, pay close attention to each student's rationalization demands, and improve the management level and work ability. In different affairs related to the vital interests of students, we should strictly follow the system, do a good job in the principles of fairness, justice, openness and transparency, establish a long-term and effective trust relationship with the students, timely resolve every bad campus network public opinion, reasonably guide students to think positively, and establish and maintain the reputation and image of the school among the students [8].

4.2 Innovation on the Content of Counsellor's Work

All education contents that cannot attract the attention of students are invalid. No college students are willing to click on the cold web page and read the old and rigid contents online, let alone learning and understanding. Therefore, college counsellors must work hard on innovate the content of network education on politics and ideology, so that students will really want to learn and enjoy learning.

How to take the advantage of rich network resources and network platforms to innovate the content of education is the key factor to promote the education reform and to give full play to the educational function. Counsellors should establish a network interaction platform that students can easily accept, timely release information related to students' life and learning, and timely respond to questions and queries raised by students, so as to establish the mutual trust relationship between students and counsellors, and establish a close and reliable identification relationship. Institutional trust is a kind of indirect trust relationship in which people agree with the rules and regulations [9]. To enhance the mutual understanding and trust between teachers and students and reduce mutual suspicion can help counsellors to carry out daily management work and strengthen ideological guidance, so as to avoid the emergence of negative public opinion on the Internet, and effectively guide the network public opinion.

Meanwhile, counsellors should also take the initiative to occupy the main position of campus network publicity. First, they should get related information from the latest information and achievements on the network, use words, images, videos and other ways to supplement the content of ideological and political education, and process it into a form that students enjoy. Secondly, the network information is complex, with various contents, different cultures, and diversified values. College students are prone to be dependent on the network. They may be addict to the Internet and neglect the normal learning and life. Counsellors should pay attention to content of college students' mental health in their daily work. They can popularize knowledge about network psychology, discover the psychological changes of students in time, carry out the mental health education in a targeted way, and cultivate the healthy and good network psychology of college students. In addition, counsellors should also popularize the knowledge of network legal system, so as to cultivate students' awareness of network legal system, enhance their ability for self-protection and restraint, and help them to establish correct network morality.

4.3 Innovation on the Method of Counsellor's Work

Based on characteristics of students, counsellors should keep an eye on the innovation of methods, and take the advantages of new media platform to realize the ideological and political education. In the network world, everyone is a news agency and a microphone. Therefore, counsellors should pay more attention to strategies and methods, and improve the scientific level of counsellors' work at the ideological level. To make the Internet ideological and political education play a positive role, we must rely on the backbone team of students who have high prestige and great influence in the Internet, and improve their own quality, moral standards and discrimination ability [10].

In the Internet age, counsellors should grasp the initiative of Internet public opinion guidance, and be good at mining and cultivating a group of backbone students who have credibility, the right of speech, and the ability to channel public sentiments. Counsellors should fully play to the positive role of the network team, strengthen mainstream public opinion, win over centrists' speech, and isolate the wrong public opinion. Through these ways, we can guide the network public opinion positively. On the other hand, counsellors can also realize equal dialogue with students through QQ, microblog, wechat and other network platforms. The anonymity and concealment of the network can help college students to open their hearts and express their true thoughts and opinions to counsellors. Counsellors can take this opportunity to timely understand the status of students and conduct targeted ideological guidance and education.

4.4 Innovation on the Position Construction of Counsellor's Work

Under the network environment, college counsellors should timely explore and expand the education function of mainstream online platforms, and master the voice and initiative of network public opinion position. Network current affairs, campus cultural life, various policy notices and other affairs are good materials for this education. Under the help of these topics, counsellors can encourage college students to analyze the content and connotation, and realize the best teaching effect. In addition, through WeChat, Microblog and other platforms, the education can be carried out in a comprehensive way and from multiple angles, which can also create a new field of education. This platform also provides counsellors with service opportunities of internet psychological consultation, employment guidance, and career planning, which can effectively stimulate students' learning enthusiasm, broaden new channels of education, and imperceptibly complete the education.

5 Conclusion

By keeping an eye on the information of online public opinions as well as practical problems and prominent contradictions exposed on the Internet, counsellors can grasp the ideological development trends of students, keep alert to possible contradictions and crises, and take corresponding effective measures in time, so as to solve potential and even real contradictions and problems in the bud.

References

1. The 45th Statistical Report on China's Internet Development in 2020: the Scale and Structure of Internet Users. http://www.cac.gov.cn/gzzt/ztzl/zt/bg/A0920010206index_1.htm
2. Lyu, C.F., Zhang, P.C.: Discussion on the Mechanism of Guiding Network Public Opinion in Colleges and Universities from the Perspective of Ideological and Political Education. The Party Building and Ideological Education in Schools, no. 497, pp. 90–92 (2015)

3. Qian, J.: On the innovation of college counsellor's working method in the era of microblog. Theory Pract. Educ. **36**, 49–51 (2011)
4. Wang, Z.B. Research on the innovation of Political Instructor's Work in the University Under the New Social Conditions, Central China Normal University (2004)
5. Li, Y.J.: On the Innovation of Work Methods of College Counsellors. The Party Building and Ideological Education in Schools: Higher Education Edition, no.2, pp. 75–76 (2008)
6. Yin, G.M.: On the characteristics and requirements of college counsellor's work in the new era. J. Cangzhou Normal Univ. **22**(1), 50–51 (2006)
7. Peng, Z.H.: On mass disturbances and internet public opinion. J. Shanghai Public Secur. Acad. **018**(001), 46–50 (2008)
8. Liu, Z.M., Liu, L.: Identification and analysis of opinion leaders in microblog network public opinion. Syst. Eng. **06**, 12–20 (2011)
9. Tang, L.F., Zhao, X.L.: Internet public opinion and the response of college ideological and political work. Heilongjiang Res. High. Educ. **04**, 70–71 (2007)
10. Deng, Y.: On the influence of internet public opinion on mass disturbances in colleges and universities. J. Chongqing Univ. Posts Telecommun. Soc. Sci. **021**(003), 123–126 (2009)

Analysis of India's Big Data Industry

Yurui Dong(✉)

Institute of South Asian Studies, Sichuan University, Chengdu, Sichuan, China
Yinxing_txl@163.com

Abstract. This paper briefly summarizes the development situation of India's big data industry, analyzes the advantages and problems in the development of India's big data industry, and puts forward corresponding countermeasures and suggestions. Generally speaking, the main characteristics of India's big data industry include, great market potential, fast development speed and backward technology level. The future development direction mainly focuses on increasing digital infrastructure investment, increasing scientific research investment, improving the market supervision system and improving the management level.

Keywords: Big data · India · Aadhaar

1 Introduction

According to the report released by Praxis Business School of analytics India magazine, the output value of data analysis and big data industry in India has reached 30 billion US dollars by 2019. In terms of market size, the data analysis industry accounts for about 21% of the total output value of the whole information technology enabled services industry in India. In terms of employees, India's big data industry has provided more than 90000 jobs for India's IT industry in 2019. According to industries, in India, the five industries with the largest number of employees are, the scientific research and technical service industry, which employs about 25% of big data practitioners; information industry, about 17%; the manufacturing industry, about 15%; the finance and insurance industry, about 9%; the retail industry, about 8%. In addition to above industries, public management, wholesale, education services, health care and social assistance are also main fields of big data practitioners [1].

2 Overview of India's Big Data Industry

2.1 General Situation of India's Big Data Industry

In the whole data analysis industry in India, more than 90% of the output value, namely about 27 billion US dollars comes from the field outside data analysis. The top five service export objects are the United States, accounting for 47% of the industry's income, about 12.7 billion US dollars; the United Kingdom, accounting for 9.6%, about 2.6 billion US dollars; Australia, accounting for 4.2%, about 1.13 billion US

J. MacIntyre et al. (Eds.): SPIoT 2020, AISC 1282, pp. 242–249, 2021.
https://doi.org/10.1007/978-3-030-62743-0_34

dollars; Canada, accounting for 3.4%, about 920 million US dollars; the Netherlands, accounting for 1.7%, about 460 million US dollars [2].

According to different industries it serves, the financial services and insurance industry (BFSI) is the largest segment of data analysis industry services in India, accounting for 36% of the total market share; it is the most critical domestic industry in India's big data analysis market, closely followed by the marketing and advertising industry, accounting for about 25% of the total market share, and the e-commerce industry, accounting for about 15%.

From the perspective of market share, local IT companies account for the largest market share of data analysis in India. Examples include TCS (Tata Consulting Services), Wipro, Genpact and tech Mahindra. These companies account for about 35% of the total market share.

Multinational companies followed local IT companies closely, accounting for about 32% of the total market share. Examples include Accenture, Cognizant and IBM.

In the third place is the internal analysis organization called "Captive analytics" under large groups, accounting for about 12% of the total market share.

In addition, there are also small professional data analysis companies known as "Boutique Analytics firms/Niche Analytics companies", accounting for about 11% of the total market share.

2.2 India's Big Data Industry Policy Represented by "Digital India" Strategy

The "Digital India" strategy was approved by Narendra Modi, the Prime Minister of India, in July 2015. The project aims to prepare India for the transformation based on knowledge, improve the level of social governance and provide better services for citizens by coordinating the actions of central government and state government simultaneously.

Digital India is a holistic and comprehensive plan with a large number of innovation and ideas. It includes nine key areas: broadband expressway, ubiquitous mobile Internet, public Internet access, e-government, e-service, open data, e-information manufacturing, IT employment and the "Early Harvest Plan".

2.3 Application of Big Data Technology in India

2.3.1 The Aadhaar Biometric Identification System

The Aadhaar biometric identification system (UIDAI), is also known as the Aadhaar project. It is a digital identity authentication system covering the entire population of India launched by the Indian government in September 2010. The system aims to collect the address, photo, fingerprint, iris and other data of residents in India by means of biotechnology and big data, and store the data in an Aadhaar ID card with a chip and 12 digit ID card number. The card is bound with the bank card and mobile phone number of the resident. As of January 2020, Aadhaar has more than 1.25 billion registered users, covering more than 90% of India's population.

The Aadhaar system is the core project of the Indian government to promote the "digital India" strategy. The system designs API (application programming interfaces), which enables Indian government and enterprises to develop applications on the basis of that database.

2.3.2 The Jan Dhan Inclusive Financial Program

Pradhan mantri Jan Dhan Yojana is an inclusive financial plan promoted by the Indian government to Indian citizens on the basis of the Aadhaar system. It aims to expand and obtain financial services such as bank accounts, remittances, credit, insurance and pensions at affordable prices. As of 2017, driven by this plan, India has opened about 285 million new bank accounts nationwide. Due to the lack of reliable credit data in a relatively large population for a long time, India's financial institutions have been limited to expand their business areas. The emergence of the Aadhaar system and the support of its affiliated electronic KYC system have solved this problem that has plagued India's financial industry for many years, reducing the bank's account opening cost by 90% [3].

2.3.3 The BharatQR Integrated Payment System

The BharatQR integrated payment system is an integrated payment system developed by NPCI (National Payments Corporation of India), MasterCard and Visa based on the Aadhaar system. The system was launched in September 2016. Using big data and other IT technologies, users can easily transfer funds from one source to another. The money transferred through BharatQR will be received directly in the user's associated bank account. The system can operate without physical infrastructure other than mobile phones, overcoming the problem that India lacks infrastructure to develop non cash payments.

3 Analysis on the Development Prospect of India's Big Data Industry

3.1 Macro Environment of India's Big Data Industry Market

3.1.1 Macroeconomic Environment

In terms of GDP, India's annual GDP (exchange rate) in 2019 was 2852.1 billion US dollars, with a growth rate of 5.3%, surpassing Britain and France to become the fifth largest economy in the world [4]. In terms of economic growth, India's GDP growth rates in 2015–2019 were 8.0%, 8.2%, 7.2%, 6.8% and 5.3% respectively, and the average growth rate in the past five years was 7.1%. Due to the huge economic scale and growth rate, India's big data industry has huge development potential. It is hopeful to become the most important big data market in the world.

3.1.2 Internet Population

The Internet population is an important basis for the development of big data industry. According to statistics, in 2019, the number of Internet users in India was about 627 million, with the Internet penetration rate of 46%. It accounted for 12% of the world's 3.8 billion Internet users, and was the second largest Internet user group in the world. In terms of growth rate, the Internet users in India are still in rapid growth, with an average annual increase of nearly 80 million Internet users. In terms of traffic data, in the first half of 2019, Indian Internet users generated 9.77 GB of data traffic per month, far exceeding the world average of 5.7 GB.

In terms of growth potential, in 2019, India's total population reached 1.353 billion, ranking second in the world. More than half of India's population was not connected to the Internet. In terms of population structure, India's young population under 30 years old is as high as 720 million, accounts for more than half of the total population. The population structure is young. In the next 15 years, the proportion of young people in India's population will continue to increase [5]. The huge size of population and large number of young people who are willing to accept new technologies provide Indian market with great potentials.

3.1.3 Digital Infrastructure

In terms of fixed broadband network, India's penetration rate of fixed broadband in 2017 was only 1.3%, which was at the world's lowest level. In terms of fixed broadband speed, according to the Speedtest index released by Okla, India's average download speed of fixed broadband in January 2020 was 41.48 Mbps, lower than the 74.32 Mbps in the world, ranking 66th globally.

In terms of mobile broadband network, in 2019, the 4G signal covered more than 90% of the population in India, but the mobile broadband download speed was only 11.58 Mbps India, lagging behind the global average download speed of 31.95 Mbps, and ranked 128th in the world [6].

In terms of mobile phone penetration rate, in 2018, India's mobile phone penetration rate was 64%, of which 24% were smart phones, 40% were non smart phones, and 35% of the population did not have mobile phones. In developing countries participating in the survey, the average ownership rate of smart phones was 45%, [7] and India ranked the bottom.

3.2 Advantages of Developing Big Data Industry in India

3.2.1 Huge Market Scale and Market Potential

As mentioned above, currently, India has the second largest Internet user group in the world, and the scale of this group is still expanding rapidly. In the foreseeable future, India will surpass China to become the country with the largest population in the world. It can be said that Indian is one of the most potential big data markets in the world.

3.2.2 Policy Support of the Government

For a long time, the Indian government has always regarded the high-tech service industry represented by IT as the focus of national economic development. In 1986, the government issued the *Policies on Export, Development and Training of* Computer *Software*, which took the IT industry as the strategic focus of India's economic development. In the 1990s, the government implemented the *IT Technology Park Plan* and the *Telecom Port Construction Plan.* In 1998, the IT industry was identified as the pillar industry of priority development. It proposed to build an information technology superpower within 10 years [8]. In 2011, India launched the project of "national network optical fiber in India" to extend broadband services to rural areas; in 2015, the strategy of "digital India" officially launched.

The government's attention and supporting policies are important reasons for the rapid development of Indian IT industry for a long time; they are also the unique advantages of India's IT industry.

3.2.3 Sufficient STEM Talent Supply

"STEM" is a combination of science, technology, engineering and mathematics. STEM talents refer to university graduates study above subjects. Big data and other related industries are high-tech industries, whose development is inseparable from STEM graduates who master relevant technologies. According to statistics, India had about 2.6 million STEM graduates in 2016, ranking second in the world [9]. Sufficient STEM talent supply is the most solid foundation for the future development of big data industry in India.

3.2.4 English and Wage Advantages

In 2019, more than 90% of the output value of India's data analysis industry (US $27 billion) came from the outsourcing sector. In terms of outsourcing objects, the top four service objects were the United States, Britain, Australia, Canada and other English speaking countries. It can be seen that the outsourcing export to English speaking countries is of great importance to India's big data industry. India has about 130 million English population [10] and sufficient STEM talent supply, which are unique advantages of India in developing its data outsourcing industry.

3.3 Problems and Risks Faced by India's Big Data Industry

3.3.1 Digital Infrastructure is Relatively Backward

Vertically, India's digital infrastructure has achieved great development in recent years; but in horizontal comparison, India's digital infrastructure level is still relatively backward. In terms of network speed, fixed broadband network coverage and the proportion of people with smart phones, India is still relatively backward even in developing countries. India performs well only in the field of mobile Internet penetration, but its mobile broadband download speed is one of the slowest in the world. The backward infrastructure is a big obstacle to the development of India's big data industry.

3.3.2 The Split of Market and Language Environment

India is a multi-ethnic country, with diversified language environment. According to statistics, there are more than 2000 recognized languages in India, of which 122 are used by more than 10000 people and 29 are used by more than 1 million people [11]. According to the constitution of India, there are currently two official languages in India: Hindi and English. Hindi is the most populous language in India, with 500 million population use it. But the number only accounts for about 45% of the total population. In addition, there are 22 national languages that can be regard as semi-official language. The split language environment actually divides the audience of IT industry into dozens of markets, which is one of the main obstacles and challenges that Indian big data companies face when developing the huge local market.

3.3.3 Low Level of Market Governance

The Indian government has been criticized for its corruption and inefficiency. Many scholars and institutions regard the low level of market management as one of the main obstacles in India's economic development. According to the 2019 edition of global governance indicators released by the World Bank, India ranked 76th, 112th and 106th in terms of government efficiency, management quality and corruption control in 2018 [12].

According to reports, more than 210 government websites in India e leaked the detailed information of citizens in the Aadhaar database; more than 110 million users' Aadhaar information was leaked by the telecom company Reliance Jio [13]. In December 2017, Airtel, the telecom giant in India, disclosed the payment accounts of 3 million users without the consent of users.

Due to the information leakage and a series of problems in the implementation process, Aadhaar has caused a series of disputes and dissatisfaction in India. On September 26, 2018, the Supreme Court of India ruled that it was illegal for private companies to access Aadhaar's biological database and stipulated that it was unconstitutional to bind Aadhaar number with bank account number. In addition, schools and educational institutions were also forbidden to ask for Aadhaar information of students. Although the result of this judgment retained the Aadhaar project, it undoubtedly hit the development of India's big data industry hard.

3.3.4 Lack of Core Technology

Although India's IT industry has developed well compared with other industries, there have been problems such as lack of core technology and low industrial added value for a long time. Its IT industry, including big data industry, has long played a role of outsourcing factories of European and American technology giants. There's no local technology giant with world influence.

4 Countermeasures and Suggestions

4.1 Strengthen the Construction of Digital Infrastructure

As mentioned above, the backwardness of digital infrastructure is the biggest obstacle to the further development of India's big data industry. In the future, continuing to invest in digital infrastructure, increasing Internet penetration, improving broadband network speed, and promoting the further popularization of smart phones will still be the primary core to promote the development of India's big data industry.

4.2 Strengthen Supervision and Improve the Governance Level

Problems represented by the Aadhaar personal information disclosure reflect that, the low level of government governance hinders the further development of India's big data industry. The Indian government needs to improve its management ability in related fields, of which the most noteworthy is the issue of personal privacy protection.

As a new industry, the birth and development of big data cannot be separated from the large quantity data collection, storage, analysis and processing. In this process, the issue of personal privacy is inevitable. If this problem cannot be properly handled, the big data industry will lead to common people's antipathy, and eventually limit the development of the big data industry.

4.3 Increase Investment in Scientific Research and Cultivate Local Big Data Enterprises

According to UNESCO, in 2018, India's research and development expenditure accounted for only 0.6% of its GDP, while China, a developing country, the research and development expenditure accounted for 2.15% of its GDP and the number for the United States was 2.79% [14]. The backwardness of scientific research ability results in low technology level of local big data enterprises; most of these companies are in the downstream of the industrial chain and can only undertake the role of outsourcing service for multinational technology giants. They cannot reach the highest value-added field in the industry.

Increasing research and development investment in science and technology is the key to improving the competitiveness of India's local big data enterprises and the overall level of the industry.

5 Conclusion

India's big data industry has the advantages of great market potential, fast development speed, policy support and sufficient talent supply. But it also has some disadvantages. In the future, through increasing digital infrastructure investment, increasing scientific research investment, improving the market supervision system and improving the management level, the big data industry in India can develop more smoothly.

References

1. Scope of Big Data in India, The Future is Bright. https://www.upgrad.com/blog/scope-of-big-data-in-india/
2. Analytics & Data Science Industry in India: Study 2019. https://analyticsindiamag.com/analytics-data-science-industry-in-india-study-2019-by-aim-praxis-business-school/
3. India's Digital Leap-The Multi-Trillion Dollar Opportunity. https://www.useit.com.cn/forum.php?mod=viewthread&tid=16806
4. List of Countries by GDP(nominal) Per Capita. https://en.wikipedia.org/wiki/List_of_countries_by_GDP_(nominal)_per_capita
5. India's Population is Likely to Surpass China's by 2028; the Increase in Young Population is a Double-Edged Sword. http://news.xinhuanet.com/world/2013-06/16/c_116161356.htm
6. India Ranks 128th in the Ookla Internet Speed Test Global Index. http://shuo.chuangyetv.com/chuangye/20200225/022523448.html

7. Pew Research Center: Smart Phone Ownership is Growing Rapidly Around the World, but Not Always Equally. https://www.pewresearch.org/global/2019/02/05/smartphone-ownership-is-growing-rapidly-around-the-world-but-not-always-equally/
8. The Analysis of Investment in India's Software Information Technology Industry; Software Outsourcing is Competitive. http://go.gdcom.gov.cn/article.php?typeid=38&contentId=9982
9. The Countries With The Most STEM Graduates. https://www.statista.com/chart/7913/the-countries-with-the-most-stem-graduates/
10. List of Countries by English-speaking Population. https://en.wikipedia.org/wiki/List_of_countries_by_English-speaking_population
11. Indian Language. https://zh.wikipedia.org/zh-hans/%E5%8D%B0%E5%BA%A6%E8%AF%AD%E8%A8%80
12. The Worldwide Governance Indicators 2019. https://info.worldbank.org/governance/wgi/
13. India is being Torn Apart by Aadhaar. http://www.ceconline.com/it/ma/8800098304/01/
14. Research and Development Spending as a Percentage of GDP. https://data.worldbank.org.cn/indicator/gb.xpd.rsdv.gd.zs

The Application of Computer Virtual Technology in Modern Sports Training

Fuquan Wang(✉)

Chongqing College of Architecture and Technology, Chongqing, China
xiaoinfo@eiwhy.com

Abstract. Computer virtual reality (VR) technology is widely used in the sports industry. It can help athletes to train, analyze the physical condition of athletes, improve the fairness of sports games, and promote the development of modern sports. This article studies the application of computer VR technology in modern sports, and discusses the impact of computer technology on modern sports industry, the application of computer VR technology in sports competition broadcasting, and the application of computer VR technology in sports training.

Keywords: Virtual reality · Computer · Physical education · Physical training

1 Introduction

With the advent of the information age, China's computer technology has developed rapidly, which has had a huge impact on all walks of life. Computer virtual reality (VR) technology can improve the sports' competition level, improve the fairness of sports competitions, and contribute to the development of the sports industry. With the computer technology application in the sports industry, it becomes possible to gradually become intelligent in the development of sports events, which can effectively improve the sports competitions level. Computer VR technology will play an important role and greatly promote the development of sports [1].

2 Overview of Virtual Technology

VR technology refers to the use of computer technology to provide learners with a perceivable three-dimensional space virtual environment. In this virtual environment, learners can perceive and communicate with each other through vision, hearing, and touch to produce an immersive feeling. As a technology aimed at enhancing human-computer interaction functions, VR technology integrates the advantages of various advanced technologies such as computer graphics, human-machine interface technology, artificial intelligence technology and visualization technology, so that learners can be immersed in VR. In the environment, stimulate learning motivation, enhance the learning experience, and realize situational learning [2, 3].

Computer technology is the core of VR technology and combines with other related disciplines to create a digital simulation environment highly similar to the real

J. MacIntyre et al. (Eds.): SPIoT 2020, AISC 1282, pp. 250–256, 2021.
https://doi.org/10.1007/978-3-030-62743-0_35

environment. In this virtual environment, users use the necessary hardware peripherals to interact the objects of the digital environment. This immersive interaction brings the user an immersive experience.

The American scientists G. Burdea and P. Coiffet put forward the "VR technology triangle" in the article VR Systems and Applications published at the 1993 World Electronics Annual Conference. "3I" characteristics [3]: interaction, immersion and imagination, as shown in Fig. 1.

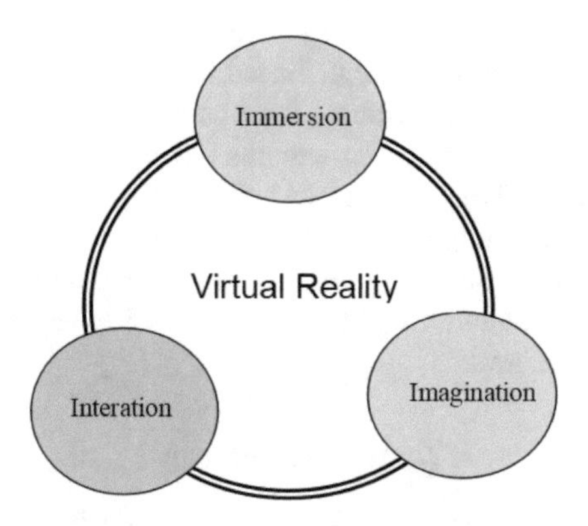

Fig. 1. The "3I" feature of VR

2.1 Interactive Features of Virtual Technology

Traditional human-computer interaction means that people interact with the computer and get feedback from the computer through the keyboard, mouse, and monitor. The interaction of VR technology refers to that participants interact with the virtual environment in a natural way with the help of a dedicated three-dimensional interactive device. Such interaction has a richer form than planar graphical interaction, as shown in Fig. 2.

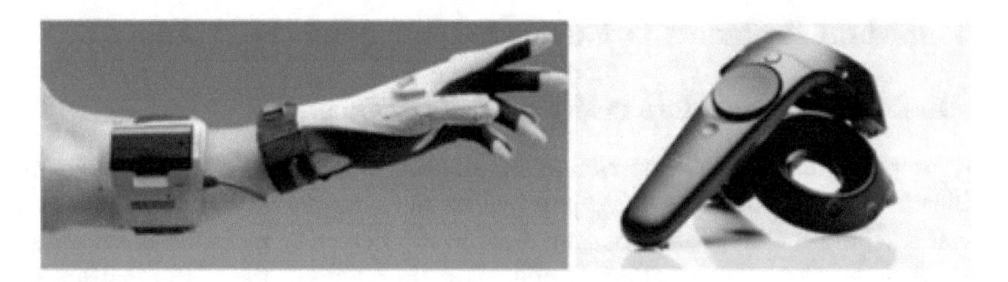

Fig. 2. Three-dimensional interactive device

In addition, interaction can also be achieved through "eye tracking". The dizziness of head-mounted display devices has always been criticized. By optimizing the "eye tracking" technology, low latency, improved image stabilization and the clarity of the current viewing angle can be achieved, which greatly enhances the user's experience. Eye tracking technology is one of the key technologies that VR devices can really put on the market. Oculus founder P. F. Luckey (P. F. Luckey) also called the technology "the heart of VR", and its importance can be seen [2].

2.2 Immersive Characteristics of Virtual Technology

Immersion, also known as presence, refers to the immersive experience that the virtual environment brings to the participants. It is considered to be an important indicator of the performance of the VR environment. Based on the sensory and human vision and hearing psychological characteristics, the computer generates realistic three-dimensional stereoscopic images. Participants appear to be in the real objective world by wearing interactive equipment, helmet displays and data gloves [4]. In general, multi-sensory and autonomy are the most important factors that determine the degree of immersion. Multi-sensory refers to the various human perception functions that an ideal VR environment should have objects in the virtual environment act according to the physics laws.

2.3 Imaginative Features of Virtual Technology

Imagination means that in a virtual environment, the user imagines the future progress of the system movement based on changes in the operating state of the system through association, reasoning, and logical judgment based on the various information obtained and his own behavior in the system. Acquire more knowledge to further understand the deep-level motion mechanism and regularity of complex systems [4].

Before the advent of VR technology, people could get inspiration to deepen their understanding of things. Unlike traditional perceptual cognition, VR technology changes people from passive to active acceptance of things [5]. From an integrated environment of both qualitative and quantitative, people actively explore information through perceptual knowledge and rational knowledge, deepen concepts and then produce cognitive new ideas and ideas.

3 The Main Problems in Current College Sports Training

3.1 Less Class Hours, More Projects, and Poor Quality

Many training items and the reduction of physical education courses are the most significant characteristics of college physical training at present. This also makes traditional teaching methods and concepts unable to effectively meet the basic requirements of society for college physical education. In addition, the traditional teaching methods pay more attention to students' sports level, fail to recognize the important impact of comprehensive quality on students, and the training method is too simple [5]. This kind of problem leads to students' learning interest is difficult to improve.

3.2 The Students Lack Interest and the Judging Criteria Are Unclear

The perfunctory problems of student learning often occur in the physical education teaching in colleges and universities, and teachers are neglected in management, so to some extent, this kind of improper learning style is also promoted. Under such a state for a long time, the comprehensive quality of students is difficult Was effectively improved [6]. At the same time, some colleges and universities do not currently formulate uniform sports standards. Teachers blindly increase their scores to take care of students' learning emotions. Under this vicious cycle, college sports training will inevitably have problems.

The computer virtual technology application in college physical training can improve the participation of students. Students can better experience the training in a environment. Meanwhile, the emergence of various new things can also enhance the interest of students in learning [6]. It will also greatly promote the improvement and development of sports training quality. In addition, the virtual technology also has the important function of information collection. Through this technology, teachers can also analyze and process the students' assessment results, which has an important impact on the objectivity and accuracy of the achievement review [7].

4 The Computer VR Technology Application in Sports Training

The computer technology application in the modern sports industry can improve the sports competition level and sports performance. Through computer technology, it can help athletes to train and judge athletes' movement standards, so as to regulate athletes' movements and improve athletes' performance. In traditional sports training, coaches provide training guidance to athletes based on their experience. The lack of professional scientific testing has prevented the training details of athletes from being correctly guided, affecting athletes' performance [7]. Computer technology can decompose the movements of athletes, monitor various parts of athletes, and can scientifically point out areas for improvement, which is conducive to athletes to improve the level of competition. There are many sports projects in China that use computer technology for training.

4.1 Make a Comparison Between Virtual Technology and Sports Technology

The key content of college physical education courses is training. Through the optimization and improvement of a certain technical action of students, the standardization of technical actions is realized. This is to improve the students' sports ability and literacy. With the overall development of computer technology in recent years, many countries have strengthened the construction of sports action simulation and simulation virtual systems [8]. This simulation system can not only effectively analyze the training of athletes, but also master the problems in technical actions, so that teachers can improve and perfect the technical actions in the process of guidance, as shown in Fig. 3.

Therefore, in the process of college sports training, the application of VR technology is very necessary [8].

In applying process of this technology, the teachers should also actively learn the technical advantages, so as to master the use of computer virtual technology, at the same time fully master the students' sports literacy requirements, and achieve targeted guidance to students [9]. At the same time, teachers can also add standard sports actions in the system construction, through the action decomposition of computer technology, so that students can have a more comprehensive and systematic grasp of prescribed actions before training.

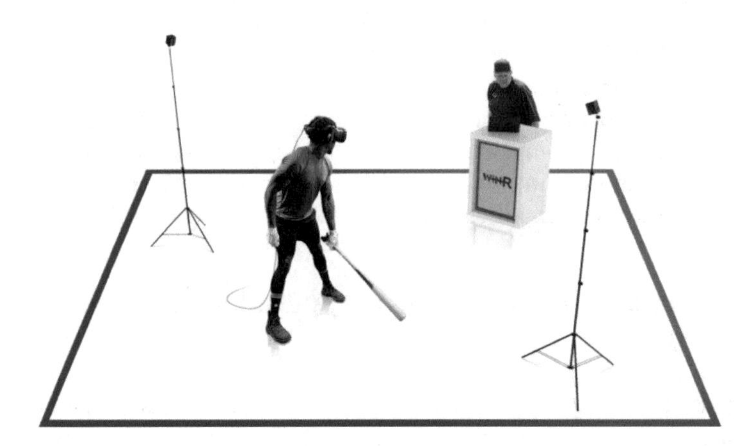

Fig. 3. Virtual technology in sports action simulation and simulation

4.2 Using VR Technology to Perform Virtual Comparison of Actions

In the development of sensor technology, its advantages are becoming more and more significant, but we should also realize that although sensor technology has been significantly developed, in the actual application process, its perception of weak signals still has some problems. Therefore, the standard actions on the computer, the computer virtualization technology must be analyzed with the impact information in student training to ensure that the difference is better found through the comparison between virtual and virtual [10].

The application of computer VR technology to carry out comparative analysis work cannot only help students achieve a comprehensive understanding of technical movements, but help students to accurately understand their own movements and clarify the distance between their training and standard movements [10]. To achieve the adjustment of the direction of teacher training guidance, optimize training efficiency and work quality.

4.3 Using Virtual Display Technology for Remote Interactive Training

Based on the interactive characteristics of VR technology, teachers should also carry out interactive training according to the actual situation of students in teaching. Current college sports training courses are mostly traditional athletics, and rarely involve advanced sports training. But the interactive advantage of VR technology can effectively improve the school training level and increase the number of love training subjects reasonably. Therefore, under a new background, students can also use VR technology to communicate in different places, and achieve a comprehensive increase enthusiasm students' interest for learning through communication and communication between different schools [11].

5 Conclusion

In summary, computer VR technology can promote the sports industry development and bring huge economic benefits to the sports industry. The impact of computer technology in modern sports industry mainly includes improving the fairness of sports competitions, as well as improving the level and performance of sports competitions. The computer VR technology application in the relay of sports games mainly includes virtual sports analysis system and computer virtual replay system. The computer VR technology application in sports training is mainly reflected in functional requirements and specific applications. Through weight-lifting training assisted decision support system, to assist athletes to scientifically carry out weight-lifting training, thereby improving weight-lifting performance.

References

1. Xie, X.: An analysis of the current situation and countermeasures of sports training in Chinese schools. Contemp. Sports Tech. **33**, 37–39 (2017). (in Chinese)
2. Wang, Z.: To study the significance, principles and countermeasures of school sports training innovation. Sports World **03**, 82–85 (2018). (in Chinese)
3. Equation: The application of virtual reality technology in the teaching of computer courses at the National Open University. Comput. Knowl. Tech. **15**, 147–148 (2017). (in Chinese)
4. Qi, H., Li, D., Liu, L.: Application of computer virtual reality technology in the digital restoration of ancient buildings. Wirel. Internet Tech. **10**, 135–136 (2017). (in Chinese)
5. Han, G., Song, Y., Zhang, S.: Research on the application of virtual reality technology based on computer vision in physical education. J. Northwest. Polytech. Univ. (Soc. Sci. Edn.) **36** (02), 92–96 (2016). (in Chinese)
6. Ye, Y., Wei, H.: Research on the application value of virtual reality technology in computer assembly and maintenance teaching. Hum. Resour. Develop. **20**, 236–239 (2017). (in Chinese)
7. Ying, L.: Application research of computer virtual reality technology in college physical education. Wirel. Internet Technol. **01**, 144–145 (2017). (in Chinese)
8. Zhang, L.: Analysis of the application of virtual reality technology in modern display art. Heilongjiang Text. **02**, 31–34 (2018). (in Chinese)

9. Bai, H., Gao, Y.: Application research of computer virtual reality technology in college physical training. J. Heilongjiang Bayi Agric. Reclam. Univ. **31**(3), 105–107 (2013). (in Chinese)
10. Wang, J., Yang, J., Sun, L.: Research on the application of computer "virtual reality" technology in college sports training. Journal of Beihua Institute of Aerospace Technology **22**(2), 56–59 (2016). (in Chinese)
11. Xu, D.: Research on the application of virtual reality technology in college physical education. J. Lanzhou Univ. Arts Sci. (Nat. Sci. Edn.) **32**(1), 120–124 (2018). (in Chinese)

The Research on the Development and Utilization of Hospital Archive Information in the Big Data Era

Dawei Yun, Zhengjun Zhou(✉), and Wei She

Hainan Vocational University of Science and Technology, Haikou, Hainan, China

Abstract. With the continuous improvement of people's health awareness, the intensity and working methods of hospital archives management have undergone great changes. What is certain is that if the hospital continues to adopt the traditional file information management model, it will inevitably cause certain obstacles to the hospital management level and development speed. To this end, this article analyzes the development and utilization of hospital archives information in the era of big data based on the impact of big data on hospital archive management, combined with the problems of hospital archive management, and aims to standardize hospitals. The usage of file information makes the hospital file management develop in the direction of digitization and informationization.

Keywords: Big data · Hospitals · Archives · Hospital archives · Information systems

1 Introduction

The progress and development of the times have produced a large amount of digital information in the society. In the face of accumulating data and information, the past archive IT service model cannot meet people's requirements of using information. In order to be able to change this problem, innovative file management has become an inevitable development. The emergence and development of big data technology provides technical support for innovative file management. With the advancement of the medical industry, a large amount of information has appeared in the hospital. How to efficiently process and use this information has become a problem that the hospital staff need to think and solve. Today, big data technology has become an important key to the solution of hospital file management due to its fast, convenient and comprehensive characteristics. To this end, the article explores the issue of hospital archives management in the big data era [1].

J. MacIntyre et al. (Eds.): SPIoT 2020, AISC 1282, pp. 257–263, 2021.
https://doi.org/10.1007/978-3-030-62743-0_36

2 Characteristics of the Big Data Era

2.1 Large Amount of Data

The most important and basic feature of the era is a large amount of data. With the popularity of the Internet, the Internet is an important part in people's lives and work. Through the Internet, many data resources are used by people every day [2].

2.2 Various Forms of Data

Data forms are expressions in the form of text, photos, audio, video, and animation. People can choose their favorite data format in their daily work and research [2].

2.3 Extensive Data Sources

With the development of the Internet, the concept of space and time has become blurred. People can almost ignore the limitations of time and space, and can access data through the network [1]. Data sources are more extensive and diverse. People can choose useful data resources from many data to increase the value of data.

2.4 Data Processing Is Rapid

The Internet contains all kinds of data. With the advancement in data processing technology, the speed of data processing has increased, and it also reflects the advantages of information management [3]. Therefore, all walks begin to use information technology, gradually introduce information management methods, and improve data processing capabilities.

3 The Era of Big Data Brings New Atmosphere to Hospital File Management

3.1 Diversified Hospital File Management Environment

The information management of hospital archives is an inevitable development. As the result of information technology, a large amount of information has appeared internally and externally in the development of hospitals, which requires the approval of hospital archives management personnel [3].

3.2 Intelligentization of Hospital File Management Information

Today, various advanced information technologies have been applied to hospital archives management, optimized the hospital archives management mechanism, management model, and management methods, and built a hospital information archive management platform, which provides the hospital with various tasks [3].

4 Problems in Hospital Archives Management Under the Background of Big Data Era

4.1 Lack of Understanding of the Importance and Necessity of File Management

The development of some hospitals still pays more attention to medical construction. They have not fully realized the importance of file management to the development and construction of hospitals, nor have they fully considered the management of medical files in their daily work. At this stage, most hospitals have a relatively simple file management model [4]. The collecting and sorting out data has invisibly wasted more human resources and complicated operations.

4.2 The Form of File Preservation Needs to be Improved Urgently

Archive preservation is an important task of archive management. If hospital archives are not kept properly, it will be difficult to use archives. In severe cases, there will be problems of file loss and misuse. But today, the original complex and rigid paper archives management was gradually replaced [4].

4.3 No Application of Massive Data to Secondary Development of Archives

The development of archives management has become more complicated, and hospital archives resources are often used for secondary development, that is, with the aid of data mining technology, data analysis technology, etc. Play the role of archive information resources to the limit. However, from the actual situation of development, it is difficult for archive managers to find the connection between the data and information of the hospital's complex archives, and they have not applied advanced technology to in-depth excavation and utilization of hospital archive resources [5].

4.4 The Quality of File Management Personnel Needs to be Improved Urgently

The archive management of hospitals requires the application of advanced technology for analysis. However, from the perspective of actual development, the archive management personnel of some hospitals often do not know how to apply technology appropriately for archive management, and are subject to traditional archive management. Under the influence of the concept, some archive managers reject the application of new technology for archive management from the bottom of their hearts, making the hospital archive management and utilization effect unsatisfactory [6].

5 The Development and Utilization of Hospital Archives Management Information Under the Big Data Era

5.1 Improve the File Management System

The informatization construction of hospital archives is a long-term and systematic work in itself. This requires that before carrying out various tasks, first establish a set of comprehensive, highly operable and highly operable documents that conform to the characteristics of the hospital. A binding file management system. At the same time, the system must comply with relevant laws and regulations and various regulations of the medical and health industry [7]. Specifically, it includes the file management system, file borrowing process and related approval system, file management personnel job responsibility system, reward and punishment system, performance evaluation system, etc., to ensure the overall standardization and scientific rationality of file management.

5.2 Carry Out Infrastructure Construction

First of all, before the hospital establishes a file management platform, it should first set up a special fund and be equipped with professional staff [7]. For the old staff, it must do a good job of training various new concepts and new technologies for the improvement and development of the file management platform. Secondly, hospitals need to promptly introduce information technology-related equipment, hardware facilities and software systems, and do a good job in the later maintenance of related equipment and facilities, reduce operating failures and unnecessary expenses, and

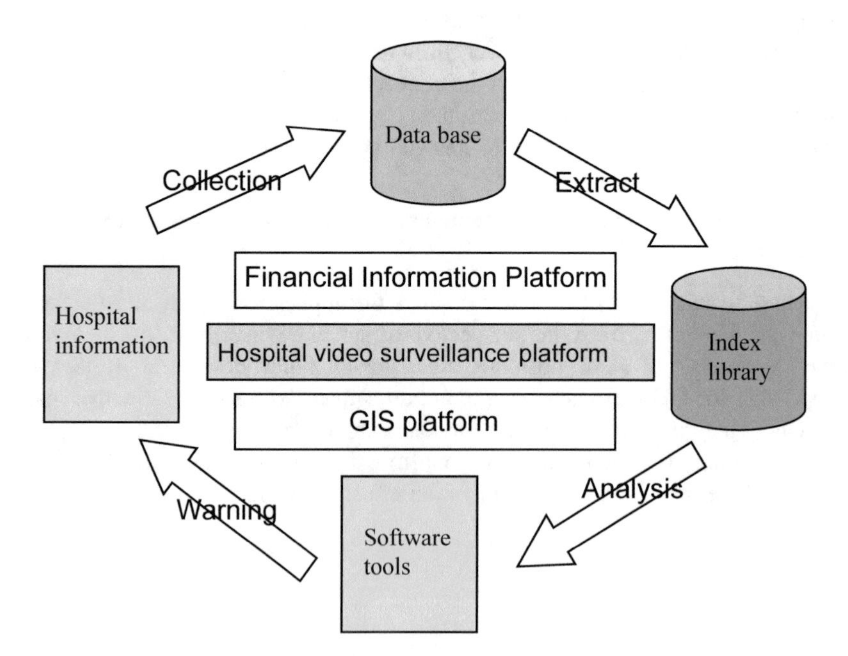

Fig. 1. Big data archives comprehensive management platform

ensure the information management of archives Smooth development; Moreover, if you want to realize the digital management of archive information, you must build a comprehensive management platform, as shown in Fig. 1. On the platform, real-time updates and efficient sharing of information can be realized. At the same time, it is also necessary to classify and manage archive information according to its source and purpose, so as to facilitate the rapid integration of data information and provide convenient conditions for people to query, retrieve and borrow archives [8].

5.3 Reasonable Use of Big Data Related Technologies

In the informatization reform of hospital archives, we should establish a big data information center first, realize the distributed processing of various data, and integrate and utilize different data information. In this process, it is necessary to start from the comprehensive characteristics of cloud computing technology and make full and reasonable use of cloud computing technology to realize the network of data information and data word management. The reason why cloud computing technology is so valued is mainly because it is important in guiding basic data construction. It can help staff quickly and smoothly complete the construction of data information center, as shown in Fig. 2. So as to ensure the effective integration of the hospital's various file information data. From the aspects of data information storage, extraction, query, etc., it jointly acts on the comprehensive management of data information to promote the overall improvement of its overall control level and management level. In addition, we must build a information management platform from the perspective of distributed storage [9]. In addition, when the hospital builds the information data management platform, it must proceed from the perspective of business growth to ensure that the various management functions of the platform can fully meet the relevant requirements for the large-scale management of business data information [10]. From the perspective of server node data management and architecture management, it is necessary to ensure

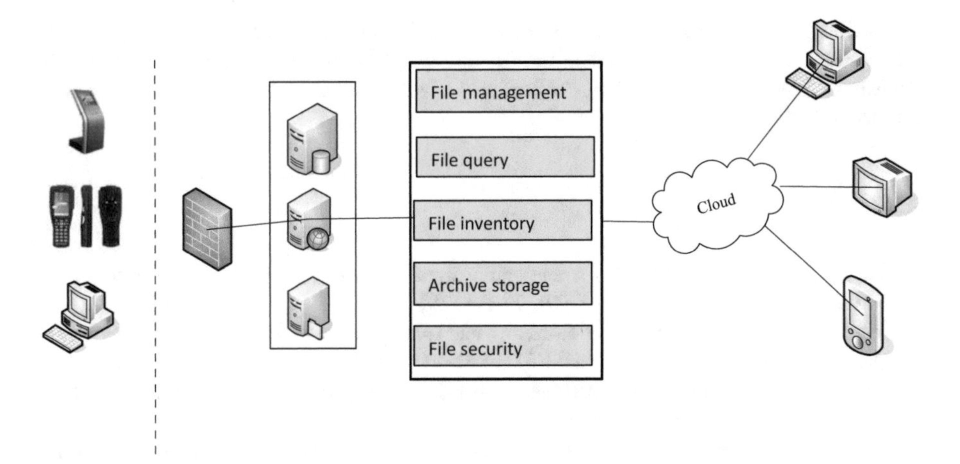

Fig. 2. Big data archive information management center system

that relevant node data information meets the relevant requirements of cloud computing control, and provide the necessary guarantee for the efficient operation of cloud computing technology and the efficient use of servers.

5.4 Keep the Files Properly

When the hospital develops file information construction, there will be two file formats: paper files and electronic files. Because in the electronic archives, only some textual information and data information can be stored [9]. In addition, the archives also carry many legally effective graphic signs or handwritings, such as official seals, signatures, etc., and these information cannot be passed through the electronic archives Store it. Therefore, it is necessary for archival managers to coordinate the relationship between the two types of files, do a good job of file sorting and classification, to ensure the authenticity and completeness of all files, and to make a one-to-one correspondence between electronic files and paper files to ensure file query, The high efficiency of borrowing and other links. When files need to be borrowed, the file management personnel must strictly follow the approval process to perform related work to avoid file loss, damage, and random tampering, and to ensure that hospital files are always properly kept [10].

6 Conclusion

Today, information technology is the primary productive force. Hospital information management can further improve the service level of the hospital, and can better meet the requirements of the masses. In the informatization construction of hospital archives, we work hard from the perspectives of awareness, investment, systems, methods, and personnel, and at the same time take various measures to build archive information and contribute to the development of the hospital.

Acknowledgement. This paper is the outcome of the study, Research on Intelligent Party Building and Key Algorithms Based on Universities, which is supported by the Foundation of the Ministry of Education for Education Projects on University-industry Collaboration. The project number is 201901012011.

References

1. Kun, L.: Try to analyze how to strengthen hospital personnel file management in the era of big data. Inside Outside Lantai **8**(09), 23–24 (2018)
2. Wei, X.: Methods of improving efficiency and quality of hospital archives management in the era of big data. Lantai World **11**(02), 140–143 (2018)
3. Jing, L.: Research on hospital archive information management strategies in the era of big data. Office Bus. **9**, 25–27 (2016)
4. Deng, S.: The influence of the era of big data on the development of hospital financial file management. China Manage. Inform. **8**(4), 112–114 (2017)

5. Ge, C.: An analysis of the informatization construction of hospital archives in the era of big data. Sci. Tech. Commun. **10**(24), 124–125 (2018)
6. Zhao, S.: Hospital personnel file management in the era of big data. Arch. Time Space **12** (06), 42–44 (2018)
7. Wang, M.: Thoughts on the information construction of hospital archives in the era of big data. Office Bus. **2**(08), 85–87 (2018)
8. Bai, Q.: The importance of establishing a sound file security system. China Manage. Inf. **12**, 33–35 (2011)
9. Song, Y.: Informatization construction of college archives in the era of big data. Inf. Comput. (Theoret. Edn.) **5**(14), 209–210 (2018)
10. Li, S., Wei, S.: Overall planning of hospital archives network management system. China Manage. Inf. Tech. **6**(12), 103–107 (2016). (in Chinese)

Integration and Optimization of College English Teaching Information Resources in the Context of Big Data

Guotai Luo(✉)

Nanchang Institute of Technology, Nanchang 330099, China
Luogt3457@126.com

Abstract. Information resources are key factors in the quality of talent training, the core support for the innovation of talent training methods, and a vital support for the improvement of teachers' teaching skills. In the context of big data, the informationization of college English education in universities relies on "massive" information resources to support the creation of a simulation learning environment that allows students to acquire information and actually use the language. However, the effect of the construction of teaching information resources is not satisfactory. Based on the analysis of related concepts, the article expounds new problems in college English teaching in the context of big data. Then, in view of these problems, the related strategies for the integration and optimization of college English teaching information resources in the context of big data are put forward, in order to make improvements of the efficiency and effectiveness of college English education.

Keywords: Big data · College English · Teaching information resources · Integration · Optimization

1 Introduction

Since 2009, "big data" is very popular in the daily life and in the academic world. At present, there is no completely unified definition of big data in the academic world, but the basic consensus that can be reached is that big data covers two basic contents, one is massive basic data; the other is the analysis and processing of these basic data. Simply, big data technology is a technology that quickly obtains useful or valuable information from a variety of vast data. Big data extends the time and the space of teaching, and has a huge impact on higher education. The combination of traditional education methods and modern education technology is more conducive to improving the level of teaching and the quality of teaching. The construction of college English teaching information resources is mainly used in learning. The effective use of information resources not only reflects the level of information resources construction, but also becomes a guarantee to improve students' ability of getting resources [1].

J. MacIntyre et al. (Eds.): SPIoT 2020, AISC 1282, pp. 264–270, 2021.
https://doi.org/10.1007/978-3-030-62743-0_37

2 Characteristics of Teaching Methods in the Context of Big Data

The basic characteristic of the era of big data is the rapid development of the Internet, and the various online education platforms, learning clients, microblogs, and WeChat. It is no longer limited to computers, but extends to more convenient and intelligent tools such as cell phones and tablets. The direct result of this development is that learners can implement online learning or download learning material anytime, anywhere, and most of these learning contents are free and readily available. In the context of big data, this method of education has been hit hard. Learning is more flexible, online learning is available anytime, anywhere, and after-school tutoring is also more personalized. Modern college education is more diversified and data-based teaching is bound to become the development trend of education [2].

3 College English Teaching's New Problems in the Context of Big Data

3.1 Insufficient Classroom Materials to Meet Teachers' Needs for Teaching Activities

College English teaching has been the focus of reform since 2003. After years of continuous renovation of teaching materials, a number of new teaching materials adapted to students' autonomous learning and multimedia application have been produced. However, these new teaching materials are still lacking when teachers start teaching activities. Especially in the foreign language world, universities are advocating a kind of flip-flop and micro-class teaching, and teachers feel the urgent need of teaching materials. Students need a variety of learning materials. Traditional English teaching uses a unified textbook. However, it takes time to compile and publish textbooks, and sometimes it is hard to meet the new needs of society. Although many teachers will add some new knowledge and content to students during the lesson preparation, the high speed and openness of information in the information technology makes teachers' efforts often inadequate.

3.2 Traditional Single Teaching Mode Cannot Effectively Cultivate Students' Comprehensive English Ability

The theory of cognitive linguistics helps us realize that the single input (such as visual or auditory input) of language learning is less effective than the simultaneous input of stereo sound, image, and situation. The traditional teaching mode is only through single-direction language input and single sensory stimulation, which is easy to form isolated images in the brain, and it is difficult to know things in many aspects, and it is not conducive to training students' comprehensive English ability.

3.3 Big Data Triggers Changes in College English Teaching Methods

Previous English education focuses on the learning of English knowledge and neglects the training of basic skills. The large amount of information,abundant resources and teaching methods are no longer limited to time and place, which has brought new opportunities and new methods for college English classes [3].

4 Contribution of Big Data to English Teaching

4.1 Obtaining the Actual Learning Needs of Students

Judging from the current situation, the information technology is constantly improved, and the reform of the education system is also continuously promoted. Reform of college English teaching should improve traditional teaching methods to meet the needs of the times. The teachers adjust traditional teaching methods and teaching concepts. When arranging English courses, the foreign languages schools should consider different grades, and consider the learning level of the learners, set up teaching methods suitable for students, and effectively guide students to master language usage [4].

During the learning foreign language knowledge, teachers must do a good job of guidance to expand students' English knowledge and enrich the channels for students to learn English. Teaching English expertise is a gradual process. When students' learning tasks are arranged, the advantages of big data technology should be used to collect and process data information. After the teachers have special information about students, they will analyze the actual needs of students' English learning knowledge. Another example is that the teacher needs to know the results of the students' daily tests and at the same time master the students' daily learning status. After obtaining some behavioral performance data, using big data technology for analysis, students' comprehensive ability tables can be obtained. The obtained information has high credibility and can provide strong support for the smooth development of teaching activities [5].

4.2 Dynamically Monitor Students' Learning

Big data technology may be used in the teaching industry, covering a wide range. Thanks to the needs of teaching improvement, different data sets can be selected. For example, to availably charge the learning status of students, the teachers need to use online teaching platforms to collect students' learning status extensively. If you need to analyze students' strengths, you should retrieve real data about the students' strengths. Big data processing technology scientifically obtain detailed data analysis diagrams. In traditional teaching activities, teachers need to have a certain understanding of student dynamics. Student performance is the core content of dynamic monitoring, but measuring students' ability based on performance alone is a relative aspect. To reform the English teaching model, it is necessary to achieve comprehensive monitoring of students' English learning to get better teaching effect. With the support of educational big data, dynamic monitoring of English teaching is realized. In this case, the teachers can monitor the learning situation at any time, and can better analyze the learning trajectory of the learners [6].

4.3 Effectively Expand Teaching Content

College English teaching activities combined with big data technology to expand teaching content. With the improvement of the technology, the ideological consciousness of university students has become more diversified, their thinking is more open, and have concerned more about popular culture. In teaching activities, English teachers should choose more content that students like. For example, when learning English songs, we must use the power of multimedia network technology to perform imitation and exercise activities to enhance learners' passion for studying foreign language. During the exercise, students will gradually develop an awareness of English language and their ability to master English grammar will be vastly perfected. In the context of big data, it is necessary to adopt diversified teaching methods to increase students' interest in learning. Taking the teaching of thank- you letters as an example, students are arranged to gather information related to thank-you letters via the Internet in order to expand students' language point reserves and record their language point. The ideas in the book complete the preliminary construction of the knowledge framework. During class discussions, students can effectively complement the knowledge framework and improve writing skills [7].

5 Integration and Optimization Strategies of College English Teaching Information Resources

5.1 Strengthening Government Guidance and Overall Planning for the Construction of Teaching Information Resources

There are many issues in the current state of foreign language information resources, which are related to the shortage of comprehensive planning. If universities want to realize the integration and common use of resources in the true sense; and reduce the repetitive construction of low-level resources, and maximize to integrate the common use of college English teaching resources, they are inseparable from the active guidance and overall planning of relevant functional governments. First of all, the relevant functional governments need to start from formulating relevant policies and conduct macro-control. At present, the need for resource integration and sharing between universities is urgent, but due to regional and status differences, it is not enough to rely on the advocacy and appeal of a certain university or individual. A more authoritative organization is also required. The relevant local government agencies as a functional department can just take on this responsibility. Under the active guidance of relevant government agencies, through changing people's concepts, formulating relevant policies, break up the situation of individual universities, establish a mechanism for the integration and sharing of teaching resources in universities, and promote the coordinated development of resources among universities. Secondly, through overall planning, focus on the superior forces; give priority to the development and construction of high-quality information resources, so as to reduce the repeated construction of low-level resources, and achieving the largest extent of openness and sharing [8].

5.2 Overall Planning of Universities

To build foreign language information resources, schools must set up a resource construction team with the participation of all parties. According to the national education and teaching resource construction specifications, it is determined the resource construction standards that meet school-based standards, and integrate planning with students (professional, level, number, etc.). The construction of information resources must be guaranteed by process monitoring and evaluation mechanisms, and various environmental monitoring mechanisms such as demonstration, selection, implementation, integration and optimization of information resources must be established. For universities to build information resources, they must make full use of the convenient conditions of the network environment, combine purchasing, downloading, and independent development, and solve them through multiple channels. On the one hand, it collects, introduces and excavates various resources scattered in the hands of various departments and teachers, including resources that have been accumulated in the traditional way (audiotapes, videotapes, slides, question banks, other materials, etc.) accumulated in the previous period, owned by other educational institutions or units or published digital media resources (CD-ROMs, databases, etc.) and self-developed educational resources (e-courses, learning websites, case libraries, multimedia resource libraries, etc.). On the other hand, it is necessary to organize specialized personnel to collect, summarize, and categorize information resources that may be provided by various educational websites or other new media, and at the same time, digitally transform or process some resources (video tapes, slides, etc.) that cannot be transmitted online. Some resources (text, pictures, audio and video, etc.) need to be organized and integrated according to certain teaching requirements to complete the interactive, multimedia, and network optimization and transformation to meet the needs of online teaching and learning [9].

5.3 Strengthen Sharing Between Universities to Avoid Waste of Resources and Low-Level Duplication of Construction

At present, the universities are still in a state of self-contained and self-closed in the construction of information resources and lack of communication between universities and sharing of inter-university foreign language information resource construction are usual. The education department shall give full play to the functions of administrative coordination and planning, play an active guiding role in regional colleges and universities' informatization cooperation, and reduce unnecessary duplicate resource development. Schools can establish information resource sharing platforms, such as multimedia courseware sharing platforms, e-book resource sharing platforms, academic report sharing platforms, demonstration course sharing platforms, learning exchange forum sharing platforms, etc., to load, release, review, retrieve, and manage information resources, instant communication, etc., to facilitate teachers and learners to check knowledge and share information resources. Through the inter-construction and common use of information resources, the integration, socialization, and all-round configuration of information resources are achieved to meet the requirements of educators and learners [10].

6 Conclusion

Rational selection and optimal utilization of information resources is an inevitable choice in modern foreign language teaching and is the core of foreign language teaching informatization. The integration and optimization of information resources for college English teaching in the context of big data is the voice of the people and the needs of society. The current lack of overall planning for resource construction, uneven resource allocation, lack of quality teaching resources, and serious repetitive construction of resources are all factors. Although the strong support of national policies and documents, the availability of basic guarantees and the urgent needs of the times have provided feasibility for the integration and sharing of foreign language teaching resources. However, this project is a systematic project, which requires the active guidance of relevant government departments, the participation of various parties in the community, and the gradual unification of standards and norms. Only in this way can the teaching information resources be reorganized, integrated and shared to form a new organic whole to better display its effectiveness, serve education and society, and work to make improvements of college English education.

Acknowledgments. This research was supported by Jiangxi Educational Science Project during the 13th Five-Year Plan Period (project number: 19YB238), and Instructional Reform Project of Jiangxi Educational Commission (project number: JXJG-15-18-7).

References

1. Chen, J., Di, Z.: The integration and optimization of foreign language information resources: a survey based on the construction of information resources in some universities. Foreign Lang. Stud. **2014**(005), 95–100 (2014)
2. Xia, F., Fang, H.: Research on the integration and optimization of foreign language information resources based on autonomous learning. Tech. Enhanced Foreign Lang. Educ. **3**(2), 47–52 (2013)
3. Shi, G., Zou, J.: Research on foreign language information in universities in China. Strategies for optimizing the application of information resources. Tech. Enhanced Foreign Lang. Educ. **5**(3), 54–58 (2013)
4. Xia, F., Chen, J.: Research on the construction and utilization of foreign language information resources in colleges and universities. Modern Educ. Tech. **22**(6), 65–69 (2012)
5. Ma, W., Hu, J.: Impact and reconstruction of international MOOCs on college English courses in China. Technol. Enhanced Foreign Lang. Educ. **2014**(03), 48–54 (2014)
6. Cai, L., Wu, W.: Thoughts on information technology as the competence of modern foreign language teachers. Technol. Enhanced Foreign Lang. Educ. **2014**(01) 49–57 (2014)
7. Wei, W.: Analysis of integration and optimization of foreign language information resources based on autonomous learning. Chin. J. Educ. **2015**(S1), 82–83 (2015)

8. Li, Z.: The impact of big data era on college English teaching and coping strategies. Contemp. Educ. Res. Teach. Pract. **2016**(07), 30 (2016)
9. Shouren, W.: Interpretation of key points in the guide to college english teaching. Foreign Lang. **2016**(03), 2–10 (2016)
10. Chen, J.: Research on MOC and foreign language teaching in the era of big data: challenges and opportunities. Technol. Enhanced Foreign Lang. Educ. **2015**(01), 2–8, 16 (2015)

Discussion on the Training of Cross-border E-commerce Application Talents Based on the Internet Era

Dandan Zeng and Jianfang Ding(✉)

School of Economics, Wuhan Donghu University, Wuhan, Hubei, China
174222038@qq.com

Abstract. At present, Internet technology develops rapidly. At the same time, CBEC (short for cross-border e-commerce) develops rapidly as a new industry. Thus, CBEC enterprises propose not only quantitative demands for related talents, but also for higher quality. As an application-oriented university, it also puts forward new and higher requirements in the training of CBEC talents. To improve the training of CBEC application talents and analyze the existing problems in this training process, it is necessary to start with entrepreneurship of teachers, the construction of curriculum system, the using of new teaching methods and school-enterprise partnership. These can improve the current situation of CBEM talent training. At the same time, these can promote the cultivation of applied talents of CBEM in the Internet.

Keywords: Internet technology · Applied talents · Cross-border e-commerce

1 Introduction

Economic globalization is a normal state at present. At the same time, the operating cycle of commodities is shrinking rapidly. On this basis, CBEC develops rapidly in China. Generally speaking, the quality of imported goods is better. It has become a new trend for consumers to buy imported goods. With data display, the scale of CBEC transactions in China reached 9.1 trillion yuan by the end of 2018, an increase of 19.5% over the previous year. It is predicted that these transactions will exceed 10 trillion yuan in 2019 in China. In this context, the demand for professionals with relevant professional quality is also growing at the same time. Meanwhile, enterprises' requirements for the quality of CBEC talents are also increasing to adapt to the rapid update of social economy and technology. This requires colleges to cultivate more application-oriented CBEC talents [1].

1.1 The Necessity of Training Applied Talents for CBEC in the Internet Era

With the application of Internet technology, countries have more frequent contacts with each other, and the trade distance is shrinking. Higher requirements for quality of life have been put forward, and the demand for quality goods has continued to increase. At

J. MacIntyre et al. (Eds.): SPIoT 2020, AISC 1282, pp. 271–276, 2021.
https://doi.org/10.1007/978-3-030-62743-0_38

the same time, our country is rich in land and materials, and the goods produced by domestic enterprises need to be sold to other countries. The development of CBEC has provided a better bridge for this. This requires professional talents to provide corresponding professional services. Therefore, the training of CBEC application talents is particularly important.

According to the relevant data analysis of the Ministry of Commerce, there are more than 200000 CBEC enterprises currently operating in China. Most of these CBEC enterprises operate well and have created more than 5000 CBEC platforms [2]. The competition between these platforms and enterprises is very fierce. In order to survive in this kind of competition, the only way for an enterprise is to develop itself and strengthen itself, which is inseparable from the participation of excellent talents. At present, these enterprises have employed many relevant professional talents from universities, but the fact shows that the lack of talents in enterprises is still serious. What they lack are professional and versatile talents with professional knowledge of CBEC and familiar with this process. In other words, CBEC talents trained by colleges and universities are different from what enterprises need. And the gap is mainly in the application. As there were few opportunities to participate in the practice and production of the corresponding enterprises, the practical application ability of college graduates is relatively weak. This requires universities to take this problem into consideration when conducting talent training, and to consider how to train application talents for CBEC.

1.2 Training for Applied CBEC Talents

At present, colleges and universities do not set up a special CBEC major. Instead, they often set up this direction in the majors of international trade, business English or e-commerce. Moreover, schools only emphasize English and e-commerce in the training objectives, and do not fully consider the differences and requirements between traditional trade and CBEC. Meanwhile, because most of the college education is based on the traditional way of education, students mainly learn theories in schools [3–5]. However, cross border e-commerce practitioners only with theoretical knowledge is not enough, and need more practical experience. Therefore, the current application-oriented CBEC training needs to strengthen the training of practical ability.

1.3 Main Content and Direction

There are some problems in the current application-oriented CBCM training. Combined with the development needs of CBCM talents in the context of Internet, this paper first analyzes the current application-oriented universities in the application-oriented CBEC personnel training problems. Then, from the aspects of curriculum system construction, use of new teaching methods, teachers' Entrepreneurship and the cooperation between schools and enterprises, this paper puts forward countermeasures and suggestions for the training of applied CBEC talents in the Internet era.

2 Problems in Training Applied Talents for CBEC

2.1 Unreasonable Theoretical Curriculum and Outdated Teaching Materials

Because of the continuous changes in social economy, many things in the field of CBEC are constantly changing. Moreover, the compilation of teaching materials often needs to go through a long process. By the time they are officially put into use, some knowledge points have been updated or are no longer applicable. If teachers do not make adjustments timely, what students learn will be inconsistent with reality. The selection of a textbook is often used in several teaching years, which has led to a disconnect between theory and reality. Thus, if teachers use outdated teaching materials to teach students, it is difficult to better integrate the new situation, and to a certain degree, students cannot adapt to their positions in time [6–8].

2.2 Single Teaching Method and Low Learning Interest of Students in Class

Though colleges and universities are conducting teaching reforms and innovation, traditional teacher teaching is the main teaching method. Students' class participation is not very high. In addition, the dullness of theoretical learning and the autonomy of college learning make students less interested in class learning, which is not good to students' mastery of theoretical knowledge related to CBEC. At present, mobile Internet is developing rapidly. There are more and more things that attract students' interest. If college teachers still use traditional teaching methods, it is obviously not suitable for the times.

2.3 The Lack of Practicality of Practice Teaching, Far from the Actual Positions

As CBEC is not a major, but a direction of international economy and trade, there are many practice teaching methods. Some colleges and universities choose case teaching to enable students to learn specific CBEC from cases, which have a deeper understanding of its practice. However, these cases are often more typical, and more uncertainty will appear in actual work; Some schools choose teaching software that simulates real transactions, such as full-process training software for international trade. Although the actual CBEC process is simulated, there are still big differences between the specific operation of enterprises and the teaching of schools. Although some colleges and universities will also send students to relevant enterprises for internships, there are still some problems. On the one hand, enterprises can only receive a limited number of students. Due to the time limit of internship, students can often participate in the most basic work, it is difficult to participate in the specific business, can play a limited role.

2.4 The Need for Improvement of Teachers' Practical Application Ability

College teachers often require higher education. Thus, the CBEC teachers generally have a master's degree, and their work track is often from school to school. They lack the practical working experience and cannot provide students with favorable practice teaching. In theoretical teaching, teachers can indeed provide professional theoretical knowledge. However, due to their lack of practical ability, it is hard to integrate the new dynamics of the industry into the class, which makes the curriculum boring. In practice teaching, they can only conduct teaching according to practical training software or cases, but some problems in the middle are hard to explain.

3 The Construction of CBEC Applied Talent Training Mode in the Internet Era

3.1 Constructing a Reasonable Curriculum System and Increasing the Proportion of Elective Courses on the Basis of Major Courses

The curriculum system has a direct impact on talent training and can reflect the training intention. This paper analyzes the differences between international economy and trade in Colleges and universities and CBEC in Application-oriented Colleges and universities, finds out the characteristics of CBEC specialty, and sets up corresponding courses in combination with the specific requirements of enterprises for talent demand. Apart from the mainstream theoretical curriculum of "E-commerce", practical training courses should be provided accordingly. Meanwhile, the international perspective of CBEC should be considered. Thus, it is necessary to increase the classification and proportion of elective courses, and incorporate language, culture, business etiquette, and world economic forms into the elective curriculum system. In this way, students can choose according to their interests and needs, increase their knowledge intake, and provide a good foundation for future relevant work.

3.2 "Playing" in Learning, and Improving Students' Enthusiasm for Learning

At present, students have more and more channels to obtain information and knowledge, and the traditional teaching style is becoming less attractive. Thus, teachers should adopt new teaching methods in classroom teaching. According to the quality requirements of CBEC practitioners, the teaching mode of "E-commerce" course can be considered to be mainly taught by teachers, and combined with the sharing of students' groups. In this way, students can understand the actual transaction in cases sharing, and because of their own participation, it can promote the classroom resonance of teachers and students. Outside the classroom, micro-classroom construction can be built. The latest news and policies can be uploaded to the corresponding platforms through short videos, and students are required to watch them in time, and then establish discussion topics [9, 10].

Apart from the teaching of teachers, college students should pay attention to the ability of independent learning in their study. Thus, while invigorating the first classroom, learning should also be integrated into other time of students, so it is particularly important to conduct the second classroom, and the subject competition is a better reflection of the second classroom. At present, students majoring in international economy and trade can participate in subject competitions such as simulation business negotiation, evaluation of foreign trade correspondence and letters, disorderly fight of trade terms, international economic law debate, international arbitration simulation, trade sandbox, financial confrontation and CBEC, etc. Students can participate in such competitions, and apply the theoretical knowledge of this industry learned in the classroom to the simulation, so as to have a better grasp of relevant knowledge. Since most of the subject competitions will simulate the actual application and form an actual trading environment, it will have a better promotion effect on students' learning interest.

3.3 Promoting the "Going Out" and "Bringing in" of Teachers and Students, and Realizing the Application of Theoretical Knowledge

The proportion of practice teaching in their talent training programs is often very high for applied colleges and universities. Considering this, apart from setting up a certain proportion of practice courses, schools often hope students to conduct production practice in enterprises. So these schools can establish partnerships with CBEC enterprises. Then these schools can regularly send students to corresponding enterprises for on-the-job internship at the stage of college education. In this way, schools can provide enterprises with qualified professionals after graduation. Due to many students in Colleges and universities, enterprises have different abilities to receive students, colleges and universities can cooperate with many enterprises and establish a two-way linkage mechanism for the needs of both parties to form a stable internship system.

In order to ensure the practical ability of teachers and its integration with the market, teachers should be encouraged to "go out" to CBEC enterprises for long-term on-the-job learning. Besides, they need to conduct research on the latest industry needs to provide solid knowledge and methods for the training of CBEC applied talents.

By introducing teachers with the working background in CBEC enterprises, schools can improve the application ability of teachers. Meanwhile, it is also possible to invite the professionals with noble ethics, good business, excellent technology and rich experience in enterprises that are cooperating with schools to offer teaching guidance. In this way, a system of joint guidance from teachers inside and outside schools can be formed, which can further achieve a good combination of theoretical teaching and practice teaching, and better train these applied talents.

3.4 Conducting Innovation and Entrepreneurship Education and Encouraging Teachers and Students to Start Businesses

Under the background of "mass entrepreneurship and innovation", teachers and students majoring in international trade and economy can be encouraged to start businesses and establish CBEC enterprises. In this way, teachers and students can accumulate certain practical experience. After the entrepreneurial results are stable, enterprises can also serve as training bases for schools, and absorb students. This not only trains teachers, but also provides new posts for students' practical education.

Acknowledgement. This work was supported by the grants from Hubei Provincial Collaborative Innovation Centre of Agricultural E-Commerce (under Construction) (Wuhan Donghu university research [2016] No. 15 Document

References

1. Ming, Z.: Thoughts on the teaching reform of cross-border e-commerce under the background of Internet+. E-commerce **09**, 67–68 (2019)
2. Min, P.: Discussion on the development of cross-border e-commerce in China. Manager' J. (9) (2017)
3. Lin, R.: Exploration on the training mode of cross-border e-commerce talents in colleges and universities in the era of "Internet+". Chin. Foreign Entrepreneurs **13**, 197 (2019)
4. Li, X.: Exploration on the training mode of cross-border e-commerce talents in private colleges in the era of "Internet+". J. Lanzhou Inst. Educ. **35**(02), 107–109 (2019)
5. Guo, F.: Exploration on the training mode of cross-border e-commerce talents in higher vocational colleges under the background of "Internet+". J. Commer. Econ (14) (2017)
6. Zhu, Q., Zhong, F.: Training of modern apprenticeship talents in vocational colleges under the background of cross-border e-commerce. Educ. Vocat. **22**, 106–111 (2019)
7. He, J., Lin, C., Lin, B., Chen, L.: Researches on "1+X" ecosphere mode of cross-border e-commerce professionals' cultivation. Educ. Forum **47**, 76–77 (2019)
8. Chai, X.: School-enterprise collaboration to build a long-term mechanism for cross-border e-commerce talents training. Think Tank Era **44**, 99–102 (2019)
9. Zhao, L.: Evaluation of cultivation conditions for cross-border e-commerce innovation and entrepreneurship talents——taking foreign trade majors in applied undergraduate colleges as an example. J. Nanchang Inst. Technol. **38**(05), 100–108 (2019)
10. Wu, C.: Training of cross-border e-commerce talents in local applied undergraduate universities. Western China Q. Educ. **5**(20), 151–153 (2019)

Level of Technology Innovation Development and Promotion Strategies of High Technology Industry in Hubei Province Based on Smart City

Li Sun[1] and Danqing Li[2(✉)]

[1] Economics Department, Wuhan Donghu University, Wuhan, China
[2] Business School, Jianghan University, Wuhan, China
31612819@qq.com

Abstract. This paper analyzes the data of 31 provinces and cities from 2008 to 2017 to illustrate the development level of input and output of technology change in high technology industry, and compares the province with other provinces, find out where you rank in the country. The study finds that there are problems of insufficient investment and low efficiency in the transformation of technological achievements in high technology industry. The methods to improve the technology change and development level of high technology industry in Hubei Province are: to build a mechanism can lead to a steady increase in research funding; to build an effective talent training mechanism; to actively promote the cooperation between industry, University and research and to Change the transformation mode to increase the number of transformation results.

Keywords: High technology industry · Technology change · Technology transfer

1 Introduction

The level of development of the high technology industry can play a leading role in future economic development. The development of high technology industry is an important strategy to seize the commanding point of a new round of economic and technological development. Under the background of implementing "innovation driven development strategy", how to seize the opportunity, define the direction, highlight the key points, act quickly, make the high technology industry bigger and stronger is of great significance to adjust the structure, improve the quality and efficiency, Promote the transformation of the industry so that it can be upgraded and developed, and to achieve a higher level of economic development [1–4]. In 2018, the innovation level of Hubei Province ranked 7th in China. There are 6500 high-tech enterprises here, making a big breakthrough compared with the previous ones. There are 12 national and 20 provincial high-tech zones. In 2018, high tech enterprises in Hubei Province organized 1315 scientific and technological achievements transformation projects. It won 26 national science and Technology Awards. Technical contract turnover reached 123.7 billion yuan, ranking among the best in China. In order to better grasp the technology

J. MacIntyre et al. (Eds.): SPIoT 2020, AISC 1282, pp. 277–282, 2021.
https://doi.org/10.1007/978-3-030-62743-0_39

change development level of high technology industry in Hubei and improve the technology change competitiveness of high technology industry in Hubei, this article makes the Comparison between different years between 2008–2017 and the Provincial comparison between Hubei Province and another provinces in the country from two aspects of technology change input and output, so as to understand the current situation of high technology industry technology change in Hubei and the gap between Hubei and developed provinces. It points out the shortcomings of high technology industry in Hubei, and puts forward some suggestions to improve the performance of high technology industry in Hubei Province.

2 Current Situation of Investment in Technology Change of High-Tech Industry in Hubei

2.1 Full Time Equivalent of R&D Personal

This index is the sum of the number of full-time personal plus the number of part-time personal converted into the number of full-time personal according to the workload. This index is internationally used, and it is a common index for scientific and technological researchers to compare their input.

(1) Provincial comparison
In 2017, this index in high technology industry in Hubei Province was 21000, slightly lower than that in 2016, ranking 10th in the country, ranking in the middle and top level. The full-time equivalent of R&D personal in high technology industry in Hubei Province is lower than that in Anhui Province, ranking second, far higher than that in Shanxi Province and Jiangxi Province. Compared with the eastern developed areas, Hubei Province has a big gap in this index. Guangdong Province ranks first in the country in this index, which is 2000057 person years. Hubei Province only accounts for one tenth of Guangdong Province [5–7].

(2) Comparison between different years
From 2008 to 2017, the full-time equivalent of R&D personal in high technology industry in Hubei Province first increased to 25642 person years in 2014, and then decreased to 21000 person years in 2017 by a small margin year by year. The overall trend is still on the rise, from 9555 person years in 2008 to 21000 person years in 2017. In 2017, the index was 2.19 times that of 2008. The growth rate between 2008 and 2017 was generally declining, although there were fluctuations in individual years.

2.2 R&D Investment

R&D investment refers to the part of the actual expenditure of the unit with scientific research activities used for internal R&D activities. Including direct expenditure for R&D project activities, management fee, service fee, capital construction expenditure and outsourcing processing fee, etc. R&D personnel take this as the basis of scientific research.

(1) Provincial comparison
In 2017, the R&D investment of high technology industry in Hubei Province was 11.366 billion yuan, with growth rate of 10%, ranking the eighth in the country and at the national leading level. Among the six provinces in Central China, Hubei Province ranks first in R&D investment in high technology industry. Compared with Shanxi Province, the R&D investment in Hubei Province is about 5 times of that in Shanxi Province. However, compared with the eastern coastal developed provinces, Hubei Province has a certain gap. Compared with Guangdong Province, which ranks first in the country, the R&D investment of high technology industry in Guangdong Province is 98.378 billion yuan, and Hubei Province is only about one tenth of it.

(2) Comparison between different years
The R&D investment of high technology industry in Hubei Province increased from 1.348 billion yuan in 2008 to 11.366 billion yuan in 2017, during which, except for the decline in 2010, it increased year by year. In this period, the year-on-year growth rate is positive except 2010, average growth rate is 32.36%, with a high growth rate.

3 Current Situation of Technology Change Output of High Technology Industry in Hubei Province

3.1 Number of Patent Applications

Patent is the abbreviation of patent right. It refers to the exclusive right granted by the patent office to the inventor and designer according to the patent law after the inventor's invention and creation have passed the examination, reflecting the situation of science and technology and design achievements with independent intellectual property rights.

(1) Provincial comparison
In 2017, the number of high technology industry patent applications in Hubei Province was 6213, ranking tenth in the country, with a growth rate of 2.8% compared with 2016; among them, the number of invention patents was 3961, accounting for 63.8% of the number of patent applications. Among the six central provinces, the total number of patent applications for high-tech industries in Hubei Province is lower than 8326 in Anhui Province. the number of high technology industry patent applications in Anhui Province is about 1.34 times that in Hubei Province. the number of high technology industry patent applications in other six provinces in Central China is lower than that in Hubei Province, the lowest one is Shanxi Province, only 281, only one in 22 of Hubei Province. Compared with the developed coastal provinces in the East, the number of high technology industry patent applications in Hubei Province is still far behind. The number of high technology industry patent applications in Guangdong Province, which ranks first in China, is 84084, seven times that of Hubei Province [8–10].

(2) Comparison between different years
From 2008 to 2017, the number of patent applications for high-tech industries in Hubei Province showed an upward trend, from 660 in 2008 to 6213 in 2017. Although there were fluctuations during this period, the overall trend was relatively obvious. Based on

2008, the growth rate was 841.36%. The growth rate in this period is positive except for 2010, with an average growth rate of 35.07%, indicating that the number of patent applications for high technology industry in Hubei Province has increased significantly.

3.2 Number of New Product Projects

(1) Provincial comparison
In 2017, the number of new product projects of high technology industry in Hubei Province was 2885, ranking 11th in China. Compared with other provinces in the central six provinces, Anhui Province has 3896 new product development projects of high technology industry, Jiangxi Province has 2987 projects, ranking sixth and tenth respectively in the country, higher than Hubei Province; other provinces have lower number of new product development projects than Hubei Province, the lowest is Shanxi Province, with only 631 new product development projects, only about one fifth of Hubei Province. Compared with the eastern coastal developed cities, the number of new product projects of high technology industry in Hubei Province is still relatively low. Compared with 32392 projects in Guangdong Province, which ranks first in China, Hubei Province is only one twelfth of Guangdong Province.

(2) Comparison between different years
The number of new product projects of high technology industry in Hubei Province increased from 1145 in 2008 to 2885 in 2017, a year-on-year increase of 1.52 times. From 2008 to 2017, the number of new product development projects of high technology industry in Hubei Province showed a trend of ups and downs, and the downward trend in 2010 was more obvious. During this period, the highest growth rate was 102.71% in 2009, and the decline rate was - 40.54% in 2010, with an average growth rate of 16.72%.

3.3 Sales of Newly Developed Products

Sales of newly developed products refers to the sales revenue realized by the enterprise in selling new products in the reporting period.

(1) Provincial comparison
In 2017, the sales of newly developed products of high technology industry in Hubei Province was 94.839 billion yuan, with growth rate of 74.69% compared with 24.006 billion yuan in 2008, ranking 14th in China. Compared with other six provinces in Central China, Hubei Province is in the middle level, lower than Henan Province, Anhui Province and Hunan Province, and higher than Shanxi Province and Jiangxi Province. The sales of newly developed products of high technology industry in Henan Province, which ranks first in Central China, is 335.689 billion yuan, 3.5 times that of Hubei Province, and sales of newly developed products of high technology industry in Shanxi Province, which ranks second, is only 17.018 billion yuan, Hubei Province One sixth of the province. Compared with the eastern coastal developed provinces, Hubei Province has a large gap in sales of newly developed products of high technology industry. Compared with Guangdong Province, which ranks first in the country, sales

of newly developed products in Guangdong Province is 1869.315 billion yuan, 19.7 times of that in Hubei Province.

(2) Comparison between different years

The sales of newly developed products of high technology industry in Hubei Province increased from 24.06 billion yuan in 2009 to 94.839 billion yuan in 2017, increase of 2.95 times. During this period, there was a continuous increase. Although the year-on-year growth rate fluctuated during this period, it was all positive, with average growth rate of 19.32%.

4 Conclusions and Suggestions

4.1 Conclusion

(1) The technology change scale of high technology industry in Hubei Province is increasing, but the total amount still needs to continue to expand. The diffusion and radiation of related industries are not strong, the driving force is not strong, and the promotion of economy is not obvious. In 2017, the number of R&D projects of high technology industry in Hubei Province was 2765, with growth of 22.67%, but only ranked 11th in the country. From 2008 to 2017, the number of R&D projects in high technology industry in Hubei Province showed an overall upward trend. Although the year-on-year growth rate fluctuated, the overall growth rate still showed a downward trend.

(2) The R&D investment of high technology industry in Hubei Province is insufficient, the ability of independent innovation is weak, the mechanism of high-end talent introduction and cultivation is not perfect, and the "bottleneck" of technology and talent is serious. From the perspective of personal input, in 2017, the full-time equivalent of R&D personal in high technology industry in Hubei Province was 21000 person years, down 1.03% compared with 2016, and reached a peak of 25600 person years in 2014 during the period from 2008 to 2017, followed by a small decline year by year.

(3) The technology change output of high technology industry in Hubei Province has been increasing, no matter the number of patent applications or the number of new project development, ranking 10th and 11th in the country, but the sales revenue of new projects ranking 14th in the country, which shows that the achievement transformation mechanism is not perfect, and the efficiency of technological development stage is higher than that of technological achievement transformation.

4.2 Countermeasures and Suggestions

(1) Through the establishment of a stable growth mechanism of investment in science and technology and a talent training mechanism, the technology change level of high technology industry in Hubei Province will be guaranteed. Personal input and fund input are the main factors influencing the output of technology change. Hubei Province should establish a stable growth mechanism of scientific and

technological fund input, continue to increase the investment in high technology industry, and it is necessary to make a plan for the training of technology change talents and improve the training system of innovative talents.

(2) We will actively promote cooperation between industry, University and research institutes, and to Change the transformation mode to increase the number of transformation results. As the transformation ability of high technology industry's technological achievements in Hubei Province is still insufficient, efforts should be made to cultivate the technology transformation market, accelerate the transformation of results into effective production capacity, increase the investment in the transformation of scientific and technological achievements. It is necessary to increase investment in this area to achieve the effect of promoting the transformation of results.

Acknowledgments. This research was supported by the soft science project of Hubei Provincial Science and Technology Department (2019ADD164) and the youth fund project of Wuhan East Lake University (2019dhsk007).

References

1. Ali, M., Park, K.: The spiral model of indigenous technology change capabilities for developing countries.In: 6th International Student Conference, pp. 106–110 (2010). Izmir, Turkey
2. Halla, B.H., Helmers, C.: Innovation and diffusion of clean/green technology: can patent commons help? Management **66**(1), 33–51 (2013)
3. Zijun, L.: Research on the technology change efficiency of high-tech industry in Jiangxi Province and its influencing factors. Nanchang University **5**, 36–39 (2018). (in Chinese)
4. Hao, L.: Evaluation of technology change efficiency of high-tech industries in Heilongjiang Province based on DEA. Economist **12**, 167–169 (2017). (in Chinese)
5. Colombo, M.G., Grilli, L., Murtinu, S.: R&D subsidies and the performance of high-tech start-ups. Econ. Lett. **1**, 97–99 (2011)
6. Weizhan, M., Chunyan, L., Xiaodong, S.: Staged analysis of innovation efficiency of China's high-tech industry: based on a three-stage DEA model. Macroecon. Res. **34**(04), 88–91 (2019). (in Chinese)
7. Xiao, P., Junling, C.: Research on technology change performance evaluation of high-tech industry in Hubei Province. Sci. Tech. Prog. countermeasures **11**, 128–132 (2013). (in Chinese)
8. Song, M.L.: Environmental regulations, staff quality, green technology, R&D efficiency, and profit in manufacturing. Technol. Forecast. Soc. Change **133**(8), 1–14 (2018)
9. Qingjin, W., Qiang, W., Xue, Z.: Research on efficiency evaluation and influencing factors of R&D activities in regional high-tech industries. Sci. Tech. Progress Countermeasures **12**, 61–62 (2018). (in Chinese)
10. Zhang, Y., Wang, J.R., Xue, Y.J., et al.: Impact of environmental regulations on green technological innovative behavior: an empirical study in China. J. Clean. Prod. **188**, 763–773 (2018)

Wisdom Media Era of Big Data in the Application of the Short Video from the Media

Zhi Li(✉)

Department of Journalism Studies, The University of Sheffield, South Yorkshire, UK
471047820@qq.com

Abstract. In today's era, with the wide application of artificial intelligence AI, the application of media is gradually approaching to intelligence, the application of big data in the era of smart media has gradually become an indispensable part of social life. The era of big data has promoted the development and progress of the society and facilitated people's life. Accordingly, new media such as short video "we media" have sprung up like mushrooms and entered the public's vision. Through the application of big data in the era of smart media, this paper analyzes the development trend of short video we media in social life, analyzes the influence of the era of big data on short video, and further reflects the general advantages of the era of smart media from the discussion of short video we media in the era of intelligent big data. This paper discusses some entertainment and convenience created by the application of the short video "we media" in the era of smart media for People's Daily life. Based on the problems and challenges encountered in the application of smart media in some fields, it puts forward specific plans for the security protection of people's personal information in the era of big data. The purpose of this paper is to try to reveal the application of smart media and big data to short video in this era and the corresponding research. Based on the above discussion, this paper combines big data analysis and relevant theoretical knowledge in the field of news media, combines intelligence with "we media", and studies the value of "we media" short videos to the social development in the era of smart media. This article research results show that the wide application of wisdom media era of big data is a trend of rapid development of today's society, in this trend, a short video from the development of the media heat continues to increase, not only make the communication between people more close, and accelerated the development of the modern intelligent society and optimize the traditional mode of transmission medium.

Keywords: Era of intelligence · Era of big data · Short video · We-Media

1 Introduction

With the rapid development of science and technology, in every field of society, the leading development of intelligence reflected in every field. Especially in the field of the spread of news media, as before Marshall McLuhan once put forward the theory

J. MacIntyre et al. (Eds.): SPIoT 2020, AISC 1282, pp. 283–289, 2021.
https://doi.org/10.1007/978-3-030-62743-0_40

connotation of "medium is the message", the birth of intelligent media era of big data technology and common application development continue to refresh the short video from the media in today's society the significance of the form and output mode, also aroused the continuous hot orgasm. We can clearly see that today's short video "we media" have been put into the era of smart media, changing the direction of development. Various performances indicate the development trend of a media. These intellectual media big data times have different effects in different fields, and have been making continuous progress and development.

Social development, economic situation and technological improvement are some important factors that cause the continuous development of short video "we media" towards intelligence [1]. With the combined efforts of these factors, the short video "we media" of smart media has become a new favorite for people to understand real-time dynamics, and it is supported by powerful technical equipment [2]. In terms of its theoretical significance to analysis of media under the background of big data era for short video from the media, this paper studies think tank under the media from the media for the use of all aspects of the social impact of the user, use a short video from the media of the new media as a powerful expression of media advantage, can help us users and researchers better analysis under the big data, since the role of the media, connotation, to distinguish the difference from traditional media division, so as to better achieve a higher cognitive [3].

Based on the environment of the media era of big data, research think media under the short video from the media to make a change on the impact of the user and, more can reflect the effect of the era of big data media, can also be convenient to know the detailed feedback of user for a short video from the media, it is widely used, but it also faces some problems [4, 5]. Based on big data in the era of smart media, this paper comprehensively understands the new application and effect of short video "we media", scientifically analyzes new opportunities and profoundly grasps the new development direction [6]. In this paper, in a short video for wisdom media era of big data from the application of comprehensive analysis of the media, put forward to promote its giving full play to the advantages of better opportunities, through specific ways to make our country engaged in intellectual media big data research and related aspects of the short video from the media personnel to improve their professional ability, make our country in the area of intelligent stepped up big time, catch up with the trend, in the news media communication is getting better and better in the development of [7]. It will lay a solid foundation for our country to build a strong socialist modern culture. Only with the application of big data in the era of smart media and the leading development of intelligence can we develop better in the whole intelligent field [8, 9].

2 Method

2.1 Core Concepts

(1) The era of smart media

Smart media: the definition of smart media is to use artificial intelligence technology to reconstruct the whole process of news information production and dissemination of media. Smart media is actually an ecosystem based on artificial intelligence, mobile Internet, big data, virtual reality and other new technologies. It consists of intelligent media, intelligent media and think tank media. Features: all-media, integrated media and intelligent media. The three stages also present different characteristics [10]. In the age of omnimedia, media fusion is a physical change, while in the stage of media fusion, it is a chemical reaction. In the age of smart media, it is a deep change at the level of genes. The composition of intelligent media is the combination with artificial AI. Intelligent media is an essential and important part of intelligent media, whose key lies in leading values. Think tank media provide the public with intellectual support [11].

(2) Short video "we media"
Short video refers to the video content that takes new media as the communication channel and is less than 5 min long. It is another emerging content communication carrier after text, picture and video. Characteristics of short video industry: low production cost, fragmentation of transmission and production; Fast propagation speed, strong social attribute; The line between producer and consumer is blurred. An important reason for the emergence of short video applications is the development of mobile Internet technology. In recent years, the short video "we media" has developed by leaps and bounds and become an indispensable part of People's Daily life.

2.2 Research Objects and Methods

This article research is the development of the era of media short video from the media as the main research object, the basis of analyzing the short video from our daily life media for social users life changes have effect and the significance, objective and real data analysis of the present wisdom media age big short video from the media in some obvious deficiencies and problems in the process of application. The research on short video "we media" is the basic starting point for the discussion and research of the full text. In the discovery of some current thorny problems and explore solutions, based on the famous academic websites in China, we can fully search and query relevant literature, and formulate feasible research routes and ideas.

As the researcher of this paper, he drew up the concrete practice plan and steps when the practical problems were investigated. The specific practice plan and steps mainly cover the discussion of professional technicians in the field of intelligent industry, the application of big data in short video "we media" in the era of intelligent media, the current problems and some solutions to these problems. Conduct professional software analysis and research on statistical products and service solutions to collect good data, so as to ensure that the research of this paper stands on an objective and neutral position, and ensure the credibility of the research results. And put forward the effective solution strategy from the Angle of specialization [12].

3 Experiment

3.1 Experimental Data Sources

In this paper, the application of big data in short video "we media" in the era of smart media is carried out based on relevant literature query and discussion of professionals in the actual industry, and relevant conclusions are taken as the verification link of the investigation. In the process of investigation and research, big data in the era of smart media is the key word to search. The objectivity of the research results is ensured by taking its reference point.

3.2 Experiment Implementation

In the process of searching literature, this study took "era of smart media big data" as the key word. After searching, it found that relevant literature had a large reference, and adopted its views and improvement measures. On the basis of critical thinking, this paper has fully absorbed the excellent results of previous studies to ensure that this article has its own characteristics.

In the actual investigation and research, the opinions of professionals on smart media and big data are strictly in accordance with the specified plan and steps to ensure the accurate recording of professionals' opinions. In order to ensure the researchability of this paper, a strict and objective analysis and description are conducted on the recorded data.

4 Discuss

4.1 Development Path of Short Video We-Media

Through experiments and actual investigations, the results show that big data plays an important and fundamental role in the development of we media in short videos in the current era of smart media. The rapid development of short video "we media" is closely related to big data and users' acceptance and use in the age of smart media. Users can get information from mobile phones in any way in any place, so that the masses have a sense of dependency in their daily habits. At the same time with the continuous progress of smart media big data. The extensive promotion of better smart media big data is also the primary factor for the surging use of short video "we media". Compared with text and text and audio, video occupies more memory and USES more traffic. In the current environment of the Internet market, a large number of Internet short video users have accumulated, and users have a new demand for short video "we media". The actual investigation results of this paper based on the specific development path of short video "we media" are shown in Table 1.

Table 1. Application development of short video we-media

Year field	Quick hand	The micro view	Shaking, audio and video	YouTube
2015	35	30	0	60
2016	60	45	20	75
2017	78	66	35	78
2018	86	79	70	85
2019	100	100	100	100

It can be found from Table 1 that the application of short videos, which are popular in Chinese society, started from traditional media such as TV videos and print ads in recent 5 years, and then gradually developed into smart media big data, such as kuaishou video, micro-video, douyin and YouTube. Therefore, it can be seen that big data in the era of smart media plays an important role in the development of China's overall network media industry.

4.2 Bottlenecks in Development

But because our country in recent five years before a short video of the media from the media developing slowly, the development of the media short video is shorter, so the development of the short video also is more, the problem of intelligent level is not enough to cause our short video applications as well as the short video control technology is not yet mature, the survey results of the main problem facing our nation today are shown in Fig. 1 below.

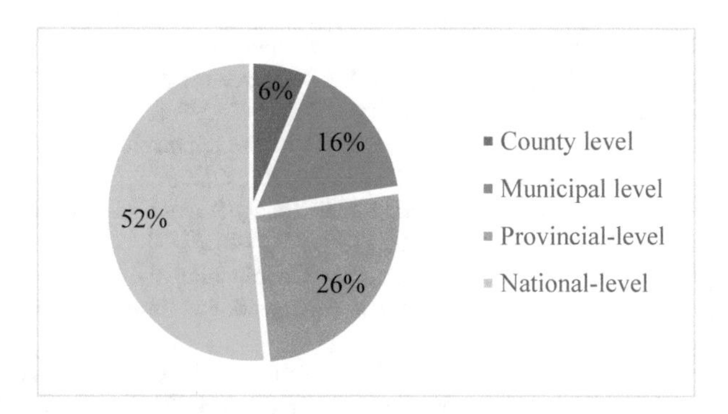

Fig. 1. Short video we media application software

We can see from the figure above that the market proportion of each short video has its own proportion. It can be clearly seen from the figure that the main part of the current market proportion is douyin APP, followed by kuaishou video. YouTube and micro-video account for a similar proportion. It can be concluded that nowadays, in daily life, people's main relaxation time is to brush up on some short videos to enrich their hobbies and add some life tips. However, due to the current short video development in many aspects of intelligent technology is not mature, the application and management of some videos is not mature, and there is a lack of video APP design talent, the application of smart media is not very skilled. Therefore, in view of these problems, we need to proceed from the reality, put forward effective improvement measures and solve the problem.

5 Conclusion

Only by constantly promote the development of modern society, in general, intellectual media age the rapid development of big data, continuously improve the ability of intelligent short video from the media, improve the technical level, to make a short video of our current society and the development of some other news media big stride forward in the direction of more advantage. For the current we are for the use of short video media problems encountered in the operation, also need to professionals in the field of artificial intelligence and the various aspects of multilateral assistance and hard work, innovation of new advanced technology actively, increase personnel training, and improve the system of each APP application management, just have to make our country striding forward steadily in the direction of the modern cultural power, and cultural development in the field of better, and better meet the needs of the people for a better life.

References

1. Maoying, W.U., Keji, H., Management, S.O., et al.: Appraising netnography: its adoption and innovation in the smart tourism era. Tour. Tribune **29**(12), 66–74 (2017)
2. Heath, A.C., Lessov-Schlaggar, C.N., Lian, M.: Research on gene-environment interplay in the era of "big data". J. Stud. Alcohol Drugs **77**(5), 681–683 (2017)
3. Dang Anrong, X., Jian, T.B.: Research progress of the application of big data in china's urban planning. China City Plann. Rev. **3**(12), 24–30 (2017)
4. Markl, V.: Breaking the chains: on declarative data analysis and data independence in the big data era. Proc. VLDB Endowment **7**(13), 1730–1733 (2018)
5. Koziri, M., Papadopoulos, P.K., Tziritas, N.: On planning the adoption of new video standards in social media networks: a general framework and its application to HEVC. Soc. Netw. Anal. Min. **7**(1), 32 (2017)
6. Meek, S., Goulding, J., Priestnall, G.: The influence of digital surface model choice on visibility- based mobile geospatial applications. Trans. GIS **17**(4), 526–543 (2017)
7. Hoecker, M., Kunze, M.: An on-demand scaling stereoscopic 3D video streaming service in the cloud. J. Cloud Comput. Adv. Syst. Appl. **2**(1), 1–10 (2018)
8. Inoue, T., Kurokawa, S.: Derivation of path of contact and tooth flank modification by minimizing transmission error on face gear. J. Adv. Mech. Des. Syst. Manufact. **116**(1), 15–22 (2019)
9. Wei, X.-S., Zhang, C.-L., Zhang, H.: Deep bimodal regression of apparent personality traits from short video sequences. IEEE Trans. Affect. Comput. **9**(3), 303–315 (2018)

10. Xiangping, W.U., Ruoqing, Q.I.U., Xiaofang, Y.A.N.G.: Influence on compliance of subcutaneous immunotherapy in patients with allergic rhinitis by We-Media management. J. Clin. Otorhinolaryngol. Head Neck Surg. **32**(53), 90–91 (2018)
11. Vo, K.A., Nguyen, T., Pham, D.: Combination of domain knowledge and deep learning for sentiment analysis of short and informal messages on social media. Int. J. Comput. Vis. Robot. **420**(221), 22–24 (2018)
12. Kendon, E.J., Blenkinsop, S., Fowler, H.J.: When will we detect changes in short-duration precipitation extremes? J. Clim. **31**(7), 41–43 (2018)

Development and Risk of Smart City

Yanping Chen and Minghui Long(✉)

Wuhan Donghu University, Hubei, China
757088198@qq.com

Abstract. Smart city is the development direction of future urban construction and the main driving force to promote sustainable urban development. In this paper, starting from the concept of intelligent city, profound interpretation of the wisdom of the city development and technical support. At the same time, in view of the wisdom urban construction existing in the information security risk, weak top-level design and problems of traditional urban management discussion, and explore the corresponding Suggestions and countermeasures, in order to promote the construction of wisdom city better and faster.

Keywords: Urban construction · Smart city · Risk

1 Introduction

With the continuous development of cities, the construction concept of smart cities should be based on digital cities and smart cities in order to deal with the increasingly serious phenomenon of "urban disease". Wisdom city through the Internet of things, cloud computing, big data, interconnected melting, intelligent technology to build a comprehensive visual system, the huge amounts of information needed for the urban operation management in the convergence, classification, analysis, can achieve real-time field operation of urban resource monitoring and tuning degree, to construct human-oriented livable city [1–3].

2 Overview of Smart Cities

2.1 Smart City Planning

Smart city planning is to determine the direction of urban development, arrange the supporting construction of urban industries and the comprehensive layout of urban industries. It is the guiding direction of urban development in a period of time and the basis of urban construction and management. Compared with the traditional urban planning, the planning of smart city firstly focuses on building a unified platform to realize the dispatching management and servitization of various information resources, thus supporting the wisdom of urban management and public service. Secondly, it is planned to set up the urban data center to build an open and Shared data system through the collection of massive data, so as to effectively improve the generation and application of supporting data used by urban managers in decision-making, and realize the

J. MacIntyre et al. (Eds.): SPIoT 2020, AISC 1282, pp. 290–294, 2021.
https://doi.org/10.1007/978-3-030-62743-0_41

pooling and sharing of urban resources and the coordination and linkage across departments. Finally, the basic network of urban intelligence composed of the communication network, the Internet and the Internet of things is planned as a whole, so as to collect and manage effectively the state data of the city's macro-economy, municipal facilities, ecological environment, public security, and people's public opinion [4–6].

2.2 Construction of Smart Cities

The focus of smart city construction is government information, city management information, industrial economy information, social people's livelihood information. Among them, government information mainly includes the construction of urban cloud computing data resource center, collaborative office center of various departments, electronic government affairs platform, online government affairs approval and other aspects. Urban management information construction includes the construction of urban intelligent management system, urban emergency management, urban public safety, urban infrastructure and other aspects. The construction of industrial economy information includes the construction of efficient enterprise resource management platform, enterprise e-commerce system, enterprise service platform and so on.

2.3 Application of Smart Cities

Through the construction of broadband multimedia information network, geographic information system and other infrastructure platforms, smart cities integrate urban information resources, provide ubiquitous public services for citizens, provide efficient processing for government public management, and provide powerful guarantee for enterprises to improve work efficiency and increase industrial capacity. At the same time, smart city brings “P2P wireless communication” into the “M2M era”, thus creating three application modes of digital administration, digital livelihood and digital industry. Among them, digital government affairs strengthen resource integration, information sharing and business coordination, promote wireless e-government projects, build mobile government affairs platform, so as to improve the governance ability and service level of the government, promote the establishment of service-oriented government.

3 Problems Existing in the Development of Smart Cities in China

3.1 Information Security System Needs to Be Improved

The hidden danger of system security mainly occurs in the construction of hardware and software. Wisdom in the process of urban construction in China at present is hardware field giant monopoly. Such as operating systems, database construction, and other areas of the software is dominated by software companies such as Microsoft. Too much of imported products, technology and solutions, combined with low localization rate make the wisdom of the city information system like a giant “black box”. The software and hardware of loopholes and even the back door to other users information blockade, exacerbated by data reveal that security hidden danger.

Data safety problems mainly come from the external network attacks, internet of things collection side, all kinds of switch and server, big data and cloud computing center and a large amount of data, all the data in the collection, transmission, exchange, storage, processing. Because the external network attack on or in the process of its own security strategy of holes can lead to leakage of dew, tampering and destruction of information data, and even lead to the chaos of common People's Daily life or cause great economic losses to the enterprise.

Hidden danger of human operation mainly comes from data management personnel and data management system. It is difficult to define the department ownership and personnel authority of the massive data generated by smart cities. At the same time, the large number of users and management departments makes it more difficult to regulate the use of data and supervise and picket.

3.2 Technical Reliability and Security Risks

The purpose of smart city construction is to provide a set of overall solutions for the future urban development, involving a large scope and a long time span. Therefore, it is difficult for the planners of smart city to rely on the development of information technology to solve all the problems in the construction process. In order to achieve the wisdom of the city in perception, need to everywhere in the city the deployment of all kinds of intelligent devices, constitute the "nervous system" of the city, but neither as nerve endings RFID tags, or as the nerve center of the data processing center, are faced with the risk of failure caused by natural or man-made factors, causing heavy to intelligence cannot be perceived, key data cannot pass, resulting in heavy big event is unable to be processed in a timely manner or early warning.

4 Suggestions for Smart Cities

4.1 Strengthen the Top-Level Design

Wisdom city construction is _ a systems engineering, the city's top layer design should be based on politics, people and companies three aspects of demand analysis, and based on the local culture, history, resource, system and their characteristics. Such as the regional economic level around the local wisdom city overall planning of strategic positioning. In the wisdom of creating distinctive city along with the local history and culture for effective protection and inheritance.

The sharing of information and resource data is also the focus of top-level design. The realization of data and information sharing in multiple fields will greatly improve the management efficiency of cities and the economic value of information resources. To this end, on the top floor of a wisdom city needs exploring in the design of information sharing between departments machine system, implement the strategy of standardization of data information. Build a unified city cloud computing data center is used to break the system of regional segmentation "information island" phenomenon, city government, resources, environment and economy in the field of information resources sharing.

4.2 Innovation of Coordination Mechanism

Smart cities are characterized by the extensive use of high and new information technology, resource sharing, intelligent response and business collaboration in many industries and fields such as e-government, urban transportation, education and medical treatment, community management, talent cultivation and enterprise incubation. However, according to the development status of smart cities in China, although traditional smart cities led by information departments can effectively integrate high and new information technologies to build urban network platforms. It is difficult to effectively integrate information resources into specific urban management business. Therefore, it is particularly important to establish a new smart city coordination and management mechanism that can coordinate the business content of all parties and fully utilize the resource advantages brought by big data and cloud platform.

In 2019, the construction of a new smart city in jiaxing adopted the management mode of "two centers", which integrates five elements of ubiquitous public service, precise and accurate urban management, transparent and efficient online government, integrated industrial economy and autonomous and controllable network system. "Big data center" as the heart into offers a wealth of information resources, while the "urban comprehensive operation management center" as a brain coordination of various resources, realize the city governance. The service of the people's livelihood, infrastructure, industrial economy, the situation of ecological environment in key areas such as operating display, analysis, management, analysis, prediction, auxiliary decision-making and business support, thus effectively improving the city overall operation efficiency and management level.

4.3 Pay Attention to Information Security

There are three main directions of dynamic prevention and information security management, construction of smart city information security system, including active prevention.

Active prevention includes the use of localized basic hardware and software products as much as possible, and it plays a decisive role in information security to build an active prevention system based on independent innovation on the basis of ensuring that the key core technology of information and the construction of information infrastructure are independent, controllable and credible. At the same time, in the construction stage of smart city system architecture, information security strategy needs to be fully planned from top to bottom, so as to achieve efficient resource sharing, interconnection and mutual access while fully controlling the information security architecture and eliminating security risks.

Passive prevention mainly includes the use of boundary protection system, network identity authentication, intrusion detection, access authentication and other security products to enhance the security of information system. At the same time, the information security disaster warning and emergency disposal system should be established, and the entire information security system should be reviewed from time to time to quickly find problems, and security hidden dangers should be quickly repaired and consolidated.

Information security management, including establishing information security propaganda system, strengthen the public information security awareness, strengthen urban wisdom relevant professional personnel training, increase information related personnel information confidential responsibility consciousness, make information security related laws and regulations, strengthen the network security management, the implementation of the accountability mechanism, fully mobilize all social forces to join and the construction of information security work, provide the good development foundation for the wisdom of the city construction. At the same time, the supervision and treatment mechanism of electronic waste and electromagnetic radiation should be established to avoid the biological risk of urban information network as much as possible.

5 Conclusion

Smart city is the deep integration of urban urbanization, industrialization and information. It is an important means to improve local productivity, improve citizens' happiness index, and promote the reform of urban management concept. It is also the core development direction of future cities. In the construction of smart cities, top-level design and overall planning, as the core of smart cities, and information security as the foundation of stable operation, have big problems and challenges to be solved. These problems are not only limited to the economic field, but also involve ecological environment, urban management and regional culture. However, these problems will be solved with the joint efforts of local governments at all levels, enterprises and public institutions, and all the citizens. Smart cities and information security construction in different regions will surely develop together and jointly promote the healthy and stable development of smart cities.

References

1. Gibson, D.V., Kozmetsky, G., Smilor, R.W.: The technopolis phenomenon: smart cities, fast systems, global networks. Behav. Sci. **383**(2), 141–143 (1993)
2. Zhao, H., Lan, X., Wu, J., et al.: Remote sensing technology helps build smart cities. Constr. Sci. Technol. (13), 30–32 (2017). (in Chinese)
3. Liu, X.: Building a smart service platform to solve the “three difficulties” of people's livelihood. China Trust Interest Commun. (3), 88–90 (2018). (in Chinese)
4. Wang, D.: Spatial and temporal information application of smart cities. Informatiz. China's Constr. (07) (2019). (in Chinese)
5. Lu, J.: Chongqing characteristic scheme of smart city. Surv. Mapp. China. (04) (2018). (in Chinese)
6. Geng, D., Li, D, Wang, D.: Top-level design concept and practice of smart city in characteristic parks – a case study of the administrative office area of a sub-center of a city. Eng. Surv. (05) (2017). (in Chinese)

Impact of Education with Chinese Feature Reform of Management Courses Based on Big Data Analysis on Employment Quality of College Students

Qinglan Luo[1(✉)] and Ying Zhao[2]

[1] College of Business Administration, Jilin Engineering Normal University, Changchun, Jilin, China
sunny203@126.com

[2] The Tourism College of Changchun University, Changchun, Jilin, China

Abstract. Big Data era to our lives has brought a revolutionary change. The rapid development of science and technology and society has put forward higher requirements for the comprehensive quality and Education with Chinese Feature education of management college students. In this case, management teachers must fully understand the current teaching style of management in universities, fully grasp the requirements and characteristics of Education with Chinese Feature education of management college students, and constantly adjust the Education with Chinese Feature teaching structure of current management courses. And the use of large data and multimedia teaching methods of political education practice teaching management courses thought. The purpose of this paper is to study the quality of employment for the students in the course of Education with Chinese Feature reform analyze large data management classes. It provides methods and solutions for education reform and management in universities Education with Chinese Feature course the big data era. This article on how to promote the proper use of large data Education with Chinese Feature education and management in universities Education with Chinese Feature curriculum reform and management of college students conducted a comprehensive study of the system. And the quality of employment data research, research results shows that the Education with Chinese Feature reform curriculum management class big data analysis of employment quality students than traditional college students the quality of employment is higher by 16.8%.

Keywords: Big data analysis · Education with Chinese Feature · College student employment · Employment quality

1 Introduction

With the rapid development of technologies such as the Internet, global data and information have exploded. Big Data era has quietly embedded in all aspects of social life, to lead a new round of changes [1–3]. With the continuous development of the Internet and the deepening of information technology, human footprints in cyberspace gradually, more and more data generation and accumulation. According to IBM, all of

J. MacIntyre et al. (Eds.): SPIoT 2020, AISC 1282, pp. 295–300, 2021.
https://doi.org/10.1007/978-3-030-62743-0_42

the data obtained in human civilization is to generate 90% in the past two years. By 2020, the global scale of the data generated will be 44 times [4] of today's data. Big data types include complex data sets, which can be video and audio information, location information, and web logs. Modern society for data processing capacity of a higher demand [5]. Management college students in the era of big data and new media are good at thinking and can use these new network resources and technologies skillfully.

In the new situation, the Education with Chinese Feature education of management college students is facing many new problems and new challenges. Among them, the supply-side reforms may be able to solve the long-term pain Education with Chinese Feature education [6–8]. The core supply-side reform is "supply-side adjustment according to demand." Big Data can solve the problem of "how much and what is" in. In this process, management courses Education with Chinese Feature teachers must be fully and comprehensively updating teaching concept of Education with Chinese Feature theory, and the introduction of modern teaching methods are highly feasible [9, 10] on the existing basis. Career Guidance Center has established an employment service for colleges and universities to promote the employment of university students' played an important role. The management of college students Education with Chinese Feature education reform services plays a key role. This article on how to properly use Big Data thinking, tools and technology to promote Education with Chinese Feature theory teaching reform and Education with Chinese Feature education of college students a comprehensive and systematic study. And the quality of employment data research, study results show that big data analysis and management courses Education with Chinese Feature quality of college students on employment reform employment quality than traditional college students is higher by 16.8%.

2 Method

With the continuous deepening of the Internet and the continuous progress of information technology, human footprints in the cyberspace have gradually formed, and more and more data are continuously generated and accumulated. The term "big data" was first coined by NASA researchers to describe the data challenge that occurred in the 1990s, the massive amount of information and data generated by supercomputers. As for the development and utilization of data resources, the United States, as a world economic, technological and military power, is still in a leading position. Since 2009, "big data" has gradually become the focus of the Internet information technology industry, and it has been heating up. 2013 can be said to be the first year of China's big data, from the generation and development of data science to the establishment of many cloud computing bases and laboratories. Big data has become the focus of global attention. "Speak with data" and "quantify everything", the world has become more three-dimensional, clearer and more scientific.

The strategy of rejuvenating the country through science and education continues to develop in our country. The combination of science and technology and education is an inevitable trend in the development of education. Make full use of the advantages of information technology to better serve education and create a more complete information environment so that everyone can fully enjoy the dividends brought by the information society. Provide personalized learning services for each student to maximize student growth. At both the international and domestic levels, the emphasis on big data resources is obvious. The combination of big data technology and education is also an inevitable development trend. The favorable conditions brought by big data to education have become a fact recognized by academia, and it is possible to use the power of data to promote the further development of personalized education. Big data itself has great value and advantages, and has promoted the transformation of personalized education from perception to touch. In a digital environment, students' performance is visible, and they no longer rely on guesswork and intuition, which is conducive to improving the personality of students.

3 Experiment

This article mainly analyzes the concept of generalized big data. Taking two graduates of management and other disciplines from a university as an example, the first session did not have management Education with Chinese Feature reform based on big data analysis, and the second session applied big data analysis to management Education with Chinese Feature reform. This article analyzes the relationship between the Education with Chinese Feature reform of management students and the quality of employment from relevant data such as the distribution of undergraduate professional disciplines, initial employment status, graduation destination, and employment regional mobility. Table 1 shows the undergraduate discipline distribution of the two fresh graduates.

Table 1. Distribution of undergraduate disciplines of two graduates

Subject	First graduate (%)	Second graduate (%)
Management	82.22	92.56
Economics	85.79	93.7
Jurisprudence	87.93	91.11
Pedagogy	82.05	85.64
Literature	85.76	90.48
Science	84.55	88.87
Engineering	91.47	94.56
Information Engineering	86.25	95.57
Civil Engineering	89.31	92.68
Automotive Engineering	84.96	88.96

Research on the employment quality of graduates still focuses on theoretical analysis. Employment quality evaluation has not yet formed a perfect system. Often, the result positioning ignores the process factors of employment quality formation, and the dimensions and data of employment quality evaluation are also subjective and objective. Therefore, it is important to construct a research with a more comprehensive graduate evaluation index system.

4 Discuss

It is not difficult to find from the annual employment quality reports of various colleges and universities that the first management university graduates have a turnover rate of about 45% after six months of employment, and a large proportion of voluntary turnover. Small personal development space, low income, poor unit development prospects, difficult colleagues or supervisors, poor working conditions, etc. have all become departure factors, forcing graduates to change companies, and to a certain extent, it has also affected job promotion. After six months of employment, the turnover rate of the second management university graduates is about 25%, which is lower than the first employment rate. Figure 1 shows a comparison of the satisfaction of Education with Chinese Feature education between two science graduates and the school.

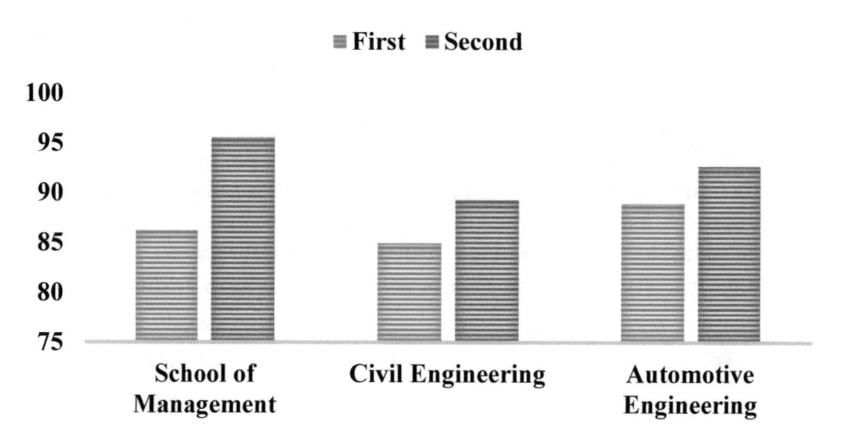

Fig. 1. Comparison of two graduates' satisfaction with school Education with Chinese Feature education

Table 2. Graduation destinations of the two fresh graduates

	First graduate (%)	Second graduate (%)	Increase situation (%)
Medical and health units	23.18	32.56	9.38
Other institutions	16.84	10.48	−6.36
State-owned enterprise	11.93	24.88	12.95
Non-state-owned enterprises	22.05	27.14	5.09
Other forms of employment	11.34	25.41	14.07
Going abroad	22.84	10.87	−11.97
Further education	15.47	27.61	12.14

From the data of the average annual growth rate of graduates in Table 2, it can be seen that the average annual growth of the employment flow of the two college students is mainly other forms of employment, medical and health units, continuing education and state-owned enterprises. The graduation situation of the two graduates is shown in Table 2.

In summary, the data comparison shows that in recent years, the number of graduates and the average annual growth rate of graduates from state-owned enterprises, medical and health units, and advanced education have all been ahead, and their performance in absorbing graduate employment and transferring employment has been strong. However, the number of graduates from higher education institutions and those going abroad and abroad is relatively small, and the average annual growth rate is also negative, indicating that they are weak in recruiting graduates.

The employment statistics of university graduates in management include the factors that affect the employment status of graduates as indicators and monitor them. They analyze the monitoring results of information and data to predict changes in the employment situation. Reflect the social and industry demand for talent. Most post-95 college graduates grew up under the careful care of their parents and elders. They are characterized by poor independence, lack of solidarity, cooperation and tolerance, and limited ability to withstand difficulties and setbacks. All of these make it difficult for internships to meet the needs of the job, and poor career stability and development ability. At the same time, private enterprises are the main channel for college graduates. From an operating cost perspective, these SMEs often require graduates to quickly compete for positions and "complete multiple tasks in one job." In addition to professional ability, comprehensive work ability is also required, which is also a good test for graduates.

5 Conclusion

In the era of big data, to do a good job of Education with Chinese Feature education for management students, we must always grasp the students' ideological situation. According to the training needs of management students, it is necessary to dig out some rich and diverse practical teaching resources. Effectively infiltrate some mainstream ideas and values into the actual course teaching process. And make full use of unstructured and semi-structured big data to fully stimulate students' learning motivation and initiative.

Acknowledgment. This work was supported by Higher Education Research Project from Jilin Provincial Association of Higher Education (Project Title: Education with Chinese Feature advancement, problems and countermeasures of professional curriculum in Jilin Province local colleges. Contract No. JGJX2019C48).

This work was supported by Vocational and Adult Education Teaching Reform Research Project from the Education Department of Jilin Province (Project Title: Research on the cultivation path of craftsman spirit of college students in private universities. Contract No. 2019ZCZ054).

References

1. Edwards Jr., D.B., Loucel, C.: The EDUCO program, impact evaluations, and the political economy of global education reform. Educ. Policy Anal. Arch. **24**(92), 1–50 (2016)
2. Daire, J., Kloster, M.O., Storeng, K.T.: Political priority for abortion law reform in Malawi: transnational and national influences. Health Hum. Rights **20**(1), 225–236 (2018)
3. Yoo, T.: A trust-based perspective on French political economy: law and corporate control in corporate governance reform. French Polit. **17**(4) (2019)
4. Scholtz, C.: The architectural metaphor and the decline of political conventions in the Supreme Court of Canada's Senate Reform Reference. Univ. Toronto Law J. **68**(4), 1–33 (2018)
5. Ugoani, J.N.N.: Political will for effective reform management and sustainable development goals achievement in Nigeria. Indep. J. Manage. Prod. **8**(3), 918 (2017)
6. Eyraud, C.: Quantification devices and political or social philosophy. Thoughts inspired by the French state accounting reform. Hist. Soc. Res./Historische Sozialforschung **41**(2), 178–195 (2016)
7. Scott, J., Holme, J.J.: The political economy of market-based educational policies: race and reform in urban school districts, 1915 to 2016. Rev. Res. Educ. **40**(1), 250–297 (2016)
8. Patil, N.A.: Reform of the Indian political system: the Gandhian alternative by A. M. Rajasekhariah K. Raghavendra Rao R. T. Jangam A. V. Kowdi. Nonlinear Dyn. **83**(4), 1919–1937 (2016)
9. Eren, K.A., Aşıcı, A.A.: Subjective well-being in an era of relentless growth: the case of Turkey between 2004 and 2014. J. Happiness Stud. **19**(2), 1–25 (2018)
10. Dyck, A.R.: Oratory and political career in the late roman republic by Henriette van der Blom (review). Class. World 110 (2017)

Test on the Combined Application of Nitrogen and Phosphate Fertilizer to Provide Intelligent Fertilization for Silage Maize Yield and Quality

Jiaxi Dong[1], Lie Zhang[2], Ruqing Cui[2], Qi Liu[1], Jin Du[1], Gaoyi Cao[1], and Xiuping Tian[1](✉)

[1] College of Agronomy and Resources and Environment, Tianjin Agricultural University, Tianjin 300384, China
15142505355@163.com, tian5918@sohu.com

[2] Tianjin Zhongtian Dadi Technology Co., Ltd., No. 2, Haitai Development Third Road, Huayuan Industrial Park, Xiqing District, Tianjin 300392, China

Abstract. In this test, 'Jinzhu 100' was used as the test material. The effects of different nitrogen and phosphorus fertilizers on maize yield index and quality index were studied through field experiments. Provide theoretical basis for Tianjin's big database of smart agriculture. Under N4P3 treatment, the number of maize rows, vane length and 100-grain weight were the highest. Under N3P2 treatment, maize grain yield was the highest. Under N3P3 treatment, the CP and starch content of maize grains were highest. Under N3P4 treatment, the starch content and CP of maize stems and leaves were the highest. Under N4P4 treatment, the ADF was highest in seeds and stem leaves. The content of NDF of seeds was the highest when treated with nitrogen fertilizer 300 kg/hm^2 and phosphorus (P_2O_5) 75 kg/hm^2. The content of NDF of stems leaves was highest under the treatment of nitrogen application of 300 kg/hm^2 and phosphorus application amount (P_2O_5) of 112.50 kg/hm^2. The rational application of different nitrogen and phosphorus fertilizers for different maizes is conducive to the improvement of its yield and quality.

Keywords: Combined application of nitrogen and phosphorus fertilizer · Maize yield · Forage quality · Smart agriculture · Maize ADF · Maize NDF

1 Introduction

Silage maize is an important roughage resource. However, our country's planting methods are mainly grain production methods [1–3]. In order to promote high yield and high quality of silage maize the targeted research on silage maize is necessary.

At present, excessive chemical fertilizers have caused serious problems and reduced fertilizer utilization [4]. Because of farmers' need to improve agricultural production, smart agriculture came into being [5–7]. Since the implementation of "agricultural water intelligence" and "metering management system", farmers' water costs in Beijing and Tianjin have been greatly saved. At the same time, in 2015, the Ministry of Agriculture included green feed maize into the national science and technology plan, and it was important to promote silage maize.

J. MacIntyre et al. (Eds.): SPIoT 2020, AISC 1282, pp. 301–307, 2021.
https://doi.org/10.1007/978-3-030-62743-0_43

In this experiment, the effects of nitrogen and phosphorus levels on yield and quality indicators were studied, and the application of nitrogen and phosphorus fertilizers to the yield and quality of silage maize was explored in order to improve the yield and quality of Tianjin silage maize and increase the fertilizer utilization rate. It provided a theoretical basis for the scientific fertilization of Tianjin silage corn and the extension of Tianjin smart agriculture.

2 Materials and Methods

2.1 Test Materials

The test was set in a fine seed farm in Jinghai District, Tianjin. The test soil is fluvo-aquic soil, and the organic matter of the cultivated layer soil is 32.26 g/kg; total nitrogen 2.15 g/kg; 1.08 g/kg; available phosphorus (P) 45.47 mg/kg; total potassium 1.03 g/kg; available potassium (K) 119.73 mg/kg; pH (H_2O) 8.2. The test crop used in this study was 'Jinzhu 100'. The test fertilizer was urea (total N $\geq$ 46%) and phosphate superphosphate ($P_2O_5 \geq 12\%$). Potassium fertilizer is potassium chloride ($K_2O \geq 50\%$).

2.2 Experimental Design

Adopt field test and random block design. The plot area is 18 m^2, each plot has 6 rows, the row length is 5 m, the row spacing is 60 cm, the plant spacing is 28 cm. 25 treatments have been set up. The experimental treatment and the amount of fertilizer applied are shown in Table 1. Each treatment was repeated 3 times. Potassium fertilizer was used as a basic fertilizer at a rate of 140.00 kg/hm^2 potassium chloride (K_2O 70.00 kg/hm^2) to each plot, and was used as the base fertilizer at one time. Nitrogen was applied twice, with 40% for base fertilizer and 60% for topdressing fertilizer. The trial was sown on May 4, 2018. The field management during the growing period is conventional field management.

Table 1. Different nitrogen and phosphorus fertilizer application rates

Trea.	Amount of N (kg/hm^2)	Amount of urea (kg/hm^2)	Amount of phosphate (kg/hm^2)	Amount of superphosphate (kg/hm^2)
N1P1	0.0	0.0	0.0	0.0
N1P2	0.0	0.0	37.5	314.5
N1P3	0.0	0.0	75.0	625.0
N1P4	0.0	0.0	112.5	937.5
N1P5	0.0	0.0	150.0	1250.0
N2P1	80.0	172.4	0.0	0.0
N2P2	80.0	172.4	37.5	314.5
N2P3	80.0	172.4	75.0	625.0
N2P4	80.0	172.4	112.5	937.5

(*continued*)

Table 1. (*continued*)

Trea.	Amount of N (kg/hm^2)	Amount of urea (kg/hm^2)	Amount of phosphate (kg/hm^2)	Amount of superphosphate (kg/hm^2)
N2P5	80.0	172.4	150.0	1250.0
N3P1	160.0	344.8	0.0	0.0
N3P2	160.0	344.8	37.5	314.5
N3P3	160.0	344.8	75.0	625.0
N3P4	160.0	344.8	112.5	937.5
N3P5	160.0	344.8	150.0	1250.0
N4P1	240.0	517.2	0.0	0.0
N4P2	240.0	517.2	37.5	314.5
N4P3	240.0	517.2	75.0	625.0
N4P4	240.0	517.2	112.5	937.5
N4P5	240.0	517.2	150.0	1250.0
N5P1	320.0	689.6	0.0	0.0
N5P2	320.0	689.6	37.5	314.5
N5P3	320.0	689.6	75.0	625.0
N5P4	320.0	689.6	112.5	937.5
N5P5	320.0	689.6	150.0	1250.0

2.3 Sampling and Measurement Methods

In the wax ripening period, 10 plants were randomly selected in the middle of the third row of each plot to determine each plant's fresh weight, and the stubble 10.00 cm to calculate the fresh material yield. The whole plant was mixed and pulverized, and the quality index was measured. In the early stage of maturity, 10 plants were randomly selected in the middle of the fourth row of each plot to determine the weight of the vanes. After being air-dried, they were weighed, and the yield traits were measured indoors. The quality indicators were determined after the grains were crushed. The two samples of the above-mentioned kernels and all plants were sieved respectively through 10-mesh and 40-mesh sieves, for the determination of crude protein, starch, neutral detergent fiber (NDF) and acid detergent fiber (ADF). Crude protein: Dumas combustion method. ADF: acid detergent method; NDF: neutral detergent method; starch content: acid hydrolysis method.

2.4 Data Processing

Data processing and analysis were performed using Excel and SPSS software.

3 Results and Analysis

3.1 Effects of Combined Application of Nitrogen and Phosphorus on Yield Characteristics of 'Jinzhu 100' Maize

From Table 2 we can see that, in all treatments, the top 3 maize vanes are N1P2 > N5P3 > N2P3, there is no significant difference among the three treatments. The top 3 of the number of kernels in the maize lined are N4P4 > N4P3 > N5P3, and the number of grains was significantly higher than other treatments. The top 3 maize vane length are N3P2 > N4P3 > N5P5, although there are no significant difference among the three treatments, the vane length of maize in these 3 treatments was significantly higher than that of others. The top 3 bald tip length of maize were N3P1 > N2P1 > N1P3. Among them, N3P1 is significantly higher than others, and there is no significant difference between N2P1 and N1P3.The top 3100-grain weight of maize are N4P3 > N3P2 > N5P3. There are no significant difference between the three treatments, but N4P3 is significantly higher than other treatments. The top 3 yield are N3P2 > N4P3 > N5P3, the three treatments are significantly higher than others. There are also significant differences between the three treatments.

In summary, the combined application of nitrogen and phosphorus had no effect on the vane row number of maize. As phosphorus can affect the synthesis of plant polysaccharides, without phosphate fertilizer can cause balding and decrease yield.

Table 2. Effect of nitrogen and phosphorus interaction on yield traits of 'Jinzhu 100'

Trea.	Number of rows (line)	Number of grains	Spike length (cm)	Bald length (cm)	100-grain weight (g)	Yield (kg/hm[2])
N1P1	14.4a	37.74d	21.55c	1.65cd	25cd	9500i
N1P2	14.8a	37.63d	22.65bc	1.69c	25.1cd	9700.8fg
N1P3	14.53a	38.73cd	21.93c	1.77bc	25.21cd	9877.36g
N1P4	14.4a	39.64bcd	21.88c	1.6de	26bc	9720.55fg
N1P5	14.33a	38.65cd	21.8c	1.4hi	25.4cd	9600.35hi
N2P1	14.42a	38.12d	21.68c	1.8b	26.8bc	9780.5fg
N2P2	14.43a	42.61ab	22.6bc	1.7c	27bc	9928.85ef
N2P3	14.67a	42.14bc	22.65bc	1.67c	28.12ab	10006.4ef
N2P4	14.5a	40.24cd	23.95a	1.66bcd	28ab	9985.5ef
N2P5	14.6a	39.84bd	22.37bc	1.68bcd	26.99bc	9980.45ef
N3P1	14.44a	38.15d	21.93c	2.12a	24.28cd	9877.36g
N3P2	14.54a	43.5a	24.65a	1.25j	29.12ab	11206.4a
N3P3	14.33a	40.99c	23.09bc	1.66cd	28.59ab	10174.28d
N3P4	14.62a	40.27cd	24.19a	1.42ghi	27.2bc	10044.28de
N3P5	14.58a	40.26cd	23.26abc	1.38hi	26.74bc	10025.42def
N4P1	14.46a	41.71bc	22.97bc	1.4hi	26.24bc	9981.65ef
N4P2	14.38a	42.49ab	24a	1.45gh	27.17bc	10012.5ef

(continued)

Table 2. (*continued*)

Trea.	Number of rows (line)	Number of grains	Spike length (cm)	Bald length (cm)	100-grain weight (g)	Yield (kg/hm$^{2)}$
N4P3	14.63a	43.97a	24.24a	1.45gh	30.01a	11044.28b
N4P4	14.5a	44.11a	23.75ab	1.5fg	27.34bc	10024.56ef
N4P5	14.3a	42.25b	21.5c	1.4hi	28ab	9995.82ef
N5P1	14.38a	42.26b	22.5bc	1.41hi	26.27bc	9562.88h
N5P2	14.52a	42.79ab	23.1bc	1.5fg	27.22bc	9865.45fg
N5P3	14.67a	43.77a	23.26bc	1.4hi	28.74abc	10325.42c
N5P4	14.4a	43.29a	23.19bc	1.35i	27.26bc	9893.55fg
N5P5	14.38a	43.09ab	24.2a	1.4hi	25.73cd	9658.47h

3.2 Effect of Combined Application of Nitrogen and Phosphorus on the Quality of 'Jinzhu 100' Maize Forage

Table 3 shows that, in all treatments, the top 3 CP of seeds are N3P3 > N2P3 > N3P2, the top 3 are significantly higher than others, but there are no significant difference among the top 3. The top 3 CP of stem leaves are N3P4 > N4P4 > N4P3. There are no significant difference among the top 3. N1P1, N5P1, and N5P5 are significantly lower than the top 3 treatments. The top 3 starch of seeds are N3P3 > N3P4 > N3P2. Among them, N3P3 is significantly higher than others, and there is no significant difference among the other treatments. The top 3 starch of stem leaves are N3P4 > N3P5 > N5P4. Although these 3 treatments are significantly higher than others, there is no significant difference among the 3 treatments. The top 3 ADF of seeds are N4P4 > N4P5 > N5P4,the three treatments are significantly higher than others, but there is no significant difference among the 3 treatments. The top 3 ADF of stem leaves are N4P4 > N4P3 > N5P4. The performance is similar to that of seeds. The top 3 NDF of seeds are N5P3 > N4P3 > N3P3, they are higher than most other treatments, but there is no significant difference among the three treatments. The top 3 NDF of stem leaves are N5P4 > N5P5 > N5P3, the performance is similar to that in the seed.

Table 3. Effect of nitrogen and phosphorus interaction on quality of 'Jinzhu 100'

Trea.	CP of seed (%)	CP of stem leaf (%)	Starch of seed (g/100 g)	Starch of stem leaf (g/100 g)	ADF of seed (%)	ADF of stem leaf (%)	NDF of seed (%)	NDF of stem leaf (%)
N1P1	8.125b	5.71b	57.3de	27.7c	1.8c	18c	20.9c	39.2c
N1P2	8.146b	6.307a	58cde	28.1bc	2.2b	18.2c	22.8b	42.2b
N1P3	8.352b	6.637a	63bcd	28.8bc	2.3ab	21ab	23.8a	42.5a
N1P4	8.246b	7.201a	59cde	29.3b	2.5ab	21.3ab	22.5b	42.9b
N1P5	8.188b	5.841b	58.4cde	29bc	2.4ab	20.7ab	22.6b	42.8b
N2P1	8.354b	5.833b	63.3bcd	28.9bc	2.1bc	18.3b	21.3bc	38.8c
N2P2	8.65ab	6.43a	64bc	29.3b	2.5ab	18.5b	23.2bc	41.8bc

(*continued*)

Table 3. (*continued*)

Trea.	CP of seed (%)	CP of stem leaf (%)	Starch of seed (g/100 g)	Starch of stem leaf (g/100 g)	ADF of seed (%)	ADF of stem leaf (%)	NDF of seed (%)	NDF of stem leaf (%)
N2P3	9.077a	6.761a	69b	30ab	2.7a	21.3ab	24.2a	42.1bc
N2P4	8.887ab	7.324a	65bc	30.5a	2.8a	21.6ab	22.9b	42.5b
N2P5	8.453b	5.964ab	64.4bc	30.2ab	2.7a	21b	23ab	42.4b
N3P1	8.38b	7.178a	69.3b	29.4ab	2.2b	19.1bc	22.3bc	41.5bc
N3P2	8.961a	7.775a	70b	29.8ab	2.6ab	19.3bc	24.2a	44.5a
N3P3	9.186a	8.105a	75a	30.5a	2.7a	22.1a	25.2a	44.8a
N3P4	8.373b	8.669a	71b	31a	2.9a	22.4a	23.9a	45.2a
N4P1	8.153b	7.051a	62.3bcd	28.7bc	2.4ab	21.2ab	22.5bc	41.6b
N4P2	8.166b	7.648a	63bcd	29.1bc	2.8a	21.4ab	24.4a	44.6a
N4P3	8.206b	7.978a	68bc	29.8ab	2.9a	24.2a	25.4a	44.9a
N4P5	8.168b	7.182a	63.4bcd	30ab	3a	23.9a	24.2a	45.2a
N5P1	8.147b	5.092b	58.3cd	29bc	2.4ab	20.8ab	22.8bc	42.1b
N5P2	8.221b	5.689b	59cd	29.4ab	2.7a	21ab	24.7a	45.1a
N5P3	8.334b	6.019a	64bc	30.1ab	2.9a	23.8a	25.7a	45.5a
N5P4	8.254b	6.583a	60cde	30.6a	3a	24.1a	24.4a	45.8a
N5P5	8.221b	5.223b	59.4cde	30.3a	2.8a	23.5ab	24.5a	45.7a

4 Conclusion and Discussion

From the above results, it can be seen that the combined application of nitrogen and phosphorus has no effect on the maize's vane rows. The highest maize row, vane length, and 100-grain weight are all under N4P3, indicating that a higher amount of nitrogen fertilizer combined with a certain amount of phosphorus fertilizer is beneficial to increase the number of kernels, vane length and 100-grain weight of maize. The highest maize grains appeared in N3P2, which indicated that excessive nitrogen fertilizer was not conducive to the accumulation of nutrients in the grains. Among 25 treatments, the combined application of nitrogen and phosphorus had a significant effect on the quality of maize. N3P2 was beneficial to maize's CP of seeds. Increasing the CP of maize, N3P4 is beneficial to increase the CP of stem leaves, and N3P3 treatment is beneficial to increase the starch of maize seeds, and N2P4 treatment was beneficial the increase of starch of maize stem leaves. Appropriate increase of nitrogen fertilizer and phosphorus fertilizer are helpful for protein accumulation. Grain protein accumulation needs to reduce the amount of phosphate fertilizer application; N4P4 increased the ADF of stem leaves. Combined application of high nitrogen and high phosphorus like N4P3 or N5P4,NDF of seeds and stem leaves accumulation, therefore, to reduce the nitrogen and phosphorus fertilizer can reduce the increase in ADF and NDF, helps to improve the quality of silage maize.

In summary, under the conditions of soil fertility in TianJin, under the condition of potassium fertilizer application, in the case of weighing maize yield and crude protein, starch and medium acid cellulose, N3P3 is recommended as the best level. The rational application of different nitrogen and phosphorus fertilizers for different maizes is conducive to the improvement of its yield and quality.

Acknowledgements. Tianjin Seed Industry Science and Technology Major Project (16ZXZ YNC00150).

References

1. Zhang, X., Mu, H., Hou, X., Yan, W., Li, P., Li, P., Su, J.: Research progress on silage corn planting and its yield and quality in China. Anim. Husbandry Feed Sci. **34**(1), 54–57+59 (2013). (in Chinese)
2. Maasdorp, B.V., Titterton, M.: Nutritional improvement of maize silage for dairying: mixed-crop silages from sole and intercropped legumes and a long-season variety of maize. 1. Biomass yield and nutritive value. Anim. Feed Sci. Technol. **69**(1) (1997)
3. Yuan, X., Guo, Q., Wang, C., Xu, Y., Wei, Y.: Effect of nitrogen, phosphorus and potassium formula fertilizer on growth and chemical composition content of Inula japonica. China J. Chin. Materia Medica **44**(15) (2019)
4. Fosu-Mensah, B.Y., Mensah, M.: The effect of phosphorus and nitrogen fertilizers on grain yield, nutrient uptake and use efficiency of two maize (Zea mays L.) varieties under rain fed condition on Haplic Lixisol in the forest Savannah transition zone of Ghana. Environ. Syst. Res. **5**(22), 1–17 (2016)
5. Agrochemicals - Fertilizers; Researchers' Work from Wageningen University Focuses on Fertilizers (Exploring nutrient management options to increase nitrogen and phosphorus use efficiencies in food production of China). Food Weekly News (2018)
6. Tong, Q., Swallow, B., Zhang, L., Zhang, J.: The roles of risk aversion and climate-smart agriculture in climate risk management: evidence from rice production in the Jianghan Plain China. Clim. Risk Manage. **26** (2019)
7. Çiftçioğlu, G.A., Kadırgan, F., Kadırgan, M.A.N., Kaynak, G.: Smart agriculture through using cost-effective and high-efficiency solar drying. Heliyon **6**(2) (2020)

Development and Innovation of Education Management Information in College with Big Data

Jie Zhao(✉)

Jinjiang College, Sichuan University, Meishan, Sichuan, China
Lubianinfo@foxmail.com

Abstract. Mining of relevant content or systems is not enough, and they have not been able to properly integrate and manage them, making relevant colleges cannot better use data information content. Therefore, this article first discusses the current information development with big data. Problems of low standardization of teaching and scientific research management, poor application technology awareness, and independent management data of various departments exist. Secondly, in response to related issues, college education has proposed innovative application approaches to information management, such as building the standardized network information platform, correctly applying big data technical teams to get high teaching quality in colleges.

Keywords: Innovation · Development · Big data

1 Introduction

With the acceleration of multimedia and communication technology, information is important indicator of the international competitiveness and modernization level of a country and region, and an important indicator of the overall national strength and economic growth level [1, 2]. Traditional teaching, the campus, classroom, library and other education and teaching environment has quietly taken place-amazing changes, education and teaching methods have changed a lot [3].

"Digital campus" is the physical manifestation of education Information. Under the guidance of theory, computer, network and multimedia information technology is integrated with education and teaching to form a new education system [4]. "Digital campus" is a complex information system. Its construction can be divided into four sections: reconstruction of education concepts, construction of network hardware, and construction of teaching systems as the core application system. Building digital campus makes the original single information mode of education management information face the integration challenge in the "campus all-in-one card" [5]. Now, the management information of education faces many new challenges. Admittedly, many colleges have made some achievements by the management information of education. However, due to technical limitations and personnel quality constraint [6].

In addition, colleges has become a problem that relevant personnel need to focus on. Based on this, this article first discusses the current information of college

J. MacIntyre et al. (Eds.): SPIoT 2020, AISC 1282, pp. 308–313, 2021.
https://doi.org/10.1007/978-3-030-62743-0_44

education with big data. Problems of low standardization of teaching and scientific research management, poor application technology awareness, and independent management data of various departments exist. Secondly, in response to related issues, college education has proposed innovative application approaches to the management information of education, such as building standardized network information platform to get high teaching quality in colleges.

2 Innovative Development of Education Management Information with Big Data

Today, results of information construction in colleges and colleges are increasingly abundant [7]. With big data, the management information of education in colleges has become more important. The management information of education in colleges requires management staff to overcome technical difficulties by using big data. At present, the following problems still exist in reform of the management information of education [8].

(1) Lack of awareness of big data application. In the practical work of higher the management information of education, there are many problems, such as the data model is not unified, the technical ability is insufficient, the use method demonstration is insufficient, the big data application research is insufficient, the fund is limited and so on. These problems affect the realization of big data application goal.
(2) Educational management data are not standardized enough. Data can promote the establishment of database only if it has unity and standardization. The work of student education and management is complicated. Colleges and universities need to rely on a variety of information application systems to do better in the management information of education.
(3) The data of each department of the college is independent. Different departments and different units apply diversified information processing systems, making the management information of education of the school flawed. Information independence has affected the establishment of a unified college database.
(4) The integration of management information of students in various majors is difficult. It is difficult for college education administrators to obtain socio-economic data.
(5) Applying big data in universities is hard. Because structure in student management data in universities is much more complicated than that of other education departments.

3 The Construction of Teaching Management Information Environment and Strategies for Innovation and Development with Big Data

3.1 A Standardized Network Information Platform

Colleges can create mobile platforms for various data and use BIG DATA technology reasonably. This is an effective way to do better of the management information of education in colleges.

Big data has three characteristics: Volume, Velocity, and Variety. Current university digital campus popularization, all kinds of the management information of education platform in all aspects of school management to application. The majority of teachers and students daily use teaching service system, including the construction of teachers' professional platform, the construction of the talent training scheme, as well as the common use of network MOOC platform. By many teachers and students handsets, PC collected in different types, different meanings of structured, semi-structured, and unstructured data. Using business system with the large-scale teacher-student user group constitutes the basis of big data; we call it big data center. With data, such massive data need to be processed on a large scale. The Velocity characteristic of big data is reflected here. Data analysis, retrieval and mining are completed through the cloud-computing platform [9]. The collected data must have a lot of junk data, or the data that is not useful to us. First, we must clean it, dig out the correlation between the data, or make certain statistics on the data to obtain a certain knowledge. The analyzed data needs to be searched by the search engine for the analysis result that the user wants. The architecture of the teaching the management information of education platform is shown in Fig. 1.

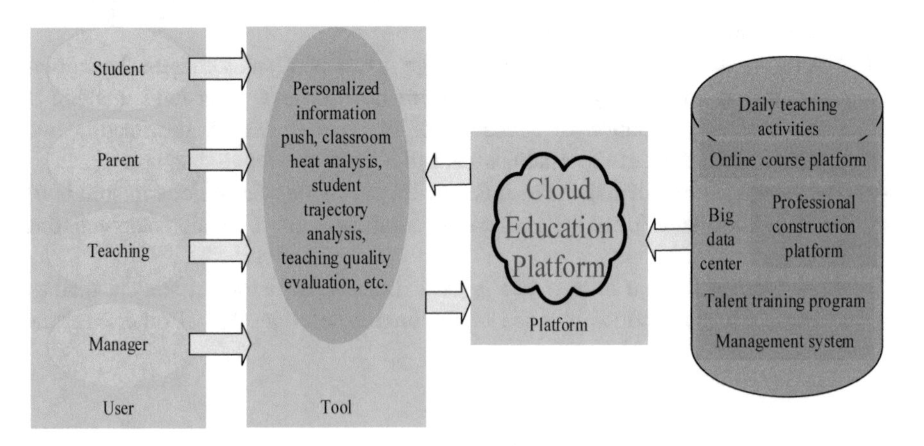

Fig. 1. Architecture of a teaching the management information of education platform by using big data

Using big data platform can help colleges and colleges to perform related tasks such as Information data storage, management, reception, research, cleaning, and calculation [10]. In addition, incomplete data or independent data should not be regarded as the auxiliary work of information management platform. Instead, universities should try their best to create big data processing platform and communication platform. In this way, college education can be applied reasonably. As shown in Fig. 2, teachers and students can make interactive through remote teaching.

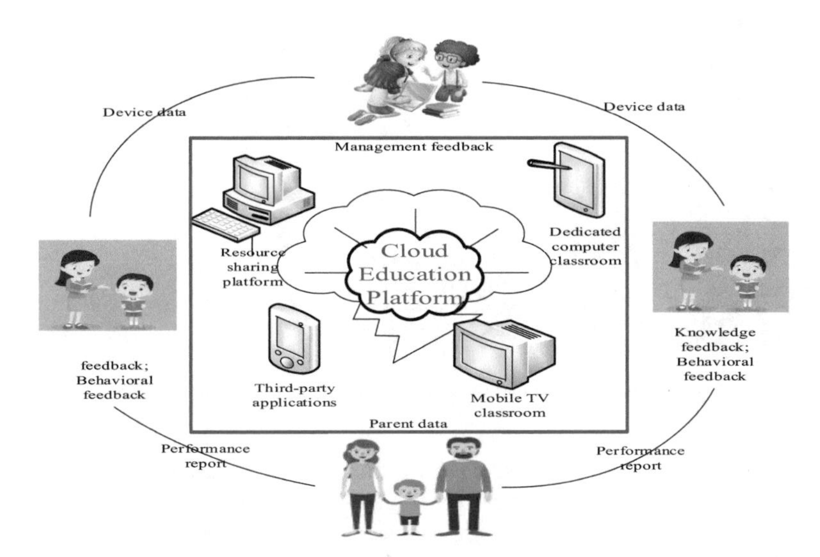

Fig. 2. Teacher-student teaching interaction under Information technology

3.2 Big Data Decision Support Services in Teaching Management

Big data provides powerful decision support in many areas such as teaching quality evaluation, teaching ability analysis, personalized curriculum analysis, learning behavior analysis, work-study demand forecasting, and student public opinion analysis. It is a big data decision support system, as shown in the Fig. 3 shown. Use cloud services to promote high-quality education resources, promote the popularization and normalization of teaching resources, integrate cloud services, digital teaching materials, and subject teaching tools to build the management information of education. Big data decision support system can be effectively explored teaching modes to improve quality, playing an irreplaceable role in promoting teaching reform [11].

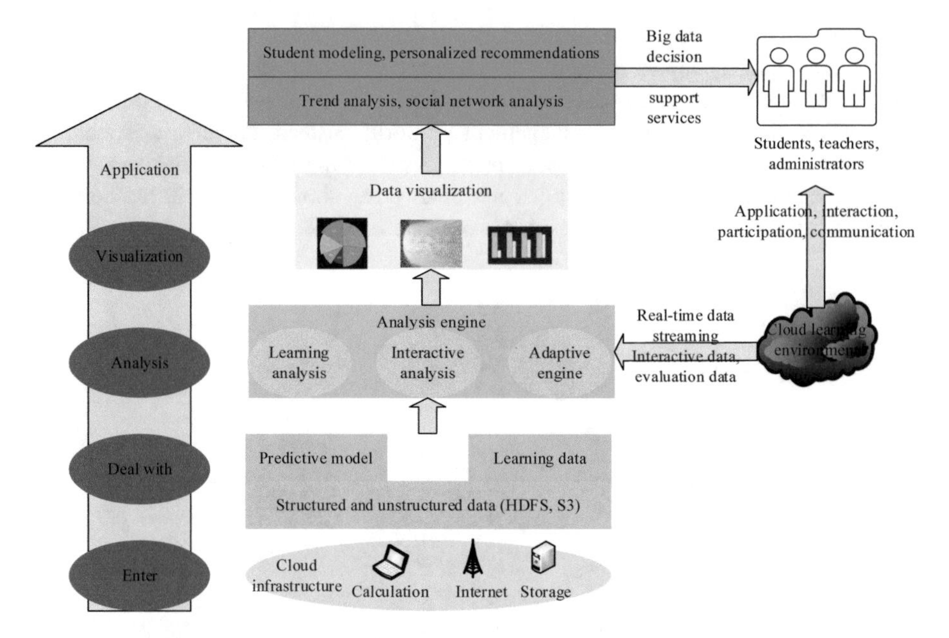

Fig. 3. Big data decision support system in teaching management

Big data based the management information of education environment provides decision support. Its service targets include college students, teachers, teaching managers, and student parents. Through big data teaching quality evaluation, teachers have opportunity to reflect on their own teaching effects, to establish an equal relationship in real classroom teaching. During the whole learning process of the teaching class, all learning behaviors of all learning objects are recorded automatically. The analysis of these records will form an evaluation, which will help students to reflect on their classroom performance, and can give students an academic warning, to warn students. Through the big data analysis system, teaching managers push personalized information to students, teachers and parents in real time. Colleges should create correct concept by applying big data technology for each organizer on campus.

4 Conclusion

When conducting big data management, colleges should reasonably use big data and encourage students to obey the management of schools. At the same time, many forms of data application systems are indispensable. This article first discusses the current information of college education with big data. Problems of low standardization of teaching and scientific research management, poor application technology awareness, and independent management data of various departments exist. Secondly, in response to related issues, college education has proposed innovative application approaches to the management information of education, such as building standardized network information platform to get high teaching quality in colleges.

References

1. Roetzel, P.G.: Information overload in the information age: a review of the literature from business administration, business psychology, and related disciplines with a bibliometric approach and framework development. Bus. Res. **12**(2), 479–522 (2019)
2. Smalley, T.N.: Information literacy in the information age: what I learned on sabbatical. Coll. Res. Libr. News **62**(7), 689–704 (2019)
3. Borba, G.S., Alves, I.M., Campagnolo, P.D.B.: How learning spaces can collaborate with student engagement and enhance student-faculty interaction in higher education. Innov. High. Educ. **45**(1), 51–63 (2020)
4. Taylor, J.L., Dockendorff, K.J., Inselman, K.: Decoding the digital campus climate for prospective LGBTQ+ community colleges students. Commun. Coll. J. Res. Pract. **42**(3), 155–170 (2018)
5. Chuan, S.: Design of digital campus office automation system based on Portlet and workflow. Int. J. Smart Home **10**(6), 321–328 (2016)
6. Aiqun, Z., An, I.T.: Capability approach to informatization construction of higer education institutions. Procedia Comput. Sci. **131**, 683–690 (2018)
7. Zhu, H., Lan, Y.: A summary of research on informatization of specialized language in the framework of Sino-Foreign cooperative education. Theory Pract. Lang. Stud. **6**(9), 1863–1868 (2016)
8. Wang, S., Zhang, T.: Research on innovation path of school ideological and political work based on large data. Cluster Comput. **22**(2), 3375–3383 (2019)
9. Li, Z., Zhu, S., Hong, H., et al.: City digital pulse: a cloud based heterogeneous data analysis platform. Multimedia Tools Appl. **76**(8), 10893–10916 (2017)
10. Senthilkumar, S.A., Rai, B.K., Meshram, A.A., et al.: Big data in healthcare management: a review of literature. Am. J. Theoret. Appl. Bus. **4**(2), 57–69 (2018)
11. Tan, X.: On reform of college English teaching based on the FiF smart learning platform. J, Lang. Teach. Res. **10**(5), 1067–1072 (2019)

Data Visualization Association Algorithm Based on Spatio-Temporal Association Big Data

Zhang Zhang[1(✉)] and Liang Tian[2]

[1] Baoji Vocational and Technical College, Gaoxin Avenue No. 239, Baoji National High-Tech Industrial Development Zone, Baoji, China
liangmeizhang@163.com
[2] Shaanxi Fenghuo Electronic Co., Ltd., Qingjiang Road No. 72, Weibin District, Baoji, Shaanxi, China

Abstract. This paper mainly studies and realizes the spatio-temporal multidimensional visualization under the big data environment. By analyzing the spatiotemporal correlation of wireless sensor networks under the random large-scale dense deployment of sensor nodes, ensure that there is no monitoring blind spot in the area, the accuracy of data sampling and the reliability of communication. By studying the Markov chain's spatial correlation data prediction algorithm, it solves the problem that nodes need to store a large amount of historical data during prediction, and the requirements for node storage space are relatively high; At the same time, the spatial correlation of the data in the network is analyzed for data prediction, which solves the problem of low prediction accuracy using the time-based correlation prediction method in the case of irregular data fluctuations.

Keywords: Big data · Spatio-temporal association · Visualization association algorithm

1 Introduction

At present, we are already in the era of big data, big data contains rich value need to use big data analysis method to obtain. Data visualization is an important means of data analysis, which can intuitively express the pattern and law of abstract data [1].

Based on the analysis of the persistence point of multi-dimensional time series data visualization in big data environment, a visualization scheme of multi-dimensional time series data based on clustering is proposed, and the visualization design and implementation of on-line monitoring data are carried out as an example. In order to solve the problem of line overlapping coverage in big data environment, the parallel coordinate clustering method is used to optimize the problem effectively.

For multi-dimensional point data, this paper uses the method of map and parallel coordinate graph association cooperation to display multiple attribute dimensions, and explores the relationship between different attribute dimensions by sorting the coordinate axis of parallel coordinates.

J. MacIntyre et al. (Eds.): SPIoT 2020, AISC 1282, pp. 314–319, 2021.
https://doi.org/10.1007/978-3-030-62743-0_45

2 Visual Framework Based on Big Data Platform

Distributed database refers to the use of high-speed computer networks to connect physically dispersed multiple data storage units to form a logically unified database . The two important characteristics of multidimensional time series data are timing and multi-dimension. Because multi-dimensional time-series is both time series data and multi-dimensional data, the visualization of multi-dimensional time series data needs to consider both the multi-dimensional properties of the data itself and the time series of the data set.

The visualization methods of multidimensional data can be divided into three categories: spatial mapping method, icon method and pixel-based visualization method. In this paper, the visualization of multidimensional data is realized based on parallel coordinate method [2].

Clustering refers to the operation of dividing a data set into multiple subsets with some similarity. The process of clustering actually completes the abstraction of the data, thus allowing analysis and visualization of the clustered collections, and directly processing large-scale data sets on the basis of clustering. The core of the clustering operation is to define the appropriate distance or similarity measure, which is related to the specific application and data set. During the clustering process, users usually need to adjust the parameters and verify the results to achieve the best results.

3 Design of Temperature and Humidity Auto-control System

In the WSNs, the sensor nodes need to be deployed in a random and large scale in order to ensure there is no blind spot in the area, the accuracy of the data sampling and the reliability of the communication. According to the famous “first law of geography” of Tobler: all the geography is related, and the correlation and the geography are inversely proportional to the spatial distance.

In WSNs, sensor nodes collect data in adjacent geographic areas. The physical parameters have continuity and similarity, so the data of adjacent nodes have spatial correlation [3, 4]. The data space correlation representation function is:

$$\rho = \begin{cases} 1 - \mathrm{d}/r_{corr} & d \leq r_{corr} \\ 0 \end{cases} \tag{1}$$

ρ represents the spatial correlation of the data between the sensor nodes, The correlation radius between sensor nodes is represented by r_{corr}.

Because the monitoring data has spatial correlation, the redundant data generated by the nodes is transmitted in the network, causing communication interference between a large number of nodes or even the nodes themselves, wasting node energy, and increasing communication delay.

In order to ensure the real-time monitoring of the environment by WSNs, it is necessary to sample the physical parameters of the region frequently. Because of the continuity and continuity of the physical phenomena in time, the data collected by the same node at the time point before and after the same node are similar, and the collected time series data have a certain degree of functional relationship. The time correlation between each sensor node depends on the sampling frequency of the node and the physical phenomena observed. The essence of the time correlation of data in wireless sensor networks is the similarity of perceived data when the sampling time interval is t.

4 Design of Temperature and Humidity Auto-control System

Most of the current data prediction algorithms only consider the time correlation of the data. In the prediction, the nodes need to store a large number of historical data, and the storage space of the nodes is required to be high. Moreover, this kind of algorithm has low prediction accuracy when the data fluctuates greatly, and is only suitable for predicting data with high time correlation. In practical application, WSNs network has the characteristics of large-scale and dense deployment of nodes, which often makes the data or data collected by adjacent nodes similar or similar, that is to say, the spatial correlation is large. Thus it can be seen that the irregular situation of data fluctuation can be solved by analyzing the spatial correlation of data in the network for data prediction. In this case, the prediction accuracy of the prediction method based on time correlation is low [5]. In this section, based on the topology of clustering network based on WSNs, the spatial correlation data prediction algorithm based on Markov chain is studied.

4.1 Workshop Air Conditioning System Model [6]

In a clustered network, generally only one cluster head and other nodes in each cluster are ordinary member nodes. The cluster head is responsible for data collection of the entire cluster, and only the cluster head can transmit data with the base station in one cluster.

The nodes in wireless sensor networks are generally divided into two categories [6]: cluster head member nodes. Different clustering protocols select cluster heads in different ways. Clustering networks have the following advantages:

(1) The redundant nodes in the cluster can sleep, which can not only ensure the full coverage of the monitoring range, but also save the energy consumption of the network, and can prolong the life cycle of the monitoring network.
(2) The cluster head can preprocess the data after collecting the monitoring data of all nodes in the cluster to remove invalid data. Reduces the amount of data transferred in the network
(3) The cluster head is responsible for communicating with the base station to avoid the communication between the member nodes and the base station. The member nodes do not need to save a large number of routing information, which reduces the storage and computing burden of the nodes.
(4) In a clustered network, the death of some nodes does not affect the overall network topology and enhances robustness. At the same time, the network has good scalability and is suitable for large-scale wireless sensor networks.

4.2 Spatial Correlation Data Prediction Algorithm Based on Markov Chain

The spatial correlation data prediction algorithm based on Markov chain is mainly suitable for clustering networks. After clustering, the adjacent graph of the Delaunay triangle of the inner cluster is constructed according to the position of the member node [7].

Delaunay triangle has the characteristics of empty circles and minimum angles. The network neighborhood graph generated using this graph has more uniform and reasonable distribution of neighbor nodes. Therefore, it is more accurate to calculate the spatial correlation between nodes based on the neighborhood graph.

The adjacent graph of Delaunay triangle is the dual graph of Voronoi graph. The nodes in the network are divided into monitoring areas according to the nearest neighbor principle, and the continuous polygons connected with the vertical bisection lines of the nodes in which the two adjacent monitoring areas are located are Voronoi graphs. Let N be the finite node set in region R^2, the Euclidean distance from any location (s1, s2) to node (n1, n2) in the node monitoring area is:

$$||n - s|| = [(n_1 - s_1)^2 + (n_1 - s_2)^2]^{\frac{1}{2}}, n \in N \tag{2}$$

The node monitoring area partition graph generated by the above method is the Voronoi graph.

4.3 Verification and Analysis of Algorithm Simulation

In order to verify the prediction effect of the algorithm, 12 wireless sensor nodes are randomly deployed, and the prediction effect of the algorithm is verified by the monitoring data of these nodes. The monitoring parameter is the relative humidity of the air, and 288 monitoring data are extracted every five minutes as a monitoring cycle [8]. Twelve sensor nodes are grouped according to the relative position of the deployment. The initial Delaunay triangle adjacent graph in the cluster is shown in Fig. 1.

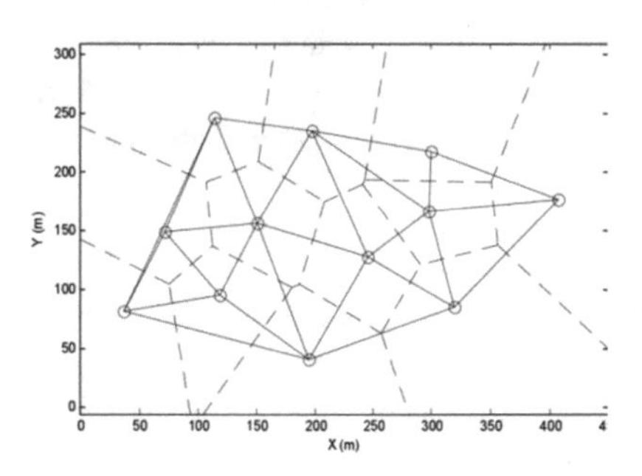

Fig. 1. Delaunay triangle of nodes in cluster figure

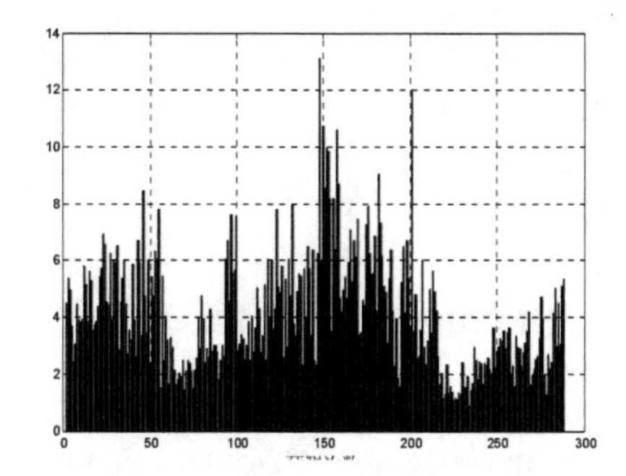

Fig. 2. Average error of all nodes in each round of prediction

The dot in the graph is the node, the dotted line is the node-centered Voronoi graph, and the real line is the adjacent graph of the Delaunay triangle of the node in the cluster.

Figure 2 shows the average prediction error of 287 rounds of data prediction using spatial correlation data prediction algorithm. The average prediction error is less than 6%, accounting for 67% of the total, and the prediction error is less than 5%, accounting for 56% of the total.

In this paper, the prediction effect of spatial correlation data prediction algorithm is compared with that of grey prediction algorithm in literature.

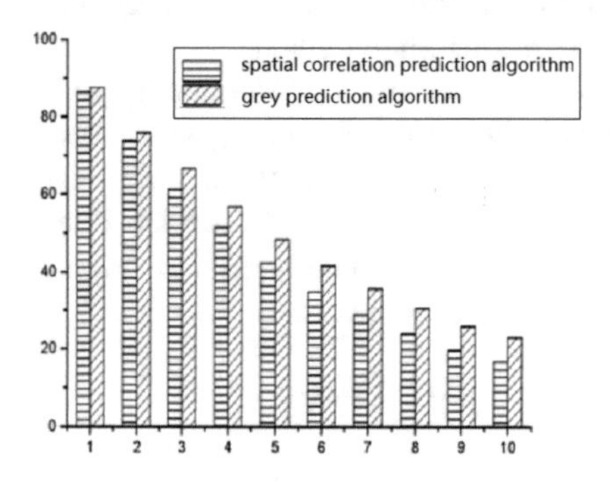

Fig. 3. Comparison of data upload between the two algorithms under different error threshold

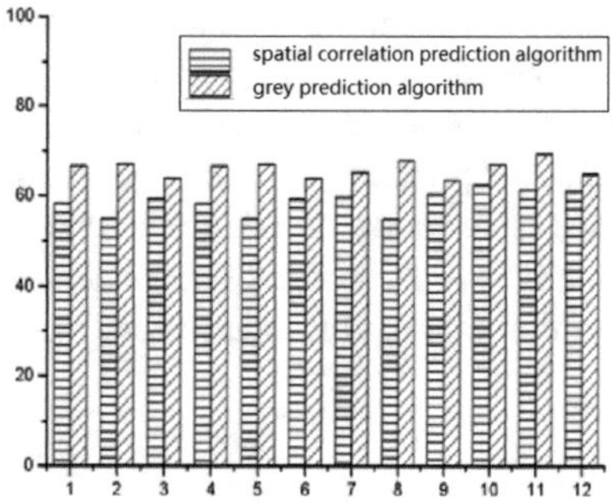

Fig. 4. Comparison of data upload volume of each node with an error threshold of 3%

Figure 3 shows the contrast using two algorithms. the error threshold ε is set to 1% to 10%, respectively. The percentage of data uploaded in the cluster. With the increasing of the error threshold, the amount of data uploaded in the cluster using the two prediction algorithms is gradually decreasing. But it can be seen from the graph that when the error threshold is 1%–10%, the amount of data upload in the cluster is lower than that in the grey prediction algorithm when using the spatial correlation data prediction algorithm [9–11].

Figure 4 shows the contrast using two algorithms. At the threshold of 3%, each node uploads to the percentage of data volume of convergent node through cluster head. By comparison, we can see that the data upload amount of each node is lower than that of grey prediction algorithm when using spatial correlation data prediction algorithm. Through the comparison of the above figure, we can see that the prediction accuracy of the spatial correlation data prediction algorithm is higher than that of the grey prediction algorithm.

5 Summary and Outlook

The data volume to be collected in the monitoring system is large. With the increase of time, the increase of the automatic monitoring station and the increase of the sampling frequency will result in the geometric growth of the environmental quality monitoring data, which is far from the storage and processing capacity of the existing monitoring system.

Based on the theoretical research on big data technology, this paper researches and designs a monitoring system based on big data technology to make full use of its ability to store and process massive data. Effectively integrate the advantages of relational database and distributed database in data query, data storage, and expansion respectively, and build an efficient, stable, scalable, and hybrid storage data center subsystem that stores and manages massive environmental quality Monitoring data. At the same time, in the data application layer of the application subsystem, functional modules such as data query, data simulation, data evaluation, over-standard alarm and trend prediction were integrated into the environmental quality monitoring system, and the preliminary development of system functions was completed, and the big data technology was verified The feasibility of integrating into the environmental quality monitoring system has carried out theoretical research and technical reserves for future research and development.

References

1. The global LBS market is expected to be 34.8 billion euros [CP/DK] by 2020, 01 Mar 2016. http://www.Techweb.com.cn/data/2015-09-08/2199235.shtml
2. Waltenegus, D., Christian, P.: Fundamentals of Wireless Sensor Networks, pp. 1–43. Tsinghua University Press, Beijing (2018)
3. Ling, W.: Research on spatial-temporal correlation data fusion algorithm for wireless sensor networks. Chongqing University (2017)
4. Sun, L., Li, J., Chen, Y.: Wireless Sensor Network. Tsinghua University Press, Beijing, 41/45 (2015)
5. Yang, J., Li, X., Chen, Y.: Group mining method based on incremental space-time trajectory big data. Comput. Res. Dev. **S2**(76), 85 (2014)
6. Liu, S., Meng, X.: Location-based social network recommendation system. J. Comput.
7. Zhai, T., Song, W., Fu, L., Shi, L.: Space-time trajectory clustering based on road network perception. Comput. Eng. Design **37**(3), 635–642 (2016)
8. Li, A., Wang, N., Wang, C., Qiu, Y.: The application of big data technology in environmental information. Appl. Comput. Syst. **24**(1), 60–64 (2015)
9. Yuan, X.R., Ren, D.H., Wang, Z.C., et al.: Dimension projection matrix/tree: Interactive subspace visual exploration and analysis of high dimensional data. IEEE Trans. Vis. Comput. Graph. **12**, 2625–2633 (2013)
10. Cao, N., Lin, Y.R., Gotz, D., et al.: Z-Glyph: visualizing outliers inmultivariate data. Inf. Vis. **17**(1), 22–40 (2018)
11. Zhou, Z., Sun, C., Le, D., et al.: Collaborative visual analytics of multi-dimensional and spatio-temporal data. J. Comput.-Aided Design Comput. Graph. **29**(12), 2245–2255 (2017). (in Chinese)

A Proposed Three-Dimensional Public Security Prevention and Control System Based on Big-Data

Guangxuan Chen[1], Weijie Qiu[2], Zhanghui Shen[2], Guangxiao Chen[3(✉)], Lei Zhang[4], and Qiang Liu[1]

[1] Zhejiang Police College, Hangzhou 310053, China
[2] Tongxiang Public Security Bureau, Tongxiang 314500, China
[3] Wenzhou Public Security Bureau, Wenzhou 325027, China
chenguangxuan@zjjcxy.cn
[4] Joint Services Academy, National Defence University, Beijing 100858, China

Abstract. The purpose of this article is to analyse the needs of three-dimensional prevention and control mechanism during the construction of local safety projects and informatization projects, to promote a new ecology of public security work in the new environment, and to promote the in-depth integrated application of new technologies in the entire public security field, so as to realize the integration of intelligence, information, command, service, and operation of the police operation mechanism, keenly aware of various types of risk hazards, and improve the level of socialization, legalization, professionalism, and intelligent of social governance of public security organs.

Keywords: Big-data · Social governance · Public security

1 Introduction

With the continuous development of mobile Internet, cloud computing, big data and artificial intelligence, and the continuous improvement of the informatization level in various fields of society, this has set new goals and requirements for safe construction projects in various places. In order to take effective measures to fundamentally control the occurrence of high-probability crimes, to achieve effective prevention and control mechanism, strengthening the construction of the public security management and control system in the region, and focusing on building a three-dimensional public security protection control network with local social governance characteristics, has become the top priority of public security organs at all levels [1–4].

Constructing a three-dimensional social security prevention and control system is an effective way to combat the crimes in modern urban governance. The orderly and stable development of regional social order not only provides a solid guarantee for the stable life of local residents, but also an important reflection of local spiritual civilization. However, as a huge social project, the construction of the system must be carried out scientifically, systematically and reasonably under the premise of the protection of laws and regulations. Only in this way can this system fully play its full role.

J. MacIntyre et al. (Eds.): SPIoT 2020, AISC 1282, pp. 320–325, 2021.
https://doi.org/10.1007/978-3-030-62743-0_46

2 Current Research

Research on social security prevention and control systems is relatively early. With the continuous deepening of research, the current research theories on crime prevention are mainly focus on crime prevention theory and related practice from the perspective of sociology of crime, criminal economics, and comprehensive disciplines.

American criminologist Weiss published a paper on crime prevention "the community and Crime Prevention" and put forward the famous three levels of prevention theory. The first one is the "primary" crime prevention model. This model is mainly to reduce the probability of crime and there is no direct relationship with the offender. It is mainly based on the crime incidents rather than the offenders with criminal motivation. The second one is the "intermediate" crime prevention model, which is mainly to change the criminals before they commit criminal activities. It belongs to the crime prevention stage. The third level crime prevention model mainly focuses on cutting off the criminal process of the perpetrators and minimizes the harm of the crime, such as the treatment and education of the perpetrators. The latter two focus on the perpetrators themselves. In his books "Evidence Based Crime Prevention" and "Preventing Crime: What Works for Children, Offenders, Victims, and Places", Farrington proposed a more comprehensive analysis of crime prevention. He divided crime prevention into four aspects: legal sanctions, developmental prevention, social prevention and situational prevention. However, in the related research process, it only conducts detailed research on the last three types, and there is very little research on crime prevention that focused legal sanctions. Therefore, it constitutes the theory of dichotomy of crime prevention strategies, which provides a broader concept than a mere general description [5–7].

Bottoms and Wiles jointly published the book "Environmental Criminology", which deeply researched and analyzed the types of crime prevention, and reclassified the current four main types of crime prevention: defensive strategies (Such as looting alarms, private police, street inspections); protection and surveillance (targeted protection against subjects who may be at risk of damage, monitoring, video reports, etc.); building new social maintenance methods (mutual cooperation between different institutions, public-private cooperation, public security measures to eliminate potential harms); prevention of criminal behaviors (planning in advance to prevent juvenile delinquency, and timely education and correction of juvenile delinquents).

Regarding the interpretation of criminal phenomena and the control of criminal behavior, Philip Zamba proposed the famous "Broken windows theory", but it did not draw any direct conclusions about the "Broken windows" effect and social security prevention and control. Some scholars believe that when social information is relatively scarce, people pay insufficient attention to the legitimacy of the incident, and there may be a phenomenon of regional crimes occurring by blindly imitating the actions of others. Therefore, the effective method for curbing crimes lies in the effective information publicity and release make people fully aware of crimes and risks, and at the same time increase their trust in the government, so that the people consciously comply with the law and improve the effect of social governance. This management method is called impression management. American scholar Honey studied the relationship between disordered management of urban communities and rising crime rates based on broken window theory, and believed that the decline and disorder of communities had a positive correlation with rising crime rates. In addition, Greenov and Alport proposed the "TAP" defense theory (Time if arrival of police). This theory believes that if law enforcement personnel can arrive at the crime scene as soon as possible through unified command and rapid response, they can deter similar crimes.

3 Approach on the Construction of Prevention and Control Mechanism Based on Big-Data

Here, we analyze the needs of three-dimensional prevention and control during the construction of local safety projects information projects, and conducts research from three aspects: business model upgrade, technical factors, and full data chain to achieve the integrated running mechanism of intelligence and information, command, service, and operations.

3.1 Business Model Upgrade

The comprehensive promotion and application of new technologies and new formats such as big data, artificial intelligence, and mobile internet have changed many mechanisms in the social environment and changed the behavioral patterns of perpetrators. This is particularly evident in public security work. Some traditional prevention and control models are no longer suitable for existing new business environments. Therefore, it is necessary to combine the local characteristics and the actual level of informatization to promote the update and iteration of the original security prevention and control model, as well as creatively promote the research of new business models [8–10].

Here, we take disciplinary repository (DR) construction as an example. The construction of the DR requires full consideration of various business needs and analysis of common data, so as to use core standard data tables as the main data source and fuse multiple data tables to form a wide table that is highly versatile, easy to expand and easy to use; The DR is a data table formed by the aggregation of topic domains. The DR can not only provide standard data services directly, but also can effectively complete the missing data and improve the accuracy of the data. The construction of the DR is the core of data governance. The high-quality DR data can effectively feed back the data tables in the standard library and improve the standard data tables. The higher the integrity of the DR, the more able to compare the omissions and weak items of each standard data table to help improve the data quality, and the higher the quality of the standard data table, the more accurate and correct the information will be. So that the DR provides services and more accurate support; Generally, if there is a high data missing rate or incorrect rate in the table, it is necessary to rectificate relevant data from the source.

The design of the DR logic model usually adopts a self-defined method, which summarizes the business objects within the scope of the requirements, classifies them hierarchically, divided into different topic domains, and then establishes the relationship between entities. From the perspective of public security, its business mainly involves the following aspects (Table 1):

Table 1. Example DR domain

DR domain definition	
Personnel	Attributes of personnel in the system
Article	Articles that related to the personnel
Address	Basic geographic area and address information, such as country, province, city, county, village, etc.
Case or incident	Incidents and cases, such as traffic accidents, crimes
Agency	Agencies involved in the incidents or cases, such as banks, gas stations, Internet cafes, hotels, offices
Track	Track library, including track of personnel in civil aviation, train, Internet, accommodation, etc., and track of items such as bayonet, MAC, RFID, etc.

3.2 Technical Factors

The technical factors for the information construction in three-dimensional prevention and control mechanism of grass-roots public security departments include the core difficulties and bottlenecks in the sensing, transmission, storage, and computing layers and technologies as load balancing, preprocessing, edge computing, etc.), and various algorithms based on the actual business model of public security (such as visual analysis, natural language understanding, public security knowledge graph, etc.).

The entire prevention and control system accesses multi-source data, including streaming data, web page data, structured data, unstructured data, semi-structured data, to achieve multi-network, multi-platform, and multi-type comprehensive aggregation of data. The data center relies on the support of multiple basic services of the big data platform to realize the fusion design of multi-source data, and builds a standard data governance system based on the actual demands of public security, and provides data basic services for police practical applications. The intelligent information center is based on the standard data system and uses NLP engine, relational engine, analysis engine, rule engine, algorithm model engine and other engines to provide data analysis and judgment, semantic recognition, relationship mining, algorithm modeling, prediction and early warning and other application services.

Figure 1 shows the Technical architecture diagram of three-dimensional public security prevention and control system.

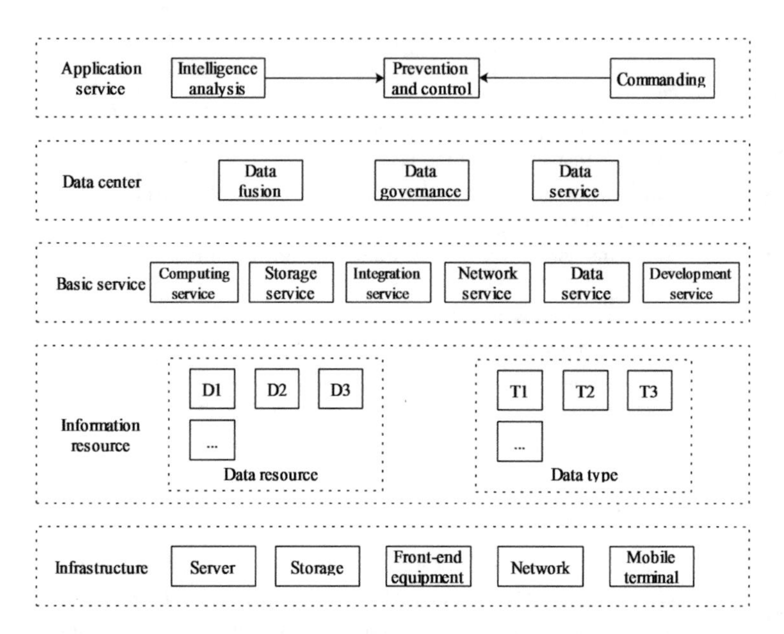

Fig. 1. Technical architecture diagram of three-dimensional public security prevention and control system

3.3 Data Chain Building

Based on local characteristics, research on full data chain based on prevention and control is the main purpose. It mainly includes: multi-dimensional source data acquisition, cleaning, "standardization", de-redundancy, classification, etc., so as to provide basic guarantee for researching business models.

The construction of a full data chain includes several key stages, namely data access, data processing, data organization, data service and data governance standard management. Among them, the data access phase needs to clarify the data sources, access methods, access implementation, data reconciliation and data aggregation; the data processing phase includes data standardization, data standard construction, data preprocessing, multi-source data fusion, and data classification; data organization includes resource library, DR, thematic library and knowledge base; data services include data query service, data comparison service, data push service, data authentication, data benchmarking service and cross-domain resource sharing; data governance standard management includes data resource directory construction, data standard system, data quality management and data operation and maintenance management.

4 Conclusion

The three-dimensional public security prevention and control system based on big data follows the concept of "platform + data + application", with cloud computing & big data platform as the core of construction, it can manage and schedule computing, storage, network, middleware, data, and business resources in a unified manner. It also can provide support for actual combat, research, testing, and teaching. The system is designed to realize the integrated police operation mechanism of intelligence and information, command, service and action, promote the in-depth integration and application of new technology in the whole field of public security, and promote the socialization, legalization, professionalism, and intelligent of social governance of public security organs, and finally to achieve the goal of three-dimensional prevention and control.

Acknowledgments. This research was supported by School-Bureau Cooperation Program of Zhejiang Police College under Grant No. 2018XJY003, and Basic Public Welfare Research Program of Zhejiang Province under Grant No. LGF19F020006.

References

1. Skidmore, M.J.: The U. S. system of social security: emphatically not a ponzi scheme. Poverty Public Policy **2** (2010)
2. Friedmann, R.R.: Effect of different policing strategies upon citizens' attitudes toward the police and their community (1985)
3. Chen, X.: Community and policing strategies: a chinese approach to crime control. Policing Soc. **12**(1) (2015). (in Chinese)
4. Stummvoll, G.: Environmental criminology and crime analysis. Crime Prev. Commun. Saf. **11**(2), 144–146 (2009)
5. Herbert, D.T., Hyde, S.W.: Environmental criminology: testing some area hypotheses. Trans. Inst. Br. Geogr. **10**(3), 259 (1985)
6. Wartell, J., Gallagher, K.: Translating environmental criminology theory into crime analysis practice. Policing **6**(4), 377 (2012)
7. Harcourt, Bernard, E: Reflecting on the subject: a critique of the social influence conception of deterrence, the broken windows theory, and order-maintenance policing New York style (1998)
8. Gau, J.M., Pratt, T.C.: Revisiting broken windows theory: examining the sources of the discriminant validity of perceived disorder and crime. J. Crim. Justice **38**(4), 758–766 (2010)
9. Ramos, J., Torgler, B.: Are academics messy? Testing the broken windows theory with a field experiment in the work environment. Rev. Law Econ. **8**(3), 563–577 (2009)
10. Livermore, C.: Unrelenting expectations: a more nuanced understanding of the broken windows theory of cultural management in urban education. Penn GSE Perspect. Urban Educ. **5**(2), 9 (2008)

Innovation of Ideological and Political Education Model in the Context of Big Data

Xiangzhen Pan(✉)

School of Management Science and Engineering,
Shandong Technology and Business University, Shandong, China
2440829015@qq.com

Abstract. Times are developing, technology is innovating, and big data thinking and methods promote the progress of human civilization, triggering all-round deep thinking and change in society. In the age of Internet communication, ideological and political education is currently confronting new chances and threats. It must advance with the times, reflect the times, grasp the laws, and be creative. Combined with the analysis of the innovation of ideological and political education course mode for college students in the Age of Big Data, the paper proposes cognition, work and assessment approaches in favor of curricular innovation based on cloud media technology, agent technology and Internet of Things technology, and comes up with a model to innovate student ideology education course.

Keywords: Big data · Cloud computing · Ideological and political education · Model · Innovation

1 Introduction

The Age of Big Data brings about a great many chances and threats to the ideological and political education in China [1]. There arises a new revolution in student methodology in the new age. This is closely associated with the advent of more flexible means, convenient carriers, diverse channels, agents, AI and cloud media technologies. Some domestic experts and scholars believe that fast and seamless network data collection can be used for the convergence and sharing of ideological and political education data. In the opinion of some researchers, the priority is to reinforce the adaptability of ideological and political education network in response to the rapid changes in network space, the technicality of Internet technology updates, and the convenience of network interaction space [2]. As shown by existing literature review in China and foreign countries, there exist numerous associations among Internet+ technology, big data technology and ideology education objects. The paper attempts to sum up the influential factors about the optimization of ideological and political education methodology and sorts out available methods to improve college student ideology education course in the Big Data Age.

J. MacIntyre et al. (Eds.): SPIoT 2020, AISC 1282, pp. 326–331, 2021.
https://doi.org/10.1007/978-3-030-62743-0_47

2 Innovation Status of Students in the Era of Big Data

2.1 The Rise of Intelligent Agents Changes Students' Models

Agent (Agent) is a term that belongs to the field of computers, also known as agent, mainly refers to intelligent objects that can sense the surrounding environment through intelligent sensors, and use the main actions of intelligent actuators for the objective environment. Agents through the search intelligent, decision-making intelligence, learning intelligence and other intelligent carriers produced by a series of intelligence carriers impact on students' ideological and political education methods. It mainly manifests in three aspects: First, it breaks the time limit. Traditional student ideological and political education is usually carried out at a fixed period of time. With the rise of intelligent agents, the course oriented towards students is aided by intelligent platforms such as WeChat, Weibo, QQ, Douyin, and short videos. "Going forward; second, breaking space constraints. Traditional models oriented towards students is often conducted in a fixed place. With the rise of intelligent agents, relying on intelligent software video teaching and intelligent network interaction can be carried out "anywhere". Third, the unified model has been broken. The agent uses intelligent analysis software, intelligent search engine, etc. to develop a personalized and humanized teaching plan for each student's ideological and political education. Under the action of agents, ideological and political education course for students has broken the traditional education methods, and has developed in the direction of convenience, efficiency and individuality.

2.2 The Popularization of Cloud Media Changes Students' Ideological and Political Education

Cloud Media (Cloud Media) refers to the collection of traditional media and fragmented media displayed in the form of Weibo, WeChat, space homepage, group, etc. Could media enjoys great advantages in its application in student ideological and political education course. Mainly manifested in: First, virtual teaching. Virtual teaching is the implementation of ideological and political education through network teaching, that is, the use of network video teaching to reproduce the classroom, through the virtual classroom to display the teaching content; Second, micro organization. Micro-organization is a form of group learning, that is, gathering students online through groups, official platforms, etc. to form a network micro-organization, and students can communicate and learn within the network group. Third, the information resource library. Information resource is a new way of students' autonomous learning, the students can take advantage of the cloud media of information resource under the relevant knowledge of student ideological and political education course of independent learning.

2.3 Integration of Homo Sapiens and Ideological Education

Homo sapiens fusion refers to the integration of artificial intelligence (AI) and people. Artificial intelligence simulates human thinking and consciousness, and integrates artificial intelligence behavior with human science. Through intelligent search, knowledge acquisition, machine learning and other carriers that cater the needs of student education, considering the promotion of ideological and political education in the student group, the education distance is greatly shortened. The main performance is as follows: first, the intelligent network teaching platform shortens the gap between the teacher and students, enabling students to obtain educational content through intelligent teaching; Second, the database intelligent retrieval system has shortened the distance between students and educational resources, making Students can quickly and accurately obtain the ideological and political education resources they need; Third, intelligent assisted robots have shortened the distance between students and learning places, so that students do not have to be fixed in a specific environment to receive counseling, but can use the help of intelligent assisted machines Under the scientific analysis of their own problems, to obtain solutions.

3 Innovative Content of Ideological and Political Education Model for Students in the Age of Big Data

3.1 Cognitive Innovation of Students' Ideological and Political Education

3.1.1 Innovation in Cloud Information Acquisition

In the Age of Big Data, cognitive approaches proper for student no longer relies solely on textbooks and classroom teachers to acquire knowledge. Instead, it combines cloud media to collect online and offline information and comprehensive port information, and then integrates them into cloud data to establish large-scale. Cloud resources library enables students to rely on cloud resources in ideological and political education and use intelligent search engines to obtain information. The innovation of information knowledge acquisition method has realized the supreme efficacy in student ideological and political education.

3.1.2 Innovation of Big Data Information Analysis Methods

In the Age of Big Data, traditional methodology for students can no longer rest on simple and fuzzy information collection and analysis methods. Big data has three salient features: focus on the correlation of things, focus on value mining, and focus on the prediction of trends [3]. Innovation of big data information analysis methods Big data, starting from data correlation, extensively collect data information required for ideological and political education, establish an information resource database, and cooperate with qualitative research based on quantitative analysis to perform parallel analysis of related data Then establish an information analysis model. This analysis method innovation guarantees the comprehensiveness of data collection for student ideological and political education course, and improves the accuracy of course statistical analysis results.

3.1.3 Innovation of Agent Information Decision

Agent decision-making method refers to the intelligent information system for information support, undertake mission in student ideological and political education course, independent analysis on the content of ideological and political education course, multi - path comparison and choice, scientific decision. A typical student's ideological and political education analysis and decision-making consists of decision points, state nodes, and end nodes. It analyzes various possibilities, determines alternatives, and selects the best. The innovation of this decision-making method makes the decision-making process concerning the informatization of student ideological and political course.

3.2 Innovations in Students' Ideological and Political Education Work

3.2.1 Theoretical Education Loads Client-Side Innovation

The client-side method of theoretical education loading mainly refers to an educational method that uses the client as a carrier to provide a new front for theoretical education. Traditional students' ideological and political education mainly relies on textbook knowledge and classroom indoctrination. In the "Internet+" era, theoretical education is not limited to traditional model, and the development of mobile clients provides a good opportunity for theoretical education [4]. Using the "Internet+" Theoretical knowledge, relying on streaming media technology, vividly displays the theoretical knowledge through online classrooms, PC clients, mobile apps, etc., which can be connected to the Internet for online browsing, or work offline when offline. Research on student ideological and political education methodology attempts to investigate the thoughts of students with the aid of customers at any time and at any place. Through applying "Internet+" technology in student ideological and political theoretical teaching, it is possible to better the teaching effects, reinforce education functions and achieve the goal of innovation and efficacy.

3.2.2 Innovation of Practice Education Mode on Cloud Platform

The practice education method of cloud platform mainly refers to the education method that provides resource service for students' ideological and political practice education with the cloud platform as the carrier. In the "Internet+" age, practical education uses the cloud platform to adopt "human-computer interaction", "virtual machine", and "virtual laboratory" methods to innovate practical education methods such as "case project driven" and "curricular modularization" under the cloud platform. The cloud platform stores students' ideological and behavioral data resources in different time and space, and integrates different types of ideological and behavioral information data into a huge database, so that data can be reconstructed, distributed, managed, and fed back through cloud technology To ensure that the data is used reasonably efficiently. To some degree, the innovation of cloud platform ideology education cuts down the cost of practice, expands the teaching staff, surmounts the limit of time and space. These advantages ensure the convenience and high efficiency of students' ideology education practice.

3.2.3 Innovation of Visual Criticism and Self-criticism

The method of visual criticism and self-criticism mainly refers to making targeted suggestions to students' shortcomings or errors with the support of visualization technology. Students make self-recognition, reflection and improvement of their shortcomings or errors based on visual criticism. The visualization technology better reflects the students' ideological problems through the complete data collection, data analysis, data presentation, and "visual message board", "visual chat room", "visual assessment" and other methods to solve the problems. The use of big data can calculate and streamline students' ideological behaviors in batches and predict their ideological tendencies, values, and work behaviors. The innovation of this method avoids the face-to-face embarrassment and achieves accurate and effective criticism and self-criticism, In order to better achieve the goal of students' ideological and political education.

3.3 Innovative Evaluation Methods for Students' Ideological and Political Education

3.3.1 Innovative Feedback Adjustment Methods for the Internet of Everything

The Internet of Everything feedback adjustment method refers to an evaluation method that externalizes feedback information into actions and actively regulates ideological and political education activities under the influence of the Internet of Everything technology that perfectly integrates people, things, and things [5]. The feedback of the Internet of Things may be shown in the two aspects as below. At first, continuous student thought-behavior communication. Organizations can continue to communicate with students through thoughts, questionnaires, conferences, parties, and other methods to maintain sensitivity to students 'thoughts and behaviors, and then use emotional analysis to promote the design of students' ideological and political education in a timely manner. Second, the arrangement and analysis of students' ideological and behavioral information. Organizations can use sensors, radio frequency identification, data retrieval tools, mobile devices, and social networks to collect relevant data that reflects the students' learning process, and filter useless or wrong data through fuzzy or cluster analysis methods. Data integration database; analyze the above ideological and behavioral data through cloud computing technology. This method innovation can better realize the comprehensive collection of student statistics, and statistical analysis accuracy, and the scientificization of feedback adjustment effect [6].

3.3.2 Innovation of Intelligent Media Cloud Cluster Information Feedback

In the era of big data, the intelligent media cloud cluster information feedback method refers to the combination of intelligent media technology and cloud information collection technology to collect information contained in student ideological and political education network survey feedback and data browsing distribution, and then use cloud technology to intelligently. The feedback information is analyzed and modularized, and the feedback information is packaged and bundled. The intelligent media cloud cluster information feedback method innovation breaks through the simple and one-sided feedback information-based information feedback method. Instead of requesting students to avoid the information cluster feedback in ideology education, the model enriches the content of course on time, and in the meantime, affords professional information support. As to the collection of statistics concerning student ideology education behaviors, the quality of data directly affects the assessment and development of student ideology education. So as to make sure of the high quality of cluster statistics collected via intelligent media cloud, it is necessary to introduce pipe network into the methodology of student ideology education course. Subject to the mode of embedded network, the paper controls student speech and act, and learning statistics, and collects student ideology education statistics in a more complete, straightforward, accurate and quality manner.

3.3.3 Innovation of Group Homo Sapiens Integrated Evaluation Method

The comprehensive evaluation method of the integration of group sapiens and humans refers to the evaluation of the student's taught group, the integration of AI and other technologies, and then comprehensively evaluates the implementation process and results of ideological and political education. It mainly combines the current status of students' ideological and political education at various stages, simulates the structure and methods of human brain thinking, fuses human thinking and consciousness, and builds a "super brain" of sapiens and human intelligence. A series of processes comprehensively evaluate student performance in ideological and political education course. This method enables a more accurate assessment of the ideological and political education of student groups. The comprehensive evaluation of the integration of group sapiens and humans is mainly to use the relational database to store, analyze, and use students 'thoughts, words and deeds, etc., to identify students' thoughts and behaviors. The higher the value realization rate of students, the higher their overall quality, and their relative preferences for ideological behavior.

References

1. Guangxue, J., Lichen, M.: A new interpretation of information literacy and educational connotation in the Internet age. China High. Educ. **10**, 44–47 (2019)
2. Shoulin, W.: Research on the innovation of ideological and political education methods in colleges and universities in the era of big data. Res. Ideol. Polit. Educ. **6**, 85–87 (2015)
3. Jingbo, X., Zhixin, W.: Research on the teaching model of ideological and political theory courses in colleges and universities in the era of big data. Heilongjiang High. Educ. Res. **5**, 152–155 (2018)
4. Zhiping, X.: Research on the implementation methods of ideological and political education for college students. China Youth Res. **10**, 25–28 (2019)
5. Fengfan, Z.: Research on the prospect of the age of artificial intelligence. Soc. Sci. Yunnan **1**, 53–58 (2020)
6. Leng, L., Shan, G.: Analysis of university network public opinion management based on big data. Int. J. Front. Eng. Tech. **1**(1), 38–47 (2019)

Communication Response of Internet Rumor Audiences Based on Big Data Analysis

Sitong Liu(✉)

Guilin University of Aerospace Technology, Guilin, China
liusitong@guat.edu.cn

Abstract. In the era of big data, the importance of data analysis has been put on the agenda, and at the same time, the authenticity of massive information has become a challenge. In the face of the fact that the number of Internet users in China has exceeded 850 million yuan and the trend of obvious growth, the research on the propagation response rule of the network rumor audiences in the era of big data is urgent. Based on the background of the era of big data, this paper USES the classic wt-sir rumor propagation model to simulate the rumor propagation mechanism in the Internet environment, and finds that the rumor propagation scale and propagation speed are significantly positively correlated with the importance and credibility of the rumor.

Keywords: Internet rumors · Big data · Audience response · Law of communication

1 Introduction

By the end of June 2019, the number of Internet users in China has reached 854 million, among which the mobile Internet users account for a large scale of 847 million [1]. In the era of big data, the form and speed of information dissemination are different from the past. Internet rumors are the product of the gradual evolution of rumors in the Internet era. With the development of social networks in the era of big data, rumors become increasingly fierce [2]. In particular, some deliberate online rumors, amplified and widely spread by social media, continue to ferment in the civil society, arouse the attention and heated discussion of all Internet users, and cause adverse effects on the normal public opinion field of the society, and seriously threaten the order of social life [3]. Compared with the rumors spread by word of mouth and three people become tigers in the past, Internet rumors have more striking influence. Like the speed of virus propagation, some Internet rumors with serious distortion and confusing the public can even cause butterfly effect in the world [4]. Under such a realistic background, scholars have begun to pay attention to the propagation characteristics and laws of Internet rumors. On the characteristics of the spread of Internet rumors understanding, some scholars from the spread of Internet rumors and influence degree, the rumors will be divided into the local and national rumor, the former refers to the community or confined to a network rumors of regionality, which is a across the network of the national common rumor, the latter is more influential to the society [5, 6]. Other scholars have made a more clear distinction on the propagation path of Internet rumors:

J. MacIntyre et al. (Eds.): SPIoT 2020, AISC 1282, pp. 332–337, 2021.
https://doi.org/10.1007/978-3-030-62743-0_48

realistic participatory network rumors refer to those "originating from reality, spreading to a certain extent and then entering the Internet, presenting a state of parallel oral communication, mobile communication and Internet communication"; However, the original Internet rumors refer to "starting from some or a number of netizens' posts, and having a certain degree of influence after many netizens' reposts and discussions, the spread of reality can only occur, or it just stays in the cyberspace and does not produce the process of real communication" [7]. In view of the great changes in rumor propagation mode in the era of mobile Internet, on the one hand, the spread of rumors is no longer limited to isolated offline interpersonal network or online Internet, and the spread of rumors between coupled networks has become more common. On the other hand, it is inaccurate to study the spread of rumors in the past by means of probability, and the differences of individual characteristics of the audience play a more important role in the spread of rumors. Therefore, based on the characteristics of information network communication under the background of big data, this paper will analyze the characteristics of network rumor audiences and the rumor propagation law.

2 Method

As a social contagion, rumor propagation is very similar to disease propagation, so most studies on rumor propagation are based on epidemic propagation model. Daley and Kenal put forward the classic DK rumor propagation model, which divides individuals into three categories: the ignorant, the disseminator and the immune [8, 9]. The ignorant person is the one who does not know the rumor, the disseminator is the one who spreads the rumor, and the immune person is the one who knows the rumor but does not spread the rumor [10]. When the ignorant come into contact with the rumor spreader, the rumor spreader spreads the rumor with the probability of the rumor spreading rate. After the ignorant come into contact with the rumor, they become immune to the rumor by seeing through the rumor or believing but not spreading the rumor, and by believing and spreading the rumor with the probability of the rumor spreader. The greater the degree of the node, the greater the possibility of believing the rumor when it comes into contact with the neighbor node spreading the rumor. Therefore, although nodes with large degrees have more opportunities to contact with rumor disseminators, their probability of believing rumors when they come into contact with rumors is lower than the average infection rate of network due to their strong anti-rumor ability. Define the following propagation rules:

A) if a healthy node s is in contact with a propagation node I, the healthy node will be converted into a propagation node with probability P.
B) when the propagation node I encounters an immune node, it will be converted into an immune node with probability Pi.
C) the rumor comes from external I and infuses into a certain point K, so that K(s) or I has no external injection, and the closed social network has no so-called rumor. Similarly, the rumor dispelling information also comes from the external injection.

3 Experiment

3.1 Simulation Parameter Value Setting

In this paper, the conversion parameter value of the model is set by questionnaire survey. The questionnaire was issued on January 5, 2020, solstice, 25, with a total of 547 copies. 499 valid questionnaires were returned. Nearly 99% of the respondents are between 18 and 39 years old, belonging to China's major Internet users. The questionnaire divided the rumors into four categories according to their importance and credibility: low in importance and weak in credibility, low in importance but strong in credibility, high in importance but weak in credibility, high in importance and strong in credibility. In each case, specific cases and actual scenarios are given for the reference of the respondents. The probability value of the role transformation of the respondents in different situations is investigated by means of scenario substitution. After filtering out some invalid samples, the average value of the data obtained from the survey is taken as the conversion parameter value. The information attention cycle of users in different situations is understood through questionnaire survey, so as to determine the value of attenuation rate a. Table 1 describes the proportion of the number of users in the rumor information attention cycle in different situations, and the expected attention cycle in each case after the unreasonable value is removed. If the expectation is 10.25, it means that when t = 10.25, the proportion (a·t) $\rightarrow$ 1 will be removed naturally. For delta (a. t) can only approach 1, so can't directly find out the value of a (a $\in$ (0, 1]), the a_1 = 0. 05, a_2 = 0.1, a_3 = 0. 15, a_4 = 0.2, a_5 = 0.95, a_6 = 1, find out first to the delta (a. t) acuity 0. 99 set up a value, to determine interest in each case decay curve.

3.2 Numerical Simulation

The final size of rumor propagation is an important parameter for rumor propagation in the network, which reflects the maximum number of points that the rumor can touch in the network. The final scale in this model describes how many points in the network are exposed to rumors, including two situations, one is transformed into a disseminator after exposure to rumors and the other is transformed into an immune person after exposure to rumors. Z is used to represent the final size. When t = infinity, the rumor propagation reaches the final size, and the final size z can be obtained as follows:

$$\dot{\theta}(\infty) = -(\omega + \gamma)\theta(\infty) + (\omega + \gamma) = 0, Z = 1 - I(\infty) = 1 - \varphi(\theta(\infty)) = 0$$

The numerical solution and simulation experiments are carried out, and the results are compared. In the simulation, each point was divided according to the state of the rumor information: the ignorant, the rumor disseminator and the rumor immune. The experimental results show that the proportion of rumor spreaders in the population shows an upward trend at the beginning, and then decreases gradually after reaching the peak, and finally stabilizes to a value. The results of the model are very close to the results of the random simulation. Network size N = 2 $\times$ 10^4, network average degree (K) = 6, rumor spread rate = 0.4, the probability of ignorant people to believe the

Table 1. Periodic table of information related communication

Pay attention to cycle	Day	Low importance & low credibility	Low importance & high credibility	High importance & low credibility	High importance & high credibility
1–2 days	1.5	0.43	0.33	0.23	0.27
3–5 days	4	0.28	0.34	0.32	0.34
1–2 weeks	10.5	0.22	0.20	0.26	0.22
15–30 days	22.5	0.06	0.06	0.26	0.18
3–6 months	135	0.04	0.03	0.05	0.06
0.5–1 year	270	0.03	0.02	0.01	0.02
1–0 years	1980	0	0	0	0.01
More over 10 years		0	0	0	0
	Except	10.33	13.68	32.79	36.08

rumor after hearing the rumor is equal to 0.6, the probability of people who believe the rumor into the rumor disseminator is equal to 0.7, the forgetting rate of rumor disseminator into the rumor immune is equal to 0.4. The random simulation experiment was carried out for 100 times, and the average value was compared with the numerical solution.

4 Discuss

In ER random networks, WS small-world network and simulation experiments on the BA scale-free network, three types of nodes are N = 1000, the total average degree of network are k = 4, parameter selection for b = 0.4, u = 0.04, beta = 0.6, theta equals 0.6, were randomly selected from 1% of the degrees of 4 I_1, I_2 two types of nodes as the initial transmission nodes, spread in the spread of node density BA network node density peak spread the largest and fastest. This is because the degree distribution of BA network follows the power-law distribution, and a few nodes in the network have very large degrees. Once the Hub node is infected, it has more opportunities to contact other nodes in the network, and the propagation speed will be significantly accelerated (Fig. 1).

Compared with the BA scale-free network, the rumors spread to a smaller extent in the real Twitter network. This is mainly because Twitter real total number of nodes in the network than the total number of nodes in the BA scale-free networks, the initial

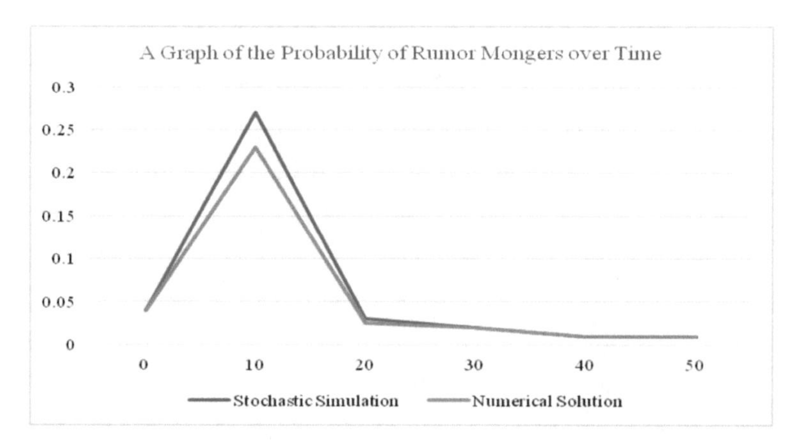

Fig. 1. The probability of rumor spreader changes with time

rumors spread under the same number of nodes, Twitter in real network, any node spreading rumours in the neighbor node number is relatively small, game, considering the effect of conformity for nodes do not spread rumors can get higher yields, node tends to keep the same choice, with most of the node that is not to spread rumours, finally the speculation in the real network Twitter than in BA scale-free network transmission range is smaller. Has never known the change of the density of S(t) can be seen that the number of users are reduced gradually, and gradually tends to be stable over time, and from the disseminator, the density of watching and rumours that I(t), W(t), t(t) can be seen in the change of the I, W, t type the number of users basic showed a trend of increase after decreases first (part I) user directly decreasing, R, RR, RT class user is on the rise, the density of the entire transmission system with the development of the time gradually tends to a stable state.

5 Conclusion

In this paper, rumor propagation nodes are only considered to be divided into two categories. In fact, the ability of rumor unknown node to resist rumor, namely its probability of being infected, is also different, so the rumor unknown node can be further subdivided. Along this line, the next step of our work is to construct a more consistent model of network rumor propagation, so as to further discover the law of network rumor propagation, and then put forward more effective Suggestions to control network rumor propagation.

Acknowledgments. This paper is supported by the Youth Fund for Humanities and Social Sciences of the Ministry of Education: “Research on Audience Communication of Internet Rumors in the Era of Big Data” (Project number: 18YJC 860021).

References

1. Fung, I.C.-H., Fu, K.-W., Chan, C.-H.: Social media's initial reaction to information and misinformation on Ebola, August 2014: facts and rumors. Public Health Rep. **131**(3), 461–473 (2018)
2. Bordia, P., Difonzo, N.: Problem solving in social interactions on the internet: rumor as social cognition. Soc. Psychol. Q. **67**(1), 33–49 (2019)
3. Chen, H., Lu, Y.K., Suen, W.: The power of whispers: a theory of rumor, communication, and revolution. Int. Econ. Rev. **57**(1), 89–116 (2018)
4. Fisher, D.R.: Rumoring theory and the internet: a framework for analyzing the grass roots. Soc. Sci. Comput. Rev. **16**(2), 158–168 (2018)
5. Sun, Z., Wang, P.P.: Big data, analytics, and intelligence: an editorial perspective. New Math. Nat. Comput. **13**(2), 75–81 (2019)
6. Bordia, P.: Studying verbal interaction on the Internet: the case of rumor transmission research. Behav. Res. Methods Instrum. Comput. **28**(2), 149–151 (2019)
7. Firth, R.: Rumor in a primitive society. J. Abnorm. Psychol. **53**(1), 122–132 (2018)
8. Goonetilleke, O., Sellis, T.K., Zhang, X., et al.: Twitter analytics: a big data management perspective. ACM SIGKDD Explor. Newslett. **16**(21), 11–20 (2018)
9. Fisher, D.R.: Rumoring theory and the internet: a framework for analyzing the grass roots. Soc. Sci. Comput. Rev. **16**(2), 158–168 (2019)
10. Rosnow, R.L.: Rumor as communication: a contextualist approach. J. Commun. **38**(1), 12–28 (2018)

Design of University Admissions Analysis System Based on Big Data

BoYu Zang(✉)

Shenyang Aerospace University, Shenyang, Liaoning, China
176921382@qq.com

Abstract. With the continuous expansion of the enrollment scale of universities in China, the number of college admissions has increased year by year, and the professional settings and student quality are also constantly changing. This has put higher requirements on the construction and management of colleges and universities. Aiming at the shortcomings of the current domestic college admissions information analysis system, the college admissions business was analyzed, and the admissions data processing and analysis design were studied in depth. In terms of methods, this article mainly builds an enrollment data model, statistics student information, and compares the scores and admission scores of most students in previous years. In terms of data processing, pre-processing of the data of each province is mainly performed, and the data of each province is merged. The standardized admissions freshmen data of each province, city, and category are combined and summarized. And through the construction of the system, the final results are obtained and classified. The system construction is introduced from three aspects: system architecture, system structure and system function modules. In the aspect of experiment, it is mainly discussed from the aspects of experimental purpose, experimental object and experimental design. Finally, the results are obtained. The data proves that the enrollment analysis system designed in this paper has certain feasibility and reliability.

Keywords: Analysis system · College admissions · Admissions data · System construction

1 Introduction

With the development and popularization of computer networks and the rapid development of information technology, it has promoted major changes in university admissions. College entrance examination is closely related to the fate of most candidates, which is a top priority for tens of millions of candidates. In order to make the college entrance examination smoothly and meet the needs of the people, colleges and universities should establish a set of admission management systems suitable for their own colleges and universities as soon as possible, improve the efficiency of admissions work, improve the service level, and provide a reliable basis for scientific management and decision-making. Image, seize competitive opportunities.

At present, the existing national online admissions system for college admissions provides reliable, standardized student data sources for admissions analysis information. However, it has the following shortcomings: the information systems of each university

J. MacIntyre et al. (Eds.): SPIoT 2020, AISC 1282, pp. 338–344, 2021.
https://doi.org/10.1007/978-3-030-62743-0_49

are independent of each other and cannot share data; the functions of post-processing data recording and printing notifications are weak; the functions are relatively simple, there are no important functions such as student registration and management, and decision analysis. Therefore, it is necessary to develop a college admissions data processing and analysis system to realize the later analysis and processing of admissions data. Otherwise, it will affect the quality and efficiency of admissions management, which will directly affect the quantity and quality of college student resources and the subsequent links of college management. Affect the development and reputation of the school.

In terms of methods, this article mainly builds an enrollment data model, statistics student information, and compares the scores and admission scores of most students in previous years. In terms of data processing, pre-processing of the data of each province is mainly performed, and the data of each province is merged. The standardized admissions freshmen data of each province, city, and category are combined and summarized. And through the construction of the system, the final results are obtained and classified. The system construction is introduced from three aspects: system architecture, system structure and system function modules. In the aspect of experiment, it is mainly discussed from the aspects of experimental purpose, experimental object and experimental design. And finally came to a conclusion.

2 Method

2.1 Admissions Data Model

The data related to the admission test scores and the subject of admission analysis include the situation of the admission institution (the nature of the institution, location, affiliation, etc.), the admission plan (including the specialty name, specialty category, specialty plan, etc.), Candidates' basic information (including attributes such as gender, ethnicity, height, vision, etc.), candidate's performance (including total scores, single subject results, and other attributes), and candidate's voluntary status (including attributes for each batch of schools, majors etc. 2. The source of candidates (including the graduated middle school, the location of the middle school, etc.) [1, 2].

2.2 Research on Data Processing and Analysis Methods

(1) Data processing

1) In the data pre-processing of each province, the structure of the original information table and the table output through the admission system of the Ministry of Education are very different from the data tables required for practical applications in colleges and universities [3]. Most fields appear as codes, and the meaning of the codes for each province is different. In order for the summarized new data table to be consistent in table structure and field meaning, each province must be registered [4]. The new original information table has been standardized. The standardized pre-processing process calls the data preprocessing table, selects and sets the category parameters of the data table to be standardized, and implements the addressing and calling of the data table.

2) Merge provincial data. Standardized data on freshmen in each province and category must be aggregated before students can be compared more specifically [5, 6]. Due to the many provinces, levels, and categories involved in the registration process, data classification is stored in different paths, and you may need to merge dozens of data tables in the end. In order to ensure the completeness and accuracy of the merged data, it is necessary to merge by category before summarizing and merging [7].

(2) Research on data analysis methods
In enrollment, it is necessary to timely count the number of students admitted by college, specialty, level, province, department, and gender, and the highest and lowest scores. You can use the SQL language to achieve flexible statistics of custom conditions [8, 9]. The main purpose of data analysis is to analyze various situations in the coming year based on student data from previous years. Therefore, it is necessary to analyze and process the data scientifically and accurately from the statistical data over the years to establish a suitable prediction model [10]. According to the difference equation theory and the least square method, a second-order difference equation mode $y_t = a_0 + a_1 y_{t-1} + a_2 y_{t-2}$ can be established. According to the model combined with the enrollment data information of previous years, the value of a_0, a_1, a_2 can be solved, and then the model is used to perform relevant predictive analysis, such as the number of applicants, regional volunteer Analysis, report rate analysis, etc. [11].

The Z-Scores method is also called the standard score method. It is a quantity that expresses the relative position of a score in the collective in units of standard deviation. The calculation formula is:

$$Z_i = \frac{X_i - \bar{X}}{S} \tag{1}$$

X_i is the original score of the i-th candidate, $\bar{X}$ is the average raw score of all admitted candidates in the same year, S is the standard deviation of the distribution of the scores of the admitted candidates for the year, and Z_i is the standard score of the i-th candidate.

2) The system uses T scores to standardize the scores. For the annual admissions situation, calculate the average score $\bar{X}$ and standard deviation S of the admission for that year; then, use the previous formula to calculate the corresponding standard score Z_i for each candidate i's total score X_i:

$$Z_i = \frac{X_i - \bar{X}}{S} \times 10 + 50 \tag{2}$$

2.3 System Architecture

According to the actual needs of the system, through the in-depth comparison of the existing infrastructure, from the perspective of technical stability, analyze and study the advantages and disadvantages and applicability of specific technologies that can be adopted by each layer. According to the breadth of support and the rationality of meeting the requirements, based on the mixed structure of C/S and B/S (that is, the presentation layer, the logic layer and the data layer), the most appropriate three-layer architecture is selected and constructed [12].

2.4 System Structure

The college admissions analysis system inevitably requires it to be able to share the candidate's electronic archive data provided by the online system in order to achieve a good link between admissions and admissions management. Candidates' electronic data can be sorted and edited in the analysis system, and provide various forms of output ports such as information query, report printing, and online information publishing. The system includes two major subsystems, a data processing subsystem with C/S structure and a recruitment analysis subsystem with B/S structure. The data processing subsystem of the C/S structure interacts with the "National Admissions Management System for Colleges and Universities" to complete the import and export of data and the printing of notices. The enrollment analysis subsystem of the B/S structure mainly completes data analysis and statistics, provides decision support and related information query.

3 Experiment

3.1 Purpose of the Experiment

It verifies whether the practical application of the college-based admissions analysis system based on big data designed in this paper is reliable and feasible. Through cooperation with a school, and applying the college admissions analysis system designed in this paper to analyze its effect.

3.2 Subjects

Regarding the enrollment of a school.

3.3 Experimental Design

This article mainly obtains some student information through cooperation with a university. Build an admissions data model, collect statistics on student information, and compare the scores and admission scores of most students in previous years. In terms of data processing, it mainly preprocesses the data of each province, combines the data of each province, and combines and summarizes the data of all provinces, cities, and categories after standardization. Through the system of the Ministry of Education, the original information tables of freshmen admitted in different provinces and cities are compared with the data tables required for practical applications in colleges and universities. In terms of data analysis, from the statistical data of the past years, the data is scientifically and accurately analyzed and processed to establish a suitable prediction model.

In terms of system, build a 3-layer architecture solution based on a mixed structure of C/S and B/S, which is the presentation layer, logic layer, and data layer. Can share the candidate's electronic archive data provided by the online system to achieve a good link between admissions and admissions management. University admissions analysis system includes two major subsystems and system maintenance modules.

4 Discussion

4.1 Standardization of Submission Results

Due to different factors such as the difficulty of the exam each year, the meaning of the same score in different years is different. For example, the difficulty of the 18-year test paper is generally small, so 447 points have just reached the score of the private undergraduate; and the 19-year test paper difficulty Generally speaking, it is more difficult. In 19 years, the score line of private undergraduates was only 425 points, and the score line of the second class was 463 points. Then the students with the same score of 460 are likely to be accepted by the second colleges. However, after we convert the test scores into standard scores by using a systematic conversion method, we can eliminate the "value" difference of the same score caused by the difference in score distribution caused by factors such as the difficulty of the exam each year, and make it directly comparable. At the same time, the original score treats the same difference in different distributions as equivalent, while the standard score converts different distributions into the same distribution (standard normal distribution), and then treats the same difference as equivalent.

Table 1 shows the average freshman submission results and the standard deviation of each year for a college in 2017–2019.

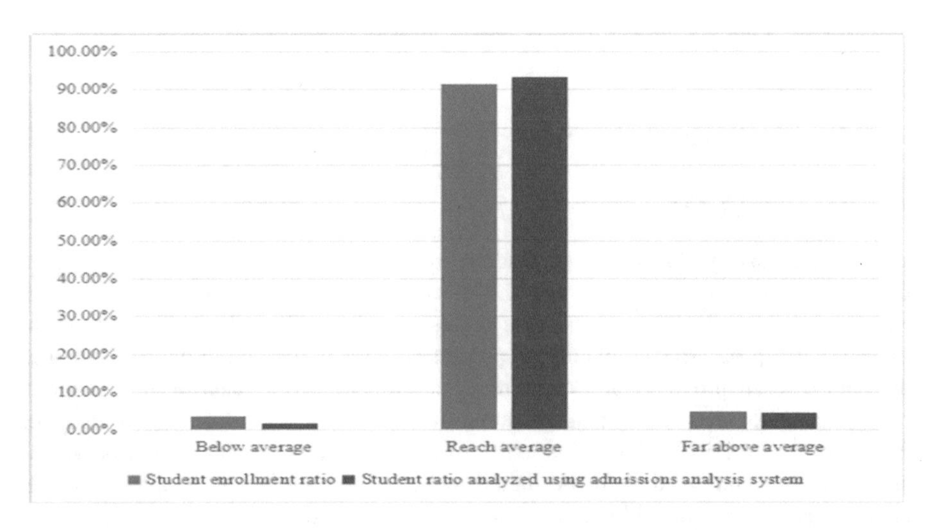

Fig. 1. Comparison of results using the admissions analysis system designed in this paper

Table 1. Means and standard deviations of submission results in different years

Admission year	Mean score	Standard deviation
2017 year	435.7	48.25
2018 year	456.8	39.63
2019 year	452.9	42.87

It can be seen from Table 1 that the average of the scores in the past three years has remained between 435 and 460, and the overall score is relatively low. The standard deviation of the scores of the transfers is maintained between 39 and 50, which reflects that the scores are relatively uniform. And has a certain effectiveness. From the comparison of the results of the admissions analysis system in Fig. 1, we can observe that the overall level of students using the admissions analysis system designed in this paper is still slightly higher than the original school. In terms of not reaching the average, the school's enrollment is about 3.7%, and students using the admissions analysis system are about 1.9%; in terms of reaching the average, the school's recruitment is about 91.6%, and students using the admissions analysis system The analysis is about 93.5%; in terms of reaching the average, the school enrollment is about 4.7%, and the students using the admissions analysis system are about 4.6%. In general, the enrollment analysis system designed in this paper is proved to be feasible and reliable through experiments.

5 Conclusions

In terms of methods, this article mainly builds an enrollment data model, statistics student information, and compares the scores and admission scores of most students in previous years. In terms of data processing, pre-processing of the data of each province is mainly performed, and the data of each province is merged. The standardized admissions freshmen data of each province, city, and category are combined and summarized. And through the construction of the system, the final results are obtained and classified. The system construction is introduced from three aspects: system architecture, system structure and system function modules. In the aspect of experiment, it is mainly discussed from the aspects of experimental purpose, experimental object and experimental design. Finally, the results are obtained. The data proves that the enrollment analysis system designed in this paper has certain feasibility and reliability.

References

1. Rocca, S.R.D.: Review: abstract state machines—a method for high-level system design and analysis. Comput. J. **47**(2), 270–271 (2018)
2. Bajić, D., Polomčić, D., Ratković, J.: Multi-criteria decision analysis for the purposes of groundwater control system design. Water Resour. Manage **31**(15), 4759–4784 (2017)
3. Wilson, A., Baker, R., Bankart, J.: Understanding variation in unplanned admissions of people aged 85 and over: a systems-based approach. BMJ Open **9**(7), e026405 (2019)

4. Stecker, M.M., Stecker, M.M., Falotico, J.: Predictive model of length of stay and discharge destination in neuroscience admissions. Surg. Neurol. Int. **8**(1), 17 (2017)
5. Lee, N.S., Whitman, N., Vakharia, N.: High-cost patients: hot-spotters don't explain the half of it. J. Gen. Intern. Med. **32**(1), 28–34 (2017)
6. King, A.J.L., Johnson, R., Cramer, H.: Community case management and unplanned hospital admissions in patients with heart failure: a systematic review and qualitative evidence synthesis. J. Adv. Nurs. **74**(7), 1463–1473 (2018)
7. Helmuth, I.G., Poulsen, A., Mølbak, K.: A national register-based study of paediatric varicella hospitalizations in Denmark 2010–2016. Epidemiol. Infect. **145**(13), 1–11 (2017)
8. Done, N., Herring, B., Tim, X.: The effects of global budget payments on hospital utilization in rural Maryland. Health Serv. Res. **54**(3), 526–536 (2019)
9. Lueckel, S.N., Teno, J.M., Stephen, A.H.: Population of patients with traumatic brain injury in skilled nursing facilities: a decade of change. J. Head Trauma Rehabil. **34**(1), 1 (2018)
10. Morawski, K., Monsen, C., Takhar, S.: A novel information retrieval tool to find hospital care team members: development and usability study. JMIR Hum. Factors **5**(2), e14 (2018)
11. Josea Kramer, B., Creekmur, B., Mitchell, M.N.: Expanding Home-based primary care to american indian reservations and other rural communities: an observational study. J. Am. Geriatr. Soc. **66**(4):A2673 (20182018)
12. Hutchison, J., Thompson, M.E., Troyer, J.: The effect of North Carolina free clinics on hospitalizations for ambulatory care sensitive conditions among the uninsured. BMC Health Serv. Res. **18**(1), 280 (2018)

Big Data on the Influence of SMEs in Export Trade Financing Costs

Jingqing Wang(✉)

School of Economics, Shanghai University, Shanghai, China
jqsun_97@163.com

Abstract. Big data is regarded as one of the digital technologies and new production factors, profoundly affects the society and economy. SMEs are often difficult to obtain financing from banks and other institutions due to information asymmetry. However, big data has three characteristics, such as diversity, high efficiency, and great potential value. It is usually used in analysis, prediction and evaluation scenarios, which plays a certain role in reducing the financing cost of SMEs. Therefore, this paper analyzes the impact mechanism of big data on the export financing cost of SMEs by describing the current situation of the application of financial big data in China. The conclusion is that big data can reduce the cost of SMEs' export financing by reducing the cost of credit assessment and risk control brought by information asymmetry, cross-border e-commerce platform and big data help banks to innovate credit model. In addition, according to the conclusion of the study, the paper also draws the Enlightenment of strengthening data sharing and cooperation among relevant departments, promoting the in-depth application of digital technologies in finance and international trade, and paying attention to the protection of data security and privacy.

Keywords: Big data · SMEs · Export financing cost

1 Introduction

Big data, AI, blockchain, IoT, cloud computing and other digital technologies are emerging technologies based on the Internet. The biggest difference between these digital technologies and the previous information technology is digitalization. Therefore, big data is closely related to all other digital technologies. In 2011, McKinsey announced that mankind has entered the age of big data in a research report, "the next frontier of big data: innovation, competition and productivity". Nowadays, data is regarded as a new factor of production. In collaboration with big data technology, data has a profound impact on society and economic activities, including international trade.

By 2020, small and medium-sized enterprises, referred to as SMEs, have accounted for more than 90% of the market entities, providing 80% of the employment, playing an increasingly important role in international trade and other economic activities, so they have become the research object of scholars. Under the framework of the new trade theory, enterprises need to pay a fixed cost to enter the export market. Only high productivity enterprises can overcome the export cost and choose export behavior

J. MacIntyre et al. (Eds.): SPIoT 2020, AISC 1282, pp. 345–351, 2021.
https://doi.org/10.1007/978-3-030-62743-0_50

[1–3]. However, because most of the entry costs of international trade need to be paid in advance, enterprises need to have sufficient liquidity [4, 5]. Therefore, financing constraints are the key factors that affect the export decision-making of enterprises. Many studies show that the main reasons why SMEs in China face greater financing constraints than large enterprises or multinational enterprises are information asymmetry, unstable operation and insufficient collateral [6–9]. Big data has the characteristics of massive data, timeliness and diversity of data, which effectively reduces the information asymmetry in the credit market and helps to reduce the threshold of export financing for SMEs [10–12]. At present, China's development is facing transformation, and the economic growth slows down, so it is urgent for SMEs to activate the market atmosphere. And solving the financing difficulties of SMEs will help further release the space for economic development. Therefore, this paper studies the impact mechanism of big data on the export financing cost of SMEs in China, and provides suggestions for big data to further reduce the financing cost of SMEs, so as to promote the development and growth of SMEs.

The next content and structure of this paper are as follows: the second part introduces the definition, characteristics and financial big data application; the third part summarizes reasons why SMEs in China are facing financing difficulties and the impact mechanism of big data on SMEs' export financing; the fourth part is the conclusion and enlightenment.

2 Big Data Overview

2.1 Big Data Concept and Characteristics

So far, there is no standard definition of big data, and scholars and research institutions have made their own definitions according to their own research. The earliest concept of big data comes from a META Research Report in 2001. In the report, 3vs model is used to describe the phenomenon of explosive growth of data [13]. The "volume" index data in the 3vs model is growing; "velocity" represents the timeliness of big data; and "variety" represents all data types including traditional structured data, semi-structured data and unstructured data. Such as, IBM and Microsoft still use 3vs-model to describe big data in the next decade [14]. Later, both Apache Hadoop (2010) and McKinsey (2011), a global consulting firm, defined big data as a data set that traditional computers could not collect, process and manage [15]. In 2011, a IDC Research Report defined big data as "big data technology describes a new generation of technology and architecture, aiming to achieve high-speed capture, discovery and/or analysis, and obtain value economically from a large number of diverse data" [16]. The definition is 4Vs model, which emphasizes the value dimension implied by big data. In addition, in early 2012, a "big data, big impact" published at Davos Forum in Switzerland defined big data as a new economic asset like monetary gold. China's definition of big data mainly focuses on that big data is an emerging technology. For example, the China big data industry ecological alliance and CCID Consultant (2019) believe that big data is a key part of the "Sensor-IOT-Cloud-BD-AI" new technology system.

The application of big data came into being when the Internet reached a stage of rapid development [10]. As a new factor of production, big data shows an exponential growth trend with the development of the Internet of Things and social media [14]. IDC, an international data company, released "the digital universe in 2020" in December 2012. The report points out that the total amount of data created and copied in the world in 2011 is about 1.8 ZB, and then it will double every two years. It is estimated that the total amount of global data will reach 40 ZB by 2020.

2.2 Financial Big Data Application

After years of development, China's financial industry has accumulated huge structured data, and various unstructured data are also growing rapidly, laying a good foundation for the financial big data application [10]. The financial big data application in China is mainly reflected in banking, securities and insurance. China's financial sector is dominated by Banks. Therefore, different from the centralized application of foreign financial big data in the investment banks and government supervision, China's financial big data is mainly used in bank credit. And bank credit is the most important external financing mode for SMEs [17]. According to a 2014 Research Report, more than 85% of the financing methods of SMEs in China are bank loans.

In the era of big data, China's banks are facing the needs of development model transformation. The bank is the place where Huge amounts of data is generated. Therefore, mining the value hidden in the massive data is the main driving force for the bank to introduce big data. The specific application is mainly reflected in four fields, namely customer management, operation management, risk management and coordination supervision [10] (Fig. 1).

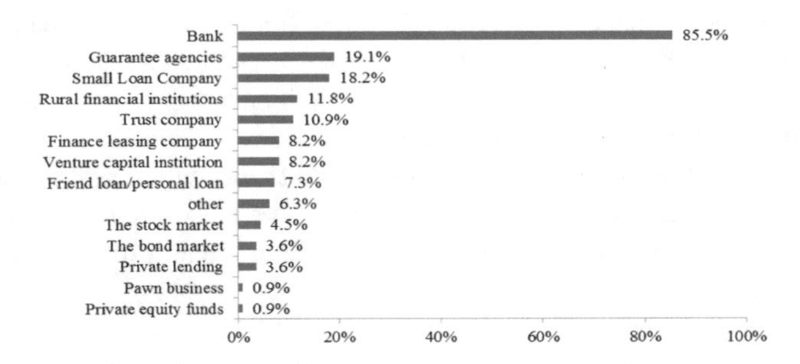

Fig. 1. Proportion of financing channels of SMEs in China in 2014. Source: https://mp.weixin.qq.com/s/k2X7jMngz1gZGP24JXvsWQ

3 Influence Mechanism of Big Data on Financing Cost of Export Trade of SMEs in China

The research shows that the reasons for the high financing cost of SMEs are mainly as follows: First, due to the lack of financial data management and the relatively high asset-liability ratio, SMEs have a low credit rating in Banks. Second, SMEs are reluctant to provide relevant services due to their small loan scale and small profit margins. Third, SMEs are unstable in operation, lack of asset mortgage, and have high risk premium. Fourth, because of the SMEs internal surplus retained funds, lack of internal financing basis. Fifth, the immature development of China's guarantee system and strict listing requirements lead to a single external financing channel for SMEs. Sixth, the bank's loan approval process is cumbersome, and SMEs are afraid to miss business opportunities and turn to private lending with higher interest rates.

The above reasons show that the high export financing cost of SMEs is mainly reflected in two aspects: high financing threshold and low financing efficiency. The essential reason for the high financing threshold of SMEs is information asymmetry, while the low financing efficiency is due to the old credit model of banks and other financial institutions for SMEs. Big data and big data technologies can solve these two problems to some extent. Therefore, this paper analyzes the impact mechanism of big data on the export financing cost of SMEs from two aspects of inclusiveness and efficiency.

3.1 Big Data Reduces the Threshold of Export Financing for SMEs

The development of the Internet not only gave birth to big data, but also created a new business model of e-commerce. The e-commerce platform is also the source of massive data, which is one of the main scenarios of big data application. SMEs carry out international trade activities with the help of cross-border e-commerce platforms such as Alibaba and dunhuang.com. While enjoying one-stop foreign trade services provided by the platform, they also leave their own information and transaction data on the platform. Through processing, mining and analyzing these historical transaction data, the platform evaluates the credit, stability and future behavior of enterprises, and provides certain trade financing services for SMEs. For example, Alibaba's fast selling platform provides SMEs with financial services in the whole international trade process [18]. Therefore, with the help of cross-border e-commerce platform, big data added new financing channels for the export of SMEs.

In the traditional mode of bank credit, the banks evaluate the credit of SMEs before credit and supervise the risk after credit. Therefore, the banks bear a lot of labor cost. SMEs' loan amount is relatively small, which leads to small profit space for banks. Therefore, banks are generally reluctant to provide financing services to SMEs. Therefore, the high labor cost and risk premium of SMEs are the direct reasons for the high financing cost of SMEs in banks. Reducing the cost of information asymmetry is the most important thing for banks to reduce the cost of export financing for SMEs. On the one hand, under the big data mode, banks use digital technologies such as big data, cloud computing and AI to batch and automate the processing of massive customers It not only improves the financing efficiency of enterprises, but also saves the labor cost, which helps to reduce the financing threshold of banks for SMEs. In addition, with the help of big data technologies, banks will be more motivated to independently design relevant financial services for SMEs with large financing demand, and provide more small-scale and convenient financing products for SMEs financing or export financing. On the other hand, Banks can also use cross-border e-commerce platforms to reduce the cost and risk of providing financing services for SMEs. When deciding whether to provide export financing services for SMEs. Banks can refer to the credit level of the enterprise provided by the cross-border e-commerce (CBEC) platform. In addition, some CBEC platforms also provide equity guarantee for enterprises on transactions, which enhances the anti-risk capability of SMEs. For example, under the guarantee service launched by Alibaba in 2015, Alibaba provides endorsement for export trade of SMEs and guarantees for financing of SMEs, which will help Banks to lower the threshold of export financing for SMEs. Therefore, big data can optimize the bank credit model.

In addition, the financial management of SMEs is not standardized, such as the lack of assets and liabilities and other information, and the confusion of business accounts. These financial management loopholes increase the difficulty of bank credit management and audit. Big data has the characteristics of automation, mass processing and intelligent decision-making, which helps to standardize the financial management and daily management of enterprises, thus reducing the financing threshold of banks for SMEs.

3.2 Big Data Improves Export Financing Efficiency of SMEs

Big data mainly improves the financing efficiency of SMEs from the perspectives of financing diversification and process digitalization. On the one hand, big data brings about the universality for the export financing of SMEs, which is mainly manifested in two aspects: increasing new financing channels, providing more diversified financial services and financing channels for the export financing of SMEs. SMEs have a greater choice when they carry out export financing. On the other hand, the reference of big data will make the credit process upgrade to digital and intelligent direction. In addition, new financing channels also use big data technology to automatically lend. For example, Alibaba's credit loan uses big data technology to automatically analyze whether it has the conditions to lend to the enterprise through obtaining the transaction data of the enterprise, and there is no manual intervention in the whole process [14].

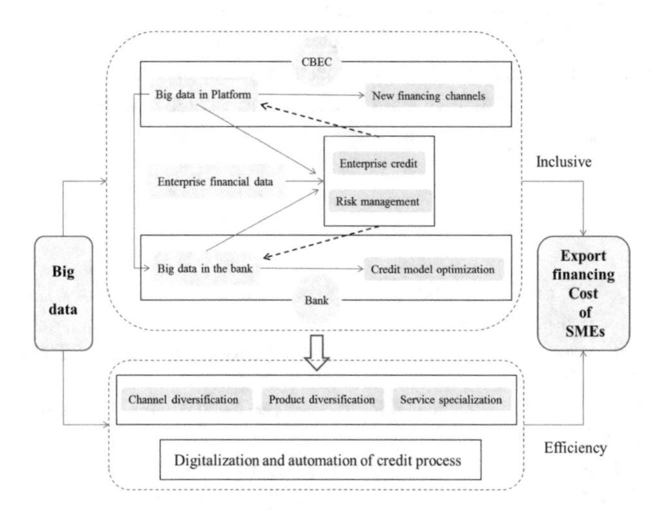

Fig. 2. Influence mechanism of big data on export financing cost of SMEs

This paper analyzes the influence mechanism of big data on the export financing cost of SMEs from the perspective of inclusive and efficiency, and then draws the impact mechanism chart. From the figure, we can find that enterprise credit assessment and risk management are the key factors for the high cost of SMEs' export financing; the inclusive nature of big data helps to improve the efficiency of SMEs' export financing; CBEC platforms play an important role in this impact mechanism (Fig. 2).

4 Conclusions and Enlightenments

4.1 Conclusions

This paper draws the following conclusions from the impact mechanism of big data on SMEs' export financing cost. First, big data mainly reduces the threshold of banks' export financing to SMEs by reducing the cost of credit assessment and risk control brought by information asymmetry. Second, CBEC can help that big data reduce the financing costs of SMEs. Third, big data helps banks innovate credit models and provide more convenient and rich financing products or services for SMEs' export financing.

4.2 Enlightenments

According to the conclusion of the study, this paper draws the following three enlightenments. First, strengthen big data cooperation among financial institutions, enterprises, e-commerce platforms and customs. The main reason for the high financing cost of SMEs is information asymmetry, and CBEC platforms, customs and enterprises are the source of data generation. Therefore, sharing data will more accurately judge the credit level of SMEs. Second, promote application of digital technologies. With the widespread application of blockchain, AI, cloud computing and other digital technologies in finance and international trade, the efficiency of big data mining value will be further improved, so as to help financial institutions to innovate more financial service products. Third, strengthen big data security and privacy protection. We should pay attention to the importance of data security and privacy protection, and avoid the extra risks and costs brought by data abuse to enterprises.

References

1. Melitz, M.J.: The impact of trade on aggregate industry productivity and intra-industry reallocations. Econometrica **71**, 1695–1725 (2003)
2. Bernard, A.B., Jensen, B.J.: Exceptional exporter performance: cause, effect, or both? J. Int. Econ. **47**, 1–25 (2004)
3. Helpman, E., Melitz, M.J., Yeaple, S.R.: Export versus FDI with heterogeneous firms. Am. Econ. Rev. **93**, 300–306 (2004)
4. Chaney, T.: Liquidity constrained exporters. J. Econ. Dyn. Control **72**, 141–154 (2016)
5. Manova, K.: Credit constraints, heterogeneous firms and international trade. Rev. Econ. Stud. **80**, 711–744 (2013)

6. Wei, S.: Credit rationing, export credit insurance and export financing of SMEs. Spec. Econ. Zone (03), 224–227 (2011). (in Chinese)
7. Nan, J.: Problems and countermeasures of international trade financing of SMEs in China. Res. Bus. Econ. **08**, 35–36 (2015). (in Chinese)
8. Cao, X., Shen, B.: How to solve the international trade financing problems of SMEs. People's Forum. **36**, 70–71 (2017). (in Chinese)
9. Guo, W.: Big data helps inclusive financial development. China Finance **12**, 34–35 (2019). (in Chinese)
10. Chen, J., Zhu, N.: Research on application of big data in financial industry. Libr. Inf. Work. **59**(S2), 203–209 (2015). (in Chinese)
11. Ding, Z., Sun, R., Xiong, Y.: Exploration of big data in small and micro financial business practice. Theory Pract. (01), 98–102 (2016). (in Chinese)
12. Song, J.: Big data and small and micro enterprise financing services. China Finance **16**, 55–56 (2018). (in Chinese)
13. Laney, D.: 3-D Data Management: Controlling Data Volume, Velocity and Variety. META Group Research Note (2001)
14. Chen, M., Mao, S., Liu, Y.: Big data: a survey. Mob. Netw. Appl. **19**(2), 171–209 (2014)
15. Manyika, J., McKinsey Global Institute, Chui, M., Brown, B., Bughin, J., Dobbs, R., Roxburgh, C., Byers, A.H.: Big Data: The Next Frontier for Innovation, Competition, and Productivity. McKinsey Global Institute (2011)
16. Gantz, J., Reinsel, D.: Extracting Value from Chaos. IDC iView, pp. 1–12 (2011)
17. CCID consultant, Big Data Industry Ecology Alliance: 2019 China Big Data Industry Development White Paper (2). China Computer Daily (008) (2019). (in Chinese)
18. Chen, J.: "New finance" in the future under the "new foreign trade" mode. Banker **10**, 54–55 (2017). (in Chinese)

Big Data and Economics Research

Wanting Zhang(✉)

School of Economics, Shanghai University, 99 Shangda Road, Baoshan District, Shanghai 200000, China
13681637472@163.com

Abstract. Big data technology is a new digital technology, which is as famous as the Internet of things and artificial intelligence in the past 20 years. It is an important technology that will profoundly change the way of human production and life. For economic researchers, the popularization and development of big data will bring subversive changes to the acquisition, collation and timeliness of data in economic research, and will play a positive role in promoting the development of economic research methods. In terms of the availability of data sources, the combination of big data technology and computer technology can make researchers easily obtain the required data, no matter the structured data we usually use or the unstructured data we don't often use. In terms of the timeliness of data acquisition, big data technology makes the timeliness of economic data possible and effectively solves the lag in economic research Post problem: in terms of the value of data acquisition, although massive data reduces the value density of data, the quality of data captured by big data technology is far higher than that of ordinary data acquisition methods.

Keywords: Big data · Economic research · Data acquisition

1 Introduction

Economics is a social science, which studies human behavior and how to allocate limited or scarce resources reasonably. In this sense, human behavior is the cornerstone of economic analysis and the basic premise of human economic activities. With the wide application of econometric analysis in economic research in recent decades, it has attracted more and more humanities research to introduce econometric analysis, because econometric analysis is an objective mathematical analysis based on fact data, while pure theoretical economic research usually takes the subjective understanding of the author as the starting point, and it is difficult to get objective and unbiased analysis conclusions [1, 2]. The Journal of econometrics, published in 1933, marks that econometrics has officially become a branch of economics. It is also known as the three basic courses of economics introduction together with microeconomics and macroeconomics.

When it comes to econometric analysis, we can't get around the problem of data selection and application. The data needed for economic research is different from the experimental data for science and engineering research. The experimental conditions of science and engineering can control the independent variables artificially. The control variable method can be used perfectly to complete the research process and clarify the

J. MacIntyre et al. (Eds.): SPIoT 2020, AISC 1282, pp. 352–358, 2021.
https://doi.org/10.1007/978-3-030-62743-0_51

relationship between the independent variables or between the independent variables and the dependent variables [3]. Economics is a human science. Researchers abstract the phenomena of human economic activities in the objective world into data, but they can't make the data change freely under a certain condition. In view of this characteristic of economic data, measurement methods such as double difference in difference and synthetic control came into being, trying to construct data under specific control conditions, so as to compare with the economic data obtained in real activities, and explore whether a certain variable has an impact on the dependent variable and the degree of influence geometry [4–6]. That is to say, in the study of economics, the immutability of data from natural economic activities is made up by the auxiliary method of measurement. In fact, in econometric theory, there are also small sample size ($n <= 30$) and large sample size ($n > 30$). In practical regression application, the latter has less statistical restrictions than the former. The more data, the more representative the research object is.

However, the arrival of big data has subverted the inertial thinking of economic data acquisition. For big data (big At present, there is no official definition of data), which generally refers to massive, high growth rate and diversified information assets, with five characteristics of huge volume, variety, speed, low value density and veracity. The first phase trade agreement between China and the United States, just signed, suddenly became uncertain because of the epidemic. It took only a few months, but before the outbreak, all the predictions of the economic situation had been made. The classics are quite different [7, 8]. With the rapid development of modern communication technology, everything in the world has become more and more changeable, and the timeliness of economic data has become more and more important. At this time, the advantages of big data technology are fully reflected.

2 Big Data Technology and Data Mining for Economic Research

At present, the data sources used by economists for economic research are mostly second-hand data, but it is difficult to obtain the first-hand data through interviews, questionnaires and other ways according to their own research needs. Although we can often read relevant articles and research reports of economics in many media focusing on economics, these can not be called "economic research" [9]. Economic Research in a narrow sense refers to the process of writing and publishing papers in journals. It is a pioneering exploration and research in a certain subdivision field of economics. It is a process of independent exploration of the author's model and knowledge of Applied Economics. It represents the wisdom crystallization of "from scratch" in an economic field and is mostly applied by full-time economic researchers. The analysis of economic phenomena and the research reports published by research institutes in the social media and WeChat official account are not the way to give the readers the economic phenomena. They are not called economic research in a strict sense [10]. They are more about transmitting economic thoughts and knowledge to the general public. The study of economics in this paper is a narrow reference.

The use and selection of data are the key elements to distinguish these two types of articles. In addition to a few pure literature review articles, the vast majority of economic research is that it is necessary to make mathematical conclusions after the data is processed by measurement, and then combine the background, purpose and economic significance of the mathematical conclusions to confirm the problems to be studied and draw research conclusions. However, the common articles on economic phenomena seldom use data, and usually put forward their own views on the current popular economic hot spots. In order to make economic research as broadly representative as possible, the amount of data needed is large and the data source must be reliable. This determines that most economic research uses second-hand data - data published by government agencies, business research institutions for or not for a specific research object [11]. Generally speaking, the ability of an independent researcher is limited, and it is difficult to obtain a large number of reliable first-hand data. Second-hand data is not only the first choice of researchers, but also helps guide researchers to adjust and explore more suitable research directions according to the data situation. At present, the second-hand data of economic research mainly come from the following sources (Table 1):

Table 1. Sources of secondary data in economic research

Data categories	The data source	Whether to pay	Note
Single country data	China – Official website of National Bureau of Statistics of China	No	Data on all aspects of the country's economy, politics, health, etc.
	China - National and provincial statistical yearbooks	No	The amount of information is so large and complex that it takes a lot of time to sort out the records
	United States – Official website of the Bureau of Economic Analysis	No	Macroeconomic operation data
World data	United Nations Conference on Trade and Development	No	More data on economic and trade exchanges between countries can be arranged in a freely selected form
	United Nations Comtrade Database	No	World economic and trade data
	International Monetary Fund	No	Economic and financial indicators such as loans and exchange rates
	OECD data	No	Economic and social data for OECD countries
	World Bank	No	Database of world economic indicators
Other data	CSMAR database	Yes	Individual country and world data are available and varied
	Wind database	Yes	Complete categories, easy to download and analysis
	The home of concession	Yes	Academic forum website, can post paid to obtain data

Some of the above data sources are related data columns on the official website of the world organization, and some are specialized data cluster websites, with different availability and practicability. But one thing in common is that these data need researchers to travel to various websites to find out whether there are the variables and corresponding indicators needed by the research, which is time-consuming and laborious. Under the big data technology, data is not limited to the official website data of data collection or organization in traditional economic research, but can capture valuable real-time data from non structural data such as social media, websites, forums, columns, application traces, etc. Big data technology is also behind the common "kill" routine of businesses. For the same product, different user accounts may be used to log in for different purchase prices. This is Taobao's calculation of user's purchasing power based on user's consumption records. In fact, it is the application of big data technology to form a three-level price discrimination against consumers and grab all consumer surplus.

Among them, the process of grabbing data needs natural language processing algorithm, which transforms the text that only records economic activities in the network into valuable economic data by imitating the transformation process of human brain. At present, the mainstream data mining tools include Google file system, BigTable, MapReduce, and the most classic Dremel, bigquery used to convert unstructured data into structural data. Because of the complexity of massive data, deep mining the causal link between economic variables is also a bright spot in the application of big data technology in economic research. The trend of variables has been preset between the economic data obtained by traditional methods. For example, the total GDP, per capita GDP and national fiscal expenditure can be found in the same data set, and it is easy to construct the causal relationship between them with traditional economic ideas.

Using big data thinking to collect data to judge the trend of macro-economy will help researchers find more potential connections between economic variables, "Li Keqiang index" is a typical case. "Li Keqiang index" is a combination of three economic indicators: new industrial power consumption, new railway freight volume and new medium and long-term loans from banks. It comes from Premier Li Li Keqiang's preference to judge the economic situation of Liaoning Province by power consumption, railway freight volume and loan granting volume when he was Secretary of Liaoning Provincial Party committee. These three indicators do not seem to directly reflect the economic situation, but in fact, they are inextricably linked with the macroeconomic situation of a region, and they can also ensure the authenticity of the data. The economist used ten years' data to compare the trend of "Li Keqiang index" with the official GDP, and found that the trend of the two is the same. Compared with the data that needs manual calculation and arrangement, these three data values are better obtained, "Li Keqiang index" is an important case in which big data thinking is typically used.

3 Big Data Technology and Timeliness of Economic Data Acquisition

Any information has time value, so does data. With the rapid development of communication technology in the 21st century, traditional data collection methods have been difficult to keep up with the changing economic environment, and can not meet the needs of people to use real-time data to create wealth. For some very abstract data which can only be calculated by researchers, it is not only very difficult to obtain, but also the economic variables which depend on subjective conjecture may be distorted. In fact, timeliness can not only be used to judge the data generated at the moment, because the economic phenomenon we have seen is always the past, and its more important significance is to accurately predict the future phenomenon, so that we can make policy adjustments to obtain the maximum benefits or avoid losses.

In addition, the application of big data technology in economic research is not only reflected in technology acquisition, but also in the transformation thinking of data acquisition. Some direct economic phenomena can be seen from indirect resource consumption and output. Li et al. (2016) used the web crawler tool to mine the information data of 271 small and medium-sized American green food and manufacturing company websites to study the impact of the micro level small and medium-sized enterprise performance and the contact with the government, industry and academia on the sales growth, established the panel regression model of the contact between the government, industry and academia, and verified that the enterprise and the government, industry Academic connections have a positive impact on sales. In addition to the above data mining tools, the relatively rudimentary python, R, MATLAB and other crawlers and statistical software can also obtain, extract and calculate big data. It is no longer monopolized by high-end technologists to engage in big data research. Most researchers can independently model research and obtain real-time first-hand data, which can greatly improve the research efficiency and prediction accuracy of economics, with strong application and practical significance.

4 The Value of Data in Big Data Technology and Economic Research

As mentioned in the introduction, big data value has the characteristics of low density (value) and huge volume. Although there is a correlation between these two characteristics, there is no absolute causal relationship. On the surface, the value density of data is inversely proportional to the total amount of data. The higher the value density of data, the smaller the total amount of data. The lower the value density of data, the larger the total amount of data. Valuable information is contained in the massive basic data. The more data, the more valuable information. But in fact, there are also cases where the value density of massive structured data is not low. For example, UnionPay, visa and other clearing organizations have massive daily transaction data, which is not only large in data volume, but also valuable.

China attaches great importance to the development of big data, and promotes the development of big data to the national strategy. In 2016, the 13th five year plan outlines the direction goals and tasks for promoting the development of big data in an all-round way; in 2017, the report of the 19th National Congress of the Communist Party of China proposes "promoting the deep integration of Internet, big data, artificial intelligence and real economy". This shows that the country has fully realized that big data technology has been closely integrated with people's production and life.

Developing big data technology is an important step to promote the development of national economy and people's livelihood.

The primary value of big data needs to build application platform to develop value. Looking at all kinds of emerging technologies, cloud computing platform, Internet of things platform and artificial intelligence platform are developing rapidly. Platform based development can not only accelerate the opening speed, but also reduce the development threshold. Small and medium-sized enterprises can use the platform to realize big data innovation in the application field of the company. Large enterprises can use the platform to construct their own big data system. Ctrip is one of them A classic example. On April 20, 2020, James Liang, chairman of Ctrip's board of directors, published an article "does working from home work? Evidence from a Chinese experience" in the famous international economic journal quarterly of economics to study whether working at home during the epidemic has a positive or negative impact on the work efficiency of employees. The highlight of this article is data collection. The data in this paper are all first-hand data collected for this experiment. James Liang collected data on the length of time and efficiency of employees working at home and in the company, and calculated the related benefits to get a top-level paper. Based on the number of more than 25000 employees in Ctrip, it can fully meet the requirements of data collection. This is a classic case of large enterprises completing economic research through their own big data system.

It should be noted that big data itself does not generate value, but its fundamental purpose is to use big data mining analysis to provide rules, knowledge and experience and other scientific basis for our decision-making, so as to objectively reduce the uncertainty of future decision-making. Therefore, the collection of big data is one aspect. How to develop and utilize it well determines the value of data utilization.

5 Conclusion

From the perspective of economic research, this paper discusses the promotion mechanism of data source, data acquisition and data value for economic research. In the academic circles with traditional data as the main research topic, the emergence of big data will be the innovation point of future economic research, and will greatly improve the efficiency and practical degree of economic research.

However, we should also note that it is not easy for a new technology to be recognized, popularized and applied to economic research by the public, which will take at least 5–10 years; however, from the actual situation, China's big data processing technology is only in its infancy, and its development is not yet mature, so we should pay attention to scientific planning and put forward suitable for China's actual situation. The situation of big data strategy and development path, forming a good big data development environment. For economic researchers, we should learn from the positive attitude of learning and acceptance and learn from the frontier economic model. We should not be too frustrated in the process of empirical research. After all, it is an immature research method to extract its essence and discard its dross.

References

1. Tao, H., Bhuiyan, M.Z.A., Rahman, M.A., Wang, G., Wang, T., Ahmed, M.M., Li, J.: Economic perspective analysis of protecting big data security and privacy. Future Gener. Comput. Syst. **98**, 660–671 (2019)
2. Anejionu, O.C.D., (Vonu) Thakuriah, P., McHugh, A., Sun, Y., McArthur, D., Mason, P., Walpole, R.: Spatial urban data system: a cloud-enabled big data infrastructure for social and economic urban analytics. Future Gener. Comput. Syst. **98**, 456–473 (2019)
3. Dembitz, Š., Gledec, G., Sokele, M.: An economic approach to big data in a minority language. Procedia Comput. Sci. **35**, 427–436 (2014)
4. Khan, R., Jammali-Blasi, A., Todkar, A., Heaney, A.: PHS61-Using big data for an economic evaluation of reducing the innappropriate prescribing of atypical antipsychotics among dementia patients. Value Health **18**(3), A257 (2015)
5. Roth, S., Schwede, P., Valentinov, V., Žažar, K. and Kaivo-oja, J.: Big data insights into social macro trends (1800–2000): a replication study. Technol. Forecast. Soc. Change **149**, 119759 (2019)
6. Vialetto, G., Noro, M.: An innovative approach to design cogeneration systems based on big data analysis and use of clustering methods. Energy Convers. Manag. **214**, 112901 (2020)
7. Aboelmaged, M., Mouakket, S.: Influencing models and determinants in big data analytics research: a bibliometric analysis. Inf. Process. Manage. 57(4), 102234 (2020). ISSN 0306–4573
8. Baig, M.I., Shuib, L., Yadegaridehkordi, E.: Big data adoption: state of the art and research challenges. Inf. Process. Manage. **56**(6), 102095 (2019). ISSN 0306–4573
9. e Camargo Fiorini, P., Seles, B.M.R.P., Jabbour, C.J.C., Mariano, E.B., de Sousa Jabbour, A.B.L.: Management theory and big data literature: from a review to a research agenda. Int. J. Inf. Manage. **43**, 112–129 (2018). ISSN 0268-4012
10. Surbakti, F.P.S., Wang, W., Indulska, M., Sadiq, S.: Factors influencing effective use of big data: a research framework. Inf. Manage. **57**(1), 103146 (2020). ISSN 0378-7206
11. Cuquet, M., Fensel, A.: The societal impact of big data: a research roadmap for Europe. Technol. Soc. **54**, 74–86 (2018). ISSN 0160-791X

Protection and Value of DaYing Zhuotong Well Based on Big Data

Jinlan Xia(✉)

Sichuan Vocational and Technical College, Suining, Sichuan, China
13651381@qq.com

Abstract. In recent years, more and more people have used big data analysis to study the protection and value of China's cultural heritage. China's historical and cultural heritage is rich, with distinctive regional characteristics and precious historical and cultural values, and its research value is extremely high. high. Among them, China's Great DaYing Zhuotong well has a relatively high level of development and has been included in the "World Cultural Heritage List." This article takes big data analysis as an entry point, and tries to emphasize the identity of DaYing Zhuotong well's cultural heritage, and studies the protection and utilization of value. Taking cultural heritage as the main research purpose, and put forward separate recommendations for the protection and utilization of Chinese DaYing Zhuotong well cultural heritage. The data analysis of my country's historical and cultural heritage, Great DaYing Zhuotong well was carried out, including the explanation of related concepts and the order of protection process. Secondly, taking DaYing Zhuotong well as a research case, analyzing its cultural heritage resources one by one, protecting the status quo and threat factors, the results show that DaYing Zhuotong well cultural heritage resources are extremely rich in types and types. And the degree of cultural inheritance is very high. The quantitative evaluation method is used to evaluate the value of resource use in different ways, and to evaluate the four indicators of resource funding status, economic growth status, environmental quality and social participation. The survey shows that 50.5% of the residents believe that among the influencing factors for the protection of DaYing Zhuotong well, the government's protection policies have the greatest impact on their protection and have a certain degree of credibility.

Keywords: Big data analysis · Daying Zhuotong well · Cultural heritage protection · Value research

1 Introduction

With the progress and development of human civilization, from the perspective of big data analysis, people are paying more and more attention to the protection of heritage. In addition, China is recognized as a large country with rich natural and cultural heritage resources in the world, and shoulders the important task of protecting world heritage [1, 2]. Among them, the British historical and cultural heritage DaYing Zhuotong well, which is an important part of Chinese heritage resources, is of great significance for the protection of China's material and intangible heritage [3]. The

J. MacIntyre et al. (Eds.): SPIoT 2020, AISC 1282, pp. 359–365, 2021.
https://doi.org/10.1007/978-3-030-62743-0_52

value of agricultural cultural heritage is very rich, but the residents living in the field of cultural heritage understand agricultural cultural heritage and believe that agricultural cultural heritage is a common thing or content in daily life, so it cannot give enough attention to its appropriate cultural connotation [4]. Through field investigations on the value of DaYing Zhuotong well and questionnaire surveys and interviews with local residents, on the one hand, they can gain a deeper understanding of local residents' wishes and true views on the protection of agricultural cultural heritage, and on the other hand, they can also strengthen their agricultural culture. Propaganda of the heritage, deepen local residents' understanding of the content of the heritage [5, 6].

As with other historical and cultural villages and towns in China, the business development of DaYing Zhuotong well also has a negative impact on local development. The commercial atmosphere and the external cultural atmosphere gradually cover up the culture of DaYing Zhuotong well. The quality of cultural products and their own commercialization are serious, and the bottom-up protection power is obviously insufficient. This has gradually led to a series of phenomena that have disappeared in tradition and culture [7, 8]. People need to look at the problem of the development and protection of the DaYing Zhuotong well from the perspective of big data analysis [9, 10].

Based on the analysis of the agricultural cultural heritage connotation, characteristics, protection status and its value composition, non-use value assessment methods, etc., this paper derives the non-use value monetary value of the agricultural cultural heritage given by the residents of the heritage site and the factors affecting the willingness to pay At the same time, explore the cooperative protection development model of agricultural cultural heritage, and provide a basis and reference for the local government to formulate protection policies. This article collects questionnaires through field research, uses Epidate software to input the data content of the questionnaire, uses Excel software to perform mathematical statistics and graphs in the questionnaire data, and uses SPS statistical software to perform multiple Logit data analysis.

2 Big Data Analysis of the Protection and Value of China's Daying Zhuotong Well

2.1 The Exploration of the Value of China's DaYing Zhuotong Well by Big Data Analysis

The era of big data has changed the methods and techniques of asset evaluation and is used to evaluate cultural enterprises on the Internet. In addition, in the era of big data, a large number of cultural heritage data companies are constantly exploring the management of Chinese cultural heritage data. For example, based on advanced concepts and big data technology, the CNFS DaYing Zhuotong well value dynamic evaluation system for the value of large stocks serving the banking industry and other financial institutions is a system tool required for dynamic evaluation. Chinese researchers have also studied the main data in the evaluation of financial assets. By comparing the three valuation methods of Great DaYing Zhuotong well, they decided to choose the cost method for evaluation, and established the valuation model of Great DaYing Zhuotong well in the era of big data.

In recent years, big data has gradually begun to be applied to the evaluation of cultural enterprises. For example, Google and some home film and television sites have conducted a lot of data analysis to predict the value of film and television projects to cultural business evaluation. Considering the income forecast in the field of cultural tourism, in addition to the traditional factors that affect the evaluation value of cultural enterprises, try to use big data to measure consumer market evaluation factors.

2.2 The Great Contribution Value of DaYing Zhuotong Well to Human Civilization

DaYing Zhuotong well is a precious cultural heritage in China. It was widely used in the Northern Song Dynasty (1041–1048). It is the earliest small-caliber drilling technology invented by mankind. It opened the prelude for mankind to explore the deep underground. Greatly improved the level of well salt production, and also discovered oil and gas resources buried deep underground, which promoted the rise of modern petroleum, chemical, aerospace, automotive and other industrial technologies, and accelerated the world's energy revolution and the development and progress of human society. Like gunpowder, papermaking, printing, and compasses, it has made a huge contribution to mankind.

2.3 DAYING ZHUOTONG WELL Cultural Heritage Value Evaluation

The British DaYing Zhuotong well entered the "World Cultural and Natural Heritage Preparatory List" due to meeting the selection criteria of (ii), (iv) and (v) of the World Cultural Heritage, combining the above for the types of ancient towns in the existing World Heritage List The result of comparison of the selection criteria for heritage projects found that the sum of the adoption rates of selection criteria (ii) and (iv) exceeded half of the total number. In general, the value of the heritage of the Great DaYing Zhuotong well is higher. In terms of the number of combinations of its selection criteria, the DaYing Zhuotong well meets a total of 3 entry criteria at the same time, which ranks in the top third of the ancient town-type world heritage, and the number is relatively rich; in addition, from the combination of selection criteria In terms of ways, the combination of the selection criteria of the DaYing Zhuotong well is consistent with the combination of the three heritages of the ancient city of Lijiang, the ancient city of Luang Prabang and the church village of Gmeldard in Lulea, which reflects its comparative advantages. High universal value. The above analysis results show that, compared with the existing ancient town-type world heritage, the cultural value type of DaYing Zhuotong well is more common and the value connotation is richer.

2.4 Reflections on the Protection Model of the Great DaYing Zhuotong Well

DaYing Zhuotong well is kept intact, with rich ancient buildings, residential houses, and more historical and cultural legends. The discovery of Cao Cao's tomb in Xigao Cave of Anfeng Township in the past few years has attracted wide attention from all walks of life. Its huge historical and cultural value and celebrity effect should have greater development value if it is jointly protected with DaYing Zhuotong well. DaYing Zhuotong well is close to the urban area of Sichuan, and is convenient to tourist sites such as the world cultural heritage Yinxu and museums. The development of DaYing Zhuotong well cannot achieve remarkable results by itself. According to the principle of joint protection, it can be integrated with the surrounding historical and cultural relics to form an Anfeng cultural protection and development circle dominated by DaYing Zhuotong well. Integrating the cultural resources of northwestern Sichuan has also expanded the popularity of the ancient cultural capital of Sichuan.

3 Experimental Data Collection

3.1 Questionnaire Design

In this study, through the payment card questionnaire, the residents of the Great DaYing Zhuotong well Heritage Area were surveyed on the willingness to pay for the protection of the Great DaYing Zhuotong well, so the design of the questionnaire is related to the success or failure of the survey. According to the 15 principles put forward by NOA and combining the characteristics of Inei Jianjing, the designed questionnaire mainly includes four parts.

3.2 Sample Collection

The scope of the DaYing Zhuotong well heritage site is in the Guanchang village area. The residents of Guanchang village all meet the sampling target of this survey. The population of Guanchang village was 214,500 at the end of 2014. According to the Scheafer sampling formula, N = 419.5 was calculated. It can be seen that the number of samples collected above 421 can meet the statistical requirements. However, Mitchel and Carson believe that there are many deviations in CVM, the sample size should be greater than the general statistical threshold, and that the estimated WTP value can be within 15% of the true WTP value, then a specific sample is being carried out When sampling, at least the number of samples to be taken in the study area should be around 60. To this end, a total of 60 questionnaires were distributed in this survey, and 30 questionnaires were distributed in the core area and non-core area of Daying DaYing Zhuotong well. In addition, in order to avoid errors caused by the choice of survey method, this survey chose the interview method with high recovery rate in the questionnaire survey.

4 Based on the Analysis of Big Data Analysis of the Protection of China Daying Zhuotong Well and Its Value Experiment

4.1 The Big Data Experiment Analysis of the Value of DaYing Zhuotong Well

This paper evaluates the non-use value of Daying DaYing Zhuotong well mainly including the existence value, selection value and heritage value of Daying DaYing

Zhuotong well. In order to be able to fully investigate the composition of non-use value, a survey was conducted on the respondents' payment reasons. In order to enable the interviewees to intuitively understand the key questions of the survey, the questionnaire describes the reasons why the respondents are willing to pay in popular language as: A: In order to enable the use of DaYing Zhuotong well and the cultural values it contains in the future; B: In order to preserve the Great DaYing Zhuotong well and its related heritage for future generations; C: In order to make the Great DaYing Zhuotong well and its related content exist forever, correspond to the Great DaYing Zhuotong well non-use value selection value, heritage value and There is value, and the statistical results are as follows (Table 1).

Table 1. Non-use value classification survey

Classification	Core area (%)	Non-core area (%)
Choice value	23.65%	38.35%
Legacy value	41.58%	32.48%
Existential value	34.97%	36.37%

The highest proportion of non-use value in the core area is heritage value, accounting for 41.58%, followed by the existence value of 34.97%, and finally the selection value of 23.65%; the highest proportion of non-core area is the selection value of 38.35%, Followed by the existence value of 36.37%, and finally the heritage value of 32.48%. Through comparative analysis, it can be seen that the residents of the core area tend to the heritage value of the heritage, hoping to protect the DaYing Zhuotong well by paying a certain amount of money, so that this valuable wealth can be left to future generations, and the existence value is also high. The non-use value distribution of the residents in the non-core area is more even, and they are mainly inclined to pay for the protection of themselves and future generations in order to have the opportunity to consume the DaYing Zhuotong well in the future.

4.2 Research on the Protection Measures of the Cultural Heritage of Great DaYing Zhuotong Well

The main body of the protection work of DaYing Zhuotong well is the residents of the heritage site, and the behavior and wishes of the residents of the heritage site are the basis for the protection and inheritance of agricultural cultural heritage. Based on the views of residents in the core and non-core areas on the factors affecting the protection of DaYing Zhuotong well, the core area statistics are shown in Fig. 1 below.

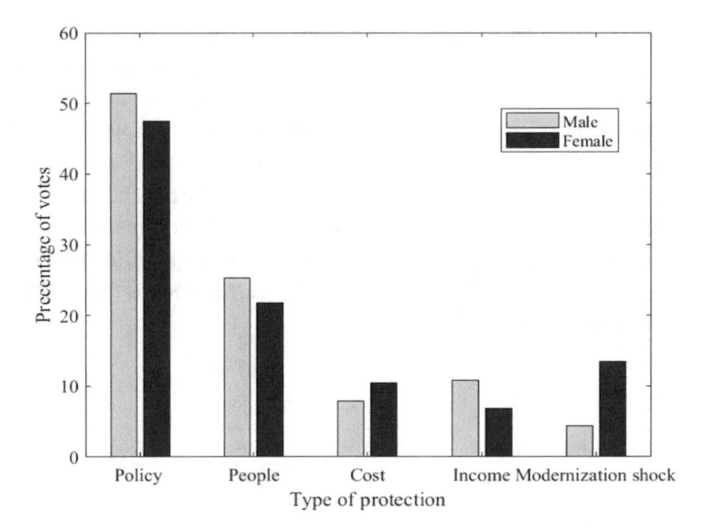

Fig. 1. Investigation and statistics of influencing factors of protection of residents in the core area

Residents of the core area believe that among the factors affecting the protection of DaYing Zhuotong well, the government's protection policy has the greatest impact on their protection, accounting for 51.5%. The payment by the state and the government echoes, and the result has a certain degree of credibility. Secondly, the protection consciousness of the public also has a greater impact on the protection of DaYing Zhuotong well, accounting for 22.3%, followed by protection gains, accounting for 8.6%, and then the impact of modernization shocks and protection costs, accounting for 7.9%, respectively and 8.1%.

5 Conclusions

Based on field investigation and questionnaire survey, this paper evaluates and analyzes the sustainable use value of DaYing Zhuotong well. Taking the sustainable use value of the world heritage as the target and the resource characteristics, economic characteristics, ecological characteristics and social characteristics as the evaluation indicators, the evaluation index system of the sustainable use value of the cultural heritage resources of DaYing Zhuotong well has been established. The comparison between two pairs and the quantitative grading of the subjective evaluation indicators finally lead to the conclusion that Dayingzhuo Jane's heritage resources are in good condition and meet the requirements of sustainable use; the economic and social use values are high and can be reasonably developed and used; the ecological environment. The utilization value is slightly lower, and it is necessary to pay attention to protection while reducing the damage of unreasonable utilization methods. In the process of development and utilization of the ancient town's heritage resources, in addition to considering the scarcity and vulnerability of the heritage resources themselves, it is also necessary to consider whether the ancient town's ecological environment carrying capacity can meet the use requirements, and on the basis of this condition, carry out a scientific and reasonable economy Development and social utilization can only take into account the needs of sustainable use of heritage resources in many aspects.

Acknowledgements. Subject: Sichuan Provincial Department of Education Humanities and Social Sciences Key Project: Research on Oral History and Culture of China DaYing Zhuotong Well (Project Number: 16SA0152).

References

1. Guo, R.: China's maritime silk road initiative and the protection of underwater cultural heritage. Int. J. Mar. Coast. Law **32**(3), 510–543 (2017)
2. Xinyao, N., Xiaogang, C.: Protection and inheritance of native culture in the landscape design of characteristic towns. J. Landscape Res. **10**(02), 90–93 (2018)
3. Li, J.: Huangmei opera cultural development research under intangible cultural heritage protection background. J. Comput. Theor. Nanosci. **13**(12), 10015–10019 (2016)
4. Hao, Z., Taeyoung, C., Huanjiong, W., et al.: The influence of cross-cultural awareness and tourist experience on authenticity, tourist satisfaction and acculturation in world cultural heritage sites of Korea. Sustainability **10**(4), 927 (2018)
5. Lei, H., Hongyu, H., Chongcheng, C.: Three-dimensional modeling of hakka earth buildings based on the laser scanned point cloud data. Remote Sensing Technol. Appl. **82**(5), 2007–2016 (2015)
6. Wang, G., Wang, M., Wang, J., et al.: Spatio-temporal characteristics of rural economic development in eastern coastal China. Sustainability **7**(2), 1542–1557 (2015)
7. Miao, J., Wang, Z., Yang, W.: Ecosystem services of Chongyi Hakka Terraces. Chin. J. Appl. Ecol. **28**(5), 1642 (2017)
8. Aretano, R., Semeraro, T., Petrosillo, I., et al.: Mapping ecological vulnerability to fire for effective conservation management of natural protected areas. Ecol. Model. **295**(Sp. Iss. SI), 163–175 (2015)
9. Yang, Y.: Implementation strategies of low-carbon tourism. Open Cybern. Systemics J. **9**(1), 2003–2007 (2015)
10. Bartolini, I., Moscato, V., Pensa, R.G., et al.: Recommending multimedia visiting paths in cultural heritage applications. Multimedia Tools Appl. **75**(7), 3813–3842 (2016)

Analysis and Research on Modern Education Technology Under "Internet+" and Big Data Technology Environment

Xiaofang Qiu(✉)

Sichuan Vocational and Technical College, Suining 629000, Sichuan, China
214838468@qq.com

Abstract. The research purpose of this article is based on the discussion and research of modern educational technology under the background of "Internet+" and big data environment. The research method used in this article is to first adopt the establishment of a mathematical model of precision education based on the era of big data to formulate accurate modern education programs and multi-stage dynamic precision education programs. Secondly, under the blessing of the "Internet+" era, after the background and data collection, the data is processed and the indicators are extracted, and a mathematical model for precision education is finally formulated. Finally, a questionnaire survey was used to conduct a sample survey on the H school, and the effective data obtained in the final survey was counted and combined with the previous two methods to analyze the analysis of "Internet+" and big data technology on modern educational technology. Experimental research results show that the emergence of Internet+ and big data greatly facilitates the development of educational work. Then on this basis, it is necessary for all parties to explore suitable modern education methods according to the characteristics of cloud Internet+ and big data, in order to improve teaching ability and cultivate excellent young talents.

Keywords: Internet+ · Big data technology · Modern education · Education analysis

1 Introduction

With the rapid development of Internet technology, the distributed processing, storage and calculation of network data information rely on a large number of resource pools, and the computer terminals of these resource pools constitute a "cloud" [1, 2]. Cloud computing technology has changed the traditional education model and brought many conveniences to modern education. In recent years, with the development and improvement of big data technology, big data has attracted the attention of various industries with a perfect computing and storage advantage, and it has also entered the field of education [3]. Through big data technology-assisted teaching and "Internet+" assisted teaching, teaching efficiency has been greatly improved. Respected by all walks of life, and thus promote the further improvement of contemporary modern education [4].

J. MacIntyre et al. (Eds.): SPIoT 2020, AISC 1282, pp. 366–373, 2021.
https://doi.org/10.1007/978-3-030-62743-0_53

"Internet+" includes not only Internet technology, but also modern information technology, such as big data, cloud computing technology and Internet of Things technology [5]. With the proposal of the Internet plus action plan, various industries in the society have increased the construction and development of informatization. Similarly, colleges and universities in order to train talents who can meet social needs, accelerate the development of education informatization, and construct a networked, digital, and intelligent teaching system. Efforts in the direction of education modernization, and eventually achieve a good learning environment for everyone to learn, learn everywhere, and always learn [6].

Teachers must be clear about the student's subject status, let students become masters of learning, guide them to spontaneously explore, and give full play to their subjective initiative [7, 8]. In the environment of "Internet+" and big data, teaching is no longer limited by time and space, and teaching is no longer limited by time and space [9]. Teachers should establish an exploratory teaching model, encourage students to establish an exploratory knowledge structure, use knowledge results and experience flexibly, and improve students' subjective participation consciousness. Under the "Internet+" environment, contemporary education changes teaching strategies, encourages students to conduct cooperative and inquiry-based learning, and improves learning motivation [10].

2 Method

2.1 Educational Technology Evaluation Model

In practical applications, the overall distribution P is often unknown, so it is impossible to evaluate the precise teaching plan by directly obtaining the conditional expectation E(Y|A, X) of the teaching effect, which requires estimation. Considering that the teaching effect is related to the teaching plan and the corresponding student characteristics and behavior performance, based on the regression model, the evaluation model for building an accurate teaching plan is as follows:

$$\mathrm{E(Y|A,X)} = \gamma^T\tilde{X} + A \cdot \left(\beta^T\tilde{X}\right)$$

In the model, $\gamma^T\tilde{X}$ partly depicts the basic influence of student characteristics and behavior on teaching effect, while $A \cdot \left(\beta^T\tilde{X}\right)$ part characterizes the influence of student characteristics and behavior on teaching effect under different teaching schemes. In fact there are:

$$\beta^T\tilde{X} = \mathrm{E(Y|A=1,X)} - \mathrm{E(Y|A=0,X)}$$

In summary, the formula for estimating the parameters of the educational technology assessment model is:

$$\widetilde{\beta_{LASSO}} = \min_{\beta}\left\{ ||y - X\beta||^2 + \gamma \sum_{j=1}^{p} |\beta_j| \right\}$$

Due to the singularity of the penalty term at the zero point, the LASSO method can make the model's partial coefficient estimate be exactly 0, based on which the coefficients of some unimportant variables (variables with smaller parameter estimates) are compressed to zero with a higher probability. For important variables with large parameter estimates, the LASSO method compresses its coefficients lightly, thereby ensuring the accuracy of parameter estimation.

2.2 Development of a Single-Stage Modern Educational Technology Program

The essence of formulating the single-order precision education technology plan is the optimal decision function d:X → A, which makes the teaching effect Y the best. In other words, looking for the optimal decision function $d^*(X)$, such that:

$$d^*(X) = arg\ \max_{d}\ \mathrm{E(Y|A = d(X), X)}$$

Therefore, to formulate a single-stage precise teaching plan, only the optimal decision function $d^*(X)$ needs to be found, so that under its precise teaching plan, E(Y| A = d(X), X) obtains the maximum value.

2.3 Cooperative Teaching

The establishment of the cloud sharing model has created great convenience for students' cooperative learning. Teachers should improve the teaching plan, formulate teaching guidelines, encourage students to work together and improve teaching efficiency. Cooperative teaching is for students to establish their own study groups. In this group, everyone should understand their responsibilities. First, they should collect data according to the division of labor, and then share the data in the cloud, and integrate effective resources to complete the learning task.

3 Experiment

3.1 Object of Study

In order to be able to more deeply analyze the analysis and research of modern educational technology under "Internet+" and big data technology, this paper selects a total of 90 students from two classes of H school for experimental investigation. Respectively their learning motivations (S1–S3), learning strategies (S4–S7), learning objectives (S8–S10), attribution (S11–S14) and curriculum factors (W1–W7) and school factors among external factors (M1–M3) Six observation variables, a total of 29 topic options. All physical items are represented by a five-level scale, "1" indicates non-compliance; “2” indicates not conforming; “3” indicates general; “4” indicates relatively conforming; “5” indicates conforming; The second part is the main body of this article.

3.2 Experimental Design

This research is for the education model learning of two classes in H school. After the learning is completed, the knowledge to be tested is tested, and then the knowledge grasp of the two classes is compared, and then the process design based on case teaching is adopted. To investigate the status quo of modern education technology public course teaching, the purpose is to discover the deficiencies or problems in the modern education technology public course teaching, so as to better provide a realistic basis for the design and implementation of the later teaching process. After the implementation of case teaching, we interviewed the students in the public education class of modern education technology, evaluated the effect and satisfaction of the case teaching implementation, and further revised and improved the implementation process.

4 Results

4.1 Analysis of Questionnaire Results

Table 1. Test of α for internal factors of learning input

	(S1–S3) Learning motivation	(S4–S7) Learning strategies	(S8–S10) Learning target	(S11–S14) Attribution	Internal factors
Number of items	3	4	2	3	12
Cronbach’s alpha	0.732	0.733	0.755	0.633	0.857
Based on standardization Cronbach’s alpha	0.752	0.781	0.758	0.321	0.832

Table 2. Test of α for school factors

Cronbach’s alpha	Cronbach’s α based on standardized projects	Number of items
0.731	0.721	7
	Beta factor test for school factors	
Cronbach’s alpha	Cronbach’s α based on standardized projects	Number of items
0.813	0.825	8

Table 3. Alpha coefficient test of the questionnaire population

Cronbach's alpha	Cronbach's α based on standardized projects	Number of items
0.819	0.912	27

According to the data in Table 1, we can see that when analyzing the internal factors that affect the learning input, the overall reliability coefficient reaches 0.857 > 0.8, indicating that the internal consistency of the internal factors is very good. At the same time, we can see that after the internal factors are processed in layers, the intrinsic reliability a coefficient of each level will be lower than the overall, but all maintain a good internal reliability.

According to the data in Table 2 and Table 3, after conducting the hierarchical analysis above, the author conducted an overall a coefficient test on the 27 options of the entire questionnaire to obtain the reliability statistics table of the total table, as shown in Table 3. It can be seen that the overall internal consistency coefficient of the questionnaire is 0.914, and the standardized internal consistency coefficient is only 0.916. This situation indicates that the internal consistency of the scale is very high, and the reliability of the scale is very good. This questionnaire can be used for the next research. After the questionnaire has undergone item analysis and reliability test, we also need to test the validity of the questionnaire.

The validity of the questionnaire refers to the validity of the questionnaire. The higher the degree of agreement between the measurement results and the content to be investigated, the higher the validity. In order to test the validity of the questionnaire, it is necessary to perform factor analysis on the questionnaire. During the test, the author selected the value of the appropriateness of sampling KMO to determine whether it is suitable for factor analysis. The purpose of factor analysis is to find the construct validity of the questionnaire. Commonly used KMO metrics are as follows: 0.9 means "excellent"; 0.8–0.9 means "good"; 0.8 means "moderate"; O.ren 0.7 means "ordinary"; below 0.5 means "unable" Accepted". The significance level of the Bartlett spherical test value is less than 0.05, which means that it is suitable for factor analysis

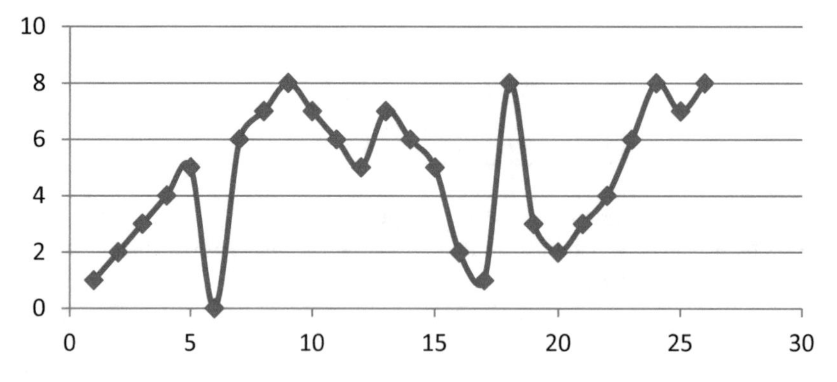

Fig. 1. Characteristic diagram of influencing factors of modern educational technology

At the same time, in order to ensure that the research is more complete and reasonable, the author has also tested the factor steep slope map of the overall questionnaire, as shown in Fig. 1. The steep slope chart here is mainly to help the author to verify whether the number of influencing factors is correct, and at the same time to check whether the questionnaire is reasonable or not. The abscissa is the title item, the number of variables, and the ordinate is the feature value. The judgment criterion of the steep slope chart inspection is to consider the factor of the sudden rise of the slope line, and delete the factor of the relatively gentle slope. You can see from the figure that after the fourth factor, the slope gradually becomes gentle, indicating that there are no special factors to extract, so in this study, it is appropriate to retain three influencing factors. This coincides with the author's previous formulation of the number of factors. When determining the influencing factors of learning investment, the author comprehensively considers three major factors as internal factors (including four observation variables, namely learning motivation, learning strategy, learning goals and attribution), curriculum factors and school factors. Study the influencing factors of times. In this way, the formulation of this questionnaire is also quite reasonable.

4.2 Characteristics of Modern Education Under the Environment of Big Data Technology

On the basis of big data technology, it integrates distributed processing and parallel processing into an integrated model. Through the support of the Internet, a large amount of network data is distributed, and a large amount of data is stored and calculated at the same time. The advantages of big data technology have also brought great help to contemporary education. Here are some examples.

First, the biggest advantage of diversified big data services is the powerful storage space and fast computing efficiency. These two points can not be surpassed by ordinary computers. Then this is mainly derived from the resource library composition of massive basic data in the cloud platform. Specifically, the resource database in big data is a massive data platform. When a user obtains certain information, after processing valid information in the resource database, the information will be transmitted to the user. In addition, big data technology also integrates. The characteristics of information storage, interception and processing provide users with unified services, and at the same time, they can also establish a terminal sharing mode, so that the information obtained by users is more accurate and fast, and the data that meets the requirements of modern education is quickly updated. Second, improving the security of data storage is one of the characteristics of big data technology and an important function in big data services.

4.3 Application Advantages of "Internet+" in Modern Education Technology

Under the new situation of social development, the demand for high-quality talents by employers continues to increase, which puts forward higher requirements for modern education. To further promote the improvement of education quality in my country, we must introduce modern "Internet+" technology into the field of education, and give full play to its channel and resource advantages. Taken together, the advantages of the application of "Internet+" technology in modern education are mainly reflected in the following aspects: First, the application of "Internet+" technology has made modern education free from the limitations of time and space. In the field of time, the memory with powerful functions in the modern "Internet+" technology can save a large amount of education and teaching information, and can process it efficiently and accurately. It can be called at any time when needed, and the efficiency of the teaching system has been greatly improved.. At the same time, "Internet+" technology can also make teaching activities get rid of the restrictions of class hours, and students can carry out learning activities through diversified teaching resources outside the classroom.

5 Conclusion

In the teaching of modern educational technology, the application of "Internet+" technology and big data technology not only promotes the reform of teaching methods and means, but also breaks the limitations of traditional education, expands the coverage of modern education, and achieves quality education. The sharing of resources has greatly promoted the development of China's education. Teachers should pay attention to the use of modern educational technology and give play to the advantages of modern educational technology. Improve the deficiencies in the previous teaching. Return the classroom to the students, and make the classroom atmosphere warmer. Secondly, this method can also be used to stimulate students' enthusiasm and interest in learning and enhance students' participation in the classroom. Teachers should pay attention to the use of modern educational technology and give full play to the advantages of modern educational technology. To sum up, it is necessary to improve the deficiencies in the previous teaching. Return the classroom to the students and make the classroom atmosphere warmer.

References

1. Bttger, T., Cuadrado, F., Tyson, G., et al.: Open connect everywhere: a glimpse at the internet ecosystem through the lens of the Netflix CDN. ACM SIGCOMM Comput. Commun. Rev. **48**(1), 1–6 (2018)
2. Xu, W., Zhou, H., Cheng, N., et al.: Internet of vehicles in big data era. IEEE/CAA J. Automatica Sinica **5**(1), 19–35 (2018)
3. Kim, S.H., Roh, M.I., Oh, M.J., et al.: Estimation of ship operational efficiency from AIS data using big data technology. Int. J. Naval Archit. Ocean Eng. **12**, 440–454 (2020)
4. Kim, M., Man, K.L., Helil, N.: Advanced Internet of Things and big data technology for smart human-care services. J. Sensors **2019**, 1–3 (2019)
5. Ekong, E.E., Adiat, Q.E., Ejemeyovwi, J.O., et al.: Harnessing big data technology to benefit effective delivery and performance maximization in pedagogy. Int. J. Civ. Eng. Technol. **10** (1), 2170–2178 (2019)
6. Chen, K., Zu, Y., Cui, Y.: Design and implementation of bilingual digital reader based on artificial intelligence and big data technology. J. Comput. Methods Sci. Eng. **2**, 1–19 (2020)

7. Kisiel, M.: The musical competence of an early years teacher the challenge a modern education. Pedagogika **27**(2), 193–205 (2018)
8. Gao, P.: Risen from chaos: the development of modern education in China, 1905-1948: dissertation summaries. Aust. Econ. History Rev. **58**(2), 187–192 (2018)
9. Gartner, G.: Modern cartography - research and education. Allgemeine Vermessungs Nachrichten **126**(4), 81 (2019)
10. Sun, Y.: Liberal arts education and the modern university. Eur. Rev. **26**(2), 1–13 (2018)

Management Innovation of Network Society Based on Big Data

Yue Zhao(✉)

Department of Economics and Management, Yunnan Technology and Business University, Kunming, Yunnan, China
375051678@qq.com

Abstract. In the process of rapid development, due to the imperfection of the relevant system structure, the network society management and the development of network information cannot keep up with the matching facilities, which is a serious threat. The management of the network society. The research purpose of this paper is to study the safety and innovation methods of network society management in the continuous development of big data and network. This article analyzes the network system, file transfer and remote call (Telnet), SQL injection and SQL vulnerabilities, DDOS attacks and other network technologies. And the analysis of network management technical measures in which VLAN technology strengthens network management, which strengthens network management through VLAN technology, and improves network security performance. Through the binding of IP address and MAC address to the computers in the entire network that use static IP access. It can solve the problems of network social management and propose innovative points and innovative methods. The experimental research results show that the establishment of an effective network social management security defense system is of great value to the security of users and network applications. Through comprehensive network social management education to improve the security awareness of network users, a complete Network security system.

Keywords: Network system · SQL injection · Big data · Network management innovation

1 Introduction

Traditional forms of communication, commerce and business operation models have generally been impacted, and anti-corruption and popular political participation using the Internet as the main channel have gradually become a reality [1, 2]. The influence of the Internet on social politics, economy, culture, and the daily life of the public is increasing, and the ways of influence are also becoming diversified. The characteristics of the contemporary Internet also determine the characteristics of information autonomy, non-real name, and low cost in the network age. This makes the objectivity of network information unable to be effectively guaranteed, and the dissemination of information has become difficult to control, and it is difficult to establish corresponding

J. MacIntyre et al. (Eds.): SPIoT 2020, AISC 1282, pp. 374–381, 2021.
https://doi.org/10.1007/978-3-030-62743-0_54

regulations. Supervision system to ensure the authenticity of information and social security [3].

The network security management system uses the advanced K-Means algorithm to build a data analysis engine that can mine potential viruses or Trojan horse genes in the data to build a highly automated network security management system [4, 5]. The system can process unstructured data resources, realize application data analysis and result report generation. The back-end logical business processing of network security management based on K-Means algorithm mainly includes two components: offline module and database. The offline module can complete the current functions of network raw data collection, preprocessing, and data mining, so it is also called network data Acquisition stage [6].

The realization of network security management should include education for computer managers and users, requiring them to consciously follow computer management and use systems, including established computer control and management systems, operation maintenance and management systems, and various data management System, personnel management system, computer room security management system, etc., but also to comply with basic daily rules, strengthen safety education for computer users, establish corresponding safety management institutions, increase penalties for computer network security violations, and even adopt legal measures Maintain computer network security [7, 8]. Let them follow a series of basic principles such as the principle of legitimate users, the principle of information utilization, the principle of resource restriction, the principle of information system, the principle of information disclosure, etc., and consciously fight crimes against the network, which is of great significance for maintaining network security, ensuring the normal operation of computers, and maintaining the security of information systems [9, 10].

2 K-Means Algorithm Optimization

K-Means algorithm is also known as k-means algorithm. The algorithm idea is roughly as follows: first randomly select k samples from the sample set as the cluster center, and calculate the distance between all samples and these k "cluster centers", for each sample, divide it to the nearest "cluster center". In the cluster where is located, calculate the new "cluster center" of each cluster for the new cluster. Based on the above description, we can roughly guess the three main points of implementing the K-Means algorithm:

(1) Selection of the number of clusters k
(2) The distance from each sample point to the "cluster center"
(3) Update the "cluster center" according to the newly divided cluster

Key points of K-Means algorithm:

(1) Selection of k value

The choice of k is generally determined according to actual needs, or the value of k is directly given when the algorithm is implemented.

(2) Measurement of distance

For a given sample $x_1^i = \{x_1^i, x_2^i, \ldots, x_n^i\}$ 与 and $x_1^f = \{x_1^f, x_2^f, \ldots, x_n^f\}$, where i, f = 1, 2…m, represents the number of samples given, and n represents the number of features. The distance measurement methods are mainly divided into the following centralized methods:

Ordered attribute distance measurement (discrete attribute {1, 2, 3} or continuous attribute): the formula of Minkowski distance algorithm is shown in (1):

$$\mathrm{dist}_{mk}\left(x_1^i, x_1^f\right) = \left(\sum\nolimits_{u=1}^{n} \left|x_u^{(i)} - x_u^{(f)}\right|^p\right)^{\frac{1}{p}} \tag{1}$$

(1) Euclidean distance, that is, Minkowski's distance formula when p = 2 is shown in (2)

$$\mathrm{dist}_{mk}\left(x_1^i, x_1^f\right) = \left\|x^i - x^f\right\|_2 = \sqrt{\sum\nolimits_{u=1}^{n} \left|x_u^{(i)} - x_u^{(f)}\right|^2} \tag{2}$$

(2) Manhattan distance, that is, when p = 1, the Minkowski distance algorithm formula is shown in (3)

$$\mathrm{dist}_{mk}\left(x_1^i, x_1^f\right) = \left\|x^i - x^f\right\|_1 = \sum\nolimits_{u=1}^{n} \left|x_u^{(i)} - x_u^{(f)}\right| \tag{3}$$

(3) The formula of distance measurement algorithm for disordered attributes is shown in (4):

VDM (Value Difference Metric):

$$\mathrm{VDM}_p\left(x_u^{(i)}, x_u^{(f)}\right) = \sum\nolimits_{z=1}^{k} \left|\frac{m_u, x_u^{(i)}, z}{m_u, x_u^{(i)}} - \frac{m_u, x_u^{(f)}, z}{m_u, x_u^{(f)}}\right|^p \tag{4}$$

Where $m_u, x_u^{(i)}$ represents the number of samples whose value is $x_u^{(f)}$ on the attribute u, $m_u, x_u^{(i)}, z$ represents the attribute u in the zth sample cluster. The above value is the number of samples of $x_u^{(f)}$, $\mathrm{VDM}_p\left(x_u^{(i)}, x_u^{(f)}\right)$ represents two discrete values $x_u^{(i)}$ VDM distance from $x_u^{(f)}$.

The execution of the K-Means algorithm starts from an initial cluster center and continues until all data points are divided into K categories, so that the distance between all data points can be reduced, which is the sum of squares. The main reason is the totality of the K-Means algorithm. The sum of squared distances can tend to decrease with the data objects contained in the cluster. For this reason, under a certain number of categories K that can be determined, the total sum of squared distances takes the minimum value.

The evaluation standard used in the algorithm experiment in this paper is accuracy. The evaluation method can analyze the degree of accurate classification. The calculation formula is as follows:

$$P(T) = \frac{\sum_c A_1(c,T)}{\sum_c A_1(c,T) + A_2(c,T)} \tag{5}$$

3 Modeling Method

3.1 Parameter Construction of Classic EM Algorithm Model

Classical EM algorithm: For a data set D containing m d-dimensional numerical records, you can set the stop threshold $\varepsilon > 0$. The steps are as follows:

For each data record in the data set D as x, the formula for calculating the probability data parameter model of h = 1,…k, belonging to the cluster is as follows:

$$w_h^t(x) = \frac{w_h^t \cdot f_h(x|\mu_h^t, \sum_h^t k)}{\sum_{i=1}^k w_i^t \cdot i(x|\mu_i^t, \sum_i^t k)} \tag{6}$$

The updated mixed model parameters are as follows:

$$w_h^{t+1} = \sum_{x \in D} w_h^t(x), \mu_h^{t+1} = \frac{\sum_{XD} w_h^t(x) \cdot x}{\sum_{XD} w_h^t(x)} \tag{7}$$

$$\sum_h^{t+1} = \frac{\sum_{x \in D} w_h^t(x)\left(x - \mu_h^{t+1}\right)\left(x - \mu_h^{t+1}\right)^T}{\sum_{xD} w_h^t(x)} \tag{8}$$

Termination condition: If $|L(\varnothing) \wedge t - L(\varnothing) \wedge (t + 1 \leq \in)|$, the estimated parameters are considered to conform to the distribution of the original data set, and the algorithm stops. Otherwise, t = t + 1, jump to step 1, iterative execution.

It is worth noting that during the steps of the algorithm model, in order to calculate the family to which the data point belongs, it is necessary to scan each record of the original data set. The calculation amount of this step is relatively large. The number of algorithm iterations is affected by initial condition parameters and data distribution. Generally speaking, the number of iterations is unpredictable, but the algorithm tends to converge. It is not only suitable for spherical Gaussian clustering, as long as the appropriate distribution function is selected, such as Poisson distribution function, polynomial distribution function, etc.), it can also be applied to clustering of other non-spherical distributed data to estimate the data distribution.

4 Case Analysis of Network Social Management

4.1 Application of Data Mining Algorithm Based on Stored FPTREE Large-Scale Data Set

Table 1. Original data set M

TID	ItemsBought
T001	A,C,E,F,J,K,L
T002	A,B,D,G
T003	E,F,H,J,L,
T004	A,B,C,D,G,M,N
T005	G,E,J,L

Table 2. Sorted data set M^E

TID	ItemsBought
T001	E,F,J,K,L,C,A
T002	G,D,B,A
T003	H,J,E,F,L
T004	C,D,B,A,M,N,G
T005	J,E,L,G

According to the application analysis situation of the data set shown in Table 1 and Table 2, it is constructed according to a similar construction process. The difference is that the single path in the middle needs to be identified and the depth-first method is used to add identification. The construction process is always top-down, Jiangsu University doctoral dissertation: large-scale data set efficient data mining algorithm research, and in the subsequent visits to build conditional subtrees are always bottom-up, so the method can be accelerated. The access efficiency plus identification.

4.2 Analysis and Research on the Characteristics of Network Social Management

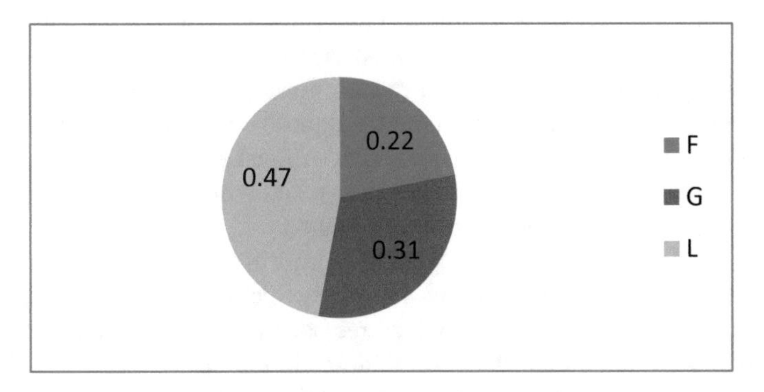

Fig. 1. The proportion of each mode

Based on the well-established network social management system, the statistical characteristics of node distribution are used to analyze the network, mine the key information in the data, and grasp the movement status of the AUV. First observe the proportion of each mode to understand the overall structure of the data. Figure 1 shows the proportion of each mode in the network, among which the M mode has the highest proportion, which shows that most of the data is in a "hold" state and the AUV operating state is relatively stable.

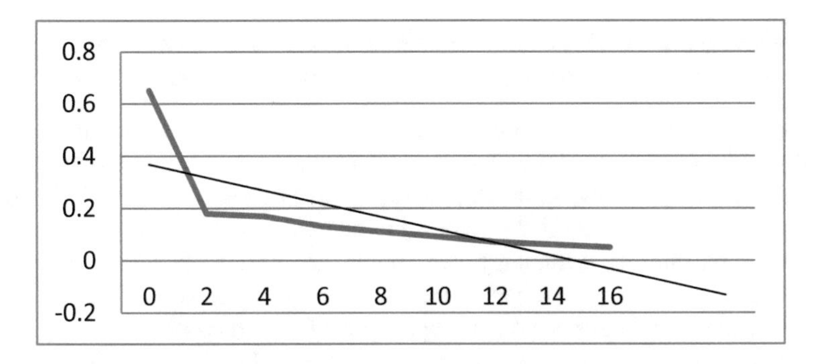

Fig. 2. Trend chart of strength analysis of each node

As shown in Fig. 2, the node is used as a unit to analyze the network using the node strength ns_i. The node strength not only considers the number of neighbors of the node, but also considers the weight of the neighboring nodes.

$$\text{ns}_i = \sum\nolimits_{j \in x} \omega ij$$

ns_i is the strength of the i-th node; ns_i represents the weight between nodes i and j. However, although the PAA algorithm smoothes the fluctuations, the effect is still not ideal. Among them, when n = 5, the network formed by the PAA algorithm reaches 27 nodes, showing all possible fluctuation modes, which means that the scale of sliding window length n = 4 is saturated for the PAA algorithm with a fragment length of 4. In the state, further increasing the fragment length will make the algorithm perform better here, but it will also smooth out more key peaks, and will further reduce the sequence dimension, which is unfavorable for other analyses.

The algorithm uses disk table storage to reduce memory usage. When traditional algorithms occupy too much memory and cannot perform mining work, disk table storage technology can be used to reduce memory usage and continue to complete mining work. Research on effective data mining algorithms for large-scale data sets is suitable for spatial performance priority. In addition, the algorithm integrates association rule mining and lightweight database, which overcomes the inefficiency and development difficulties of file system-based algorithms. The experimental results show that this algorithm is an effective disk-based association rule mining algorithm when the storage space is limited.

Among them, for fast storage and reconstruction, we designed a novel data structure, modified the multi-pointer connection in the original algorithm implementation to a single-pointer connection, reducing the number of pointers in the original data structure. Since it only needs to be traversed in order, such a change is feasible. And it can reduce the memory usage, reduce the workload of rebuilding the pointer, and make it possible to store and rebuild quickly.

4.3 Ensuring the Development of Normative Laws for Network Social Management and Forming a New Management Pattern

First of all, it is necessary to conduct in-depth research on Internet technology, accurately grasp the operating rules and management functions of the Internet, vigorously control bad behavior in the Internet, and solve the phenomenon of online violence. Government departments must do a good job of guidance and demonstration. In the development of Internet technology, we should not use the excessive influence of administrative power to prevent some officials from trading power and money, create a clean and healthy environment for the operation of the Internet, and ensure that the public has a good way to actively express their wishes. In addition, it is necessary to summarize the government's experience in anti-corruption operations and further study the rules of network operation to ensure the stability of network operations. Government departments should encourage the public to express their opinions with the help of online platforms and actively participate in social management. The government and the public should use the Internet to strengthen communication and answer various questions in the process of public participation in social management.

5 Conclusion

This article is an exploratory research, which helps to understand the important research significance of model innovation in network social management in the era of big data, contributes to the spirit of "strengthening network social management", and can also help China's network social management innovation to provide An exploratory study of feasibility. This paper analyzes the risks and challenges of big data application, puts forward the research on the construction of big data security protection system, and conducts detailed research on each module of big data security protection system. Use a variety of methods to ensure the security of the enterprise-level big data platform, and finally build a big data security management and control platform to protect the privacy of sensitive data, reduce corporate legal risks, regulate the operation process of the big data platform, and ensure the security of the open source system.

References

1. Wen, Z., Hu, S., Clercq, D.D., et al.: Design, implementation, and evaluation of an Internet of Things (IoT) network system for restaurant food waste management. Waste Manag. **73** (MAR), 26–38 (2018)
2. Krundyshev, V.M.: Preparing datasets for training in a neural network system of intrusion detection in industrial systems. Autom. Control Comput. Sci. **53**(8), 1012–1016 (2019)
3. Mukherjee, A., Deb, P., De, D., et al.: WmA-MiFN: a weighted majority and auction game based green ultra-dense micro-femtocell network system. IEEE Syst. J. **PP**(99), 1–11 (2019)
4. Wang, Y., Kung, L.A., Byrd, T.A.: Big data analytics: understanding its capabilities and potential benefits for healthcare organizations. Technol. Forecast. Soc. Change **126**(JAN), 3–13 (2018)
5. Lee, C.Y., Chong, H.Y., Liao, P.C., et al.: Critical review of social network analysis applications in complex project management. J. Manage. Eng. **34**(2), 04017061:1–04017061:15 (2018)
6. Jiao, H., Wang, Y., Liu, M.: The effect of the social network of the top management team on innovation in cultural and creative industries: A study based on knowledge network embedding. J. Chin. Hum. Resour. Manage. **10**(1/2), 4–18 (2019)
7. Morris, R.L., Caroline, S.: Critical moments in long-term condition management: a longitudinal qualitative social network study. Chronic Illn. **14**(2), 119–134 (2018)
8. Ming-Fu, H., Ching-Chiang, Y., Sin-Jin, L., et al.: Integrating dynamic Malmquist DEA and social network computing for advanced management decisions. J. Intell. Fuzzy Syst. **35**(1), 1–11 (2018)
9. Francia, P., Iannone, G., Santosuosso, U., et al.: Daily physical activity performed and egocentric social network map analysis in the management of young patients with type 1 diabetes. Diab. Technol. Ther. **22**(1), 184–185 (2020)
10. Herrera, R.F., José-Manuel, M., Ramírez, C.S., et al.: Interaction between project management processes: a social network analysis. Int. J. Proj. Organ. Manage. **12**(2), 133–148 (2020)

Short Term Load Forecasting Model of Building Power System with Demand Side Response Based on Big Data of Electrical Power

Xiang Fang(✉), Yi Wang, Lin Xia, Xuan Yang, and Yibo Lai

State Grid Hangzhou Power Supply Company, Hangzhou 310000, Zhejiang, China
fangxiang0330@126.com

Abstract. With the development of the times, modern people's life is more and more inseparable from electric energy, which leads to the rapid growth of the pace of power marketization, rapid development at the same time, users' demand for related service quality is also increasing. Because of all kinds of reasons, the development direction of distribution enterprises is moving towards convenience, safety, flexibility and stability. In the daily management of the building, to ensure the uninterrupted power supply is an important factor affecting the service quality of the building. Whenever there is a power outage, it will cause unpleasant impact and experience to users and tourists, bring fright to employees, and cause unnecessary trouble to daily management. Therefore, this paper introduces the demand side response, which is based on an important interactive mode for the active distribution of power grid. According to the implementation of a series of incentive mechanisms, guide users to actively cooperate with the operation of the distribution network management. In recent years, according to the corresponding measures for energy consumption, adjusting the energy consumption mode can prevent the random fluctuation caused by intermittent renewable energy, distribute the demand side resources on demand, control the distributed power supply more effectively and operate the energy storage equipment more flexibly. According to the analysis and summary of the demand response of the system, this paper discusses the examples of short-term load forecasting and demand side response, which shows that the experiment of load forecasting model in this paper has high accuracy.

Keywords: Smart grid · Demand response · Short-term load forecasting · Hierarchical coordination

1 Introduction

For the new ultra short term prediction, the former main purpose is in the field of real-time security analysis, automatic generation control system\economic planning [1]. Generally speaking, the short-term forecast of the total load of lines or multiple clients running at the same time is from the system aspect [2]. In recent years, the emerging system demand [3] and multi energy simultaneous control technology [4] have raised

J. MacIntyre et al. (Eds.): SPIoT 2020, AISC 1282, pp. 382–389, 2021.
https://doi.org/10.1007/978-3-030-62743-0_55

new expectations for the technology demand of ultra short term load forecasting. First of all, in terms of the time required for the operation of the technology, it is hoped that the prediction of the project can be achieved in just a few minutes; second, in terms of the requirements of system operation, it is necessary to provide different services for different types of users, so the prediction of different types of users must be done [5]. From the current detection methods of ultra short term load forecasting, artificial intelligence and wavelet analysis are the main methods. Although the prediction period of the above method is short, its accuracy is much lower than the usual prediction results.

With the reform of China's power market system and the rapid development of new energy technology, China's power grid is facing various uncertain challenges [6]. The principle of demand side response is to distribute the power grid through an interaction. According to the implementation of a series of incentive mechanisms, guide users to actively cooperate with the operation of the distribution network management [7]. In recent years, according to the corresponding measures for energy consumption, adjusting the energy consumption mode can prevent the random fluctuation caused by intermittent renewable energy, distribute the demand side resources on demand, control the distributed power supply more effectively and operate the energy storage equipment more flexibly. The traditional distribution network will not generate electricity according to the user's demand, so there is demand response, which changes the operation mode [8], to a large extent, it solves or alleviates the problem that the system is unable to automatically distribute power or power shortage, improves the stability of power supply demand, and it has practical significance to study the short-term load combination forecasting model considering demand side response Meaning. In the current stage of rapid development of the power industry, load forecasting plays an important role. If we trace the evolution of load forecasting, we can find that its development process has gone through a long period of time, and in this process, the accuracy of load forecasting is gradually improving, but due to the complexity of the internal structure of the power system, And the demand for low consumption and high efficiency load forecasting in the whole power industry is gradually increasing, so a more accurate forecasting method is needed.

This paper studies the short-term power load forecasting model of building power system based on demand side response. The example shows that the load forecasting model proposed in this paper has high accuracy.

2 Method

2.1 Demand Side Response

Demand side response (DR) is an important interactive unit in automatic power dispatch. According to the implementation of a series of incentive mechanisms, guide users to actively cooperate with the operation of the distribution network management. While responding to the power consumption and adapting to the power consumption mode, making corresponding measures and adjusting the energy consumption mode according to the energy consumption can prevent the random fluctuation caused by intermittent renewable energy and make rational use of the demand side resources. The traditional distribution network can not actively monitor whether the power is sufficient, so there is demand response, which changes the operation mode, to a large extent solves or alleviates the problem that the system cannot automatically distribute power or power is insufficient, improves the stability of power supply demand, and meets the needs of the building as shown in Fig. 1.

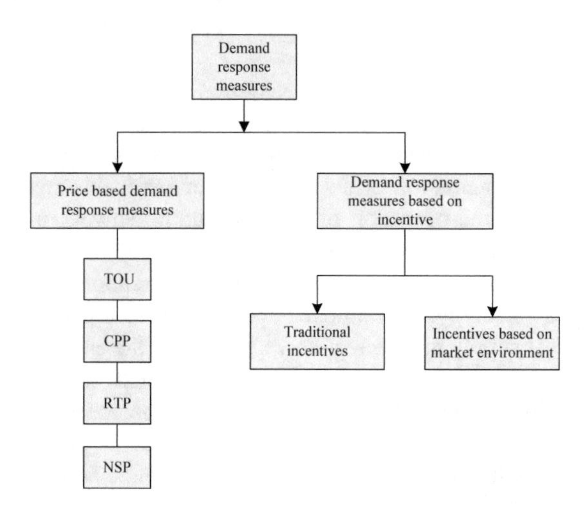

Fig. 1. Schematic diagram of demand side response process represented by building

In smart grid, the influencing factors of demand response are more complex and uncertain. Under the power market mode, demand side response technology has developed rapidly. The demand side response of power adjusts the power price according to the demand of market power, which can not only change the traditional power control behavior, but also improve the user's electricity experience.

2.2 Load Form Based on Fuzzy Clustering

The main steps of this experiment are: set the maximum clustering number n that may appear; when the clustering number is $2 \sim n$, the validity function of the clustering result of the budget fuzzy c-means; find the corresponding K value, which is the minimum value of the validity function, which is the best clustering number, and then select the average value of the function to get the typical negative of the users Load curve.

When the fuzzy c-means clustering algorithm is used to cluster the load curve, the clustering number C should be given in advance. In many cases, it is difficult to select the value, because it is difficult to obtain the value, which further leads to the inaccuracy of fuzzy partition. Therefore, when it is difficult to correctly select the load data, the efficiency index can be used to search the optimal number of clusters.

3 Experiment

3.1 Thinking of Short-Term Load Forecasting Model Based on Demand Response

(1) Select similar days;
(2) The results selected after the prediction of the two models are added, and the data obtained is the preliminary prediction results;
(3) Achieve multi-level coordination of prediction results.

The schematic diagram of the model framework is shown in Fig. 2.

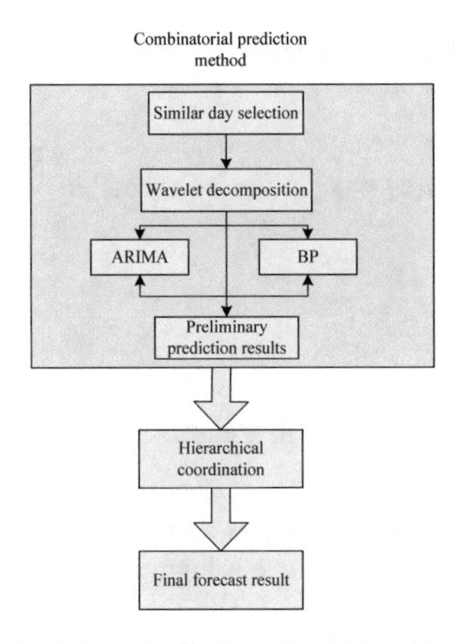

Fig. 2. Schematic diagram of model architecture

3.2 Data Source

In the process of load forecasting in this study, historical date data is used as the basis of short-term load forecasting. Because there is a large difference between some specific historical days and the predicted load characteristics in this experiment, the accuracy of the predicted results may be reduced, so we need to screen similar days for these historical days. According to the fuzzy rules, the factors closely related to the change of load shape are assigned to the daily eigenvector. Average daily eigenvectors for all days in each category. By Euclidean distance, the feature vector of forecast day is used to judge which kind of similar day the forecast day belongs to.

When determining the daily eigenvector, we should first determine the main factors that affect the historical load. The principles to be considered are shown in Table 1.

Table 1. Conversion value principles of main influencing factors

–	1	2	3	4
Type (D)	Sunday	Monday to Saturday	–	–
Maximum temperature	<0 °C	0–10 °C	10–20 °C	20–30 °C
Minimum temperature	<0 °C	0–10 °C	10–20 °C	20–30 °C
Rainfall	No rain	Light rain	Moderate rain	Heavy rain

So the daily eigenvector is:

$$X = (D, T_{\max}, T_{\min}, R) \tag{1}$$

In load forecasting, if the load sequence is discrete, the discrete wavelet transform function is:

$$W_{\psi}f(a,b) = |a|^{-1/2}\Delta t\sum_{k=1}^{n} f(k\Delta t)\overline{\Psi(\frac{k\Delta t - b}{a})} \tag{2}$$

4 Discussion

4.1 Load Forecasting Analysis with Combined Forecasting Model

When building the neural network of load forecasting based on fuzzy algorithm, the influence of time, voltage type and other factors is added. The neural network of load forecasting is divided into two kinds, one is high-voltage load forecasting, the other is low-voltage load forecasting.

In the high-voltage load forecasting neural network, the classification can be divided into two forms according to the different levels of high-voltage. One is to add corresponding nodes in the neural network model, take the voltage as the input variable of the model, and analyze the curve that affects the load; the other is to calculate the parameters of the associated high-voltage type, and build the matching model. In the load forecasting neural network, the fixed high voltage is set to 0, and the non fixed high voltage is set to 1. The factors that affect the load forecasting are integrated into the model, and the voltage of the week before the forecasting date is added to the neural network, and the corresponding voltage parameters are used as the input nodes of the neural network. Randomly select the historical load data of a certain day, then analyze the data, use the fuzzy c-means clustering method to select the similar days of historical load, use the effectiveness index to calculate the clustering number, and find that when the clustering number is 2, the clustering effect is the best. The clustering method proposed in this paper can be used to select similar days of forecast days.

In the low-voltage load forecasting neural network, the input nodes and output nodes are 15 and 1 respectively. The prediction results of neural network are mainly related to the load value in the first three years of the prediction year, the voltage type and average voltage in the first two years. Because the voltage types may be different every day in the prediction year, the input parameter information is used to select the voltage type that appears the most times in the year, And set it to standard voltage.

A large amount of historical data is needed as the basis for building the load forecasting neural network, but the corresponding parameter information needs to be obtained through the power collector and other equipment. Therefore, due to the influence of the equipment itself, the data may have a large error, so the acquired load data must be processed before the construction, Select different processing methods for different data types. Data types can be divided into error data, missing data and error data. Error data is input into neural network according to a certain range of values, using effective filtering means and according to these conditions. The source of the missing data is extracted from the original parameters and the corresponding filtering links. If the actual missing data is less, it can be filled through the linear interpolation method. The filling formula is:

$$T_{n+j} = \frac{T_{n+i} - T_n}{i} \times j(0<j<i) \tag{3}$$

Finally, in order to prevent the phenomenon of neuron saturation in the constructed neural network, the corresponding processing operations should be carried out for the load data before the actual training, so as to solve other negative effects caused by the difference of the original data. The final prediction result is shown in Fig. 3, and the prediction error is 7%. The prediction error of a single model is 10%, which shows that the method in this paper improves the prediction accuracy.

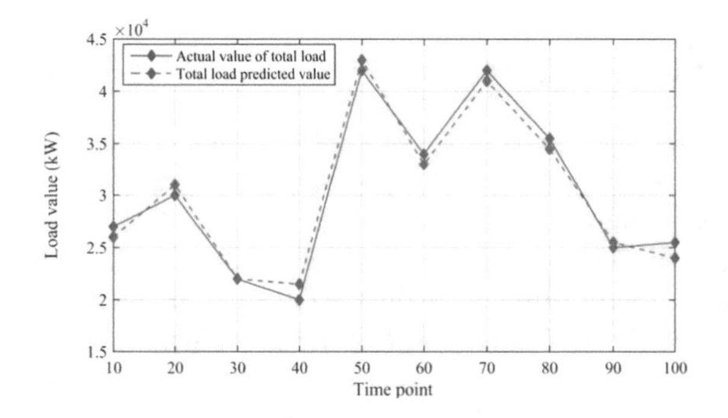

Fig. 3. Regional short-term load prediction results

4.2 Hierarchical Coordination Analysis of Load Forecasting Results

Using the method proposed above, the experiment designed in this paper estimates the two groups of experimental groups respectively, and then compares the prediction curve obtained from two groups of users and the total prediction curve of the total curve. It is not difficult to see the imbalance between the user group and the total load group, which will make the power supply company make some changes. The prediction results of this method are shown in Fig. 4.

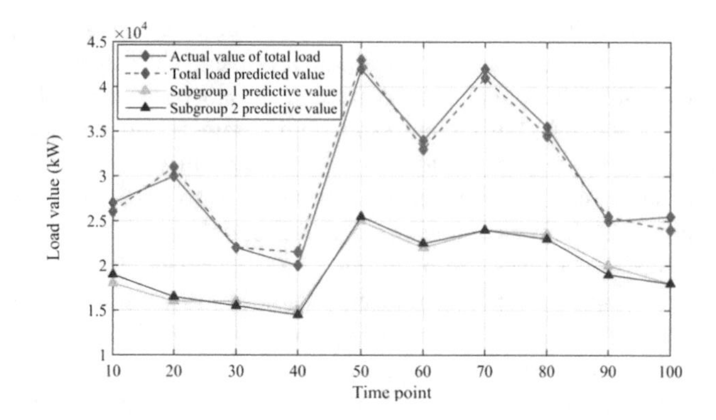

Fig. 4. Load prediction values at all levels after coordination

It can be seen from the figure above that the error of the calculation results before and after coordination is 7%, 6%, 12% and 7%, 6% and 13% respectively. According to the results, it can be concluded that the conclusions of the coordination method used in this experiment are similar to the theoretical results, and the accuracy of the prediction results has been improved to a certain extent.

On the basis of combined forecasting, the combined forecasting model proposed in this paper uses state estimation to adjust the inconsistency of regional prediction and its sub regional prediction results. The results show that the load forecasting model proposed in this paper has high accuracy.

5 Conclusion

With the development of smart grid demand response, people put forward higher requirements for smart grid load shape analysis and short-term load forecasting, especially for more refined short-term load forecasting. In this paper, a high-precision short-term load combination forecasting and hierarchical coordination model are proposed. The fuzzy c-means method is used to extract the similar days of historical load days, and the validity index is used to ensure the similarity of the extracted similar days. On this basis, wavelet analysis is used to decompose the historical similar days, and different features of the decomposed subsequences are used to predict respectively, which improves the accuracy of load forecasting. In the aspect of short-term load forecasting, the widespread application of distributed generation has a certain impact on the power consumption behavior of the user side. Therefore, considering the demand side distributed power consumption prediction can become the next research direction.

References

1. Morello, R., Mukhopadhyay, S.C., Liu, Z., et al.: Advances on sensing technologies for smart cities and power grids: a review. Sensors J. IEEE **17**(23), 7596–7610 (2017)
2. Khalifa, T., Shaban, K.B., Abdrabou, A., et al.: Modelling and performance analysis of TCP variants for data collection in smart power grids. Comput. Commun. **103**(MAY1), 39–48 (2017)
3. He, X., Ho, D.W.C., Huang, T., et al.: Second-order continuous-time algorithms for economic power dispatch in smart grids. IEEE Trans. Syst. Man Cybern **48**(9), 1482–1492 (2018)
4. Le, Y., He, J.: Practical data-driven analysis on stochastic property of power assumption in smart grids. ICIC Express Lett. **11**(1), 95–100 (2017)
5. Ibrahim, M.M., Danbala, A.A., Ismail, M.: Towards attaining reliable and efficient green cloud computing using micro-smart grids to power internet data center. J. Comput. Commun. **07**(7), 195–205 (2019)
6. Christopher, F., Stefan, N., David, D., et al.: Project 'power-to-heat in smart grids'–a multi-objective approach for a maximised value of flexibilities in grids. Cired Open Access Proc. J. **2017**(1), 2842–2845 (2017)
7. Reddy, N.V.R.S., Krishna, P.V.: Context aware power management in smart grids using load balancing approach. Int. J. Adv. Intell. Paradigms **11**(3–4), 335–347 (2018)
8. Bindra, A.: Securing the power grid: protecting smart grids and connected power systems from cyberattacks. IEEE Power Electron. Mag. **4**(3), 20–27 (2017)

Big Data in Art Design

Jian Xu(✉)

Shanxi Vocational and Technical College of Finance and Trade,
Department of Engineering Art Design, Taiyuan 030031, Shanxi, China
57208741@qq.com

Abstract. With the rapid development of the information age, big data is gradually applied to all walks of life. In the field of education, big data has brought strong technical support to the construction of databases and scientific research. The application of big data to the humanities and social sciences, especially in the art design discipline, is particularly effective, and can further promote the sustainable and healthy development of art design teaching. The purpose of this article is to study the application of big data in art design. This paper firstly improves the traditional Apriori algorithm, proposes the Apriori algorithm based on Hadoop, and introduces the idea of the improved Apriori algorithm. Then introduced the workflow of distributed parallel computing framework. The experimental results show that when the number of pictures is very large, the use of distributed retrieval can obviously improve the retrieval efficiency. When the data set size is 54 KB, 3.97 MB, 33.87 MB, the image retrieval time is only 100 ms, 150 ms and 1000 ms.

Keywords: Big data · Apriori improved algorithm · Art design · Distributed parallel retrieval

1 Introduction

Nowadays, the application of big data technology has penetrated all fields of society. After time test, all walks of life no longer doubt the ability of big data, from just a vague concept to practical applications, from structured data analysis to unstructured data analysis and semi-structured data analysis, big data technology is constantly moving forward [1, 2]. According to IBM estimates, there are 2.5 one thousand and six bytes of data generated every day. In the past two years, 90% of such big data has been generated in the world [3]. This is an unbelievable number, but unfortunately, despite such huge data, the amount of familiar information people get is very small [4].

Art can intuitively show the height and standard of a country at the level of education and spiritual civilization. To strengthen the cultivation of art talent education, on the one hand, by strengthening the overall artistic level and professional ability of art students, let these talents use the means of art to create an atmosphere of artistic civilization [5, 6]. On the other hand, as China gradually enters the ranks of major technological nations, technological progress also requires art packaging and beautification in order to truly demonstrate China's comprehensive promotion of science, technology, culture, and education from inside and out [7, 8]. Use the resources provided by today's big data era to build an art professional network practice platform

J. MacIntyre et al. (Eds.): SPIoT 2020, AISC 1282, pp. 390–396, 2021.
https://doi.org/10.1007/978-3-030-62743-0_56

based on the Internet and big data. By improving the platform from manpower to resources to art blending, talent promotion, and complete equipment, it is a modern education system of art students provide a comprehensive and diversified art education platform that can effectively meet their personal and social dual education needs, thereby cultivating more outstanding art professionals and comprehensive talents to inject fresh blood into the development of China's innovative science and technology [9, 10].

This paper firstly improves the traditional Apriori algorithm, proposes the Apriori algorithm based on Hadoop, and introduces the idea of the improved Apriori algorithm. Then introduced the workflow of distributed parallel computing framework. The experimental results show that when the number of pictures is very large, the use of distributed retrieval can obviously improve the retrieval efficiency.

2 Big Data-Related Technologies

2.1 Association Rules

Let $I = \{i_1, i_2, \ldots, i_m\}$ be a set, where items is shortened to I, Transaction is shortened to T, T is shortened to D, and $T \in I$. Each T has a unique identifier, such as a transaction number, called TID. Let A be A set of I, if $A \in T$, then T contains A.

Support: The ratio of the amount of consumption of a good contained in a transaction set D to the total amount of consumption in D is called support for that good.

$$\sup = P(A \cup B) = \frac{C(A \cup B)}{C(D)} \quad (1)$$

Confidence as formula (2).

$$con = P(A \mid B) = \frac{P(A \cup B)}{P(A)} = \frac{C(A \cup B)}{C(A)} \quad (2)$$

Where, sup is support, con is confidence, and C is count.

2.2 Improved Algorithm for Apriori Based on Hadoop

Disadvantages of Apriori algorithm: Many candidate item sets with infrequent item sets can also be generated when generating candidate item sets. At this time, if these candidate items can be identified before generating candidate frequent item sets, the database transaction set can be reduced when scanning the database. Some sets of transaction items have the same items in different order, and if you can somehow combine or convert these items into the same item set, it simplifies the transaction set.

The design idea of improved algorithm for Apriori based on Hadoop (H-Apriori algorithm) is as follows: The user specifies a map() function, through which the key/value pair is processed and a series of intermediate result sets are generated, and finally the reduce() function is used to merge all intermediate result sets with the same key, thus the occurrence times of corresponding frequent item sets are obtained. Taking a top10 or more sets yields a frequent itemset, which greatly reduces the number of database transactions.

2.3 Distributed Parallel Retrieval

MapReduce is a relatively easy to use distributed computing model, which adopts the idea of "splitting first and then combining" to process parallel computing. When MapReduce processes tasks, it first divides the input data into several databases according to certain rules. A data block corresponds to a calculation task, which is called map task. Then each map task is assigned to each slave node for execution. Then, through reduce task, the operation results on each slave node are integrated and merged to get the final result.

3 Experimental Design of Art Design Based on Big Data

3.1 Data Acquisition

In the experiment, the time needed to retrieve images with different number of nodes was tested. Six different Numbers of pictures (50,000, 100,000, 150,000, 200,000, 250,000 and 300,000) were used for experiment and analysis at one node, two nodes and three nodes respectively.

In order to verify the time performance of the improved h-apriori algorithm, the AssociationsSP and two files of Frequent Item set Mining Dataset Repository, retail and accidents (a classic Dataset for association rule study) were used as experimental data sets. The attributes of the dataset are shown in Table 1.

Table 1. Data set attributes

Data set	Size	Record
AssociationsSP	54 KB	1001
Retail	3.97 M	88161
Accidents	33.87 MB	340184

3.2 Experimental Environment

The system is run on Hadoop platform, and the foundation of the experiment is to build Hadoop cluster. In this study, three ordinary computers were used to build a Hadoop distributed system. One of them ran NameNode and JobTracker as Master nodes, and the other two ran DataNode and TaskTracker as Slave nodes. All nodes were connected through Ethernet.

4 Discussion of Experimental Results of Art Design Based on Big Data

4.1 Experimental Results and Discussion

(1) Analysis of performance test results of H-Apriori algorithm

The time needed to obtain frequent item sets in the original Apriori algorithm is matched with the time needed to obtain frequent item sets in the H-Apriori algorithm, as shown in Table 2 and Fig. 1.

Table 2. Performance comparison of Apriori and H-Apriori algorithms under different data sets

Dataset	H-Apriori	Apriori
AssociationsSP	100	105
Retail	150	1000
Accidents	1000	16000

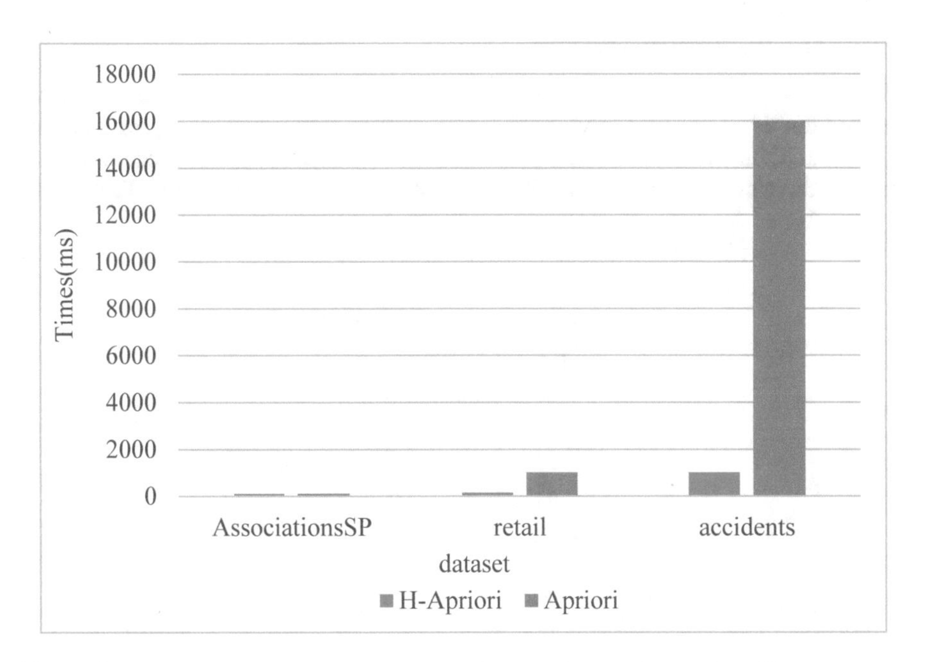

Fig. 1. Performance comparison of Apriori and H-Apriori algorithms under different data sets

As can be seen from Fig. 1, with the increase of file data sets, the traditional Apriori algorithm takes more time to calculate frequent item sets, and the job time required by map and Reduce phases to complete tasks in Hadoop distributed cluster environment is also more. But overall, the improved h-Apriori algorithm takes much less time to process data sets than the traditional Apriori algorithm. When the original file on 54 KB size of data sets, Apriori algorithm and H-Apriori algorithm basic same time needed for calculation of frequent itemsets, H-Apriori algorithm does not significantly improve the running time, this is because the Hadoop cluster to analyze the data size, determine whether to need to break up, in addition, according to the reduce task set how many files into the original data and the partition of the data copied to the corresponding the reduce task, complete computing tasks assigned to the DataNode, map and reduce tasks need time to calculate respectively. Sometimes, in the mix-wash phase, the time of data mix-wash takes longer than the calculation time, which is determined by the network bandwidth at that time, the speed of CPU processing, and the amount of data generated in the Map and Reduce phases and the time of data generation. However, as the file data set increases, the acceleration ratio between H-Apriori algorithm and Apriori algorithm increases significantly, because the time cost and split file cost assigned to the DataNode working node are very small relative to the calculation task.

(2) Image retrieval performance analysis

When the number of images is small, the more nodes the image retrieval time is more. As the number of images grows, the advantage of multiple nodes becomes apparent. Therefore, when the number of images is very large, distributed retrieval can obviously improve the efficiency of retrieval. The experimental results are shown in Table 3 and Fig. 2.

Table 3. Image retrieval performance test results

	1 node	2 nodes	3 nodes
5	30	35	40
10	45	55	50
15	65	60	55
20	85	70	65
25	100	90	70
30	160	100	80

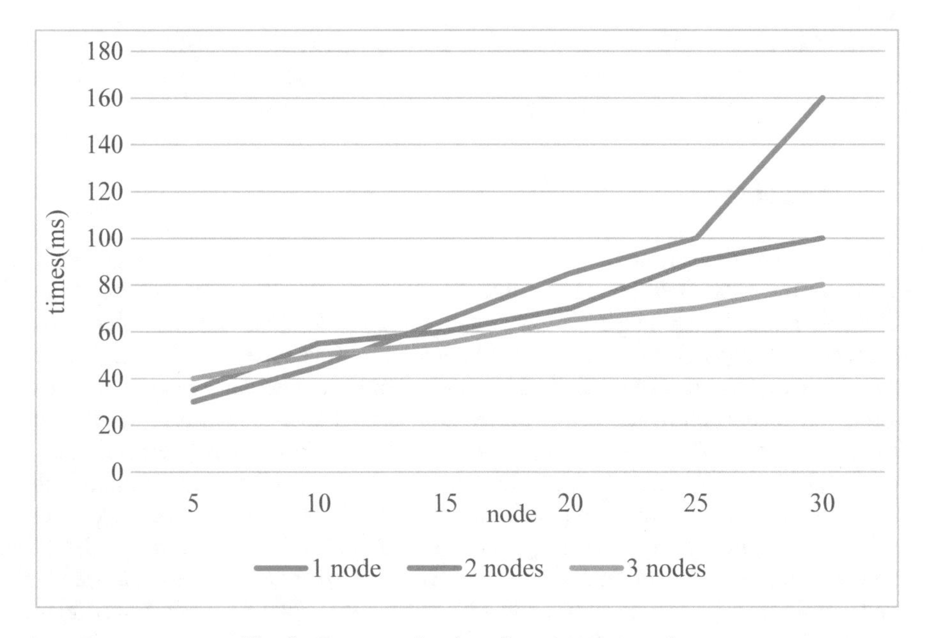

Fig. 2. Image retrieval performance test results

4.2 Suggestions for Artistic Design Creation in the Context of Big Data

(1) Infrastructure of big data analysis needs to be cultivated

At present, big data analysis is not yet mature in China, and big data is still at the enlightenment stage in the field of artistic creation in China, which is mainly reflected in the lack of data sources and the lack of authenticity of data. This shows that our country does not have an integrated analysis for large data systems, although some in China art website development faster, but not art field to a macro view of big data analysis and research of these sites, the real artistic creation can be used for large data analysis of rare, big data did not play a substantial role, most of these data collected by the site is all garbage data, it took a lot of resources to collect data, but the availability of data is not high, this also is a worth highly focus on the core issue of art workers. Data source is just on the one hand, the second in the work in the big data technology industry environment, our country has not set up a perfect institution and technology does not pass, in big data analysis industry, engaged in the enterprise is not standard, not professional, all data collected by the lack of accuracy, timeliness, get the conclusion of natural lack of accuracy.

(2) It is necessary to pay attention to the artistic value of design

Big data is supposed to bring the field of artistic creation from simple to diversified, and then close to interactive. In the art industry, the analysis of big data can achieve unprecedented success in design work. However, nowadays, big data technology is only at the enlightenment stage in the field of artistic creation in China. Incomplete data analysis and unqualified technology lead to more simplification in the field of artistic creation. Artistic creation blindly caters to the preferences of audiences and the investment direction of investors, thus losing its due artistic value.

5 Conclusion

To ensure the advancement and timeliness of art design, it is often inseparable from the support of big data. Only by making full use of big data can we give full play to our subjective initiative, stimulate people's vitality and wisdom, and maintain the unique charm of art design discipline. Due to the limitation of time, the work completed in this paper is only a preliminary result. In the future research and learning process, it will be carried out in the following aspects: increasing the number of nodes in the Hadoop distributed system. This paper mainly tests the efficiency of storage and retrieval from 1 node, 2 nodes and 3 nodes. In the future, multiple nodes can be used to improve the efficiency of system storage and retrieval. In addition, the intermediate output and results in the process of mass image storage and retrieval are all saved in HDFS, requiring frequent disk IO read and write operations, so it cannot meet the needs of image retrieval applications with short execution time. How to properly use cloud computing and the Spark platform based on memory computing to reduce the execution time will be the focus of future work.

References

1. Kong, W.: Digital media art design based on human-computer interaction technology in the background of big data. Revista de la Facultad de Ingenieria **32**(14), 485–489 (2017)
2. Heecheol, Y., Wonjae, S., Jungwoo, L.: Private information retrieval for secure distributed storage systems. IEEE Trans. Inf. Foren. Secur. **13**(12), 2953–2964 (2018)
3. Baoli, H., Ling, C., Xiaoxue, T.: Knowledge based collection selection for distributed information retrieval. Inf. Process. Manag. **54**(1), 116–128 (2018)
4. Gao, G., Li, R., Xu, Z.: Mimir: a term-distributed retrieval system for secret documents. Int. J. Inf. Commun. Technol. **12**(1/2), 209 (2018)
5. Tajeddine, R., Gnilke, O.W., Rouayheb, S.E.: Private information retrieval from MDS coded data in distributed storage systems. IEEE Trans. Inf. Theory **64**(11), 7081–7093 (2018)
6. Vadivazhagan, K., Karthikeyan, M.: Mining frequent link sets from web log using apriori algorithm. J. Comput. Theor. Nanosci. **16**(4), 1395–1401 (2019)
7. Guo, W., Zuo, X., Yu, J., et al.: Method for mid-long-term prediction of landslides movements based on optimized apriori algorithm. Appl. Sci. **9**(18), 3819 (2019)
8. Luhanga, E.T., Yonah, Z.O., Nyambo, D.G.: Characteristics of smallholder dairy farms by association rules mining based on apriori algorithm. Int. J. Soc. Syst. Sci. **11**(2), 99 (2019)
9. Hong, J., Tamakloe, R., Park, D.: Discovering insightful rules among truck crash characteristics using apriori algorithm. J. Adv. Transp. **2020**(2), 1–16 (2020)
10. Zhou, Y.: Design and implementation of book recommendation management system based on improved apriori algorithm. Intell. Inf. Manag. **12**(3), 75–87 (2020)

User Interface Code Automatic Generation Technology Based on Big Data

Chunling Li[1(✉)] and Ben Niu[2]

[1] School of Electronics and Internet of Things, Chongqing College of Electronic Engineering, Chongqing 400000, China
LiChunling0202@163.com
[2] Chongqing Research Institute, ZTE Corporation, Chongqing 400000, China

Abstract. With the wide application of Web applications in many fields, the efficiency of development and the quality of code requirements are also increasing. In the development of the project, especially in the application of B/S structure, there are great similarities between the code structures of each module, and the code writing has become a repetitive work. In order to improve the efficiency of code writing, this paper starts from the user interface code generation technology and combines with the background of big data to study the way of code writing automatic generation. This article first introduces the knowledge of automatic code generation. Then based on MDA, JAVA, JS, HTML code and other abstract to XML, can effectively save a lot of application development time, and effectively unify the application interface and operation mode. The definition of the language is mainly to design the foreground JSP page, which simplifies the preparation of JSP pages. The experimental results show that the automatic code generation saves 21.2% of the cost time and effectively improves the efficiency of user interface code writing.

Keywords: Big data era · User interface · Code automatic generation · Code writing

1 Introduction

Different application platforms have different development languages. For the same graphical user interface, tedious development process is required to realize cross-platform, which consumes a lot of work energy of developers and reduces their time to realize the actual functions and logic of the software [1, 2]. At the same time, with the development of large-scale parallel computing, big data, deep learning algorithm and human brain chip, as well as the reduction of computing costs, artificial intelligence technology has made rapid progress [3, 4].

Code generation is the primary way to make high-performance software reusable. Effects are indispensable in code generators, whether reporting failures or inserting statements and protections. Yukiyoshi Kameyama has shown that unrestricted effects interact with the generated bindings in undesirable ways, resulting in unbound variables or, worse, accidentally bound variables. These subtleties prevent domain experts from using and extending generators [5]. Stenzel Kurt gave us a formal calculus for the

J. MacIntyre et al. (Eds.): SPIoT 2020, AISC 1282, pp. 397–403, 2021.
https://doi.org/10.1007/978-3-030-62743-0_57

operational QVT. Stenzel, Kurt also provides a provably correct Java code generation framework. The framework USES the metamodel of the Java abstract syntax tree as the target for QVT transformation [6]. This problem is exacerbated by the emergence of a new generation of programming languages for large-scale, dynamic, and distributed systems, as most existing ADLs do not capture the features of this language. In this context, Everton Calvacante investigates the generation of source code in the Go programming language from the architectural description of the ADM language, as they are all based on calculus process algebra. Everton Calvacante also defines communication between adM-ADL and Go elements [7].

The code automatic generation model improves the efficiency of developers and eliminates the need to repeat tedious development work for the user interface of multiple software platforms, thus reducing the burden of developers [8, 9]. Referring to the idea of neural network framework, this paper proposes a whole set of code automatic generation technology based on neural network, conducts research on the cluster search technology in the testing process, and the experimental results prove its feasibility [10].

2 Big Data and Code Generation Technology

2.1 Big Data

There are many kinds of data to be processed in big data applications. From the perspective of data generation, it could be log output from the company's business, observation from research institutions, and user data from network applications. In terms of storage formats, it can be text in character format, audio in binary format, video, encrypted information that may have been compressed, etc. From the perspective of data organization, it may be structured data with fixed format or unstructured data with scattered information.

2.2 Automatic Code Generation Technology

On the basis of NET development platform, a MIS oriented software automation development framework based on database and code generation engine is used to realize Web automation development under the three-layer architecture, so as to improve the development efficiency, reduce the development cost and flexibly respond to the constant changes of user requirements in practical application. The framework consists of four parts: database conceptual model, common code template, code generation engine and automatically generated three-tier architecture code.

As the most commonly used nonlinear activation function in artificial neural networks, the mathematical form of Sigmoid function is as follows:

$$f(x) = \frac{1}{1 + \exp(-x)} \tag{1}$$

According to the above expression, it can map the input value to a numeric range of (0, 1). Applied in neural network, the process of output value changing from 0 to 1 shows that it well fits the state of neurons from inactive to active in biology. Deform the function in the following form:

$$\tanh(x) = \frac{\exp(x) + \exp(-x)}{\exp(x) - \exp(-x)} \tag{2}$$

3 Design of the Experiment

3.1 Experimental Background

The rapid growth of the Web is changing the landscape of the software industry, and the trend for the next few years will be browser-based applications. Web applications have gradually entered into every aspect of our daily life, from portal sites to search sites, from online forums to online shopping malls, Web applications are changing people's lives with a large number of exciting advantages. With the rapid development of software development technology, reusable, extensible and well-tested software components are becoming more and more popular among developers. This means that developers can devote enough energy to analysis and building business-level logical applications to get rid of simple code work.

3.2 Experimental Design

GUI file first after image preprocessing to get high-dimensional sparse image feature vector, and then use this article innovative application since the encoder based on the deconvolution network coding image features extraction of low-dimensional dense: after a DSL code in the file character results after word embedded model for dense term vectors, then use GRU helped network extraction from word feature vector, the above two vector characteristics after cascade is a coding model in this paper.

The coding model in this paper is similar to the pix2Code network structure, which is mainly composed of two parts, namely the visual model and the language model. Among them, the self-encoder based on the deconvolution network and the full connection layer constitute the visual model in this paper, and the word embedding layer and the GRU network constitute the language model in this paper. The experimental results are shown in Table 1.

Table 1. Experimental results

The attribute name	If required	The default value	Constraints
id	Y	No	Must be unique within the same module, and the home page must be indexed
title	Y	No	The page title
init Op	N	No	Multiple SQL; Delimit, where SQL is no longer supported? Expression, if parameters need to be passed, follow the system variable definition
Params	N	No	–

4 Automatic Generation of User Interface Code for Big Data

4.1 Analysis of the Application of User Interface Code Automatic Generation Technology Based on Big Data

As shown in Fig. 1, when the code automatic generator runs, the corresponding code file is generated. When a new application app is created, the index. jsp and Struts.xml files will be generated first, and corresponding Java packages will be generated, including Action, Service, DAO, domain, etc., and then the application will be written into the data table. For visual model, using the deconvolution deconvolution in the network layer is able to restore the original image, and the encoder output among the best feature extraction results can be used as input features, using GUI design training innovative application since the encoder, based on the deconvolution network using deconvolution network as well as the excellent characteristics of the encoder, make the network after training to maximize the extraction of features in the image. When the

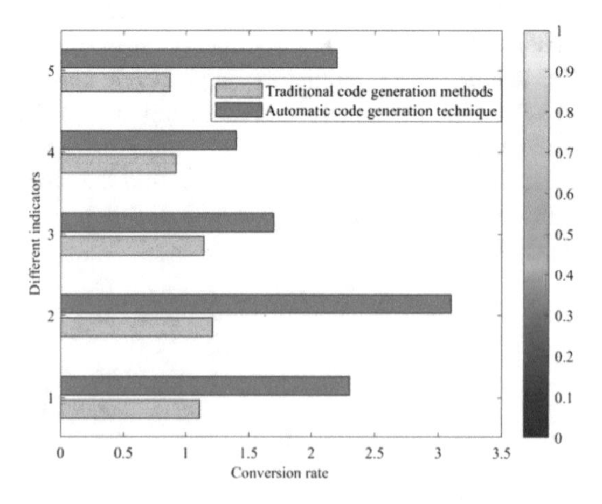

Fig. 1. Comparison of conversion rates between traditional code generation methods and automatic code generation techniques

auto-encoder training is completed, the multi-dimensional feature vectors of the auto-encoder output are reduced through the full connection layer to obtain the low-dimensional feature vectors of the visual model, so as to facilitate the subsequent cascade with other low-dimensional feature vectors.

As shown in Fig. 2, the model designed in this paper transforms from the original $256 \times 256 \times 3$ to $32 \times 32 \times 128$, which is the network part of the original visual model in Pix2Code that removes the full connection layer. This part is taken out separately in this paper to connect with the subsequent deconvolution network. Among them, the operation of the deconvolution network is basically a mirror version of the convolutional network with multiple deconvolution layers and up sampling layers. The purpose of convolutional network is to reduce the number of activated neurons through pooling, i.e., the down-sampling operation. Unlike convolutional network, the purpose of deconvolution network is to expand the scale of activated neurons through the combination of deconvolution operations. Using convolution neural network in this paper, the visual model unsupervised feature extraction, it is mainly composed of two parts of the convolution and deconvolution, among them, the convolution network is mainly used to convert the input image into multi-dimension feature vectors of feature extractor, and deconvolution network as a kind of image generators, based on the extracted from the convolution network characteristic vector, the value of the final output is the same as the input image size.

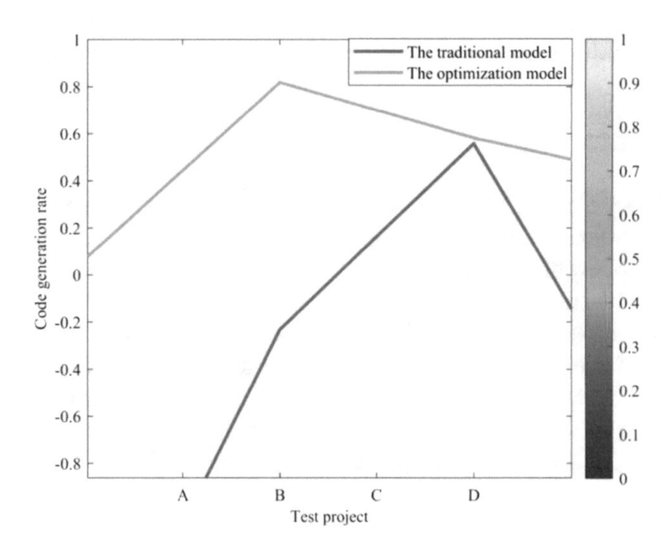

Fig. 2. Code generation efficiency of traditional and optimized models

Such applications are still CRUD operations in nature, but the interface is different from the standard interface rules. In this case, it is recommended to use the include tag, place special pages in external files, and then include them in the appropriate location, so that the application will be built with the contents of these files unoverwritten. Such as report design function module. In this case, the configuration files in the Resource directory, including the Struts and Spring configuration files, may need to manually specify actions and beans, and the tool has determined that it will not overwrite the handwritten content, so you can safely use it. However, for JAVA code, as JAVA source files are not structured files like XML, they can only be overwritten. Therefore, if there is manual JAVA code, it is recommended to write it to another JAVA class instead of directly modifying it in the automatically generated JAVA class.

4.2 Suggestions on the Application of User Interface Code Automatic Generation Technology Based on Big Data

In user page setup, the message/service bus is carried as a message and consists of two parts: the header and the message style. The header mainly includes the public information of each type of message, and the function of the message is analyzed from the header information. The header contains the length of the message style, message type, event identification, source node, source process, target node, target process, and other information. The header is the public information that must be carried by each message, and the style of the message depends on the application. Newspaper style is the specific data information sent. Quote style according to different application function definition of each data volume, application system before sending a message to serialize data body, receive a message through the deserialization again after get the result of the request, so that each application needs to provide serialization and deserialization of the style of interface, the interface application scenarios are different, but the function is to serialization and deserialization of input parameters, the results as output parameter.

5 Conclusions

In order to better the user interface code automatic generation technology, combining with the background of big data, based on the coding, try to use DSL symbol is transformed into low dimensional continuous value at the same time, further, the words of similar symbols in vector space is mapped to a similar position, so as to meet the needs of the original model serialization on one hand, fundamentally solves the problem of limited dimension vector, in order to further extract GUI rich characteristics in image corresponds to the DSL code provides a possibility. Experimental results show that the word embedding model can improve the accuracy of the model on the basis of the original, which proves the feasibility of the method.

References

1. Erdi, P., Huhn, Z.: Special issue: Hippocampal theta rhythms from a computational perspective: code generation, mood regulation and navigation. Neural Netw. **18**(9), 1202–1211 (2005)
2. Benouda, H., Essbai, R., Azizi, M., et al.: Modeling and code generation of android applications using Acceleo. Int. J. Softw. Eng. Appl. **10**(3), 83–94 (2016)

3. Syriani, E., Luhunu, L., Sahraoui, H.: Systematic mapping study of template-based code generation. Comput. Lang. Syst. Struct. **52**, 43–62 (2017)
4. Possatto, M.A., Lucredio, D.: Automatically propagating changes from reference implementations to code generation templates. Inf. Softw. Technol. **67**, 65–78 (2015)
5. Kameyama, Y., Kiselyov, O., Shan, C.C.: Combinators for impure yet hygienic code generation. Sci. Comput. Program. **112**, Part 2, 120–144 (2015)
6. Stenzel, K., Moebius, N., Reif, W.: Formal verification of QVT transformations for code generation. Softw. Syst. Model. **14**(2), 981–1002 (2015)
7. Calvacante, E., Oquendo, F., Batista, T.: Architecture-based code generation: from π-ADL architecture descriptions to implementations in the go language. In: Lecture Notes in Computer Science, vol. 8627, no. 14, 130–145 (2015)
8. Tan, W.J., Tang, W.T., Goh, R.S.M., et al.: A code generation framework for targeting optimized library calls for multiple platforms. IEEE Trans. Parallel Distrib. Syst. **26**(7), 1789–1799 (2015)
9. Basu, P., Williams, S., Van Straalen, B., et al.: Compiler-based code generation and autotuning for geometric multigrid on GPU-accelerated supercomputers. Parallel Comput. **64**, 50–64 (2017)
10. Song, L., Di, L., Fan, L., et al.: CPSiCGF: a code generation framework for CPS integration modeling. Microprocess. Microsyst. **39**(8), 1234–1244 (2015)

Cultural Heritage Research Model Based on Big Data

Jialin Li[1(✉)], Shuo Wang[2], and Yujuan Yan[2]

[1] Changchun Institute of Land Surveying and Mapping, Changchun, Jilin, China
15876031@QQ.com

[2] Jilin Engineering Normal University, Changchun, Jilin, China

Abstract. With the advent of the era of big data, digital protection of cultural heritage presents new opportunities and challenges. At present, the digital protection and utilization of cultural heritage is a new topic for relevant practitioners in China. From the perspective of big data, this paper tries to sort out the background and necessity of the cultural heritage research model under the background of big data. Based on the characteristics of big data, the digital protection of cultural heritage is considered. Then, in this paper, through investigating the city A cultural heritage research, with A corresponding data of cultural heritage as A sample, will be based on the GIS platform is introduced into the construction of the technical tools and analytical knowledge system heritage research space visual characteristics, to build the relationship between the cultural heritage space visual and quantitative analysis, the corresponding relationship between, according to the results of the analysis of the cultural heritage efficiency increased by 23.11%. Finally, this paper puts forward the suggestion of establishing standard operation platform in the background of big data.

Keywords: GIS platform · Big data background · Cultural heritage · Research model construction

1 Introduction

Architectural cultural heritage is the historical gene and cultural element of a city, and its protection and utilization is one of the important research fields of architecture [1, 2]. So far, digital technologies such as 3D laser scanning, BIM information model, GIS geographic information, VR virtual reality have been widely applied in this field, while the application of the new generation of artificial intelligence technology in this field has just started and needs to be further developed [3, 4].

Through the application of information and communication technology, the preservation and promotion of cultural heritage around the world has become one of the most important research topics and has a wide application prospect. Bartolini Ilaria presents a generic recommendation framework that manages heterogeneous multimedia data from multiple Web repositories and provides users with context-aware recommendation technology to support intelligent multimedia services. A dynamic access path for a given environment. Through case study, the specific application of the system in

J. MacIntyre et al. (Eds.): SPIoT 2020, AISC 1282, pp. 404–411, 2021.
https://doi.org/10.1007/978-3-030-62743-0_58

the field of cultural heritage is proposed [5]. Cultural heritage sites are threatened by a variety of natural and man-made factors. Innovative and cost-effective tools for systematically monitoring landscapes and CH sites are needed to protect them. Agapiou A proposed A multidisciplinary approach based on remote sensing technology and geographic information system (GIS) analysis to assess the overall risk in the Paphos district (Cyprus) [6]. Nakamura, Satoshi proposed in a semi-closed environment threat assessment of air pollution on cultural heritage of the examples of synthesis method, which in Florence, Italy Vicky monitoring campaigns in the palace is used as a case study, conducted a wide range of research projects, main purpose is to get on the air quality in the courtyard and micro climate condition of the first batch of quantitative data, and where possible, to determine the main cause of degradation and puts forward appropriate protection strategy [7].

The importance of cultural heritage is self-evident. Based on the background of big data, this paper constructs a RESEARCH model of cultural heritage based on GIS, discusses the applicable boundary of the model, and USES SD method to verify the results [8, 9]. At the same time, the data collection, optimization and analysis methods are fully verified on the technical level, and are applied to the spatial analysis of A culture in this city to verify the feasibility and applicability of the proposed method [10].

2 Big Data and Cultural Heritage Model

2.1 Status Quo of Digital Protection of Cultural Heritage

With the background of globalization and modernization, culture has been impacted unprecedentedly, and cultural heritage has gradually lost its original soil and social environment. Digital protection, cultural heritage is the digital information technology is applied to the ethnic and folk cultural heritage in the rescue and protection, with the aid of digital photography, 3d information acquisition, virtual reality, multimedia and broadband network technology, build a comprehensive digital system based on computer network, so as to realize the protection of cultural heritage, inherit and carry forward. It can be seen that the ways of cultural heritage digital protection include digital collection and storage, database management, digital virtual, digital publishing, digital entertainment and so on. However, the digital protection in China is mainly based on the primary information collection led by the government or driven by projects, and the application means are a little single, which needs to be further strengthened.

2.2 Construction of Cultural Heritage Research Model

Traditional GIS can only present the X, Y, 3d GIS extends the vertical axial (Z), the calculation of the overall data have Z axis, highly vision analysis and calculation of obstacles of elevation, Angle of view analysis at the same time with a flat Angle and vertical Angle data, visibility analysis and stadia approximation space syntax, but increased the obstacles with elevation attribute calculation, greatly improved the data precision and accuracy. In GIS system, the nonlinear distortion needs to be corrected in time, so that the camera parameters can get higher precision. The establishment of camera nonlinear model is also critical to the accuracy and speed of camera calibration:

$$\sigma_x(x, y) = k_1 x\left(x^2 + y^2\right) + \left(p_1\left(3x^2 + y^2\right) + 2p_2 xy\right) \quad (1)$$

$$\sigma_y(x, y) = k_2 y\left(x^2 + y^2\right) + \left(p_1\left(3x^2 + y^2\right) + 2p_1 xy\right) \quad (2)$$

3 Design of the Experiment

3.1 Experimental Background

In recent years, the main driving force for the development of digital heritage protection, UNESCO has been actively promoting digital means for the research, archiving, analysis and display of human cultural heritage for many years. It has become an important issue that needs to be actively addressed in the current field of cultural heritage digitization to build a theoretical system of heritage digitization for different research and protection needs and to lay solid professional support such as method tools, data models and technical processes. Therefore, the development of informatization constitutes an important technical support for the protection of architectural cultural heritage, and it is necessary to conduct a holistic research from the perspective of methodology and construct the research paradigm of digital heritage protection.

3.2 Experimental Design

Taking A cultural heritage area of this city as the experimental sample, the image and data of A cultural heritage area are collected on the spot. The GIS based workflow requires the photogrammetric results to generate A clear grid object conforming to THE GIS usage standards, and then it is imported into ArcGIS through Arc GIS3D Analys Text Ension module for data analysis. Other relevant data attached to the 3D model should also be imported and correlated in ArcGIS to form a property parameter library that is easy to manage. Using 3d editing tools in Arc Scene, polygons, polylines, and points, elements such as doors and Windows can be digitized directly to the model, which can then be used to simplify the representation of the model for spatial analysis in GIS. The experimental results are shown in Table 1.

Table 1. Experimental results

Regional division	The ranking	The environment	The atmosphere	The landscape
A	1	78.21	83.11	80.16
C	2	87.99	72.16	78.69
B	3	82.14	57.21	72.23
D	4	72.11	72.43	68.32
E	5	68.32	78.21	67.37

4 The Construction of Cultural Heritage Research Model

4.1 Analysis of Cultural Heritage Research Model Based on Big Data

As shown in Fig. 1, this paper adopts IDW interpolation method to judge the historical accumulation of the city. IDW interpolation is a weighted distance average function, which is a method to judge the historical age of non-cultural relics and historic sites according to the time of construction of all cultural relics and historic sites. In this interpolation result, the size of any raster value is between the maximum value and the minimum value of the sampled data, which is relatively stable. IDW interpolation can use different power exponents, the higher the power exponent, the smaller the influence of distant points, that is, the higher the degree of local influence. The results show that the oldest areas do not coincide exactly with concentrations of existing heritage sites. On the one hand, the development of primitive settlements took a long time, during which human civilization was not yet developed and there were few material and cultural heritages left. On the other hand, there may be historical deposits buried underground that have not yet been discovered. The analysis of historical deposits can be used as a reference for delineating underground burial areas.

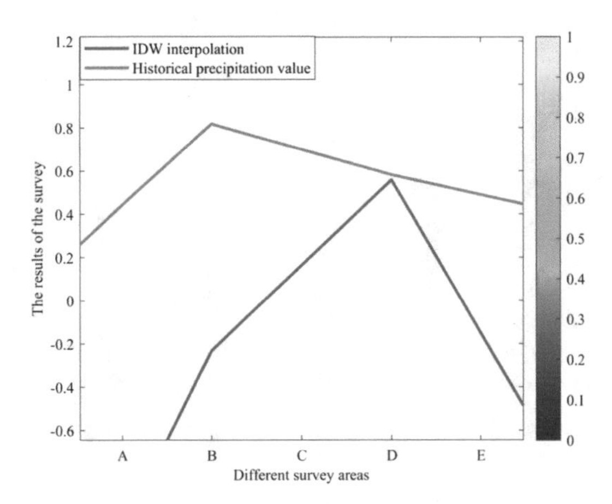

Fig. 1. Distribution of historical precipitation value of cultural heritage in this city

Location degree refers to the convenience of getting to a place. Location degree is related to transportation, geographical location and land price potential, which has great influence on the protection and utilization of cultural heritage. The location degree analysis of A cultural heritage in this city is carried out. Firstly, the geographical location analysis is carried out. Many popular points with concentrated passengers and convenient transportation are selected as the city center, such as the core square, railway station and airport, etc., and the distance between cultural relics and historic sites and the city center is calculated. Since there are multiple objects in the center element, the command to obtain the minimum distance should be adopted to select the distance from the nearest city center. Rate the distances on a scale of 5 to as far as 1. On the basis of geographical location analysis, traffic accessibility is analyzed. The route near A cultural heritage in this city is divided into road network grading map. The road system is introduced into GIS to calculate the linear distance between the heritage resources and the main and secondary trunk roads, and the shortest distance between each heritage resources and the main and secondary trunk roads is obtained. These distances are also divided into 5 intervals according to their magnitude and scored respectively. Weighted and superposed analysis shows that yellow represents good location, convenient transportation and intensive heritage resources, and the more green, the more disadvantaged it is. It can be intuitively seen that among all the cultural heritage resources, the central city is the best, and some villages and towns are again. Combining the location and value of cultural relics and historic sites can provide reference for the development sequence of heritage protection and utilization.

As shown in Fig. 2, as the result of the current investigation of cultural heritage, the current status analysis chart is needed for easy analysis. The traditional method is the use of CAD, Photoshop and other software, thematic separate painting. It takes a lot of work and time. The accuracy rate is low due to the reliance on researchers' recall and

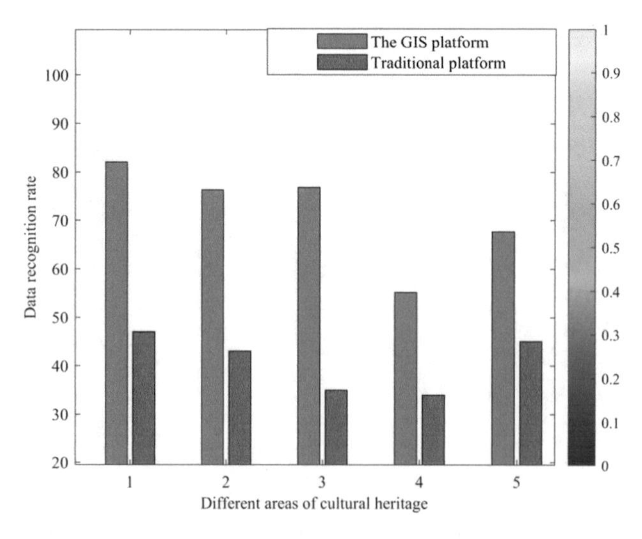

Fig. 2. Efficiency comparison between traditional methods and GIS platform for A cultural heritage

supplement. And the application of information technology, in the early information collection and storage based on the completion of the database can be used to quickly output the thematic analysis chart. Take the thematic map of current architectural features as an example, open the property sheet of A cultural heritage historical block in GIS, double-click the building features column, and make the property sheet rank by building features. Select the building feature of the same level and export it to a separate layer. Set the colors of the six layers respectively. Select the legend type and generate it automatically. The process is similar for other thematic layers.

A large number of images and attribute data obtained in the survey need to be timely integrated and input into the database after the survey. In the database, graphic data, attribute data and picture data can be associated, that is, click on the current building plane to quickly call up relevant attribute data and photos. The relevant steps are as follows: First, the building or lot should be numbered in CAD, which is consistent with the number in the survey. In the CAD plan, the building is divided into blocks with single or courtyard as the unit, and the name of the block is named as the building number. The numbering is generally divided into blocks and named from left to right and top to bottom, such as A001. A is the block number and 001 is the building number. The number is unique and does not repeat. The CAD after the chunking is imported into THE GIS, and a column of numbered information is automatically generated in the property sheet. Secondly, input the contents of the survey form into Excel, and import the excel contents into GIS by connecting the Excel number column with the GIS number column, so as to realize the association between the attribute contents and the plane.

4.2 Suggestions on Cultural Heritage Research Model Based on Big Data

Demand mining the value of data, highlight the relevance of data. Data does not exist in isolation, but has a complex internal relationship with other data, and has a close logic. In platforms, tree diagrams, node diagrams, and parent-child hierarchy diagrams are often used to represent internal associations between data. This enables the user to start from one data and query all relevant data, and through visualization, it is very easy to grasp the internal relations between the data. In this respect, the GIS system platform is definitely ahead of the Scuderia platform, thanks to a more powerful backend infrastructure framework.

It is required to enhance the artistic presentation power of data visualization to meet the sensory needs of users. In addition to meeting functional requirements for actual use, platform construction must also meet certain aesthetic requirements, which is conducive to the improvement of user experience and the establishment of a good man-machine relationship. At the same time, different from ordinary engineering projects, cultural heritage itself is full of profound cultural and artistic deposits. While meeting the functional requirements, it also requires the artistic expression of data visualization. Since Scuderia platform has been used for 3d display of cultural heritage before, it has higher requirements on artistry. In addition to the platform itself, which has good feedback on visual experience, users can also design artistic user interface and graphic display mode at will, enhancing the role of artistry in the process of data visualization.

Users' control over data is required to be strengthened. Whether zooming in, zooming out, roaming, or displaying, disappearing, or hierarchical browsing, keep the data loaded and updated in real time. All feedback from the platform should be fast and accurate. For data analysis, the platform needs to be able to make enough effective choices. In this respect, the GIS system platform is superior to the Scuderia platform. Besides this, the graphic transformation of data analysis process and results is also one of the important connotations of data visualization. Although the form of the data is transformed through graph transformation, the kernel of the data is not transformed, and the data can also be analyzed using statistical and computer science analysis methods such as cluster analysis, factor analysis, correlation analysis, regression analysis, A/B testing and data mining.

5 Conclusions

In order to better the construction of a cultural heritage research model, this article embarks from the background of big data, based on GIS technology, comprehensive visual area, visual Angle, distance, skyline, shadow and volumetric analysis "passive" quantitative methods, and put it into two levels of architectural space and cultural space comprehensive experiments. This paper takes A cultural heritage area of This city as the sample for experiment, and the results show that this model can effectively protect cultural heritage digitally. How to effectively manage "big data" based on the platform and make use of the advantages of visualization, and how to deeply dig data value is the focus and direction of future research work.

Acknowledgements. The research projects of Jilin Engineering Normal University—Key Technologies of Virtual Reality Fusion and Interactive Operation for Intelligent Learning in Network Environment.

References

1. Lagomarsino, S., Cattari, S.: PERPETUATE guidelines for seismic performance-based assessment of cultural heritage masonry structures. Bull. Earthq. Eng. **13**(1), 13–47 (2015)
2. Pechlaner, H.: Cultural heritage and destination management in the Mediterranean. Thunderbird Int. Bus. Rev. **42**(4), 409–426 (2015)
3. Picard, D., Gosselin, P.H., Gaspard, M.C.: Challenges in content-based image indexing of cultural heritage collections. IEEE Signal Process. Mag. **32**(4), 95–102 (2015)
4. Guzman, P.C., Roders, A.R.P., Colenbrander, B.J.F.: Measuring links between cultural heritage management and sustainable urban development: an overview of global monitoring tools. Cities **60**(PT.A), 192–201 (2017)
5. Bartolini, I., Moscato, V., Pensa, R.G., et al.: Recommending multimedia visiting paths in cultural heritage applications. Multimedia Tools Appl. **75**(7), 3813–3842 (2016)
6. Agapiou, A., Lysandrou, V., Alexakis, D.D., et al.: Cultural heritage management and monitoring using remote sensing data and GIS: the case study of Paphos area, Cyprus. Comput. Environ. Urban Syst. **54**, 230–239 (2015)

7. Nakamura, S., Fukuda, M., Issaka, R.N., et al.: An integrated approach to assess air pollution threats to cultural heritage in a semi-confined environment: the case study of Michelozzo's Courtyard in Florence (Italy). Nutr. Cycl. Agroecosyst. **106**(1), 47–59 (2016)
8. Vojinovic, Z., Hammond, M., Golub, D., et al.: Holistic approach to flood risk assessment in areas with cultural heritage: a practical application in Ayutthaya, Thailand. Nat. Hazards **81** (1), 1–28 (2016)
9. Chianese, A., Piccialli, F., Valente, I.: Smart environments and Cultural Heritage: a novel approach to create intelligent cultural spaces. J. Location Based Serv. **9**(3), 209–234 (2015)
10. Bertolini, L., Carsana, M., Gastaldi, M., et al.: Corrosion assessment and restoration strategies of reinforced concrete buildings of the cultural heritage. Mater. Corros. **62**(2), 146–154 (2015)

Changes in Enterprise Human Resource Management in the Context of Big Data

Jun Wang(✉)

Department of Human Resource Management, Dalian Neusoft University of Information, A7, 8 Ruanjianyuan Road, Dalian 116023, Liaoning, China
wangjun_xg2@neusoft.edu.cn

Abstract. As big data gradually becomes the focus of all walks of life, its role in enterprise human resource management cannot be ignored. The rapid development of Internet technology has also enabled companies to collect a large amount of data and process it effectively. Based on the background of big data, this article analyzes the changes in the six modules of human resource management, human resource planning, recruitment and allocation, training and development, performance management, salary management, and employee relations. Development provides theoretical guidance.

Keywords: Big data · Human resource management · Enterprise information

1 Big Data Overview

1.1 Big Data Definition

In the context of big data, how to optimize human resource management and how to better find the talents needed by enterprises in accordance with the development of the times, and develop and retain talents, are important topics that must be considered in the development of human resource management [1].

Summarizing the opinions of scholars, the meaning of big data can be divided into the following four levels: First, "big data" refers to the use of technical tools and processing modes that are beyond conventional. The data has a high growth rate and global quantification, and requires stronger Information assets of decision-making power and insight. Second, the basic characteristics of "big data" can be summarized in four points: massive data, high-speed processing capabilities, diversified data types, and flexible processing methods. Third, big data has dual attributes, including technical attributes and social attributes. Technical attribute refers to the efficient, advanced and systematic analysis of massive data. The method used is professional and requires strong analytical ability; social attribute refers to its great influence on the market, organization, government, society, etc. The big impact reflects the regularity of the development of the historical process of human society. Fourth, "big data" is a new way of thinking and data wisdom of big data. It represents a new era, leads a new trend of The Times, and has important theoretical guiding significance for the optimization of human resource management in enterprises today.

J. MacIntyre et al. (Eds.): SPIoT 2020, AISC 1282, pp. 412–418, 2021.
https://doi.org/10.1007/978-3-030-62743-0_59

1.2 Big Data Characteristics

First, the volume of data is huge. From TB to PB; Second, there are various types of data, such as web logs, videos, pictures, geographic location information and so on. Third, the value density is low. In video, for example, there may be useful data for only a second or two during continuous monitoring; Fourth, speed of processing. This point is also different from the traditional data mining technology. The Internet of Things, cloud computing, mobile Internet, Internet of cars, mobile phones, tablets, PCS and various sensors all over the world are all sources of data or ways of carrying it [2].

2 Limitations of Traditional Human Resource Management

2.1 Management Philosophy Is Relatively Limited

Traditional human resource management does not pay much attention to people, and the importance of related work is not recognized. In many cases, human resource management is even equivalent to logistics work. Such misunderstandings have also led to enterprises not paying enough attention to the planning of human resource management work, and insufficient training of relevant staff, their professional level is relatively low, and the advantages of human resource informatization cannot be highlighted [3, 4].

2.2 Data Records Are Not Detailed and Comprehensive Enough

Traditional human resource management has many problems and deficiencies in data recording. The work efficiency is often low and the recorded information is cumbersome and unorganized. Sometimes, there may even be cumbersome information and incomplete records. Due to the incomplete and insufficiently detailed information records, many actual data about employees cannot be fully obtained, and the analysis results based on the data analysis cannot reflect the real problems of the employees, which greatly reduces the pertinence and effectiveness of management [3, 4].

2.3 Human Resource Planning Does Not Match the Actual Situation

Avery important task in human resource planning is to predict the demand and supply of human resources. Human resource demand forecasting is to predict the number and quality of employees needed by the company in a certain period in the future according to the requirements of the development of the company, and then determine the personnel supplement plan to implement the education and training program. Manpower supply forecasting, also known as staffing forecast, is another key link in human resource forecasting. Only after forecasting the number of personnel and comparing it with the demand for personnel can various specific plans be made. In traditional human resource planning, managers often rely on experience or brainstorming to make decisions. They lack the analysis and summary of data related to human resource planning, and they do not trust or recognize the data. As a result, human resource planning is often inconsistent with the actual situation. The phenomenon of balance seriously even leads to the loss of employees and market changes in the enterprise, which affects the long-term strategy of enterprise development [5].

2.4 Recruitment Data Utilization Rate Is Not High

Traditional human resource management has already used network platforms and methods when recruiting, such as Liepin.com, 58.com, 51job.com, Zhaopin.com, etc., which provide companies with a large amount of candidate data. For applicants, they can obtain more recruitment information through the Internet, which broadens the source of information channels; for companies, they can search the candidate database for personnel matching the company's position for screening and hiring. However, online recruitment is still not the main recruitment channel of enterprises. Enterprises still rely on offline on-site recruitment. Online recruitment is often used as a supplementary form, which makes the utilization rate of recruitment data on online platforms not high. In addition, online recruitment also has the problem of redundancy of personnel information and difficulty in judging the authenticity of the information [5].

2.5 Staff Training Is Not Effective

As one of the modules of human resource management, employee training and development has received more and more attention from enterprises, which is also a reflection of the increasing emphasis on talents by enterprises. However, the training and development of employees in many companies are often mere formalities. The training needs analysis is not comprehensive and specific enough, the training implementation lacks supervision and control, and the training evaluation system is not perfect enough to form a cyclically improved training system. In fact, the mining of employee training data will have a very beneficial impact on employee training and development work [5].

2.6 Performance Pay Is Not Systematic Enough

Performance management and salary and benefits management are two very important modules in human resource management. Traditional human resource management lacks big data mining and analytical thinking in these two modules. On the one hand, when formulating systems, the performance compensation system often fails to keep up with the pace of enterprise development, and there is a certain lag; on the other hand, performance compensation data, especially performance data at all levels, companies often don't know how to use it after collecting, On the contrary, the evaluation of employees is not accurate enough, and it may not be possible to objectively and comprehensively evaluate when the relevant work is summarized.

2.7 Not Paying Enough Attention to Employee Relationship Management

Employee relationship management includes labor contract management, communication management, corporate culture construction, service and support, employee emotion management, etc. Due to the many aspects involved in employee relationship management, especially the management of labor contracts and disputes, a large amount of management information and data will be generated. If big data and other information methods are used for mining and analysis, it will be an improvement and perfection of employee relationship management. Important basis. However, traditional human resource management often ignores these data and information, ignoring the potential management value, which makes related work costly, low efficiency, and difficult to carry out.

3 Big Data Types in Enterprise Human Resource Management

In the context of big data, enterprise human resource management is gradually moving towards data, including data on people, positions, training resources, incentive resources, etc. The specific content is as follows.

3.1 Basic Data-Employee History Reproduction

Basic data refers to the original data of employees, including their personal information, such as age, academic information, professional skills, personal expertise, practical experience, position, working years, turnover frequency, and other information that truly reflects the overall quality of the individual. With the characteristics and capabilities, it can provide some objective reference standards for the human resource management department. These records, as important data for companies to analyze employees, reflect the employees' previous growth experience relatively truly, and companies can also use this data to make a comprehensive analysis of the employees' previous comprehensive qualities [6, 7].

3.2 Competence Data-Employee Development Feedback

Competence data refers to the data on the changes of employees' personal abilities before and after the job, including the comprehensive situation of the employee's personal development, such as training experience (including pre-job and post-job training), training assessment, problem-solving efficiency, participation in competitions, rewards and punishments Situation, position changes, etc. When the company collects data on the ability of the survey, it includes the assessment of new employees' past abilities and the assessment of changes in the abilities of old employees. It includes both horizontal comparative analysis between employees and vertical development analysis of the employees themselves, which helps the company to understand more comprehensively. The development of employees, to understand the personal capabilities of employees, so as to provide targeted guidance for their own development [6, 7].

3.3 Efficiency Data-Work Effectiveness Test

Efficiency data refers to data on the completion of work tasks of employees, including the comprehensive performance of employees at work, such as work task completion efficiency, completion time of different types of work, completion time of individual tasks, error rate, failure rate, etc. Through the collected efficiency data, companies can make objective and true evaluations of the working conditions of employees at different levels. At the same time, according to the objective situation, analyze the deep-seated reasons that cause the difference in employee efficiency to formulate a reasonable human resource training plan, and carry out employee ability training in a hierarchical and targeted manner to improve employee work efficiency [6, 7].

3.4 Potential Data-Employee Future Analysis

Potential data refers to the data on the analysis of the employee's career planning, including the employee's future development direction and potential, such as work efficiency improvement rate, income increase, promotion frequency, performance improvement rate, etc. Through the analysis of employee potential data, the company is conducive to further discovering outstanding talents and developing their career potential, thereby promoting the progress and development of the company itself. At the same time, the analysis of potential data is helpful for companies to better formulate corporate human resource training plans, teach students in accordance with their aptitude, truly give full play to the capabilities of each employee, form corporate cohesion, and continuously improve corporate competitive advantages [6, 7].

4 Human Resource Management Changes in the Context of Big Data

4.1 Big Data and Human Resource Planning

In the big data environment, companies can collect a large amount of internal and external data information, combined with traditional subjective and qualitative methods, to analyze the real data situation of each employee more comprehensively, and to gain insight into the basic data, ability data, and efficiency of employees Data and potential data, combined with the current employee's personal career development plan and the company's personnel turnover in recent years, comprehensively analyze the company's future personnel needs, and make reasonable plans and arrangements for internal promotion and external recruitment based on the facts As the premise, based on data, make enterprise human resource planning more reasonable [8–11].

4.2 Big Data and Recruitment

After applying big data technology, collect data on various factors such as personality, work ability, career tendency, academic qualifications and experience of candidates through the Internet, and use the computer language of big data to deeply analyze and explore the internal driving force and comprehensive quality of candidates, To have a more comprehensive and profound grasp of the comprehensive situation of the applicants, based on the analysis of big data information, to carry out intelligent personnel screening and evaluation, and to innovate the recruitment plan. Based on the analysis of the comprehensive information of the applicants, to realize the data analysis of job matching, In order to improve the subjective defects of traditional recruitment, to achieve a comprehensive analysis of subjective and objective combination, to make recruitment more accurate and reliable [8, 12–14].

4.3 Big Data and Training and Development

The application of big data in training and development can be analyzed from the perspectives of the company and employees. From the company side, we can use big data analysis to design more effective training courses and systems, and at the same time monitor and collect various data of employees in the training process. Through in-depth data mining, we can objectively and scientifically analyze the training effects of employees and employees. For training problems, continuously improve and perfect the company's training and development. From the perspective of employees, under the big data training system, they can make full use of their fragmented time through the network to carry out fragmented learning, and independently choose the training courses of interest to strengthen and improve [8, 12, 14].

4.4 Big Data and Performance Compensation

In the era of big data, performance appraisal can use big data informatization to make the appraisal more specific and targeted. Big data can be used to establish personal performance portraits of employees, which can be used as the basis for evaluation, collect various performance information of employees, and conduct comprehensive information-based performance evaluation of employees. Salary management can use big data information mining technology to do a good job in the analysis and comparison of external salaries in the same industry, and then design a more reasonable and competitive salary system based on the actual situation of the company and job needs, and it can also better analyze the company The internal salary situation makes the company's salary distribution more fair and balanced [8, 12–14].

4.5 Big Data and Employee Relations

Big data era, can a comprehensive collection of employee relations, complex data analysis from the result of data analysis and mining problems of employee relationship management, analysis and summary of employee needs, timely solve the related issues of employee relationship management, truly achieve "people-oriented" thought, make employees feel warm, caring and harmonious corporate culture.

5 Conclusions

In the era of big data, human resource management must follow the trend, comprehensively collect and organize various information and data from all aspects, and then use big data for in-depth data mining, in order to analyze, guide and plan the company's development more objectively, combining tradition The human resource management method realizes the combination of subjective and objective, and realizes a more comprehensive digital information human resource management.

Acknowledgements. 2018-2019 Liaoning Science and Technology Innovation Science and Technology Think Tank Project, Liaoning Province Science and Technology Talents Innovation and Entrepreneurship Ecological Environment, project number LNKX2018-2019C37.

The key project of the Social Science Planning Fund Project of Liaoning Province in 2018, The construction of talent ecological environment in Liaoning Province based on digital transformation, project number L18ARK001.

References

1. Xin, M.: Some thoughts on the reform of enterprise human resource management in the era of big data. Int. Public Relat. **05**, 143+145 (2020). (in Chinese)
2. Yun, W.L.: Human resource management reform strategy in the era of big data. J. Sci. Technol. Econ. **28**(16), 201+200 (2020). (in Chinese)
3. Liao, Y.: Thoughts on the reform of enterprise human resource management in the era of big data. Hum. Resour. Dev. **12**, 82–83 (2020). (in Chinese)
4. Li, C.: Discussion on enterprise human resource management reform in the big data era. Mod. Mark. (Late Issue) **05**, 184–185 (2020). (in Chinese)
5. Pei, F., Zhen, Z.: Innovative understanding of enterprise human resource management in the era of big data. Mod. Bus. **11**, 36–38 (2020). (in Chinese)
6. Wang, Q., Zhu, X.: Innovative thinking on enterprise human resource management in the era of big data. J. Shenyang Univ. Technol. (Soc. Sci. Ed.) **03**, 255–259 (2015). (in Chinese)
7. Liu, Z., Hao, N.: Changes in human resource management in the era of big data. Talent **05**, 373 (2015). (in Chinese)
8. Wang, Z.: A preliminary study on enterprise human resource management innovation in the big data era. Shangxun **17**, 183–184 (2020). (in Chinese)
9. Ting, L.: The impact of "big data" on human resource management. Shang **15**, 50 (2015). (in Chinese)
10. Zhang, R.: Thoughts on the reform of human resource management by big data. Hum. Resour. Dev. **14**, 35–49 (2015). (in Chinese)
11. Liu, Y.: The innovative application and potential dilemma of big data in human resource management. Hum. Resour. Manag. **6**, 11–12 (2015). (in Chinese)
12. Zhu, L.: Overview of human resource management application research in the context of big data. Times Econ. Trade **12**, 65–66 (2020). (in Chinese)
13. Tong, F.: The application of big data in human resource management. Digit. Commun. World **05**, 251 (2020). (in Chinese)
14. Shen, J.: Innovative analysis of human resource management application based on the background of big data. World Labor Soci. Secur. **12**, 7 (2020). (in Chinese)

The Research on Big Data Promoting Innovation and Development of Inclusive Finance

Wangsong Xie(✉)

Business School, Wuxi Taihu University, Wuxi, Jiangsu, China
862028100@qq.com

Abstract. With the rapid development of technologies such as the Internet, big data and cloud computing, new changes have also occurred in the financial industry. Big data technology plays an important role in accelerating financial development and promoting financial innovation. This article aims to provide a new way of thinking for the innovative development of inclusive finance by studying the innovative development model of promoting inclusive finance under the background of big data. This article investigates the current situation of inclusive finance by conducting a questionnaire survey on vulnerable groups such as small and micro enterprises, farmers, urban low-income people, and poor people in our city. It is concluded that 78.70% believe that small and micro enterprises have difficulties in financing, slow financing, expensive financing, etc., and insufficient financial services; 80.40% think that the development of small and micro enterprises lacks the support of financial resources, and put forward measures to promote the innovative development of inclusive finance. The research in this paper helps to use big data technology to effectively reduce the cost of traditional inclusive financial services and better realize the business sustainability of inclusive financial development.

Keywords: Big data · Inclusive finance · Financial services · Innovative development

1 Introduction

Under the influence of modern informationization development mode, the development of all industries are affected by the era of big data and change, the opportunity for them at the same time also brought many challenges [1]. In this mode of development, the financial industry is also influenced by a lot, a lot of business development makes the development of financial industry market more vast. In order to keep pace with the development trend of the times, only innovative development models can meet the needs of the era of big data and further develop internet finance [2, 3]. With the continuous innovation of technology, financial providers use mobile Internet, big data, cloud computing, artificial intelligence and other technologies to provide users with better services. Internet financial advantages under the background of big data, which brings a lot of echo in the society, the financial industry, the Internet has become a very promising under the influence of the industry [4, 5].

J. MacIntyre et al. (Eds.): SPIoT 2020, AISC 1282, pp. 419–425, 2021.
https://doi.org/10.1007/978-3-030-62743-0_60

In recent years, the rapid development of Internet finance has led to the transformation of traditional financial industry. Big data has changed traditional financial products and services, and exerted a significant impact on financial service system, organizational structure, management mode and development mode [6]. Big data technology can greatly promote the development of Internet finance, promote the innovation of development mode of Internet finance, and provide data support and technical support for the innovative development of internet finance. The close combination of Internet finance industry and big data technology is of great significance for effectively promoting the sound development of Internet finance [7, 8]. Big data technology has become the technical support and guarantee for the rapid development of Internet finance, especially playing an important role in financial risk control and precise positioning of target consumer groups. Big data technology for Internet business model innovation and perfecting laid a solid foundation, the development of financial innovation has become the future of the internet [9, 10].

Based on big data and Internet finance, this paper focuses on how to promote the innovative development of inclusive finance under the background of big data. Through the questionnaire survey of small and micro enterprises, farmers, urban low-income groups, poor groups and other vulnerable groups, this paper understands the current situation of inclusive finance, and puts forward targeted countermeasures to promote the innovative development of inclusive finance. In order to make financial inclusion more people.

2 Big Data and Inclusive Finance

2.1 Large Data

This is an era of information, will produce a lot of data in today's life, we are also more and more began to research using these data, and on further the concept of "big data". In some data, big data is also called "massive data", mainly referring to a kind of more diversified data information. In general, under the premise of new thinking mode, big data USES the function of data analysis and integrates data, data processing technology and data application, making it have many characteristics. In general, in the social informatization unceasingly development today, some equipment is more and more intelligent, which produced a huge amount of data such as mobile phone and computer. These data have excellent timeliness and processing ability, which makes big data more suitable for modern development. What is more prominent is that big data shows many excellent performance in the financial industry. In addition, the Internet is a very free and open virtual space. A lot of Spaces generate a lot of data all the time, which are intertwined with each other. As a result, the tasks of big data processing and analysis are more and more complex, so big data also has a very diversified situation. Big data has a very strong ability to collect and contains a large amount of data, so its contents are very extensive. Almost all information of users will be recorded in big data. Once the information is leaked out, it will have a great impact on users.

2.2 Inclusive Finance

Inclusive finance refers to the provision of appropriate and effective financial services at an affordable cost for all social strata and groups in need of financial services. Mainly to meet the financial needs of small and micro businesses, farmers, urban low-income groups, poor people, the disabled, the elderly and other vulnerable groups easily excluded by formal finance. Due to the education level, age, and the influence of living environment, they can't correctly understand pratt & whitney financial availability, convenience and universality. Traditional inclusive finance is often limited by physical branches, equipment and personnel arrangement. "To serve enterprises and residents, it must reach users through offline branches". Even if their pay a high cost, financial services or group will still exist lots of blank region, consumers don't have access to high quality financial services.

3 Experimental Methods

Using means such as mobile internet and big data, combined with inclusive finance, to provide customers with efficient, convenient and high-quality financial service quality, but there are still many difficulties. This paper mainly adopts literature analysis method and questionnaire survey method. Through collecting and reading literatures, books, relevant policy documents issued by the country and various text materials on the development of inclusive finance driven by big data, I sorted out and analyzed. Through the questionnaire survey of small and micro enterprises, farmers, urban low-income groups, poor groups and other vulnerable groups, the current situation of inclusive finance is understood, and the main problems in the development of inclusive finance in our city are analyzed. There are 100 small and micro enterprises participating in the questionnaire survey, 300 people from farmers, low-income groups, poor groups and other vulnerable groups. Through telephone survey and questionnaire distribution, 354 valid questionnaires were collected, and the questionnaire efficiency was 88.5%. Based on the analysis of the current situation and problems of inclusive finance in Zhuhai, this paper proposes innovative development countermeasures to promote inclusive finance in order to benefit more people and better realize the development of inclusive finance.

4 Results Analysis and Innovative Development Strategies

4.1 Survey Results and Analysis

Table 1. Current status of inclusive financial development in our city

The problem we are facing	Percentage
Small and micro enterprises have difficulties in financing, slow financing, expensive financing, and insufficient financial services	78.70%
Farmers and rural enterprises have low loan acquisition rates	55.70%
There is a problem with the information connection between the client manager and the client	62.80%
The development of small and micro enterprises lacks the support of financial resources	80.40%
People's financial awareness is weak, and financial literacy is not high	57.40%

Through telephone and questionnaire surveys of some small and micro enterprises, farmers, urban low-income people, poor people and other groups in our city, the collected questionnaires are summarized and analyzed, as shown in Table 1 and Fig. 1.

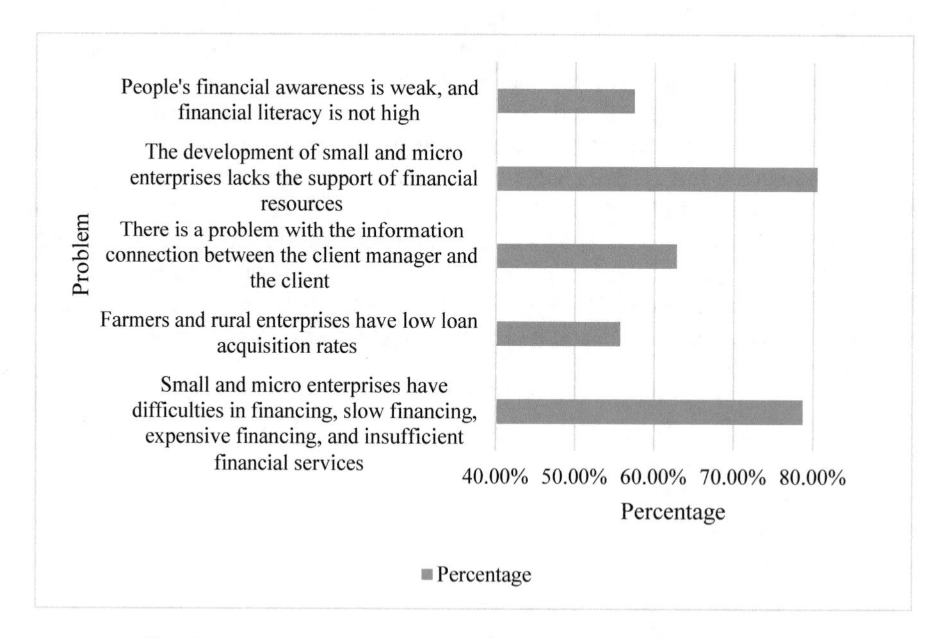

Fig. 1. The status quo of inclusive financial development in our city

It can be seen from the survey that 78.70% believe that small and micro enterprises have difficulties in financing, slow financing, expensive financing, and insufficient financial services; 80.40% think that the development of small and micro enterprises lacks the support of financial resources; It may be that the city's inclusive finance is also subject to low-efficiency data analysis, risk pricing, process approval and other technical measures in terms of risk control, which is insufficient to support small and micro enterprises. Therefore, it is very urgent to solve the financing problem of small and micro enterprises at present. Starting from my country's inclusive financial system, it is difficult to improve the financing of small and micro enterprises and other related issues to drive the economic development of enterprises and society. 55.70% believe that the loan acquisition rate of farmers and rural enterprises is low; 62.80% believe that there is a problem with the information connection between the client manager and the client; 57.40% believe that people are weak in financial awareness and financial literacy is not high; It may be that inclusive finance cannot provide different forms of services according to the people of different development levels in different regions. Especially in areas with extremely strong geographical restrictions, the lack of own resources and the imperfect financial infrastructure cause customers' financial needs to be unrealized, and these areas are often the main venues for inclusive financial development.

4.2 Promote the Development Strategy of Inclusive Financial Innovation

(1) Create a new inclusive finance model for online and offline interaction

The bank combines a wealth of physical outlets and an integrated service platform for online finance, and uses the three core means of openness, mobile and big data to create a new model of online and offline interaction, providing customers with inclusive, efficient, convenient and safe online finance Services, while supporting the development of the real economy. On the one hand, banks continue to promote the transformation of outlets to achieve intelligent, networked, interactive and integrated outlets. On the other hand, on the basis of an open platform, the bank's services are integrated into the mobile Internet, into the business ecology, and into the pan-finance service field. By strengthening the combination of financial and non-financial products, the integration of finance and related industries has been achieved, and the goal of "every customer has a mobile bank" is steadily advancing. Practicing Pratt & Whitney Finance focuses on solving three issues: "Where are customers", "What do customers need" and "How to provide solutions for customers", so as to effectively achieve communication and mutual benefit with customers. With "customer experience first" as the center, build a product and service system. Focus on launching inclusive financial products and services on the Internet, including Huimin Finance, Micro Finance, and overseas finance. Huimin Finance takes the service of people's livelihood and the convenience of the people as the starting point, focusing on "clothing (medicine), food, housing, transportation, beauty, entertainment, and play", and vigorously develops Huimin Finance based on the Internet model, covering "paying fees" and "helping health", "benefit education", "benefit protection", "benefit financial management", "benefit consumption", "benefit agricultural assistance" and so on. Through the rational use of big data analysis technology, we actively explore new small and micro financial service models, and provide efficient and convenient online financing services to small and micro merchants (enterprises) around the community (online) to solve the problem of "difficult financing and expensive financing". By taking advantage of big data, developing online and offline integrated financing services, enabling small and micro customers to apply online and obtain loans, breaking the constraints of loan time and space, and providing convenience for financing. Through extensive cooperation with third-party platforms, the use of big data risk control technology to provide batch online services, provide a new model of unsecured, unsecured, full-process online, efficient and convenient network financing. Take financial inclusive as the leading idea, meet the needs of small and medium-sized enterprises and individuals for funds, and act as an intermediary in the financial market.

(2) Innovate inclusive financial services grid and develop characteristic poverty alleviation paths

The gridding of financial services not only formed the gridding model of inclusive finance, but also helped the development of targeted poverty alleviation with the help of the inclusive finance model. The effective combination of financial development models and policies can actively promote the precise positioning and development of poverty alleviation work. Based on the analysis of different social structures and people's living structures in my country, the poverty-causing factors are more diversified. The development of inclusive finance under the grid service model can provide different forms of inclusive financial projects to solve the poverty-causing problem caused by various factors. Focus on building a financial poverty alleviation model with regional characteristics, and effectively promote the efficiency of financial inclusion. Based on reality, establish rural finance to support the development model of e-commerce, through banking financial institutions relying on professional financial services, drawing on the Internet financial development concept, creating a local characteristic "e-commerce poverty alleviation" model, and innovating rural inclusive financial development methods. The grid development of financial services can solve the problem of convenient wide coverage of financial services. Convenience and wide coverage in the development of traditional inclusive financial services are inherently conflicting issues. There will not be any form of convenience if they want to popularize widely. However, the grid model of financial services can achieve wide coverage and convenience. Ensure that financial services can take into account remote areas and solve the last mile of financial services. The grid service will form fixed-point grid sites in the financial service area. Even people in remote areas can use grid sites to realize financial needs, so that financial services can really benefit the people.

5 Conclusion

In short, in the era of big data as the background, Internet finance needs to keep up with the times and continue to innovate and develop. Conform to current trends, inclusive future development. Because this is a very systematic project, we can proceed from many aspects, use the technology of big data, make full use of the cutting-edge technology of data analysis, use social networking as a platform, master data information, and make our competitiveness competitive. improve. Improve financial efficiency, better benefit more people, make due contributions to serving the real economy, and jointly build a benign and sustainable inclusive financial business model. Understand what the user thinks, know what the user needs, prevent the user from preventing, and continue to provide users with "temperature" inclusive financial services.

References

1. Arthur, A.K.N.: Financial innovation and its governance: cases of two major innovations in the financial sector. Financ. Innov. **3**(1), 10 (2017)
2. Tsonchev, R., Tonova, S.: Financial innovation and the new regulation. Economics **73**(1), 28 (2017)
3. Bernier, M., Plouffe, M.: Financial innovation, economic growth, and the consequences of macroprudential policies. Res. Econ. **73**(2), 162–173 (2019)
4. Hernandez, R.J., Liu, P., Shao, Y.: Financial innovation and aggregate risk sharing. Theoret. Econ. Lett. **08**(11), 2182–2198 (2018)
5. Boulanger, P.P., Gagnon, C.: Financial innovation and institutional voices in the canadian press: a look at the roaring 2000s. J. Bus. Commun. **55**(3), 383–405 (2018)
6. Md, Q., Wei, J.: Nexus between financial innovation and economic growth in South Asia: evidence from ARDL and nonlinear ARDL approaches. Financ. Innov. **4**(1), 1–19 (2018)
7. Khraisha, T., Arthur, K.: Can we have a general theory of financial innovation processes? A conceptual review. Financ. Innov. **4**(1), 4 (2018)
8. Horsch, A., Richter, S.: Climate change driving financial innovation: the case of green bonds. J. Struct. Finan. **23**(1), 79–90 (2017)
9. Singh, S.K.: Currency demand stability in the presence of seasonality and endogenous financial innovation: evidence from India. MPRA Paper **9**(2), 122–139 (2016)
10. Ouma, S.A., Odongo, T.M., Were, M.: Mobile financial services and financial inclusion: is it a boon for savings mobilization? Rev. Dev. Finan. **7**(1), 29–35 (2017)

Spatial Econometric Analysis of House Prices in Guangdong Province Based on Big Data

Yibing Lai[1(✉)] and Qinghui Sun[2]

[1] Department of Basic Courses Teaching and Research, Guangdong Institute of Technology, Zhaoqing, Guangdong, China
1282976924@qq.com

[2] College of Construction, Guangdong Institute of Technology, Zhaoqing, Guangdong, China

Abstract. The price of houses has a great deal to do with the local economy, which influences house price to a great extent. Based on the theory of Economic Convergence, Based on big data and spatial econometric model theory, Based on big data and spatial econometric model theory, the convergence analysis of ESDA in recent ten years is conducted in Guangdong Province. The results show that there is spatial correlation but absolute β convergence is not significant, the empirical results of this paper can provide some reference for relevant departments to understand the relationship between housing prices in Guangdong Province.

Keywords: House prices · Big data · Exploratory spatial data analysis · Spatial econometric models

1 Introduction

With the marketization of housing, a series of problems have arisen, Housing has always been a hot topic of People's Daily attention and discussion, as well as the focus and difficulty of the government's Macroeconomic regulation and control.

In recent years, the overall economic level of Guangdong Province has grown rapidly, but the economic development of the various regions in the province is not balanced, The level of development in developed areas such as Shenzhen and Guangzhou is much higher than that in other areas, guangdong Province has become one of the provinces with the most prominent regional economic disparity in China. Because economic development largely affects housing prices, Therefore, the economic level of Guangdong province is high in the region housing prices are also high.

Based on the close relationship between housing price and economy, this paper applies the theory of economic convergence to the study of housing price, and combines the analysis of big data to study whether there is spatial correlation between housing prices in various cities in Guangdong Province, it is of great significance to discuss whether the growth of house price has convergence or not.

J. MacIntyre et al. (Eds.): SPIoT 2020, AISC 1282, pp. 426–432, 2021.
https://doi.org/10.1007/978-3-030-62743-0_61

2 Introduction to Theory

2.1 Global Correlation Analysis

Global *Moran' I* was first proposed by Moran [1] in 1948 and is defined as follows:

$$I = \frac{\sum_{i=1}^{n}\sum_{j=1}^{n} W_{ij}(Y_i - \bar{Y})(Y_j - \bar{Y})}{S^2 \sum_{i=1}^{n}\sum_{j=1}^{n} W_{ij}} \tag{1}$$

Where I is the index Moran, Among them $S^2 = \frac{1}{n}\sum_{i=1}^{n}(Y_i - \bar{Y})^2$, $\bar{Y} = \frac{1}{n}\sum_{i=1}^{n} Y_i$, Y_i represents observations for each region, n is all the locales, W_{ij} is the normalized spatial weight Matrix [2]. *Moran' I* values range from [−1,1], Greater Than 0 indicates that there is a positive correlation in space, Greater Than 0 indicates that there is a positive correlation in space, and less than 0 indicates that there is no spatial correlation. $E(I)$ is the expected value of index *Moran' I*.

2.2 Local Correlation Analysis

Local *Moran' I*, defined as follows:

$$I_{\mathrm{i}} = \frac{n^2}{\sum_i \sum_j w_{ij}} * \frac{(x_i - \bar{x})\sum_j w_{ij}(x_j - \bar{x})}{\sum_j (x_j - \bar{x})^2} \tag{2}$$

Where W_{ij} is the space weight, the other symbols mean the same as above.

2.3 Space Weight Matrix

The Space Weight Matrix is defined as [3]:

$$W = \begin{bmatrix} w_{11} & w_{12} & \cdots & w_{1n} \\ w_{21} & w_{22} & \cdots & w_{2n} \\ \vdots & \vdots & \vdots & \vdots \\ w_{n1} & w_{n2} & \cdots & w_{nn} \end{bmatrix} \tag{3}$$

Among them

$$w_{ij} = \begin{cases} 1 & \text{当区域}\ i\ \text{和区域}\ j\ \text{相邻接} \\ 0 & \text{其他} \end{cases} \tag{4}$$

If two regions are spatially adjacent, then $W_{ij} = 1$ indicates that there is significant spatial correlation between the two regions; if two regions are not adjacent to each other, then $W_{ij} = 0$ indicates that there is no spatial correlation between the two regions.

2.4 Econometric Model [4]

(1) the following ordinary least squares models (OLS) were used without considering the spatial effect:

$$\frac{1}{T-t}\log\frac{y_{iT}}{y_{it}} = \alpha + \beta \log y_{it} + \mu_{it} \tag{6}$$

(2) considering the β convergence of spatial effect, the spatial lag model (SLM) is adopted to analyze the β convergence:

$$\frac{1}{T-t}\log\frac{y_{iT}}{y_{it}} = \alpha + \rho W \log\frac{y_{iT}}{y_{it}} - \beta \log y_{it} + \mu_{it} \tag{7}$$

(3) the β convergence considering the spatial effect is analyzed by the following spatial Error Model (SEM):

$$\frac{1}{T-t}\log\frac{y_{iT}}{y_{it}} = \alpha - \beta \log y_{it} + \gamma W \varepsilon + \mu_{it} \tag{8}$$

Where α is the constant term and β is the coefficient of the regression equation. i: The price of a house in Block i, y_{it}: Initial value of house price, y_{iT}: The late value of the house price, β: The convergence rate of regional house prices, T: In the late stages of housing prices, t : Initial stage of house price, μ_{it}: Random factor. The selection of the model is based on the Lagrange multiplier, and Anselin proposes the following criteria [5–7].

3 Exploratory Spatial Data Analysis (ESDA)

3.1 Data Source and Processing

In this study, 21 cities in Guangdong Province from 2011 to 2020 a total of 10 years of average price data, data mainly from the "statistical yearbook of Guangdong Province" and home security and other websites. In order to reduce the influence of Heteroscedasticity, the data were standardized, and the data of regression analysis of econometric model were logarithmically processed [8].

3.2 Quartile Map of House Prices in Guangdong Province

In order to get a better understanding of the housing price growth and spatial distribution of the cities in Guangdong Province in the past decade, we have made a quartile map of the housing price spatial distribution in Guangdong Province.

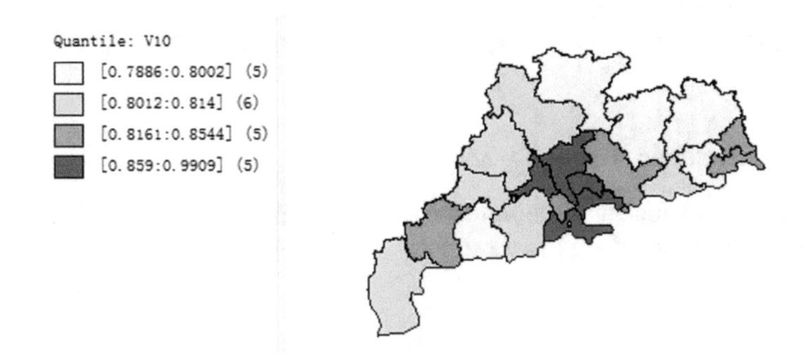

Fig. 1. A spatial quartile map of housing prices in Guangdong province in 2019

According to the depth of color, the housing prices in prefecture-level cities in Guangdong province are divided into four levels: high, higher, lower and low. The deepest color represents the highest price, the lightest color is the lowest price. As can be seen from Fig. 1, Shenzhen, Guangzhou, Foshan, Dongguan and Zhuhai are the five cities with the deepest colors. They are all relatively developed areas in Guangdong Province and are also the cities with the highest housing prices in Guangdong Province.

3.3 Autocorrelation Test in Global Space

The global spatial autocorrelation test is made on the house price data of 21 prefecture-level cities in Guangdong Province from 2011 to 2020. Table 2 shows the empirical results.

Table 1. House price autocorrelation test in Guangdong Province

Year	*Moran'I*	Mean	S.d	P	Z-Value
2011	0.3204	−0.0427	0.1426	0.01	2.5468
2012	0.2495	−0.0476	0.1363	0.023	2.1805
2013	0.239	−0.0542	0.135	0.029	2.1723
2014	0.2978	−0.0507	0.1343	0.015	2.5949
2015	0.2934	−0.0576	0.1226	0.013	2.864
2016	0.358	−0.0476	0.1334	0.007	3.0402
2017	0.4338	−0.0502	0.1366	0.001	3.5427
2018	0.4379	−0.0496	0.1342	0.001	3.633
2019	0.4485	−0.063	0.1328	0.001	3.8514
2020	0.4434	−0.0547	0.1323	0.001	3.7645

As Table 1 shows, the *Moran'I* index of house prices in Guangdong Province passed the 5% significance level from 2011 to 2016, and passed the 1% significance level from 2017 to 2020, and the Z value is more significant than 3.5. Using Monte Carlo simulation test, 999 random repeated permutations are selected to get the S.d value of the *Moran'I* index of house price in Guangdong Province between 0.1226 and 0.1426, which further guarantees the stability of the result. The results show that housing prices in 21 prefecture-level cities in Guangdong Province are spatially correlated.

3.4 Autocorrelation Test in Local Space

In order to further understand the situation of the spatial agglomeration effect in cities at various levels in Guangdong Province, local spatial autocorrelation analysis is needed.

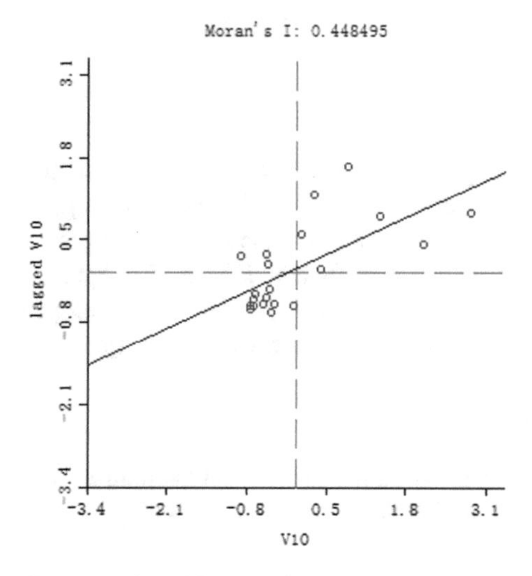

Fig. 2. *Moran'I* scatter plot of house prices in Guangdong Province in 2019

From the scatter plot in Fig. 2, the spatial correlation model of House prices in 21 prefecture-level cities in Guangdong Province is obtained. Among them, HH (33.3%) and LL (52.3%) account for a total of 85.6%. From this we can see that the prices of houses in most areas of Guangdong Province mainly show positive and negative spatial correlation, and the housing prices in local cities have an obvious agglomeration effect.

As shown in Fig. 3, the gray areas are not significant, and the color areas are significant at the level of 1% to 5%. There were 4 areas with high concentration, 2 areas with low concentration, and no cities with high concentration and high concentration. This is related to the high economic level of Guangdong Province and the mutual influence of the economic and technological development of the adjacent areas, thus forming the pattern of high concentration.

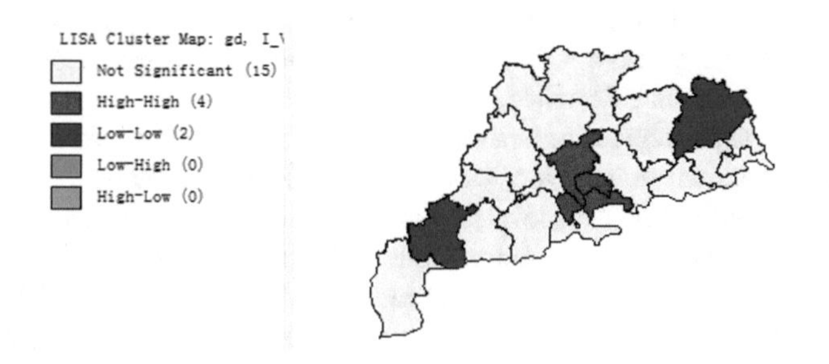

Fig. 3. LISA cluster map of house prices in Guangdong Province in 2019

4 Guangdong Province Housing Price Spatial Measurement Analysis

Table 2. Spatial correlation test results

Terms of inspection	MI/DF	VALUE	PROB
Moran's I (error)	0.1072	1.2013	0.2296
Lagrange multiplier (lag)	1	4.2263	0.0398
Robust LM (lag)	1	10.2867	0.0013
Lagrange multiplier (error)	1	0.4333	0.5104
Robust LM (error)	1	6.4937	0.0108
Lagrange multiplier (SARMA)	2	10.7200	0.0047

Table 3. The regression results of different models of house price in Guangdong Province

	OLS	SLM	SEM
α	−0.2409775*	−0.1623882*	−0.0600
	0.0326	0.0581	0.5479
β	0.2878101*	0.1825571**	0.1202971**
	0.0793	0.0024	0.0039
ρ		0.5062631**	
		0.0084	
γ			0.514874*
			0.0109
R^2	0.3165	0.4935	0.4169
LIK	49.9350	52.2421	50.7301
AIC	−95.8699	−98.4842	−97.4602
SC	−93.7809	−95.3506	−95.3712

Note: The mean and mean significant levels of 10% and 1%, respectively.

Table 2 shows the test results of model residuals, *Moran'I* is positive, this indicates that there is a spatial positive correlation between the housing prices of all prefecture-level cities in Guangdong Province.

Table 3 shows the regression results of different models of house price in different cities of Guangdong Province. OLS Model is significant at 10% level, SLM model and SEM model are significant at 1% level, the decision coefficients of SLM model and SEM model were higher than OLS regression model, which confirmed the existence of spatial autocorrelation. At the same time, several indexes of goodness of fit of the evaluation model are combined:The results of R^2, LIK, AIC and SC show that SLM model is a better fitting model. However, the β coefficients are all positive, which shows that there is no obvious trend of absolute β convergence in cities of Guangdong province.

5 Conclusion

The results of exploratory spatial data analysis show that there is spatial correlation between housing prices of cities in Guangdong Province. By analyzing and comparing the results of three regression models, it is concluded that SLM model is a better fit model. However, the β coefficients are all positive, which shows that there is no obvious trend of absolute β convergence in cities of Guangdong province. This and in recent years, the real estate market overheated development, some areas higher speculation, house prices continue to rise and other factors, thus breaking the absolute convergence trend of β [4].

Acknowledgements. This work is supported by the university. The fund is from the 2018 Guangdong Institute of Technology "Innovation Strong School Project" Science and Technology Project: GKJ2018015 regional economic research project of Guangdong Province based on spatial econometric analysis.

References

1. Moran, P.A.P.: The interpretation of statistical maps. J. Roy. Stat. Soc. Ser. B (Methodol.) **10** (2), 243–251 (1948)
2. Anselin, L.: Local indicators of spatial association-LISA. Geograph. Anal. **27**(2), 93–115 (1995)
3. Yang, S., Jurphy, R.D.: A review of space metrology theory and its application. Stat. Decis. Mak. **06**, 39–42 (2020). (in Chinese)
4. Lai, Y.: Empirical analysis of annual electricity consumption in Guangdong province based on spatial measurement. Sci. Technol. Econ. Inner Mongolia **08**, 80–83 (2018). (in Chinese)
5. Kang, S.: Analysis and research on the spatial econometrics of Chinese economic growth. Value Eng. **29**(091), 216–217 (2019). (in Chinese)
6. Drake, L.: Testing for convergence between UK regional house prices. Reg. Stud. **04**, 357–366 (1995)
7. Meen, G.: Regional house prices and the ripple effect: a new interpretation. Hous. Stud. **06**, 733–753 (1999)
8. Wenbin, W.: Spatial econometric analysis of regional housing price convergence in China. Stat. Decis. Making **01**, 130–132 (2012). (in Chinese)

State Grid Yichang Power Supply Company's Young Employee Training Plan Based on "Big Data Profile"

Xianjun Liu(✉), ChengChen Ji, Yi Zhang, and Sisi Yang

State Grid Hubei Electric Power Co., Ltd., Yichang Power Supply Company, Yichang 44300, Hubei, China
Liuxianjun2019@163.com

Abstract. In the 21st century, with the advent of the training model for young employees based on the "big data profile" and the development of information technology, modern enterprises have brought both opportunities and challenges. To build an excellent modern company, State Grid Yichang Power Supply Company must implement the "people-oriented" and "talent-powered enterprise" management ideas, and strengthen the training of young employees to achieve the healthy growth of new young employees and the common development of the company. This paper uses the "one person, three files" data of young employees to construct an analysis model to portray the "personal portraits" and "organizational portraits" of young employees and experts, and develop software to automatically generate training maps, and build on the basis of training maps. A two-level organization system, implementation of a number of assessment and incentive measures, and preparation of a "training navigation manual" to enhance the control and effectiveness of training management work, and to stimulate the internal driving force of young employees to participate in training. This article studies the "State Grid Yichang Power Supply Company" young staff training plan and implements a new training model, but the purpose is to modify and supplement this basic framework, and finally design a relatively complete training model. Including effective training system, internal training mechanism, triple training incentive mechanism. This article is based on the principle of "being a person first, doing things later", innovating training methods, extending training time limits, vigorously implementing a professional mentor system, and planning training projects with innovative ideas and methods. Research shows that managers and young employees respectively believe that about 73% and 64% of companies have formulated corresponding training plans before the implementation of training, accounting for the vast majority of the total number of companies.

Keywords: Young employees · Training projects · Planning and implementation

J. MacIntyre et al. (Eds.): SPIoT 2020, AISC 1282, pp. 433–439, 2021.
https://doi.org/10.1007/978-3-030-62743-0_62

1 Introduction

"Portraits" of individuals and organizations. Create "three files per person" for new young employees and experts and collect relevant data, sort them to obtain the scale of analysis, extract key factors, and then visualize the "personal portraits" of new young employees and experts. Taking "professional development portrait" as an example, post-development portraits, professional development portraits, etc. can be formed. Then merge and analyze the "personal portraits", and finally assemble the "personal portraits" into scattered scatter plots, and further form "organization portraits" based on the scatter plots.

Training as a research topic is first carried out in the fields of psychology and scientific management. Subsequently, with the development of management science theory, training theory roughly experienced three development stages: training in the traditional theory period, training in the behavioral science period, and training in the system theory period. In the traditional theory period, training focuses on the development of personal skills and attitudes, and less consideration of the relationship between individuals and others, or between individuals and groups; training in the theoretical stage of behavioral sciences, in addition to continuing the traditional theoretical period, emphasizes the development of personal skills and attitudes In addition, more attention is paid to the relationship between young employees and others; after the 1960s, training theory entered the period of system theory. The most important basic assumption of system theory is the openness of the system to the external environment, which means that the organization is regarded as an open system, and special attention is paid to the adaptation and communication between systems.

Based on the theories and research results of modern enterprise management, enterprise young employee training, and human resource management at home and abroad, this paper conducts a more comprehensive and systematic study on how to strengthen the training of young employees of State Grid Yichang Power Supply Company. Based on the actual situation of the existing young employees of State Grid Yichang Power Supply Company and the "Eleventh Five-Year" development plan, this paper designs a training program for young employees to ensure the sustainable development of State Grid Yichang Power Supply Company. The research in this article not only provides specific guidance for the training work of State Grid Yichang Power Supply Company, but also provides a reference for similar companies.

2 Young Employee Training Plan Based on "Big Data Profile"

2.1 Fine Portrait Formation Training Map

Refine big data comparative analysis model. According to the practical experience of big data analysis and based on the factor scale of "one person, three files", an "organization profile" covering all factors is formed one by one, and the "organization profile" is coded to extract a comparative analysis model. The continuous analysis and application of subsequent data has laid the technical foundation [1, 2].

Build the foundation of analysis technology. Based on the above comparison analysis model, comprehensively considering factors such as economy, applicability, safety, etc., the stand-alone software of "one person, three files" automatic analysis is developed in a targeted manner, and functions such as one-click data import analysis and automatic report generation are realized. The basic raw data of the "one person, three files" of existing young employees is recoded according to the corresponding rules to meet the needs of software analysis.

2.2 Basic Ideas for the Design of the Training Program for Young Employees of State Grid Yichang Power Supply Company

With the continuous development of training theory and the deepening of practical training, various training methods have become more and more mature. In company training, the application and selection of training methods are an important part of determining the impact of training [3, 4]. Currently, the training methods commonly used by companies are: classroom teaching method, seminar method, audio-visual method, game training method, job training method, case analysis method, role method, inspection method, independent study method and training method.

(1) Introduction to various training methods and scope of application
1) Classroom teaching method

Classroom teaching means that trainers systematically impart knowledge, concepts and skills to trainees through language expression, blackboard writing and other auxiliary teaching tools. So far, this method is the most used training method in enterprise training [5, 6]. This method is often used in the training of management knowledge, product knowledge, marketing knowledge, accounting knowledge, and job management.

2) Research method
The seminar method means that the trainer can effectively organize the trainees to discuss work issues or problems in groups and reach common conclusions so that the trainees can communicate and discuss with each other during the seminar. Ways to improve learners' awareness and competition. This method is mainly suitable for the training of leadership skills, decision-making strategies and business negotiation skills.

3) Audiovisual method
The so-called audio-visual method refers to the use of audio-visual materials such as slides, movies, video tapes, and computers for training, mainly for the training of new employees. Audiovisual methods are widely used to improve the communication skills, interview skills and customer service skills of new employees, and are also widely used to describe how to complete certain work processes. In general, audio-visual methods are rarely used alone, usually in combination with other methods (such as classroom teaching). It is generally considered to be an auxiliary teaching method [7, 8]. However, with the development of modern technology, it plays an increasingly important role in the training process, gradually making it an independent and effective training method.

(2) Training program design for professional and technical personnel

Professional technical personnel refer to professional technical business workers engaged in power transformation, power transmission, power distribution, relay protection, dispatching and communication in the State Grid Yichang Power Supply Company. They all have junior professional titles and are safe and reliable operation of the Changchun power grid. Guarantee power.

1) Training objectives

Provide professional and technical personnel with sufficient professional knowledge and technical level to ensure the safe operation and reliable power supply of the company's power supply equipment and facilities, and reserve talents for the company's sustainable development

2) Training plan

Regular training is conducted once or twice a year, including job adaptability training, job safety production training, etc.; job qualification training is conducted from time to time, mainly to master the knowledge and technology required for the job, and the prerequisite requirements for the job; for key training objects Training in career planning is also required.

(3) Training content and methods

1) Through training, build a team of "one specialization and multiple abilities", and cultivate young employees with superb skills, outstanding professional skills, and standardized professional literacy; to achieve a comprehensive quality of young employees during the "Eleventh Five-Year Plan" period improve. The training mainly adopts training methods such as audio-visual method, work coaching method and independent learning [9, 10].
2) All employees at the production site will evaluate their professional skills and apply a professional qualification certificate system. It is necessary to adopt a combination of centralized, decentralized, self-study, and guidance, and actively organize various positions for vocational skills training to create conditions for participating in vocational skills assessment and obtaining vocational qualification certificates. During the "Eleventh Five-Year Plan" period, all employees must be certified. Education mainly adopts training methods such as classroom teaching and independent learning.

3 Experimental Research on Young Employee Training Plan Based on "Big Data Profile"

3.1 Document Method

Through the collection and analysis of various existing related literature materials, we have a preliminary understanding of domestic and foreign research related research results, select valuable information to provide theoretical basis for ourselves, grasp the latest research trends, avoid repeated research and reflect the original Innovation of research. Through reading and sorting out relevant research results on the training of young employees of enterprises, analyzing and summarizing from the perspective of theory and reality, and further improving personal related research.

3.2 Questionnaire Survey Method

Through the business questionnaire survey, the survey learned and checked first-hand information about the training status of new employees. Through the statistical analysis and classification of the data, the problems existing in the training of young employees of SMEs are discussed. Analyze, think and discuss and propose objective countermeasures.

4 Experimental Analysis of Young Employee Training Plan Based on "Big Data Profile"

4.1 Analysis of the Company's Training for Young Employees

Through the understanding of young employees, it is believed that the main purpose of training is also three aspects: First, young employees believe that training can enhance their own value and meet their own future development needs; second, it can meet the needs of future development of the company. Most companies are not willing to carry out training due to their subjective intentions. The main constraints are lack of necessary funds and high work intensity, which is difficult to arrange for a while. They are also worried that the training will not be effective in the short term, and young employees cannot serve the company for a long time. Managers choose young employees. One aspect of training. The survey results are shown in Table 1.

Table 1. Reasons for training

Manager	Employee	Reason
75.3%	67.3%	The knowledge and ability of employees cannot meet the requirements of the current job position
45.3%	41.6%	The need to continuously improve enterprise production efficiency and product (service) quality
72.6%	59.8%	The needs of the company's long-term development in the future
43.7%	44.3%	The need for enterprise technological innovation and product upgrading
48.3%	54.3%	The needs of employees' future work development
43.2%	54.3%	The need to comprehensively improve the overall quality of employees
69.7%	34.2%	The need for employees to better integrate into the corporate culture

The survey results show that some business managers said that their companies have carried out training for young employees, and some young employees said that they have received corporate training. Among them, front-line young employees and new young employees are the main target of corporate training. This article has conducted an in-depth analysis and understanding of the majority of corporate managers and young employees who believe that companies conduct training and do not conduct training. According to the table, most managers believe that the main purpose of training is three aspects: First, to make young employees competent through training The job requirements meet the basic development and operation of the company; the second is to train young employees who can adapt to new opportunities and challenges in the future for the long-term development of the company; the third is to cultivate the sense of belonging of young employees and make them recognize the company Culture makes the team cohesive.

Most companies formulate training plans in advance when implementing training. From the analysis of the graph, managers and young employees respectively believe that about 73% and 64% of companies have formulated corresponding training plans before the implementation of training, accounting for the total number of companies. The vast majority.

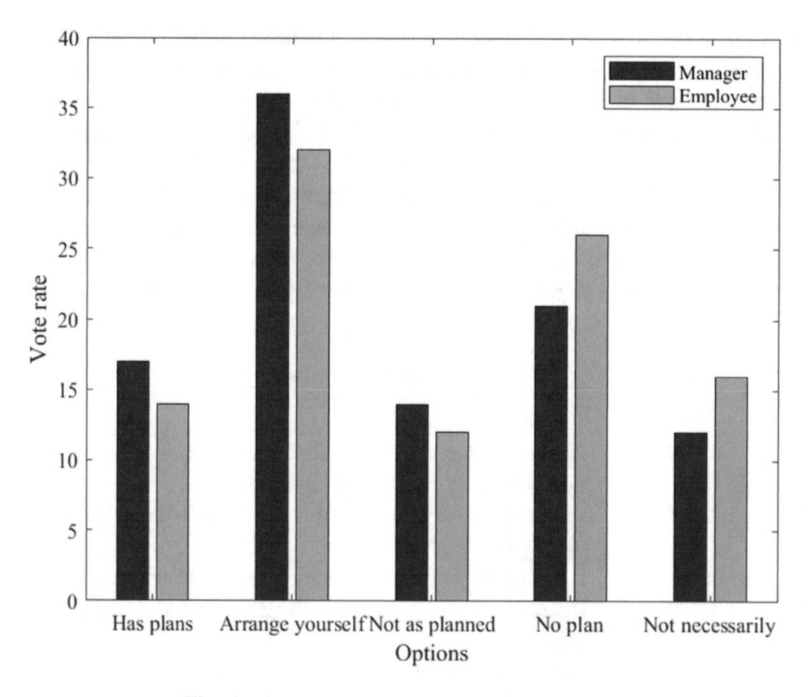

Fig. 1. Experimental analysis of training plan

As shown in Fig. 1, the companies that can actually implement the plan in full account for only 23% and 19% of all statistics. Most companies will adjust their training plans according to the actual situation. In other words, it shows that the implementation of the young employee training program is unsuccessful for most SMEs.

5 Conclusions

In the selection of corporate training objects, most managers and young employees have chosen to train frontline personnel and new recruits, while the training for management cadres is relatively small. As the managers and operators of enterprises, management cadres shoulder more important functions than ordinary young employees. Their management level, innovation ability, professional skills and other aspects directly affect every young employee. When companies choose managers Usually recruited directly for use, without paying attention to its future development and recognition of the company's culture, it is easy to cause managers to have no sense of belonging and have certain worries about their future development, which in turn affects the management team under their jurisdiction. At the same time, it was found in the survey that the faculty of the enterprise often chose the management cadres of the enterprise itself, which led to the management cadres assuming the training function of the enterprise, while the enterprise itself ignored the training of management cadres.

References

1. Liberty pumps launches employee ownership plan. Pump Ind. Anal. **7**, 12–13 (2015). https://doi.org/10.1016/S1359-6128(15)30250-0
2. Liberty pumps launches employee ownership plan. World Pumps **9**, 4–9 (2015). https://www.worldpumps.com/waste-wastewater/news/liberty-pumps-launches-employee-ownership-plan/
3. Valentin, M.A., Valentin, C.C., Nafukho, F.M.: The engagement continuum model using corporate social responsibility as an intervention for sustained employee engagement. Eur. J. Train. Dev. **39**(3), 182–202 (2015)

3. Valentin, M.A., Valentin, C.C., Nafukho, F.M.: The engagement continuum model using corporate social responsibility as an intervention for sustained employee engagement. Eur. J. Train. Dev. **39**(3), 182–202 (2015)

4. Ismiyanti, F., Mahadwartha, P.A.: Does employee stock ownership plan matter? An empirical note. Soc. Sci. Electron. Publ. **14**(3–2), 381–388 (2018)
5. Tariq, S., Jan, F.A., Ahmad, M.S.: Green employee empowerment: a systematic literature review on state-of-art in green human resource management. Qual. Quant. **50**(1), 237–269 (2016)
6. Thomas, T.N., Leander-Griffith, M., Harp, V., et al.: Influences of preparedness knowledge and beliefs on household disaster preparedness. MMWR Morb. Mortal. Wkly Rep. **64**(35), 965–971 (2015)
7. Jordan, D.W., Cotter, J.J.: Association between employee earnings and consumer-directed health plan choices. J. Healthc. Manage./Am. Coll. Healthc. Executives **61**(6), 420–434 (2016)
8. Sung, S.Y., Choi, J.N.: Effects of training and development on employee outcomes and firm innovative performance: Moderating roles of voluntary participation and evaluation. Hum. Resour. Manage. **57**(6), 1339–1353 (2018)
9. Guan, X., Frenkel, S.: How perceptions of training impact employee performance: evidence from two Chinese manufacturing firms. Pers. Rev. **48**(1), 163–183 (2019)
10. Riles, L.: Four benefits of ongoing employee training. Water Well J. **71**(7), 10–11 (2017)

Framework of Smart Regulation System for Large-Scale Power Grid Based on Big Data and Artificial Intelligence

Xinlei Cai(✉), Yanlin Cui, Kai Dong, Zijie Meng, and Yuan Pan

Electric Power Dispatching and Control Center of Guangdong Power Grid, Guangzhou 510600, China
517665114@qq.com

Abstract. In order to effectively improve the operating environment of large-scale power grid and effectively monitor the safety of power grid, this paper analyzes the concept of smart regulation system for large-scale power grid based on big data and artificial intelligence (AI) in the new era environment of smart operation, then analyzes problems existing in the current dispatching system of large-scale power grid in China, as well as the specific needs of smart dispatch system in the design process. And the corresponding smart control system is designed to effectively optimize and improve the operation environment of power grid, and upgrade and improve the core technology of smart dispatching system of power grid.

Keywords: Big data · AI · Power grid environment · Smart control system · Design framework

1 Introduction

With the development of society and the progress of the times, people have more and more demand for electricity in their daily work and life. The traditional power grid dispatching system has been unable to meet the current social production and life dispatch needs for electricity. In order to effectively improve the dispatching ability of power grid system, the scale of power grid is expanding, and the smart regulation of power grid is born. The smart system of grid control effectively meets user's demand for electricity and conform to the development trend of power grid. The smart regulation system of power grid mainly relies on big data and AI [1]. In order to make the operation of regulation system of large-scale power grid more stable and perfect, AI and big data need to be introduced when designing the relevant control system framework. Due to the high computing power of AI, the smart system of large-scale power grid can be effectively optimized [2].

J. MacIntyre et al. (Eds.): SPIoT 2020, AISC 1282, pp. 440–445, 2021.
https://doi.org/10.1007/978-3-030-62743-0_63

2 Concept of Smart Regulation System of Large-Scale Power Grid

The smart control system of large-scale power grid consists of the power system and the smart grid. Among them, the power system mainly includes power generation, power transmission, voltage transformation and power distribution. The smart grid is mainly based on big data and AI combined with cloud computing functions to improve the power system to form a smart grid. The smart control system of large-scale power grid not only improves and optimizes power technology of China, but also enables power technology to be effectively reformed. By combining big data and AI, the power grid system and cloud computing are integrated with each other, breaking through the traditional power supply scheduling system. It provides an effective reform plan and theoretical and technical foundation for the power system [3]. In addition, the intelligence of smart grid lies in its ability to integrate modern computer technology, power technology and smart devices through big data and cloud computing to form a highly integrated platform based on big data and AI. It enables the power system to have the ability of automatic monitoring and controlling. By combining big data and cloud computing, the power system can feed back relevant data and return the data to the terminal, so that engineers can monitor power grid. For instance, the smart grid can monitor the power consumption in a short time and optimize power supply of the system by combining smart control system to ensure the normal power supply. It can carry out self-monitoring, alarm and fault maintenance, and realize automatic control and automatic adjustment through feedback of big data [4]. Secondly, information integration capabilities of Internet are used to integrate information and resources through big data and cloud computing platforms to improve and optimize the smart control system of large-scale power grid [5].

3 Status of Smart Control System of Large-Scale Power Grid

In the traditional process, most power regulation needs to rely on engineers' rich work suggestions and artificial calculation. And power regulation is implemented according to the data fed back by regulation center, leading to the lack of regulation ability in the regulation process. In order to make the power system better complete the regulation and optimize its function, the smart regulation is introduced into the system of large power grid. The control center in the smart system controls the power system dispatch process [6]. With the optimization and popularization of smart control system, combination of big data and AI in the smart control of large-scale power grid improve the application of power grid effectively. Through big data and cloud computing platform, the analysis and modeling can be carried out according to the power consumption of users that can be predicted and evaluated to strengthen the load capacity of large power grid and balance the grid effectively. In addition, by introducing big data and AI into the smart control system of grid, it can effectively simulate the dispatching in the operation process of power grid, improve the self inspection ability and monitoring ability of power grid, and reduce obstacles in the actual dispatching process [7].

Therefore, in the future grid dispatching system, the smart control system of large grid based on big data and AI will be widely used.

4 Design Framework of Intelligent Regulation System of Large-Scale Power Grid Based on Big Data and AI

In the design of smart control system of large power grid, it is necessary to establish a certain model, and design the control system of power grid based on big data and AI by means of physical modeling. And in the design process, in order to improve the stable operation efficiency of power grid, it needs to meet the design principles of instant measurement, instant resolution and instant control [8]. Therefore, this paper mainly constructs the scheduling framework of big data and AI effectively from two aspects of function design and system framework design (Fig. 1).

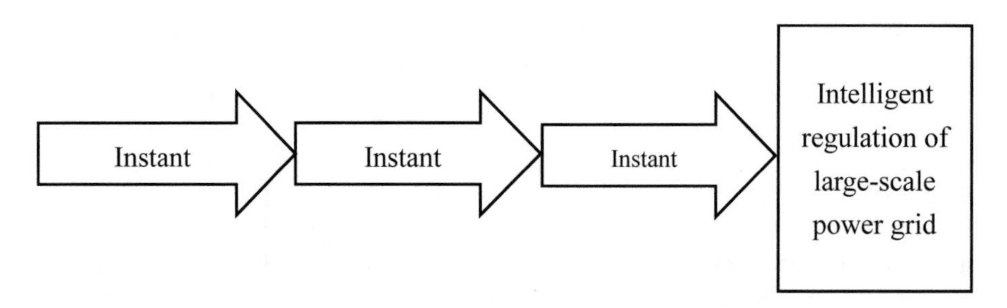

Fig. 1. Design principles of intelligent regulation system

4.1 Functional Design

In order to effectively design the scheduling control framework based on AI, it is necessary to design its functions. In the process of designing related functions, it needs to be based on big data and the storage capacity of data cloud. During the design process, a large amount of historical data is calculated to select appropriate samples. And a complete and diverse scheduling control system based on AI is designed by establishing a physical model. In this way, the application ability of scheduling control system is effectively promoted. In the design process, simulator is used to model based on the feedback data samples and elements, so that its operation results are close to the actual situation [9]. In the design process, its hardware functions mainly include central processing and graphics processing. By improving the central processing capacity and image processing capacity of intelligent control system of power grid, it can effectively improve the processing and learning ability of the intelligent control system for data samples. In the process of working, the intelligent control system is mainly operated and controlled by the control system. Through real-time control of business links and continuous feedback of grid operation data, a certain control experience is formed. In addition, in the framework design of intelligent control system of large-scale power grid, it is also necessary to optimize the intelligent function of power grid including the storage of a number of power grid operation data, so that the intelligent regulation system can carry out equipment measurement and fault alarm feedback according to these data, and effectively store its operation log and fault handling plan, so as to provide data reference for maintenance and automatic fault repair. At the same time, in order to effectively improve the decision-making ability o intelligent control equipment, it also needs to design an intelligent learning system based on the existing physical model in the functional design to improve the algorithm application ability of AI and its multi-functional processing ability. For example, when the dispatching control system of power grid has an operating failure, its intelligent learning system can prioritize the failures in order to avoid the impact of failure on the operation of the entire power grid, and provide corresponding solutions according to the importance of failures, so as to repair itself with more priority processing principle, and the self-checking system can check the potential failures [10] (Fig. 2).

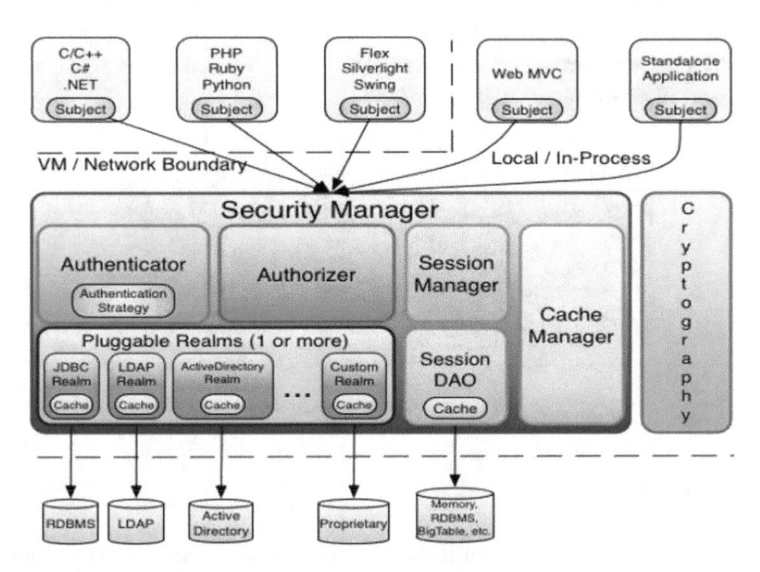

Fig. 2. System platform architecture

4.2 System Framework Design

When designing the framework of smart control system of large power grid, we need to refer to a lot of data models, and design with a variety of calculation methods. Among them, in the process of system framework design, such mixed framework as CPU and GPU can be used to calculate to effectively reduce the model technology time, so that the smart control system of power grid can have more opportunities for learning and use, and a powerful computing cluster can reduce the training and learning time. Furthermore, various structures and unstructured data can be used to refer to word-of-mouth operation data, external operation environment data, and operation management and monitoring data. And these data can be integrated into a big data network, and be collected, classified, stored and optimized with the use of cloud computing platform.

And relevant data needs to be stored according to different data sampling methods, for example, with the mode of structure type of the data and frequency of data collection [11]. Secondly, in the process of integrating text and log, the language processing system is an smart control system for scheduling and operation. The design process is transformed into knowledge base and rule base. In the design of algorithm engine framework, different algorithms need to be designed according to different types and processes of operation, for instance, cluster analysis or knowledge map calculation are used in the calculation of similar engines. In addition, in the design of business scenario framework, the main package is to sense the situation, such as forecasting the load of power grid through big data and AI, forecasting and assessing related faults, and analyzing user's electricity situation through big data. It is completed by the main combination of physical modeling and construction method to store and analyze big data. And during the design of smart decision-making framework, it needs to design the overload, failure and maintenance of equipment to provide effective auxiliary decision-making. This scheme is mainly based on the simulation and prediction of physical model, and provides strategies for emergencies by combining relevant scheduling processes and disposal plans [12] (Fig. 3).

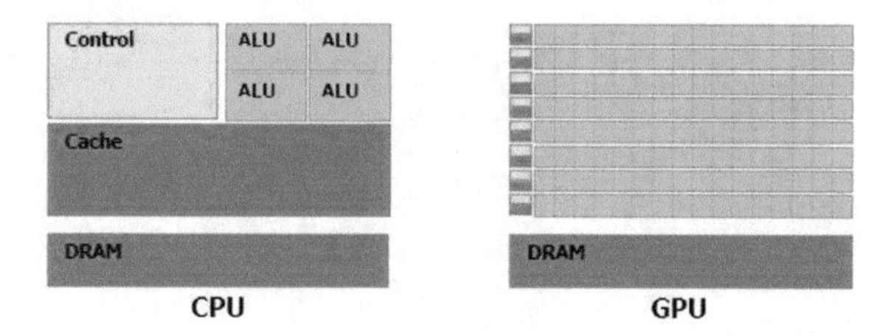

Fig. 3. Key technologies

5 Conclusion

To sum up, the design principles of instant measurement, instant resolution and instant control need to be met in the design of control framework of smart control system of large power grid. Through the functional design and system framework design, the smart control system of large power grid can be effectively optimized, so that it can effectively carry out smart control by integrating big data and AI together with cloud computing functions, perfect the operation environment of power grid, and further improve the core technology of smart control system of power grid.

References

1. Li, J., Yan, Q., Wu, Q.: Intelligent control system framework of large power grid based on big data and artificial intelligence. Commun. Power Technol. **37**(03), 5–7 (2020)
2. Liu, D., Li, B., Shao, G., et al.: Intelligent control system framework of large power grid based on big data and artificial intelligence. Electr. Power Inf. Commun. Technol. **17**(03), 14–21 (2019)

3. Pan, P., Tian, D.: Big data analysis of equipment monitoring based on smart grid regulation and control technology support system. Sci. Technol. **33**, 66 (2018)
4. Xing, Y., Lang, Y., Li, Q.: Discussion on centralized operation and maintenance mode of multilevel smart grid regulation system. Power Syst. Protect. Control **46**(15), 142–148 (2018)
5. Tong, C., Wu, J., Zhan, W., Gao, Q.: Big data analysis of equipment monitoring based on smart grid regulation and control technology support system. Rural Electr. **06**, 10–13 (2018)
6. Ren, X.: Research and application of on-line monitoring and intelligent diagnosis of operation state of power grid control automation system. Power Syst. Protect. Control **46**(11), 156–161 (2018)
7. Li, W., Yang, H.: Development of intelligent grid regulation automation assessment simulation system. Commun. Power Technol. **34**(05), 229–230 (2017)
8. Jia, K.: Basic data detection and identification of smart grid control technology support system. North China Electric Power University (Beijing) (2017)
9. Lan, C.: Analysis of abnormal intelligent management analysis and automatic inspection technology of power grid control technology support system. Fuzhou University (2016)
10. Wei, W., Jin, Y., Zhao, Y.: Key technologies and standards for training and simulation of training personnel for smart grid dispatching control system. Smart Grid **4**(06), 626–630 (2016)
11. Hu, H., Zhang, G., Wang, M., Ren, T.: Design and implementation of power grid intelligent monitoring system based on integration of regulation and control. Inner Mongolia Electr. Power Technol. **32**(04), 24–27+45 (2014)
12. Xie, Y., Wu, X., Xiang, Z., Zhang, Z.: Regional grid intelligent monitoring and auxiliary decision-making system based on integration of regulation and control. Zhejiang Electr. Power **33**(02), 59–62+65 (2014)

"21-st Digital Maritime Silk Road" Based on Big Data and Cloud Computing Technology Facing Opportunities and Challenges–From Digital Trade Perspective

Shengnan Zheng(✉)

Department of Economics, Shanghai University, Shanghai, China
zsnjysh@163.com

Abstract. The "21-st Digital Maritime Silk Road" based on big data and cloud computing technology is a new proposition and mission given to China by the digital age. It is of great importance for China, developing countries, even all countries. Taking digital trade as the starting point, analyzing the opportunities and challenges the countries and regions along the "Belt and Road" would encounter from following perspectives, including digital infrastructure, digital payment, logistics system construction, and international digital trade rule-making. On the basis of accurately grasping the current condition of digital trade among countries and regions along the "Belt and Road", the paper put forward suggestions, such as all-round assistance to narrow the digital divide, start with e-commerce, "Connecting people" contribute to "Unimpeded Trade" and so on, to promote the "21-st Digital Maritime Silk Road" construction.

Keywords: 21-st Digital Maritime Silk Road · Big data · Digital trade

1 Introduction

In 2013, during his visits to Kazakhstan and Indonesia, the idea of the Silk Road Economic Belt and the 21-st Century Maritime Silk Road was proposed by President Xi firstly, and his thought drew much attention from other places. The core of the Belt and Road (B&R) is Wutong, while the core of Wutong is Unimpeded Trade. The advent of the Digital Era has given birth to Digital Trade, a new form of trade, which bring new vitality to traditional trade and new impetus to the B&R. Under the new background, President Xi put forward "the 21-st Century Digital Maritime Silk Road" at The First Belt and Road Forum for International Cooperation in 2017, which advocated that Chinese Digital Economic should benefit more countries and regions along the B&R through digital trade. With this political background, this paper analyzed the opportunities and challenges we would face during the process from the perspectives of digital infrastructure, digital payment, logistics, and new international trade rules, and made several suggestions for the "21-st Century Digital Maritime Silk Road".

J. MacIntyre et al. (Eds.): SPIoT 2020, AISC 1282, pp. 446–451, 2021.
https://doi.org/10.1007/978-3-030-62743-0_64

2 Status of Digital Trade Among China and Other Countries and Zones Along the "Belt and Road"

Currently, as for the definition of digital trade, there is still no unanimous opinion. According to report from China Academy of Information and Communications Technology, digital trade is a kind of new business activities in this new age, during which the information and communication technology play an important role, involving three different levels of products, (1) the goods. (2) digital products, including software, music, etc. (3) digital service, including telecommunications, Internet, etc. This section will analyze the digital trade among corresponding countries and regions in the zone of B&R from the three aspects [1].

Traditional trade refers to exchange of goods in the real world with currency as the medium. With the popularity of the Internet, e-commerce is getting more acceptance. Unlike traditional trade, it is not limited by the time and space and greatly improves the trade efficiency [2]. According to bilateral trade figures and cross-border e-commerce connectivity index of China and 65 corresponding major countries and regions. On the whole, the higher the cross-border e-commerce connectivity index is with China, the higher the bilateral trade volume will be. The average cross-border e-commerce connectivity index of the top 10 countries ranked by bilateral trade volume is 7.55. The average cross-border e-commerce connectivity index for the bottom 10 is 2.72. Furthermore, we find that the top 10 countries ranked by bilateral trade volume are mainly Southeast Asia countries. At the same time, the top 10 countries ranked by cross-border e-commerce connectivity index are mainly countries of Eastern Europe [3]. There are some obvious differences between the two group countries, indicating that cross-border e-commerce still have a great space in the regions along the B&R.

At present, we can only know digital products and digital service through relevant reports and news. In 2014, Ant Financial began to move towards the country and region along the Belt and Road, and successfully helps more than 100 SaaS software in China successfully go abroad. In 2019, China-Singapore Internet data channels will open, with the aim of strengthening China-ASEAN cooperation on the field about Artificial Intelligence. It is obvious that, as the digital technology development in countries and regions along the B&R is relatively backward, therefore, in digital products and digital services, China tend to play the role of export countries while other countries are importers. There is still a very large development space for China and these countries.

3 The Opportunities for the "21-st Century Digital Maritime Silk Road"

3.1 Countries and Regions Along the B&R Attach Great Importance About the Fourth Industrial Revolution

Big Data, Cloud Computing slowly opened the Fourth Industrial Revolution. In order to catch this rare chance, all the countries or regions have expressed the willingness to cooperate with China and jointly realize the economic development. By August 2019,

China has signed memorandums of understanding on e-commerce cooperation with 19 countries and established e-commerce cooperation mechanisms, including Colombia, Italy, Panama, Argentina, Iceland, Rwanda, UAE, Kuwait, Russia, Kazakhstan, Austria, Hungary, Estonia, Cambodia, Australia, Brazil, Vietnam, New Zealand and Chile. These countries and regions are willing to trust China and to reform themselves. Policy Communication is the basis for Unimpeded Trade. With the joint promotion of leaders from various countries and regions, the 21-st Century Digital Maritime Silk Road will certainly usher in good prospects.

3.2 The Effect of Facility Connectivity Begin to Show

"Facility Connectivity" is the basis and premise for "Unimpeded Trade". After the "Belt and Road" initiative, China has signed infrastructure construction cooperation documents with a number of countries. Furthermore, these documents are in an orderly implementation, and has begun to show results. In terms of CR railways, by April 2019, 14,691 lines had been opened, connecting 51 cities in 15 countries along the route [4]. As for Network communications, by the end of 2017, China's three major network operators as the main information and communications enterprises, has helped 12 countries along the route finish 34 land cables and 6 sea cables [5]. Up to now, China has reached cooperation with most countries and regions along the B&R in the sea, land, air, network, which has laid a good foundation for the 21-st Century Digital Maritime Silk Road.

3.3 The Construction of International Digital Trade Rules Is Imperative

In recent years, digital trade has developed rapidly, but the establishment of international digital trade rules has been relatively slow, or even stalled. The main reason is that there are different claims about digital trade from developed countries and developing countries. Developed countries, represented by the United States, Europe and Japan, with the developed digital technology and digital trade level, require to establish completely relaxed and free cross-border digital circulation rules. At the same time, most developing countries represented by China and Russia hold the opposite view, and they believe that out of consideration of national security and national interests, cross-border data should be classified, or even refused to connect to the outside network, to prevent the outflow of data involving national security and interests [6]. At present, the more notable negotiation about digital trade rules are TTP, TTIP and TISA, in which the TPP signed in 2016 involves more detailed digital trade rules. But the Trump Administration withdrew from the TPP in 2017, and some provisions infringe on U.S. and European interests, so the TPP negotiations on international digital trade rules have stalled for a while. But this condition may be an opportunity for China and the countries surrounding the B&R. On one hand, countries along the B&R can conclude bilateral digital trade provisions to push the international digital trade rules. On the other hand, as a major member of the TPP, China can speak on behalf of the developing countries and promote to establish international digital trade rules that are in interests of developing countries.

4 The Challenges for the "21-st Century Digital Maritime Silk Road"

4.1 A Digital Divide Among Countries and Regions Along the "Belt and Road"

Digital economy is the hotbed of digital trade, so the level of digital economic level determines the level of digital trade. According to the Global Digital Economic Development Index, published jointly by the Ali Research and KPMG, there is a short-term insurmountable digital divide among China and countries and regions surrounding the Belt and Road [7]. As shown in Fig. 1, the figure shows the digital economic development index, which is a composite index consisting of five primary indicators (digital infrastructure, digital consumers, digital business ecology, digital public service and digital education and scientific research levels) and 16 secondary indicators, reflecting the digital economic development level of a country or region. China's digital economic development index is 0.718, much larger than other countries and regions. The scores of about 40% of the countries and regions sit 0.3–0.4, and 7 countries and regions ranked behind 150. The huge digital gap between China and other countries means that the first step to construct the "21-st Century Digital Maritime Silk Road" will be pretty tough.

Fig. 1. The digital economic development index

4.2 Digital Payments Are in Trouble

Digital payment is a critical part of digital trade. Without secure and convenient payment methods, consumers will lose confidence in digital trade. In countries and regions along the Belt and Road, digital payments are in such an awkward condition. On the one hand, due to the backwardness of the information and communication industry, most countries don't have sound credit guarantee mechanism, corresponding policies and regulations, and legal cooperation mechanism, it is not possible to effectively protect the rights and interests of consumers, resulting in a considerable number of consumers "dare not use" [8]. For example, in South Africa, locals prefer physical shopping because of a lack of confidence in digital payments [9]. On the other hand, backward digital infrastructure will lead to a variety of payment problems, such as frequent network failures, limited service scope, thus bring various inconveniences to consumers, resulting in consumers "cannot use" or "unwilling to use." Therefore, creating a safe, fast and convenient digital payment environment has become a major challenge in building the "21-st Century Digital Maritime Silk Road".

5 Proposals to Promote the "21-St Century Digital Maritime Silk Road"

5.1 All-Round Help to Bridge the Digital Divide

The digital divide is the primary problem when we construct the Silk Road. Firstly, China should increase investment in countries and regions surrounding the B&R zones, especially in digital infrastructure investment, and encourage outstanding private enterprises to go abroad. At the same time, the host countries should actively optimize the local investment environment in order to form a benign interaction. Secondly, it is necessary to strengthen the exchange of digital talents. leaders can sign relevant documents to encourage the exchange of talents, universities should roll out some preferential policies for studying abroad, to encourage the students from these countries and regions to study in China, especially to learn the digital economy-related knowledge. Finally, with the expansion of friends circle of the "Belt and Road", more and more developed countries will become the main force. It will be advisable for China to encourage developed countries to help developing countries, in order to reduce the pressure of China.

5.2 Cross-Border E-commerce as a Breakthrough

At present, cross-border e-commerce is the main body of digital trade among these countries, so China should take cross-border e-commerce as a breakthrough, unblock cross-border e-commerce channels, establish cross-border digital payment and credit system, which will contribute to the progress of digital trade. In order to promote the development of cross-border e-commerce, on one hand, government should help train a number of leading enterprises in the digital industry [10], in addition to continuing to support Dun Huang, Ali, BAT and so on, encourage the formation of a number of digital trade-related outstanding enterprises. On the other hand, China should speed up the construction of Digital Free Trade Zones and Digital Free Trade Ports, so as to open the "21-st Century Digital Maritime Silk Road".

References

1. China academy of information and communications technology: digital trade and development impact report (2019). (in Chinese)
2. China academy of information and communications technology: china digital economic development report (2020). (in Chinese)

3. Ali research: eWTP to help the belt and road–Alibaba's economic practices (2017). (in Chinese)
4. Du, Z., Wu, P., Pan, J.: China's cross-border logistics collaboration with countries along the Belt and Road. Pract. Foreign Econ. Relat. Trade **07**, 89–92 (2019). (in Chinese)
5. Chen, B.: Digital maritime silk road information infrastructure research. Defense Sci. Technol. Ind. **3**, 34–36 (2020). (in Chinese)
6. Lan, Q., Dou, K.: Connotation evolution, development trend and China strategy of US-Europe-Japan digital trade. Int. Trade **006**, 46–53 (2019). (in Chinese)
7. Ali research, KPMG: meet the new wave of the global digital economy (2018). (in Chinese)
8. Yan, Z.: China-ASEAN co-builds digital economy the belt and road core area. China Econ. Times **005**, 87–96 (2019). (in Chinese)
9. Huang, Y.: China and Africa build the digital silk road: opportunities, challenges, and path choices. China Int. Stud. **4**, 156–164 (2019). (in Chinese)
10. Xia, J.: The origin of digital trade, international experience and development strategy. J. Beijing Technol. Bus. Univ. (Soc. Sci.) **33**(05), 1–10 (2018). (in Chinese)

Outdoor Sports Industry Platform Construction Based on Big Data Technology

Yuyan Wang(✉)

Department of Physical Education, Huazhong Agricultural University, Wuhan, Hubei, China
wangduanyuyan@163.com

Abstract. The development of my country's sports big data industry is still in the preliminary stage of development, but its application prospects are broad and it has extremely high R&D value. Through literature, logical analysis and other research methods, and on the basis of reviewing relevant research on the application of big data technology in the sports industry, a platform architecture for the outdoor sports industry of big data technology is constructed. After the outdoor sports industry data platform is built, the application of big data analysis can achieve the following expected results: First, it can be opened to provincial mountaineering associations and outdoor sports clubs, and provide guidance for them; Second, provide future sports brands for related sports Product development suggestions to achieve precise marketing; third, a precise positioning system for outdoor sports rescue can be established to provide technical support for emergency rescue.

Keywords: Outdoor sports · Big data · Sports industry

1 Introduction

Since the "National Fitness Campaign" rose to the national strategy, how to build a sports power has become the focus of attention in the political and academic circles. The "Outline for Building a Powerful Sports Country" proposes that my country should accelerate the deep integration of the Internet, big data, artificial intelligence and the real sports economy. As the main industry of the national fitness program, the sports industry will emerge with many innovative development models with the rapid development of information technology in the new era. Big data is a special strategic resource that can penetrate into various industries and promote the transformation of traditional and existing industries to informatization and advanced technology. The application of big data technology in the field of sports promotes the emergence of new formats of sports big data. Compared with the existing sports industry, the sports big data industry is more conducive to the realization of the goal of a sports power. In recent years, research on the sports industry has mainly focused on development logic, dynamic mechanism and upgrade paths, development opportunities and challenges, etc., while research on the sports industry structure has mostly been carried out from the perspectives of quantitative optimization and layout policies. Since the concept of outdoor sports began to enter China in the 1980s, after nearly 40 years of development,

J. MacIntyre et al. (Eds.): SPIoT 2020, AISC 1282, pp. 452–459, 2021.
https://doi.org/10.1007/978-3-030-62743-0_65

nowadays, with the improvement of China's economic level and the transformation of consumption structure, as well as the promotion of government policies, outdoor sports have ushered in new developments [1]. Obviously, the pace of the leisure age is moving step by step into people's lives, and outdoor sports, as a branch of the leisure industry, has huge development potential. Through combing, it is found that in the past 10 years, the government has issued a number of guidance documents related to the sports and health industry, with the 2014 State Council document No. 46, "Several Opinions of the State Council on Accelerating the Development of the Sports Industry and Promoting Sports Consumption" as the dividing line, China has raised national fitness to the national strategic level, actively promotes the development of the sports industry, and strongly supports the development of fitness running, cycling, mountain climbing and other outdoor leisure projects [2]. Local governments have actively responded to and issued policies to develop sports and leisure as keywords. In 2016 Document No. 77 "Guiding Opinions of the General Office of the State Council on Accelerating the Development of the Fitness and Leisure Industry", the focus is on the development of outdoor sports. At present, my country's outdoor sports market has the main characteristics of high growth, and has formed a parallel pattern of professional niche and popularization. But how to use big data technology to collect, organize, and analyze the data generated by outdoor sports has become an urgent problem for us [3].

2 Overview of the Application Research of Big Data Technology in the Sports Industry

2.1 Definition of Sports Big Data Industry Concept

This research further clarifies the definition of the sports big data industry based on the theory of industrial economics, that is, the sum of economic activities that provide various sports products and services to the society based on big data science and technology [4]. Among them, science and technology based on big data mainly include three aspects: big data resources, big data technology, and big data thinking mode. (1) Big data resources: it can be divided into athlete data, non-athlete data and sports media data. Athlete data refers to the data collected by professional intelligent recording equipment on the changes of various indicators of professional athletes during training and competitions. Non-athlete data refers to the data that non-professional athletes take the initiative to upload or automatically record changes in various indicators of the body during daily exercise and fitness due to health considerations. Sports media data refers to sports news, information, events and other data published and disseminated through the Internet. (2) Big data technology: refers to software development and hardware manufacturing technologies involving the capture, storage, management, processing, and mining of sports-related data. (3) Big data thinking model: refers to following the "data-technology-application" thinking model, using big data analysis technology to mine the value of sports big data and then apply it to sports products and services. Compared with the traditional sports industry, the sports big data industry pays more attention to the use of data to create value, so that sports organizations can rely on massive data and emerging technologies in management, competition, and training activities, and enable related activities to be better carried out. Data resources will present comprehensive, specific and accurate characteristics, and the acquisition of sports data will no longer rely on manual records, questionnaire surveys and other methods. Instead, smart terminals can be used to collect various "natural data" anytime and anywhere, avoiding traditional methods. The resulting loss of data is missing, untrue, and inaccurate.

2.2 Application Fields and Value of Sports Big Data

Cui Jiujian believes that integrating the development of the sports industry with big data and using big data analysis techniques can create greater value for my country's sports [5]. Liu Qi and He Xinsen proposed that we need to establish a competitive sports big data platform to adapt to the development trend of the unclear boundaries of competitive sports in the era; at the same time, we should also adapt to the economy of the times by establishing a joint mechanism of economic sports big data [6]. The coordinated development trend of the sports system operation mode; finally, we should also adapt to the diversified trend of cooperation between the subjects of the competitive sports system in the era by cultivating a team of competitive sports big data talents. Wang Qi and Yan Xiaoyan believe that the application of big data in the selection, training and competition of competitive sports should be encouraged, and the establishment of theoretical basis should also be forward-looking; construct a sports information collection system that conforms to the development trend of big data, and also this system should be applied to the practice of sports; it should be used to break the path and direction of traditional sports research, and strive to achieve a leap from "measurement paradigm" to "calculation paradigm"; the right choice for China's sports path today is to promote the coordinated development and research of big data and innovation research, we should strive to achieve this. At the same time, we should also strengthen innovation drive and improve the awareness of information security [4]. Xu Yun and Zhang Hui believe that data mining must give full play to its potential value in game technical and tactical analysis, national fitness monitoring, and sports monitoring data [11]. At present, there is still a lot of research work that we urgently need to carry out. There are mainly the following three aspects: (1) Establishing a platform for sports big data; (2) Sharing and generalization of mining tools; (3) Intersection and integration with other technologies. Lei Hui and others explained the basic ideas of visual analysis of sports data: analysis based on statistics, movement and clustering, feature detection, etc.; and looked forward to the future development direction.

2.3 Research Significance of Big Data in Outdoor Sports Industry

The study found that the current research materials mainly focus on simple theoretical analysis, while the establishment and application of databases in sports-related fields are relatively rare, which also causes partial or biased results. However, it is worthy of recognition that some domestic experts and scholars have begun to introduce data mining technology into sports projects. Although the current research is not in-depth enough, it is important for us to study sports data mining in the future and truly apply data mining technology to sports technology. It is of great significance to make tactics and make decisions on sports development, and to promote the development of data mining technology in the field of sports. The above research has conducted research and discussion on the characteristics of big data, the value of big data mining in the sports industry, the choice of big data analysis path and application practice, etc., and provides good enlightenment for the research of outdoor sports industry platform architecture based on big data technology. At present, there are not many related researches on outdoor sports big data. This research has important theoretical and application value.

3 Outdoor Sports Industry Platform Architecture Based on Big Data Technology

As our country has a large population and a wide area, and outdoor sports not only involve many projects, but also divided by sea, land and air areas, it can be divided into land outdoor sports, coastal outdoor sports and low-altitude outdoor sports. Only land outdoor sports can be divided into alpine outdoor sports, low-altitude outdoor sports, plain outdoor sports according to altitude, and ice and snow outdoor sports according to season. These outdoor sports are under the jurisdiction of the China Mountaineering Association under the State Sports General Administration [7]. However, most of the currently qualified outdoor sports are dependent on tourist attractions, tourist attractions are managed by the Tourism Bureau, outdoor sports are managed by the Sports Bureau, and low-altitude outdoor sports like paragliding involve the management of the Aviation Committee, resulting in many management departments. Coordination and management is difficult, the amount of data is large and scattered, and the frequency of occurrence is frequent. The general statistical methods are slow, and it is difficult to meet the requirements of the mountaineering association and other management departments to dynamically grasp the outdoor sports market, timely formulate relevant policies and measures, and promptly guide local associations and civil organizations to develop outdoor sports meet the needs of outdoor sports enthusiasts. Although domestic and foreign scholars have begun to try the application research of big data technology in the sports industry, and some research results have been applied to sports events, research in the field of outdoor sports is relatively rare. Therefore, build an outdoor sports industry platform based on big data technology (Fig. 1), carry out forecasts of future project development in the outdoor industry, formulate plans for the corresponding data construction of the mountaineering center, and provide practical and feasible policy recommendations and related data support, grasp the preferences of outdoor sports enthusiasts in time, and ensure that they can safely engage in outdoor sports, so that the work of the mountaineering management center will enter the forefront of the entire industry. It has important theoretical and practical significance to promote the development of China's outdoor sports industry.

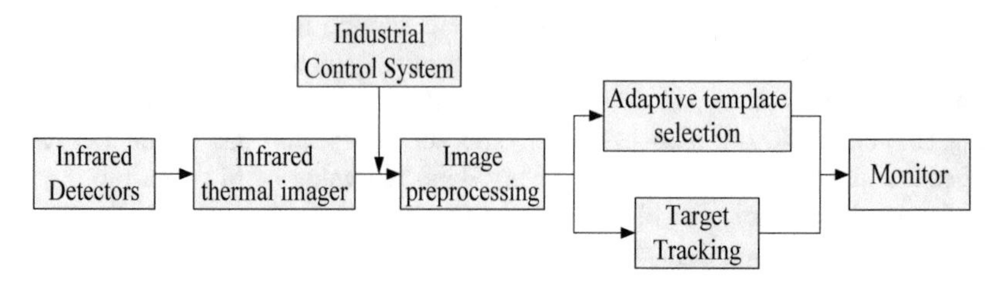

Fig. 1. Flow chart of outdoor sports target tracking platform

3.1 Data Source Acquisition

First-hand data collection, including statistical data from the National Tourism Administration, the Aviation Association has approved the opening of regional data that can engage in low-altitude outdoor sports such as paragliders, gliders, and hot air balloons, and those who have received driving certificates for paragliders, gliders, and hot air balloons. Information data, as well as water sports data from the Water Sports Management Center of the National Sports Administration, statistical data from outdoor manufacturers and suppliers of the Textile Association, and data reported by various mountaineering associations and outdoor clubs under the China Mountaineering Association. Data obtained from a sample survey on the current status of sports development [8]. At the same time, it forms a resource sharing pool with the "population database", "corporate database", "spatial geographic database", "macroeconomic database", and "cultural resource database" in various places, resulting in a multiplier effect with half the effort. Foreign outdoor sports second-hand data collection methods are becoming more and more abundant. For example, the International Citizen Sports Federation, established in 1968, has experienced nearly half a century of development. Its organizational system, trail standard operating mechanism and certification system have become increasingly mature and perfect. A lot of hiking data can be obtained from it. The Outdoor Sports Foundation of the United States conducts surveys on participation in outdoor sports every year, and Japan also releases survey reports on mountaineering every year. These reports have important reference value for us to timely understand the development status of outdoor sports in foreign countries, outdoor sports facilities standards and certification systems. Domestic second-hand data can be obtained from Luye Outdoor Network, Outdoor Travel Forum, 8264.com Outdoor Information Network and other outdoor sports websites, as well as large-scale forums and related WeChat public accounts, Weibo, Taobao and other e-commerce platforms and other channels to organize related event participants, outdoor sports brand suppliers, to build an outdoor sports database.

3.2 Data Organization

The mountaineering management center needs to collect a huge amount of data when establishing a database. At present, the mainstream technology in the world that processes massive amounts of data is the Hadoop technology, which was developed by the Apache Foundation and also includes those developed based on the Apache Foundation. Users can develop distributed programs. The special feature is that this development process does not require a detailed understanding of the distributed bottom layer. The high-speed computing and storage need to be realized through the full use of the power of the cluster. This software framework can realize distributed processing of large amounts of data, and the data processing is also very time-sensitive, reliable and scalable. Hadoop is open source, so the software cost of the project will be greatly reduced.

3.3 Data Security

The data center platform of the mountaineering center will collect a large amount of mountaineering enthusiasts' data. These data will involve many personal privacy and certain commercial secrets, and the importance of data security needs to be taken into consideration [9]. Big data security is a very broad field, which can include: the security of big data systems, the security (encryption) and privacy protection of the data itself, the security and privacy issues brought about by big data applications, and the application of big data technology in the security field.

3.4 Data Mining

After the big data platform of the Mountain Sports Management Center obtains a large amount of customer data, only by analyzing the data can the data generate value. Data mining technology is a means and method to realize the value of data. In recent years, people from all walks of life have also begun to focus on data mining, mainly because for the current research field, the extensive use of large amounts of data is very necessary, which is conducive to the realization of the research from data to information and knowledge [10]. Conversion, data mining is produced along with the naturalization of information technology. We can understand that data mining is the refinement and extraction of large amounts of data, or the extraction and mining of knowledge. The knowledge discovery process in the database is shown in Fig. 2 below, which consists of the following steps: First, data cleaning is designed to eliminate noise or inconsistent data; second, data integration, so that multiple data sources can be combined; third, data selection, extract data related to analysis tasks from the database; fourth, data transformation, data transformation or unified into a form suitable for mining; fifth, data mining, use intelligent methods to extract data patterns; sixth, pattern evaluation, according to a certain degree of interest measurement, identify the really interesting patterns that provide knowledge; seventh, knowledge representation, use visualization and knowledge representation technology to provide users with the knowledge to be mined.

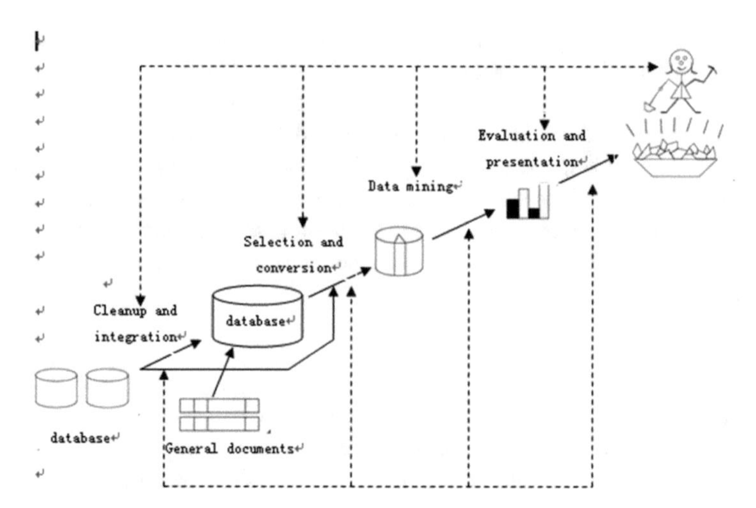

Fig. 2. Knowledge discovery process based on data mining technology

4 Conclusion

The construction of a big data platform for the outdoor sports industry will have the most authoritative outdoor sports data in the country, which can be opened to the provincial mountaineering associations and provide appropriate industry forecasts for the development of outdoor mountaineering sports projects and for the future construction of mountaineering associations. Provide guidance and authoritative suggestions. It can provide future product construction suggestions for related sports brands, and can charge a certain consulting fee every year to prepare for its precise marketing. A precise positioning system for outdoor sports rescue can be established to provide technical support for emergency rescue and break the situation of irregular outdoor sports.

Acknowledgements. This work was supported by the Fundamental Research Funds for the Central Universities (2662019FW014).

References

1. Si, H.: The development characteristics and trend analysis of outdoor sports research in China. Guangzhou Inst. Phys. Educ. **2**, 48–56 (2018)
2. Jiang, G.: Survey on the development of outdoor sports clubs in my country. Sports Cult. Guide **4**, 40–43 (2014)
3. Huang, B.: Analysis of the development status of outdoor recreational sports in my country. Sports World **3**, 49–51 (2020)
4. Wang, Q.: Opportunities and challenges faced by my country's sports development in the era of big data. Sports Sci. **1**, 75–80 (2016)
5. Cui, J.: Analysis of the development strategy of the sports industry under the background of big data. Electron. Test. **24**, 47–48 (2015)

6. Liu, Q.: Development strategy of competitive sports system in the era of big data. J. Capital Inst. Phys. Educ. **3**, 156–159 (2015)
7. Li, Q.: Analysis and optimization of the system structure of the sports big data industry. Sports Sci. Technol. **1**, 100–105 (2020)
8. Zhou, L.: Application management of outdoor fitness equipment for all people based on big data of Internet of Things. J. Shazhou Vocat. Inst. Technol. **1**, 17–20 (2019)
9. Urakov, A.L., Ammer, K., Dementiev, V.B., et al.: The contribution of infrared thermal imaging to designing a "winter rifle"-an observational study. Thermol. Int. **29**(1), 40–46 (2019)
10. Kobelkova, I.V., Martinchik, A.N., Keshabyants, E.E., et al.: An analysis of the diet of members of the Russian national men's water polo team during the sports training camps. Voprosypitaniia **88**(2), 50–57 (2019)
11. Zhang, H.: Application of data mining in the field of sports. J. Wuhan Inst. Phys. Educ. **11**, 27–30 (2012)

Exploration on the Application of Personalized Teaching System Based on Big Data

Jie Cai(✉)

School of Marxism, Shandong Polytechnic College, Jining, Shandong, China
992237086@qq.com

Abstract. Information technology promotes personalized learning. The article revolves around the issue of a personalized teaching system based on big data. It conducts in-depth analysis of educational big data, personalized teaching, and related ideas through the review of relevant literature. Under the guidance of theory, humanistic theory, and blended learning theory, a personalized teaching concept based on big data is proposed. The study proposes that a personalized teaching system based on big data is of great help to teaching, and can promote the change of learners from a "passive receiver" of knowledge to an "active builder"; it can promote the future of education to be truly digital and intelligent the development of personalized and personalized directions.

Keywords: Big data · Personalized teaching · Teaching

1 Introduction

With the development of the global informatization wave, education has entered a new historical period, and the development of personalized education has become the consensus of the international community [1]. The extensive application of personalized education can make every student's development conform to his own characteristics, and it is a manifestation of humanization and fundamentalism. In the current classroom teaching, although each teacher has also contacted and used multimedia teaching tools, most of them are still teacher-centered. The multimedia equipment used is only a teaching tool for teachers to transfer knowledge to students, and does not really change the teaching. The model is still in a one-way teaching model in which teachers speak and students passively accept, and students' personalities cannot be developed. In classroom teaching and subject teaching and research, there are still some core and important teaching problems that cannot be broken through and solved by traditional teaching and research methods. Teachers lack quantitative data support in subject teaching and research, subject teaching and research rely on experience; students lack specific and timely help in learning; standardized and standardized unified exercises are just a means to take care of most students. Parents do not understand the learning process of students in class, nor can they provide targeted help. Heavy extracurricular tutoring has increased the pressure on students, caused students' fatigue, and even affected their physical and mental health. Based on this, this paper designs and analyzes the concept of personalized teaching based on big data, and builds a personalized teaching system based on big data, in order to achieve personalized

J. MacIntyre et al. (Eds.): SPIoT 2020, AISC 1282, pp. 460–466, 2021.
https://doi.org/10.1007/978-3-030-62743-0_66

"teaching" for teachers and personalized "learning" for students, and ultimately promote students personality development [2].

2 Theoretical Analysis of Personalized Teaching Based on Big Data

Big data technology provides technical support for personalized teaching [3]. Through the mining and analysis of learner characteristics, it predicts and intervenes learners' learning behavior, making it possible to teach students in accordance with their aptitude. First, personalized teaching based on big data. Through the process of tagging the knowledge points of all resources such as courseware, materials, homework, test papers, teaching videos, etc. in the whole teaching process, the knowledge points corresponding to the resources can be accurately located, which is convenient for teachers to learn carry out accurate resource push and targeted guidance. Secondly, it collects data from the whole teaching process such as teacher preparation, student preview, teacher classroom teaching, classroom interactive tests, after-school homework exercises, examination evaluation, etc., to obtain all the teaching digital chemistry data of teachers and students. Analyze and process the collected academic data, and establish a learning model for each student through cloud computing, semantic web, visualization and other technologies. Teachers, parents, and students themselves can effectively realize personalized teaching through the analysis of the model. Precision teaching driven by management and data intelligence. Through the labeling of knowledge points and the analysis and processing of academic data, the learning characteristics of students are given, and personalized resources are pushed according to the degree of mastery of each knowledge point by students, helping students to accurately locate weak points and improve students' learning effects. Big data is not only a huge amount of data, but more importantly, it emphasizes the whole process, non-sampling, and full sample data. With the help of big data analysis and processing technology, it can comprehensively record the learning phenomenon of students and analyze the characteristics of each student. Discover the personality behind the commonality. The purpose based on big data is to focus on the micro performance of each student by collecting the data of the student's learning process, gain insight into the real student, understand the student's learning path, the development of subject literacy at different stages, and the status of each knowledge point. In order to achieve targeted teaching in accordance with their aptitude [4].

3 Big Data Supports Personalized Teaching

Compared with traditional personalized teaching, personalized teaching under the background of educational big data opens up new ideas and ways. Big data provides technical support for personalized teaching. Personalized teaching under the background of big data is based on learners [5]. As the center, full consideration of learners' learning characteristics is a change of traditional learning methods. With the help of data mining technology, learning analysis technology and academic big data collection technology, all real-time data can be integrated and collected, and presented in a visual way, which optimizes the traditional education process and provides scientific innovation for personalized teaching guarantee and support.

3.1 Changes in Teaching Mode

Through the collection of big data of academic conditions, with the help of data analysis and prediction functions, the traditional "one size fits all" teaching model is changed, which is very beneficial to the development of personalized education [6]. On the one hand, big data technology has promoted the innovative application of learning models such as autonomous learning and blended learning. It can achieve a comprehensive upgrade of pre-class preview, lectures in class, and consolidation and improvement after class. The data of the learning process is presented in front of students, making learners become the master of learning, and provide scientific analysis and suggestions for learners' learning behaviors, so that learners understand themselves better; on the other hand, the teaching content is determined by teachers based on the data of usual tests or examinations. Adjust and change the corresponding teaching strategies and teaching methods. The teacher analyzes the cognitive situation of the students through the presented students' academic data, and then controls the teaching.

3.2 Push of Personalized Learning Resources

The use of big data technology can record in detail the learner's whole-process data such as quizzes, unit tests, homework, exams, etc. The collected test, homework, exam, each question and each item of data are labeled as knowledge points, and the collected data is analyzed to understand the students' mastery of each subject, each unit, and even each knowledge point. Since each student has a different degree of mastery of a knowledge point, the resources pushed by each student are also different according to the degree of mastery of the knowledge point [7]. Through personalized learning resource recommendation, help students use the recommended resources to understand themselves, Improve yourself. For example, every student who scored 86 points is considered an excellent student, but using big data analysis, it can be found that the knowledge points behind the same 86 points of two people are different. It is precisely by this that when assigning homework. The resources pushed to the two learners are also different. Big data can record students' exam reports, homework reports, and test reports in detail. Teachers can understand each student's learning situation by viewing the reports, and use the presentation of knowledge points to clearly find out the students' deficiencies in their knowledge. Students send exercises of the same type that match the wrong knowledge points to further consolidate their knowledge.

3.3 Early Warning and Decision-Making

Big data technology can effectively support learning early warning and help teachers make decisions. The use of big data technology to organize and visualize relevant information data and dynamic data generated during the learning process, combined with teachers' teaching experience and understanding of students, can modify and intervene students' learning behaviors, learning methods, and learning trajectories [8]. It understands its own errors and shortcomings, thereby improving students' learning ability and performance.

Using the big data collection technology of learning situation to track and record the learning behavior of students, you can discover problems in learning in time, formulate intervention strategies in time, and provide personalized learning early warning to prevent precipitation learning and ensure the quality of learning [9]. The use of big data technology can also set the minimum task compliance value for each learning module and task, and track and record the learning behavior of students. If the minimum task compliance value is lower than the set minimum value, teachers need to intervene in learning. Teachers can list unqualified students as the objects that need intervention based on the results of the process evaluation. Process evaluation includes pre-class preparation, after-class test, unit test, homework completion, mid-term test, etc., each item is set completion time and performance evaluation standards, those who are not completed within the time or those who fail to pass the performance can implement early warning and supervision of learning behavior.

In addition to the teacher's personalized intervention, students can monitor and control themselves by watching their own learning progress bar, and can also compare with other students' learning progress bars to motivate themselves to study hard, improve learning efficiency, and quickly complete learning tasks, so that students can quickly enter the learning state and learning process, to ensure the durability and continuity of learning, and improve learners' learning perseverance.

3.4 Changes in Evaluation Methods

The traditional "empirical" educational evaluation method is changing to a new "data doctrine" educational evaluation method. The most important concept of the new evaluation method is to use "data to speak". The application of big data technology in education makes the evaluation method of learning not only based on the final academic performance and learning results as the entire evaluation basis. More attention should be paid to a series of activities such as learners' grasp of learning content, knowledge transfer ability and emotional attitude shown in the learning process. The combination of formative evaluation and summative evaluation makes the traditional meaning of excellent students no longer exist, and the recognition of learners' learning behavior can promote learners' personalized learning.

4 Personalized Teaching Design Content Based on Big Data

4.1 Individualized Student Model

Using big data analysis and processing technology, it is possible to construct a student model composed of students' preview status, classroom performance, homework status, mastery of various knowledge points, interactive communication status and other elements. Teachers and parents can analyze each element of the student model. Provide students with personalized learning strategies. Students find their own shortcomings and promote self-improvement by observing their own learning models.

4.2 Personalized Learning Resources

The design of personalized learning resources based on big data is conducive to meeting the learning characteristics and diversity of different students, individual learning needs, and promoting the development of students' individualization. The design of personalized learning resources based on big data includes the following two aspects: building a knowledge system tree and establishing a media space based on the knowledge system tree.

The knowledge system tree is the core of personalized learning resources and the basis for the labeling of knowledge points. The knowledge system tree is composed of knowledge points and the relationship between knowledge points [10]. The hierarchical processing of knowledge points becomes the key to the formation of the knowledge system tree. In the process of designing personalized learning resources in this paper, knowledge points are divided into different levels by subject-grade-knowledge block-knowledge points. In the concept of personalized teaching based on big data, personalized learning resources are composed of related subjects. The learning content of a subject is divided into different grades, and the learning content of a grade is divided into different knowledge blocks, like this one divide layer by layer until it is divided into atomic knowledge points that cannot be subdivided. The hierarchical structure diagram is shown in Fig. 1.

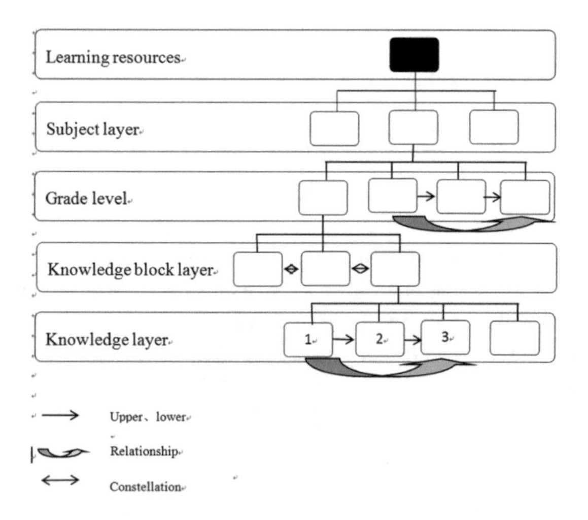

Fig. 1. Hierarchical structure diagram of learning resource knowledge system tree

On the basis of constructing the knowledge system tree, different types of digital teaching media need to be used to present knowledge points, form digital learning resources, and determine a rich media space [11]. These digital learning resources include texts, courseware, cases, videos, and audios. Teachers can push different learning resources according to their teaching needs. Students can choose learning resources according to their own learning styles. Each knowledge point can be used.

Presented with various forms of digital resources, many knowledge points constitute a media space for personalized teaching based on big data. In the process of pushing to students, it is necessary to choose the appropriate resource presentation form according to the different types of knowledge points: concept and theorem-based knowledge points can push text-type resources to students to help students master concepts and understand theorems; complex skill-based knowledge points you can push PPT type resources of courseware to students; the applied knowledge points, key and difficult points in the chapters should be made into videos and animation resources to be pushed to students to help students establish a logical knowledge system.

On the basis of constructing the knowledge system tree, different types of digital teaching media need to be used to present knowledge points, form digital learning resources, and determine a rich media space [12]. These digital learning resources include texts, courseware, cases, videos, tutorials and audios. Teachers can push different learning resources according to their teaching needs. Students can choose learning resources according to their own learning style. Knowledge points can be presented in various forms of digital resources, and many knowledge points constitute a media space for personalized teaching based on big data. In the process of pushing to students, it is necessary to choose the appropriate resource presentation form according to the different types of knowledge points: concept and theorem-based knowledge points can push text-type resources to students to help students master concepts and understand theorems; complex skill-based knowledge points you can push PPT type resources of courseware to students; the applied knowledge points, key and difficult points in the chapters should be made into videos and animation resources to be pushed to students to help students establish a logical knowledge system.

5 Conclusions

With the development of society and the advancement of science and technology, the information age has come. Therefore, it is necessary to provide suitable education for each student, and promote each student to actively and lively develop into the main theme of education in the information age. In the past, the one-size-fits-all education model could not meet the individual development of students and the society's demand for diversified talent training. Traditional teaching is in urgent need of change, and personalized teaching has attracted much attention. With the emergence of technologies such as cloud computing, the Internet of things, and educational big data, personalized teaching has received technical support. Applying emerging technologies to the teaching process, to help teachers carry out personalized teaching, stimulate students' interest in learning, meet the individual needs of students, and break through traditional teaching methods. The personalized teaching system based on big data can well realize personalized teaching, enhance the pertinence and initiative of students' learning, improve students' interest in learning and classroom efficiency, improve classroom teaching effects, and promote the improvement of students' comprehensive ability and promotion teachers' professional growth will play a proactive role.

References

1. Chen, J.: Model construction for continuous promotion of education informatization oriented to big data. China Audio-visual Educ. **6**, 52–57 (2019)
2. Li, R.: An analysis of personalized teaching from the perspective of role theory. Teach. Manage. **18**, 73–76 (2017)
3. Yang, X.: Development of Educational Big Data: Connotation, Value and Challenge. Mod. Distance Educ. **1**, 50–61 (2016)
4. Fan, Y.: Educational evaluation model and its paradigm construction in the era of big data. Chin. Soc. Sci. **12**, 139–155 (2019)
5. Sun, H.: The core technology, application status and development trend of education big data. Distance Educ. J. **5**, 41–49 (2016)
6. Li, X.: Flipped classroom precision teaching model under the formative evaluation of big data. J. Hangzhou Dianzi Univ. **3**, 64–68 (2020)
7. Zhou, J.: Analysis of Teaching Formative Evaluation Strategy Based on Big Data Features. J. Guangxi Radio TV Univ. **2**, 38–41 (2018)
8. Jiang, Q.: Research on personalized adaptive learning-the new normal of digital learning in the era of big data. China Audio-visual Educ. **2**, 25–32 (2016)
9. Kim, J.R.: English learning analytics and its application. Asia Pac. J. Multimedia Serv. Convergent Art Humanit. Sociol. **6**(9), 321–330 (2016)
10. Wu, F.: Study on the design of learning outcome prediction framework based on the analysis of learner's personality behavior. China Audio-visual Educ. **1**, 41–48 (2016)
11. Mo, L.: A preliminary study of student portraits based on the analysis of college student status data. Mod. Inf. Technol. **6**, 32–33 (2018)
12. Ren, H.: Accurate teaching based on big data: generation path and realization conditions. Heilongjiang High. Educ. Res. **9**, 165–168 (2017)

A Big-Data Based Study on the Construction of Beforehand Alarming System for Public Health Emergency

Qiaofeng Gao(✉) and Yucong You

Guangzhou College of Technology and Business, Guangzhou, China
472253732@qq.com

Abstract. At present, numerous types of public health incidents are quite inevitable for our human society, especially in recent years, the frequent occurrence of public health incidents has caused huge losses that are irreversible to the society. In this context, it is quite necessary to strengthen Construction of Beforehand Alarming System for Public Health Emergency. This article mainly introduces the role of big data and public health emergency beforehand alarming system, and launches an in-depth analysis of the construction of the beforehand alarming system of public health emergency with big data era.

Keywords: Big data · Public health emergency · Beforehand alarming system

1 Introduction

Establishment of a scientific and systematic beforehand alarming system for public health events helps China more efficiently confront health crises, yet from the current point of view, compared with the current beforehand alarming system and mechanism for public health emergency. Some progress has been made; there still exist certain unfavorable factors in its actual application process that restrict the efficient use of its application value and affect our country's ability to, accordingly, respond to public health emergency. As a result, it is quite necessary to strengthen the degree of heed to it, as well as taking corresponding optimization and polishing measures.

2 The Role of Big Data and Public Health Emergency Alarming System

2.1 Big Data

Big data serves as an emerging term and means in the current Internet cyber era. It mainly refers to a data collection that analyzes and manages Internet data on the basis of making full use of database software means. By means of the application of analysis technology and high-speed capture technology in big data, it is possible to extract high-value information accurately from quite a large number of data and realize efficient applications in all respects. Currently, our country mainly adopts big data thinking and

J. MacIntyre et al. (Eds.): SPIoT 2020, AISC 1282, pp. 467–472, 2021.
https://doi.org/10.1007/978-3-030-62743-0_67

big data technology in the domain of public health emergency. As far as big data technology is concerned, the core content is nothing but prediction. Application of this technology is not like some command robots that must input corresponding command actions provided they wish to perform related actions. Big data analyzes massive amounts of relevant data based upon advanced mathematical algorithms, and then makes corresponding comprehensive judgments based on the actual results of data analysis to obtain beforehand alarming results. This is a positive promotion for efficient response to public health emergencies Role [1, 2, 3].

In public health emergencies, they mainly include some infectious disease outbreaks. Beforehand detection, isolation in advance, and pre- treatment must be done for such incidents. Only in this way can the harm of public health incidents be minimized. In the medical industry at present, big data technology has been widely used. Evidently, for asthma patients, corresponding smart sensors can be installed on their respirators, so that doctors can dynamically follow the patient's daily routine in real time. Activity data and respiratory change data, and then judge those environments that will adversely affect the patient's respiratory system.

2.2 Public Health Emergency System

When a public health emergency occurs, a sound and complete public health emergency beforehand alarming system can help people find public health crises more quickly, and take proper handling and response measures as soon as possible to alleviate the public health crises as much as possible. At the same time, a decent and sound public health emergency system is able to carry out risk assessments of internal risk factors for public health emergencies that are difficult to predict, and make full use of health big data technology to complete the intensity and scope of public health crisis outbreaks. In this way, we can achieve corresponding emergency preparedness, including formulating emergency plans and preparing emergency health and medical supplies. Under such a complete public health emergency system, the public will not panic too much, and it has a good guarantee for social stability [4].

3 Construction of a Beforehand Alarming System for Public Health Emergency Based on Big Data

3.1 Emergency Health Response System

The composition of health emergency agencies in my country is very rich, including blood collection and supply agencies, health supervision agencies, medical research and teaching agencies, disease control agencies, medical treatment agencies, and health administrative departments at all levels. The above-mentioned agencies have their own functions in the process of emergent health emergencies, as well as the actual situation of the current health emergency management and the stage at which they are in, and take corresponding emergency treatment measures. Starting from the actual setting of the traditional emergency procedures in the past, the health emergency can be divided into the following steps, including emergency preparedness, monitoring and beforehand alarming, emergency response, and summary evaluation. Judging from the process of health emergency decision-making and handling in China at this stage, after the occurrence of related emergencies, emergency workers need to fully understand the details of the incident and conduct scientific verification, and give feedback immediately and provide feedback. It is on the record. Proceeding from the specific situation of time, strengthen the interconnection and intercommunication with the health emergency linkage department and various governments, carry out a comprehensive evaluation and consultation, and select appropriate emergency measures on the basis of formulating relevant principles. At this stage, the expert database and knowledge base of the local emergency command and decision system should be activated and managed as soon as possible. Through the application of information statistical analysis and spatial display analysis functions, and fully integrated with the evaluation opinions of experts and relevant personnel, the leadership the decision-making carried out is assisted. In response to some particularly major incidents, a national emergency command center was established, and the national health and family planning emergency command system was activated in time. In the process of handling, it is necessary to dynamically track the actual development trend of the incident, and provide timely feedback on the relevant implementation and the effectiveness of the measures taken. The health administrative department needs to promptly terminate the incident from the relevant real-time tracking situation. Determine, and at the same time strengthen the archiving of later data and later evaluation after the end of the time, and strengthen the understanding of the importance of accumulated knowledge and experience [5, 6].

3.2 Emergency Command and Decision System Planning

From the current point of view, the efficiency of government affairs in countries around the world has been greatly improved in the context of rapid development of informatization. All countries are facing the opportunities and challenges brought by the information technology revolution. For health informatization and emergency command and decision-making system is one of its ultimate subsystems. The establishment of a unified, authoritative and efficient emergency organization and command system will play a positive role in improving the efficiency of health emergency. The emergency command and decision-making system is mainly to strengthen the full tracking, support and decision-making of crisis response through the application of information technology and modern networks. The system can realize the determination of the best response measures in the shortest time and issue them as soon as possible. Efficiently unite its various agencies to jointly respond to the health crisis. Under normal circumstances, complete the daily management of health emergency and the standardization of related emergency management business processes with high quality. In an emergency state, it can complete the processing, analysis, transmission, storage and collection of information. At the same time, it also has the functions of commanding and dispatching, assisting decision-making, and selecting and starting the plan, thereby efficiently realizing emergency management in the information system Combination of peace and war. In terms of the structure of the emergency command

and decision-making system, my country mainly divides the original emergency command platform for public emergencies into a three-level emergency command and decision-making system, which mainly includes the national, provincial, and prefecture levels. Horizontally, it is divided into linkage departments, emergency command and decision systems of various departments, and other comprehensive information systems. In the overall structure of the system, emergency command and decision-making can be divided into operation management system, safety assurance system, standard specification system, application software system, and network and technology implementation. In order to maximize the probability of occurrence of the phenomenon of "information chimney" and "information islands", the local-level system network layer should realize the reserved interface with various information systems and higher-level health emergency systems [7].

3.3 Establishing a Sharing Mechanism for Healthy Big Data-Driven Decision-Making

During the process of emergency response to public health emergencies, falling into the quagmire of dogmatism and empiricism is not suggested, while following the principle of seeking truth from facts, observing objective laws to carry out emergency response work, and avoiding subjective assumptions and blind decision-making; build a scientific decision-making sharing mechanism based on big data. For this reason, relevant departments and staff should strive to meet the following requirements:

First off, we must focus on selecting high-end talents with strong professional capabilities and comprehensive qualities, forming a health big data analysis team, and assisting relevant departments in setting up a command center in the face of public health emergencies. In this process, it is necessary to rely on artificial intelligence methods and use current advanced algorithms to reasonably predict and accurately analyze public health events. To ensure that all decisions made in response to public health incidents must be based on data, a scientific beforehand alarming and decision-making mechanism can be constructed [8].

Second, the beforehand alarming work carried out on it should give full play to the role of the health big data sharing mechanism. Relevant departments should strive to achieve information and data sharing. During the construction of beforehand alarming mechanisms, it is necessary to ensure that public health management departments, public security and transportation departments, commercial enterprises, market supervision departments, and material management departments can exchange information and data with each other. By virtue of using this means, it is definitely possible to ensure that big data can flow smoothly among various departments, thereby constructing a scientific linkage mechanism, thereby improving the efficiency of handling public health incidents and realizing its standardized operations.

Last, relevant departments need to hire professional teams to rely on their health big data information maintenance capabilities to provide technical assistance in handling emergencies, and collect, save and effectively analyze public health data in the first time. Increase the rate of decision sharing and strengthen the effect of decision sharing.

Finally, there is a need to strengthen cooperation among various departments. Generally, information and data related to public health incidents are often scattered in various fields such as public travel, social media, and telecommunication services. Relevant departments need to achieve a high degree of integration of these relatively scattered data through the scientific use of health big data, so that it can be used efficiently. By strengthening cooperation, we can prevent the spread of public health incidents and minimize the losses caused by them [9].

3.4 Give Full Play to the Role of Non-governmental Organizations

As the environment deteriorates off, public health incidents have unfortunately become a normalized problem. In addition, in the self-media era, the development trend of multiple subjects is becoming more and more obvious. Therefore, NGOs are likely to become an essential part of the construction of public health incident beforehand alarming system.

In the beforehand alarming of public health incidents, the more important link is the acquisition of crisis information. At present, due to the mixed content on the Internet and the extremely complex causes of public health incidents, traditional analysis and processing methods can no longer meet the needs, and big data must be used. Technology provides technical support for public health event beforehand alarming. At present, some non-governmental organizations that focus on technology research and development are more advanced in big data technology. They can quickly obtain useful data from the massive information of the Internet through Internet of Things technology, search engine technology, and data mining technology, and process these data. Make it useful information. It can provide timely beforehand alarming information for the decision-making layer to take corresponding countermeasures in the first time. For such non-governmental organizations, we must actively cooperate with them and give full play to their role [10].

To give full play to the role of NGOs in the beforehand alarming of public health events, we can start from the following three aspects: First off, NGOs must actively establish effective communication with other NGOs, and even with government organizations, in a timely manner. Build a data sharing platform to exchange and share relevant information in the first place to improve the speed and ability to respond to public health incidents. Second, non-governmental organizations should actively participate in social welfare. As we all know, in the past when health incidents occur, the public often blindly grab certain commodities, leading to chaos in the market. Such derivative crises often lead to further deterioration of public health incidents. To solve this problem, the assistance of non-governmental organizations is needed. In the community, in order to improve efficiency, big data analysis should also be used. Through this technology, more typical communities should be accurately selected, and knowledge about public health incidents can be preached, so that the masses can not only master the response to public health incidents, but also Data technology has also been recognized to further improve the level of awareness; third, in the beforehand alarming stage of public health incidents, some criminals often maliciously spread rumors, causing serious social chaos. In this regard, NGOs should actively use the Data technology guides public opinion, accurately pushes the correct information to the public, and avoids the negative impact of the spread of malicious speech. At the same time, big data technology can also be used to accurately find the source of these rumors and provide these data to law enforcement agencies. The conduct of rumors by law-breakers is subject to due sanctions [11].

4 Conclusion

A conclusion may be drawn and introduced based on the above analysis that a scientific and complete beforehand alarming system for public health emergencies can effectively function to enhance our country's response to public health emergencies and play a positive role in promoting the stable development of the society. Therefore, relevant departments are highly anticipated to strengthen the application of big data and efficiently realize the comparison and analysis of relevant public health event data, so as to reduce various losses caused by public health events.

Acknowledgements. Fund: 2020 Foshan Social Science Planning Project: A Big-Data Based Study on the Path of Foshan's Response to Public Health Emergency (Project Number: 2020-GJ031).

References

1. Li, J.: Construction of an beforehand alarming system for public health emergencies based on big data. All Circles **10**, 169–176 (2020) (in Chinese)
2. Han, A., Isaacson, A., Muennig, P.: The promise of big data for precision population health management in the US. Publ. Health **185**, 110–116 (2020)
3. Ding, D., Del Pozo Cruz, B., Green, M.A., Bauman, A.E.: Is the COVID-19 lockdown nudging people to be more active: a big data analysis. Br. J. Sports Med.**54**, 1183–1184 (2020)
4. Husnayain, A., Fuad, A., Su, E.C.Y.: Applications of Google search trends for risk communication in infectious disease management: a case study of the COVID-19 outbreak in Taiwan. Int. J. Infect. Dis. **95**, 221–223 (2020)
5. Effenberger, M., Kronbichler, A., Shin, J.I., et al.: Association of the COVID-19 pandemic with internet search volumes: a Google trendsTM analysis. Int. J. Infect. Dis. **95**, 192–197 (2020)
6. Liu, M., Caputi, T.L., Dredze, M., et al.: Internet searches for unproven COVID-19 therapies in the United States. JAMA Intern. Med. **180**, 1116–1118 (2020)
7. Wood, W., Neal, D.T.: Healthy through habit: interventions for initiating & maintaining health behavior change. Behav. Sci. Policy **2**, 71–83 (2016)
8. Ding, D., Nguyen, B., Learnihan, V., et al.: Moving to an active lifestyle? A systematic review of the effects of residential relocation on walking, physical activity and travel behaviour. Br. J. Sports Med. **52**, 789–799 (2018)
9. Bian, W., Leng, F.: Research on the overseas public health strategy information system for emergency decision-making of major public health emergencies—taking response to the new crown pneumonia epidemic as an example. Books Inf. **2**, 13–18 (2020) (in Chinese)
10. Wang, W.: The intelligence mechanism and system for responding to public health emergencies in my country. Books Inf. **1**, 15–26 (2020) (in Chinese)
11. Wang, F.: Improving the ability of social risk information perception and perfecting the beforehand alarming mechanism of public health emergencies. Libr. Inf. Knowl. **2**, 4–6 (2020). (in Chinese)

Internet of Things Information Construction Combined with Big Data in Cloud Computing Environment

Jiani Hong(✉)

Department of Economics, Shanghai University, No.20, Chengzhong Road, Jiading District, Shanghai, China
1530203900@qq.com

Abstract. With the advent of the cloud computing era, many convenient conditions have been proposed for people to work and live. This shows the importance of cloud technology development and application. On the basis of the traditional structure of the IoT, this paper constructs the system architecture IoT, which is composed of There are three levels. The design of the perception control layer adopts the edge computing technology to complete the network data collection. The virtualization technology is used to set up the virtual server in the network layer to speed up the network data transmission. Use big data technology to effectively analyze the data transmitted, and provide service application to the cloud computing service layer, so as to complete the Internet of things informatization construction combined with big data. Within the scope of cloud computing. Finally, results show that the bit rate after these pathways change reaches 486 basis points, while the bit rate of the previous physical network construction is only 248 basis points, indicating it has a relatively high level of informatization.

Keywords: Cloud computing · Big data · Internet of Things · Informatization

1 Introduction

The concept of cloud computing in China is: cloud computing integrates a large number of network resources and storage resources into one computer through the Internet, realizes network resource sharing, and provides dynamic and virtualized network resources through large-scale distributed parallel computing to provide network resources User service [1]. Cloud computing mainly involves two technologies, one is virtualization technology, and the other is distributed parallel technology. Among them, virtualization technology is to virtualize and abstract physical resources such as hard disks, memory, and storage of servers, and convert them into logical resources that can be dynamically managed, thereby improving the utilization of network resources; distributed parallel technology is the use of a large number of computer software. The hardware resources are distributed on different IP addresses through the Internet for integration and collaborative operation, providing network users with distributed a lot of data storage and management [2]. In the cloud computing environment, the informatization construction of the IoT puts forward new construction

J. MacIntyre et al. (Eds.): SPIoT 2020, AISC 1282, pp. 473–478, 2021.
https://doi.org/10.1007/978-3-030-62743-0_68

requirements. The traditional Internet of Things has a low degree of informatization and cannot satisfy the development requests of the cloud computing environment. For this reason, a cloud computing environment combined with big data is proposed. The construction of networked informatization will promote the IoT development and improve the level of informatization of the IoT.

2 Internet of Things Information Construction Combined with Big Data in a Cloud Computing Environment

The IoT uses networks to establish connections and exchanges between things. The Internet is used as a connection tool to extend the collection of information to the things, and the things can exchange information with each other [3]. Since the Internet of Things covers a wide range, from the perspective of layering, it is mainly divided into application layer, perception layer, and application layer.

There are a lot of sensing information on the sensing layer, which is mainly responsible for collecting object information and is the source of information acquisition; Network layer has strict transmission requirements, multimedia network must meet the requirements of interaction and real-time; The application layer is composed of a variety of terminal application programs, which intelligently control objects through data processing and analysis. Based on this, the method of processing smart grid big data on account of cloud computing software platform can not only realize the unified management, intelligent storage, data processing, data analysis, data visualization and other applications of smart grid big data, but also increase the encryption of data and greatly improve the privacy effect of user data. The following figure shows the structure of the Internet [4] (Fig. 1).

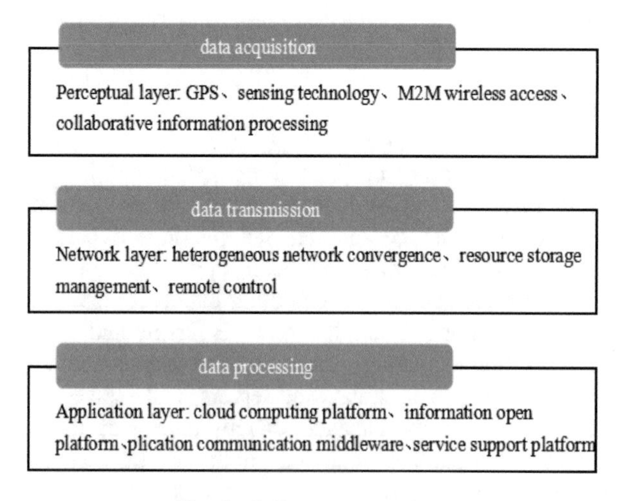

Fig. 1. IoT structure system

2.1 IoT Information Architecture Design

This time, using the core concepts of cloud computing and the Internet of Things, as well as the traditional hierarchical design concept, using the "cloud network side" model, combined with big data in the cloud computing environment, the Internet of Things information architecture is designed, as shown in Fig. 2.

Perception control layer	Sensor→Wired network→GPS IC Card→Wireless base station→ RFID
Network transport layer	VPN→ LAN → Internet Optical fiber ring network: Primary substation, Secondary substation
Cloud computing service layer	Software services:ERP,CRM, website, Electronic Commerce Application service: Storage application, software application and collaborative application

Fig. 2. IoT information architecture combined with big data

In the existing technology, the construction of new information sensing contact system, effectively solve a series of existing problems, so as to design a safe, reliable, digital and information-based Internet of things system. The whole system is operated and managed by three parts, Coordinate the reception and processing of information [5]. With the help of comprehensive perception, computer virtualization and other features, it integrates big data data data to realize information connection across people and things, and truly realizes the informatization construction of the Internet of things.

2.2 Perception Control Layer Design

The perception control layer is mainly responsible for the collection of multi-dimensional perception node data. The node is the most basic organization in the Internet. It can be a front-end perception device or a front-end system divided by function. Because the data resources of the Internet of Things are complex and diverse, the collection and transmission cycles of structured data and semi-structured data are different. Different types of data vary in structure, data collection period, network transmission speed, transmission period, etc. [6]. Edge computing terminals are selected as the transition link between sensors and cloud computing service platforms. Edge computing has strong computing capabilities, It can complete the collection of Internet of Things data, intelligently extract effective information from the collected data, effectively share the computing pressure of the cloud center, make the Internet of things more efficient and useful. The edge computing is used to obtain data from the perception layer of the Internet of things, and combined with the characteristics of the data, it is flexibly deployed in the control layer, and the same type of data is allocated to the corresponding Internet nodes to assure that the deployed data is close to the Internet nodes in the calculation and transmission, so as to realize the design of Perception layer and its control [7].

2.3 Construction of the Network Transmission Layer

The network transmission layer is an independent and autonomous organization in the Internet of Things, with important data storage and processing functions. He is the central point of the whole system, responsible for the connection of the upper and lower levels. The convergence of data from the edge nodes upwards, and the task of "gathering edges to domains" is completed; the network data for analysis is provided to the cloud computing service layer downwards on demand to complete the task of "data to the cloud". As a "midfield" organization, the network transport layer realizes the collection of IoT data to the cloud computing service layer level by level on demand. The construction of the network transmission layer adopts virtualization technology, which uses virtualization technology to build a part of virtual machines in networks such as local area networks, the Internet, and optical fiber ring networks to increase network transmission speed and provide a smooth network environment for data transmission [8]. At the same time, virtualization technology is used to flexibly deploy functions to receive and process the data sent from the perception control layer, and uniformly transform the data into structured data, which is distributed to nearby virtual machines. The network transmission layer uses virtualization technology to first complete the data reception work, then process the data, and finally transmit the processed data to the other layers [9], so that the cloud computing service layer can concentrate its computing resources on big data analysis and data mining.

2.4 Realizing the Information Construction of the Iot

Through the establishment of big data drive architecture, the smart grid big data drive is realized. The classification model is established by data mining algorithm, which makes multiple different types of smart grid database output different target data according to user demand, shortens the time of users using data, and improves the data processing efficiency. The classified data is stored permanently through the blockchain system, and the users in the network fulcrum of the blockchain realize data sharing through the generated key, which effectively improves the confidentiality and security of the data. The scheme of this design is novel, which introduces the blockchain technology into the application and processing of smart grid big data, which provides certain technical reference for the application of smart grid big data, and provides important technical support for the next step of smart grid big data processing [10].

3 Experiments

In the context of cloud computing, combined with big data to carry out the construction of the IoP, in order to prove the validity of the research results. The experiment uses an Internet of Things as the experimental object. This experiment uses 5 servers and several ports to build an experimental environment. The specific parameter configuration is: the terminal is divided into a virtual server and a backup server. The memory is 128 GB and 64 GB, and the IP address is 11.26.51.110 and 11.26.51.110, the hard disk is 16 G, the server types are FTP server and FTP server respectively. 2500 pieces of effective data were transmitted to the informatization construction and the original IoT respectively, and the effective data analysis rate of the IoT before and after the informatization construction was compared. The experimental results are shown in Fig. 3.

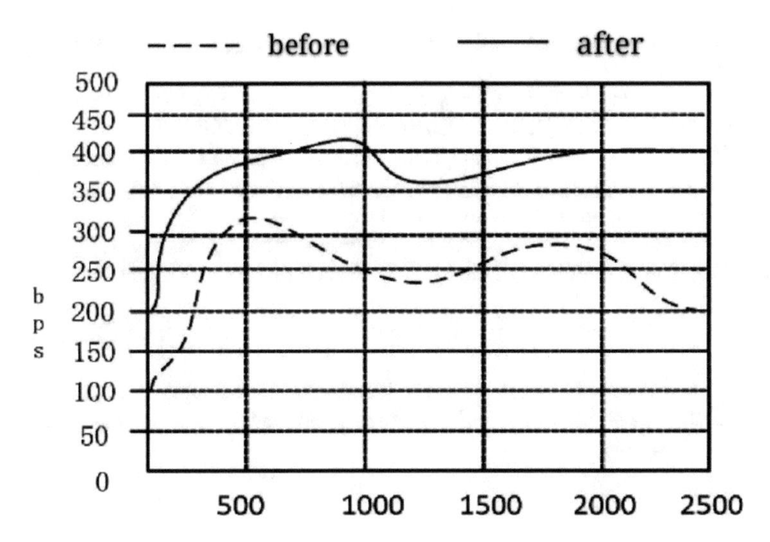

Fig. 3. IoT information architecture combined with big data

The bit rate is the basis for judging the data analysis rate of the IoT. The higher the bit rate, the faster the data analysis rate of the IoT. It can be seen from Fig. 3 that the bit rate of the Internet of Things after construction is 486 bps, while the bit rate of the physical network before construction is only 248 bps, which is much lower than before construction data analysis rate.

4 Conclusion

In the context of cloud computing, the application of big data data to the informatization construction of the Internet of Things has effectively improved the informatization level of the Internet of Things. It is combined with information technology to adapt to the needs of modern development, efficient and rapid implementation of transactions, intelligent connection between people and the network, as a new technology, widely used in various fields to promote the development of society.

References

1. Doan, C.H., Emami, S., Sobel, D.A., et al.: Design considerations for 60 GHz CMOS radios. IEEECommun. Mag. **42**(12), 132–140 (2004)
2. Razavi, B.: A 60-GHz CMOS receiver front-end. IEEE J. Solid-State Circuits **41**(1), 17–22 (2006)

3. Gaucher, B.: Completely integrated 60GHz ISM band front end chip set and test results. IEEE **802**, 10–12 (2006). 15-06-0003-00003c
4. Zhang, F., Gao, Z., Ye, Q.: Construction of cloud platform for personalized information services in digital library based on cloud computing data processing technology. Autom. Control Comput. Sci. **49**(6), 373–379 (2015)
5. Shi, W., Cao, J., et al.: Edge computing: vision and challenges. IEEE Internet Things J **3**(5), 637–646 (2016)
6. Satyanarayanan, M., Bahl, P., Cceres, R., Davies, N.: The case for VM-based cloudlets in mobile computing. IEEE Pervasive Comput **8**(4), 14–23 (2009)
7. Aazam, M., Huh, E.: Fog computing micro datacenter based dynamic resource estimation and pricing model for IoT. In: IEEE 29th International Conference on Advanced Information Networking and Applications (AINA 2019), pp. 687–694 (2015)
8. Bonomi, F., Milito, R., Zhu, J., Addepalli, S.: Fog computing and its role in the Internet of Things. In: The First Edition of the MCC Workshop on Mobile Cloud Computing, pp. 13–16 (2012)
9. Schmidhuber, J.: Deep learning in neural networks: an overview. Neural Netw. **61**, 85–117 (2015)
10. Redmon, J., Divvala, S., et al.: You only look once: unified, realtime object detection. In: 2016 IEEE Conference on Computer Vision and Pattern Recognition (CVPR 2016), pp. 779–788 (2016)

Research on the Development and Management of Electric Energy Under the Background of Big Data

Zhen Dong[1(✉)], Yanqin Wang[2], Yong Wang[1], Ning Xu[2], and Dongchao Wang[2]

[1] State Grid Hebei Electric Power Company, Shijiazhuang 050000, Hebei, China
dongzhen197603@163.com, wangyong198104@163.com
[2] State Grid Hebei Electric Power Company Economic Research Institute, Shijiazhuang 050000, Hebei, China
wangyanqin2020@163.com, xuning197311@163.com, jyy_wangdc@163.com

Abstract. The report of the 19th National Congress of the Communist Party of China pointed out that "socialism with Chinese characteristics has entered a new era", and China's economy has changed from a high-speed development to a high-quality development stage. The adjustment of power energy structure and market-oriented reform have been pushed forward in depth. One belt, one road, is the traditional cost management mode. The first is that the traditional power project cost above the budget quota has been unable to meet the needs of the full market competition. Two, the new technologies change rapidly in complex power engineering. At the same time, the overseas electricity market is broader. This puts forward new requirements for the cost management of electric power engineering. At this time, the management based on cost big data will bring more convenience to the cost management.

Keywords: Big data environment · Power engineering · Cost management

1 Introduction

The report of the 19th National Congress of the Communist Party of China pointed out that "socialism with Chinese characteristics has entered a new era", and China's economy has changed from a high-speed development to a high-quality development stage. The adjustment of power energy structure and market-oriented reform have been pushed forward in depth. One belt, one road, the traditional cost management mode will face the following problems: first, the traditional power engineering cost technology based on the budget quota has been unable to meet the needs of the full market competition; two, the new technologies in complex power engineering change rapidly; meanwhile, under the strategic background of "one belt and one road", the overseas power market is broader. New requirements are put forward for the cost management of electric power engineering. At this time, the management based on cost big data will bring more cost management work facilitate [1].

J. MacIntyre et al. (Eds.): SPIoT 2020, AISC 1282, pp. 479–483, 2021.
https://doi.org/10.1007/978-3-030-62743-0_69

2 Current Situation of Power Engineering Cost Management Based on Big Data Application

2.1 Data Acquisition

At present, it is very difficult to obtain the cost data of electric power projects. The efficiency of manual collection or filling in data is low and the accuracy is difficult to be guaranteed. The cost personnel lack the channels and methods to obtain sufficient external data, and it is difficult to obtain market information in time. Although the electric power industry has a relatively perfect industry quota release mechanism, it is difficult to avoid the ill considered and can not meet the requirements of new technology. The cost results based on this are often difficult to reflect the actual situation of the project. This not only affects the accuracy of the budget preparation of the design unit, but also makes the preparation unit of bidding price limit and the settlement audit unit unable to determine the cost consistent with the actual, so it is difficult to form a full market competition mechanism.

2.2 Data Processing

At present, the increasingly mature cloud cost applied in the construction industry has not been widely used in the power industry, which makes the power industry almost have no effective means and methods to deal with engineering information. When the external data such as the market price of power materials fluctuate, especially the data that may affect the project cost during the construction period are lack of reliable analysis methods, which will cause the unscientific management of power project cost; at the same time, the information update cannot be realized in time, so it is difficult to realize the refined control of power project cost [2]. This will not only increase the cost of the construction enterprise, but also make it difficult for the owner to control the investment reasonably.

3 Application Value of Big Data in Power Engineering Cost

From the current application of cloud cost technology based on Internet and big data technology in the construction industry, the application of the above technology will greatly promote the quality of cost management of owners, design, construction, supervision and other main bodies involved in the construction of electric power projects.

3.1 Employer

In China's power engineering investment entities, whether power generation projects or power grid projects, the owner unit is usually an energy group with economic strength, and each group contains many investment entities. Different investment entities have different types and scales of power engineering price data, which means that the data can be divided into four levels: Project, enterprise, group and industry Acquisition and processing. If the owner controls the investment of a certain scale of electric power project, it can control the investment of new project more scientifically by comparing with the previous project cost. If it is a different type of project, we can learn from the past data of the internal and external of the group, estimate the investment amount accurately and reasonably, and realize the refined management of power engineering cost [3].

3.2 Designer

Enterprises mainly engaged in electric power design often undertake a lot of the same or similar project budget. However, due to the different projects undertaken by different technical and economic personnel, it is often unable to grasp the project cost information undertaken by other colleagues in time. In the case of similar projects, if the technical and economic personnel independently prepare new budget estimates and budgets, it is not only not conducive to the improvement of work efficiency, but also for the same problems may be corrected from time to time, affecting the quality of work. The application of enterprise level big data can provide technical and economic personnel with reference to similar project cost information in time. At the same time, use the Internet or database to match the big data of material and equipment prices in time and accurately, improve the accuracy of estimate and budget, and complete the cost work efficiently. Using the data of engineering cost between enterprises or group level, we can monitor the cost index dynamically and control the cost more reasonably.

3.3 Construction Unit

Construction enterprises need cost management based on big data. Power engineering often involves a large number of building materials, and these materials will fluctuate due to the length of the construction period. For example, the change of steel price which is widely used in substation or transmission line often has a great impact on the settlement work. The big data can collect all kinds of data related to the cost of electric power engineering in time, analyze the historical engineering cost data through accurate summary, so that the construction unit can grasp or reasonably predict the price changes of bidding quotation or materials during the construction period, and the construction unit can more convenient to the project cost composition, influencing factors, various materials and technologies. Optimize the construction progress, take corresponding measures to avoid price risk and ensure the realization of profits.

3.4 Construction Control Unit

Using big data, electric power engineering supervision enterprises can track the real-time changes of network quotation composed of similar project cost and key cost in industry database; using database to collect a large number of index data related to bidding quotation, transaction contract, cost control objective, external information, construction quality and safety, weather change of project location, etc. In the process of construction, the electric power engineering supervision unit can realize the real-time comparison and control of the node cost of similar projects, the project node cost budget and the actual node cost in the database. In the environment of big data, real-time monitoring and feedback are carried out for every program and every node of project implementation through big data, so as to effectively control the cost of power engineering under the condition of ensuring safety and quality.

Big data has a broad application prospect in power engineering cost. The cost of electric power engineering based on big data can better realize the sharing of cost data, which can not only improve the efficiency of current cost work, but also provide a new solution to the difficulties and bottlenecks in cost work. With the development of Power Engineering Intelligence and information management intelligence, big data will bring a revolution to power cost in the near future.

4 The Realization of Sharing Platform of Electric Power Engineering Cost Information Resources

Establish a unified data standard structure. The research of big data requires a unified and standardized data structure, which provides convenience for later data analysis and data mining. Under the double track system of fixed price and list price, it is the key and difficult point to establish the whole process standard and unified data structure from feasibility study, preliminary design to bidding and settlement. From the perspective of pricing activities and total project construction cost, it can be seen that the pricing activity mode of power grid project is closely related to the elements of "quantity, price and cost" after the composition of engineering pricing cost is gradually decomposed. Therefore, the data standard of electric power engineering can consider the unified calculation rules of engineering quantity and the unified cost composition, so as to realize the double track system of pricing system. The difference between fixed price and list price is only in "price", one is the government's leading price, the other is the market competitive price. At the same time, enterprises are encouraged to calculate and prepare enterprise quotas in accordance with their own development characteristics on the basis of industry quotas. Enterprise quotas will also provide reference for the continuous improvement of industry quotas and the level of quotas closer to the market, and promote the common development of enterprise quotas and industry quotas. Unify the cost information data exchange standard of engineering and transaction stage, and realize interconnection and data information collaborative exchange of all stages.

Data mining and analysis application. Data mining is divided into three steps: data cleaning, data mining, data mining results analysis and application. At present, the algorithm of data mining has developed well. The main algorithms are classification algorithm, clustering algorithm, association algorithm, sequence mining and so on. Some data analysis companies have developed mature business data mining tools, but the cost data of electric power projects usually involves business secrets or even state secrets, which is not conducive to the information security of enterprises and governments, so it is necessary to actively develop data mining tools with independent intellectual property rights, and strengthen the research of Engineering cost information technology.

Results analysis and application is the key link of data mining. Through the construction of the platform, improve the cost information service ability, build the project cost prediction model, carry out the dynamic monitoring of the project cost, improve the monitoring sensitivity of the comprehensive index of the project cost, labor, materials and other indexes, and establish the market analysis, multi-party linkage and rapid response management mechanism. Strengthen the comprehensive development and utilization of the market price information, cost index, project case information and other types of professional cost information, and predict the regional, company and industry trends of electric power investment in combination with the market situation to form a diversified information application service system.

Digital technology represented by the Internet is accelerating the deep integration with various fields of economy and society. Big data technology provides various technical support for construction project decision-making. With the development and application of big data, gradually realize the layer by layer integration of management data, business data and industry market data, and achieve the precise control of project investment and project cost. State Grid Zhejiang Electric Power Co., Ltd. Briefly analyzes the current situation of electric power engineering cost information, puts forward the idea of building the electric power engineering cost information resource sharing platform. Through cloud computing technology and big data processing technology, it provides a unified open platform for the electric power industry to realize the integration, sharing and interaction of electric power engineering cost information resources, Provide timely and accurate information services for all parties involved in construction, and provide data support for national, industrial and enterprise research and decision-making.

5 Conclusion

The price, cost index, index and other information and data structure of each factor of production in electric power engineering are complex.

With large amount of data and strong timeliness, it is difficult to collect and analyze information, and it has been in a state of “rich data and lack of information” for a long time. Based on big data processing technology, State Grid Zhejiang Electric Power Co., Ltd. Constructs a sharing platform of power engineering cost information resources, which provides reference and reference for the interactive application of engineering cost information in the power industry.

References

1. Li, X.: On the cost management and control of electric power engineering. Build. Mater. Decoration **17**, 133–134 (2018)
2. Qiao, F.: Cost management and control strategy of electric power engineering. Account. Study **10**, 201 (2018)
3. Zhang, X.: Discussion on the whole process cost management and control of power engineering construction. Low Carbon World **03**, 321–322 (2018)

Secure Big Data Computing Based on Trusted Computing and Key Management

Yang Mei(✉) and Fancheng Fu

School of Computer Information Engineering,
Nanchang Institute of Technology, Nanchang 330044, China
maynni@163.com, wangleil8291826361@126.com

Abstract. MapReduce system is widely used under the cloud computing architecture, and plays an important role in finance, medical health, scientific research, transportation and energy. Big data computing shows its incomparable convenience and quickness. But at the same time, more and more people begin to care about the security of big data platform, and worry about their own privacy security. Due to the sensitivity of the data involved in the above fields, users are exposed to great threats to privacy and security all the time. Big data platform may cause huge disaster due to system intrusion, data replacement and wrong result of calculation. The current solutions mainly focus on the encryption process before data transmission and storage, and the decryption process during data calculation, but fundamentally, these solutions can not prevent user data from being leaked by big data processing program, nor can they guarantee the accuracy of calculation results. A security oriented MapReduce architecture is proposed to provide a secure and reliable computing architecture by integrating big data processing framework, key management system and trusted computing module. Big data processing architecture controls the whole life cycle of cloud computing platform from creation, operation to resource recovery, and is the cornerstone of the whole architecture. The key management system ensures the security of data encryption and decryption by providing a complete set of key verification and authority authentication. Trusted computing architecture is the ultimate scheme to provide security guarantee for the whole architecture. Through the module authentication at the operating system level and the user-defined white list authority control, the security authentication of the whole system module is carried out. Through the above three modules, user data can be encrypted and decrypted before the big data platform is processed. Trusted computing can protect the security status of the big data platform in real time. The security oriented MapReduce architecture provides continuous and strong security guarantee for user data and calculation results.

Keywords: Trusted cloud computing · Big data security · Trusted big data architecture · Trusted computing · Key management · Secure data computing

1 Introduction

For large-scale cloud computing systems, MapReduce provides an extremely successful example in simplifying parallel data processing processes [1]. It is widely used in finance, health care, scientific research, transportation, energy and many other fields.

J. MacIntyre et al. (Eds.): SPIoT 2020, AISC 1282, pp. 484–489, 2021.
https://doi.org/10.1007/978-3-030-62743-0_70

While greatly improving the efficiency, the implementation complexity of the algorithm is controlled within a controllable range. But at the same time, the security of these platforms attracts a wide range of attention, especially when dealing with high-risk and sensitive data. For example, Taobao knows our shopping habits, Tencent knows our friends' contact information, Amazon and Dangdang know our reading habits, and Google. Baidu knows our retrieval habits. Even after a large number of harmless data are collected, personal privacy will be exposed. Once upon a time, data leakage led to users' privacy being cracked, and data tampering led to analysis results deviation, which resulted in serious consequences. Various data leakage incidents emerged in endlessly, and personal data of users on the network were common. Thus, security and privacy issues are the focus of growing threats and vulnerabilities.

It is very complex for big data computing platform to manipulate data for processing. However, broadly speaking, the ultimate goal is basically the same. It can memorize the user data, increase the readability and visualization of data, and simplify the complex interaction. Then, considering the security of big data platform, key management system emerges as the times require. We introduce it to solve the problem of encryption and decryption. As we all know, a complete key management system is the core machine of cloud computing system data protection.

System. With the help of the key management system, the expansion and implementation of the system will reduce the difficulty and complexity of encryption and decryption. The design of key management system directly affects whether the whole system ecology can meet the security requirements of users, and the risk and extra cost can be reduced to a particularly low level. After the trust guarantee of storage layer and data layer, the trust degree of the whole big data computing platform will be included in the evaluation. Trusted computing architecture provides hardware based security protection and establishes a consistent guarantee based on machine authentication. Under the protection of trusted computing, computers will continue to run in the desired way. By calculating the unique value of each running software and distinguishing each software according to the trust program, the credibility of the whole system can be measured. Under the comprehensive protection of big data framework, key management system and trusted computing architecture, trust and reliability can be generally accepted.

2 Overview of Trusted Computing

Trusted computing is first proposed by trusted computing Organization (TCG), which includes the protection of trade secrets and system security authentication, strong consistency of user identification, establishment of machine-based identification and network consistency, and allowing secure computing under open standards and definitions. Trust detection architecture of trusted computing organization measures the trusted program of application by hashing all loaded applications into TPM (trusted platform module). CRTM (the core root of trust for measurement) is the first code executed when the system starts. By measuring bootloader, CRTM helps build a chain of trust. After loading the kernel, this trust chain connects all modules and their configurable files to form a network structure in series [2].

Trusted computing loads all executable modules into the platform configuration registers (PCRs) for trust verification. The specific method is to replace the PCR value with the hash value of the original value and the hash value of the latest loaded executable file. All values loaded into the PCR are immutable and will not be changed until the system is rebooted. Generally speaking, attribute based verification is used in the verification mechanism of trusted platform. SML (srored measurement log, also called ml) records the detailed data of these measurements and the necessary metadata of executable files, which represents the current consistency state of the platform. WL (white list) records the expected value of the verifier, which means that the executable program that meets the expectation will appear in the white list. The difference between ml and WL can explain many problems in trusted verification. TPM is used to detect hash values in PCRs and protect key signatures by TPM such as AIK (attention identity key). AIK is similar to privacy CA third-party signature authentication, and is bound with a certain trust chain key to ensure that - an AIK will only be used for a certain TPM, that is, a certain system or platform. In this way, AIK cannot be forged or misused. After signing, the set of attributes, such as the authentication certificate, is sent to the verifier. According to the hash value recorded in ml, the software running on the platform can be easily distinguished. This process requires WL collected from the original data, that is, the data collection absorbed from the software and hardware levels during the trust chain establishment process.

3 Framework Design

Big data computing architecture is the cornerstone of the whole system, providing comprehensive data conversion, flexible resource allocation and load balancing resource optimization. Through distributed parallel computing and effective cloud storage, big data can be deeply and quickly analyzed and extracted. All the next operation of the whole system benefits from the perfect construction of data computing platform. Big data computing architecture is responsible for the incubation, startup and operation of the entire processing platform. Stop, expand and shrink, can quickly control the remote day mark machine. Running MapReduce task on a daily standard machine is actually a distributed computing task. By dividing the large task into small parts, each node can run and process part of the task and complete the whole task. Hadoop relies on the hardware based distributed file system HDFS and computing framework MapReduce. The basic processing process is as follows [3].

The input file is partitioned according to a certain standard. Different nodes process different fragmentation tasks, and each group of nodes undertakes the function of storage and calculation at the same time. This process is called map, and the task of slicing is called map task. In the calculation stage, the partitioned task is resolved into key value pairs according to certain rules, and the map task is executed. Each group of key value pairs is calculated to get a new key value pair.

As the name implies, the key management system is the management and control of the key life cycle, including the whole process of key creation, modification and destruction, involving key server, cryptography protocol, user program and other related protocols. Every time the user data is encrypted, the key management system will open the interface, store the decryption key and the attribute set of the machine, and manage the key value pairs with the agreed cryptosystem. When the data processing platform wants to decrypt user data, it must obtain the decryption key from the key management system. Key management connects the data processing platform and sensitive information to provide basic security in the form of the first barrier.

The specific process of key management is as follows: key. Jid is the key number obtained from the computing platform virtual machine, VM. ID is the number of computing platform virtual machine, property. Set. With_Sign is the set of virtual machine attributes after signing obtained from trusted computing module, key_ Entity is the key entity obtained by key number. Property is returned when the key signature is verified using the signature certificate and property collection_Set to determine whether the signature is valid. If it is invalid, an error is returned directly. After the signature is verified successfully, whether the attribute set is matched is detected. Key management verification algorithm (Table 1).

Table 1. Key management verification algorithm.

Algorithm 1 key management verification
key_ lid- key_ id_ from _wm
vm_ lid←-vm id. fom. w
property. set _with sign←ger. _wm propery. fom. OAT(vm id)
key_ enityt ger key. by. idkey. Jd)
propenty. se1←werify. signanureqproperty. se1 with sign. key enit: CERT)
if properny. set = false then
retun_ wrong. stahus_ to. OA7I)
return false (Verification Failed)
else if matchproperty. set, key. _entityproperty. set) = false then
retum wrong. status_ 10_ OA7U)
return false (Verification Failed)
else
retum_ right staus. 10. OAT()
return key. entity.key (Success)
end if

4 Trusted Computing Framework

Trusted computing standard architecture is based on cryptography, with chip as the root, motherboard as the platform and software as the core. The network will transfer trust, including independent password, active control, dual system architecture, active measurement, ternary and three-layer point-to-point trusted access, etc. The first stage is the encryption scheme of trusted computing platform, which determines the encryption mechanism, including encryption algorithm, key management and certificate configuration. In the second stage, four main standards are proposed, namely chip, motherboard, software and network, which are run by trusted computing. In the third stage, four supporting standards are proposed, namely, trusted computing specification architecture, trusted server platform specification, trusted storage specification and trusted computer trustworthiness evaluation specification. The fourth stage has formed a trusted computing system, including a complete set of facilities, applications, office services, class protection and so on (Fig. 1).

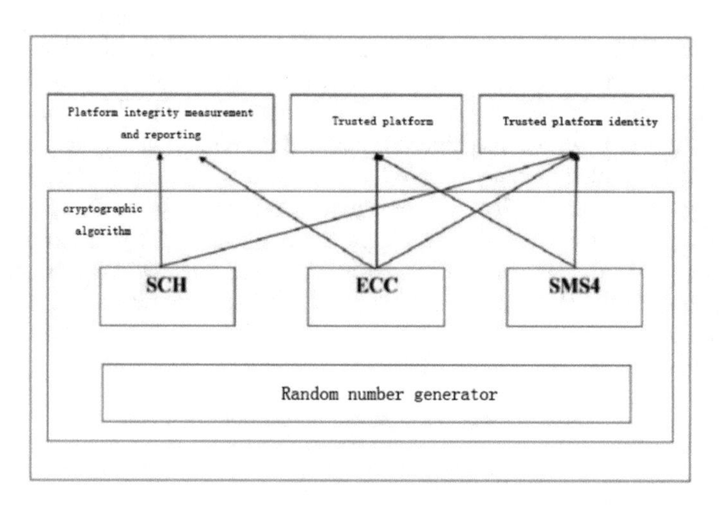

Fig. 1. Trusted computing architecture

The scheme is compatible with dual certificate system, conforms to China digital certificate management system, simplifies key transmission protocol, and is convenient for project implementation and large-scale application. Trustworthy infrastructure software innovations include a dual system architecture that enables independent hardware and software systems to run simultaneously with existing systems, and to measure and monitor the performance of hardware and software in existing computer systems. The trusted computing platform software system includes the host software system and the trusted software library; the host software system, namely the trusted computing platform, realizes part of the trusted software foundation of the normal system functions, including the trusted software components of the trusted computing platform, to achieve the overall credibility. Cloud computing platform, which is based on the existing information system and infrastructure, is a new business operation mode. Its distributed, shared, heterogeneous and autonomous modes have not changed. Software service is actually an autonomous mode. In the era of cloud computing, users enjoy the convenience brought by services, and the introduction of outsourcing services may bring new security risks. The key applications and data of users on cloud computing service platform, confidentiality, integrity, availability, reliability, audibility, controllability, etc. are all the concerns of users, and the core problem is reputation. In the era of cloud computing, we can provide better, cheaper and more comprehensive security services for users on the basis of trust, which fully reflects the characteristics of cloud computing service aggregation, openness, low cost and high efficiency.

5 Conclusions

A security oriented MapReduce architecture. In the computing platform provided by the big data processing framework, the encryption and decryption process provided by the key management system, and the remote trust verification provided by the trusted computing architecture, we have demonstrated comprehensive security protection on the platform. The whole system runs on the openstack infrastructure, multiple modules cooperate and unify, and has the characteristics of high cohesion and low coupling. We implement the big data processing architecture through Sahara, the key management system with Barbican, and trust verification by oat, providing the security verification of virtual machines. When the whole system starts, all the programs from the kernel are included in ml, and the whitelist is inserted into the database at this time. When the computing platform wants to process the user data and requests the decryption key from Barbican, Barbican will entrust oat to verify the trust of the virtual machine of the computing platform. Only virtual machines that meet the trust requirements have permission to obtain and decrypt users. After the functional evaluation and performance evaluation of the experimental prototype, we show a safe and reliable system that can resist most of the attacks. At the same time, we also show a very satisfactory answer in terms of performance loss.

Acknowledgements. The Design of Computer Network and Maintenance Experiment System Based on VR. Subject No.: GJJ180998.

References

1. Lu, Y., Ye, X.: Key management issues and challenges in cloud service environment. Comput. Sci. **03** (2017)
2. Hu, Q., Li, Z., Yan, X.: Research on cryptograph database key management scheme under hierarchical access control. Comput. Sci. Explor. **06** (2017)
3. Li, H., Fan, B., Li, W.: Research and improvement of a trusted virtual platform construction method. Inf. Netw. Secur. **01** (2015)

Data Mining and Statistical Modelling for the Secure IoT

The Research on Student Data Analysis System Based on Data Mining

Guozhang Li and Xue Wang(✉)

Hainan Vocational University of Science and Technology, Haikou, Hainan, China
xuewininfo@eiwhy.com

Abstract. Due to the advanced modern computers and storage technologies, more and more data are generated and recorded every day. These complex data cover all aspects of student learning, life and management, and are a hidden resource for the school. However, due to the mass data and the intricate information channels, a large amount of data is not taken seriously, which is regarded as "junk' and ignored. The expansion of the scale of universities and the development of information technology pose new challenges to student management, classroom teaching and employment. Using big data in the education field to predict and judge student behavior in a timely manner can provide decision-making assistance for universities in mental health analysis, teaching quality evaluation, and student employment.

Keywords: Big data · Data mining · Student data · Data analysis

1 Introduction

Big data is a relatively general concept, which is more of an era background and digital platform, including data optimization, distribution and management. In the actual operation of the platform, the potential information of the data itself is difficult to achieve accurate query, which requires in-depth mining or optimization of data mining technology, and data mining technology will follow. Nowadays, various industries in China are beginning to infiltrate big data technology. Big data analysis has become the mainstream of industry development, and it is also an important means for current enterprises to break the bottleneck of development. Therefore, the single data information analysis system in the past has begun to be phased out, and new data mining technology has become student data analysis in the main trend [1].

2 Analysis of Data Mining Technology in the Era of Big Data

2.1 Data Mining

Data mining technology was first proposed in the early 1990s. This technology is more oriented towards artificial intelligence research in the commercial field. Today, the value of data mining is more obvious. In actual application, it can be used to effectively

J. MacIntyre et al. (Eds.): SPIoT 2020, AISC 1282, pp. 493–498, 2021.
https://doi.org/10.1007/978-3-030-62743-0_71

learn the situation of the product itself [2]. It can also optimize data in a large amount, and truly provide an important reference for enterprise development. In the current development of this technique, the previous method of finding information from simple and clear data has gradually changed to extracting valuable information from fuzzy and complex data. This in itself is a brand new technological breakthrough. The realization of this technological breakthrough requires more support such as Internet technology, information technology, and cloud computing technology [1].

2.2 Cluster Analysis

Usually in the data mining process, cluster analysis is an important data processing technique. Through clustering analysis technology, it is possible to reorganize transactions that are difficult to understand effectively, and promote them to be presented in a more visualized state. Combining the different properties of the data, they can be divided into different groups [2]. The whole is a kind of Effective data analysis process. Clustering analysis technology can classify relatively large data, thereby extracting the required information resources. However, in actual application, this technology is significantly different from the traditional data classification processing method. Its own advantage is mainly that it can effectively group the information data of fuzzy objects. The current cluster analysis methods are mainly divided into hard clustering and fuzzy clustering. Among them, the hard clustering method is more suitable for the data information, while the fuzzy clustering is mainly to achieve the classification effect by dividing the fuzzy data [2]. On the whole, the two classification methods of cluster analysis have obvious differences, but the goals that can be achieved are basically the same, and data classification can be achieved.

2.3 Characteristic Data Analysis Method

The data analysis method can integrate the overall data analysis, and then extend to the characteristic analysis level, thus discovering the useful value Data information. Therefore, this analysis technology itself has the characteristics of fast and comprehensive, can effectively deal with most data resource analysis, and has become the main research direction of most related researchers. In practical applications, relevant designers usually propose a variety of different characteristic data analysis methods. For example, artificial neural networks can be used for data collection, or corresponding neural networks can be built in the data terminal to collect available information in this way. Content. It is also possible to directly use the corresponding genetic algorithm in deep analysis, and to reorganize and select the huge data information [3]. We need to fully use the corresponding visualization technology to mine and collect relevant data, so the practicality of data mining technology is constantly increasing.

3 The Overall Design of Student Data Analysis System Based on Data Mining

The construction of teaching management system is the foundation of school management. The teaching quality management system has various sub-systems [4]. Therefore, finding out the lack of the current internal teaching quality assessment system and conducting a comprehensive review, establishing the correct system direction, establishing the concept of teaching management modernization, and constructing a reasonable teaching quality management system can improve the teaching and talent training quality.

At this stage, all colleges and universities are more or less using teaching management and evaluation systems to assist teaching management, but few schools can comprehensively analyze, evaluate, and share the feedback information generated by each system to form guidance with a certain theoretical height. Sexual opinions help teaching management personnel adjust various systems in time to effectively improve teaching quality. In view of this situation, this research group starts from the perspective of educational data mining, collects and analyzes educators, learners, managers and other data to build a multi-dimensional teaching quality management system [5]. The overall design idea of the basic framework is as follows, as shown in Fig. 1:

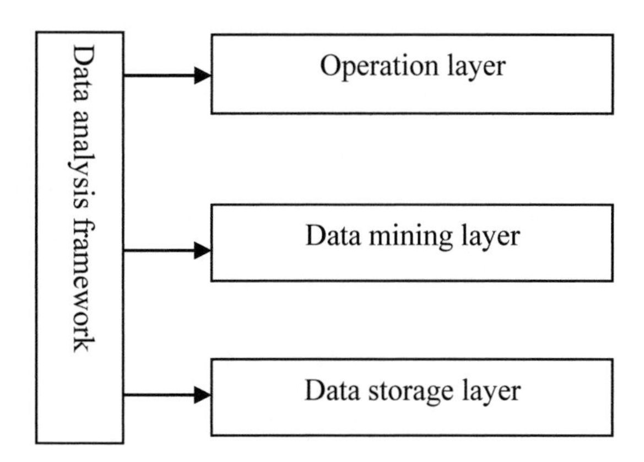

Fig. 1. Overall framework of data analysis system

1) Operational level: Users input relevant teaching quality evaluation data at the operational level, select relevant data sources and set mining parameters, etc., and display various information results generated by data mining at this level.
2) Data mining layer: The main function is to mine the database according to the mining parameters set by the user, and then generate association rules, which are displayed on the operation layer interface.
3) Data storage layer: It mainly stores all kinds of data and provides operating data to all layers of the system according to user instructions.

4 Big Data Mining in the Construction of College Student Data Analysis System

4.1 Collect Daily Behavior Data of College Students

The operation process of this work is very complicated, and the data sources are different, so data preprocessing should be done to optimize the effect of data integration. Data collection mainly includes initial login information, usage record information, and iterative update information. For initial login information, that is, college students make learning goals and plans, college students' learning styles and hobbies, all such information is contained in the database; for usage record information, college students course learning is carried out on the basis of filling in personal information [6], and student browsing records and IP addresses are automatically stored, the system is shown as Fig. 2. In addition, college student interaction data, registration information, and test scores will also be stored in the database; for iterative update information, due to college students' learning goals and hobbies dynamic changes, then the database system will also change accordingly, smoothly realize the storage of new information, and then recommend suitable learning resources to college students [7].

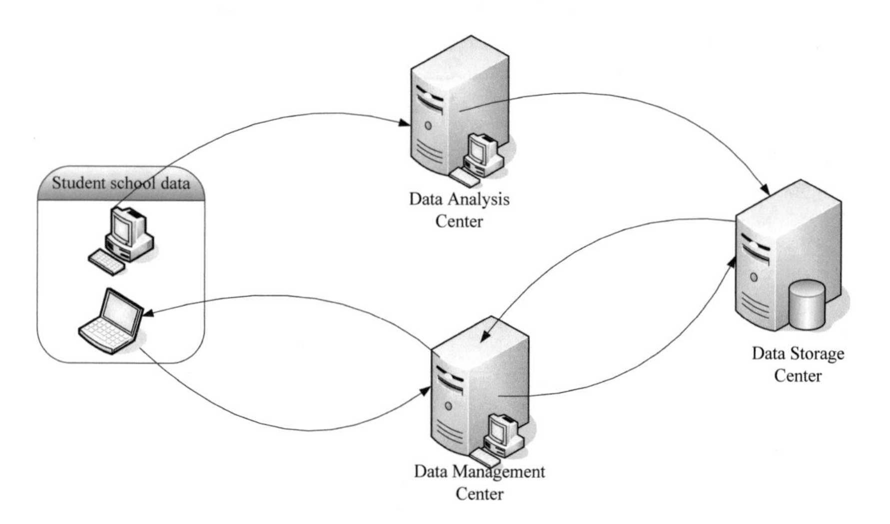

Fig. 2. Collection and analysis system of student data

4.2 Store and Manage the Daily Behavior Data of College Students

Today, if you want to efficiently store and manage massive amounts of data, you must filter out value information, and if necessary, use storage technology and feasible strategies to complete the storage and management of differentiated standards, structures, and real-time requirements [8].

First, data encryption. In the process of storing the daily behavior data of college students, to avoid the phenomenon of network hackers and realize the effective management of data and information within the security range, the school introduces advanced encryption technology and uses encrypted information to realize the conversion of content and code. Only the college students have the right Decoding, which can improve the effectiveness of data transmission. Secondly, warehouse storage. Big data management is more difficult. Manage big data through warehouse storage to realize the purpose of reasonable planning and orderly calling of big data in the storage center [9]. This method has strong applicability. If considered from a long-term perspective, the warehouse storage method should be appropriately improved. Finally, cloud backup. Nowadays, the storage and management of daily behavioral data of college students is gradually moving away from physical machines and turning to the digital field. Due to the rapid growth of big data, the existing physical machines and warehouses are limited in capacity, which provides a broad space for cloud backup. At present, the scope of cloud storage services is gradually expanding, and the efficiency of cloud computing applications is greatly improved [10]. Even if there are network attacks, the cloud can be defended by data migration.

4.3 Analyze the Daily Behavior of College Students

In the process of analyzing students' behavior, the R language is used cleverly to complete data and information analysis tasks, and the analysis results are visually presented in the form of reports, which improves the practicability of the personalized recommendation platform for high-quality educational resources as a whole. During this period, comprehensive use of data statistics methods, machine learning, and data mining techniques, through the method of model establishment, to deeply explore the relationship between college students' learning content, learning behavior, and employment opinions, and then determine the employment direction of college students, and make it based on this Individualized talent training program [11]. Finally, rationally allocate high-quality educational resources to jointly complete the goal of training outstanding talents.

Based on the R language to construct a data model of college students' learning performance and personality characteristics, the specific steps are: collecting individual performance data → inputting data into the database → constructing a special database → constructing a special data model → using data algorithms to analyze performance data → issuing individual character features. Next, analyze the similarity of the types of performance, and classify the data of similar personality characteristics into the same group of databases, and sort out the relationship between learning performance data and personality characteristics. Use regression analysis to construct mathematical models, and use least squares method to obtain the relationship between learning performance data and personality features [12]. The BP neural network method can comprehensively analyze the learning performance data and personality characteristics, and the data classification task is completed by the decision tree method. In the end, the personality of college students can be displayed intuitively.

Finally, teachers can purposefully guide college students and help college students make employment and entrepreneurial plans. Therefore, students can find a suitable position in society after graduation and make positive contributions to social development and national prosperity [12]. At the same time, college students can realize their self-worth.

5 Conclusion

To sum up, the student data analysis system meets the needs of current society, and at the same time it can guide the smooth development of education and teaching work, which can expand the application scope of information technology and improve data mining to a certain extent. It is hoped that school educators and information technology researchers can use this as a reference, attach great importance to data mining, and actively use the technology in student data management to deepen education and teaching reform.

Acknowledgement. Project: This paper is the outcome of the study, Research on the Integrated Application of the Internet + Smart Campus Big Data Cloud Platform, which is supported by the Foundation of the Department of Higher Education in the Ministry of Education for the Second Batch of Education Projects on University-industry Collaboration in 2019. The project focuses on the reform of teaching content and curriculum system; the project number is 201902160018.

References

1. Lu, S.: Data mining technology and application in the era of big data. Digit. World **4**(2), 44–47 (2017) (in Chinese)
2. Liu, C.: Data mining technology and application in the era of big data. Think Tank Era **9**(3), 1–3 (2019) (in Chinese)
3. Sun, Y., Yu, F.: Research progress of personalized services based on big data in China. Mod. Inf. **8**(02), 171–174 (2018) (in Chinese)
4. Liu, Q., Chen, E., Zhu, T.: Research on educational data mining technology for online smart learning. Pattern Recogn. Artif. Intell. **9**(01), 77–90 (2018) (in Chinese)
5. Liu, Y., He, S., Xiong, T.: Library smart service from the perspective of big data mining. Mod. Inf. **7**(11), 81–84 (2017) (in Chinese)
6. Chen, D., Zhan, Y.: Application analysis of deep learning technology in education big data mining. Audio-visual Educ. Res. **7**(2), 68–71 (2019) (in Chinese)
7. Zhao, Q.: Research on teaching quality evaluation system based on data mining. Wireless Internet Technol. **6**(13), 86–87 (2019) (in Chinese)
8. Fu, Y.: Analysis of the results of students' autonomous learning based on educational data mining. Inf. Comput. **7**(16), 219–220 (2019) (in Chinese)
9. Pei, Q.: Research on higher vocational teaching quality evaluation system based on data mining method. Modern. Educ. **8**(37), 285–286 (2017) (in Chinese)
10. Wang, P., Qiu, T.: The construction of a professional teaching quality evaluation system in colleges and universities. Eng. Educ. **9**(3), 240–246 (2018) (in Chinese)
11. Zhao, Q.: Research on teaching quality evaluation system based on data mining. Wireless Internet Technol. **11**(13), 86–87 (2019) (in Chinese)
12. Pei, Q.: Research on higher vocational teaching quality evaluation system based on data mining method. Modern. Educ. **5**(37), 285–286 (2017) (in Chinese)

The Application and Research of Data Mining Technology in Enterprise Service

Yetong Wang and Kongduo Xing(✉)

Hainan Vocational University of Science and Technology, Haikou, Hainan, China
kyummail086@eiwhy.com

Abstract. The Internet has evolved into a complex of information transmission. Modern computer and information technology as well as the growing network data bring us into the era of big data. Over the years, people have used data mining techniques to analyze and integrate massive amounts of data, discover laws and knowledge from massive amounts of data, and dig out valuable information. This article starts with the research of data mining technology, combing with the features of different fields, analyzes the application of data mining technology in different network platforms. And by the value of data mining in enterprise services to improve business operations, perfect data services on the enterprise platform, and maximize the data mining positive role.

Keywords: Data mining · Big data technology · Enterprise service · Data model

1 Introduction

The Internet has evolved into a complex of information transmission. Both individuals, companies, platforms, etc. frequently use the Internet, such as personal communications, corporate propaganda, and media platform communication all rely on it. Network data continues to grow. Social networks, e-commerce platforms, financial platforms, education platforms, music platforms, medical platforms and other platforms in different fields, various service tools, etc., all provide and process a large amount of network data [1], and at the same time, they also serve as data sources. For example, the daily transaction data generated by hundreds of millions of users on an e-commerce platform is about 20 TB, and the data on an education platform such as course resources also exceeds 100 TB. Data is transmitted via the Internet and is updated every moment, gradually forming a large amount of more complex information. The Internet provides people with many applications and services [1].

2 Data Mining Technology

Data mining is not only the collection of certain types of data, but also the process of mining various valuable information such as trends. Data mining technology is the integration, statistics and analysis of a large amount of data, and in-depth analysis

J. MacIntyre et al. (Eds.): SPIoT 2020, AISC 1282, pp. 499–504, 2021.
https://doi.org/10.1007/978-3-030-62743-0_72

through machine learning, artificial intelligence or data mining methods, discovering laws and knowledge and applying them in various fields [2].

There are many data mining methods, such as genetic algorithms, rough set methods, decision tree methods, and neural network methods. You can usually choose one for modeling or multiple for comparison and verification. Generally, the steps of data mining are: analyzing problems, selecting and preprocessing data, creating and debugging data models, mining data, and maintaining data mining models and evaluation results [2]. Take the education platform data mining as an example, the general steps of analyzing data mining are shown in Fig. 1:

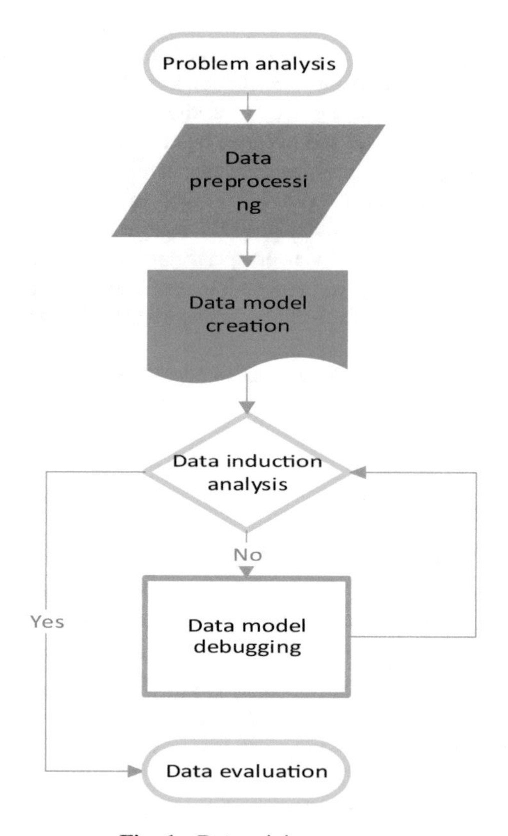

Fig. 1. Data mining process

3 The Data Mining Application in Enterprises

Many platforms in different industries have used data mining technology, and they have effectively improved the quality of platform services by using data mining results. In the Internet, common e-commerce, finance, education, music, health care and other platforms use data mining technology to mine and integrate data information, and then integrate beneficial data results into different applications [3]. The results of data mining not only come from everyone, but also related to the future development of individuals, enterprises, and society [3]. We need to play to the social value of data mining, improve people's lives, improve platform data services, and maximize the data mining positive role.

3.1 Application in E-Commerce Enterprises

The popularity of the Internet has subverted many traditional physical industries, online shopping has become a fashion, and China's marketing market has made breakthrough progress [4]. The computer data mining technology application in e-commerce is mainly reflected in the preferences of users. Most platforms use this technology to collect huge data, mine the data and analyze the products that users' prefer. Provide enterprises with intelligent data analysis, optimize the operation of e-commerce platform, improve marketing methods, and enhance marketing services [5]. The specific operations are shown in Fig. 2.

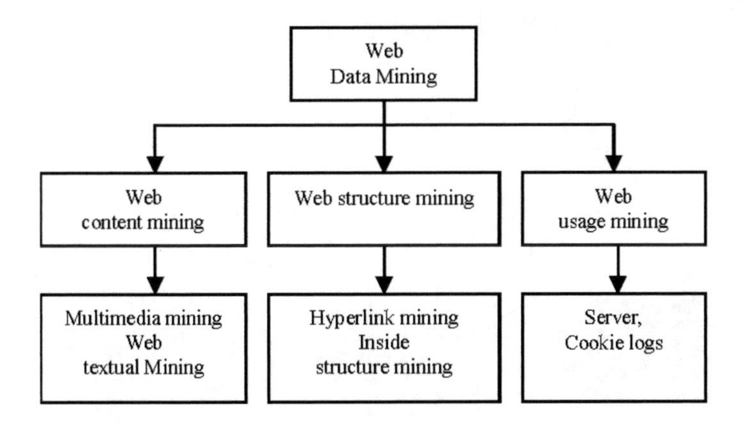

Fig. 2. Application of data mining technology in e-commerce

3.2 The Application of Data Mining Technology in Financial Enterprises

The banking industry has realized the important role of big data strategy in its operation management, customer marketing and product optimization, as well as the profound influence on the future development direction of the bank [6]. In-depth mining of the big data, its value is huge, thereby promoting the innovation of bank products, services and management.

3.3 Application in the Field of Education

Big data has changed the traditional classroom teaching model, from "small classrooms" to "world lecture halls," and tens of thousands of course teaching contents are presented on the Internet. Online education has reached hundreds of millions of hits.

Among the knowledge resources in the network platform, big data makes education development have a broader space [7].

In the past, the learning characteristics of learners dissipated in classrooms, white study rooms and other places, and many data were difficult to collect [5]. Now using big data technology, these data are stored on the Internet or downloaded locally, and repeated research. For teachers, data mining technology can understand the degree of students' knowledge mastery through students' network behavior, and can also plan the important and difficult points, to improve the teaching quality, as shown in Fig. 3. In addition, for courses and lesson plans with high click-through rates and good evaluations, the reasons for their popularity are unearthed [7]. A lot of learners, teachers, and school information on the Internet are worthy of deep mining.

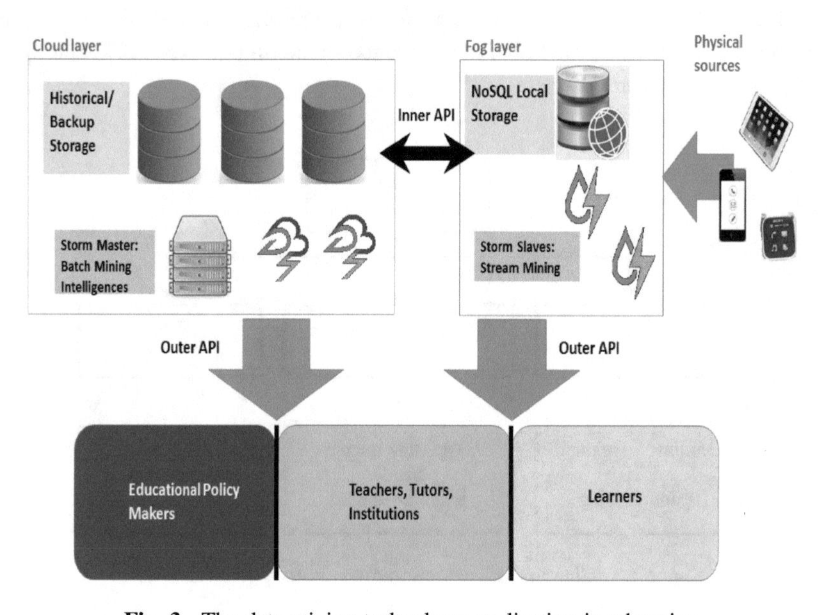

Fig. 3. The data mining technology application in education

3.4 Application in the Field of Music

Audio and video on the Internet increase dramatically today, and massive music data relies on network platforms [4]. The music network platform not only quickly integrates multi-cultural and multi-form music resources, but also gives feedback on the preferences and ideas of audience users [8]. For major music platforms, the amount of music resources and user experience are important indicators of platform advantages. Through the mining of music data, on the one hand, the current popular music resources can be analyzed and included on the platform to increase the amount of resources. On the other hand, combining user feedback information such as music listening data, it has developed a humanized recommendation function, such as recommending similar songs and similar artists on a certain platform, greatly improving user experience [9].

3.5 Medical Applications

With the popularization of medical information, hospital management is gradually modernized. Nowadays, in the hospital system, data on patient files, medical device management, medicines, etc. are all entered into the system. These data, such as patient structure and consultation time, are integrated and mined [10].

In medical work, the diagnosis methods of workers are also changing. In big data era, the data and information contained in human diseases can be consulted. Nowadays, workers can use data mining technology to conduct comprehensive research on the information in the database through various data such as the patient's medical history, the clinical symptoms that appear, and the medical history of similar patients [11]. It can not only dig out meaningful diagnosis rules, assist in disease diagnosis, improve the efficiency of doctors' diagnosis and treatment, and alleviate doctor-patient conflicts, but also can effectively warn patients of diseases and recommend treatment measures, as shown in Fig. 4.

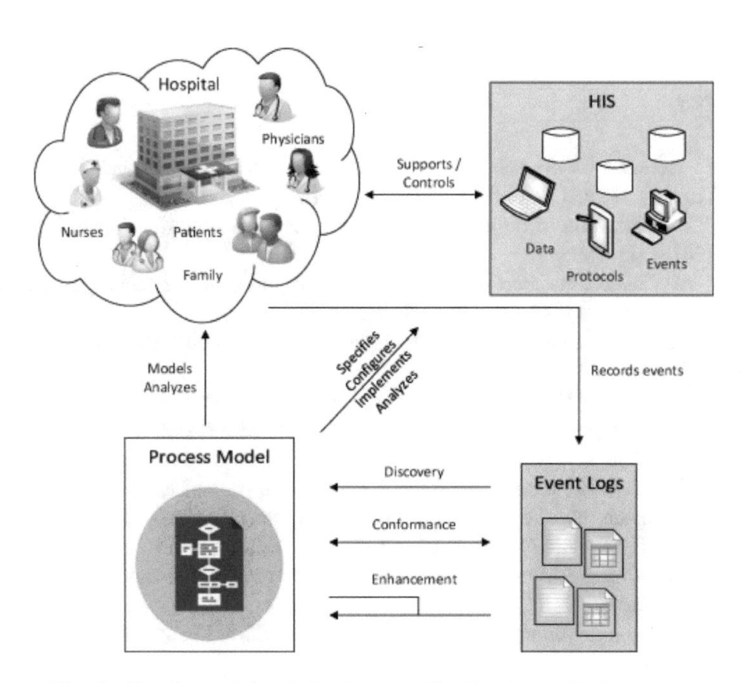

Fig. 4. The data mining technology application in medical treatment

4 Conclusion

All in all, under the situation of increasing information, it is very necessary to promote the processing of network information and data through data mining technology. This article analyzes the data mining technology application in network platforms from its research manpower, combined the data with different fields. People use data mining technology to dig out data that is conducive to the development of different fields from massive data to achieve the effect of improving social governance and increasing the utilization rate of data platforms, and apply them in various fields. We can give play to the social value of data mining enterprise services, improve business operations, improve enterprise data services, and maximize the positive role of data mining in enterprises.

Acknowledgments. This paper is the outcome of the study, Hyperspectral Remote Sensing Image Processing Based on Adaptive Differential Evolution Nuclear Extreme Learning Machine and Application in Offshore Environment Detection in Yangpu Port of Hainan, which is supported by the Foundation of Hainan Provincial Department of Science and Technology for Science and Technology Cooperation Projects in 2018. The project belongs to the Key Research and Development Program for Science and Technology Cooperation; the project number is ZDYF2018234. It is also the outcome of the study, Application Research of Humanoid Learning Algorithm Model based on Network Text Analysis of Tourism Image Perception in Hainan Free Trade Zone, which is supported by the Foundation of Education Department of Hainan Province for Scientific Research Projects in Colleges and Universities in 2019. The project number is Hnky 2019-105; this project belongs to General Projects of Scientific Research.

References

1. Yan, W.: Research on the application of CRM in digital publishing based on data mining. Xi'an Technol. Univ. **17**, 89–92 (2018) (in Chinese)
2. Li, X.: The application of big data mining in banking business. Era Financ. Technol. **8**(5), 15–19 (2017) (in Chinese)
3. He, G.: Application research of data mining in the field of computer network security. Technol. Market **23**(8), 13–15 (2016) (in Chinese)
4. Wang, Y.: Data mining technology and application based on big data. Digit. Technol. Appl. **9**(4), 112–115 (2018) (in Chinese)
5. Li, F.: An analysis of the application value of Web data mining technology in e-commerce. Electron. Technol. Softw. Eng. **14**(02), 195–198 (2016) (in Chinese)
6. Liu, Z.: Big data analysis and mining technology and its decision-making application research. Sci. Technol. Innovation **7**(23), 84–85 (2019) (in Chinese)
7. Yu, H.: Research on the application of data mining in bank customer relationship management. Hefei Univ. Technol. **13**, 52–55 (2010) (in Chinese)
8. Lv, Q.: Predicting music trends based on machine learning. Lanzhou Univ. **11**, 104–106 (2017) (in Chinese)
9. Xing, B.: Emotion-driven music data mining and retrieval. J. Zhejiang Univ. **2**(08), 210–213 (2013) (in Chinese)
10. Tian, Z., Yu, L., Xiong, B.: Application and research of data mining in marketing. J. Baicheng Normal Univ. **6**(02), 24–26 (2017) (in Chinese)
11. Ren, F., Liu, S.: Data mining technology is widely used in medical information. Chin. J. Multimedia Netw. Teach. **3**(16), 8–10 (2019) (in Chinese)

Influencing Factors of Undergraduates Using App Learning Based on TAM Model – Taking MOOC App as an Example

Lili Chen(✉)

Oxbridg College, Kunming University of Science and Technology, Kunming 650000, Yunnan, China
519980950@qq.com

Abstract. In recent years, more and more undergraduates begin to use app learning based on the TAM model. Among them, the emergence of MOOC is a good opportunity for educational reform. Because of the convenience, low cost, lifetime and pertinence, MOOC software played a great role in educational reform. This article reviews the related content of TAM model and learning efficiency, and builds MOOC learning behavior model of college students. Through the survey, a total of 255 valid questionnaires were collected. The relevant data were analyzed, and the model was hypothesized. Through the model conclusion, we find: The behavior of college students using MOOCs will be affected by behavioral intentions, and behavioral intentions will be affected by learning motivation, perceived usefulness, and attitudes. The attitude to use will be determined by college students' perception of MOOC's usefulness and perceived ease of use. Perceived ease of use is affected by cognitive load, and perceived usefulness is affected by learning effects and motivation.

Keywords: MOOC learning behavior · Technology acceptance model (TAM) · Learning efficiency · Influencing factors

1 Introduction

MOOC, a novel learning form, has existed in China for a few years. Previous studies have conducted a large number of analyses and investigations on the effects of MOOC learning, and lack of research on MOOC learning efficiency. In addition, on the surface, the literature on learning efficiency is very rich, but the concept of learning efficiency is still very difficult to measure [1].

Based on TAM and learning efficiency, a model of college students'MOOC learning behavior is constructed, and a structural equation model (SEM) analysis is performed through this model. Finally, the influencing factors of college students' MOOC learning behavior are explored [2]. The perceived ease of use of MOOC learning is reflected in whether users can easily find the content they want and whether they can use MOOC easily. Perceived ease of use was first proposed by Davis in the mid-1989 technology acceptance model, and it was proposed that the use of the perceived ease of use variable to partially measure attitudes to use was unprecedented [3]. According to theoretical analysis, if college students can find accurate solutions quickly

J. MacIntyre et al. (Eds.): SPIoT 2020, AISC 1282, pp. 505–512, 2021.
https://doi.org/10.1007/978-3-030-62743-0_73

in MOOC when they encounter learning difficulties, it helps to enhance college students' positive attitude towards MOOC learning. And whether you can easily find the content you want on the MOOC platform is very important for MOOC learners to continue to use MOOC for learning [4–6].

Based on the existing research, this article aims to explore the influencing factors of college students' MOOC learning behaviors in view of the high drop-out rate and low completion rate. Firstly, it explores the influencing factors of college students' MOOC learning efficiency, and then explores the learning behavior of college students' MOOC learners. This article reviews the previous research of MOOC, then sorts out and analyzes the related content of learning efficiency and TAM. Finally, according to the research situation in this article, the specific theories and content will be explained. Studies have shown that the perceived usefulness of MOOC measures whether the content found by college students on the MOOC platform can be sufficiently effective. Based on the test results, this paper believes that the perceived usefulness of MOOCs has a positive impact on learners' attitudes to use [7, 8].

2 Proposed Method

2.1 Influencing Factors of Undergraduates' App Learning Based on TAM Model

According to the setting of the TAM model: perceived ease of use and perceived usefulness are two key variables in TAM. According to the logic of this article, perception is affected by several influencing factors of learning efficiency, namely learning motivation, learning effect, learning time cost, and cognitive load. The following is a detailed analysis of why the above model is constructed and the hypothetical relationship between variables.

(1) The relationship among cognitive load, learning time cost, and perceived ease of use

The learning costs of students can generally be summarized as the time invested and energy spent, and the energy spent is the cognitive load. MOOC learning is different from ordinary classroom learning. First, the arbitrariness of learning time and place. MOOC learning can be carried out anytime, anywhere. This form of learning can make better use of 24 h a day, so from this perspective, MOOC learning can save learning time. Second, from the perspective of the learning process, MOOC learning is different from the traditional learning method of listening and memorizing. In the past, if you took notes in the traditional classroom, if the frequency of taking notes did not keep up with the frequency of the teacher's speech, it is likely that you need to spend more time after school to organize, when using MOOC to study, the effective teaching time of students will be greatly increased, because students can pause or look back at any place they do not understand at any time, and although the actual teaching time will also increase, the overall classroom teaching efficiency will be greatly improved, and the students' harvest will also increase greatly.

(2) The relationship among learning effect, learning motivation and perceived usefulness

The learning effect of MOOCs has been questioned by many people. However, according to research, MOOC's learning effect is good, but the measurement and composition of specific learning effects need to be explored. Previous domestic research paid more attention to the indicator of learning completion rate to measure learning effects. It is more effective to measure the learning effect of MOOC from learning process and learning results. Specifically, MOOC learning must start from the first class, and it can only be completed after a series of processes such as classroom learning, homework, unit tests and final tests. Therefore, the MOOC learning process includes learning about MOOC content and participation in MOOC activities, and the MOOC learning results include homework completion results and exam results. Therefore, this article considers that the learning effect of MOOC refers to the sum of the effects achieved by the learning process of MOOC and the results of the exams and homework assignments.

Employment motivation will also have an impact on the learning of college students. With the current increase in the number of college graduates year by year, there will be huge pressure on employment。 College students in the school will learn about the employment difficulties that will come in the future from teachers, parents, and social channels. Therefore, during the school, college students will have the choice to maximize their own employment strength.

This paper believes that the above-mentioned secondary school motivation and employment motivation can be combined into future development motivation, that is, academic motivation and employment motivation are for the future development, so it can be measured by the variable of future development.

The variable of perceived usefulness is concerned with whether the content found by college students on the MOOC platform can have sufficient utility, which will determine its usefulness of perceived MOOC. As a MOOC platform, it is very important to provide effective and convenient MOOC content. Since the beginning of building the MOOC platform, a number of MOOC courses have appeared, but whether the current MOOC courses can meet the diverse needs of learners is still unknown. However, it is certain that the richness of the course greatly determines the usefulness of the learner's perception of MOOC; and the same course, the teaching form, teaching content and time are also very different, only those who can maximize The MOOC curriculum that meets the needs of students can maximize the perceived usefulness of learners and thus increase their attitude toward use.

(3) The relationship between perceived usefulness, perceived ease of use, and attitude toward use

Cognitive attitude refers to a person's very clear, precise and calm estimation of the results that behavior can produce, and see if it is beneficial to his future development. People are self-interested, so cognitive attitude is worthy of the fact that individuals will pay more attention to whether their behaviors are beneficial to themselves. As long as an individual believes that using the system or performing an action will have a beneficial effect on himself, he will tend to perform the action or use the system.

On the other hand, emotional attitude is the perceptual part of attitude, and it is the specific emotional perception of individual behavior. People are not completely rational. On the issue of individual attitude, the individual will find a balance between rationality and emotion. Emotional attitude is the most emotional part of individual attitude. If an individual behavior feels comfortable and beneficial, the individual's attitude towards the behavior will be more positive. Therefore, we will use attitude to analyze attitude from two dimensions of cognition and emotion. In this study, MOOC learners' use attitudes will be divided into two parts: cognitive attitudes and use attitudes. Individuals' cognitive attitudes and use attitudes will be examined to evaluate their use attitudes.

(4) Relationships between attitudes, behavioral intentions, and learning behaviors

Behavioral intent is the most predictive of behavior, and it represents the subjective possibility of a person performing an action. If a person has a strong intention for a behavior, then he will have a greater possibility and a stronger motivation to implement the behavior. Behavioral intention is an individual's belief in the implementation of a specific behavior, which measures the possibility of behavior.

In addition, perceived usefulness refers to how college students feel about the utility of MOOCs. The higher the perceived usefulness, the more it will have a positive effect on behavioral intentions. And several characteristics of learning motivation, that is, family expectation motivation, academic motivation and employment motivation also have a certain effect on behavioral intention. With regard to the variable of learning behavior, this article decides to use the evaluation of learning results, the amount of learning content, the amount of learning time, and the quality of learning effects to judge based on the results of previous studies.

2.2 Definition and Measurement of Variables

(1) Cognitive load

In this model, cognitive load refers to the total amount of cognitive resources that an individual spends during the entire MOOC learning process. It is mainly measured in terms of psychological effort and task difficulty.

(2) Cost of learning time

The learning time cost refers to the time investment of college students in MOOC learning. It is mainly divided into three parts: course search time, course study time, and time spent after class.

(3) Learning effect

The learning effect refers to the increase in the knowledge and ability obtained by students during the course of MOOC learning and after the completion of MOOC learning.

(4) Learning motivation

Learning motivation refers to an individual's perception of others' expectations of themselves.

(5) Perceived ease of use

Perceptual ease of use refers to the ease with which learners perceive the use of the MOOC platform.

(6) Perceived usefulness

Perceived usefulness refers to the extent to which learners perceive the value of the MOOC platform.

3 Experiments

3.1 Questionnaire Development and Data Collection

This study considers the questionnaire method to be the most suitable way for this study for the following reasons: first, the learner's own basic situation and learning background are unique; second, the types of MOOC platforms and MOOC courses that learners may be exposed to Diverse; Third, no valid secondary data is currently available. Fourth, the current past research is not deep enough, and some variables need to be redefined.

3.2 Data Collection

With regard to the selection of research subjects, the MOOC learning behavior of college students was selected as the research theme, based on the following reasons:

(1) College students are more receptive. According to the MOOC learner survey of Huohu.com, more than half of the MOOC learners are college students. Therefore, it is representative to study college students' MOOC learning behavior;
(2) The study of MOOC is mainly completed by self-study. As a group of college students who are about to enter or have reached adulthood, their self-learning ability is relatively strong and they are more inclined to use MOOC;
(3) College students who have received higher education are more capable of accepting new things and are more willing to try new learning methods.

In terms of data collection, due to the network characteristics of MOOC learning, this article collects the questionnaires through targeted delivery of questionnaire star sample services. On the one hand, the reliability of the sample can be ensured to a certain extent, and on the other hand, the authenticity of the data can be improved as much as possible through control of answer time and network IP control. A total of 300 responses were received after the questionnaire data was collected. The screening process of the invalid questionnaires on the collected content is as follows: first, incomplete questionnaires; second, questionnaires suspected of being fake; third, for questionnaires with regular options, such as the answer option is checked as Questionnaires such as 1, 2, 3,

1, 2, and 3 are considered invalid questionnaires. Since this questionnaire was collected through the questionnaire sample service, no data was missing.

After excluding invalid questionnaires according to the above methods, a total of 255 valid questionnaires were collected in this survey, with a questionnaire effective rate of 72.6%.

4 Discussion

4.1 Reliability Test

In general exploratory studies, Cranbach's a coefficient is above 0.6, and benchmark studies are above 0.8. Under normal circumstances, Cranbach's a coefficient is above 0.6, which is considered to have high reliability. According to the analysis, the a coefficient of this study is 0.866. The reliability test of specific latent variables is shown in Table 1:

Table 1. Reliability tests

Latent variable	Cronbach's Alpha
Overall reliability	0.846
Cognitive load	0.725
Cost of learning time	0.733
Learning result	0.803
Learning motivation	0.798
Perceived ease of use	0.758
Perceived usefulness	0.816

As shown in Table 1 above, the results show that the a coefficients of all the latent variables mentioned above are within the acceptable range. This shows that the data collected in this research has good reliability and reliable internal consistency.

4.2 Descriptive Statistical Analysis

Sort out the collected content. From the perspective of gender, grade, MOOC platform, time length of learning, professional category, and individual professional performance, some contents are as follows: 115 of the 255 questionnaires were male, accounting for 45.1%, 140 women, accounting for 54.9% of the total. In the surveyed grades, freshmen accounted for approximately 27.6%, sophomores accounted for approximately 36.7%, sophomores accounted for approximately 27.6%, and seniors accounted for approximately 8.1%. Therefore, we can learn that sophomores and juniors are more inclined to study MOOCs. As shown in Fig. 1 below.

As shown in Fig. 1 above, in terms of the use of the MOOC platform, the MOOC platforms used by Chinese universities MOOC, Curtain Net, Super Star Curriculum, and MOOC College are the most used. From the time of participating in MOOC study,

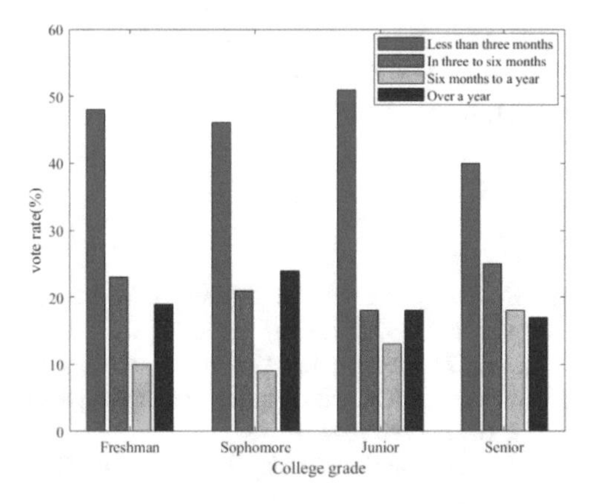

Fig. 1. MOOC learning model statistics

48% of people have used MOOC for less than three months, 23% of people have studied MOOC in three to six months, 10% of people have studied MOOC for half a year to one year, and participated in MOOC People who have studied for more than one year account for 19%.

In addition, from the perspective of the majors of college students participating in MOOC studies, liberal arts students accounted for 37%, science students accounted for 33%, and engineering students accounted for 30%. From the perspective of their personal professional performance, the top 30% of students have a professional score of 46.7%, and the professional scores of between 30% and 60% are 24.5%, so we can see that the higher the professional scores of college students. The more inclined to learn MOOC.

4.3 Research Conclusions and Recommendations

(1) Research shows that the more interesting mobile learning is, the more learners like to use mobile learning, and whether a person feels interesting during the learning process depends on whether the difficulty they feel is comparable to their own knowledge and skills.

(2) Because social influence also has a significant impact on learners' willingness to use mobile learning. If learners experience the convenience of mobile learning in the process of mobile learning, Superiority, they are likely to recommend their classmates and friends to use it.

5 Conclusions

On the TAM model, the integration model of the learning efficiency model and technology acceptance model adopted by this institute is used to explore the influencing factors of college students' MOOC learning behavior. This research is only a preliminary attempt. For the study of college students' MOOC learning behavior, there may be many important variables that cannot be considered. All of this looks forward to further research in the future.

References

1. Chee, K.N., Ibrahim, H.: Designing mobile learning communication aid as an android app. Adv. Sci. Lett. **22**(12), 4023–4027 (2016)
2. Rose, S., Shah, B.J., Onken, J.: Bridging the G-APP: continuous professional development for gastroenterologists: replacing MOC with a model for lifelong learning and accountability. Gastroenterology **13**(11), 1872–1892 (2015)
3. Yanik, E., Sezgin, T.M.: Active learning for sketch recognition. Comput. Graph. **52**(C), 93–105 (2015)
4. Zhang, X., Chen, B., Liu, H.: Infinite max-margin factor analysis via data augmentation. Pattern Recogn. **52**(C), 17–32 (2015)
5. Deshpande, S., Chahande, J., Rathi, A.: Mobile learning app: a novel method to teach clinical decision making in prosthodontics. Educ. Health **30**(1), 31 (2017)
6. Paiz, F., Bonin, E.A., Cavazzola, L.T.: Surgical learning application (app) for smartphones and tablets: a potential tool for laparoscopic surgery teaching courses. Surg. Innovation **23**(1), 106 (2015)
7. Khaddage, F., Müller, W., Flintoff, K.: Advancing mobile learning in formal and informal settings via mobile app technology: where to from here, and how? Educ. Technol. Soc. **19**(3), 16–26 (2016)
8. Sharples, M., Aristeidou, M., Villasclaras-Fernández, E.: The sense-it app: a smartphone sensor toolkit for citizen inquiry learning. Int. J. Mob. Blended Learn. **9**(2), 16–38 (2017)

Malicious JavaScript Detection Method Based on Multilayer Perceptron

Zhenyuan Guo(✉)

School of Software, Zhengzhou University, Zhengzhou 450002, Henan, China
relwayg@stu.zzu.edu.cn

Abstract. JavaScript-based attacks have become the major threat to web security. Because of complexities and obfuscation of JavaScript code, it is even more difficult to manually generate digital signatures, thus the traditional anti-virus scanners can hardly detect malicious JavaScript effectively. In the paper, the characteristics of JavaScript code and extract parts of features to distinguish benign from malicious were analyzed, then we propose a method based on multilayer perceptron to detect malicious JavaScript code. This method can effectively detect the JavaScript-based attacks. Experimental results indicate that the average classification precision achieves 98.8% and the Average rate of false positives about 3%. The various performance indicators surpass the other machine learning-based methods.

Keywords: Malicious JavaScript · Static analysis · Machine learning · Multilayer perceptron

1 Introduction

Malicious JavaScript code has become the major threat to the Internet web security. Millions of users are exposed to malicious JavaScript attacks on the Internet every day. As a scripting language, JavaScript is used in Web development, but because Java-Script allows you to programmatically scan and exploit vulnerabilities in the browser, it also provides a perfect platform for these attacks. A piece of malicious JavaScript code can redirect the page viewed by the user to a malicious site, thereby downloading and installing some malicious software [1]. Compared with the others' network attacks, malicious JavaScript codes will evade the detection of conventional antivirus software through the code obfuscations technology [2]. As JavaScript is an interpreted language, it does not need the browser to compile and runs directly. This nature of JavaScript enables attackers to obfuscate their malicious code more effectively. JavaScript code can be executed silently when a web page is accessed by the user using the browser. Hackers also try to lure users to their malicious sites, for example by creating links with popular, attractive titles to entice them to click through. As for the traditional blacklist technology [3], if the attacker frequently changes his IP address and URL [4], this method will be completely ineffective, and the continuous updating of the blacklist will also require more system overhead.

Based on the execution state of malicious JavaScript, the existing detection methods are divided into static analysis and the dynamic analysis. Static analysis is the

J. MacIntyre et al. (Eds.): SPIoT 2020, AISC 1282, pp. 513–520, 2021.
https://doi.org/10.1007/978-3-030-62743-0_74

practice of not executing code files and analyzing code features and code structures to identify malicious JavaScript code. For example, in literature [5], the author USES the method of malicious signature matching to detect malicious JavaScript. The disadvantage of this method is that it is difficult to detect unknown attacks. Dynamic analysis refers to run the script code in a controlled environment, through code to be inspected after the execution of the state, for example in the literature [6], the author by monitoring system status, network connection state, as well as the system registry file to detect abnormal process, the method is due to the malicious JavaScript code to run directly, it will increase the risk of system is under attack, and the test efficiency is also a problem.

Since malicious JavaScript code will have some specific characteristics, this paper extracts the characteristics that can distinguish between malicious and normal code through static code analysis, and then classifies them through the multi-layer perceptron algorithm in machine learning to detect malicious JavaScript codes. The work is shown as the follow: 1) Static analysis of malicious JavaScript code features, extract the relevant distinguishing features. 2) Based on these characteristics, a classification algorithm based on multi-layer perceptron to identify malicious JavaScript code was proposed. 3) The method was characterized on a specific data set. 4) Compared with the related machine learning methods.

The rest contents of this article are listed as follows: section two analyzes malicious and benign JavaScript code and extracts some important features. In the third section, the algorithm principle of multilayer perceptron is given. The fourth section gives the experimental results. The fifth section summarizes the whole paper.

2 Feature Extraction of Javascript

For text feature extraction, the simplest method is to use the unary or binary grammar model [7]. This method is very effective in natural language processing, such as naive Bayes or SVM in the application of spam filtering. However, for JavaScript code, the use of this method will not only produce high-dimensional feature vectors to increase the difficulty of calculation, but also produce a lot of meaningless features. Furthermore, the syntax model ignores the syntax structure of JavaScript, and some distinguishing features will be ignored, such as the large number of non-standard encoded strings or hexadecimal numeric variables in some malicious JavaScript code, which will increase the length of the script file. Therefore, we abandoned the grammar model, carried out feature analysis based on JavaScript feature elements [7].

This paper extracts a useful feature set to vectorize malicious and normal code samples, and finally classified them through a classifier. Microsoft's JavaScript language reference contains all the JavaScript feature elements, including methods, attributes, functions, and other keywords. The most common means of concealment is the use of obfuscation techniques, which are used to make a program more difficult to understand by changing the composition of characters. Techniques can be expressed into the following four types:

1) Confusion between space character randomization and annotation randomization. By randomly inserting space, TAB, line break, page change and carriage return characters in JavaScript, the binary stream representation of the code is changed to a great extent without changing the code semantics, which can avoid many detection techniques of opportunity content matching.
2) Confusion between randomization of variable names and rewriting of function Pointers. In JavaScript code, variable and function names are often replaced by strings that do not have a specified meaning, which makes the code less readable. Because the structure does not change, the detection difficulty is increased.
3) String confusion. String obfuscation is one of the most popular obfuscation techniques. There are two ways to achieve string recombination: string splitting and character encoding. In JavaScript code, strings are often encoded in hexadecimal or double-byte encoding standard (Unicode) to hide the meaning of strings, and special functions such as eval(), escape(), unescape(), document.write(), string. charcode() and so on are used to reorganize and encode strings.
4) Numerical confusion. Numerical obfuscation is another commonly used obfuscation technique. In some malicious code, the attacker to access to user resources, in the code will be used to some specific memory address, which in some detection mechanism will be the use of these memory address identified as suspicious or malicious operation, so the attacker in order to better hide their intentions will use numerical confusion technology to cover the use of these specific address.
5) Confusion of logical structure. On the premise of not changing the semantics, attackers will change the logical structure of JavaScript code by changing its execution process, making code analysis more complex and not conducive to detection.

In different ways, the combination of eval() and unescape() functions is often used for code obfuscation, replacing the obfuscating code with a malicious JavaScript snippet by means of character substitution, inserting the symbol "%" at the appropriate place in the obfuscation code with a for loop, and converting the hexadecimal code to the actual string using the unescape() function.

Table 1. The percentage of malicious and normal JavaScript from the code made up of different functions

Functions	eval ()	unescape ()	Combining with function of eval () and unescape()	XMLHttpRequeset
Malicious JavaScript	8%	8%	8%	3%
Normal JavaScript	2%	27%	0%	90%

From the Table 1 we can see, the single function does not make distinction between malicious and the normal sample, whereas a combination of them makes a clear distinction between malicious and normal samples. If the JavaScript code communicates with other servers through the XMLHttpRequest object in the DOM (document object model), it is likely to be malicious because some information about the user, including the user name and password, will be passed to the target server. By analyzing the JavaScript code, we chose four important features: 1) JavaScript code basic statistics attribute characteristics. (2) URL redirect feature. 3) Characteristics of the attack process. 4) Obfuscation features. Finally, the JavaScript code sample is parsed through the JavaScript debugger tool to obtain a vectorized sample set. Through analyzing, we select features set including: numbers of substring(), numbers of fromCharcode(), numbers of eval(), numbers of unescape(), composite numbers of eval() and unescape(), numbers of setTimeout(), numbers of document.write(), and numbers of creat.Element(), numbers of escape(), numbers of link(), numbers of exec(), numbers of iframe, the average length of the string, script file size, numbers of CreatObject(), numbers of ActiveXObject(), numbers of parseInt(), numbers of fromcharcode(), hex or octal numbers, and numbers of tag string in JavaScript code, etc.

3 Methods

3.1 Algorithm Principle of Multi-layer Perceptron

A typical multilayer perceptron structure is shown in Fig. 1. The multi-layer perceptron in single hidden layer is a function mapping process [8], shown as below formula (1).

$$\mathbf{y} = f(x) + z \tag{1}$$

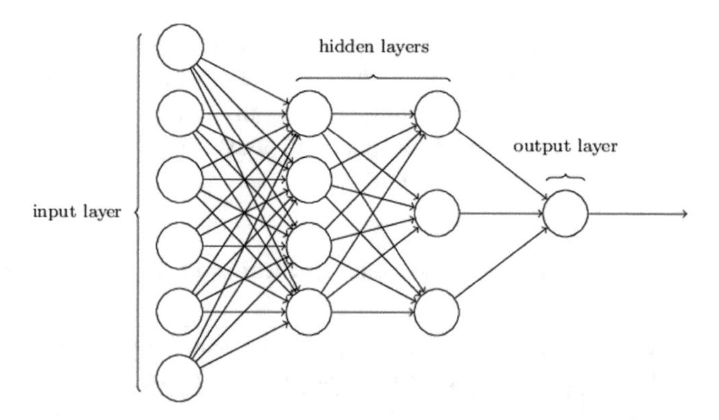

Fig. 1. The structure diagram of multi-layer perceptron

In formula (1), *D* is the sizes of the input vector of *x*, and L is the size of the output vector *f*(*x*). The concrete implementation of the network structure is shown as follows:

$$h(x) = s(W^{(1)}x + b^{(1)}) \tag{2}$$

$$f(x) = s(W^{(2)}h(x) + b^{(2)}) = s(W^{(2)}(s(W^{(1)}x + b^{(1)}) + b^{(2)})) \tag{3}$$

Where, $h(x)$ are output values of the hidden layer node, $W^{(1)} \in R^{D \times D_h}$ is the weight vector between input layers and hidden layers, $W^{(2)} \in R^{D_h \times D_l}$ is the weight vector between hidden layers and output layers, $b^{(1)}$ and $b^{(2)}$ respectively is the bias vector. Nonlinear activation function $s()$ is hyperbolic tangent function or *Sigmoid* function. In this paper, we choose a hyperbolic target function as an activation function, whose form is:

$$tanh(x) = \frac{e^x - e^{-x}}{e^x + e^{-x}} \tag{4}$$

Then we use stochastic gradient descent to send the parameters $\{W^{(1)}, W^{(2)}, b^{(1)}, b^{(2)}\}$ in the solution network. Supposed n samples of data set, then the loss function of the whole is:

$$L(x, y) = \frac{1}{n}\sum_{i=1}^{n}\left(\frac{1}{2}(h(x_i) - y_i)^2\right) \tag{5}$$

The gradient descent method updates the parameters according to the following rules:

$$W = W - \alpha\frac{\partial L(x, y)}{\partial W}, \quad b = b - \frac{\partial L(x, y)}{\partial b} \tag{6}$$

Where, α is the learning rate, we use traditional back-propagation algorithms to calculate the value [9].

3.2 Classification Model Design

The training and sample testing of the model in this paper are conducted in the environment of WEKA [9], and what WEKA needs is the ARFF file format. According to the selected characteristics, the JavaScript sample set is firstly represented by vectomization, then the sample set after vectomization is stored in the Mysql database, and the data set is transferred into the CSV format by PhpMyAdmin [10], then the CSV format file is converted into the ARFF format file we need, and finally the algorithm provided in WEKA is applied for experiments.

4 Experimental Results

In order to verify the performance of the multilayer perceptron in the detection of obfuscating malicious code, we consider the accuracy, computation speed, robustness, scalability and interpretability. In the field of malicious code detection, evaluation indexes are often used to evaluate the performance of classifiers in the accuracy. Precision, Recall, FN and FP were used.

4.1 Data Set Description

The experimental data set consists of 1500 malicious JavaScript and 1500 benign JavaScript. The data set was divided into three equal parts, with two-thirds of the samples used for the training set and one-third for the test set.

4.2 Multi-layer Perceptron Model Architecture

For the multi-layer perceptron network with three-layer structure, numbers of nodes in input layers are 30, and numbers of nodes in output layers are 2, so numbers of nodes in the hidden layer and numbers of training iterations need to be determined. For the selection of network parameters, there is no method to set parameters at present, and the optimal parameter configuration can only be found through continuous experiments. First, numbers of training iterations were randomly selected to be 30 and fixed unchanged, and influences of hidden layer nodes on the classification results were tested. The number of hidden layer nodes was selected from 3 to 30. When numbers of hidden layer's nodes increase from 3 to 12, classification error rate decreases, and when numbers of nodes exceed 12, the classification error rate will increases. The optimal choice of the model is the number of hidden layer nodes of 12, and the lowest classification error rate is 5.3%.

We measured the effects of the different training iterations on the classification results. Fixed network node number of hidden layers is 12, training iterations selected from 20 to 100. The classification error rates will decrease with the increase of the number of training iterations. When the number of iterations reaches 75, the classification error rate will decrease to 1.8%. However, as the number of training iterations continues to increase, the classification error rate does not decrease significantly, but the system cost and operation time increase exponentially. Therefore, it can be seen from the above that the classification effect of the model is optimal when numbers of nodes in the hidden layers are 12 and numbers of training iterations is 75.

4.3 Model Comparison

In order to reflect the advantages of multi-layer perceptron in classification effect, it was compared with supervised classification model SVM [11]. Among them, both the training set and the test set are the same data set, and the experimental results are the average values after 10 independent experiments. Table 2 shows the classification accuracy of different classifiers. Not only can it be seen that the classification accuracy of these four classifiers is more than 95%, which proves that feature selection is very effective, but the classification accuracy of MLP of multi-layer perceptron is completely superior to other classifiers. Table 2 respectively express the classification result of these three classifiers under ten continuous tests.

Table 2. Different classification performance result of these four classifiers

Classifiers	Precision	FN	FP
MLP	98.8%	3.6%	1.6%
Naïve Bayes	96.3%	15.4%	1.1%
SVM	97.5%	9.1%	0.7%

From the Table 2 we can see, the classification accuracy of classifier MLP, Naïve Bayes, and SVM is 98.8%, 96.3%, and 97.5%. The average false rate based on the measurement models of multi-layer perceptron is only about 3%, and the average false alarm rate of SVM is about 8%. The experimental results show that the features extracted from the static JavaScript code analysis are significantly helpful to the classification results. In the comparison process of four different models, the detection model based on multi-layer perceptron not only achieves the highest classification accuracy and true positive detection rate, but also the lowest average false positive rate. Therefore, for malicious JavaScript detection in this paper, the multilayer perceptron is the optimal model.

5 Conclusions

Based on the static analysis of JavaScript code, this paper proposes the main features that distinguish malicious from normal JavaScript code, and then establishes the classification model of multi-layer perceptron based on these features. Experimental results shown model paper proposed can identify malicious JavaScript code well and has a low false alarm rate.

References

1. Schutt, K., Kloft, M., Bikadorov, A., Rieck, K.: Early detection of malicious behavior in JavaScript code. In: Proceeding of ACM Workshop on Security and Artificial Intelligence, pp. 15–24 (2012)
2. Wenpeng, Q.U., Zhao, L., Deng, X.: Multi-feature detection identification and analysis of obfuscated malicious JavaScript code. Intell. Comput. Appl. **8**(4), 47–52 (2018). (in Chinese)
3. Zhang, H.L., Zou, W., Han, X.H.: Drive-by-download mechanisms and defenses. Ruanjian Xuebao/Jonmal Softw. **24**(4), 843–858 (2013). (in Chinese)
4. Tun, W.: Research of webpage redirection spam technology based on Javascript. Comput. Digit. Eng. **40**(3), 86–88 (2012)
5. Kashyap, V., Dewey, K., Kuefner, E., Wagner, J.: JSAI: a static analysis platform for JavaScript. In: ACM SIGSOFT International Symposium on FSE, pp. 121–132, November 2014
6. Egele, M., Scholte, T., Kirda, E., Kruegel, C.: A survey on automated dynamic malware-analysis techniques and tools. ACM Comput. Surv. **44**(2), 1–42 (2012)
7. Wressnegger, C., Schwenk, G., Arp, D., Rieck, K.: A close look on n-grams in intrusion detection: anomaly detection vs. classification. In: Proceeding of ACM Workshop on Security and Artificial Intelligence, pp. 67–76 (2013)

8. Adlakha, P., Subramanium, P.: Detection system using multilayer perceptron based on artificial neural network. Int. J. Adv. Res. Comput. Sci. Softw. Eng. 910–916 (2013)
9. Markov, Z., Russell, I.: An introduction to the WEKA data mining system. ACM SIGCSE Bull. **38**(3), 367–368 (2006)
10. Fakheraldien, M.A.I., Zain, J.M., Sulaiman, N.: An efficient middleware for storing and querying XML data in relational database management system. J. Comput. Sci. **7**(2), 314–319 (2011)
11. Raikwal, J.S., Saxena, K.: Performance evaluation of SVM and K-nearest neighbor algorithm over medical data set. Int. J. Comput. Appl. **50**(14), 35–39 (2012)

Computational Simulation of Magnetic Field of FC-Mold

Fei Li(✉), Ying Chen, and Wencheng Liu

College of Mechanical Engineering, Jilin Engineering Normal University, No. 3050 Kaixuan Road, Changchun, China
fayeleeneu@qq.com

Abstract. Applying the computed software ANSYS, a 3D mathematical model of magnetic field inside the mold with FC-Mold was bewrited, in order to achieve the magnetic induction results inside the mold with the FC-Mold. The distribution features of magnetic induction results of liquid steel inside the mold with the FC-Mold was resolved and discussed. The explanation of simulation results was that the FC-Mold equipment can form a stabilizing magnetic field, the magnetic field inside the liquid steel had adequate intensity, this valid magnetic field engendered by the FC-Mold was imploded in the overlay region of magnetic pole mainly. The valid magnetic field enshrouded the region of total mold width. In the external part of the overlay region of magnetic pole, the magnetic induction intensity was slightly smaller. The magnetic intensity distribution was basically even toward the width and thickness directions of mold. Toward the direction of mold height, the magnetic intensity was appeared parabolic shape in the overlay region of magnetic pole, the two peaks of magnetic intensity appeared in the center of the overlay region of magnetic pole, and then it gradually declined from the mold top toward the mold bottom outside the overlay region of magnetic pole.

Keywords: Magnetic intensity · Electromagnetic brake · Casting mold · Simulation

1 Introduction

Electromagnetic brake technology is one of the modern electromagnetic continuous casting technologies. Electromagnetic fluid flow control technique is able to control the flow of liquid steel in mold effectively, remove inclusions, and purify molten steel [1–5]. Especially in slab continuous casting, the technology has achieved good metallurgical effects. With the development of electromagnetic brake technology, FC-Mold has been widely used in factories at home and abroad, it can control fluid flow by arranging magnetic fields at critical locations to reduce or avoid defects in the continuous casting process [6–10]. Thereby, FC-Mold reduce or avoid defects generated during continuous casting.

FC-Mold have upper and lower pairs of magnetic poles. The magnetic poles of FC-Mold have a larger coverage area in the width direction of mold. If the magnetic fields of the upper and lower magnetic poles are not well matched, may cause non-melting of slag and non-removal of non-metallic inclusions.

J. MacIntyre et al. (Eds.): SPIoT 2020, AISC 1282, pp. 521–526, 2021.
https://doi.org/10.1007/978-3-030-62743-0_75

The current intensity will directly affect the magnetic field distribution in mold with FC-Mold, so it affects the brake effect of the main jet of molten steel directly. Which can affect the flow pattern in mold. Therefore, the magnetic field inside the mold is studied in this paper. It is a great significance to provide the theoretical basics for discussing metallurgical effects with FC-Mold mold.

2 Mathematical Model

2.1 Governing Equations of Magnetic Field

Since the magnetic field engendered by the electromagnetic brake is a constant DC magnetic field, and the liquid conductivity inside the mold is high, and the free charge is relatively short in the mold, the Maxwell equations for calculating the magnetic field inside the mold are as follows.

$$\nabla \cdot B = 0 \tag{1}$$

$$\nabla \times H = J \tag{2}$$

$$\nabla \times E = 0 \tag{3}$$

$$\nabla \cdot D = 0 \tag{4}$$

In Eqs. (1)–(4), B denotes the magnetic induction intensity, H denotes the strength of magnetic field, J denotes the c density of conduction current, E denotes the strength of electric field, D denotes the electric induction density.

2.2 Calculated Grid of Electromagnetic Field

The 3D model was built by ANSYS software, the model included air, iron core, coils, mold and liquid steel. The width of mold is 1450 mm, and the thickness of mold is 230 mm. The grid diagram is shown in Fig. 1.

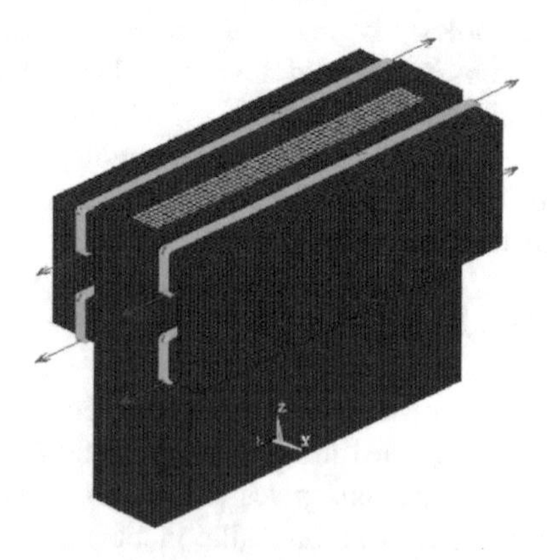

Fig. 1. Finite element mesh diagram of calculated electromagnetic field of FC-Mold

2.3 Solution Method

First, the basic governing equations were determined. And then the 3D model was built. Completion of the previous two steps, set the magnetic induction intensity to zero at the unlimited region of the electromagnetic brake equipment. We used the mean of finite element method to calculate the magnetic field inside the mold. If the calculated results meet the set convergence criteria, the calculated program completed automatically.

3 Calculation Results and Discussion

Figure 2 shows the magnetic intensity distribution of the liquid steel (half volume and whole volume) inside the mold when the lower coil was energized at 850 A and the upper coil was energized at 425 A during the FC-Mold electromagnetic braking process.

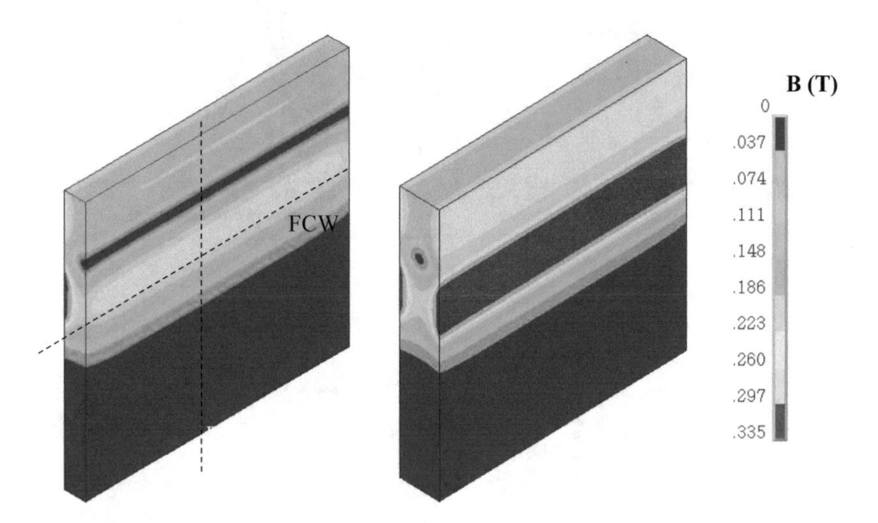

Fig. 2. Distribution of magnetic induction intensity inside mold with FC-Mold

In the Fig. 2, the magnetic induction intensity generated by the FC-Mold electromagnetic brake was imploded in the covered regions of magnetic poles primarily, and the lower magnetic induced intensity was rapidly attenuated at the lower edge of the lower magnetic poles. Since the lower coil current was larger than the upper coil current, the magnetic induction intensity was larger in covered region of lower magnetic pole. And a region with the weakest magnetic induction intensity was formed between the upper and lower magnetic pole covered regions. In the direction of mold thickness, since the casting billet surface was close to the magnetic poles and the center of the casting slab was far away from the magnetic poles, in the liquid steel the magnetic induction intensity gradually decreased from the surface of slab to the center of the slab.

In order to analyze the magnetic intensity distribution of the liquid steel in the width direction, in the height direction and thickness direction of mold, the point of intersection between the straight line FCW and the straight line FCH in Fig. 2 was marked, and it was the midpoint of the straight line FCW. Then make a vertical line of the two lines (FCW and FCH) through the intersection, and its length value was the mold thickness, which was denoted as FCT. The magnetic intensity distribution of the straight lines (FCW, FCH and FCT) was shown in Fig. 3 respectively. The diversification of magnetic induction intensity in three directions of mold can be analyzed from Fig. 3.

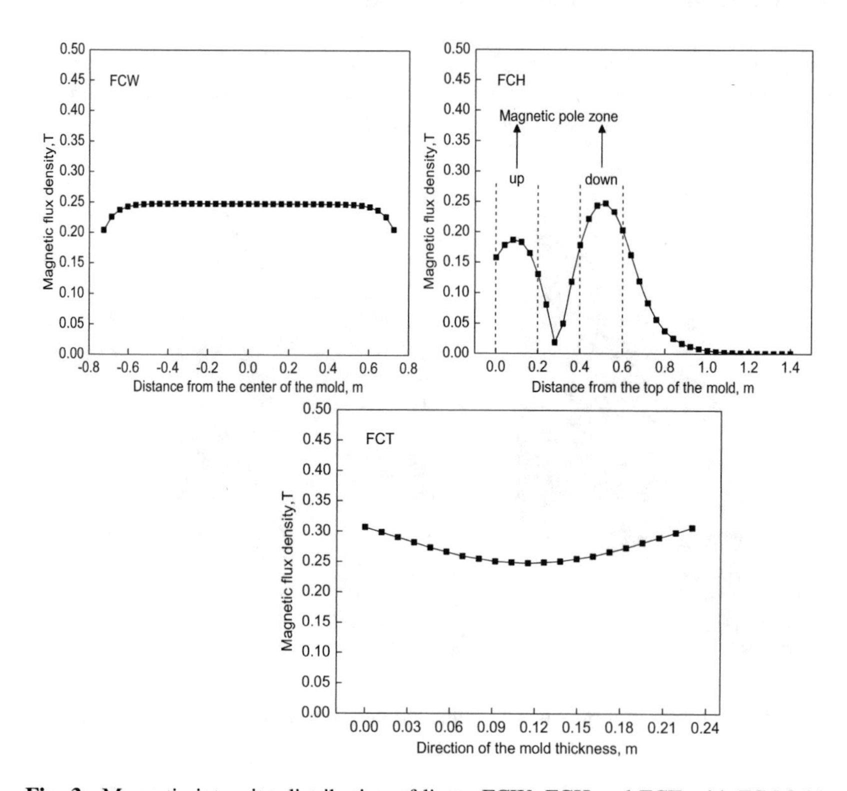

Fig. 3. Magnetic intensity distribution of linear FCW, FCH and FCT with FC-Mold

As can be drawn from Fig. 3 that the magnetic intensity distribution of the three feature lines was identical to the results displayed in Fig. 2. The respective characteristics are as follows.

In the direction of mold width (FCW), the magnetic intensity distribution was substantially even. At the edge of magnetic pole coverage position, it is reduced due to the leakage magnetic induction intensity.

In the direction of mold height (FCH), the magnetic intensity was mainly imploded in the covered regions of upper magnetic pole and lower magnetic pole. In the middle regions of the magnetic poles, there was two peaks of the magnetic intensity appeared and the magnetic induction intensity in the lower magnetic pole covered region was larger. In the external part of the covered region of lower magnetic pole, the magnetic intensity declined gradually from the mold top toward the mold bottom, it decreased to a certain value between the upper and lower magnetic poles and then gradually increased, and the magnetic induced intensity increased to a second peak in the middle of the upper magnetic pole coverage region.

In the thickness direction of mold (FCT), the distribution of the magnetic induction intensity was an arc that was slightly curved upward, and at the center of the thickness it was the smallest. From the center of the thickness to left edge or right edge of the slab, the magnetic induction intensity strengthened gradually, and it at the two edges had the same value.

4 Conclusions

The FC-Mold equipment can form a stabilizing magnetic field, the magnetic field had adequate intensity inside the liquid steel. The valid magnetic field engendered by the FC-Mold was imploded in the covered region of magnetic poles mainly. The valid magnetic field enshrouded the region of total mold width. In the external part of the covered regions, the magnetic field was slightly smaller. The intensity of magnetic field dramatic decreased far from the covered regions of magnetic poles. The magnetic intensity distribution was basically even toward the direction of mold width and the direction of mold thickness. Toward the direction of mold height, the intensity of magnetic field was appeared parabolic shape in the covered regions of magnetic poles. There were two peaks of magnetic intensity appeared in the center of the covered regions of magnetic poles. And then it gradually declined from the mold top toward the mold bottom outside the covered regions of magnetic poles.

Acknowledgments. This research was supported by Doctoral Research Funding Project of Jilin Engineering Normal University (Grant No. BSKJ201810) and Program for Innovative Research Team of Jilin Engineering Normal University.

References

1. Cukierski, K., Thomas, B.G.: Flow control with local electromagnetic braking in continuous casting of steel slabs. Metall. Mater. Trans. B **39**(1), 94–107 (2008)
2. Harada, H., Takeuchi, E., Zeze, M., Ishii, T.: New sequential casting of different grade of steel with a level DC magnetic field. Tetsu-to-Hagane **86**(4), 278–284 (2000)
3. Idogawa, A., Sugizawa, M., Takeuchi, S., Sorimachi, K., Fujii, T.: Control of molten steel flow in continuous casting mold by two static magnetic fields imposed on whole width. Mater. Sci. Eng. A **173**(1–2), 293–297 (1993)
4. Timmel, K., Eckert, S., Gerbeth, G.: Experimental investigation of the flow in a continuous-casting mold under the influence of a transverse, direct current magnetic field. Metall. Mater. Trans. B **42**(1), 68–80 (2011)
5. Chaudhary, R., Thomas, B.G., Vanka, S.P.: Effect of electromagnetic ruler braking (EMBr) on transient turbulent flow in continuous slab casting using large eddy simulations. Metall. Mater. Trans. B **43**(3), 532–553 (2012)

6. Singh, R., Thomas, B.G., Vanka, S.P.: Effects of a magnetic field on turbulent flow in the mold region of a steel caster. Metall. Mater. Trans. B **44**(5), 1201–1221 (2013)
7. Ji, C.B., Li, J.S., Tang, H.Y., Yang, S.F.: Effect of EMBr on flow in slab continuous casting mold and evaluation using nail dipping measurement. Steel Res. Int. **84**(3), 259–268 (2013)
8. Chen, Z.H., Wang, E.G., Zhang, X.W., He, J.C.: Study on the effect of electromagnetic brake control on liquid flow in continuous casting slab mold. Res. Iron Steel **33**(5), 11–14 (2005). (in Chinese)
9. Yamamura, H., Toh, T., Harada, H., Takeuchi, E., Ishii, T.: Effect of magnetic field conditions on the electromagnetic braking efficiency. ISIJ Int. **41**(10), 1236–1244 (2001)
10. Li, F., Wang, E.G., Feng, M.J., Li, Z.: Simulation research of flow field in continuous casting mold with vertical electromagnetic brake. ISIJ Int. **55**(4), 814–820 (2015)

Construction of Simulated Seismic Experiment (Experience) Platform

Chen Wu(✉), Tingting Guo, Hao Zhang, Le Yang, and Jinlong Song

Shandong Seismological Bureau, Jinan 250014, Shandong, China
wuchen800911@163.com

Abstract. At present, domestic vibration platforms used to simulate earthquakes can be roughly divided into two kinds according to their uses. One is for scientific research (experiment) purposes; the other is for experience (popular science) purposes. Because of their different uses, the technical requirements, technical specifications and technological treatment of the two vibration platforms are quite different. With the increasing attention paid to popular science propaganda of earthquake prevention and disaster reduction from the national level to the public, the departments of earthquake prevention and disaster reduction need to increase their investment in the work of popular science propaganda while doing their work well. Then the simulated seismic platform which can combine experiment and experience will have broad application prospects in the seismic industry. However, this research is still too few in the seismic industry at present. This is the case. In this paper, the background, construction scheme and theoretical technology of the research are introduced, and the basic principles and key technologies of the platform are analyzed and studied. Several key issues of the construction project, such as load and acceleration, are also discussed in the light of the construction requirements of the platform in Shandong Earthquake Emergency Rescue Training Base.

Keywords: Acceleration · The maximal displacement · Degrees of freedom of movement · Frequency · The load · Static\dynamic output

1 Introduction

In order to better develop the cause of earthquake prevention and disaster mitigation in Shandong Province, our bureau began the construction of the earthquake emergency rescue training base in Shandong Province in 2013. The platform of simulated earthquake experiment (experience) is located on the underground floor of the comprehensive training building of the training base. The purpose of the platform project is twofold. Firstly, the seismic performance of building components and building models is tested by simulated earthquake [1]. Thus the corresponding data are obtained for performance analysis [2]. The second is to make the trainees (visitors) feel the real earthquake situation to the greatest extent. The platform should be able to simulate several real earthquakes in history, such as Tangshan earthquake, Wenchuan earthquake, Osaka-Shenzhen earthquake, Japan, and so on, and combine simulation scenarios to achieve the best experience effect, so that the experiencer can have a deeper and intuitive understanding. The operating system of the platform requires openness,

J. MacIntyre et al. (Eds.): SPIoT 2020, AISC 1282, pp. 527–535, 2021.
https://doi.org/10.1007/978-3-030-62743-0_76

that is, if other (newly occurring) seismic waveforms and ground motion parameters are input, the platform can also simulate the seismic effect [3]. Aiming at the design background, requirement and advanced construction scheme of the new simulated shaking table, as well as the technical advancement of the industry, this paper gives a comprehensive introduction to the simulated seismic experimental platform.

2 Key Parameters of Vibration Platform

The main parameters of the shaking table include: load, size, frequency, displacement, degree of freedom and acceleration, etc. Any change of these parameters will directly affect the investment cost of the whole shaking table and the scale of related supporting systems [4]. Therefore, we should build a moderate scale shaking table according to the actual needs. Of course, if there are special needs, we can also expand the construction of the shaking table system appropriately Scale up.

2.1 Mesa Size

If the shaking table system is based on civil buildings, the larger the platform is, the smaller the specimen shrinkage is, and the experimental accuracy can be improved. At the same time, as the total mass of the test piece and the table increases, the output of the vibration intensifier and the capacity of valves and pumps will be increased. The direct consequence is to increase the cost of the whole system [5]. Therefore, the size of shaking table should be determined according to the object of shaking table. If large equipment is used as experimental object, the size of shaking table can be selected from 2 m * 2 m to 3 m * 3 m. If civil structure is used as experimental object, the size of shaking table can be selected from 4 m * 4 m to 6 m * 6 m. Except for a few special shaking table with high frequency, the most large shaking table is made of steel material box. Type structure. The drum type natural frequency of the steel platform is generally low, and must be controlled more than three times of the working frequency in design. Its mass is generally about two-thirds of the load.

2.2 Mesa Load

The load of the platform is the same as the size of the platform, which directly determines the input cost of the system. It is very important to choose the reasonable load of the mesa. It is generally used in the shaking table of large equipment. It is economical to use 5 * 104 N to 10 * 104 N. It is suitable for the mesa of civil construction to use 20 * 104 N to 50 * 104 N.

2.3 DOF of Motion

The number of vibration absorbers, power supply, pump source, valve and sample acquisition channels and control channels will be affected by the degree of freedom of motion of the shaking table, so it is an important factor affecting the cost. Generally speaking, if it is a shaking table for the purpose of large-scale equipment experiment, choose a horizontal and a vertical two-degree-of-freedom, as long as the specimen rotates 90° in the horizontal direction of the table, another horizontal-degree-of-freedom test can be carried out; if it is a shaking table for the purpose of civil construction, it needs the specimen to reflect the effect of three-dimensional seismic components. It can not be replaced by one-dimensional and two-dimensional vibration [6]. Only three-dimensional motion can reproduce the three-dimensional records of earthquakes at the same time. Therefore, three-dimensional vibration is generally used in large-scale shaking table for civil buildings. Since the increase of freedom of movement will greatly affect the cost, we should pay special attention to the balance between the two.

2.4 Acceleration, Velocity and Displacement of Motion

Acceleration is a very important index of vibration platform, because the output value of the vibrator is directly determined by the magnitude of acceleration, and acceleration can also indirectly affect the capacity of pumps, valves and power supply. According to the relevant experimental specifications, if the shaking table for the purpose of civil building experiment can be determined by the size and load of the table, then the required acceleration can be calculated according to the similarity theory. If the shaking table for the purpose of large-scale equipment experiment is used, the acceleration of horizontal excitation should be 4.5 g < 5.0 g, and the acceleration of vertical excitation should meet the relevant requirements of 3G < 3.5 g. Usually in the case of no-load (or small-tonnage specimens), the main horizontal acceleration of the shaking table can reach 4.5 g or more if the size of the table is more than 4 m * 4 m, and the vertical acceleration is more than 3.0 g. If the shaking table is full-loaded, the horizontal acceleration of the table can reach 1.2 g or more than 0.8 g.

The size of servo valve, pump source and power supply capacity is largely affected by the speed of the shaking table. Usually, the floor spectrum experiment method is adopted and used in the shaking table of large-scale equipment. The maximum speed of the table does not need much speed. The maximum speed is usually determined by sweeping-beat wave experiment method. If the horizontal velocity reaches 1.5 m/s and the vertical velocity reaches 1.0 m/s, it can usually meet the industry experimental specifications.

Frequency Range
Generally speaking, for large-scale equipment test shaking table, the frequency range should be 0–50 Hz, while for civil construction shaking table, the frequency can be calculated according to the scaling ratio of the structure and the frequency range of the seismic signal used. In the case of full load, in order to meet the requirements of large-scale equipment test for 4 m * 4 m shaking table, the upper limit frequency of 50 Hz can usually meet the requirements.

3 Construction Scheme of Vibration Platform

3.1 Main Technical Specifications

According to our actual needs and site status, we adopt a three-dimensional and six-degree-of-freedom vibration platform. Its main body surface size is designed to be

2×3 m, and then a 4×5 m transition platform is set on it. The main body of the platform and the extension platform are fixed by bolts. The size of 4×5 m can basically meet our needs for experiment and experience.

The main technical specifications are as follows:

(1) Load: 2000 kg;
(2) Frequency range: 40 Hz;
(3) Maximum acceleration: vertical: 10 m/s^2; horizontal: 10 m/s^2;
(4) Maximum speed: 0.8 m/s;
(5) Maximum displacement: +150 mm;
(6) Platform size: 3000 mm * 2000 mm (length * width).

3.2 System Composition

The system composition and structure sketch of the shaking table for simulated seismic test and the installation mechanism sketch of the transition platform and building specimens are shown in Fig. 1. It mainly consists of the following parts: bearing platform, transition platform, horizontal X-direction actuator, horizontal Y-direction actuator, vertical Z-direction actuator, hydraulic source and hydraulic distributor, servo control system, hydraulic source control system, connecting cable, hydraulic connection pipeline [9].

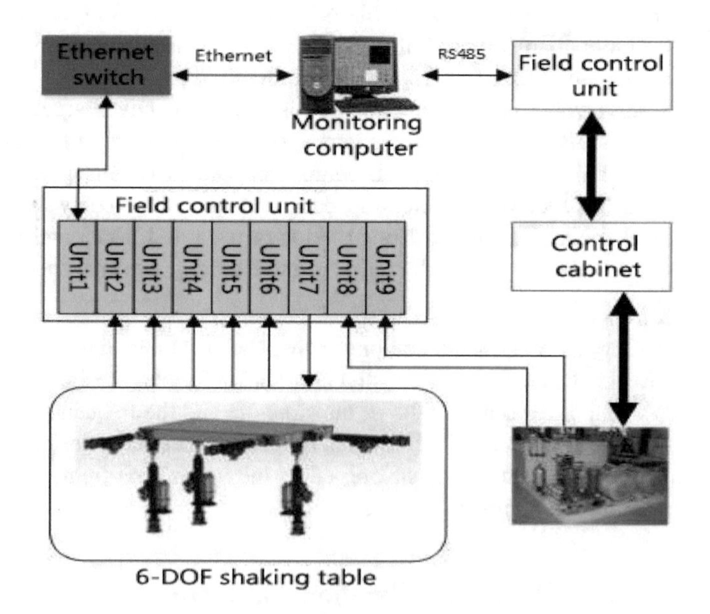

Fig. 1. Structural sketch of shaking table

3.3 System Principle

The shaking table consists of three horizontal servo actuator components and three vertical servo actuator components. Under the drive of the control system and the hydraulic source system, the shaking table vibrates according to the prescribed law. In order to achieve acceleration control, improve the stability of the control system and expand the system bandwidth, the system uses three-state control and pole placement technology to further improve the control accuracy of the system. The main function of the servo control system is to convert the DOF driving signal of the table into the driving signal of each single system (exciter). The motion of the platform is realized by controlling the motion of the exciter [10]. The driving signal is corrected by iteration operation to further reduce the difference between the system response signal and the desired signal. The control precision can be improved. The outer loop of the servo control system is added with vibration control to form a real closed-loop control of vibration [8]. If the frequency response function of the servo system is represented, the iterative compensation principle of vibration control can be expressed as shown in Fig. 2.

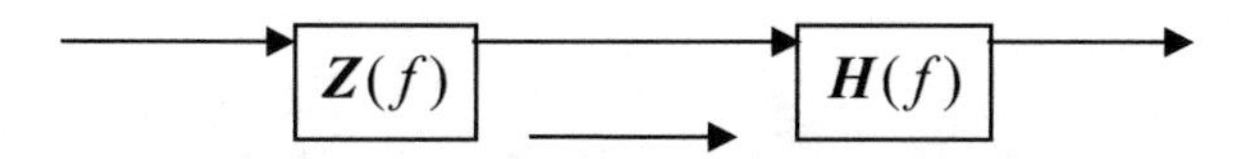

Fig. 2. Iterative compensation principle for vibration control

In the figure, the value of the expected signal corrected by the system impedance is taken as the input signal of the servo system, which represents the system impedance. The inverse of the frequency response function is the system impedance. If there are: then the output of the servo system is the expected signal [7].

4 Design Ideas and Technical Approaches

4.1 Platform Design

Because the platform needs to drive the loads connected to the upper surface to complete the movement of various postures. Therefore, the following issues should be considered in the design process (Fig. 3):

(1) The frequency of the first mode of vibration of the platform should be at least 80 Hz;
(2) The form and position tolerances of the platform, such as the parallelism error between the upper surface and the vertical actuator mounting surface, and the parallelism and verticality tolerances between the actuator mounting planes [11];
(3) The stiffness of the whole platform in all directions and the strength of the connection between the platform and the hydraulic actuator;
(4) Installation mode of the test piece and platform.

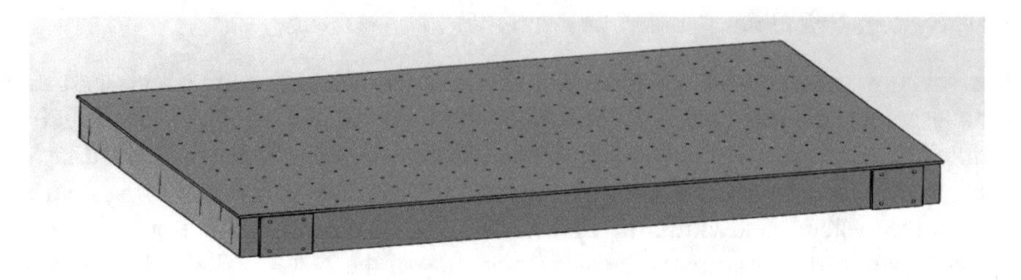

Fig. 3. Platform

4.2 Transition Platform

The transition platform is welded with steel plate. The bolt holes are machined on the upper plane to connect the building specimens. The holes of the bolts are machined on the lower plane. The structural sketches of the transition platform and the vibration platform are shown in Fig. 4.

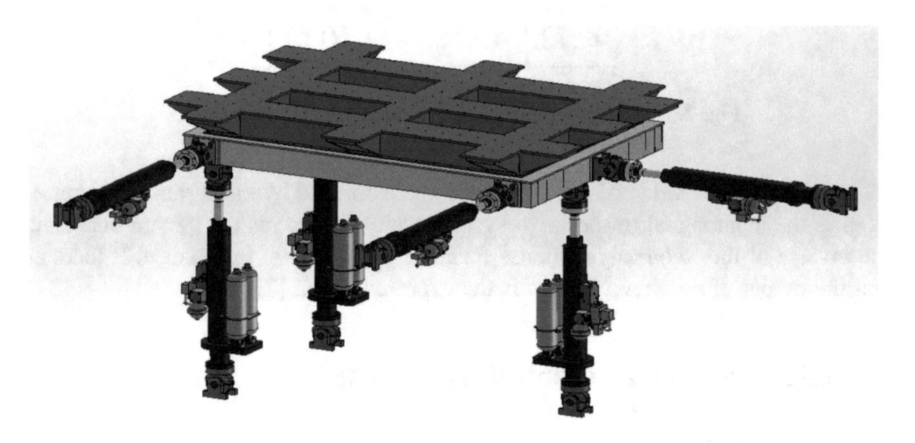

Fig. 4. Installation diagram of transition platform and vibration platform

The natural frequencies of the transition platform are: first order: 80.87 Hz, second order: 134.02 Hz, third order: 185.37 Hz. The following figure is still represented by the first natural frequency.

4.3 Horizontal X-Directional Servo Actuator

4.3.1 Actuator Parameters and Configuration

There is a horizontal X-direction actuator with a static output of 53.6 KN and a dynamic output of 44 KN, which can realize the vibration of platform and load with an acceleration of 1 g in the X-direction [12]. The horizontal X-direction actuator is shown in Fig. 5. Each horizontal actuator includes a hydraulic cylinder, a two-stage electro-hydraulic servo valve, a displacement sensor and an acceleration sensor.

Fig. 5. Structural diagram of horizontal X-direction actuator

The maximum function curve of the horizontal X-direction actuator is shown in Fig. 6. The full load mass is 2000 kg, the maximum amplitude (+150 mm) is achieved in the low frequency band (0.1 Hz ~ 0.85 Hz), the maximum speed is 0.8 m/s in the medium frequency band (0.85 Hz ~ 2 Hz) [12], and the maximum acceleration is 10 m/s 2 in the high frequency band (2 Hz ~ 40 Hz).

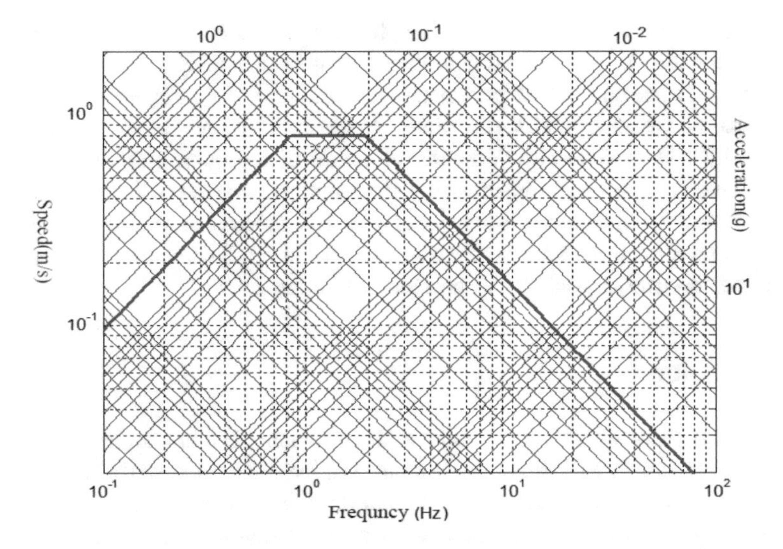

Fig. 6. Level X to the actuator biggest function curve

According to the above X-direction load parameters and actuator parameters, draw the X-direction load matching curve, as shown in Fig. 7. Obviously, the actuator design fully meets the load vibration force and speed requirements.

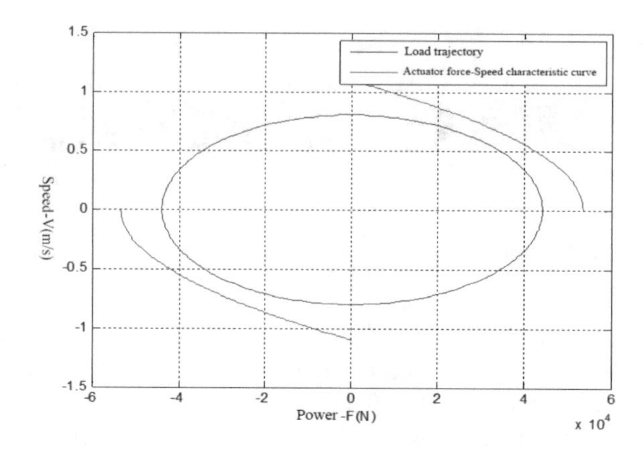

Fig. 7. 15 X to matching load curve

The composition and working principle of horizontal Y-direction actuator and vertical Z-direction actuator are exactly the same as horizontal X-direction actuator, but the technical index and output parameters are different [14]. For example, the static output force of the horizontal Y-direction actuator is 29.5 KN and the dynamic output force is 23.5 KN, which can realize the vibration of the platform and the load with acceleration of 1 G in the Y-direction; the static output force of the Z-direction actuator is 53.6 KN and the dynamic output force is 42.0 KN, which can realize the vibration of the platform and the load with acceleration of 2 G in the Z-direction, and balance the overturning moment caused by the horizontal vibration.

5 Conclusion

In this system, there are only a few units at the provincial and bureau level to construct the platform of simulated seismic experiment. We have started to study and understand this aspect very little. So we have carried out a lot of investigation and research work in the early stage of project construction to lay a foundation for the follow-up work. On this basis, we put forward the means of realizing the key parameters and the preliminary construction plan. In the project construction, we also improve the construction plan to achieve the ultimate design purpose. At the same time, we hope that our research results and experience can better serve our cause of earthquake prevention and disaster reduction, and can play a reference role for brothers provincial bureaus to carry out such research and construction.

References

1. El-Hasnony, I.M., Barakat, S., Elhoseny, M., Mostafa, R.R.: Improved feature selection model for big data analytics. IEEE Access **8**(1), 66989–67004 (2020). https://doi.org/10.1109/access.2020.2986232

2. Chen, W., Peng, W., et al.: High availability research of Shandong seismic data center. Comput. Appl. Res. **32**(Supplement), 19–21 (2015)
3. Krishnaraj, N., Elhoseny, M., Lydia, E.L., Shankar, K., ALDabbas, O.: An efficient radix trie-based semantic visual indexing model for large-scale image retrieval in cloud environment. Softw. Pract. Exp. (2020)
4. Murugan, B.S., Elhoseny, M., Shankar, K., Uthayakum, J.: Region-based scalable smart system for anomaly detection in pedestrian walkways. Comput. Electr. Eng. **75**, 146–160 (2019)
5. The determination of main parameters and technical and economic analysis of Fangzhong, large-scale simulated seismic shaking table. World Seismic Eng. **17**(4), 135–138 (2011)
6. Tian, L., et al.: System integration design of automotive multi-axle shaking table test bed. Mach. Manuf. Autom. **41**(1), 19–22 (2012)
7. Guan, G., Cong, D., et al.: Research on recurrence iteration algorithm of random vibration power spectrum. Seismic Eng. Eng. Vibr. **26**(6), 71–76 (2006)
8. Guan, G., Cong, D., et al.: 6-DOF random vibration control algorithm. J. Mech. Eng. **44**(9), 215–219 (2008)
9. Han, J., Zhang, L., et al.: The development and control technology of multi-degree-of-freedom shaking table. Hydraul. Pneumatic **10**(1), 1–6 (2014)
10. Wu, C.: GPON fiber technology in Shandong earthquake information network application. Earthq. Defense Technol. **8**(1), 1–111 (2013)
11. Wu, C., et al.: Shandong Seismic data center high availability research. Appl. Res. Comput. **32**(Suppl), 19–20 (2015)
12. Xin, G., et al.: Earthquake emergency platform and emergency communication system research. Digit. Commun. World **19**(7), 54–57 (2010)
13. Xiong, H.: Earthquake disaster rescue scene emergency communication research and design. Chengdu University of Technology, Chengdu (2013)
14. Zhang, Y.-F., et al.: Design of communication system for emergency rescue at earthquake site. Chin. J. Disaster Prev. Technol. **12**(20), 111–115 (2005)

Social E-Commerce and Marketing Strategy Based on Data Analysis

Qian Tang(✉) and Zhen Li

Beijing Institute of Technology, Zhuhai Campus, Zhuhai 519088, Guangdong, China
tangq4300@126.com

Abstract. As a derivative and transformation of traditional e-commerce, social e-commerce mine business information and user relationships through the networks, establish some models to provide personalized e-commerce services for consumers, which has become an important development direction. This paper summarizes the definition of social e-commerce through reference review, and compares the characteristics of social and traditional e-commerce, and deepens the understanding of social e-commerce business mode. Based on the analysis of the data of the article pushed by Jingdong WeChat public account, mining the marketing strategy of social e-commerce, which has certain significance for the actual social e-commerce operation.

Keywords: Social e-commerce · Data mining · Marketing

1 Introduction

Social e-commerce is one of the important forms to promote the development of e-commerce. Due to the influences of social media, the social e-commerce variability and innovation are more active than traditional e-commerce. Chinese government vigorously encourage the development of social network innovation to develop e-commerce business models. On July 6, 2018, the "Social E-Commerce Business Standards" was approved by the Ministry of Commerce as the only social e-commerce regulatory document in China. The aim is to establish a business strategy for the healthy development of the industry at a critical stage of social e-commerce. In recent years, the social e-commerce business model has been continuously innovated, and the number of social e-commerce participants has grown rapidly.

According to "Social E-Commerce Industry White Paper 2017", the number of social e-commerce Chinese users was 92 million in 2015, and increased to 162 million in 2016. It grows to 223 million in 2017, and be estimated will grow to 310 million in 2018 [1]. The rapid growth of social e-commerce participation is mainly due to extensive social interaction and interpersonal communication based on online social tools. According to the "China Social E-Commerce Industry Market Research Report of 2018 Q1", 30.4% of Chinese social e-commerce users surveyed said they would share their experience in purchasing goods on online social platforms, however the proportion of users who said they would not share on social platforms was only 10.3% [2]. Social e-commerce uses a social platform based on the mobile Internet to mine

J. MacIntyre et al. (Eds.): SPIoT 2020, AISC 1282, pp. 536–541, 2021.
https://doi.org/10.1007/978-3-030-62743-0_77

content and establish user relationships to form a new business model for e-commerce to provide consumers with personalized services.

2 The Social E-Commerce Definition

The academic definition of social e-commerce originated from Jascanu & Nicolau [3]. They described social e-commerce as a kind of "social media" based on network interconnection. The function of this kind of media is to provide an equal communication platform and trading market for merchants and consumers. Wang Nan [4], Li Yunming, etc. [5] believe that social e-commerce is the use of social relations and interaction to engage the e-commerce transaction. Zong Ganjin [6], Zhu Xiaodong, etc. [7] believe that social e-commerce is a business model for improving product sales or customized services through the formation of interest graphs in the context of multi-dimensional social platform information integration. "Social E-Commerce Industry White Paper 2017" points out that social e-commerce applies social elements such as attention, sharing, communication, discussion and interaction to the e-commerce process through the social function of network and e-commerce platform [1]. "The China Social E-Commerce Development Special Analysis 2017" pointed out that social e-commerce is a derivative model of the original e-commerce. It is primarily based on interpersonal networks and describes user images and needs through various communication ways such as social media and splatforms, and then establish a trading method with social trust as the core [8]. Fisher S. [9] believes that social e-commerce is mainly composed of six major parts: social media, reviews, social shopping, advertising, recommendations, forums and communities.

The company uses six components as a platform for communication with customers, thus forming a new way of customer relationship management. "Social E-commerce Business Standard" [10] also defines the social e-commerce clearly that is based on the interpersonal relationship network, using Internet social tools to engage business activities, including the whole process of e-commerce such as information display, payment settlement, and logistics.

As can be seen from the above references, social e-commerce is the use of Internet social tools, social media and social relationships for social interaction and communication to achieve business transactions. Due to the inclusion of social elements in it, the transaction process is distinguished from the offline store and the traditional online shopping transaction process.

3 The Characteristics of Social E-Commerce

Although China's e-commerce market has matured, the high cost of acquiring customers has become the bottleneck in the development of traditional e-commerce. E-commerce relies on social flow platforms and acquaintance networks to effectively get customers. Social e-commerce includes shop selection and product comparison, etc. before purchase, communication and interaction with e-commerce sellers through Instant Messaging during purchase, evaluation and shopping sharing after purchase.

Compared with traditional e-commerce, social e-commerce pays more attention to users and content, which makes it show great vitality in product promotion, data management and so on.

1. Traditional e-commerce focuses on the comment section. It hopes that the purchase action will be driven by users who do not know each other and cannot determine the authenticity. The effectiveness of communication is limited. Based on good communication and interactivity, social e-commerce can better push products directly and conveniently, and form a wide range of communication effects to drive consumption.
2. Users in traditional e-commerce consumption are often active search and get comments; while social e-commerce is more focused on user sharing and active communication, makes more users passively obtain relevant information.
3. Traditional electronic e-commerce needs to search, view comments, compare orders through the PC or mobile phone, which will cost users a higher time. Social e-commerce uses the fragmentation time to put marketing into the social circle and promote products invisibly, so it has a full-time 24-h "shopping". Users pay for fragmentation time to complete a series of actions such as accepting guidance, making consumption, and sharing content (Fig. 1).

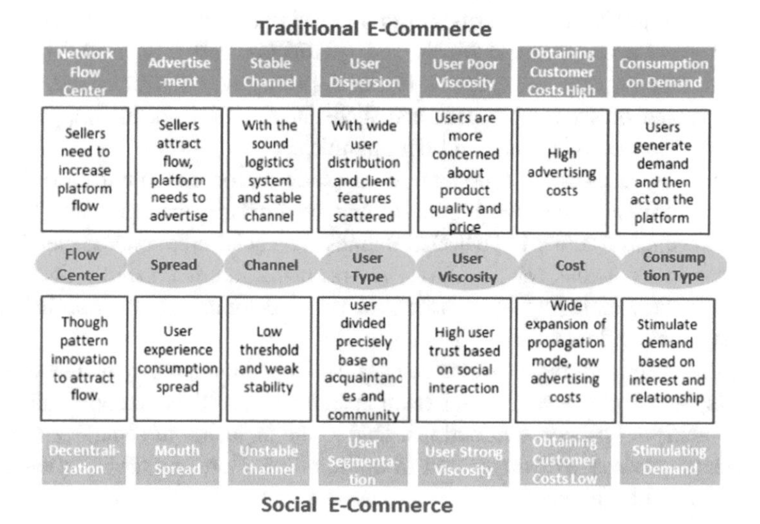

Fig. 1. The comparison of traditional e-commerce and social e-commerce characteristics

4 The Classification of Social E-Commerce Business Model

"Social E-commerce Business Standard" divides the social e-commerce model into three main forms include community e-commerce, socialization e-commerce and traditional e-commerce corporate social media. The community e-commerce mainly includes forms such as platform-operated, agency-retailed, group-buying, video-lived and content-fan. "China Social E-Commerce Industry Market Research Report of 2018

Q1" divides the social e-commerce business model into three types: social content type, social sharing type, and social retail type, while "Social E-Commerce Industry White Paper 2017" divides the business model into five types: the shopping guide type, the group-buying type, the community type, the crowd distribution type, and the comprehensive type. Compare the above two classification methods, combined with the enterprise case analysis, summarize the types as follows.

1. Social content e-commerce is similar to shopping guide type which respect with star, net red, anchor and other opinion leaders create content through social media platforms such as WeChat public account, Weibo, live broadcast platform to attract users, and then recommend products to them. Consumers buy their recommended products based on their trust in stars and so on. The establishment and maintenance of the social relationship depends on high quality, content.
2. Social-sharing e-commerce is similar to the group-buying type. It mainly spreads through social relationships and shopping sharing to attract users to buy. All of them are based on existing social relationships, which mean social activities first and followed by business activities.
3. Social retail e-commerce is similar to crowd-distributed e-commerce, mainly recruiting individuals to become distributors, and individual distributors obtain sales commissions by forwarding shopping information to social media.
4. Community social e-commerce mainly conducts online community interactive sales based on interests and hobbies.
5. Comprehensive type is a combination of the above modes.

5 Data Mining Application for Social E-Commerce

Data mining is an information processing method that integrates multiple technologies such as databases, artificial intelligence, and statistics. Social e-commerce covers a variety of business data, including marketing, shopping transactions, social advertising, reviews, customer relationship management, etc., including various types of text, audio, video, and images in social networking platforms or e-commerce platforms. It is massive, complex, diverse. These data are not independent, but have a certain relationship. Through scientific and effective mining methods and analysis processes to create more business value.

Jingdong (JD) is a self-employed e-commerce enterprise that is well-known in China. In 2014, the strong alliance between Jingdong and TengXun made more companies realize the importance of social-based WeChat marketing in the era of mobile internet. From the WeChat platform to build a platform for Jingdong shopping, create series of marketing activities based on WeChat's huge user base. Therefore, we analyze marketing strategy data from Jingdong WeChat public account.

Taking the data of the Jingdong WeChat public account from 42 days period, the public terrace account pushed a total of 78 articles to the users, and the average daily number of posts articles was 1.86, these articles were read by the users a total of 645,035 times, the average reading volume per article was 8276.68, the cumulative thumb-up number were 3,864, and the average number of per article was 49.54 (Figs. 2 and 3).

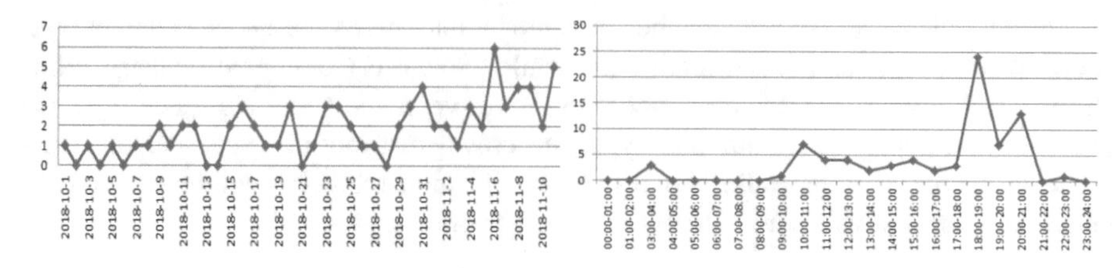

Fig. 2. The number of posted articles per day and per time period

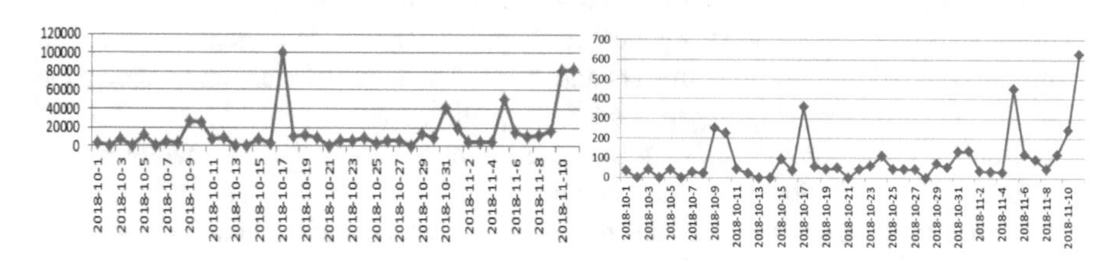

Fig. 3. The number of posted articles readings and thumb-up per day

By collecting and analyzing the marketing data of the JingDong WeChat public account, we found some operating rules and strategies based on the WeChat public account.

(1) Excessive WeChat content push frequency will cause users to be bored, thus losing the value of marketing. The daily delivery of no more than 2 messages is within the user's tolerance. JingDong WeChat public account both avoids the negative impact of the crazy push mode, at the same time maximizes the promotion of company brands.
(2) By analyzing the number of posts articles in 42 days, we found that the average number in November was much larger than that in October, especially in the five days before "double eleventh", the number reached the peak of the statistical period. It is 2–3 times the average value. This is also in full compliance with the current situation of the double eleven e-commerce war.
(3) By analyzing the number of posts articles in each one hour period of a day, it can be seen that JingDong also considers the selection of the content spread time period. Most people are busy during the day, the user does not have time to read much content. The effect of promotion is limited. In the period from 11:00 to 13:00 and 19:00–21:00, the people who worked hard for a period to find ways to relax and decompression, and the appropriate amount of articles can be promoted.
4) By analyzing the number of readings and thumb-up in 42 days, we find that the two are positively correlated. The more readings, the more thumb-up, and the readings number appear cyclical fluctuations of 6–7 days, generally peaking at weekends. It indicates that the time the user reads is closely related to the individual's free time.

Through the above data analysis, we can get the social-based WeChat public account marketing strategy. First, we should consider the fragmentation principle of time, fully grasp the user's management of fragmentation time, and push the content at precise timing to produce good marketing. The more reasonable time is selected as the two time periods of 11:00–13:00 and 19:00–21:00. Secondly, from the daily quantity analysis, it is not the optimal marketing strategy that push content more than 2 times per day. The number of posts articles can be appropriately increased during holidays. Take different strategy according to the specific situation. Finally, from the daily number of readings and thumb-up, the appropriate time to push a reasonable number of content will help improve the reading, and special days such as weekends or holiday, the effect of the activity will be better.

6 Conclusion

Social e-commerce is an effective integration of traditional e-commerce and social media. The development of its business model indicates that e-commerce will be evolved with cross-border integration and multi-channel in the future, the marketing strategies and methods of social e-commerce will also change.

References

1. Jascanu & Nicolau marketing strategy analysis based on social media. In: 2014 International Conference on Mechatronics, Electronic, Industrial and Control Engineering (MEIC 2014). Atlantis Press (2014)
2. Nan, W.: Social E-commerce: monetizing social media, Hamburg, Germany. Syzygy Deutschland Gmbh **38**(4), 88–99 (2012)
3. Li, Y., et al.: Deriving value from social commerce networks. J. Mark. Res. **47**(2), 215–228 (2010)
4. Ganjin, Z.: Social commerce: looking back and forward. Proc. Am. Soc. Inf. Sci. Technol. **48**(1), 1–10 (2011)
5. Zhu, X., et al.: Research on social commerce in web 2.0 environment. In: 2011 International Conference on E-business and E-Government (ICEE) (2011)
6. Fisher, S.: Social commerce camp-killer social commerce experience (2010)
7. Fayyad, U., Piatetsky-Shapiro, G., Smyth, P.: From data mining to knowledge discovery in databases. AI Mag. **17**(3), 37–54 (1996)
8. Kim, J., et al.: Segmenting the market of West Australian senior tour fists basing an artificial neural network. Tour. Manag. **24**(1), 25–34 (2003)
9. Pawlak, Z.: Rough sets. Int. J. Comput. Inf. Sci. **11**, 341–356 (1982)
10. Morwitz, G.V., Schmittlein, D.: Using segmentation to improve sales forecasts based on purchase intent: which "indenders" actually buy. J. Mark. Res. **29**, 391–405 (1992)

Innovation Research and Practice of Studio Teaching Mode in the Context of Smart City

Dai Chen(✉) and Fei Li

Academy of Art and Design, Anhui Technical College of Mechanical and Electrical Engineering, Wuhu 241000, Anhui, China
0121000377@ahcme.edu.cn

Abstract. The major of architectural animation and model making is a very practical subject. Students need to master the principles of indoor and outdoor design, software operation skills, so that they can be competent for such application-oriented posts as architectural animation, indoor and outdoor design. With the continuous improvement of the level of science and technology, the concept of smart city is deeply rooted in the people's hearts, and people are more and more fond of intelligent modern design. Based on this, the architectural animation major of our college actively discusses the studio teaching mode, pays attention to the integration of students' professional knowledge and skills, highlights the cultivation of students' innovation ability, practical operation ability, project reporting ability, improves their employment competitiveness, and shortens the gap between enterprises and schools.

Keywords: Smart city · Architectural animation and model making · Studio teaching

1 Introduction

With the improvement of the level of science and technology, the concept of smart city based on technology comes into being. Smart city not only emphasizes the application of technology, but also emphasizes the participation of people. The major of architectural animation and model making is mainly engaged in architectural visual design, digital construction design, BIM design, architectural modeling and other work in design and architectural enterprises, which has certain requirements for technology and talents. Therefore, under the background of smart city, strengthening the innovation research on the teaching mode of the professional studio is of great significance to improve students' professional skills and innovation ability. At the same time, under the background of smart city, strengthening the innovation of studio teaching mode to improve the practical ability of students, can also greatly improve the job matching degree of students. With the introduction of the studio teaching mode of Anhui Institute of mechanical and electrical technology (hereinafter referred to as our college), facing the needs of professional posts, optimizing the curriculum structure, integrating teaching resources, improving the innovation and practice ability of professional talents, and ensuring the integration of courses. Using the studio to optimize the project teaching,

J. MacIntyre et al. (Eds.): SPIoT 2020, AISC 1282, pp. 542–548, 2021.
https://doi.org/10.1007/978-3-030-62743-0_78

according to the characteristics of different levels of students, teach students according to their aptitude, and cultivate students' enthusiasm and innovation ability [1].

2 Analysis of the Current Situation of Architectural Animation and Model Making

At present, the current situation of the major of architectural animation and model making in China is not optimistic, mainly due to the lack of effective connection with employment opportunities, and it is difficult to guarantee graduates' employment. Specifically: first of all, there is a lack of practical projects, engaged in idle theorizing, and students cannot understand the core professional skills in depth; Secondly, we should pay attention to the explanation of theoretical knowledge. The students mainly listen to the lessons and lack the training of practical operation ability; The third is the lack of new practice projects, the low participation of students, and the inability to cultivate students' innovation ability; The fourth is the lack of practice of the simulation report course, which results in the students' weak language organization and expression ability, weak communication ability, unable to effectively communicate with customers, and lack of ability to negotiate orders at their graduation; The fifth is that teachers' evaluation is too subjective, students' innovation ability is not brought into play, personal comprehensive quality is not improved, and evaluation mechanism is not perfect; Sixth, teachers lack of enterprise training experience, no practical project support, too rigid classroom teaching, which affects students' innovation ability and practical ability [2]. As a new course, architectural animation and model making make students adapt to the development of the times, so they are chosen by more and more students. And gradually presents the trend of growth, the specific data is shown in Table 1:

Table 1. Enrollment of architecture animation and model making major in 2016–2019 (Ten thousand units)

2016	2017	2018	2019
23	27	32	39

3 Studio Teaching Mode

The studio teaching mode of the architectural animation industry is mainly to cultivate high-quality skilled talents according to the industrialization management. Studio teaching mode: establish a cartography center, operate and manage in an commercialized mode, and undertake social service projects. Under the cartography center, there are studios for architectural animation, effect performance, construction drawing deepening, architectural decoration, etc., which operate in a project-oriented way. According to the direction of interest, students choose the corresponding studio, and participate in practical project learning in the studio. Guided by the social service projects undertaken, the studio adopts the project task driven teaching method to enable students to master the core skills of the post, such as the ability of Surveying and mapping on the project site, the ability of scheme space design, the ability of three-dimensional expression of architectural animation, the ability of deepening the design of construction drawings, the ability of engineering budget and construction management, through "doing while learning, learning by doing". In the studio, students are designers, teachers are managers of enterprises, and build an integrated platform for teaching, learning and doing between schools and enterprises [3].

The advantages of establishing studio teaching mode are as follows:

(1) Posts ability recognition, role exchange, students from the probation designer, internship designer, designer assistant, designer, master designer, design director and other professional posts all-round exercise;
(2) Introduce the real project of enterprise, realize the effective connection between classroom teaching and employment real exercise;
(3) The whole process of building animation project leading, task allocation, team cooperation, scheme discussion, simulation report is used to realize project-based learning, highlight the awareness of innovation and practice, and exercise comprehensive ability;
(4) The high-intensity work, the real working environment of the enterprise, exercise the students' ability to resist pressure, and improve the employment adaptability;
(5) The production of the project adopts the way of team cooperation to stimulate students' interest in learning, improve team cooperation ability and strong sense of competition.

4 Studio Teaching Mode Innovation Path of Architectural Animation and Model Making Professional of Anhui Technical College of Mechanical and Electrical Engineering

(1) Develop the talent training plan of the studio

The leaders of art and design college led the teachers of professional courses to visit 15 famous universities and 18 enterprises in China, interviewed 12 outstanding graduates, based on the employment direction and positions of architectural animation professionals, the personnel training plan is revised according to different training directions and positions. For example, the major of architectural animation and model making is subdivided into building outdoor scene animation, building interior animation and interior design major; interior design expands to the direction of project site management and project supervision, and expands professional employment positions; the work positions are subdivided into architectural animation designer, interior designer, draftsman and space showed planner, with the specific proportion as follows in Fig. 1:

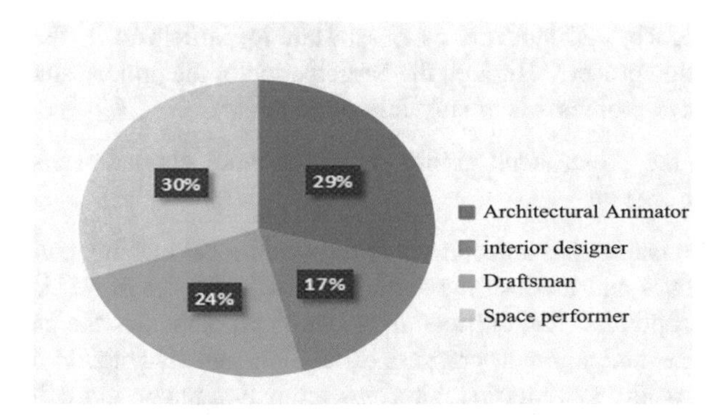

Fig. 1. Proportion distribution of architectural animation and model making professional posts

(2) Establish a platform for school enterprise cooperation demonstration training center

In order to promote the implementation of studio teaching mode, our institute and perfect space decoration & HD Design Co., Ltd. jointly build a mapping center, which is oriented to cultivate high-quality skilled talents, strengthen social service ability, actively adapt to the needs of regional economy and enterprise talents, and timely and effectively adjust the curriculum system. Relying on the drawing center and its subordinate studios, the management mode of operation mechanism is set up, and the talent cultivation mode of "project teaching, practical training, studio system" is implemented, which is committed to building a highland of high-quality skilled talents in the architectural animation industry. At present, our institute has architectural animation, effect performance, construction drawing deepening, architectural decoration studio, etc. These studios exercise the practical ability in posts of students and improve their competitiveness in employment [4].

(3) Curriculum system and teaching content reform

Deepen the professional teaching guilding committee and constantly improve the talent training program. According to the progressive teaching mode of studio stage, the overall course is divided into three stages. In each stage, it is divided into several modules according to its needs. The modules are divided into different topics according to the actual progress of the project. From the preliminary measurement, scheme preliminary design, in accordance with the industry technical specifications, the theoretical basis courses and project practice courses are integrated. Drawing deepening, engineering budget, construction process flow and other work steps aim at students' ability to achieve professional posts, and realize the curriculum system of combining professional skills, post skills and professional skills. According to the situation of classified enrollment, we should refine and adjust each topic in each module in time to meet the needs of different levels of students [5].

The implementation of "project-based" teaching mode, each studio tutor to undertake social service projects as the guide, the business projects undertaken to apply to classroom teaching, studio tutor as the corporate design director to manage and guide students to practice; and students as quasi staff to participate in the actual project design, production process. Through the participation of the project, students' ability to analyze and solve problems is greatly improved [6].

(4) Highlight the professional ability of architectural animation and optimize the curriculum system

Focusing on the major of architectural painting and model making, combined with the teaching objectives and tasks of the studio, it subdivides the professional abilities of different posts, optimizes the curriculum system, and embodies the characteristics of mutual promotion and common progress of teachers and students. In the construction of professional ability system of architectural animation major, our college subdivides the course structure according to the post ability, optimizes the course setting system, and forms the three-level course structure of professional basic ability, professional core ability and professional comprehensive ability. See Table 2 for details:

Table 2. Curriculum structure of "three levels"

Post ability	Curriculum structure	Curriculum setting	
Basic vocational ability,	Basic vocational level,	Public basic courses,	Career planning, ideological and moral cultivation, physical education, English
		Professional general courses	Design sketch and color, composition basis, architectural photography, design history, etc.
Professional core competence	Professional core level	Professional module course	Architectural hand drawing performance course, design drawing, effect drawing performance, sketch master (SU) software, etc.
		Major core courses	Architectural animation design, space design, space model production, script and lens, post production, etc.
Vocational comprehensive ability	Comprehensive vocational skill level	Practical training project course	Enterprise practical training project course, architectural animation design project course, project case design
		Elective development courses	Animation advertising creativity, graphic processing, simulation report and drill, engineering project management, etc.

(5) Emphasize the project practice process

In the process of studio teaching, the tutor needs to strengthen the professional characteristics and create a comfortable and beautiful working atmosphere of the enterprise according to the professional direction and in the transformation of studio environment. Highlight corporate culture, institutionalized management, and emphasize the authenticity, sharing and openness of studio item practice training. For example, in the teaching of home space design project, real business project design cases are introduced, from the early house type measurement, plan proposal map drawing, effect drawing performance, construction drawing deepening, project report, project budget to project implementation. Students doing while learning, learning by doing, and experience the achievement sense of project completion.

(6) "Substituting competition for practice" to improve students' competitiveness

Through participating in various professional competitions, students broaden their horizons, improve their design skills and level, form a good brand awareness, and bring a sound development momentum to the major. In 2013, it won 2 first prizes, 3 s prizes and 2 third prizes in the first college digital art works competition; 1 s prize and 1 third prize in the national vocational college skills competition landscape design project; 2 first prizes, 6 s prizes and 4 third prizes in Anhui Vocational College skills competition landscape design project; There are 2 s prizes in the architectural decoration technology application of Anhui Vocational College skill competition.

(7) Pay attention to the cultivation of "double-teacher type" teachers and improve the level of teaching staff

In the process of the implementation of the studio talent training mode, we should strengthen the construction of "double-teacher type" teaching staff, require teachers to go to the front line of the enterprise in stages, design commercial projects according to the standards of enterprise designers, and improve the design practice ability. The school introduces enterprise engineers as part-time teachers through various policies. In the course of curriculum implementation, centering on the professional development characteristics of each studio, the school takes the backbone teachers as the main body and integrates industry experts such as enterprise engineers or design directors to participate in studio teaching, so as to promote the common growth of teachers and students [7].

5 Conclusion

In the process of implementing the studio teaching mode, the architectural animation and model making major of our college gives full play to the resource advantages of schools, enterprises and society, deepens the cooperation mode between schools and enterprises, builds a practical education base with enterprises, and jointly promotes practical teaching, which embodies the characteristics of professional school running. Students can participate in practical projects, exercise skills, improve professional ability and core competitiveness of employment through the studio practical education platform.

Acknowledgements. This work was supported by Anhui quality engineering project school enterprise cooperation demonstration training center "Perfect space & HD Design Co., Ltd. school enterprise building decoration cooperation demonstration training center" of Anhui Technical College of Mechanical and Electrical Engineering (Project No.: 2019xqsxzx73).

References

1. Liu, Y.: Research on talent training mode of animation master studio. Time Educ. (15) (2017)
2. Lai, Y.: Research on studio teaching mode based on the combination of school and enterprise project orientation – taking animation design major as an example. Packag. World (2) (2019)
3. Pang, J.: Research on studio-based teaching mode – taking Suzhou Gaobo animation major as an example. Art Educ. Res. (17), 3–5 (2018)
4. Zhang, C.: Research on the application of studio teaching mode in industrial design major of colleges and universities. J. Shenyang Univ. (11), 2–3 (2018). (in Chinese)
5. Song, Y.: Application of studio mode in industrial design teaching. Bus. Mod. (11), 2–4 (2019)
6. Xia, X., Wang, Z.: Discussion on the construction of famous teacher's studio of character image design in higher vocational colleges. Art Educ. Res. (03), 3–5 (2018)
7. Chen, J.: Reform practice and exploration of animation design and production professional training mode based on "famous teacher studio". Labor Soc. Secur. World (02), 2–4 (2019)

Effects of Combined Application of Nitrogen and Phosphorus Fertilizers on Photosynthetic Characteristics of Silage Corn, Laying the Foundation of Agricultural Information

Xinqian Shi[1], Lie Zhang[2], Ruqing Cui[2], Qi Liu[1], Jin Du[1], Gaoyi Cao[1], and Chunyang Xiang[1(✉)]

[1] College of Agronomy and Resources and Environment, Tianjin Agricultural College, Tianjin 300384, China
xiang5918@sina.com
[2] Tianjin Zhongtian Dadi Technology Co., Ltd., Tianjin 300384, China

Abstract. At this stage, it is undeniable that smart agriculture has become an inevitable development direction of agricultural production. The most basic part of smart agriculture is the collection and arrangement of big data. This paper discusses the effects of combined application of nitrogen and phosphorus fertilizers on photosynthetic characteristics and dry matter yield of silage corn, and uses silage corn varieties "Jinzhu 100" to provide theoretical support for big data in smart agriculture. In this experiment, 25 kinds of mixed treatments were set up, and the relevant indexes of corn were measured. Among the 25 treatments, N4P4 (nitrogen application 240.00 kg/hm^2, phosphorus application P_2O_5 112.50 kg/hm^2) and N4P3 (nitrogen application 240.00 kg/hm^2, phosphorus application P_2O_5 75.00 kg/hm^2) applied to "Jinzhu 100" plant height, leaf nitrogen and chlorophyll accumulation promoted the effects most significantly. N4P3 (nitrogen application 240.00 kg/hm^2, phosphorus application P_2O_5 575.00 kg/hm^2) had the most significant effect on increasing the dry matter weight of "Jinzhu 100". Among the 25 treatments, N4P4 treatment had the best photosynthetic physiology and growth promotion effect of "Jinzhu 100".

Keywords: Nitrogen and phosphorus combined application · Silage corn · Photosynthetic characteristics · Dry matter production · Smart agriculture

1 Introduction

Silage corn has been widely used to raise livestock and other aspects, and the demand in China is also increasing year by year. Silage corn has higher biological yield, fiber quality and greenness [1]. The yield of silage corn is 4–5 times that of ordinary corn [2], feeding cows with silage corn can not only increase the milk production significantly, but also reduce the use of concentrated feed during feeding [3]. Crop with strong photosynthesis can accumulate large amounts of organic matter, most of the yield of corn comes from the accumulation of photosynthesis, the strength of photosynthetic characteristics is the main factor that determines the yield of corn [4], so the

J. MacIntyre et al. (Eds.): SPIoT 2020, AISC 1282, pp. 549–555, 2021.
https://doi.org/10.1007/978-3-030-62743-0_79

study on photosynthetic characteristics of corn silage is essential. Studies have shown that appropriate increase of nitrogen fertilizer can increase chlorophyll content in corn leaves and increase leaf photosynthetic rate [5], promote dry matter accumulation and increase crop yield [6]. Phosphate fertilizer has an important role in increasing crop yield, increasing dry matter weight in stems and leaves, and increasing photosynthesis in panicle leaves [7]. Related reports on the effects of nitrogen and phosphorus interactions on corn yield and dry matter also indicate that, in certain range, yield of corn increased with increasing nitrogen and phosphorus fertilizer application rates [8]. In recent years, most of studies on crop photosynthetic characteristics and yield are limited the interaction of nitrogen fertilizer or phosphate fertilizer or tillage measures with nitrogen and other factors (such as tillage measures, light, etc.) on photosynthetic characteristics, little research has been done on the effects of combined application of nitrogen and phosphorus fertilizers on crop photosynthetic characteristics, but there are little research has been done on the effects of combined application of nitrogen and phosphorus fertilizers on crop photosynthetic characteristics. And with the rapid development of the Internet of Things and big data, people's attention to smart agriculture has also greatly increased. Smart agriculture can not only solve the problems brought by the extensive agricultural production model, but also combine technologies to achieve the maximum grain output under the condition of moderate application of chemical fertilizers [9, 10]. In this test, a new variety of "Jinzhu 100" silage corn in this area was used, to study the effects of different levels of nitrogen and phosphorus fertilizers on photosynthetic characteristics and dry matter, to provide scientific guidance for the amount of nitrogen and phosphorus fertilizer applied to local silage corn, to provide new data and scientific theoretical support for fertilizer interaction for the development of smart agriculture.

2 Test Materials and Methods

2.1 Situation in the Experimental Field

The test was set up in the experimental field of Jinhai District, Tianjin. The test soil was fluvo-aquic soil, and the cultivated layer soil was found to contain organic matter 32.26 g/kg, total nitrogen 2.15 g/kg, alkaline hydrolyzed nitrogen 169.14 mg/kg, total phosphorus (P) 1.08 g/kg, available phosphorus (P) 45.47 mg/kg, total potassium 1.03 g/kg, available potassium (K) 119.73 mg/kg, pH (H_2O) 8.2.

2.2 Experimental Design

The corn variety tested in this test was Tianjin "Jinzhu 100", one of the main silage corn varieties in Tianjin. Field plot experiment was adopted and randomized block design was adopted. The applied amount of nitrogen and phosphorus ratio is shown in Table 1. In the test, potassium fertilizer (potassium chloride, $K_2O \geq 50\%$) was applied to each plot as a base fertilizer at a rate of 140.00 kg/hm^2 potassium chloride (K_2O 70.00 kg/hm^2), and phosphate fertilizer (phosphate fertilizer was superphosphate, $P_2O_5 \geq 12\%$) was also applied as a base fertilizer at one time. Nitrogen fertilizer (urea, total N $\geq$ 46%) was divided into two applications. Among them, base fertilizer accounted for 40%, and topdressing 60% was applied at the jointing stage of corn.

Table 1. Different nitrogen and phosphorus fertilizer application rates

Treatment	Amount of N (kg/hm^2)	Amount of Urea (kg/hm^2)	Amount of phosphate (kg/hm^2)	Amount of superphosphate (kg/hm^2)
N1P1	0.0	0.0	0.0	0.0
N1P2	0.0	0.0	37.5	314.5
N1P3	0.0	0.0	75.0	625.0
N1P4	0.0	0.0	112.5	937.5
N1P5	0.0	0.0	150.0	1250.0
N2P1	80.0	172.4	0.0	0.0
N2P2	80.0	172.4	37.5	314.5
N2P3	80.0	172.4	75.0	625.0
N2P4	80.0	172.4	112.5	937.5
N2P5	80.0	172.4	150.0	1250.0
N3P1	160.0	344.8	0.0	0.0
N3P2	160.0	344.8	37.5	314.5
N3P3	160.0	344.8	75.0	625.0
N3P4	160.0	344.8	112.5	937.5
N3P5	160.0	344.8	150.0	1250.0
N4P1	240.0	517.2	0.0	0.0
N4P2	240.0	517.2	37.5	314.5
N4P4	240.0	517.2	112.5	937.5
N4P5	240.0	517.2	150.0	1250.0
N5P1	320.0	689.6	0.0	0.0
N5P2	320.0	689.6	37.5	314.5
N5P3	320.0	689.6	75.0	625.0
N5P4	320.0	689.6	112.5	937.5
N5P5	320.0	689.6	150.0	1250.0

The trial was seeded on May 4, 2018. The field management during the growing period was in accordance with the conventional field management.

2.3 Measurement Items and Methods

2.3.1 Determination of Growth Physiological Indicators

Plant height: Five representative plants were picked during the harvest period, and averaged after measuring with a tape measure. Chlorophyll content and leaf nitrogen content: The chlorophyll content and leaf nitrogen content of panicle leaves were determined with a plant nutritional tester (TYS-3 N) at the silking stage. Panicle leaf photosynthetic index: A portable photosynthesis instrument was used to select three panicle leaf leaves for measurement at around ten o'clock in the morning when the tasseling period was sunny.

2.3.2 Determination of Dry Matter Yield of Silage Corn

After the plants were taken during the wax ripening period, they were killed in an oven and dried to constant weight to calculate the dry matter yield.

3 Results and Analysis

3.1 Effects of Combined Application of Nitrogen and Phosphorus on Chlorophyll Content, Leaf Nitrogen Content and Plant Height of "Jinzhu 100" Corn Leaves

It can be seen from Table 2 that the top 2 chlorophyll contents of the leaves, nitrogen content of corn leaves and corn plant heights were N4P4 and N4P3, among them, it showed that the combined application of medium and high amounts of nitrogen and phosphorus can help the accumulation of chlorophyll and nitrogen in leaves. It was also known from Table 2 of the three indicators under N1P1 treatment were the lowest, indicating that the application of nitrogen and phosphorus fertilizers were not conducive to the accumulation of chlorophyll and nitrogen in corn. According to the analysis of the above test results, it was found that N4P4 and N4P3 were beneficial to the growth of corn and the accumulation of nitrogen and chlorophyll in leaves.

Table 2. Differences in leaf chlorophyll et al. at nitrogen and phosphorus interaction of "Jinzhu 100" corn

Treatment	Chlorophyll content (SPAD)	N content (mg/kg)	Plant height (cm)
N1P1	37.53 k	2.68 k	252.82 k
N1P2	42.18 gh	3.42 ij	259.81 j
N1P3	40.73ij	3.37j	269.45 i
N1P4	43.25 fg	3.68 k	275.74 efg
N1P5	40.94 i	3.47 hij	261.91 j
N2P1	40.40 ij	3.37 j	261.93 j
N2P2	43.04 fg	3.70 bcdef	271.08 ghi
N2P3	44.21 def	3.74 abcdef	277.89 def
N2P4	44.20 def	3.69 bcdef	278.13 def
N2P5	43.13 fg	3.68 bcdef	270.70 hi
N3P1	40.48 ij	3.40 ij	271.67 ghi
N3P2	44.10 def	3.68 bcdef	277.44 def
N3P3	44.95 cde	3.78 abcde	278.11 def
N3P4	46.15 bc	3.77 abcde	283.44 bc
N3P5	43.97 def	3.64 efgh	276.78 def
N4P1	43.80 ef	3.57 fghi	274.20 fgh

(*continued*)

Table 2. (*continued*)

Treatment	Chlorophyll content (SPAD)	N content (mg/kg)	Plant height (cm)
N4P2	43.97 def	3.76 abcde	276.92 def
N4P3	47.29 ab	3.89 a	286.78 ab
N4P4	47.62 a	3.86 a	289.49 a
N4P5	41.38 hi	3.49 ghij	268.48 i
N5P1	39.73 j	3.46 ij	262.30 j
N5P2	44.12 def	3.67 cdefg	274.91 efgh
N5P3	45.99 c	3.74 abcdef	281.34 cd
N5P4	43.46 f	3.66 defg	274.78 efgh
N5P5	41.23 hi	3.45 ij	268.00 i

3.2 Effects of Combined Application of Nitrogen and Phosphorus on Dry Matter Weight and Water Use Efficiency of "Jinzhu 100" Corn

According to the analysis in Table 3, The top 2 leaf transpiration rate were N4P4 and N2P1, the top 2 net photosynthetic rate, stomatal conductance, water use efficiency and total dry matter weight were N4P4 and N4P3, this showed that the combined application of medium and high nitrogen and phosphorus fertilizers had a significant effect on the total dry matter weight of corn and can improve the net photosynthetic rate of leaves and promote photosynthesis of corn. According to the analysis of the above test results, it was concluded that N4P4 and N4P3 were beneficial to corn, leaf photosynthesis and transpiration increased the total dry matter weight of corn.

Table 3. Differences in water use efficiency et al. at nitrogen and phosphorus interaction of "Jinzhu 100" corn

Treatment	Net photosynthetic rate (μmol/m^2/s)	Transpiration rate (mmol/m^2/s)	Stomatal conductance (mmol/m^2/s)	Water use efficiency (%)	Total dry matter weight (kg/hm^2)
N1P1	16.06 l	1.58 m	15.67 j	10.68 d	49850 f
N1P2	26.31 c	2.52 ijk	26 h	11.84 d	50004.5 f
N1P3	27.44 ef	2.01 l	20.67 i	10.91 d	52200 ef
N1P4	20.97 k	1.63 m	16.67 j	9.04 f	52028.6 ef
N1P5	24.13 e	2.71 ghi	26.33 h	10.39 f	51246.5 ef
N2P1	25.98 e	4.96 a	31.00 def	7.73 l	50285 ef
N2P2	25.66 fg	3.41 d	30.33 ef	7.56 j	51600 ef
N2P3	26.11 h	4.11 b	26.00 h	7.31 kl	53800 d
N2P4	26.04 k	3.09 e	32.00 de	7.23 kl	54000 de
N2P5	23.24 fgh	2.73 ghi	28.67 fg	8.17 h	52800 ef
N3P1	23.29 ijk	3.05 e	30.00 ef	8.22 i	50900 d
N3P2	31.93 b	2.89 efg	31.00 def	13.12b	54600 d
N3P3	29.37 b	4.31 a	33.00 b	11.44 c	54400 d

(*continued*)

Table 3. (*continued*)

Treatment	Net photosynthetic rate (μmol/m^2/s)	Transpiration rate (mmol/m^2/s)	Stomatal conductance (mmol/m^2/s)	Water use efficiency (%)	Total dry matter weight (kg/hm^2)
N3P4	27.87 d	3.46 d	36.00 b	8.3 hi	51500 ef
N3P5	28.77 d	2.75 fghi	29.00 fg	9.79 d	51200 ef
N4P1	25.19 hij	2.81 fgh	33.00 b	8.52 jk	55800 cd
N4P2	22.46 k	2.45 jk	25.67 h	8.62 fg	56200 c
N4P3	34.02 a	4.12 b	40.00 a	14.82 a	68000 a
N4P4	35.40 a	4.97 a	39.33 a	15.11 a	67200 a
N4P5	24.80 k	2.5 ijk	26.00 h	7.89 jk	51800 ef
N5P1	22.6ghi	2.6 hij	28.67 fg	8.03 hi	50400 ef
N5P2	27.43 d	3.07 e	27.33 gh	8.39 i	59400 b
N5P3	28.63 b	2.33 k	27.00 gh	11.04 c	62000 b
N5P4	23.77 fg	2.7 ghi	28.67 fg	9.35 g	58800 bc
N5P5	25.43 fg	2.66 ghij	26.00 h	9.58 g	55400 cd

4 Conclusion

(1) In summary, under the N4P4 treatment (N: 240.00 kg/hm^2, P:112.50 kg/hm^2) the chlorophyll content and plant height of corn leaves were highest, and under the N4P3 treatment (N: 240.00 kg/hm^2, P:75.00 kg/hm^2) the nitrogen content of corn leaves was highest and the effect on leaf nitrogen accumulation was most significant. Without N and P treatment, the corn's 3 indicators were lowest.

(2) Under the N4P3 treatment (N: 240.00 kg/hm^2, P:75.00 kg/hm^2) the indicators such as transpiration rate reached the maximum. The reasonable combination of nitrogen and phosphorus fertilizers for different corns were beneficial to the increase of photosynthesis and dry matter weight of corn.

Acknowledgements. This work was supported by Tianjin Seed Industry Science and Technology Major Project 16ZXZYNC00150.

References

1. Zhao, F., Ding, Y., Zhang, J., Zhang, F.Q., Xia, L.K., Tang, B.J.: Planting prospect and cultivation technique of silage maize. J. Seed Ind. Guide (10), 21–23 (2018). (in Chinese)
2. Su, T.Z., Hou, L.X., Zhang, Y.Q., Wang, F.M., Peng, L.: Physiological characteristics of population and its response to density in high-yield silage maize. Acta Agriculturae Boreali-Sinica 34(02), 132–137 (2019). (in Chinese)
3. Wei, X.D.: Study on economic benefits of silage corn and its application. Country Technol. (24), 81–82 (2018). (in Chinese)
4. Liu, J.Z., Yuan, J.C., Zhou, K., Liang, R., Zhang, H.X., Ren, J., Cai, H.G.: Effect of different nitrogen fertilizer application rates on photosynthetic characteristic and yield of spring maize. J. Maize Sci. **27**(05), 151–157 (2019). (in Chinese)

5. Sun, N.X., Zong, X.F., Wang, S.G.: Effects of Nitrogen Supply on Photosynthetic Traits of Maize Journal of Southwest University (Natural Science Edition) (03), 389–392+396 (2005). (in Chinese)
6. Gu, Y., Hu, W.H., Xu, B.J., Wang, S.Y., Wu, C.S.: Effects of nitrogen on photosynthetic characteristics and enzyme activity of nitrogen metabolism in maize under-mulch-drip irrigation. Acta Ecologica Sinica **33**(23), 7399–7407 (2013). (in Chinese)
7. Wang, J.K., Li, S.Y., Zhang, X.D., Wei, D., Chi, F.Q.: Spatial and temporal variability of soil quality in typical black soil area in Northeast China in 20 years. Chin. J. Eco-Agriculture (01), 19–24 (2007). (in Chinese)
8. Zeng, J.X., Wen, X.C., Raza, M.A., Chen, G.P., Chen, C., Peng, X., MA, Y.W., Li, L., Guan, S.C., Yang, W.Y., Wang, X.C.: Effects of combined applications of nitrogen and phosphorus on interspecies interaction, yield, and dry matter accumulation and translocation in maize in a maize-soybean relay intercropping system. Acta Prataculturae Sinica **26**(07), 166–176 (2017). (in Chinese)
9. Fang, Y., Sun, G., Jin, D.D., Chen, Y.F., Du, J., Wu, S.H., Wang, X.C., Wang, W., Liu, C.: Main technical fields and development trends of modern agricultural IoT. Agric. Technol. **40** (02), 1–2 (2020). (in Chinese)
10. Wang, J.F.: Research on the development trend of big data in the era of smart agriculture. J. Tech. Econ. Manag. (02), 124–128 (2020). (in Chinese)

Imagination of Virtual Technology Application in Sports Field

Kunlun Wei[1(✉)] and Haiyan Hao[2]

[1] Department of Physical Education, Tianjin University of Technology and Education, Tianjin 300222, China
hhy632@163.com
[2] Weishan Road Middle School, Hexi District, Tianjin 300222, China

Abstract. The rapid development of virtual reality technology, artificial intelligence, 5G transmission technology and Internet has accelerated social changes, and brought great changes to social development. If it can be widely applied in the field of sports, it can not only make up for the lack of fitness exercise space and technical training, but also reduce the physical injury caused by innovative and difficult sports movements when using VR technology to create high, new and difficult sports technologies.

Keywords: Virtual reality technology · Immersion · Realistic experience · Sports field

In recent years, VR technology, AR technology, and holographic projection technology, as representatives of virtual reality and virtual imaging technology, have been increasingly applied to all aspects of social changes. It integrates many disciplines, such as computer, web multimedia, bionics, education and psychology, with the help of camera, sensing device and virtual technology founded by a memory medium by computer hardware and software, and integration of artificial intelligence, which can make the person indulge with virtual interactions, and in which participants also can experience the similarity to real visual, hearing, touch, smell and so on by the real-time interaction aiding with language, gestures, movements and other behaviors.

1 Virtual Technology Features

1.1 VR's Technical Characteristics

1.1.1 Dynamic Interactivity

Participants come into the virtual environment with the help of VR glasses, helmet display, data gloves or the photoelectric dynamic capture system of peripherals, so that people's self-will can interact with the "entity" of the virtual world in real life with language, posture and other skills, as if all these are real.

J. MacIntyre et al. (Eds.): SPIoT 2020, AISC 1282, pp. 556–561, 2021.
https://doi.org/10.1007/978-3-030-62743-0_80

1.1.2 Sense of Immersion

Putting on the VR peripherals, you would come into a dreamlike world where the characters and scenes are so realistic and the sound effects are so shocking in the highly three-dimensional images, you can learn to play basketball with Yao Ming, play badminton in the same field with Chen Dan, and interact with many celebrities autonomously and learn in different aspects, feeling the sense of tracking time or space under the guidance of your own vision, hearing, touch and even sense of smell [1].

1.1.3 Creativity

The person, who gets into the virtual world, can accept all kinds of information according to his or her own sense, and make unlimited interaction with the system through the logic, reasoning thinking process, to put those he or she hasn't grasped or verified ideas into creative action, and guide the formation of the real world of sports technology or game tactics, which be able to greatly increase motor skills, reduce the exercise intensity, improve the new action and promote the creation of a new technology.

1.2 Augmented Reality Technical Characteristics

1.2.1 Augmented Reality

Augmented reality technology is to use modern computers and multimedia technology to classify the recorded images and audio data into three-dimensional images according to the project and requirements, and transfer the Live images captured by the camera to the computer, and then use professional projection or display equipment to show, making the real environment and virtual image complement each other and form a complete event, and enhance the sense of reality.

1.2.2 Real-time Interactive

When AR technology is being applied in sports, it also needs the professional motion capture system supported by high-speed cameras to transfer the captured image data to the computer in time to merge, stack and synthesize the new graphic output of the virtual image preset by the computer so as to complete the interaction between the real scene and the virtual image.

1.2.3 3D Registration

3D Registration also known as three-dimensional positioning. It can be used in AR technology to solve the problem of "where am I" during real-time interaction between moving people and virtual scene, among which "I" refers to the moving people equivalent to the camera or eyes, "in" refers to the location, "which" refers to the map built in the virtual environment. On the other hand, the participants needs to find your motion track in the original map (virtual environment), and virtual image and real scene are 3d images, so we call such a process 3d registration.

1.3 Holographic Projection Technology Features

1.3.1 The Advanced Nature

The holographic projection technology is developed based on the inherited VR technology and AR technology. It uses a new type of photosensitive medium to replace the traditional photosensitive film to record holograms, and applies digital technology and holographic film technology to simulate optical diffraction to achieve image reproduction. It gets rid of the reliance on helmets, 3D glasses, various sensors, etc., and realizes that objects can display their true appearance in the air.

1.3.2 Sense of Reality

The holographic projection device is small, light and suitable for independent placement or hanging, and its thin and transparent holographic film that is not limited by the venue and equipment can be competent for most of the sports scene, participants do not need to wear any equipment, they can play freely without any restrictions. For example, on a basketball court, basketball enthusiasts want to play a game, but they can't play because of insufficient staff, but in a holographic projection environment, you can set the game mode. There are penalties and winning or losing according to the rules of the game, thereby enhancing the participants' sense of participation, intimacy and reality.

2 To Realize the Analysis and Application of VR Technology

2.1 The Similarities and Differences of VR Technologies

VR technology application has more than 60 years' history since the 1940s and 1950s. Virtual technology mainly includes: VR technology (virtual world), AR technology (real world + digital information), MR technology (real world + virtual world + digital information) and the holographic projection (real world + virtual world + digital information + virtual-real game) [2]. The advent of the era of holographic projection spun off wearing virtual sensing device for inconvenience the shackles of movement, by which the dynamic macroscopic among observation, speech recognition, identification, false or true interactive is becoming true, sports fans not only can exercise, but also can enjoy 360-degree observation simulation of virtual technology falling in love with it. They do not feel lonely any more.

2.2 Implementation Principles of Virtual Reality

2.2.1 Application of Interference and Diffraction Principles

The process of obtaining a hologram is produced by superimposing a diffused object beam and a reference beam formed by laser irradiation through special equipment. The optical wave phase and amplitude are spatially converted into interfering waves of varying strong and weak phases, thereby the interference, the contrast and interval between the stripes record all the information of the object's light waves on the negative film, and then obtain it through processing procedures such as development and fixing, and then use the principle of diffraction to reproduce the object's light wave information [3]. The light wave information of each point recorded on the picture is subjected to multiple exposures to reproduce the stereoscopic image process under the irradiation of a coherent laser.

2.2.2 Principles and Methods of Obtaining Holographic Projection Dynamic Material

The application of virtual reality technology in the field of sports mainly reflects the dynamic behavior of people. Therefore, most of the produced sports materials are dynamic, like sports technology and sports tactics, most of them are made by motion capture methods as follows: (1) Mechanical Motion Capture. (2) Electromagnetic Motion Capture. (3) Motion Capture Based on Inertial System. (4) Optical Motion Capture Technology. This method does not need to wear complex equipment. It only needs to put reflective labels on the important joint positions of the human body, tracking all reflective markers during the movement through multiple cameras, and then use computer vision technology to analyze and process the shooting. The image sequence of 3D is based on the position of the reflective marker in a plurality of sets of two-bit image sequences to restore three-dimensional human motion data [4] (Fig. 1).

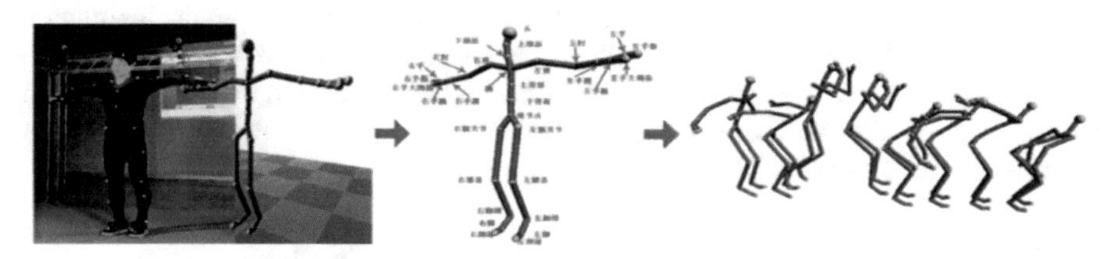

Fig. 1. Optical motion capture system

In virtual reality technologies, VR, AR, MR technology and holographic projection all need certain hardware and software, by which the acquired images can be repaired, segmented edited and stored manually, then through the intelligence and the facilities of virtual reality technology makes images reproduce and completes human-to-image interaction as Fig. 2 describes.

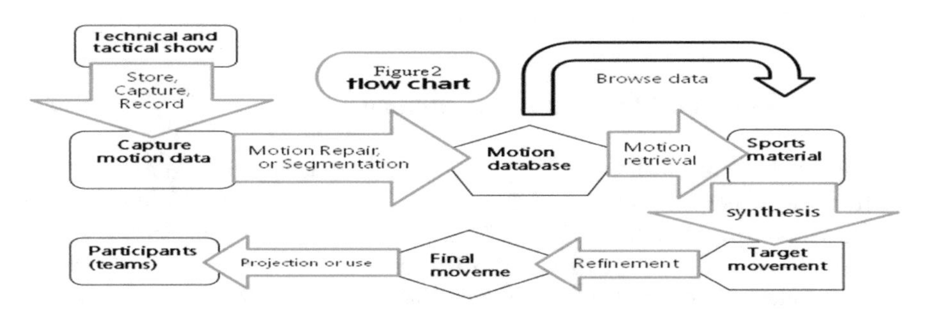

Fig. 2. Implement image reproduction flowchart

After the motion data is generated by the motion data capture system, it is mainly stored in the computer in the format of ASF&AMC file. The ASF file stores the skeleton data of human body, and defines the skeleton model of human body [5].

2.3 Application of Virtual Reality Technology in Sports

The anchor institutions of the 26th World University Summer Games have equipped a large number of equipment including high-speed, ultra-high-speed cameras and underwater cameras and used the key technologies such as 3d graphics generation, Chroma Key synthesis and camera tracking in order to provide high-quality image transmission for the majority of audiences when making public signals for swimming games, which has fully received the affirmation and high attention of the industry [6].

The Guangdong Hongyuan Club of the CBA League uses VR glasses to conduct players' offensive and defensive tactical exercises and the simplest shooting training. The KAT WALK virtual treadmill released by KATVR company can simulate the virtual environment, so that users can experience the running experience of the outdoor real environment and bring a better immersive experience [7] "Science and Technology Virtual Stadium" was set up in Beijing to virtual exercises for table tennis, badminton, racing and other sports. For example, you could wear sensors to learn technology, race across the screen, and interact with famous players full of interest, daring and smooth, not affected by the weather and the venue, through which you could achieve an effective physical exercise [8].

2.4 Research and Application of Virtual Technology in China

It has been more than 60 years since the emergence of virtual technology. In recent years, with the development of artificial intelligence, it has promoted its rapid and huge changes, which has brought great impact on the development of human society and people's lives. A large number of outstanding research institutions have also appeared in China, such as, Department of Intelligent vision Information Processing, Peking University, etc. Representatives of outstanding domestic virtual reality companies include Beijing Storm Technology Co. Ltd. Shenzhen Longitude Technology Co., Ltd., etc. At present, it is mainly used in the military, manufacturing, construction, urban planning, entertainment and other fields [9].

3 Conclusions and Optimization Strategies

3.1 Though the virtual reality technology brings people a happy experience, it also has the following deficiencies affecting the public's experience and wide-scale popularity.

(1) Virtual technology wearables are not universal. (2) The consumption cost of purchasing hardware is expensive. (3) There are too few professional employees. (4) The content of sports material is scarce, and difficulty of the production will be much higher and more complicated. (5) The wearable sensing device has a large obstacle to human movement and is prone to syncope, and the interaction experience is poor.

3.2 Although virtual reality technology has made great progress, it still a considerable distance to fully meet the needs of society, and we should also strive to achieve the following. (1) Unify industry standards. In addition to core technologies, the product accessories produced by different manufacturers should be universal to reduce consumers' additional economic burden. (2) Speed up technology research and development, improve product quality, and bring consumers a perfect experience to increase social identity. (3) R&D teams, manufacturing companies and higher physical education institutions should cooperate to create more scientific virtual sports products to meet the sports needs of different levels and different populations. (4) Cultivate more professional and technical personnel to fulfill technical research and development, market growth needs, and after-sales service requirements. (5) Increase the investment in virtual reality artificial intelligence in public stadiums to meet the desire of the general public for sports technology learning and the demand for physical fitness.

Acknowledgments. This paper is the research result of the "13th Five-Year Plan" of Tianjin Education Sciences Planning. Project Number: VE1052.

References

1. Dai, L., Li, X.: Research progress on application of virtual reality (VR) technology in sports field. Neijiang Technol. **9**, 25–26 (2019)
2. Xu, D.: Research on application of virtual reality technology in college physical education. J. Lanzhou Univ. Arts Sci.
3. Li, M.Y.: Design of two-wheeled mobile control robot with holographic projection. New Industrialization
4. Xia, G.: Analysis and reuse of human motion capture data. 6, 1–3 (2017)
5. Du, Z.: Research on analysis method of human motion capture data. 3 (2015)
6. Zhao, G.: Application cases of virtual sports system, radio and TV technology. **9**, 76–79 (2012)
7. Wu, L.: Research on the application of virtual reality technology in sports field, Fujian sports science. **10**(5), 12–14 (2017)
8. Li, W.: Application prospects of virtual reality technology in libraries of sports institutes, sports. **2**(108), 104–105 (2015)
9. Chen, W.: Research and design of urban community planning system based on augmented reality. Huazhong University of Science and Technology (2007)

Effect of Time Slot Search on DAG Scheduling Strategy in Heterogeneous Clusters

Lumei Du[1], Yanzhao Jiang[2], and Yangyan Du[3](✉)

[1] College of Electronic Information and Automation, Tianjin University of Science and Technology, Tianjin, China
[2] College of Computer and Communication Engineering, University of Science and Technology Beijing, Beijing, China
[3] Zhengzhou Public Security Bureau, Zhengzhou, China
185490728@qq.com

Abstract. For the improvement of the scheduling performance of the directed acyclic graph (DAG) system of heterogeneous cluster applications, this paper repeatedly pushes and fills the application tasks in the initial solution of the scheduling scheme, and improves the scheduling effect of the initial solution, At the same time, it keeps the high efficiency of the list scheduling algorithm. Finally, the simulation experiments in the random graph generator to create various DAGs application task scheduling models show that the proposed algorithm has significant advantages in critical path scheduling length performance indicators.

Keywords: Time slot search · Heterogeneous cluster · Directed acyclic graph · Scheduling strategy

1 Overview

As for the applications of directed acyclic graph, the task scheduling problem is a basic problem of distributed and parallel computation [1]. This issue refers to a basically NP hard problem and the optimal solution is limited to the problem's limited case. Heuristic task scheduling algorithms in this field has been extensively studied. This paper puts forward a new critical directed search algorithm complying with the time-slot stacking strategy [2]. This algorithm is feasible especially to achieve DAG scheduling in heterogeneous clustering mechanisms. Through using the "fill" operation to eliminate the communicating demands between nodes linked via slow network connections, the above problems can be solved, so as to improve the algorithm performance [3].

2 The Definition of Problem Model

Applications can make use of the DAG model $G = (T, E)$ to represent [4]. In the formula, E represents N_E edges and T denotes N_T tasks. Respective edge indicates a priority constraint relationship, revealing the task t_i required to be executed prior to the

J. MacIntyre et al. (Eds.): SPIoT 2020, AISC 1282, pp. 562–567, 2021.
https://doi.org/10.1007/978-3-030-62743-0_81

task t_j. The edge $e_{ij}(1 \leq i,j \leq N_T)$ has a weight ε_{ij} that represents the number of data for undergoing the transfer from p_i to, and the p_j task t_i exhibits a weight $\tau_i(1 \leq i \leq N_T)$ that represents the executing period required by the task t_i on the baseline processor stage.

The target heterogeneous clustering mechanism is able to be indicated in a hierarchical manner in a not directed graph $HC = (C, L_W)$. In the formula, $C = \{C_k | 1 \leq k \leq N_C\}$ represents a clusters group and $L_W = \{l_{C_iC_j} | C_i, C_j \in C\}$ represents a WAN link between a group of clusters. Respective cluster $C_k = (P(k), L(k))$ covers several processors $P(k)$ and several local area network (LAN) links $L(k)$ employed in terms of communication between the processors in $P(k)$.

In order to conduct task scheduling, heterogeneous clustering systems are able to be expressed as non-hierarchical graph models $NHC = (P, L)$. In the formula, [12, 13]:

$$\begin{cases} P = \bigcup_{k=1}^{N_C} P(k) = \{p_1, p_2, \cdots, p_{N_p}\} \\ L = L_W \bigcup \left(\bigcup_{k=1}^{N_C} L(k)\right) = \{l_{p_ip_j} | p_i, p_j \in P\} \end{cases} \quad (1)$$

This particular link represents that, if a link $l_{p_ip_j}$ is given and if p_i, p_j belongs to the kth cluster, this link represents the LAN connection in $L(k)$; if not, the link corresponds to the WAN connection in L_W. Respective processor p_q exhibits a (relative) calculated power factor w_q; performing tasks t_i on the processor p_q requires a unit of time $\tau_i \cdot w_q$. Similarly, Similarly, set λ_{p_q,p_r} as the reciprocal of the bandwidth of p_q and p_r connection; thus, ε_{ij} is sent from p_q to p_r in time unit $\varepsilon_{ij} \cdot \lambda_{p_q,p_r}$. For a heterogeneous cluster model NHC and a DAG model G, the issue solved here refers to finding a prominent schedule, set as the mapping from G to NHC. The solution quality.

3 The Task Scheduling Algorithm Complying with Time Slot Stack

3.1 The Description of the Algorithm

In the proposed task scheduling algorithm complying with the time slot stack, two steps can be made use of to obtain a high-quality solution: (1) Use the baseline algorithm m to generate a baseline schedule [5]. Recently, the developed algorithms (e.g., HEFT or HCPT) [13, 14] are capable of efficiently generating effective task scheduling for the majority of target computing scenarios, which are able to act as a baseline algorithm. (2) Improve the baseline scheduling through a critical searching process. Its essence is to coordinate the searching process, which tries to optimize the current optimal solution in an iterative manner through using the task scheduling problem and the prior knowledge of the target computing environment.

The proposed algorithm adopts two techniques to improve baseline scheduling: (1) searching the vital path pertaining to the current schedule and trying to down-regulate the completion time through moving tasks to other processors, reducing the initiation period of key task allocation. (2) include finding all the feasible time slots in a given schedule (free time gaps between two specified tasks in processor time) and checking whether it is possible to insert a set of tasks into these time slots [6]. If successfully inserting tasks into time slots can observably reduce completion time (period for performing all tasks in a DAG use) [7]. The mentioned two technologies are both pushing and filling processes, which are performed iteratively till the search solution is not optimized further. Algorithm 1 gives a formal description of the time-slot stack process for task scheduling.

Algorithm 1: Formal description of time slot stack process
1. Make use of the existing algorithms (list-based heuristics) to generate a scheduling plan ψ.
2. Repeat
3. Make use of the pushing process to obtain a new scheduling scheme ψ';
4. Make use of the filling process to obtain a new scheduling scheme ψ'';
5. Select $\rightarrow \psi$ the best schedule in $\{\psi, \psi', \psi''\}$;
6. Until K step iterates $\rightarrow \psi$ doesn't change;
7. Return ψ (currently the best scheduling scheme)

Initially, the algorithm uses a baseline algorithm (eg, a list-based heuristic) to obtain a baseline schedule. Next, it repeats the timeslot stacking process till the quality of the solution doesn't get optimized during the K-step iteration. K denotes the only control parameter in the algorithm (usually given as a small value, e.g., K = 4 or 5). During the termination of algorithm, it will return to the optimal schedule identified currently. In the algorithm, ψ' is the ψ'_1 and ψ'_2 that are defined in functions 1–2.

In terms of the definition of trigger task and target task set, the time slot stack operation attempts for decreasing the scheduling length by reallocating tasks. In the mentioned operating processes, the incentive to reassign such a task t_i is called a trigger task, and a group of tasks, such as $R = \left[t_{j_l} :: t_{j_m}\right]$, is called a target task set on the same processor that triggers task reallocation. In terms of all the push-pull operations, the target task set R and task t_i need to be triggered.

3.2 Push Operation

For a target task and a trigger task t_i in R, a target operation relates to trigger the task on the task to another processor, termed a push operating process. Accordingly, the push operating process is able to be expressed below [8]:

$$Push\_op(t_i, R \rightarrow P) \equiv Best_{t_k \in R, p_j^k \in P}\left(Reassign\left(\{t_k\} \rightarrow p_j^k\right)\right) \tag{2}$$

In the formula, p_j^k is a processor that has not been specified, and the task t_k is to be appointed to the processor. Under the scheduling of ψ_k, $Best_k(\psi_k)$ is the scheduling which has the smallest completion time in the scheduling. If the target task is located before the trigger task t_i on the identical processor, and not any priority relation is identified between the trigger task and the target task [9], the push operation is capable of pushing the target task to reduce the execution level of the trigger task t_i. Accordingly, the trigger task can be performed in advance, which is shown in Fig. 1 as below:

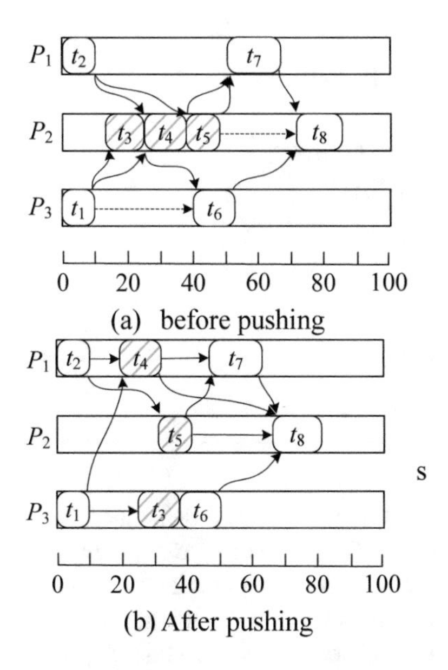

Fig. 1. Push operation example

In the example, suppose the trigger task is t_5 and the target task set R covers t_3 and t_4. Because of t_5 pushes t_3 and t_4, the two target tasks are reappointed to any processor (among p_1, p_2 and p_3) producing the optimal result schedule. Figure 3c indicates the push results. The tasks of t_3 and t_4 are moved to and separately, which led to a reduction in maximum completion time.

For performing a push operating process, a target task set should be made, which can be completed by the prerequisites between the trigger task t_i and the target task below, required to meet the following conditions: (1) $t_k$and t_i are independent to each other; (2) $\phi(t_k) = \phi(t_i)$; (3) $t_k \prec t_i$. The push operating process aims at performing a triggered task t_i earlier through the push operation of the target task away from the current processor. Thus, the third condition is introduced. The 3 conditions above above all satisfy the task between t_i and $Bits_k(t_i)$ bits. Accordingly, $Bits_k(t_i)$ acts as the matching target set R in terms of a given task t_i in a push operating process. The algorithm process is shown in Algorithm 2 as below.

Algorithm 2: Push operation
Procedure $Push\_op(t_c, Bits_k(t_c) \to P)$
1. $R \leftarrow Bits_k(t_c)$;
2. while $R \neq \varnothing$ do
3. Select tasks from;
4. Candidate processor in $P' \leftarrow$;
5. repeat do
6. Candidate processor p_j in P';
7. $\psi' \leftarrow Reassign(\{t_r\} \to p_j \vert \psi)$;
8. if ψ' acceptable, then
9. $\psi \leftarrow \psi'$;
10. $P' \leftarrow P' - \{p_j\}$;
11. until ψ' acceptable;
12. $R \leftarrow R - \{t_r\}$;
13. endwhile

In the worst case, all newly generated schedules cannot be accepted, and the algorithm has $O(|R||P||T|\log|T|)$ time complexity. Nevertheless, as the new scheduling is acceptable in the inner loop, time complexity tends to be less on average.

It is necessary to reassign all the tasks to the processor in P. For considerable processors in a representative heterogeneous clustering scenario, limiting the set of processors to be taken into account is expected. For this reason [10], in terms of a provided target task, only the following processors are taken into account as candidate processors, which is expressed as P': (1) the processor that assigns at least one task; (2) among the processors there is not the fastest task assignment in each cluster processor (with the lowest w_q).

The candidate processors are selected in this way for the following reasons: (1) For tasks distributed to a set processor, performing the rest tasks on such processor can minimize communicating delays. (2) Another feasible processor probably exists, on which no tasks are already appointed. When computing and communication costs are taken into account, the fastest processors in respective cluster in the set of available processors are likely to contribute to certain tasks as well. Through this operation, the time complexity of the push operation turns into $O(|R||P'||T|\log|T|)$.

4 Conclusion

In allusion to the heterogeneous cluster system, a novel static scheduling algorithm on the basis of directed search strategy is put forward, which is called slot stack algorithm. The algorithm seeks an optimized solution according to the initial solution yielded from the baseline algorithm. Through pushing and filling tasks in a set scheduling plan in an iterative manner, the effect of trying to optimize the initial solution is achieved. In particular, such algorithm is good at promoting the initial solution yielded by list scheduling algorithms (e.g., HCPT or HEFT), since the slot stack operation restores the main disadvantages of the list scheduling algorithm. Finally, the feasibility of the developed algorithm is verified through the experimental simulation process. The next research will mainly focus on the application system development and algorithm theory analysis so as to achieve a further development of the proposed algorithm.

References

1. Miner, G.E., Starr, M.L., Hurst, L.R., et al.: Deleting the DAG kinase Dgk1 augments yeast vacuole fusion through increased Ypt7 activity and altered membrane fluidity. Traffic **18**(5), 315–329 (2017)
2. Luo, M., Cai, M., Zhang, J., et al.: Functional divergence and origin of the DAG-like gene family in plants. Sci. Rep. **7**(1), 5688 (2017)
3. Khaleghzadeh, H., Deldari, H., Reddy, R., et al.: Hierarchical multicore thread mapping via estimation of remote communication. J. Super-Comput. **74**(3), 1321–1340 (2018)
4. Han, S.W.: The involvement of central attention in visual search is determined by task demands. Atten. Percept. Psychophys. **79**(3), 726–737 (2017)
5. Li, B., Franzon, P.D.: Machine learning in physical design. In: IEEE 25th Conference on Electrical Performance of Electronic Packaging and Systems (EPEPS), pp. 147–150 (2016). https://doi.org/10.1109/EPEPS.2016.7835438
6. Lv, Z., Li, X., Wang, W., Zhang, B., Hu, J., Feng, S.: Government affairs service platform for smart city. Future Gener. Comput. Syst. **81**, 443–451 (2018). https://doi.org/10.1109/EPEPS.2016.7835438
7. Zeng, L., Guo, X.P., Zhang, G.A., Chen, H.X.: Semiconductivities of passive films formed on stainless steel bend under erosion-corrosion conditions. Corros. Sci. **144**, 258–265 (2018)
8. Lv, Z., Xiu, W.: Interaction of edge-cloud computing based on SDN and NFV for next generation IoT. IEEE Internet Things J. (2019). https://doi.org/10.1109/JIOT.2019.2942719
9. Xia, Z., Kiratitanavit, W., Yu, S., Kumar, J., Mosurkal, R., Nagarajan, R.: Fire retardants from renewable resources. In: Advanced Green Composites, pp. 275–320 (2018). https://doi.org/10.1002/9781119323327.ch11
10. Zhao, Y., Xiao, S.: Sparse multiband signal acquisition receiver with co-prime sampling. IEEE Access **6**, 25261–25269 (2018)

"Internet+" in the Marketing Mode of Electric Power Electricity Value-Added Services to Users

Liangfeng Jin[1(✉)], Zhengguo Wang[1], Qi Ding[1], Ran Shen[1], and Huan Liu[2]

[1] Power Science Research Institute of State Grid Zhejiang Electric Power Co., Ltd., Hangzhou 31200, China
nnn88nnn@126.com
[2] State Grid Zhejiang Electric Power Co., Ltd., Hangzhou 312000, China

Abstract. With the wide use of mobile Internet, it brings great convenience to our production and life. With the advent of the information age, the application upgrading and breakthrough of artificial intelligence and big data, the changes of communication, digital and network technology are mostly attributed to the development of the network era based on computer technology. The purpose of this paper is to explore the value added service in the electricity marketing mode under the background of "Internet+". Through the "Internet+" power of cross boundary products, the unity of traditional power industry and Internet technology and information technology will be combined. In this paper, we will use the research method of specific analysis to compare the data and come to a conclusion. The results of this paper show that Power Grid Corp will fully grasp its own advantages and explore potential value of end-users in the "Internet+" mode. Starting from the marketing theory of Power Grid Corp and user relationship theory, this paper analyzes the current situation and problems of the power grid and end-user marketing management mode, and puts forward the core value orientation requirements of the "Internet+" and the essence of the new Internet transformation "Internet+" mode.

Keywords: "Internet plus" · Power marketing mode · Data fusion · Value added service

1 Introduction

With the deepening of China's power system reform, all power grid companies and power selling enterprises are also carrying out cost accounting, and also derive various energy-saving services. All enterprises continue to carry out market-oriented business strategies and participate in competitive power selling business. Because of this, the sale of electricity, one of the core businesses of power supply enterprises, has received unprecedented attention from the society and customers. At the same time, the value-added service of traditional power marketing mode has also reached the time of updating and upgrading. If electric power enterprises want to continue to expand the market, the key is whether they can do a good job in selling electricity. Therefore, it is

J. MacIntyre et al. (Eds.): SPIoT 2020, AISC 1282, pp. 568–573, 2021.
https://doi.org/10.1007/978-3-030-62743-0_82

self-evident that the importance of marketing strategy research. As the focus of basic public utilities, electric power industry is related to the people's livelihood. In this case, the research on the marketing strategy of electricity sales is not only of great significance to the development of power supply enterprises, but also of great social significance.

At present, the "Internet+" has become a major trend in the development of the world's power grid. "Internet+" is the [1, 2] developed by the traditional power industry using the Internet information technology. The "Internet+ power grid" is a new electricity market relationship management mode based on the Internet, which plays the role of digitalized information and the interactive performance of the Internet media. It aims at narrowing the distance between Power Grid Corp and the end users more conveniently and achieving the profit goal more reliably. [3, 4] As a local monopoly company integrating transmission, distribution and sale of electricity, power grid company has mastered the first-hand information of users, which can achieve a more close management effect with the end users through the Internet [5, 6]. However, in the actual operation process, the number of users is very large, and the data information is even massive. If the management and service of the grid company can not keep up, it is easy to cause confusion, resulting in the loss of users. In the research of related fields of domestic and foreign scholars, customer relationship management is a hot topic explored and discussed in the current stage of the business community, and also one of the most attractive research contents in the academic community [7]. At this stage, the phased strategy of customer relationship management is put forward. The research on customer relationship management in China is a little later than that in foreign countries. At the end of 1980s, the term "satisfaction management" was paid attention and discussed by the academic community [8, 9]. At present, major enterprises pay more attention to users' personal feelings, and relevant theoretical research is also increasing. Domestic scholars point out that there is a "win-win" situation between mobile Internet and "Internet+". If we can effectively implement the concept of "Internet+ Mobile", we will produce unexpected scenes and build a multi screen, cross platform, super user experience [10].

Power companies are not only for profit, but also to promote economic growth and serve the society. Many enterprises put forward plans to build smart grid, which has great significance and far-reaching impact on social development. As the terminal of the power industry chain, the user is an important part of the power system. For the time being, taking the "Internet+" mode for user management is imminent. Only when we begin to use the Internet to establish a sound user data management system and coordinate the relationship with users, can Power Grid Corp compete with other energy companies in a more powerful and better way.

2 Method

In the "Internet+" mode of business, Power Grid Corp generates massive amounts of data in daily operations, which can form a transaction database through Internet servers. Through the use of clustering method to analyze the market in detail, a series of data transformation is carried out, which is automatically collected and stored in the access log of the system for quantitative analysis. Combined with the use traces of fixed user groups, we can further observe and understand the end users, and then develop more targeted and efficient marketing programs to enhance the competitiveness of Power Grid Corp. The difference and commonality of each sample data can be better understood by aggregate analysis.

Let X = $\{x_1, x_2..., x_n\}$ Where Xi is the first sample. i = 1, 2, 3..., n, Cluster can be said to divide x into K clusters G_j, $J = 1, 2, 3, ...k$,

$$G_j \leq X, J = 1, 2, 3, \ldots k \tag{1}$$

$$G_j = \{x_{j1}, x_{j2}, x_{jn_e}\} \tag{2}$$

In the data processing of cluster analysis, there are often different dimensions and units of the collected customer data attributes. When calculating the distance between objects, we must first deal with the correlation of them – convert them to the same unit of measurement, standardize the data types, that is, map the value fields of different attributes in the samples to the same interval, so that they are in the same distance Each sample attribute plays an equal role in the measurement of distance.

3 Experiment

3.1 Experimental Data Source

In this experiment, we study the situation of some regions and cities in our country. These objects are selected from the official website, data library, residential users, domestic and foreign enterprise cooperation and economic factors. The comprehensive consideration of these factors is conducive to the representativeness and typicality of experimental data.

3.2 Experiment Implementation

Because of the strong dependence of the K-Means algorithm on the initial clustering centroid selection of Power Grid Corp terminal users under the "Internet+" mode, it directly affects the effectiveness of clustering results. It first determines an initial division based on the initial clustering centroid of the end users of the grid company, and then optimizes the initial division of the end users. Therefore, the clustering results of K-means algorithm have a strong dependence on the selection of the clustering centroid of the end users of the grid company. Therefore, in order to better solve this problem, the author will judge according to the actual life experience and data statistics, and make the initial center of mass in the first selected categories or clusters to the greatest extent, so as to ensure the relative rationality of clustering results. Therefore, for the end users of Power Grid Corp in the power industry, the most important index of Power Grid Corp's contribution to the "Internet+" mode is the length of the browsing time and the frequency of operation as the main characteristics of future clusters. According to the characteristics and practical experience of the power industry under the "Internet+" mode, the Power Grid Corp end user group will be divided into star users and cash cows. There are four kinds of users: users, low-end users and growing users, i.e. k = 4. By determining a reasonable initial center point and optimizing the k-means algorithm, we can get a better clustering result.

4 Discussion

4.1 Data Visualization

In the questionnaire survey, a total of 200 questionnaires were sent out, 185 of which were recovered, 180 of which were valid, the effective recovery rate was 90%, and the experimental results were valid. The basic information of the respondents is shown in Table 1.

Table 1. Distribution of sample characteristics

Content	Classification items	Number	Proportion of survey results
Age	18–28	21	11.7%
	29–40	65	36.1%
	More than 40	94	52.2%
Education	Undergraduate and below	85	47.2%
	Bachelor degree or above	95	52.8%
Occupation	Employees of state-owned enterprises	125	69.4%
	Private employees	55	30.6%
Average yearly earnings	Less than 20000 Yuan	46	25.6%
	2–5	30	16.7%
	5–10	70	38.9%
	Over 100000 Yuan	34	18.8%

4.2 Analysis and Discussion

Starting from the electricity trading market, there will inevitably be many problems and difficulties. We need to speed up theoretical research and gradually transition to a perfect and reasonable trading market. Improve the interconnection, communication and data sharing between electric power trading platform and dispatching, metering, marketing and other technical systems, so as to meet the needs of "day clearing and month closing" in the day ahead market. Another is to timely carry out the construction of spot market and real-time balance market. As shown in Fig. 1, at present, the market share is dominated by the state-owned power grid, so it is very important for the state to use financial means to improve the value-added service of user electricity.

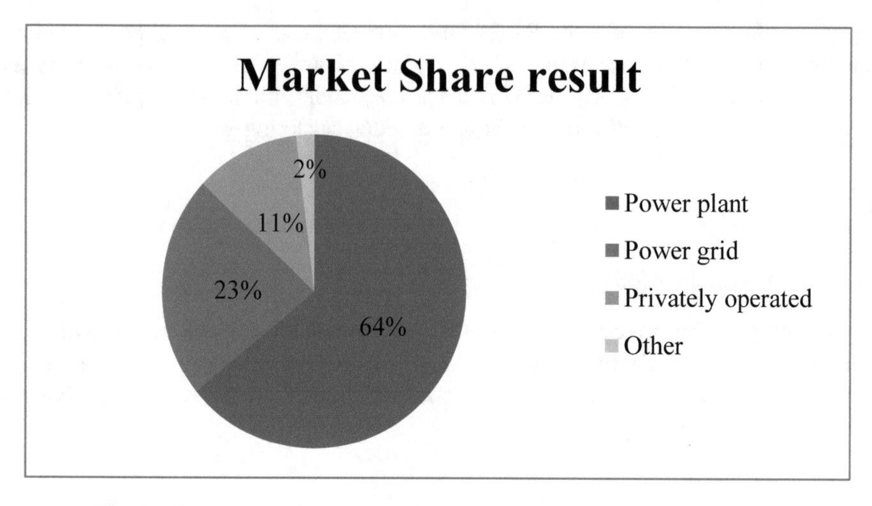

Fig. 1. Proportion of properties of existing electricity selling companies

5 Conclusion

In recent years, with the deepening of economic globalization, China's economy has changed from high-speed to high-quality development. Under this social background, the marketing mode reform follows. Under the guidance of the relevant policies of the national government, this round of electric power marketing mode is in the pilot stage. In the continuous development of the electric power market, it will gradually form a competitive, scientific and orderly electric power market system. At the same time, power grid enterprises should seize the development opportunities brought by the reform, improve the value-added service projects for users, and improve the marketing mode with the help of the Internet.

References

1. Jiangming, X., Lou, Z., Ye, J.: Incoherently pumped high-power linearlypolarized single-mode random fiber laser: Experimental investigations and theoretical prospects. Opt. Express **25**(5), 5609 (2017)
2. Gui, M., Wu, Z., Gong, B.-G.: Value-added service investment decision of B2C platform in competition. Kongzhi yu Juece/Control and Decision **34**(2), 395–405 (2019)
3. Kato, T., Manabe, Y., Funabashi, T.: A study on influence of ramp event of aggregated power output of photovoltaic power generation on electric power system frequency. Electr. Eng. Jpn. **202**(3), 11–21 (2018)
4. Xie, M., Xiong, J., Ji, X.: Two-stage compensation algorithm for dynamic economic dispatch of power grid considering correlation of multiple wind farms. Dianli Xitong Zidonghua/Autom. Electr. Power Syst. **41**(7), 44–53 (2017)
5. Sharov, V.: Application of a modal approach in solving the static stability problem for electric power systems. Therm. Eng. **64**(13), 971–981 (2017)

6. Cheon, S., Nam, K.H.: Pedaling torque sensor-less power assist control of an electric bike via model-based impedance control. Int. J. Automot. Technol. **18**(2), 327–333 (2017)
7. Refaat, S.S., Abu-Rub, H., Sanfilippo, A.P.: Impact of grid-tied large-scale photovoltaic system on dynamic voltage stability of electric power grids. IET Renew. Power Gener. **12** (2), 157–164 (2017)
8. Li, C., Li, X., Tian, S.: Challenges and prospects of risk transmission in deep fusion of electric power and information for energy internet. Dianli Xitong Zidonghua/Autom. Electr. Power Syst. **41**(11), 17–25 (2017)
9. Cárdenas, J.J., Romeral, L., Garcia, A.: Load forecasting framework of electricity consumptions for an intelligent energy management system in the user-side. Expert Syst. Appl. **39**(5), 5557–5565 (2018)
10. Pilli-Sihvola, K., Aatola, P., Ollikainen, M.: Climate change and electricity consumption—witnessing increasing or decreasing use and costs. Energy Policy **38**(5), 2409–2419 (2017)

Influencing Factors of Rural Entrepreneurship by University Students in Qiannan Under the Background of Informatization

Qinglou Cao[1,2(✉)] and Fangfang Chen[3]

[1] School of Economics and Management, Qiannan Normal University for Nationalities, Duyun, China
caoqinglou@126.com
[2] Rural Development Institute, Chinese Academy of Social Sciences Combine Guizhou Academy of Social Sciences, Guiyang, China
[3] School of Chinese, Kunsan National University, Kunsan, South Korea

Abstract. Informatization management has become the mainstream nowadays with the rapid development of network and information technology. Under the background of information, we take the study area, Qiannan Prefecture, Guizhou Province, and select the research object, the students from Qiannan Normal University for Nationalities to research the effect of university students to start up business in the rural area from himself or herself, business environment, business support and so on, furthermore, put forward the policy proposal of perfecting the support system, innovating the financing mode to solve the bottleneck of start-up financing, strengthen the training management, enhance the recognition degree, and perfect the construction of rural infrastructure.

Keywords: Informatization · Rural business · Influencing factor · Countermeasure research

1 Introduction

The social development of human being and life have been further improved from the beginning of the first industrial revolution, represented by mechanization, to electrification, automation, and the fourth industrial revolution, represented by intelligence, hitherto. The new technological revolution, represented by Internet industrialization, industrial intelligence and industrial integration, AI, clean energy, unmanned control technology, quantum information technology, VR and biotechnology, have developed into each field, and the information revolution of agricultural industrialization is developing rapidly [1].

Since reform and opening-up in China, business startups is the catalyst for socioeconomic development, a direct channel for employment, and a necessary way of stabilizing people's livelihood and market economy development. The paper takes the study area, Qiannan Prefecture, Guizhou Province, and selects the research object, the students from Qiannan Normal university for Nationalities to research the effect of university students to start up business in the rural area from himself or herself,

J. MacIntyre et al. (Eds.): SPIoT 2020, AISC 1282, pp. 574–581, 2021.
https://doi.org/10.1007/978-3-030-62743-0_83

business environment, business support and so on, combined with the new era of the current national informatization, furthermore, put forward solutions to help university students of Qiannan for innovation and starting up business, develop rural talents, and lay the foundation for the optimization strategy of overcoming poverty [2].

2 Present Situation on University Students' Rural Business in Qiannan

We take the study area, Qiannan Prefecture, and select the research object, the students from Qiannan Normal University for Nationalities to synthetically analyze. The students of Qiannan Normal University for Nationalities are 13000 persons. According to the data of 522 questionnaires about 90% of rural university students, it is inferred that the proportion of university students from rural area in Qiannan Normal University for Nationalities is large.

2.1 Structural Analysis of University Students' Business in Qiannan Normal University for Nationalities

2.1.1 Data Sources

With the support of student organizations, such as university league committee, teachers, students youth entrepreneurship center, students' union and so on, it has good representation and universality according to 600 questionnaires, issued to 17 secondary colleges of Qiannan Normal University for Nationalities, and 522 valid questionnaires, involving each major, grade, and university.

2.1.2 Structural Analysis of University Students' Entrepreneurship

As shown in Table 1, the results reflect the overall low entrepreneurial passion of university students in Qiannan Normal university for Nationalities. Through cross-over analysis, the students who are starting a business and have entrepreneurial ideas is counted and calculated. Among them, about 5.6% of sophomore students are starting a business only, and the proportion of junior and senior entrepreneurs is 8% and 7.4%, the other is 3%, and the proportion of people who do not start a business is more than 90%. Students who have ideas to start a business (including a little and a lot of ideas) account for about 70% of the total, especially the senior students who full of passion and strong entrepreneurial intention.

Table 1. University students' entrepreneurship conditions

Type	In the process of	None	Total
Junior	5 (5.62%)	84 (94.38%)	89
Sophomore year	12 (5.61%)	202 (94.39%)	214
Junior year	9 (7.96%)	104 (92.04%)	113
Senior year	7 (7.45%)	87 (92.55%)	94
Others	3 (25%)	9 (75%)	12

2.2 Entrepreneurship Motivation Analysis of Students in Qiannan Normal University for Nationalities

Compared with the employment phenomenon of graduates from rural to urban areas and inland to coastal areas during a long period of time in the early period of reform and opening up in China, the market and opportunities in rural areas are gradually reflected, more and more university students devote themselves to returning home to start a business with the development of Internet information and the strategy of rural revitalization [3]. The results show that more than 70% of 522 university students have had the desire to start a business. 52.7% of the university students who are starting a business and are interested in starting a business want to go to the vast countryside to start a business. According to the field visit investigation, and combined with the field situation analysis of Qiannan Normal University for Nationalities, its entrepreneurial motivation can be divided into the following points:

2.2.1 Guizhou Big Data Development Provides the Basis

Guizhou Province has opened the "7+N" cloud construction of the first provincial government and enterprise data overall storage, sharing, opening, development and utilization of cloud service platform (Yunshang Guizhou). In 2018, Guizhou Province deeply implemented the big data strategic action, and the telecommunications industry maintained a high growth trend. From January to November 2018, the operating income of Internet and related services, software and information technology services in the province increased by 75.8% and 21.5% respectively from the same period last year. The gross of telecom business reached 219.117 billion yuan in the whole year, with an increase of 165.5% over the previous year, and the growth rate was 19.3% points faster than 2017.

2.2.2 Homesick

Nowadays, university students are usually far away from their hometown during their school years, and usually seldom go home. After graduation, they have less chance to go home if they look for a job. Under the circumstances, 40% university students who intend to start a business in rural areas intend to return home to start a business to take care of home easily.

2.2.3 The Personal Development

Rural areas have unique entrepreneurial resources. Schumpeter points out that entrepreneurial motivation is an opportunity to create profit-generating resources by revolutionizing the integration of entrepreneurial resources to meet market demand and thus create social value [4].

2.2.4 External Environment Pressure

In recent years, university students have been increased considerably, and this situation continues every year, resulting in the unemployment of structured knowledge in education. More and more university graduates are difficult to find suitable jobs in the urban, the pressure and pace of life that the urban gives university students are also increasing, so returning home to start a business has become one of the choices of many university students.

2.2.5 Policy Factor

Over the years, national policies have strongly supported rural development. With the implementation of the policy of "agriculture, rural areas, and rural residents", and the strategy of rural revitalization, China has continuously increased the support for development funds and policies in rural areas, especially for university students returning home to start a business, which is reflected in the aspects of funds, management, technology, policy and so on, and has greatly reduced the threshold and difficulties of university students' enterprises.

3 Problems and Reasons of Rural Enterprises for University Students in Qiannan Prefecture

Although the current social and national policies attach great importance to rural development, and encourage university students and graduates to devote themselves to rural self-employment, all kinds of entrepreneurship competition is growing vigorously, but the status quo of self-employment of university students is still not optimistic. Through big data analysis, problems and reasons of rural enterprises for university students in Qiannan Prefecture are analyzed by taking the case of the students of Qiannan Normal university for Nationalities.

3.1 Entrepreneurial Practical Ability Still Needs to be Strengthened

According to the statistics of the Ministry of Education, at present, the problem of employment difficulty for university graduates in our country has appeared in a wide range. The government encourages students to start businesses every year, but less than 1% of all university graduates are truly self-employed, of which, and people who select to start a business in rural areas is less.

Compared with foreign universitys and universities, such as the United States, the proportion of students' self-employment is as high as 20%–30%, which shows that the entrepreneurial consciousness of university graduates in China is not enough.

3.2 Low Overall Success Rate

According to the data of Employment Report Chinese Undergraduates, 2018, issued by the Max Research Institute and the Chinese Academy of Social Sciences, 6.2% of university students started their own businesses after three years of graduation, and the overall survival rate of entrepreneurship decreased. Combined with research and interview, the overall national data development and current entrepreneurship situation of university students of Qiannan Normal university for Nationalities, although rural entrepreneurial opportunities are relatively large, blind participation without good opportunities can easily lead to failure.

3.3 Insufficient Entrepreneurial Resources Integration

The undergraduate entrepreneur is the entrepreneur who represents the youth generation, they are young and energetic. Compared with social entrepreneurs, the resources accumulated by university students are less connected and it is difficult to integrate the resources needed for entrepreneurship [6]. For university entrepreneurs, the resources they have are far from satisfying the development of new enterprises. 89.5% of the university students willing to start a business in Qiannan Normal university for Nationalities thought they lacked funds, and 72.9% thought they lacked technology and relevant entrepreneurial knowledge.

3.4 Insufficient Entrepreneurial Resources Integration

In addition to policy system and university education factors, the traditional social prejudice, rural backward infrastructure and uncontrollable climate have virtually increased the hardships and difficulties of entrepreneurship. The current social environment in rural areas is not conducive to the creation of the foundation of rural entrepreneurship, but also to a certain extent affects the enthusiasm of university students to rural entrepreneurship. Although the policy provides a lot of support policies for university students' rural entrepreneurship, it is difficult to make university students really get strong support because of the unclear functions of government departments in grass-roots areas and the procrastination of their work efficiency; due to their own interpersonal relationship and complex natural environment, some rural areas are extremely exclusive, which seriously affect the healthy and steady development of university students' rural entrepreneurship. In addition, the rural areas are located in remote areas, the construction of transportation is not perfect, the ability of information communication is difficult, the popularization of telecommunication network infrastructure is poor, and the severe infrastructure environment affects the willingness of university students to start a business in rural areas.

The survey data shows that 70% are willing to go to the countryside to start a business, but less than 10% are really put into practice. The grim reality tells us that we must pay attention to the construction of infrastructure in rural areas and the cultivation of entrepreneurial culture in rural areas.

4 Countermeasures and Suggestions for University Students' Rural Entrepreneurship

4.1 Improving the Policy Support System

When university students start a business to rural areas, colleges and universities should actively contact local governments to make overall planning, lead entrepreneurial university students to the rural areas, give full play to the mobility of human factors, and form a joint force with local farmers.

First, the local government at the grass-roots should introduce and adjust the system and policies to guide university students to start their own businesses in rural areas according to local conditions, and provide convenience for university students to start their own businesses in the aspects of platform development, public resources, industrial matching, venture capital and government services. Second, the local civil servant at the grass-roots give full play to the role of convergence, actively communicate with rural entrepreneurial university students, actively cooperate to provide all kinds of market information resources, and guide rural entrepreneurial university students to start and develop smoothly and smoothly. Third, coordination of rural technical experts or experienced entrepreneurs with innovative practices are invited to provide the necessary technical support and experience guidance for the development of start-up enterprises to university students, and promote the healthy development of the whole entrepreneurship value chain of university students to rural entrepreneurship.

4.2 Innovate Financing Model, and Solve Entrepreneurial Financing Bottlenecks

The biggest difficulty for university students to start a business in rural areas is lack of funds, 89.5% of university students are contrived, the biggest problem of starting a business is lack of funds. colleges and universities should combine the efforts of the government and social industries to provide a platform for university students to start a business and create conditions for entrepreneurship.

First, coordinate with departments, such as banks, financial institutions, etc., to increase the amount of loans for rural entrepreneurial university students, and simplify the loan application procedures, reduce the threshold of guarantee, shorten the processing time, and provide necessary financial support for entrepreneurial university students. Second, innovative cooperation model. build a platform to encourage successful entrepreneurs, outstanding entrepreneurs, well-known experts and scholars for university students to rural entrepreneurship guidance and advice and cooperative development. Third, encourage diversified financing models. Strengthen propaganda, guide the society to pay attention to and support the rural entrepreneurship of university students, take advantage of the technology and knowledge of university students, combine with local land, funds and existing resources, and combine with "three changes reform" and financing in many ways to solve the problem of funds [7].

4.3 Innovate Financing Model, and Solve Entrepreneurial Financing Bottlenecks

Colleges and universities, as the talent highland of university students, should pay attention to, guide and help university students to devote themselves to rural areas for entrepreneurship, and strengthen the management, and attach great importance to the subject education of university students' innovative entrepreneurship, especially strengthen the efforts to guide university students to devote themselves to rural development.

First, help university students to establish a correct world outlook and philosophy, fully combine individual ideals with social needs, guide rural entrepreneurship, serve the strategic development of rural revitalization, and realize the unity of self-worth and social value. Second, according to the demand of rural entrepreneurship and scientific and technological talents, combining with the characteristics and advantages of schools, we should innovate and reform the talent training program, increase the construction of rural technology and entrepreneurship theory curriculum, combine the training of professional talents with the demand of rural construction, and improve the level of colleges and universities serving the local and rural areas. Third, in conjunction with local enterprises and rural government departments, we should jointly develop the practice base for university students to practice in rural entrepreneurship, and provide all-round and three-dimensional services for university students to provide training on innovative entrepreneurship topics, interpretation of various national policies, rational demonstration of various entrepreneurial projects, relevant supporting technical support and follow-up and evaluation guidance. Fourth, set up a typical case of university students' rural entrepreneurial achievements, build a platform for sharing achievements, share successful experience of entrepreneurship, and then improve the success rate of university students' rural entrepreneurship.

4.4 Improving Rural Infrastructure

The reports of the 19th National Congress of the Communist Party of China of the Party puts forward the strategy of rural revitalization, which ushered in a historic new opportunity for the development of rural areas in China, especially the construction of rural infrastructure. In the new age, the condition of rural infrastructure in our country has improved obviously, but the facilities of hydropower and garbage pollution treatment still have outstanding short board, which restricts the rural development to a certain extent.

According to the data, 80% of university students now express their willingness to start a business in rural areas with wide opportunities, but more hope that rural areas have relatively perfect infrastructure. First, improve rural infrastructure and provide solid hardware conditions for entrepreneurial university students and rural local entrepreneurs. Second, actively enhance the soft environment. On the one hand, actively publicize the national strategy of rural revitalization, and improve the mass base; On the other hand, develop rural cultural resources, maximize the utilization of resources, and play an ecological effect. Third, guide farmers to change their traditional thinking, the new professional farmers have a better strategic layout and talent reserve, and actively integrate into the new changes of university students' rural entrepreneurship the new professional farmers have a better strategic layout and talent reserve, and actively integrate into the new changes of university students' rural entrepreneurship [8].

Acknowledgments. This work was financially supported by Guizhou Provincial Social Science (2019qn007) and Guizhou Provincial Federation of Social Sciences (LHKT2019YB07) fund.

References

1. Enrong, P., Zhiyan, S., Liang, G.: Intelligent integration and reflexive capital reorganization – dynamic analysis of the new industrial revolution in the age of artificial intelligence. Res. Dialect. Nat. **36**(2), 42–47 (2020)
2. Jiaqi, X., Xiaoxiao, L.: The present situation and training strategy of university students' rural entrepreneurship ability. Further Educ. Res. **2016**(1), 31–33 (2016)
3. Jun, Q., Tingjun, W.: Research on rural entrepreneurship of university students from the perspective of rural revitalization strategy. Educ. Vocat. **2019**(1), 67–71 (2019)
4. Jiangquan, H.: Identification of rural entrepreneurship opportunities for university students in China. Technol. Manage. Res. **36**(9), 192–197 (2016)
5. Xiwen, Y.: Research on quality requirements and policy support for rural university students returning home in the new age. Agric. Econ. **2019**(7), 101–103 (2019)
6. Jinglu, G.: New opportunities for university students to start a business in rural areas. Chin. Talent **2013**(15), 28–29 (2013)
7. Yiming, P.: Research on problems and countermeasures of university students' entrepreneurship in rural areas. Chin. Adult Educ. **2017**(5), 62–65 (2017)
8. Yufei, Z.: The faced dilemma and countermeasures of university students' self-employment–based on rural entrepreneurship perspective. Hunan Soc. Sci. **2010**(4), 180–182 (2010)

Transformation of Local Government Administrative Management Concept Based on Smart City Construction

Ya Dong[1,2] and Huanping Zhu[1,2(✉)]

[1] Haojing College of Shaanxi University of Science and Technology, Shaanxi, China
answerdong3@163.com
[2] Xixian New Area of Xi'an, 712000 Shaanxi, China

Abstract. With the changes of the times and the development of science and technology, science and technology have penetrated into every aspect of our lives. The emergence of smart cities has made China's urban landscape change with each passing day. In the context of smart city construction, the management concept of local governments has also changed. This article explores the change of local government administrative management concepts under the background of smart city construction. This article uses the literature analysis method to summarize the trajectory of administrative changes in local government in China, analyzes the current problems and causes of local government administration, and puts forward some related suggestions. The research in this paper finds that in the context of smart city construction, there are many problems in the administration of local governments in Shanghai. These problems are caused by a variety of reasons. Therefore, we must keep up with the times and improve related systems to improve. Local government administration.

Keywords: Smart city · Local government administration and management concept change · Urban construction

1 Introduction

The Party Central Committee and the State Council have attached great importance to the construction of new smart cities, and explicitly requested that the construction of new smart cities be accelerated. At present, China's new type of smart city has entered a period of rapid development, with a large construction volume, which cannot be sustained solely on the basis of government financial investment. To break the "bottleneck" constraint, we must change the model that used to rely mainly on government-led construction and actively promote institutional reforms and policies Innovate, establish scientific and standardized investment and financing mechanisms, and guide and encourage social capital to participate more and better in the construction of China's new smart cities.

With the emergence of smart cities, the public's perception of the relationship between local governments at all levels and the way government is governed has undergone a series of changes [1, 2]. Local government administration is a manifestation

J. MacIntyre et al. (Eds.): SPIoT 2020, AISC 1282, pp. 582–590, 2021.
https://doi.org/10.1007/978-3-030-62743-0_84

of the specific methods and means by which local governments perform their functions. Since local governments can directly contact the people, the choice of their administrative management methods is not only related to the efficiency of local government administration and office, but also to the public's evaluation of the government and Trust [3, 4]. Traditional administrative management methods have now exposed many problems, and administrative reform and innovation, especially in the reform of government agencies and management concepts, have moved towards a more humane development, moving from single hard management to good governance. Therefore, it is of great significance to explore the change of local government administrative management concepts in the context of smart city construction.

This article first analyzes and analyzes the main characteristics of smart cities. Through literature analysis combined with comparative analysis and other methods, it reviews the historical trajectory of the evolution of local government governance in China, and analyzes the problems and problems of local government administration in China. The reason and put forward some related suggestions.

2 Proposed Method

2.1 The Main Characteristics of Smart Cities

(1) Comprehensive perception

Smart cities can integrate all kinds of information in the city, with the support of big data, collect information through the Internet, comprehensively grasp every corner of the city, and then form the Internet of Things [5]. In the daily operation of the city, digitalization and intelligence are realized. Various smart devices in the city together form a huge network to capture various information in the city at all times. Then the data is aggregated to the big data center and cloud computing analysis and processing center to intelligently integrate the data.

(2) Extensive interconnection

The construction of smart cities requires the use of broadband network technology and wireless communication technology to achieve the extensive interconnection of urban information. The basis for its further development is that there is a comprehensive system and a large amount of data information that can be shared. This information can be shared on the network. With continuous sharing, the economic benefits brought by it are getting larger and larger [6]. At the same time, the interconnection, intercommunication and interaction of information resources can be achieved to promote the realization of intelligent feedback applications. Sharing and linkage The various departments are linked together to achieve cross-departmental information sharing and resource integration, break down data barriers between departments, and achieve unified management and interoperability of data, thereby improving government administrative efficiency, and then quickly performing social management and facilitating residents' lives.

(3) Highly integrated
Unlike traditional cities, smart cities can unify the independent parts of the original city into a system through information technology. Under this system, the original management methods and people's production and lifestyle will undergo huge changes [7]. Because the system is highly integrated and achieves seamless connection between urban systems, the government can make smarter, faster and easier decision management in city management, operation and maintenance, and citizens can get high-quality services quickly and satisfactorily in their daily work.

(4) Green ecology
As the saying goes, Jinshan Yinshan is not as good as green water and green mountains. Smart cities are naturally inseparable from green sustainable development. People's life cannot be separated from the natural environment, and harmonious coexistence between man and nature is an important way for sustainable development [8]. Nowadays, urbanization is developing faster and faster, and people are beginning to concentrate in cities, which has further caused environmental damage. The smart city can monitor the environment in different places in the city in real time based on intelligent equipment, such as sensors, etc. Once problems occur, it can be dealt with in a timely manner to achieve sustainable development of the smart city. Sustainable development to achieve the coordinated development of economy, population, resources and environment is the construction goal of smart cities. This is the concept of sustainable development. Green development is an inevitable choice to follow the path of sustainable development.

(5) People-oriented
People-oriented is the basic point of smart city [9]. As the most basic unit in a city, all means related to urban development are to protect the interests of the people. Regardless of whether the government revises its policies or builds public facilities, it must be closely surrounded by citizens to meet the standards of citizens' production and living, in order to achieve the ultimate goal of sustainable development of smart cities. Serving the people. The essence of serving the people is people-oriented. It requires that the interests of the people be put first. It combines the formulation of policies, the construction of public infrastructure and the needs of urban residents, responds to public demands, and truly serves the people. People and improve people's sense of gain.

(6) Innovation promotion
Innovation is the vitality of a city, and it is also the core force for the development of a smart city. To realize the smartness of a city, new blood must be injected. Today, with the rapid changes in science and technology and the advancement of knowledge, the extensive economic growth mode is no longer suitable for the development of the times, and it is necessary to rely on knowledge, technology and talents to improve the quality and competitiveness of urban development. Through scientific and technological innovation, institutional innovation, management innovation, model innovation, etc., tapping the city's potential, creating new products and providing new services, which is conducive to achieving a win-win situation for the government, enterprises and individuals.

2.2 The Historical Track of the Evolution of Local Government Governance in China

The mode of local government governance in China is not sticking to the rules, and there is no change. With the continuous deepening of China's institutional reform, the environment of the government department system is constantly changing, and the goals of local governments have also changed greatly from before the reform [10]. With the change of the times, local governments should work harder to find a governance model suitable for the times. Local governments have become more and more important in the process of governance reform [11, 12]. While the local government is based on the development of the local economy, the government's decentralization reform of the society has made the people's power more and more, and the forces of all parties are slowly penetrating into the social strata, and led by the local government to govern the local government together. Local Government Governance Process Since the reform and opening up, China's economy has developed by leaps and bounds, and great changes have been made in the reform of the economic system and the functions of local governments. The governance structure of local governments is also constantly changing.

1) Vertical change
From the beginning of the founding of the People's Republic of China in 1949, local governments made major adjustments to China's old hierarchical structure, and organized regional divisions of provinces and cities as a whole, followed by regions, provinces, counties, and townships. It is a large district, and a special commissioner's office is set up, which is set in the middle of provinces and counties as a provincial dispatch agency, and a district office is also established in the county and countryside as a county-level dispatch agency. However, as the pace of system reform is getting faster and faster, everyone slowly discovers that the governance structure of the city-administered counties still has great shortcomings, so they are considered to be replaced by provincial-managed counties. The eleventh five-year report also mentioned the need to reduce levels. Along the way, we can see that the structural characteristics of local governments have changed in two parts: from 49 years to reform and opening up, the governance structure of the government has basically remained unchanged. The decisive factor is the planned economy. All economic, political, and social rights are under the control of the central government. Local governments have basically no rights at all. Subordinate governments must obey their superiors. They have no autonomy. The superior governments are in charge of everything. In view of the timely and innovative factual verification of reform and opening up, the situation of similar power concentration has changed, and the superiors have also tried to decentralize power to lower levels, but this is only a part and not all decentralization. As the state distributes some power to local governments, local governments at the provincial level also delegate power to lower levels to promote the comprehensive and common development of the provinces and municipalities. After the local governments have some of their own rights, they have enabled the economic development of the entire society. Great progress has been made; on the other hand, higher-level governments

still have to control lower-level governments, and its means mainly include finance, administration, and personnel, and this subordinate relationship of subordinates has not changed.

2) Horizontal changes
Since the founding of the People's Republic of China in 49 years, great changes have taken place in the entire economic system, and the planned economy has slowly transformed into a market economy. As the economic system has changed, and in order to adapt to this change, the political system has also changed. A major factor is the horizontal local government governance structure. Under the traditional planned economy system, the ties between local governments were mainly vertical relationships, and there were few horizontal relationships, which led to a representative honeycomb structure. The factors that cause this situation can be seen from two aspects: one is subjective needs, the central government has all the rights and one body, the locality is completely under the jurisdiction of the central government, and the local government itself is not an independent subject. It is impractical to cooperate with other local governments. Then there are objective conditions. The local conditions and resources are small. The main factor is that the central government attributed most of the main resources to itself. This also prevents local governments from having large projects, which makes local governments interact with each other. Cooperation has become scarce. Due to the existence of these factors, there was basically no connection between local governments at that time, and some individual governments sought cooperation under the unified deployment of the central government. Since the reform and opening up, the market economy system has been gradually formed. The ability to make the market play a role in resource allocation is its core, used to improve the efficiency of the use of limited resources

3) Changes in government governance
In the early days of the founding of the People's Republic of China, the central government held power, and the source of power for local governments was the central government, while the source of power for subordinates depended on superiors. This was a one-to-one connection. Because of this connection, the governance mechanism is very monotonous. In terms of organizational structure, there are two main reasons for the presence of the same organization at the same level. One is to facilitate the transmission and execution of superior instructions by governments at all levels, and the other is to facilitate the handling of more complicated matters when the power is highly concentrated in the central government.. In the initial period of reform and opening up, the local governance mechanism did not differ substantially from the previous period. China's local governance mechanism transitioned from a "single type" mechanism to a "stress type" during this period. In order to achieve and exceed the tasks assigned by superiors, governments at all levels operate by continuously assigning tasks to subordinates. The next level of government is under the pressure of their superiors to complete the tasks assigned by their superiors in various ways. Therefore, in the case of assigning tasks layer by layer, the establishment of a hierarchical counterpart organization is an inevitable choice.

3 Experiments

Research Method

(1) Comparative analysis.

This paper studies the local governance issues of various countries and regions through the network and books, summarizes the advantages and disadvantages, and compares and analyzes the actual situation with our country to provide a powerful reference for our country's local governance work.

(2) Literature analysis method.

This paper has collected a wide range of domestic and foreign data and research results on local government governance, carried out sorting and analysis, and formed its own theoretical system, and combined with the actualities of local governance in China, put forward related suggestions.

(3) Theory guides practice

Based on the combination of theory and practice, this paper comprehensively studies the trajectory of the change of local government's administrative management concept in the context of smart city construction, finds out the problems and causes by analyzing the current situation, and then finds out effective countermeasures.

4 Discussion

4.1 Problems and Causes of Local Government Administration Under the Background of Smart City Construction

In the context of smart city construction, there are many problems in local government management in China, and the reasons are various. The problems and causes of local government administration in the context of smart city construction are analyzed. The results are shown in Table 1.

From Table 1, we can see that under the construction of smart cities, there are still many problems in China's local administration. The causes of these problems are various. To change these status quo, we cannot solve them overnight. Things need us to proceed from reality and explore a more suitable route to complete the change of local government administrative management concepts in the context of smart city construction.

4.2 Probe into the Change of Local Government Administrative Management Concept Under the Background of Smart City Construction

In order to improve the change of local government administrative management concept in the context of smart city construction, it is necessary to improve local administrative management. How to improve local government management? This article sorted related materials through various literature searches, and the results are shown in Fig. 1.

Table 1. Analysis of problems and causes of local government administration in the context of smart city construction

The problems and reasons of local government administration	Analysis of the problems in the administration of local government
The local government has a relatively backward concept of management innovation	The concept of management personnel has not kept up with the pace of the times, and the concept is old
With weak sense of rule of law and strong color of rule of man	The supervision mechanism of government administration fails to play its expected role
The role of the government is too much and there is a poor sense of efficiency	There is a conflict of interest between departments, which affects the implementation of administrative policies
The government has the phenomenon of official standard and lacks a strong sense of service	Lack of humanistic care in government administration

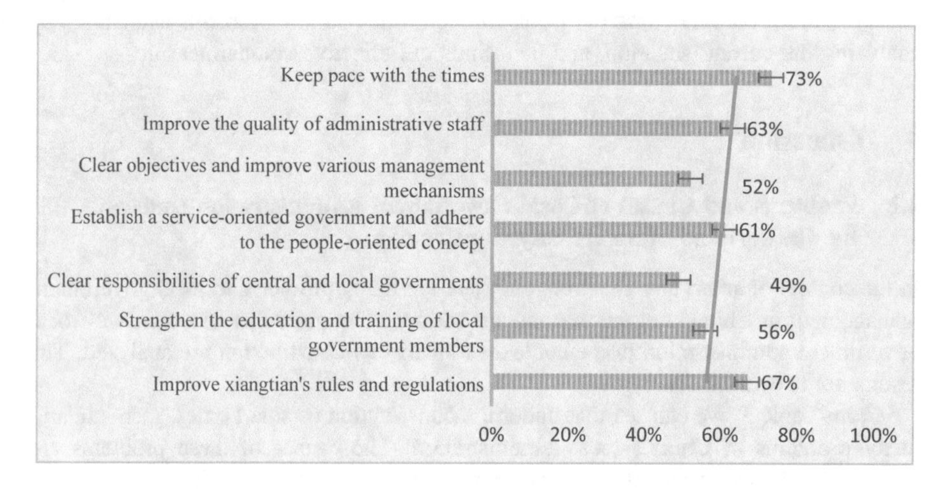

Fig. 1. Ways to improve local administration

As can be seen from Fig. 1, in order to improve the local government's administrative management in the context of smart city construction, it is necessary to improve the relevant rules and regulations, strengthen the education and training of members of local government departments, and achieve clear separation of powers and responsibilities. Clarify the responsibilities of the central and local governments, establish a service-oriented government, and adhere to the people-oriented concept, clear goals, improve various management mechanisms, and improve the quality of administrative staff. Finally, in the context of smart city construction, it is even more important to keep pace with the times, and Keeping pace with the times can improve the administrative management of local governments.

5 Conclusions

The emergence of smart cities has brought convenience to people's lives. However, there are still more or less cruxes in the administration of local governments in our country, which has also caused a series of contradictions and conflicts, which has caused a great stir in society. To this end, this article explores the change of local government administrative management concepts in the context of smart city construction. The research in this article finds that there are still many problems in the local government administration in China at present. These problems are not caused by a single reason. There are many factors that make it difficult to improve the administration of local governments. However, we must firmly believe that The construction process of local administration is long and the ending is wonderful. Our local government should pay full attention to the development of administrative management. Local government management must be innovative. It is an inevitable choice for social development. Government management innovation and perfection of local administrative management are An indispensable part of the cause of building socialism with our own characteristics is our unremitting and arduous task from the beginning and a glorious mission entrusted to us by history.

References

1. Libing, W., Wang, J., Kumar, N.: Secure public data auditing scheme for cloud storage in smart city. Pers. Ubiquit. Comput. **21**(5), 1–14 (2017)
2. Yao, Z., Xia, X., Zhou, C.: Smart construction of integrated CNTs/$Li_4Ti_5O_{12}$ core/shell arrays with superior high-rate performance for application in Lithium-Ion batteries. Adv. Sci. **5**(3), 1700786 (2018)
3. Ma, D., Li, G., Wang, L.: Rapid reconstruction of a three-dimensional mesh model based on oblique images in the Internet of Things. IEEE Access **6**, 61686–61699 (2018)
4. Zhang, M., Wang, H., Wang, Y.: Progress and prospect of urban geological survey in China. Northwes. Geol. **51**(4), 1–9 (2018)
5. Lou, Y., Duan, L., Wang, S.: Front-end smart visual sensing and back-end intelligent analysis: a unified infrastructure for economizing the visual system of city brain. IEEE J. Sel. Areas Commun. **37**, 1489–1503 (2019)
6. Wang, Y., Wang, X., Guan, F.: The beneficial evaluation of the healthy city construction in China. Iran. J. Public Health **46**(6), 843–847 (2017)
7. Yandra, A., Nasution, S.R., Harsini, Wardi, J.: A closer look on ineffectiveness in Riau mainland expenditure: local government budget case. IOP Conf. **156**(1), 012058 (2018)
8. Bolaji, S., Campbell-Evans, G., Gray, J.R.: The perils of bureaucratic complexity: education for all in Nigeria. Int. J. Educ. Adm. Policy Stud. **9**(1), 1–9 (2017)
9. Sun, T.-T., Ma, X.-H., Li, X.-X.: Discussion on research status and development ideas of biopotency for Chinese materia medica. Chin. Tradit. Herbal Drugs **48**(9), 1906–1911 (2017)
10. Paynter, S., Voinovich, G.V.: Empowering the public-private partnership: the future of America's local government (Athens, OH: Ohio University Press, 2017). 97 pp. $19.95. ISBN: 9780821422663. Pub. Adm. Rev. **79**(4), 614–615 (2019)

11. Drew, J., Grant, B.: Means, motive, and opportunity – local government data distortion in a high-stakes environment. Aust. J. Pub. Admin. **76**(2), 237–250 (2017)
12. Sadushi, M., Shatku, S.: The code of the administrative procedures according to the principle of the power separation and balancing in the central and local government bodies in Albania. Academicus Int. Sci. J. **16**(16), 133–140 (2017)

Analysis and Simulation of Oxidation Causes During Steam Turbine Oil Use Based on Fenton Oxidation

Yimin Chen(✉)

State Grid Shandong Electric Power Research Institute, Jinan 250003, China
738715729@qq.com

Abstract. Steam turbine oil mainly plays the role of lubrication, heat dissipation and speed regulation in the steam turbine unit. The viscous deposits generated during the operation of steam turbine oil can affect the heat dissipation and lubrication of the equipment. In the long-term oxidation process, due to the influence of oxygen, heat, water, impurities and other factors, degradation products will be produced. The characteristic of Fenton oxidation method is that it has super strong oxidizing ability and can be deeply oxidized and degraded. The Fenton oxidation method is mainly based on the reaction time and temperature and the pH value between the adsorbent and the adsorption phase. In order to adjust the additive formulation and prevent the turbine oil from turning green during use, it needs to be evaluated by appropriate experimental methods. The purpose of this article is to use the Fenton oxidation method to analyze the causes of oxidation during steam turbine oil use. Through the analysis of pH value-color change by Fenton oxidation method, the structure and source of the coloring matter are inferred. In this paper, three steam turbine oils are selected for physical and chemical performance and failure analysis. The experimental results show that the use of Fenton's reagent can accelerate the color and chemical structure of the oxidation products during the oxidation process similar to the color-forming substances in the used steam turbine oil.

Keywords: Fenton reagent · Fenton oxidation · Steam turbine oil · Oxidation process

1 Introduction

Turbine oil is in a forced circulation state for a long time during operation, and it is difficult to gradually age, and it will not be affected by oxygen, heat, water vapor, impurities and other degradation products [1, 2]. Some are soluble in oil, making the color of the oil darker, while others are insoluble in oil, producing a thick deposit, namely oil sludge, which is very harmful [3]. The sludge deposited in the equipment will affect the heat dissipation, lubrication, and even cause the speed control mechanism to be stuck and other undesirable phenomena. The oxidation of steam turbine oil causes the acid value of the steam turbine oil to increase, which changes the chemical environment of certain additives and reduces the function [4]. The chemical reaction with the friction surface metal and the oxygen in the environment constantly consumes

J. MacIntyre et al. (Eds.): SPIoT 2020, AISC 1282, pp. 591–598, 2021.
https://doi.org/10.1007/978-3-030-62743-0_85

the additives, resulting in extreme pressure. The wear-resistance and friction-reducing functions reduce the boundary lubrication [5, 6]. Long service life is one of the most important characteristics of steam turbine oil.

The increase in metal abrasive debris leads to three-body abrasive wear and also plays a catalytic role in the oxidation of steam turbine oil. External liquid and solid pollutants can accelerate the oxidation of steam turbine oil and the failure of additives [7]. The combined effect of these four factors will lead to steam turbine oil failure. Fenton's reagents are based on ferrous ions and undergo a series of free radical reactions to generate hydroxyl radicals (·OH). Through the paramagnetic resonance (EPR) method to further understand the characteristics of the oxidant fragments generated in the Fenton reaction, the (·OH) characteristic signal can be obtained [8, 9]. Based on this, the mechanism of high-energy free radicals and the production principle of oxidants are obtained [10, 11]. The generated (·OH) electrode has high potential and strong oxidation performance, and can oxidize some substances that are difficult to be oxidized by general oxidants.

The color of the turbine oil turns green during use. Through analysis, it is found that the color is derived from nitrogen-containing additives and has a certain relationship with the pH of other additives. In order to adjust the additive formulation and prevent the turbine oil from turning green during use, it needs to be evaluated by appropriate experimental methods. In this study, an oxidation study of the oil using Fenton's reagent showed that the oxidation products were the same as the oxidation products produced during the operation of the oil.

2 Basic Introduction to Fenton Oxidation

2.1 Fenton Oxidation

The principle of Fenton oxidation is the mixed reaction of Fenton system with complex colored sunlight or monochromatic ultraviolet radiation. This principle is based on the common Fenton method, and a photo-excited chemical reaction takes place to obtain more hydroxyl radicals (·OH), thereby improving the recovery efficiency of ferrous ions and reducing the content of pollutants. In recent years, based on the Fenton oxidation method, the common Fenton oxidation technology is used in combination with other technologies, such as changing the production method of H_2O_2 and improving the state of the catalyst to reduce the loss of the catalyst. Or to achieve the recycling of iron salt catalyst and reduce the secondary pollution of the reaction has become one of the important directions of many scholars. In the Fenton oxidation method, Fe^{3+} can be reduced to in various ways. The reduction pathway includes Fe^{3+} and H_2O_2 in the system. A reaction similar to Fenton is reduced to. This reaction needs to undergo a two-step conversion process. Fe^{3+} first generates H_2O_2. Then compound $[Fe^{III}(HO_2)]^{2+}$. Subsequent conversion from $[Fe^{III}(HO_2)]^{2+}$ to Fe^{2+} and hydroperoxy radical $(HO_2\cdot)$[23], the reaction equation is as follows.

$$Fe^{3+} + H_2O_2 \rightarrow Fe^{2+} + HO_2 \cdot + H^+ \tag{1}$$

$$Fe^{3+} + H_2O_2 \rightarrow [Fe^{III}(HO_2)]^{2+} + H^+ \quad (2)$$

$$[Fe^{III}(HO_2)]^{2+} \rightarrow Fe^{2+} + HO_2 \cdot \quad (3)$$

In an electrochemical system, Fe^{3+} can directly generate Fe^{2+} through cathode reduction. Therefore, the combination of the ordinary Fenton oxidation system and the electrochemical reduction of Fe^{3+} system is of great significance to reduce the oxidation of gasoline engine oil in the subsequent treatment process.

2.2 Influencing Factors of Cathode Electrochemical Reduction of Fe^{3+}

The electrochemical reduction effect of Fe^{3+} on the cathode is affected by many factors. The DC power supply provides energy for Fenton's electrical reactions. When the power supply voltage is too low, the current density in the system is small, and the reaction intensity of reducing Fe^{3+} at the cathode is weak, and it cannot even happen normally. When the voltage continues to increase, it will not only cause an increase in energy consumption, but also increase the side reactions in the system, and cannot effectively reduce Fe^{3+} at the cathode. The electrode material plays a very important role in the electronic Fenton oxidation process. The electrode material mentioned here generally refers to the cathode material. The type of cathode material and its conductivity will affect the occurrence of Fenton reaction. The positive electrode material with good performance can reduce the occurrence of side reactions and save energy consumption and cost. In addition, the electrode material should have corrosion resistance and high mechanical strength, which is also necessary for experimental research and industrial applications to avoid damage to the reaction process due to corrosion or external impact. When the distance between the cathode and the cathode is large, the resistance generated in the solution increases, resulting in a decrease in current density, which affects the mass transfer rate of the reaction and reduces the reaction rate. When the distance between the two electrodes decreases, the resistance decreases and the conductivity of the solution increases. The mass rate becomes faster, the electrons move faster, and the redox capacity that occurs at the two poles also increases. However, the distance between the poles should not be too small. Too small a distance will increase the difficulty of processing the reaction vessel, increase the difficulty of industrial application, and ensure the stability of the reaction. Therefore, the distance between the electrodes should be controlled within an appropriate range to ensure the stability of the reaction.

3 Oxidation Experiment Materials During the Use of Steam Turbine Oil by Fenton Oxidation Method

New turbine oil (NTO) and used green turbine oil (UTO) are provided by a lubricant company. The oil sample CKB is added with a mass fraction of 0.1% dialkylaniline, 0.1% 2,6-di-tert-butyl-p-cresol, 0.05% alkenyl succinic acid, 0.05% N,N-di-n-butylaminomethylene The oils of triphenyltriazole are of analytical grade. There are many physical and chemical indexes for steam turbine oil. When performing physical and chemical performance tests, it is impossible to analyze all items and select all the typical physical and chemical indexes for testing different varieties of lubricants. Therefore, for particle size, select kinematic viscosity, moisture and liquid phase corrosion and other indicators for analysis. The experimental oils are the three types of steam turbine oils A, B, and C that are commercially available. The basic performance is shown in Table 1.

Table 1. Properties of experimental turbine oil

Project	Colour	Moisture	Graininess	Kinematic viscosity	Liquid phase corrosion	Exterior
A	Orange	22.56	5	28.9	No rust	Transparent
B	Orange	19.67	6	25.1	No rust	Transparent
C	Orange	21.54	9	33.2	No rust	Transparent

Put 30 mL of oil sample and 30 mL of n-hexane in a separatory funnel and mix. Add 30 mL of extraction solvent, shake the mixture vigorously for at least 1 min to make it thoroughly mixed, and let stand until the lower layer solution is transparent. Put the transparent lower layer solution into another separating funnel, add 10 mL Fenton, shake vigorously, and let stand until the lower layer solution is transparent. The extract was washed again twice in the same manner, and the lower solution was transferred to a 100 mL round bottom flask.

4 Analysis and Conclusion of Oxidation Cause During Steam Turbine Oil Use by Fenton Oxidation Method

Take a 30 mL colorimetric tube, take 2 mL of the sample to be tested, add 4 mL of hydrochloric acid, then add 0.4 mL of titanium reagent, and add water to bring the volume to 30 mL. After shaking, directly measure the absorbance and measure the H_2O_2 concentration at a wavelength of 400 nm. Experiments show that the conventional index of steam turbine oil is basically within the normal range. Due to the high frequency of steam turbine oil particle size testing, over-sized particle size testing is relatively timely. After discovery, continuous tracking tests are conducted to quickly reduce the steam engine oil particle size to the qualified standard. In order to fully understand the operating status of the steam turbine oil and analyze the severity of the steam turbine, the rotating oxygen bomb values of A, B, and C turbine steam turbine oils were tested to track the changing values of the rotating oxygen bomb. It can be seen from Table 2 that the content of various metal elements in steam turbine oil is low, which is close to the content data of metal elements in ordinary color turbine oil after operation in the literature. Therefore, it is not the metal salt that turns the turbine oil green.

Table 2. Elemental content of oil products

Element name	Content/(mg·kg^{-1})		
	A	B	C
Fe	0.16	0.15	0.17
Cu	1.81	1.92	1.89
Zn	0.76	0.62	0.69
Mg	0.66	0.67	0.68
Ni	0.02	0.04	0.03
Ba	26.00	27.00	27.02

Some substances will show different colors under different acid and alkali conditions; some substances will show different colors under their oxidation and reduction states, such as the redox indicator in redox titration. Diphenylamine is one of the earliest redox indicators. Now it is generally replaced by sodium diphenylamine sulfonate which is easily soluble in water, but the discoloration mechanism is the same and the discoloration is similar. Compared with the new steam turbine oil, the diameter of the old steam turbine oil is increasing, indicating that its anti-wear performance has deteriorated a lot. But its wear scar diameter is smaller than that of base oil. It shows that although the waste engine oil is scrapped, its anti-wear performance has not been exhausted. The waste engine oil still contains some additives, and these additives still exert a certain anti-wear effect. Figure 1 shows the relationship between the amount of sludge produced by the three steam turbines and the retention rate of the rotating oxygen bomb.

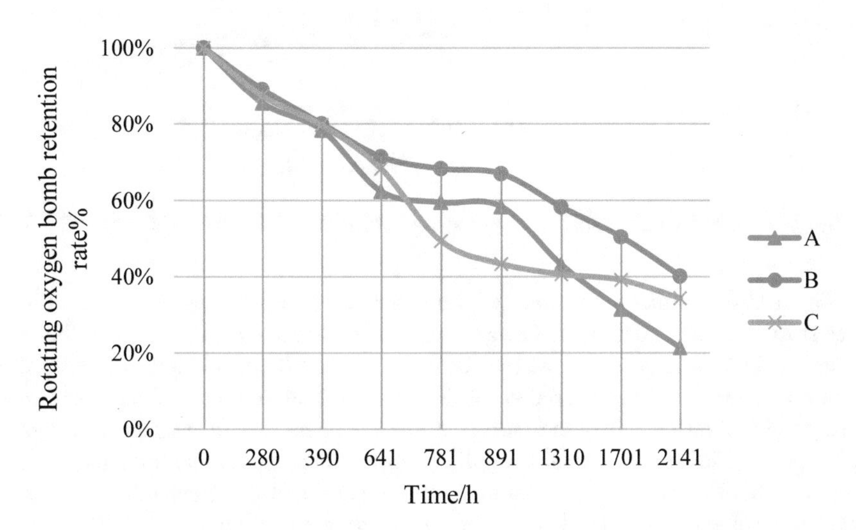

Fig. 1. Relationship between steam turbine oil oxidation life and the retention rate of a rotating oxygen bomb

It can be seen from Fig. 1 that as the retention rate of the rotary oxygen bomb decreases, the amount of sludge produced by turbine oil increases. When the retention rate of the rotary oxygen bomb was reduced from 100% to 85%, the amount of sludge generated rapidly increased to 800 μg/g, the amount of sludge generated remained above 1000 μg/g, and the retention rate of the rotary oxygen bomb accounted for 25% the amount of sludge produced is 958 μg/g.

Using the difference in the polarities of the chromophore and hydrocarbon and the difference in their solubility in solvents with different polarities, the chromophore is separated from the oil by an extraction separation method. The infrared spectrum of the steam turbine oil extract also has absorption peaks for azo and quinone groups. Azo and quinone groups are not found in the base oil and additives of steam turbine oils. Many pigments and indicators contain azo or quinone groups. The three turbine oils are oxidized by Fenton's reagent, and the oxidation products NTO-SO and UTO have the same color. The visible light absorption spectrum can show the absorption of liquid at various wavelengths, avoiding the difficulty of color description. Figure 2 shows the comparison of the visible spectrum analysis of the three turbine oils UTO and NTO-SO.

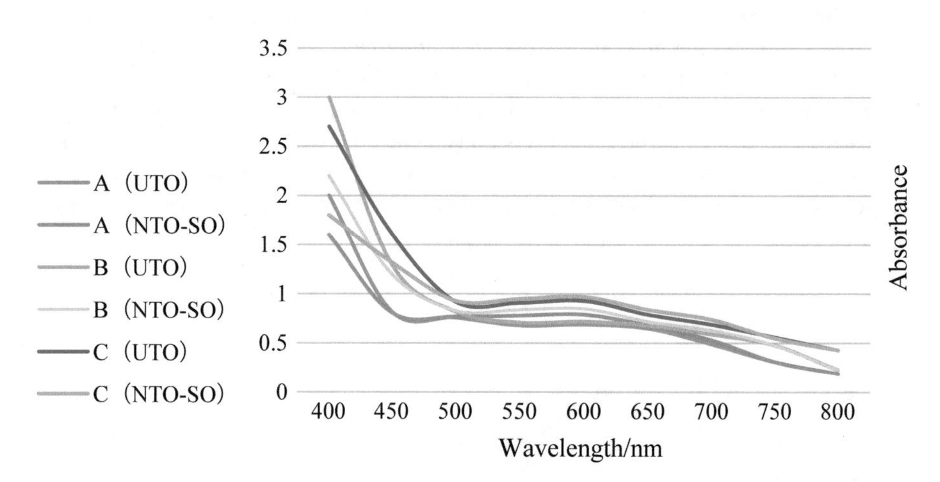

Fig. 2. Comparison of visible spectrum analysis of three turbine oils UTO and NTO-SO

Figure 2 shows that the visible spectrum curves of the two heavier materials in the three steam turbines are similar. The strength is not high, indicating that there should be green and blue substances in the oil. Therefore, it can be concluded that these three turbine oils have similar material color distributions and all contain blue-green materials. When extracted with other solvents, the oil phase is still dark green or green. Since the used steam turbine oil is usually yellow-brown, the oil that turns the oil green should be blue or blue-violet. Therefore, it is determined that methanol water as the extraction solvent can relatively completely extract the color change of the color-changing turbine oil. Therefore, it can be speculated that the coloring material in the steam turbine oil is a pigment pigment containing an azo or quinone group.

5 Conclusion

This article analyzes the reasons for the reduction in the oxidation resistance of turbine lubricants. When selecting lubricants, antioxidants must be selected reasonably to ensure the quality of the lubricants. In order to ensure the normal use of the steam turbine, the relevant personnel should regularly check the rotating oxygen bomb value of the steam turbine lubricant. Using Fenton oxidation method and elemental analysis method, the pH value of steam turbine oil extract was studied. Under certain acid and alkali conditions, the substance appears blue or blue-violet, which is the main reason why the turbine oil appears green after use. This article explains the effect of free radical scavenger antioxidants on steam turbine oil from one aspect, and speculates on the synergistic antioxidant mechanism of shielding phenolic aromatic amines at low temperatures. The analysis of acid-base discoloration, infrared spectroscopy, thin layer chromatography, and visible spectroscopy showed that the structure of the chromogenic material in the simulated oxidized oil is similar to that of the green oil in the field, indicating that the simulated oxidation experiment can quickly reflect the new turbine after use. The discoloration of the original oil formula.

References

1. Xu, J., Kong, F., Song, S., Cao, Q., Huang, T., Cui, Y.: Effect of fenton pre-oxidation on mobilization of nutrients and efficient subsequent bioremediation of crude oil-contaminated soil. Chemosphere **180**, 1–10 (2017)
2. Xu, J., Du, J., Li, L., Zhang, Q., Chen, Z.: Fast-stimulating bioremediation of macro crude oil in soils using matching fenton pre-oxidation. Chemosphere **252**, 126622 (2020)
3. Salari, M., Rakhshandehroo, G.R., Nikoo, M.R.: Degradation of ciprofloxacin antibiotic by homogeneous fenton oxidation hybrid AHP-promethee method, optimization, biodegradability improvement and identification of oxidized by-products. Chemosphere **206**, 157–167 (2018)
4. Ramesh, K., Balakrishnan, M., Vigneshkumar, B., Manju, A., Kalaiselvi, K.: Removal of colour and chemical oxygen demand from textile effluent by fenton oxidation method. Curr. Sci. **113**(11), 2112–2119 (2017)
5. Santanaraj, J., Sajab, M.S., Mohammad, A.W., Harun, S., Kaco, H.: Enhanced delignification of oil palm empty fruit bunch fibers with in situ fenton-oxidation. BioResources **12** (3), 5223–5235 (2017)
6. Anis, M., Haydar, S.: Heterogeneous fenton oxidation of caffeine using zeolite-supported iron nanoparticles. Arab. J. Sci. Eng. **44**(1), 315–328 (2019)
7. Liu, D., Gu, G., Wu, B., Wang, C., Chen, X.: Degradation of isopropyl ethylthionocarbamate from aqueous solution by fenton oxidation: RSM optimization, mechanisms, and kinetic analysis. Desalin. Water Treat. **132**, 362 (2018)
8. Cui, J., Wang, X., Zhang, J., Qiu, X., Wang, D., Zhao, Y., et al.: Disilicate-assisted iron electrolysis for sequential fenton-oxidation and coagulation of aqueous contaminants. Environ. Sci. Technol. **51**(14), 8077–8084 (2017)
9. Reyes, I.A., Patiño, F., Flores, M.U., Narayanan, J., Calderón, H., Pandiyan, T.: Use of ligand-based iron complexes for phenol degradation by fenton modified process. J. Mex. Chem. Soc. **57**(57), 96–104 (2017)

10. Thomas, M., Barbusinski, K., Kalemba, K., Piskorz, P.J., Kozik, V., Bak, A.: Optimization of the fenton oxidation of synthetic textile wastewater using response surface methodology. Fibres Text. East. Eur. **25**(6), 108–113 (2017)
11. Hao, X., Peng, H., Xu, P., He, M., Dou, B.: Production of H_2 by steam reforming in schizochytrium algae oil of cell disruption and extraction via ultrasound method. Int. J. Hydrogen Energy **44**(30), 15779–15786 (2019)

Optimization of Berth Shore Bridge Allocation Based on Simulated Annealing Algorithm

Zhichao Xu(✉) and Yi Guan

Department of Information Management, Dalian Neusoft University of Information, Dalian, Liaoning, China
xuzhichao@neusoft.edu.cn

Abstract. In recent years, with the rapid development of the world commercial trade, the importance of maritime transportation has become increasingly prominent. The port plays a pivotal role in it. Port operations are aimed at optimizing overall efficiency, rationally allocating berths, shore bridges and other resources, and improving port throughput. The issue of berth shore bridge allocation is particularly important. The traditional berth shore bridge allocation mechanism uses the first-come-first-served static scheduling method. With the increasingly complex port environment, it will reduce the port's operating efficiency and cannot better meet the needs of port development. Thence, the establishment of a new berth shore bridge allocation mechanism is the key to ensuring efficient operation of the port. Taking the container loading and unloading operations as the background, the berth and shore bridge allocation mechanism in container operations is analyzed. Based on some assumptions, the berth and shore bridge allocation model is established, and the simulated annealing algorithm is used to obtain the berth and shore bridge allocation that makes the ship in Dalian Port the shortest time The scheme optimizes the traditional berth shore bridge allocation mechanism.

Keywords: Port · Berth and shore bridge allocation · Simulated annealing algorithm

1 Introduction

With the development of world commercial trade, maritime transport plays a pivotal role. As a hub of maritime transportation, the port provides a platform for ship operations. Port to berth, shore bridge resource planning and scheduling, to ensure the orderly operation of ships. However, due to the complexity of the environment, it is impossible to better adapt to changes in the environment based on planning and scheduling alone, resulting in chaos in port operations and reducing the utilization of port resources. In order to better enhance the environmental adaptability of port operations, increase the port's resource utilization rate, and improve port throughput, terminals often adopt a first-come-first-served static dispatching method to allocate resources such as ships and quayside berths. However, with the unceasing development of the port economy, the throughput of the port is getting larger and larger, and the traditional berth shore bridge allocation method cannot satisfy the demand of port

J. MacIntyre et al. (Eds.): SPIoT 2020, AISC 1282, pp. 599–606, 2021.
https://doi.org/10.1007/978-3-030-62743-0_86

development. Therefore, a new algorithm is needed to optimize the berth shore bridge allocation mechanism. Thence, it is of great importance to research the optimization of berth shore bridge allocation based on simulated annealing algorithm.

At present, scholars at home and abroad have done a lot of in-depth research on the optimization of berth shore bridge allocation. Tang Shuang researched the optimal allocation of port container berths and shore bridge resources in 2011. Sun Bin et al. studied the berth and shore-bridge joint scheduling for a robust response strategy in 2013 [1]. Christian Bierwirth et al. investigated the follow-up problems of container terminal berth allocation and crane scheduling in 2015 [2]. Xu Wandong carried out research on berth-shore-bridge joint scheduling optimization based on compound cost in 2019 [3].

2 Berth Quayside Allocation Mechanism Based on Container Handling

The container terminal is an operation place for docking ships and loading and unloading containers in the port. Container handling is one of the main operations of container terminals. Its products are not physical objects, but labor and services provided for container transportation and the product form is the displacement of space for container cargo. These displacements must be targeted and suitable for the "effective displacement" required by the transport activity itself. The wharf needs to provide berths, use machinery and manpower to move the containers through a series of business activities. These business activities mainly include: unloading and loading.

Uploading: Refers to the activity of unloading the container from the ship and putting it in the yard.

Loading: Refers to the process of transporting containers from the yard to a given location on the ship according to the corresponding sequence.

2.1 Berth

The berth refers to a place where a ship is docked at the port in order to perform loading and unloading and has a long coastline of several hundred meters. The berth size and water flow are not uniform, they are made according to the size of the ship. According to the length of the berth shoreline and the different management methods, berths can be divided into: individual berths and continuous berths. Individual berths are usually not long, and only one large boat or two small boats can be put down. Continuous berths are usually not short, and can accommodate multiple ships. The berth usage is higher than the former, but there are many factors to be considered when allocating resources.

2.2 Shore Bridge

Modern container operations generally use shore bridges to perform ship loading and unloading operations. The shore bridge can walk around the parallel rails of the dock's shore wall line, but cannot cross, and the prescribed safety distance must be ensured when performing the operation. During loading and unloading, the arm frame is tiled, and the movement of the vehicle on the shore arm frame in the horizontal direction and the vertical movement of the vehicle tool can reach the box-to-vehicle and vehicle-to-ship directions.

3 Influencing Factors of Berth Shore Bridge Allocation

Generally speaking, the problem of berth scheduling is to reasonably arrange the docking position and docking sequence of the container terminal ships, so that the total arrival time of ships arriving at the port within a certain period of time is the shortest, which improves the port's operating efficiency. Under normal circumstances, the primary consideration in arranging berths is the allocation of shore bridges. Once the shore bridges are assigned to ships, the docking position and the length of operation of the ship are actually roughly determined. Therefore, berth plans and shore bridge plans are often inseparable.

3.1 Arrival and Departure Time of the Ship

Affected by the changing weather at sea and business lines, ships arriving at the port have an estimated arrival time t1 and a planned departure time t2. The pier is allocated to its berths and shore bridges for loading and unloading operations within a certain period of time based on the ship's estimated berthing time and planned departure time. The assigned work period must be between t1 and t2. If it exceeds the t2 period, it will affect the subsequent transportation and operation of the ship.

3.2 Ship Loading and Unloading Capacity

Usually the planner will arrange the shore bridge according to the following standards: when the ship's loading and unloading capacity does not exceed 150 boxes, arrange 1 shore bridge for its service; when the ship handling capacity is 151–500 boxes, arrange 2 shore bridges; when the volume is 501–700 boxes, arrange three shore bridges; when the ship's loading and unloading capacity is 701–1000 boxes, arrange four shore bridges; when the ship handing capacity is 1001–2000 boxes, arrange five shore bridges; when the loading and unloading volume exceeds 2000, 6 shore bridges are arranged to serve the ship[4, 5].

4 Optimized Solution of Berth Shore Bridge Allocation Based on Simulated Annealing Algorithm

The problem of berth shore bridge allocation can be regarded as a multi-objective optimization problem [7]. In the study, the ship's berthing operation was independent, and the ships operating in the port within a certain time range were regarded as the research objectives. The shore bridge berths were allocated to these arriving ships in order to optimize the ship's berthing sequence, berth and determine the reasonable operation shore bridge. The number makes the ship's total time in the port the least, the utilization rate of the shore bridge is the highest, and finally the overall comprehensive benefit of the terminal is reached. Considering only the discrete berth allocation, the ship cannot berth across multiple areas and can only be located where it is arranged. Ships are moored and assigned to shore bridges at the same time. The shore bridges of adjacent berths can be dispatched flexibly.

4.1 Assumption

According to the port container operation is composed of berths, shore bridges, ships, etc., the following assumptions are proposed:

(1) Only consider the ship's discrete berthing mode; (2) The ship can only be served by the terminal after arriving at the port; (3) The ship operation time is determined by the shore bridge configuration; (4) The physical conditions of the berth meet the requirements of ships arriving at the port; (5) The number of shore bridges allocated to a single ship must not exceed the maximum number of shore bridges allowed for operation. (6) The operation volume allocated to the shore bridge per hour must not exceed the work efficiency of the shore bridge.

4.2 Symbol Design

(1) i represents the ship's number. (2) C represents shore bridge assembly. (3) $V0$ represents the efficiency of shore bridge operation. (4) Tai represents the arrival time of ship i. (5) Ei represents the loading and unloading volume of ship i. (6) Cmi represents the maximum acceptable shore bridge number of ship i. (7) Lvi represents the length of ship i, including the length reserved for horizontal safety. (8) l represents the berth spacing. (9) Tsi represents the starting operation time of ship i. (10) Tfi represents the departure time of ship i.

4.3 Objective Function

$$F1 = \min \sum_{i=1}^{V} (Tfi - Tai) \tag{1}$$

4.4 Algorithm Solution

The model is solved by using simulated annealing algorithm in the paper. The steps to solve the function of the simulated annealing algorithm are shown in Fig. 1.

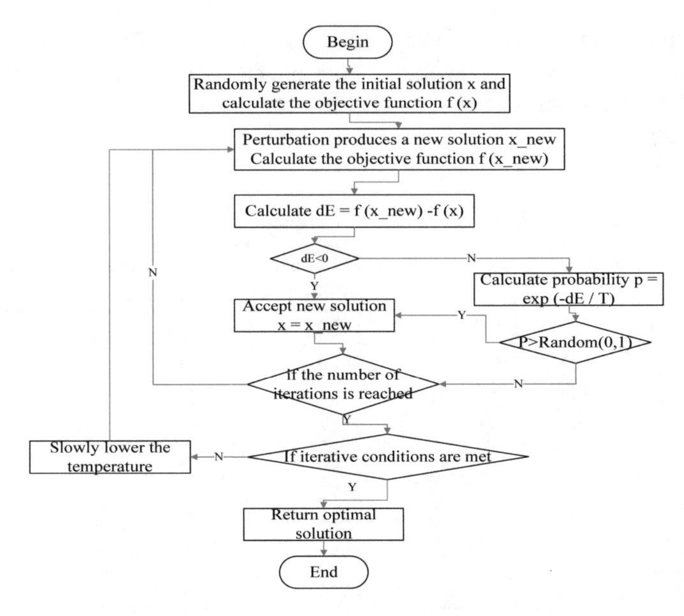

Fig. 1. Description of simulated annealing algorithm

(1) Given the initial condition, set a logical annealing rule. Generate a random initial solution x and calculate the objective function f (x). (2) The perturbation produces a new solution x_new, and calculate dE = f (x_new) − f (x). (3) If dE < 0, accept the new solution x_new, otherwise accept x_new with probability p = exp (−dE/T), the specific method is to generate a random number of 0–1, if p > this random number, then accept x_new, Otherwise it refuses and stays in state x. (4) Ditto procedure 2 and 3 until the system reaches equilibrium. (5) Reduce the temperature according to the rules given in the first step, and re-execute steps 2–4 at the new temperature until T = 0 or reach a certain predetermined low temperature [6–8].

5 Simulation

Dalian Port is the most important synthesis foreign trade port in the Northeast China [9]. It was built in 1989 and currently has more than 100 modern professional berths, including more than 70 berths above 10,000 tons. Dalian Port is an oil transportation center in Northeast Asia which has China's largest and most advanced 300,000-ton base oil terminal and 300,000-ton ore terminal [10]. It is mainly engaged in the loading, unloading, storage and transportation of base oil, refined oil and liquid chemical products and is Asia's most advanced bulk liquid chemical product transshipment base and China's largest sea passenger/car ro-ro transportation port. Its cargo throughput has reached 450 million tons now.

As shown in Table 1, take the actual sample data of Dalian Port for a certain period as an example. A container terminal has a total length of 1200 meters, with 4 berths and a berth spacing of 300 meters. The terminal is equipped with 12 shore bridges,

The shore bridge operating efficiency V0 is 30 boxes/hour. From a certain moment onwards, we will select the basic data of 10 container ships arriving one after another, covering the time of arrival, berth, number of operating boxes, and maximum number of shore bridges.

Table 1. Basic data of container ships arriving

Ship number i	Arrival time Tai	Ship length LVi	Container Volume Ei	Maximum number of shore bridges Cmi
1	6.17 01:00	260	1433	6
2	6.17 01:00	235	1635	6
3	6.17 02:00	160	501	4
4	6.17 03:00	280	2261	7
5	6.17 03:00	155	1000	4
6	6.17 04:00	240	2091	6
7	6.17 11:00	200	621	5
8	6.17 14:00	300	1547	7
9	6.17 17:00	212	1559	6
10	6.17 21:00	244	1683	6

Since the shore bridge usually moves horizontally along the shore rail, the order between the shore bridges cannot be changed, so planners generally use the following allocation method.

(1) Sort the vessels to be berthed in order of arrival. (2) Select the ship for arrangement based on the principle of first come, first served. According to this method, Table 2 shows the arrival and departure times and working hours of each ship. The sum of all ships' time in port is the objective function. The result is accurate to 0.5 h.

Table 2. Timetable for initial allocation of ships in port

Ship number	Arrival time Tai	Ship length LVi	Container Volume Ei	Number of shore bridges	Departure time Tfi	Time in port (hours)
1	6.17 01:00	260	1433	6	6.17 09:00	8
2	6.17 01:00	235	1635	6	6.17 10:30	9.5
3	6.17 02:00	160	501	4	6.17 13:30	11.5
4	6.17 03:00	280	2261	7	6.17 21:00	18
5	6.17 03:00	155	1000	4	6.17 22:00	19
6	6.17 04:00	240	2091	6	6.18 09:00	29
7	6.17 11:00	200	621	5	6.18 03:30	16.5
8	6.17 14:00	300	1547	7	6.18 13:30	23.5
9	6.17 17:00	212	1559	6	6.18 18:30	25.5
10	6.17 21:00	244	1683	6	6.18 23:00	26

In the initial plan, the total vessel residence time F1 = 186.5 h. Obviously this solution has a lot of room for optimization. Now use Matlab software and simulated annealing algorithm to solve the model:

In the program, the basic parameters are as follows:

Objective function: the total time the ship is in port F1. Initial temperature: 1000. Temperature decay function: This article adopts the most commonly used temperature update method, delta = 0.99. Markov chain length:3000. Evolutionary algebra:1000. Algorithm stop criterion: when the program optimization times reach evolutionary algebra, it terminates.

The result after 1000 iterations of the program is shown in Fig. 2.

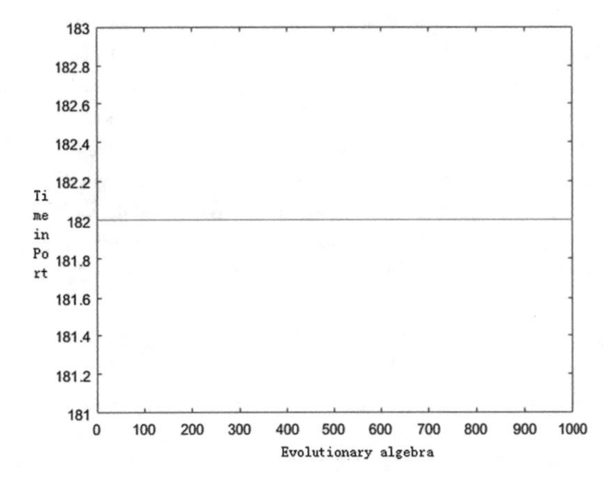

Fig. 2. F1 change process

The optimal value is 182 h. That is, after using the simulated annealing algorithm, the ship's total time in port was shortened by 4.5 h, which improved the port's operating efficiency.

6 Conclusions

This paper presents a simulated annealing algorithm and applies it to the assignment of berths and bridges to ports. In this paper, based on the traditional port container operation, the simulated annealing algorithm is applied to the berth shore bridge allocation, the traditional berth shore bridge allocation mechanism is improved, and the optimal solution is obtained. Finally, the comparative simulation experiments show that the algorithm proposed in this paper has high practicability and effectiveness, and prove that this method is beneficial to improve the operation efficiency of the entire port and then increase the throughput of the port.

References

1. Bin, S., Junqing, S., Qiushuang, C.: Joint scheduling of berth and shore bridge based on robust reactive strategy. Syst. Eng. Theor. Pract. **33**(4), 1076–1083 (2013). (in Chinese)
2. Bierwirth, C., Meisel, F.: A follow-up survey of berth allocation and quay crane scheduling problems in container terminals. Eur. J. Oper. Res. **244**(3), 675–689 (2015)
3. Wandong, X., Guiyun, L., Ciyun, W., Luyao, L.: Optimization method for berth-shore bridge joint scheduling of container terminal based on compound cost. J. Ningbo Univ. (Sci. Technol. Ed.) **32**(06), 87–91 (2019). (in Chinese)
4. Hongtao, H., Xiazhong, C., Zhen, L., Chengle, M., Xiaotian, Z.: The Joint quay crane scheduling and block allocation problem in container terminals. IMA J. Manage. Math. **30** (1), 51–75 (2019)
5. Mohammadi, M., Forghani, K.: Solving a stochastic berth allocation problem using a hybrid sequence pair-based simulated annealing algorithm. Eng. Optim. **51**(10), 1810–1828 (2019)
6. Manavizadeh, N., Hosseini, N., Rabbani, M., Jolai, F.: A Simulated Annealing algorithm for a mixed model assembly U-line balancing type-I problem considering human efficiency and Just-In-Time approach. Comput. Ind. Eng. **64**(2), 669–685 (2013)
7. Tufano, A., Accorsi, R., Manzini, R.: A simulated annealing algorithm for the allocation of production resources in the food catering industry. Br. Food J. **122**(7), 2139–2158 (2020)
8. Yang, L., Xinyu, L., Jianhui, M.: Large-scale permutation flow shop scheduling based on improved simulated annealing algorithm. Comput. Integr. Manuf. **26**(2), 366–375 (2020). (in Chinese)
9. Wawrzyniak, J., Drozdowski, M., Sanlaville, É.: Selecting algorithms for large berth allocation problems. Eur. J. Oper. Res. **283**(3), 844–862 (2020)
10. Sheikholeslami, A., Mardani, M., Ayazi, E., Arefkhani, H.: A dynamic and discrete berth allocation problem in container terminals considering tide effects. Iran. J. Sci. Technol. Trans. Civ. Eng. **44**(1), 369–376 (2019). https://doi.org/10.1007/s40996-019-00239-1

Information Security Terminal Architecture of Mobile Internet of Things Under the Background of Big Data

Jing Chen(✉)

School of Business and Information Technology, Yunnan Land and Resources Vocational College, Kunming 652501, Yunnan, China
wjhwan18@126.com

Abstract. In order to better study the information security (IS) of mobile Internet of Things, traditional methods can no longer process the big data of Internet of Things due to its characteristics of large volume, multiple types and fast speed. This paper proposes the mobile Internet of Things IS terminal architecture based on big data analysis, and gives the architecture of the whole system. Combined with block chain technology, Spark computing system, Hadoop technology, etc., the solution of system big data storage (DS) and terminal acquisition is presented. The results show that the defense efficiency of terminals is improved by 21.6%, effectively solving the problem of IS terminal protection of the Internet of Things.

Keywords: Mobile Internet of Things · Big data background · Security terminal · Big data storage

1 Introduction

While the Internet of Things (IOT) is receiving global attention, it is also suffering security threats and attacks from all over the world. In order to further promote the development and implementation of IOT technology, ensuring the security of IOT technology is undoubtedly an important and basic guarantee to reduce users' worries [1, 2].

Aafaf OUADDAH has conducted extensive research on different access control solutions in the IOT in its target, Model, Architecture and Mechanism (OM-AM) approach [3, 4]. Aafaf OUADDAH emphasizes the advantages and disadvantages of traditional access control models and protocols from the perspective of IOT, as well as the recent access control models and protocols [5]. Aiming at network IS, Xiaofeng Lu focused on the property of privacy and proposed the privacy IS Classification (PISC) model. PISC divides privacy into four security categories, each with its own security objectives. According to the data from the search engine, Xiaofeng Lu got 53 categories of privacy. Lavrova D. S proposed an ontological model for IOT applications, which provides a detailed representation of the relationships and interrelationships between system elements at different levels of abstraction and in different levels of detail. The ontology model allows understanding of the technical aspects of developing

J. MacIntyre et al. (Eds.): SPIoT 2020, AISC 1282, pp. 607–613, 2021.
https://doi.org/10.1007/978-3-030-62743-0_87

a security information and event management (SIEM) system for detecting and analyzing IOT security events [7].

Combined with the background of big data, this paper studies the IS terminal architecture of mobile IOT, and realizes reliable storage through two parts: block chain data structure and consensus algorithm. Considering the real-time performance of the IOT, Raft consensus algorithm was adopted to ensure the integrity of stored data and the difficulty of further modification through the hash pointer in the block chain data structure [8, 9]. Finally, after testing, the new scheme has improved the confidentiality of IS terminals of mobile IOT [10].

2 Mobile IOT IS Terminal

2.1 Big Data

These massive data are not a normal data set for analyzing and managing the relationship between enterprises and customers. But from a technical point of view, the integration of big data and cloud computing is as inseparable as the two sides of a coin. Complex big data will inevitably make it impossible for enterprises to effectively use a single distributed computer for comprehensive data processing, so a new distributed computing system architecture must be adopted in terms of structure. Its main technical feature is the comprehensive mining of massive enterprise data, but it must adopt technologies such as distributed processing relying on cloud computing, distributed database, cloud storage and/or cloud virtualization server.

2.2 Mobile IOT IS Terminal Architecture

There are many different ways to implement PoS, one of which is a hybrid mode that takes advantage of the balance to adjust the difficulty:

$$H(B, t) \leq balance * target \tag{1}$$

R is the current connector margin, S is the current supply of Smart Token, P is the current price of Smart Token, Smart Token = SP, then:

$$Connector = FSP, \quad P = \frac{R}{SF} \tag{2}$$

3 Design of the Experiment

3.1 Experimental Background

The expansion of mobile network bandwidth provides more powerful network support for the application of the IOT information industry, and the popularization of the IOT will be further expanded in depth and breadth. The mobile network will be the most suitable network for integration with the IOT. With the full security architecture implemented, IoT devices will be able to securely log into the service center for authorization, query, and other operations without manually setting a password, and the password for each symmetrical IoT session will be variable and unique. Both the block chain technology ensures that the storage is difficult to modify, and since the private key of each IoT device is only stored locally and Raft is the corresponding public key, this prevents user privacy violations and enhances the security of IoT systems.

3.2 Experimental Design

Taking a IoT enterprise data as sample, in the security architecture topology deployment is completed, we need to design a good key generation and distribution of the source file, source files, block Raft consensus agreement chain structure source file deployed on the application layer of NS - 3, namely the Applications in NS - 3 modules directory, and registered in Applications modules directory wscript file added. H header file and. Cc file, in order to run the program to find the new source files, and other related modules. In the security architecture, all nodes except AP nodes deploy additional key related files in the application layer, and AP nodes only route communication data. The Raft cluster node also requires the application layer to deploy the Raft consensus protocol file and the blockchain data structure file. The experimental results are shown in Table 1.

Table 1. Experimental results

The number of deals	Approximate size of block	Path size (Bytes)	Path size (hash number)	Processing speed (ms)
16	4 KB	118	2	2.11
512	128 KB	208	6	0.92
2048	512 KB	313	10	1.91
65535	16 MB	427	15	2.01

4 Big Data and Mobile IOT Security Design

4.1 Analysis of Mobile IOT IS Terminal Architecture Based on Big Data

As shown in Fig. 1, if a network communication failure occurred within the Raft cluster, the whole Raft cluster was divided into two subsets. At this time, the leader could not successfully submit the data uploaded by the client because he could not access most of the follower nodes. Since most of the follower clusters do not have a leader, they will re-elect a candidate as the leader, and then the leader will act as the representative of communication with the outside world. If the client requires to submit the new IoT public key data, the new leader will reach a consensus with most of the floor nodes according to the Raft workflow. If this is a successful network failure repair, then the original leader node. In the lost connection stage of FlowAR, any update of the old Lader node cannot be counted as confirmation and will be rolled back to force the data consistency with the new leader node.

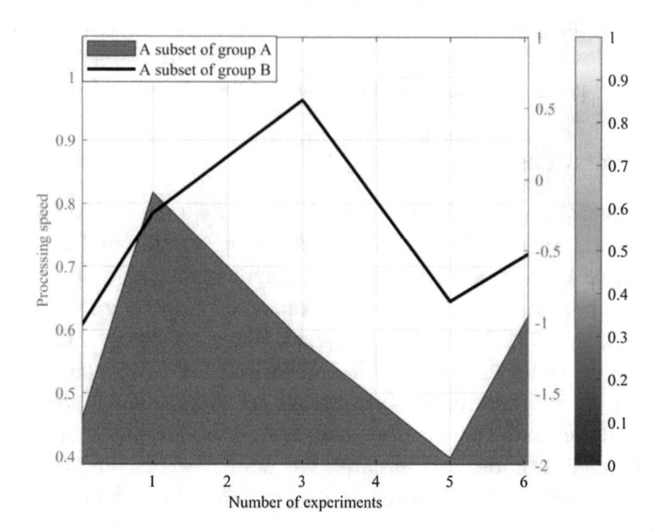

Fig. 1. Comparison of data simulation results of two subsets

Industrial PC3 Node0.Org2.example.com block storage file is deleted or incomplete, nor does it affect the normal work of the DS of the whole IOT network. The data is restored during the index synchronization process when the node restarts. If the node fails to stop, index synchronization will also be carried out after startup to ensure the node has the latest blockchain state. As shown in Fig. 5, PC3 enters the data recovery process after restart. During this period, due to the inconsistency between the local block chain state and the global block chain state, storage service will not be provided until the data recovery is completed and the local block chain state is synchronized. Therefore, compared with the traditional storage technology, the distributed storage technology based on block structure not only has better security because of the introduction of block chain structure, but also the malicious molecules must control multiple storage devices at the same time to achieve their goals, greatly increasing the cost and difficulty of their attacks.

As shown in Fig. 2, at the time of the simulation time is 3 s, IoT can be seen from the N53 log component in the console output, in design, the leader of Raft cluster nodes after upload IoT public key data obtained, did not immediately to other followers node broadcasts this data, but a 150300 ms, under the Raft cluster random cycle heartbeats consensus process, the IoT public key data included in the msg appendentries tleader type packets. The follower node stores the packet in the receiving cache, and then makes a vote response after comparing the block chain update degree of the leader node and the local node. If the majority of follower nodes are voted by the leader node, it means that the Raft cluster network supports the new IoT public key data store and a consensus is reached. Then the IoT's corresponding public key data and device ID number will be submitted to the local blockchain data structure of the leader node.

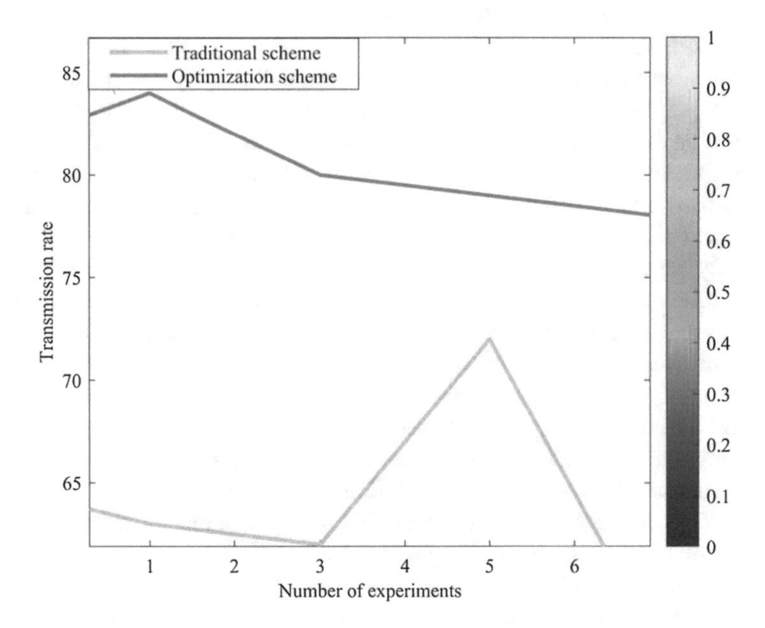

Fig. 2. Comparison of data processing rates of simulation schemes

In this design, to form an encryption consensus on the public key cache data between a Raft node cluster, it may be required to submit all corresponding public key cache data of each Follower node to their respective local public key block chain during the consensus formation process of the heartbeat data detector of the next node. After reaching a consensus with the whole network users, the Leader wireless node device agent shall make timely response to the corresponding wireless device agent of the whole network. Considering the potential risk of network IS in wireless intelligent technology, the key wireless intelligent terminal equipment should avoid directly using the network technology as the remote wireless communication basic network. If some wireless smart terminal devices need to use wireless network public network technology to transmit data in network services, they should first enable Anapn network services provided by wireless carriers free of charge. In addition, it is recommended that users adopt an end-to-end user password protection mechanism to ensure absolute confidentiality, integrity and freshness of data transmitted by users on wireless and public networks. When choosing the wireless network private network such as McWimax, McWill, and 230 MHz private network, that is, the wireless network should adopt the security management mechanism to protect the integrity and network confidentiality through access security authentication, access control, transmission of confidential information, etc. Above the whole secure network architecture of the wireless communication system database encryption and decryption, storage system data encryption block double-chain technology for specific tests.

4.2 Suggestions on IS Terminal Architecture of Mobile IOT Under the Background of Big Data

First of all, the identity of the two parties of communication is authenticated before communication. RFID authentication and key agreement protocol simplifies the public value transfer between two communication parties, will convey it to the authentication server to, even if the server was compromised, so it could pass the back of the program, because the backend authentication server, and not directly to store the user's password, but the store validation yuan, however, verify the yuan does not store confidential information. So this process, user information confidentiality performance greatly improved. This is known as defending against server leaks.

Secondly, since the sharing and transmission of user privacy information between communication parties cannot be carried out directly in plaintext, certain information encryption protection measures must be taken for communication parties to prevent the disclosure of user privacy and other information. This is the need to effectively resist the attack of the intermediary, intended to prevent interception of users' private information and replace it with error messages malicious others. This means that the system recognizes and automatically generates wrong user sessions and keys, which realizes the untraceability of user privacy information and reduces the security risks and use risks. The new Rfid key agreement bidirectional authentication protocol can effectively solve the problem of user privacy security information protection in THE Rfid bidirectional authentication system. The bidirectional authentication protocol also has the characteristics of low risk, low cost storage, low communication and convenient operation. It can effectively resist the attack of server information disclosure, the attack of the intermediary and the user online information guessing attack that the communication party cannot easily detect. It can provide bidirectional authentication, strong security of user session and key, untraceability and forward security. Now has become the general public know this kind of bidirectional information authentication, the face and face recognition authentication technology is relatively familiar, this is also the future bidirectional authentication development trend direction.

5 Conclusions

First of all, based on the IOT terminal IS architecture in the process of information transmission vulnerable to security threats, the design realizes the algorithm of IOT based on large data IS defensive scheme, in the protection of the IOT from illegal attacks at the same time improve the system robustness and fault tolerance performance, greatly enhanced the security of the system, and effective utilization of the network system of the redundant force, reduce the waste of resources. Then, the DS technology of the IOT based on block structure further ensures the privacy of the data of the IOT from another dimension, and the distributed DS structure improves the security and anti-attack capability of the storage system.

References

1. Wang, M., Perera, C., Jayaraman, P.P.: City data fusion: sensor data fusion in the Internet of Things. Int. J. Distrib. Syst. Technol. **7**(1), 15–36 (2015)
2. O'Sullivan, A.C., Thierer, A.D.: Projecting the growth and economic impact of the Internet of Things. Social Science Electronic Publishing **39**(10), 40–46 (2015)
3. Cartlidge, E.: The Internet of Things: from hype to reality. Opt. Photonics News **28**(9), 26 (2017)
4. Akpakwu, G., Silva, B., Hancke, G.P.: A survey on 5G networks for the Internet of Things: communication technologies and challenges. IEEE Access **5**(12), 3619–3647 (2017)
5. Ouaddah, A., Mousannif, H., Elkalam, A.A., et al.: Access control in the Internet of Things: big challenges and new opportunities. Comput. Netw. **112**(12), 237–262 (2016)
6. Lu Xiaofeng, Q., Zhaowei, L.Q., et al.: Privacy information security classification for Internet of Things based on internet data. Int. J. Distrib. Sens. Netw. **2015**(6), 1–8 (2015)
7. Lavrova, D.S., Vasil'ev, Y.S.: An ontological model of the domain of applications for the Internet of Things in analyzing information security. Autom. Control Comput. Sci. **51**(8), 817–823 (2017)
8. Hong, S., Park, S., Park, L.W., et al.: An analysis of security systems for electronic information for establishing secure internet of things environments: focusing on research trends in the security field in South Korea. Fut. Gener. Comput. Syst. **82**, 769–782 (2017)
9. Ying, Z., Jiezhuo, L.: Information security transmission technology in internet of things control system. Int. J. Online Eng. **14**(6), 177 (2018)
10. Yoneda, T.: Information security technologies in the age of the internet of things (IoT). Mitsubishi Electr. Adv. **157**(1), 17–19 (2017)

The Evolution of Network Platform Ecology from the Perspective of We Media

Ruixi Zhang(✉)

School of Journalism and Information Communication,
Huazhong University of Science and Technology, Wuhan, Hubei, China
zhangruixi320@163.com

Abstract. The network platform ecology can be divided into three levels: macro ecology, meso ecology and micro ecology. As a subsystem in this platform, we media platform has transitioned from the stage of social media communication with information feedback and interaction among people to the stage of artificial intelligence with interaction among machines. In recent years, they even extend and embedded into various life of people. The mature of we media is the symbol of the ecological perfection of the network platform and its growth route reflects the relationship between the platform and users, transforming from being preemptive for users to refinement operation, from use and satisfaction to payment satisfaction, from overall maintaining users to screening users, which shows the mature profit mode of China's network platform, the highlight of content and service value and the improvement of the platform's status. Meanwhile, it maps the evolution of the network platform ecology in China.

Keywords: We media · Network platform · We media platform

1 Introduction

In recent years, the rapid development of we media has become one of the most noteworthy phenomena in China's network ecology. According to the *Statistical Report on Internet Development in China* in December 2019, cell phone netizens spent the longest time using Apps for instant messaging, accounting for 14.8%, followed by network video, short video, and network audio (Fig. 1) [1]. Most of these platforms are mainly we media content from micro-content, short video to webcast, etc. The rise of we media reflects the flow of capital, technology and competent professionals from traditionally professional production sites to we media platforms. Thereafter, we media platforms have developed rapidly with obvious stages which reflects the evolution of the network platform ecology in China.

J. MacIntyre et al. (Eds.): SPIoT 2020, AISC 1282, pp. 614–620, 2021.
https://doi.org/10.1007/978-3-030-62743-0_88

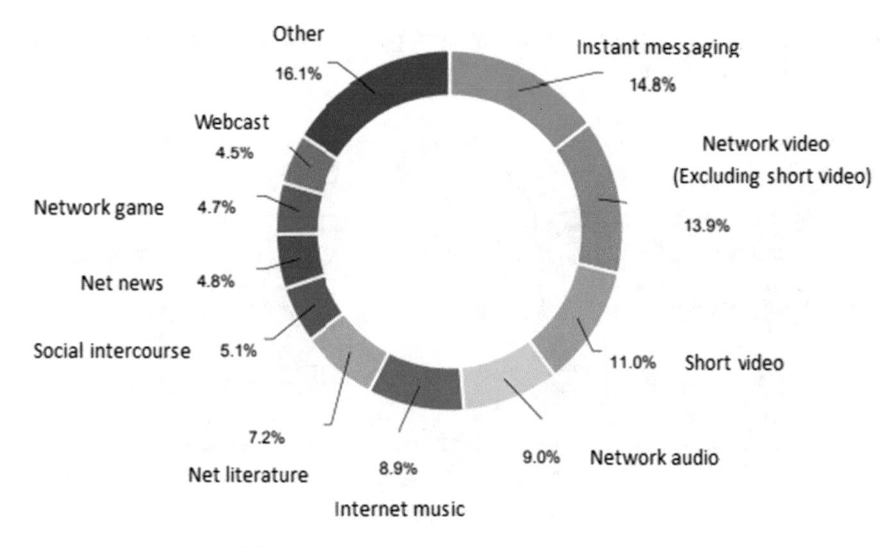

Fig. 1. Proportion of time on various APPs

2 The History and Current Situation of We Media Platform

We media is an approach to figure out how ordinary people provide and share their real lives and their news, strengthened by digital technology and connected with the global knowledge system [2]. According to the network development, we media platform can be divided into three stages: Web1.0, Web2.0 and Web3.0. Web1.0 mainly covers the content publishing, aggregation and search for we media. The second stage, Web2.0, users can share and interact, while the stage Web3.0 enables personalized recommendation, actively connecting the information with people via technology In these three stages, the role of the network changes from information provider to an information platform which can understand and supply the needs of users [3]. We media platform also transits from the stage of social media communication of human information feedback and interaction to the artificial intelligence stage of machine interaction [3]. At different stages of we media development, there is a superposition relationship, the coexistence of portal Web we media content in Web1.0, social we media content in Web2.0 and recommended we media content based on algorithm in Web3.0 [4].

We media content in each period is bound to have the technical characteristics of the mainstream media in this period, resulting in the text form with the characteristics of the times [5]. The early we media content, in printing age, is an extension of thoughts in cyber space and then our country's we media business has entered into a new era, ushering in the expansion of content distribution channels, diversification and specialization of forms. We media platforms include BBS forums popular in early years, later Blog sites and current Microblog and Wechat. The different characteristics of platforms, like Microblog, Wechat, Blog and Aggregation Clients, constitute the we media ecology (Fig. 2). Combined with the ranking of AiMedia and Qingbo Index, the most active we media platforms are Wechat public account, Sina Weibo, Toutiao, Himalaya Fm and Zhihu.

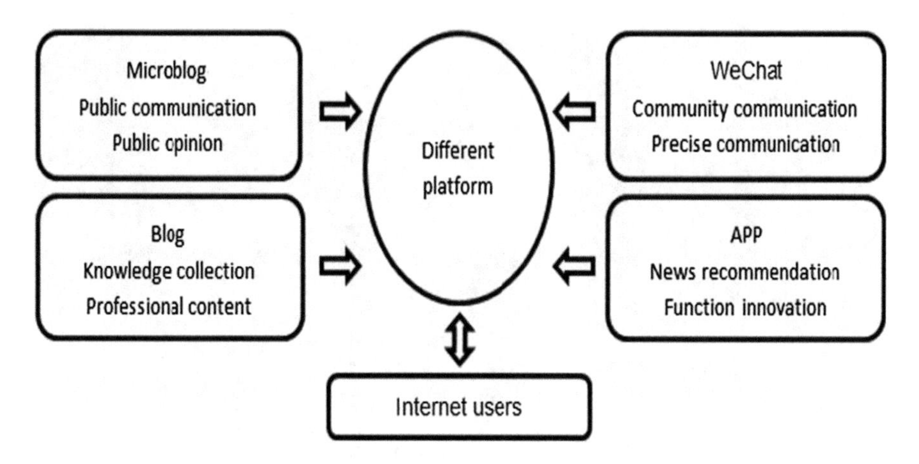

Fig. 2. Different characteristics of we media platform

We media platform can be seen as a subsystem in the network platform ecosystem. The network platform ecology can be divided into three levels: macro ecology, meso ecology and micro ecology, which refers to the external environment, composed of several elements like politics, economy, culture, and technology, the interaction between platforms and interaction between elements within the network platform, respectively [6]. With the change of external environment, the relationship between network platforms also change from learning and imitation to competition and cooperation and then to the current integration and development. Since the birth of we media platform, it has become an infrastructure in another sense with its growth mapping out the evolution network platform ecology in China [7].

3 Transformation of the Relationship Between Platform and Users

3.1 From Being Preemptive for Users to Refinement Operation

Since the birth of we media platform, the major platforms have been in a "horse race enclosure" for a long time, using free or low-cost services to increase the stickiness of users [8]. Under the current background of "flow is king", users who come to the horse race circle are also easily attracted by other platforms. In this case, more and more we media platforms have launched membership payment services, which not only make some funds return quickly, but also screen out some users for "fine operation". Network platforms, from attracting users to precisely farming products strategic tilt, is conducive to its value enhancement [9].

There is no eternal free mode in any industry. Just as the fast development of online car-hailing entered the fast lane, this novel industry has launched a "subsidy war". Giant capital drove a large number of competitors through "horse race enclosure", and after completing the division of market share, it has returned to the state of refinement operation. The relationship between several major we media platforms and users has basically turn to "refinement operation", while some new we media platforms are still being preemptive for users, trying to expand the market and seek living space.

3.2 From Use and Satisfaction to Payment Satisfaction

Compared with the state of subsidized use or even free use in "horse race enclosure", in recent years, including we media platform, more and more internet products have added paid functions, and the coexistence of "free use platform" and "paid satisfaction function" has become the norm. The upgrade of paid function is inseparable with the upgrade of user consumption concepts and policy support. As online products have taken roots in the lives of netizens, more and more netizens can accept payment for online functions and services. At the same time, the introduction of relevant policies and regulations such as "Management Regulations for Online Payment Services of Non-Bank Payment Institutions" provides a guarantee for the online payment market.

As the improvement of network platform ecology, online knowledge payment and service payment have gradually become a trend. After grabbing users, the network products begin to "cultivate" users with content and service quality, and gradually turn from "use and satisfaction" to "payment satisfaction" with many functions needing to pay for use. In this trend, high-quality content on the network pays more and more attention to copyright protection, and the membership model of major platforms is gradually flourishing.

3.3 From Overall Maintaining Users to Screening Users

In the Internet era, the traditional two-dimensional economic relationship between enterprises and customers has shifted to three-dimensional economic relationship among enterprises, users and customers. The mode for we media platform is platforms, users and customers, which will gradually screen and realize the relationship between the platforms and users through the willingness to pay. The mature of payment mode helps the platform screen and classify users, and at the same time, it is also preparing for the second half of the "Internet+".

At present, many network products are implementing a strategy of free basic functions and partial function charging, so as not to lose too many users while screening paying users, and at the same time ensure the external effect of network and continuous derivative innovation. Under the new network platform ecology, the strategy of "overall maintaining users" can no longer maximize the value. The new strategy for network products is to gradually realize the stratification of users through invisible conditions to screen users. With the maturity of the market, it becomes more and more difficult to "rob new users", therefore cultivating the old users becomes more important. At present, many network platforms use big data technology to provide individual operating mode with the most matching content through technical force, regardless whether users are willing to pay or not. While screening users, they also pursue the maximization of the number of users.

4 The Gradual Perfection of Network Platform Ecology

4.1 Gradual Stability of Industry Pattern

Screening users from the we media platform, intensively designing content and service, and moving towards the membership payment model not only reflect the general law of the development of the Internet industry, but also show normal business logic in the context of a stable industry pattern. In recent years, the standardization and formalization of we media platform has also laid the foundation for its transformation of profit model. In order to enhance competitiveness, Internet enterprises generally provide a free model at the beginning, but when users become habitual and dependent, they will gradually promote the profit mode like Taobao defeating competitor Ebay with previous free strategy. The improvement of people's acceptance of we media payment products reflects that it is on longer a savage growth state of "horse race enclosure", and the pattern of we media industry has gradually stabilized.

4.2 The Highlight of Content and Service Value

Since we media platform has undergone a phased transition from content aggregation, content sharing to personalized recommendation, the volume of content is becoming increasingly difficult to estimate. Compared with the huge volume of we media content, high-quality content and services are becoming more scarce, while the high cost of a large number of free services makes the operating expense of platform continue to increase. Free is the best marketing tool, but the overflow of free will bring about a decline of product and service quality, the high cost of enterprises, the growth hinderance of the industry. In current network ecology, the prosperity of the payment mode helps to build a healthier network content ecology.

The transformation of user demand is an important part of the ecological evolution of the network platform. Currently, the demands of users have a tendency from quantity to quality, single to multiple, which matches the growth of content volume, but does not match with the low proportion of high-quality content. The report at 19th CPC National Congress pointed out that the main contradictions in our society have been transformed into the contradiction between the people's growing need for a better life and the uneven and inadequate development. The same is true in the field of online content. The absolute number of content production is already very large, but high-quality content without sufficient supply can't meet the needs of users. In this case, the improvement of the content and service quality of the we media platform becomes the focus.

4.3 Technical Gestell and Mutual Construction of People

Heidegger's concept of "Technical Gestell" believes that when people design tools, people design the way of existence and the way people exist is reconstructed by the embedded media. Penetrating in many life scenes, we media platform, between social platform and a private place, has changed people's ways of behaviors and thinking. At present, the absorption of users' fragmented life scenes is an important means to enhance the scales of users. The process of competing for the we media market is the process in which all aspects of users' daily life are continuously penetrated by we media content from scenic spots, clothing, food to lifestyles. The popularity of various online celebrities displays that the content of we media has profoundly affected people's living habits and ways of thinking. The network platform has already become an indispensable part of people's lives, and we media content is embedded into every corner of the network platform from social networking and sharing scenes, which further accelerates the mutual construction of network technology and people.

4.4 The Redistribution of Value of Internet

Platform, user and government constitute the interactive game triangle. At present, the power of the interactive game triangle is changing and the status of the platform is improving. The "China We Media Development Report" stated that the we media industry's ecological territory has expanded and entered the stage of mass growth. With the popularization of we media, its platform has become an "infrastructure" run by enterprises, and peoples work and life are increasingly inseparable from the platform. Novel coronavirus in 2020 forces the sale of a large number of commodities to be transferred online for a time, and the "influencer marketing" become more and more prosperous, accelerating the integration of we media content with commercial advertising (Fig. 3).

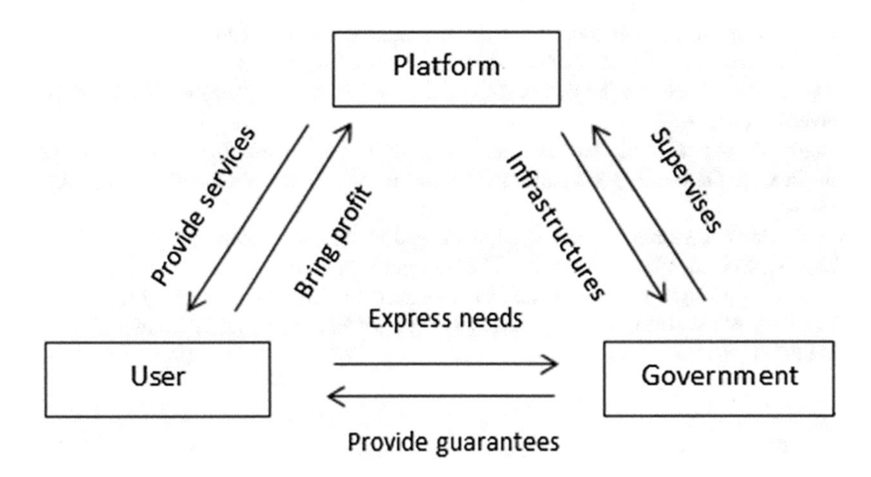

Fig. 3. The interactive game triangle constituted by platform, user and government

In the era of web2.0, more people engage in the valuable and creative work, and web3.0 requires the redistribution of Internet value. A group of we media content creators such as Li Ziqi and Papi Sauce who conform to the trend of times and the aesthetic needs of the public have entered the public vision and exerted international influence. The network platform is not only firmly embedded in people's lives, but also screens out a number of content and service producers in the new era to promote the redistribution of Internet value.

New forms of we media platform is constantly emerging, and the content of we media has been growing exponentially, but the quantification of content also brings about the difficulty of information screening, which widely exists on the network platform. Although personalized recommendation technology is used in the era of web3.0 to "top" the information most needed by individual users, algorithmic technology combined with capital power often doesn't present content most needed by users, but the content most needed by capital. The volume of online content is becoming more and more incalculable, but people generally feel that it is becoming more and more difficult to obtain valuable information. The network platform ecology is constantly maturing and improving, but at the same time, issues such as privacy leakage, unclear copyrights, and information redundancy are becoming more prominent. In this case, the network platform can attract potential customers and maintain market share by improving its own content and service quality, so as to meet the new challenges at AI stage.

References

1. Deng, X.: We media: the latest stage and characteristics of new media development. Discovery (2002)
2. CNNIC: Statistical report on China Internet development. China Internet Network Information Center, Beijing (2020)
3. Peng, L., Hanhui, H.: The transition from enterprise to platform ecosystem: Mechanism and path. Sci. Technol. Progress Countermeasures **33**(10), 1–5 (2016)
4. Zhou, L.: Research on Evolution Mechanism of Network Platform. Zhejiang Normal University (2013)
5. Hongfei, Z., Ning, L.: Six business models of we media. Friends Editors **12**, 41–45 (2015)
6. Lianghua, G.: On Heidegger's philosophy of technology. Dialectics Nat. Commun. **04**, 19–25 (1992)
7. Wu, Y.: Game Research on Traditional Media and We Media. Heilongjiang University (2012)
8. Wang, Y., Cui, C., Gao, S., Qian, X.: The leap from "internet celebrity" to "internet celebrity economy" – taking papi sauce as an example. Modern Econ. Inf. **08**, 370–371 (2016)
9. Chengwen, W.: Welcome to the golden age of content entrepreneurship. China Publishing **23**, 5–8 (2017)

Multimodal and Multidimensional Interactive Teaching Mode of Comprehensive English Under the Background of Big Data

Yongjia Duan(✉)

Department of Applied Foreign Language, Chengdu Neusoft University, Sichuan, China
duanyongjia@nsu.edu.cn

Abstract. As the core basic course of English major, Comprehensive English also needs corresponding transformation and innovation in the context of big data. Multimodal and multidimensional interactive teaching has the conditionality of depending on the sustainable development of modern technology. Therefore, taking the innovation of Comprehensive English teaching as the starting point, making full use of all kinds of information technology in big data era, fully exploring the multidimensional interactive channels between teachers and students in the background of big data and developing multimodal interactive teaching, can further meet the needs of Comprehensive English teaching and students' learning in the context of big data. This paper mainly explores the multimodal and multidimensional interactive teaching mode of Comprehensive English in big data. Through multimodal sensory integration, the creation of multimodal interactive practice activities, the improvement of teaching methods and means, and the creation of diversified evaluation system, a more scientific and complete multimodal multidimensional interactive mode is established to further promote the Comprehensive English multi-modal interactive teaching coverage and coverage quality, and finally improve the teaching effect.

Keywords: Big data · Comprehensive English · Multimodal interaction · Multidimensional interaction · Teaching effect

1 Introduction

Modality means the sense. Multimodal teaching refers to the teaching method of mobilizing the five senses and the whole body, which aims to break away from the traditional English teaching method and focus on cultivating students' intercultural communication ability, English practical ability and English cultural literacy [1]. The big data network learning platform has brought a lot of convenience to our English learners, but it is also a brand new challenge for teaching activities. As a new mode of modern education, multimodal teaching has the important value of transforming quality education, cultivating humanistic talents and promoting social development [2]. It's very necessary to grasp the "interactive" concept of multimodal interactive teaching by making the best of the Internet to further improve the teaching efficiency of Comprehensive English teaching.

J. MacIntyre et al. (Eds.): SPIoT 2020, AISC 1282, pp. 621–628, 2021.
https://doi.org/10.1007/978-3-030-62743-0_89

Under the "Internet plus" environment, many teaching cloud platforms have emerged under the background of big data. The multimodal and multidimensional interactive teaching mode also comes into being. This teaching mode mainly refers to that teachers, relying on the learning platform, social tools and free and high-quality teaching resources provided by the network information technology, take the students' multi-sensory experience as the information processing and cognitive means by using multidimensional ways for the interaction between teachers and students through a variety of instant communication tools, so as to form a variety of interactive teaching structures among teachers and students, students and students, students and network resources [3]. Therefore, it is feasible and valuable to explore the innovative development of multimodal and multidimensional interactive teaching mode in the background of big data.

2 The Current Problems of Comprehensive English Teaching Mode Under the Background of Big Data

Multimodal utterance refers to the phenomenon of communication through language, image, sound, action and other means and symbolic resources by using hearing, vision, touch and other senses [4]. The interaction between teachers and students is very important for the classroom teaching of Comprehensive English, which is also the fundamental reason that although artificial intelligence is widely used now, it can't replace teachers' comprehensive and efficient interactive teaching that touches the soul [5]. Zheng Jinzhou of East China Normal University put forward five misunderstandings of interactive teaching: interactive teaching is not a one-way teaching [6]. It can be seen that there are still many problems in the teaching of Comprehensive English.

2.1 Simplification of Classroom Model and Lack of Educational Technology Support

Some underdeveloped areas and rural areas have poor educational equipment, which cannot meet the technical requirements of multimodal and multidimensional interactive education. In addition, although most colleges and universities have introduced multimedia assisted teaching, most English classes still adopt the traditional education mode, focusing on teachers' teaching, and students passively accept knowledge. PPT is only a tool to simplify teachers' blackboard writing, and it does not really provide assistance for interactive teaching. Some teachers did not make full use of multimedia technology and did not keep up with the pace of big data information age.

2.2 Imperfection of Modern Information Technology Literacy of Teachers and Students

In the multimodal and multidimensional interactive teaching mode under the background of big data, the application of information technology and big data is the basic premise and important guarantee for the realization of that mode of Comprehensive

English, which requires relevant students and teachers to have certain information technology literacy. However, at present, students and teachers are still lack of certain information literacy, including English teachers who are not able to master high-level information skills and are not familiar with the new teaching platform, which leads to poor teaching in multimodal interactive teaching mode under the background of big data [7]. In addition, because most teachers have adapted to the current relatively traditional teaching mode, teachers will have certain resistance to modern information technology, unwilling to innovate and change, which is not conducive to the innovation of teaching methods and means, let alone the formation of multimodal and multidimensional interactive teaching mode.

2.3 Limitation of the Traditional Teaching Mode to the Multimodal Interactive Teaching Environment Under the Background of Big Data

The multimodal and multidimensional interactive teaching mode of Comprehensive English requires a specific teaching environment. However, many college English teaching is only carried out in the traditional classroom. The class size and seat arrangement are usually relatively fixed. This leads to the weakening of Comprehensive English multimodal and multidimensional interactive teaching mode, which cannot truly reflect the actual teaching effect, or it will become more formalistic. For example, in the traditional education model, students have narrow communication channels and cannot consult teachers in time when they encounter problems. Students' listening and speaking ability is not fully practiced, which leads to the inefficiency of students' review and preview after class, which also causes the situation that students fail to hand in their homework or their homework accuracy is extremely low.

As mentioned above, Comprehensive English, as the main core course of English majors, also needs to keep pace with the times. It is necessary to make the most of information technology in the classroom to form the integration goal of class and after-class with big data and Internet, by combining English teaching with computer and other auxiliary means and apps into the classroom and after class, so as to better realize the teaching knowledge and interest. The participation of various media can greatly change the single traditional teaching mode, and can better mobilize students' hearing, vision, touch and even smell, to stimulate students' multiple reading ability and language application ability; through the network to assist teachers to carry out teaching activities such as teaching test, question answering and discussion, the traditional classroom and the Internet can be organically combined "Multi directional interaction", which can lay a good foundation for the improvement of students' Comprehensive English ability [8]; the multidimensional and multidirectional interaction methods such as teacher-student interaction, student-student interaction and student-computer interaction can also combine the traditional teaching mode with the latest cloud technology and network technology, so as to fully utilize network resources and promote the diversification of teaching forms, and eventually to form a more scientific multimodal and multidimensional interactive mode.

3 The Construction of Multimodal and Multidimensional Interactive Mode in Comprehensive English

3.1 Teaching Ideas

The concept of "multimodal teaching" came into being in the background of highly developed modern information technology. Human beings have entered the era of hypertext and can carry out extensive information interaction activities through the Internet [9]. Modality is a special way to express information with a certain medium. It interacts with the external environment through the sense of touch, hearing and vision. The interaction with more than three senses is called multimodality. The interaction between normal people is multimodal [10]. Constructivist learning theory holds that learning is the process of students constructing their own knowledge. On the contrary, they should actively construct the meaning of information. Compared with the traditional teaching, the use of network cloud resources and cloud platform to create multimodal and multi-dimensional interactive teaching mode has significant advantages.

3.2 Construction of Multimodal and Multidimensional Interactive Mode

In the process of multimodal interactive teaching, learners' multiple abilities are stimulated and highlighted. In the process of learning, students perceive, decode and store knowledge through multimodal channel, and then output it to form a positive interaction between teachers and learners, so as to realize the deep conversion and internalization of knowledge [11]. Comprehensive English teaching can be divided into pre-class, in-class and after-class. In each stage, task-based and project-based teaching can be used to promote teaching with the help of the media Internet and big data cloud platform. Various teaching materials and means are used to stimulate different senses of students, and diversified multidimensional interactive methods are used to construct multimodal and multidimensional Comprehensive English teaching. Details are shown in the Table 1 below:

Table 1. The teaching process and interactive mode used in Comprehensive English

Teaching stage	Teaching contents	Interactive mode
Pre-class	Preview before class	Teacher-student interaction + student-student interaction (Online + offline)
	Brainstorming before class	Teacher-student interaction + student-student interaction (Online + offline)
	MOOC	Human-computer interaction (online)
In-class	Classroom teaching	Teacher questioning and evaluation interaction + student-student discussion and mutual evaluation interaction + teacher-student discussion interaction + student-student summary interaction (offline)
	Network light live broadcast + Live broadcast	Teacher-student interaction + student-student interaction (online)
	In-class test	Teacher-student interaction (offline)
After-class	After-class discussion	Human-computer interaction + student-student interaction (online + offline)
	After-class review	Human-computer interaction + teacher-student interaction + student-student interaction (online + offline)
	After-class exercises and tests	Human-computer interaction + teacher-student interaction + student-student interaction (online + offline)

3.3 Implementation Approaches

Comprehensive English teachers can choose a variety of materials according to the curriculum objectives and classroom objectives, such as audio, video, visual symbols and foreign language website intervention, etc., so as to form a multimodal and multidimensional interactive mode between teachers and students.

3.3.1 The Integration of Multimodal Senses to Realize Multidimensional Interaction Between Teachers and Students

In multimodal interactive teaching, multimodal sensory fusion is the foundation of successful interaction. In the process of arousing students' five senses, teachers often use individual internal transfer method and external stimulation method to guide interactive teaching. On the one hand, based on the "audio visual touching multimodal teaching method", teachers can enrich PPT, Teaching plan and carry out outdoor study activities to mobilize students' information and visual nervous system. English song with multi rhythm and speech with multi rate are used to enrich the auditory system. Games and situational simulation are used to enhance students' dynamic tactile sensitivity. The synesthesia effect of taste and smell is stimulated through three kinds of sensory systems, namely visual, listening and touching, which can be operated and perceived practically. On the other hand, based on the value of nonverbal symbol mode under the social symbol theory, teachers use a large number of nonverbal symbols such as images, audio, video and so on, in order to subconsciously construct students' English cultural literacy and improve their English comprehensive ability.

Under the mobilization of various symbols, students' multimodal senses can be constantly moved in the situational context and life, and the integration of students' five senses is constantly improved, and the level of feedback degree (i.e. interactivity) of the scene is also improved. In the context of big data, these interactions not only stay at the classroom level, but also radiate to the after-class. Through the carrier of pre-class tasks and after-class tasks, the network, cloud platform and MOOCS are used as the media to construct the human-computer interaction mode, teacher-student interaction mode and student-student interaction mode.

3.3.2 Creating Multimodal Interactive Practice Activities

English literacy is mainly reflected in students' communication and practical application. Teachers can use big data to create multimodal interactive practice activities, and cultivate students' English communicative competence through practical training in pairs and groups. For example, in the construction of situation of multi-media mini-play performance, in addition to simple language expression, but with the help of costume, not only do expressions, postures and other non-verbal factors serve as a foil for the specific meaning that language express, but also some perception in the scene, such as sound, picture, color, object, etc., are important to make the teaching content specific, visualized and expressive In order to build a multimodal interactive context between teachers and students.

3.3.3 Taking Big Data as an Opportunity to Make the Best of Cloud Technology to Improve Teaching Methods and Means

In the process of multimodal and multi-dimensional interactive language teaching, teachers use some non-verbal elements to stimulate students' interest in language learning and improve students' subjective initiative and participation. For example, teachers can design questions and activities that interest students, then make much of multimedia technology to promote students' participation in activities, link relevant videos of network resources, stimulate students' sense organs by using image mode to assist them to cognize the content related to the theme, so as to promote students' active participation; Online courses can also make the most of the functions of the teaching platform, such as the sharing and editing function of Tencent documents, so that students can answer and edit questions together at the same time. Interactive software, such as chat area, QQ learning group, Tencent classroom, etc., can also be used in the class so that students can send their own views and answers to the questions in real-time in the form of "bullet-screen comments" on the teacher's teaching display screen for teachers and students to share. In the multidimensional interactive environment, students' autonomous learning ability has been vastly improved by thinking and summarizing, and teachers use multimedia technology to combine auditory mode, visual mode and other modes, so as to involve more students in different activities in the classroom and improve their confidence in language learning and communication.

3.3.4 Establishing Diversified Evaluation System

Teaching evaluation plays an important role in improving teaching quality. The multimodal and multidimensional interactive teaching mode under the Internet environment abandons the single evaluation mode of "final paper determines performance". It adopts the multimodal evaluation method, including teachers' and students' evaluation, which effectively integrates the students' learning attitude, learning interest, learning motivation and other emotional elements into the multidimensional evaluation system. The evaluation mainly focuses on the whole learning process, so as to ensure that the evaluation system can reflect the real learning situation of students. Then through the data analysis of the evaluation results, teachers put forward plans for students' problems in teaching, provide further guidance to students through learning platform or social tools, and put forward some learning suggestions and methods individually and specifically. In addition, the feedback in the evaluation, such as students' learning attitude and emotional changes, will also be reflected in the teachers' subsequent teaching design and teaching content selection, so as to better assist personalized teaching.

4 Conclusion

Under the background of big data, through the Internet and cloud resources, the implementation of multimodal and multidimensional interactive teaching mode in the Comprehensive English is very central and essential in breaking the limitations of teaching in time and space, and it plays significant role in broadening students' learning time and learning ways. It can effectively introduce high-quality resources to teach students in classroom and with the help of convenient learning platform, foreign language classroom teaching and students' autonomous learning can be effectively combined. Teachers can present videos, images and micro lessons related to English teaching content in front of students and students can learn anytime and anywhere. And through smart phones, network and other devices, ubiquitous learning of human-computer interaction, teacher-student interaction and student-student interaction can be realized. In the future teaching practice, we teachers still need to constantly improve the multimodal and multidimensional interactive teaching mode of Comprehensive English and the function of teaching platform, expand teaching resources, refine the supervision of students' learning process and the research and classification of learning strategies, so as to form a more innovative and complete teaching mode, finally to complete the comprehensive education objective.

References

1. Yang, L.: Analysis of multimodal interactive teaching mode of college english under the background of big data. J. Heilongjiang Univ. Technol. (Comprehensive Edition)**11**, 123–126 (2019). (in Chinese)
2. Yilang, T., Ying, L.: Research on multimodal interactive teaching of college english in the information age. Overseas Engl. **22**, 139–140 (2019). (in Chinese)

3. Manqian, C., Jing, Y., Zhiying, J.: Research on the multi-mode interactive english teaching under the informationized environment in higher vocational colleges. J. Harbin Vocat. Tech. College **3**, 164–166 (2018). (in Chinese)
4. Delu, Z.: On a synthetic theoretical framework for multimodal discourse analysis. Foreign Lang. China **1**, 24–25 (2009). (in Chinese)
5. Kou, P.: Research on the interactive mode of college English teaching under the "Internet Plus". J. Heilongjiang Inst. Tech. (6), 71–72 (2020). (in Chinese)
6. Zheng, J.: Interactive Teaching. Fujian Education Press (2005). (in Chinese)
7. Wang, T.: On cultivating college students' intercultural communicative competence by multimodal teaching mode in college english. Western China Qual. Educ. **10**(5), 87+89 (2019). (in Chinese)
8. Zheng, H.: Research on interactive teaching mode under mobile internet environment in higher vocational colleges. Think Tank Era **5**, 181 (2019). (in Chinese)
9. Luo, Y.: Multimodal dimensional interactions teaching model in english listening teaching. J. Hubei Adult Educ. Inst. **9**,101–102(2011)
10. Yueguo, G.: On multimedia learning and multimodal learning. Technol. Enhanced Foreign Lang. Educ. **114**(4), 3–4 (2007). (in Chinese)
11. Ning, B.: Construction of multimodal interactive teaching mode of "internet plus business english". Overseas Engl. **5**, 51–53 (2017). (in Chinese)

Logistics Development Under the Background of Internet

YiWei Chen[1(✉)], YuQing Wang[1], ZhengXing Yang[1], YuXin Huang[2], and Xin Luo[1]

[1] School of Management, Xi Hua University, Chengdu 610039, Sichuan, China
cywzl685@163.com
[2] School of Civil Engineering, Architecture and Environment, Xi Hua University, Chengdu 610039, Sichuan, China

Abstract. With the rise of the Internet, Internet technology is playing a revolutionary role in many industries and fields, and logistics industry is also developing rapidly due to Internet technology. In recent years, the development of logistics has closely followed the pace of the development of Internet technology, promoting the emergence of many new logistics branches, and many logistics enterprises are also making efforts to realize intelligent logistics under the Internet. The emergence of modern information technologies such as "Internet+" and big data enables logistics to connect products in production, storage, logistics and transportation, and realize information exchange and sharing and precise circulation in each link, thus transforming the development of traditional logistics into a modern integrated logistics in the supply chain. This article through the analysis of the Internet technology innovation application in intelligent logistics, e-commerce logistics, logistics development trend of the Internet, on the basis of relevant theoretical basis and research status and research some key bottleneck problem, and on the Internet under the background of logistics development main problems were analyzed, and corresponding solutions are put forward, in order to promote domestic logistics can seize the opportunities under the Internet development in the future challenge coordinated development.

Keywords: Internet · Logistics development · Forecast

1 The Introduction

Since the Internet entered the commercial market in the 1990s, it has developed rapidly and played an indescribably important role in the development of global market informatization. Logistics plays an important role in connecting all parts of the economic market. With the fierce competition among enterprises, the importance of cost control, information reliability sharing and other issues has been highlighted, and has been widely concerned by the international logistics industry [1].

The development of China's domestic logistics industry has skipped over some stages of steady development. In China, logistics facilities and equipment, logistics management and operation, science and technology and other aspects have relatively

J. MacIntyre et al. (Eds.): SPIoT 2020, AISC 1282, pp. 629–634, 2021.
https://doi.org/10.1007/978-3-030-62743-0_90

complete development. Due to its late formation and weak practical foundation of professional theory, despite the broad prospect of intelligent logistics development, there are still many challenges.

By analyzing the cases of intelligent logistics and e-commerce logistics, this paper points out several significant problems existing in the development of logistics under the background of the Internet, and puts forward several corresponding solutions and optimization directions by combining theoretical knowledge and practical application. According to these problems, China's domestic logistics needs to start from the construction of intelligent logistics ecosystem and the formulation of national guidance policies, so as to better develop logistics in the context of the Internet.

2 The Development of the Logistics Industry Brought by Internet Technology

Internet technology to benefit from technology integration innovation, innovation is the essence of the Internet. Under the background of integrated innovation brought by the Internet, logistics should be reformed in three general directions: technological progress, efficiency improvement and organizational innovation [2]. The development of smart logistics and new logistics management mode are obvious in the current development trend of domestic logistics.

Fig. 1. Intelligent logistics

2.1 The Internet Improves the Innovation and Development Level of Logistics,Take Intelligent Logistics as an Example

In the context of the coming of a new round of global scientific and technological revolution, the demand for consumption-oriented logistics is growing rapidly, the overall logistics efficiency of the society is improving rapidly, and the transformation of enterprises is facing greater opportunities. In the logistics industry, more enterprises are aiming at realizing transformation through smart logistics [3]. Products in the enterprise procurement, production, warehousing, distribution, sales each link, wisdom logistics can provide a public information integration platform for data backup information sharing, as well as deployment of logistics network planning, on the basis of large data cloud computing technology for data analysis and make full use of various social resources, set up the Internet platform to develop accurate transport system, automatic intelligent warehouse system, convenient and efficient distribution planning system mechanism, realize the innovation in the Internet based on the construction of efficient logistics operation.

2.2 The Internet Promotes the Development of E-Commerce, and E-Commerce Drives E-Commerce Logistics

With the rise of the Internet in the 1990s, e-commerce has become an important part of the network economy. The Internet promotes the gradual development and progress of the network economy. Among them, e-commerce logistics and e-commerce are mutually beneficial and develop each other.

As an important part of e-commerce, logistics system plays a role of bridging the physical circulation between businesses and consumers. With the comprehensive development of e-commerce, the standards of e-commerce logistics in all walks of life are gradually improved, which brings the logistics industry to think about the problems such as innovative operation, developing new models, and how to make efficient use of modern science and technology [4]. E-commerce activities have brought new development opportunities to the logistics industry and created a new development direction. E-commerce driven by the development of e-commerce logistics has a very high practical value and theoretical role. Combined with the role of the Internet, modern logistics has more room for growth and development direction.

3 Development Bottlenecks in the Logistics Industry Under the Background of the Internet

3.1 Low Efficiency of Logistics Operation

Due to the influence of various factors such as historical accumulation, institutional cost and enterprise logistics management level, the overall operation efficiency of domestic logistics is low, which results in a large waste of material cost, time cost and opportunity cost. In under the action of Internet technology, the logistics industry can effectively reduce the corresponding cost, logistics operation efficiency is improved, but the society as a whole logistics costs than the gross national product (GNP) is still large, in addition there are many enterprises inventory, supply chain product flow velocity is slow, low operation level of money and material brings the problems such as huge pressure on enterprises, human resources in these problems, need to be further combined with the opportunities brought by the Internet and realize all logistics link closely integrated with the Internet technology to improve logistics operation efficiency as a whole [5].

3.2 The Intellectualization and Standardization of Logistics Are Relatively Low

Standardization of smart technology popularization are wisdom logistics base, modernizing the Internet under the wisdom logistics management an important means and necessary conditions, as the prerequisite of the logistics industry to realize the high efficiency operation, perfect the standard system to realize intelligence logistics, reduce logistics cost and improve service quality and operation efficiency plays an important role [6]. Logistics activities involve many enterprises in the supply chain, ranging from procurement and production to sorting, warehousing and transportation and distribution. Only by making full use of the information technology provided in the Internet era and carrying out confidence reform and innovation, can an efficient logistics operation system be realized. At the same time in the Internet developing background, logistics operation of the vehicle scheduling management, inventory, warehousing management, information software system and other important links need to implement a complete, so the whole logistics operation system, the need to develop from the system Angle and comply with the relevant standards, in order to better the combination of the Internet, data and other information technology to achieve intelligent logistics, wisdom.

3.3 The Degree of Logistics Integration Is Low

Modern logistics has the development idea of systematic control and maintenance in many aspects. It integrates the flow of production, supply, transportation, sales, storage and related information in economic activities into an integrated system. With the emergence and development of supply chain management and operation, logistics integration is particularly important. Logistics informatization is an important support for the realization of logistics integration. Under the background of the vigorous development of Internet technology, the level of logistics informatization has been improved to some extent, but there are still problems such as excess costs caused by information asymmetry [7]. Modern logistics technology is widely used in logistics activities. Modern technology and facilities can greatly improve the efficiency of logistics activities. Nevertheless, domestic logistics technology level still has a great room for improvement. As a service-oriented industry, with the increasingly complex social division of labor, the production and operation of logistics technology and management requirements are relatively strict, but so far, logistics service socialization has not covered the society. Three main factors lead to the low degree of logistics integration [8, 9].

4 Analysis of Logistics Industry Development Issues in the Context of the Internet and Relevant Supporting Measures

During the period when Internet technology revolutionized China's economy, coupled with the progress promoted by big data, cloud computing and other information technologies, the logistics industry has made extensive progress. At the same time, more logistics enterprises choose to seize the opportunity brought by the Internet to reform and develop in the direction of intelligent logistics. Internet drives the development of logistics industry in small quantity reduction, batch, grasp the personality, strong elastic characteristics, in order to better meet the enormous market demand, because is still in developing stage, the logistics development under the Internet also has low efficiency of logistics operation, logistics, intelligent and standardization of each link popularity is low, and low level of logistics integration bottleneck problem [9]. In this regard, this paper combined with the technology brought by the Internet technology and the analysis of the status quo of domestic logistics put forward the following countermeasures.

4.1 Build an Intelligent Logistics Ecosystem

From a single logistics enterprise, the wisdom of the first to build the enterprise internal logistics ecosystem, with wisdom, electrical business logistics and reverse logistics in the field of intelligent experience to reach the company internal logistics, on the basis of wisdom from a single enterprise logistics across the enterprise logistics park of wisdom, and then integrated into regional logistics wisdom, expanded from every level layers from the bottom upgrade, to build logistics ecosystem to optimize the logistics cost is high, the intelligent wisdom the low popularity [10].

4.2 Strengthen the Construction Standards, Guidelines and Policies of Smart Logistics

China's logistics industry, scattered, chaotic, complex significant problems each region should be combined with the national logistics industry policy guidance to make the corresponding regional logistics industry policy guidance, strengthen the enterprise to participate in social wisdom logistics operation management level, to achieve the purpose of promoting regional logistics index standardization, well implement policy leadership.

4.3 Building an Internet Logistics Information Sharing Platform

Modern logistics gradually incline to supply chain integration, dominated by the state, all walks of life to participate in enterprise environment, logistics is the bridge to maintain each link, in order to achieve the overall system integration, need to implement relevant in each link and the link joint connection, in this requires at every link connecting information consistency, set each link of effective resources, improve the overall competitiveness of supply chain logistics system [11].

5 Conclusions

Relevant national policies put forward to promote the in-depth integration of the Internet, big data, cloud computing and other information technologies with logistics and promote the transformation of the logistics industry, which is the "supply-side reform" of the logistics industry. Logistics can be intelligent, informatization, automation, systematic, bring more benefit for the modern social market. In the future development trend of Logistics in China, enterprises' active participation, strong policy support, upgrading of related technologies, strengthening of social resources integration ability, complying with the needs of The Times, and realizing the improvement and transformation of intelligent logistics in the future are needed.

References

1. Sun, X., et al.: Research on information sharing mechanism of agile logistics management in the context of "internet+". Inf. Sci. **05**, 160–162 (2017). (in Chinese)
2. Chen, X.: Marine transport efficiency evaluation of cross-border e-commerce logistics based on analytic hierarchy process. J. Coastal Res. **94**(sp1), 682 (2019)
3. He, L.: Development trend of china's smart logistics. China Circulation Econ. **6**, 126–134 (2017). (in Chinese)
4. Jerzy, K., Kijewska, K.: Smart logistics in the development of smart cities. Transp. Res. Procedia **39**, 201–211 (2019)
5. Hui, W.: Research on the construction and countermeasures of logistics information platform in the era of big data. Inf. Record. Mater. **020**(005), 109–111 (2019). (in Chinese)
6. Zheng, K. Zhang, Z., Song, B.: E-commerce logistics distribution mode in big-data context: A case analysis of JD.COM. Ind. Mark. Manage. **86**, 154–162 (2020)
7. Barenji, A.V., et al.: Intelligent E-commerce logistics platform using hybrid agent based approach. Transp. Res. Part E: Logs Transp. Rev. **126**, 15–31 (2019)
8. Xinyu, G.: Empirical Study on Working Mechanism of Logistics Industry in Improving Efficiency of Manufacturing Industry: With Transaction Costs as Mediating Factor. Logs Technology (2017). S. O. Economics, and N. University
9. Gong, Yu, et al.: Logistics innovation in China: The lens of chinese daoism. Sustainability **11**, 2 (2019)
10. Dai, J., et al.: Service innovation of cold chain logistics service providers: A multiple-case study in China. Ind. Mark. Manage. **3**, 85–92 (2019)
11. Xinyue, W.: Research on development problems and countermeasures of smart logistics in China. Railway Transp. Econ. **039**(004), 37–41 (2017). (in Chinese)

Model Analysis of Bowl Buckle High Support Formwork Construction Technology Based on Computer Simulation Technology

Xuan Shi(✉)

Construction Engineering College, Tongling University, Tongling 244000, Anhui, China
1052904552@qq.com

Abstract. As people's requirements for building functionality and artistry continue to increase, various super high-rise and long-span structural forms have gradually emerged. In order to meet the construction requirements of such buildings, the application of high-formwork structural systems is becoming more and more widespread. However, at the same time, the number of high-profile safety accidents is increasing day by day, bringing huge economic losses and casualties to society. The purpose of this paper is to analyze the model of the bowl buckle high support formwork construction technology based on computer simulation technology. This article first analyzes the basic structure and performance characteristics of the bowl buckle high support formwork, and then introduces the analysis model of the stability of the tall formwork support system. The experimental part uses SAP2000 software to simulate the stability of the high support template. The high support system adopts the rigid connection between the vertical rod and the vertical rod. The other rods adopt a semi-rigid connection to establish the model. The ultimate bearing capacity of the support mold under vertical and horizontal loads provides guidance and reference for the design of similar bowl button high support mold systems in the future. In this paper, through experiments on this model, the vertical ultimate bearing capacity is 86.6 KN and the horizontal ultimate load is 28.5 KN. The experimental results show that the ultimate load of the model designed in this paper is closer to the actual situation.

Keywords: Computer simulation technology · Large template support · Bowl buckle scaffolding · Support stability · Model analysis

1 Introduction

In recent years, with the rapid development of China's economic construction and the increasing scale of urban and rural construction, China's construction industry is also developing at an unprecedented speed. There are more and more urban elevated roads, viaducts and large-span structural buildings. The functional requirements of buildings are increasingly demanding in terms of space structure and height. The common requirements of this type of building are the large span of the building structure, the

J. MacIntyre et al. (Eds.): SPIoT 2020, AISC 1282, pp. 635–642, 2021.
https://doi.org/10.1007/978-3-030-62743-0_91

high strength of the building structure, the large load bearing capacity of the building structure and the use of cast-in-situ reinforced concrete structures.

Because the tall formwork support system is very different from the ordinary formwork system in terms of structure selection, form composition and load, the high formwork structure is subjected to large loads and members during the construction process. There are many and people do not understand the system [1, 2]. This also led to frequent occurrence of high formwork accidents on the construction site, which seriously affected the safety of construction workers and the performance of the building during construction, resulting in a large number of casualties and economic losses, and brought a particularly negative impact on society [3, 4]. Therefore, the research on the reliability and safety of high supporting molds has received strong attention from the society [5].

In this paper, SAP2000 software is used to simulate the stability of the high support template. The high support mold system adopts the rigid connection between the vertical rod and the vertical rod. The other rods adopt a semi-rigid connection to establish the model. Ultimate bearing capacity of the mold under vertical and horizontal loads. In this paper, through experiments on the performance of the model, the vertical ultimate load capacity and horizontal ultimate load of the model are 86.6 KN and 28.5 KN, respectively.

2 Method

2.1 Design and Construction of Bowl Buckle High Support Formwork

(1) Basic structure

In the bowl buckle type high support mold frame, the bowl buckle joint is very important and is one of its main components, including: one, the lower bowl buckle, which is located at a position below the pole and is welded to the Location. Secondly, the upper bowl buckle surrounds the surface of the pole and can slide up and down along it. Third, the limit pin is installed on the surface of the pole, through which the upper bowl buckle can be locked. Fourth, the joint is located on the horizontal rod for fixing the horizontal crossbar [6, 7].

When assembling the various members of this support frame, you should follow the corresponding steps: first, connect the lower bowl buckle with the horizontal rod joint so that the latter is placed in the former; second, connect the upper bowl buckle with the crossbar The joint is connected, the former needs to buckle the latter through another component, this component is the limit pin; third, the upper bowl buckle is turned, the direction followed is clockwise, the tool used is a small hammer, wait for it to turn after a certain degree, its spiral surface can firmly bear against another component-the limit pin; Fourth, the vertical rod is connected with another component, this component is the horizontal rod to form the corresponding frame. Through the joint position of the rod, the vertical pole can be connected to one or more horizontal poles, up to four, and they can all rotate freely in the plane without angle restrictions, which is more flexible [8, 9].

The components of the bowl buckle high support formwork mainly include:

The vertical rod can withstand vertical forces. It is the main force receiving part of this force in the entire supporting formwork. The upper load formed during construction is transmitted to the supporting formwork base through it and then to the supporting formwork base foundation.

The ejector rod, which belongs to the previous component, the vertical rod, is located at the top part of this component and can be connected to the adjustable ejector. Because of it, the support frame can be freely adjusted to various heights.

Horizontal bars can be divided into two types according to different directions, one is a vertical horizontal bar, and the other is a horizontal horizontal bar; through it, the vertical bars can be connected to form a frame as a whole; In the process of horizontal load, or in the process of constructing scaffolding, it is a main force-bearing part. If this kind of rod is located at the bottom of the bracket, then it is often called a sweeping rod.

Scissors can be divided into two types according to different directions, one is horizontal scissor and the other is vertical scissor. Because of it, the supporting frame can have stronger rigidity and stronger integrity.

The base is located at the bottom of the pole, another component of the supporting frame. With it, the frame can avoid sinking, and the load of the pole can also be transmitted to the foundation through it.

(2) Performance characteristics

The characteristics of this type of formwork can be reflected as follows:

(1) There are multiple functions, which can be flexibly and easily spliced and installed according to the actual situation, so as to obtain a supporting mold frame that can meet various requirements.
(2) Whether it is during installation or removal, the whole process is simple and convenient, so the work efficiency is stronger. Compared with the fastener type bracket, the construction of this type of mold base has one less process, that is, the fastener bolts need not be twisted, and only a simple treatment with a hammer is required, and the efficiency is increased about three times.
(3) The position of the bowl buckle node has a relatively large bearing capacity. In this type of mold base, the lower bowl buckle is welded, so it can be firmly connected to the pole, so it can cope with the shear force well; and the limit pin of the other component is also very strong. The ground is connected to the vertical pole, so the upper bowl buckle can also be firmly fixed on the vertical pole, which can cope with the tensile force.
(4) The overall strength is stronger, because in this type of supporting frame, after the related rods are connected, they can reach the state of relative axis.
(5) Easy and simple material extraction. This type of formwork is mainly made of steel pipe, so there is no need to specially select materials, just use the fastener type bracket.
(6) Avoid waste and reduce costs. In this formwork, the rod accessories are all connected, so it is very convenient to use, there is no need to worry about some parts disappearing, and the fastener type bracket is not, so its cost is more likely to increase.
(7) Whether it is maintenance, management, or transportation, you only need to do simple operations.

2.2 Analytical Model for Stability of Tall Template Support System

(1) Shelving model
The formation of the shelving model is inseparable from a study, the object of this research is the riser-type building structure group column [10]. This model is constructed through the related achievements of its stability research. In this model, the formwork bracket belongs to a multilayer rack, and its upper and lower ends are hinged to each other; therefore, when analyzing the stability of the bracket, it can be simplified, that is, it is viewed as It is a hinged equivalent column at both ends. Its stiffness and calculated length are related to the bracket, which are the total value of the single column stiffness and the column height of the latter.

(2) Multi-node continuous strut model
For a formwork support frame, the vertical bar set in it is a continuous pressure bar in the frame. There are many support points that can be supported. A horizontal member, a component in the frame, can be used as its lateral direction. Support point, and the other component scissor support can also do so; these two components can also play an adjustment function, the object involved is the vertical rod axial force, so that if the internal force formed by a vertical rod can trigger a strong axis Force can cause large deformation, then these internal forces can be transmitted to other poles with the aid of a horizontal rod, which can prevent buckling of the pole, which can make the support frame more stable.

3 Experiment

3.1 Model Establishment

In this project, the distance between the poles of the bowl-shaped high supporting molds is 1.2 m, and the horizontal and vertical distances of the horizontal members are 0.9 m. In the establishment of the finite element model, for the convenience of operation, the simplified bowl buckle high mold was 18 m in length, 9 m in width, and 12 m in height. The material used for the steel tube supporting frame is Q235 carbon steel, and the specification is $\phi 48 \times 3$ mm.

3.2 Experimental Tools

SAP2000 is a large-scale general-purpose commercial finite element software. The predecessor of SAP2000 program is SAP. Among all the existing SAP series products, its advancedness and maturity rank first; it has very strong calculation and analysis functions. In this regard, the industry has formed a consensus, whether it is in static and dynamic analysis, or in other analysis, it has performed very well and is perfect.

3.3 Stability Analysis

The support formwork is a kind of rod structure. The basic process of the stability analysis (buckling analysis) of the rod structure using SAPZ000 is: building the model:

the same as the general structural analysis process, pay attention to the selection of materials, the setting of semi-rigid nodes, and application of load and establishment of analysis case type. For the axial compression rod system in this article, it belongs to the first type of stability problem. The Bucking option in the analysis case is directly selected, and the critical load size and deformation under each load application state are calculated and analyzed. The critical load value is the product of the given load and the yield factor. The yield factor should be between 0 and 1. When it is greater than 1, it indicates that the load can be increased. When it is less than 0, it indicates that the load will yield when it is applied in the opposite direction.

4 Discussion

4.1 Experimental Results and Analysis

This paper uses SAP2000 software to analyze the stability of the support structure. They are vertical load simulation and horizontal load simulation. The simulation results are shown in Table 1 and Fig. 1:

Table 1. Experimental results

	Output case	StepNum	ScaleFactor	Ultimate bearing capacity
Vertical load	Buck	1	0.086609	86.6 KN
	Buck	2	0.086756	
	Buck	3	0.088737	
Horizontal load	Buck	1	0.028546	28.5 KN
	Buck	2	0.030483	
	Buck	3	0.037976	

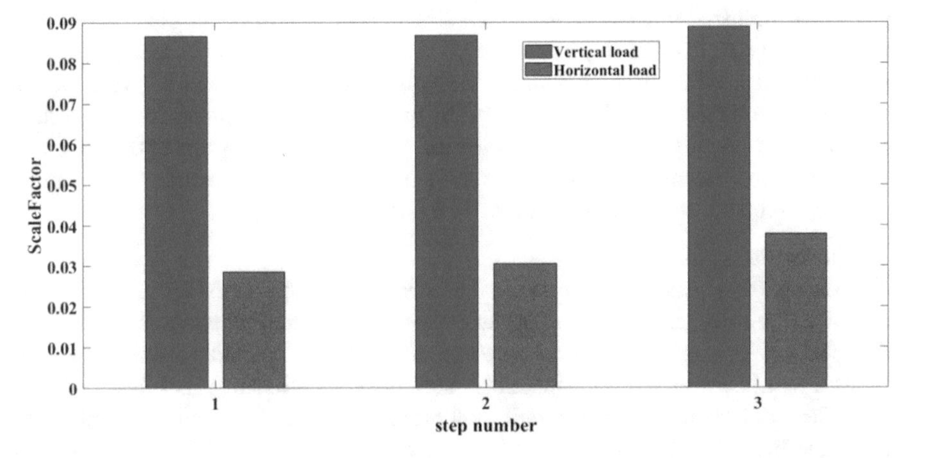

Fig. 1. Experimental results

4.2 Suggestions for Improving Construction Site Safety

(1) Construction measures in rainy season
Before entering the rainy season, pay attention to the layout of the site, the layout of water and electricity, and set up temporary drainage ditches at the site to ensure the smooth drainage of the site. According to the actual conditions of the site construction and the principle of convenience of construction, reasonable arrangements shall be made for the works performed during the rainy season. Before the rainy season, the project manager organized the engineering department, safety department and administrative department to check the construction preparations and site conditions of the rainy season, check whether the drainage pipes and water collection wells are closed, and ensure that the drainage pipes and water collection wells can be used normally. The inspection includes the drainage of various places, and detailed inspection of various mechanical equipment (welders, cables, etc.) during the construction in the rainy season. Carefully check the distribution boxes, door boxes, cables, wires and other temporary supports used at the site, and take maintenance measures in a timely manner. Ensure that the electrical equipment circuit protection is intact, the insulation equipment is grounded, access protection and high reliability. In order to prevent rain, wind, humidity, and flood, please take necessary measures. Hand-held power tools must have a leak-proof device, which must be safe and reliable. Thoroughly clean up the mechanical equipment and electrical equipment used, fully do the work of preventing rain, moisture, and flood, and timely respond to failures during maintenance.

(2) Prevention of collapse accidents
In the operation of the model, the designated person supervises. When deviation occurs, the construction must be stopped immediately, the operator should be evacuated from the scene, and the operation should be carried out after eliminating the danger. When pouring concrete, if you find that the zipper is slipping or the pole is deformed abnormally, please report it to the relevant personnel immediately. Then use the 10t jack prepared in advance to put the sliding part back to the original position to strengthen the rod. In order to prevent quality accidents and occasional collapse caused by continuous sinking. Before pouring the concrete, it is necessary to wait for the strength of the concrete of the lower support bed to meet the requirements, and then completely remove the lower support of the full hall frame before pouring the concrete. When removing the mold, it is necessary to properly fix the movable mold frame, fixing rod, support, etc. to avoid accidental injury.

(3) Prevention of fall accidents
All aerial workers must learn the safety knowledge and safe operating procedures of aerial work, and workers must accept the complete signature procedure in accordance with special safety technology disclosure and related rules. Operators at height need to submit a certificate. High-level operators must undergo a health check before being hired. The construction site project unit shall provide the operators with qualified safety helmets, safety belts and other necessary safety protection tools, and the operators shall

properly wear and use them in accordance with regulations. Safety belts worn by staff members must be inspected and certified before use. The buckle point of the seat belt is high or low. The buckle must be hung above the waist. Temporary guard rails are required around the frame-supported floor. The guardrail must be firm and reliable, and the height of the guardrail must be above 1.2 m.

5 Conclusion

To sum up, the application of the bowl buckle high-support formwork construction technology has greatly promoted the development of China's construction industry and is of great significance to the long-term development of China's construction industry. The bowl buckle high support mold system has the characteristics of diversity, complexity and high risk. Before construction, a special construction plan for a high-supporting frame needs to be prepared and verified by experts. Material verification needs to be done before construction. The construction process management and strict management need to be strengthened. Make safety and technical measures and emergency plans to avoid accidents such as mold frame deformation, displacement, and collapse during the construction of high-support molds, resulting in casualties and economic losses.

Acknowledgements. This work was supported by Tongling University School-level Scientific Research Project (2018tlxytwh9).

References

1. Duncavage, E., Advani, R.H., Agosti, S., et al.: Template for reporting results of biomarker testing of specimens from patients with diffuse large B-cell lymphoma, not otherwise specified. Arch. Pathol. Lab. Med. **138**(5), 595–601 (2017)
2. Chen, G., Liu, Y., Goetz, R.: αKlotho is a non-enzymatic molecular scaffold for FGF23 hormone signaling. Nature **553**(7689), 461–466 (2018)
3. Gobbi, A., Scotti, C., Karnatzikos, G.: One-step surgery with multipotent stem cells and Hyaluronan-based scaffold for the treatment of full-thickness chondral defects of the knee in patients older than 45 years. Knee Surg. Sports Traumatol. Arthrosc. Official J. Esska **25**(8), 2494–2501 (2017)
4. Toyota, T., Morimoto, T., Shiomi, H.: Very late scaffold thrombosis of bioresorbable vascular scaffold. JACC Cardiovasc. Interv. **10**(1), 27–37 (2017)
5. Jin, Y., Liu, C., Chai, W.: Self-supporting nanoclay as internal scaffold material for direct printing of soft hydrogel composite structures in air. ACS Appl. Mater. Interfaces **9**(20), 17456 (2017)
6. Mann, M.D.: Increasing student motivation to learn by making computer game technology more engaging: measurable outcomes that determine success. Contemp. Issues Educ. Res. **10**(2), 117 (2017)
7. Athron, P., Balazs, C., Fowlie, A.: Model-independent analysis of the DAMPE excess. J. High Energy Phys. **2018**(2), 121 (2017)

8. Anadol, R., Dimitriadis, Z., Polimeni, A., et al.: Bioresorbable everolimus-eluting vascular scaffold for patients presenting with non STelevation-acute coronary syndrome: a three-years follow-up1. Clin. Hemorheol. Microcirc. **69**(1), 1–6 (2018)
9. Liu, X., Chen, W., Zhang, C.: Co-seeding human endothelial cells with hiPSC-derived mesenchymal stem cells on calcium phosphate scaffold enhances osteogenesis and vascularization in rats. Tissue Eng. Part A **23**(11), 546 (2017)
10. Shan, Z., Lin, X., Wang, S., et al.: An injectable nucleus pulposus cell-modified decellularized scaffold: biocompatible material for prevention of disc degeneration. Oncotarget **8**(25), 40276–40288 (2017)

Vulnerability Discovery and Security Protection Based on Web Application

Hui Yuan(✉), Jie Xu, Liang Dong, Lei Zheng, Shan Yang, Bo Jin, RongTao Liao, ZhiYong Zha, HaoHua Meng, GuoRu Deng, Yan Zhuang, Shuang Qiu, and Ning Xu

Information and Communication Company of Hubei State Grid Corporation, No. 197 Xu Dong Street, Wuhan, Hubei, China
882148@qq.com

Abstract. With the rapid development of computers and the Internet, WEB applications have been increasingly used in people's lives, which has brought more hidden dangers. More and more attackers are focusing on WEB applications, using various attack methods to carry out malicious activities. In order to use web applications safely and protect the privacy of individuals or companies, this article has conducted research on web application-based vulnerability discovery and security protection, mainly for web application firewall technology, vulnerability detection technology, and data encryption Technology has made an in-depth understanding and application. Then this paper uses SVM-based multi-protocol abnormal traffic detection experiments and experimental analysis to further study vulnerability discovery and security protection. The accuracy of linear kernel function, polynomial kernel function and kernel radial basis function are respectively: 85.91%, 87.60%, 91.77%, which shows that the flow abnormality simulation experiment based on multiple protocols of support vector machine is very effective and successful Detected If a copper leak is detected in the system, it can detect vulnerabilities in web applications and issue warnings in advance to implement security protection functions.

Keywords: Web application · Vulnerability discovery · Security protection · Vulnerability detection technology

1 Introduction

Due to the low cost of the WEB architecture and the excellent ability to handle applications, more and more enterprises or governments are gradually migrating traditional application methods to the WEB-based architecture; due to the open application of WEB, it is more restricted by the protective wall. Small, to provide hackers with better access opportunities. And now there are new changes in the WEB server and application layer architecture. The current WEB service can not only realize the function of a typical dynamic page, but can even communicate with the database and use the database to dynamically generate content. These WEB services are bundled with applications, servers, etc., providing more opportunities for attacks. Many attacks take advantage of security vulnerabilities in the current WEB architecture, including

J. MacIntyre et al. (Eds.): SPIoT 2020, AISC 1282, pp. 643–648, 2021.
https://doi.org/10.1007/978-3-030-62743-0_92

modifying page content, embedding malicious code, guessing database passwords, accessing sensitive information, and so on [1].

This article studies web application vulnerabilities and related methods to achieve security protection. The working principle of WEB applications, firewall technology, vulnerability detection technology, encryption methods and other technologies, and focuses on how to achieve security protection. Now the mainstream security protection is through encryption methods. Realize it.

2 Web Application Vulnerability Discovery and Security Protection Research Methods

2.1 Working Principle of WEB Application

For modern WEB applications, they are all interactive and driven by a database. They usually consist of a back-end database and WEB pages. In addition to some information content, WEB pages also contain scripts written in specific languages. These scripts are server scripts. During the WEB interaction, they search the database based on the information parameters submitted by the user and present the specified information to the user. The traditional database-based WEB application is a three-tier system structure [2, 3]. The three-tier structure actually adds an intermediate structure between the client and the database, so business rules, data operations, and WEB verification are performed at the intermediate level [4].

2.2 Firewall Technology

A firewall is a system or a group of systems configured between the internal network and the external network to implement access control policies. It is the only entry point for information between different networks or network security domains, and can control the data flow into and out of the network according to established security policies. And he has strong resistance. Simple firewall functions can be applied to routers, and complex functions can be implemented from hosts or even subnets. The purpose is to create a security checkpoint between the internal network and the external network to allow, deny or redirect the data flow through the firewall. Check and monitor the internal and external network communication of the internal network [5, 6]. However, the firewall is not a panacea and must be used together with other security measures to ensure the security of the system (Fig. 1).

2.3 Vulnerability Detection Technology

The key part of the vulnerability detection system using vulnerability library technology is the vulnerability library it uses, that is, the standard vulnerability library formed on the basis of the research and analysis of security vulnerabilities, attack cases and actual security experience. Then the corresponding vulnerability detection rules are constructed on the basis of the vulnerability database, and the detection program adopts the rule matching technology to automatically perform the vulnerability detection work

[7, 8]. In the detection system using the vulnerability database, the central controller sends a detection request to the detection server, and after receiving the response, it matches the information in the vulnerability database accordingly. If a vulnerability is found, relevant information such as possible harm, location, etc.

In a vulnerability detection system, the execution time required to detect a vulnerability on a server is:

$$t(v_{ik}) = p_{ik} + \mu_i h_{ik}(\mu_i > 0) \tag{1}$$

Vulnerability detection system based on the vulnerability database, the vulnerability database is the core of the system, it saves the characteristics of common security vulnerabilities, corresponding processing measures, etc., and provides the main basis for the detection system. Therefore, the amount of information in the vulnerability database is an important factor to measure the strength of a vulnerability detection tool, and its size determines the number of security vulnerabilities that the vulnerability detection system can find [9, 10]. And the accuracy of the vulnerability feature description is directly related to the accuracy of the detection results, which is the main work of developing the vulnerability detection system.

2.4 Data Encryption

Encryption technologies are generally divided into two categories: "symmetric" and "asymmetric".

Asymmetric encryption means that encryption and decryption do not use the same key. There are usually two keys, called "public key" and "private key". The two must be used in pairs, otherwise the encrypted file cannot be opened. The "public key" here means that it can be released to the outside world, but the "private key" cannot and will only be known by the holder. Its advantage lies in this, because if the symmetric encryption method is to transmit encrypted files on the network, no matter which method is used, it is difficult to tell the other party's key to the other party. The asymmetric encryption method has two keys and can display the "public key", so don't be afraid that others will know. The receiver only needs to use his own private key when decrypting, which works well. Avoid key transmission security issues.

Use a variable-length key with a maximum length of 448 bits, and it runs very fast.

MD5: To be precise, it is not an encryption algorithm. It can only be considered as a digestion algorithm. MD5 processes the input information into 512-bit data packets, and each data packet is split into 16 32-bit data packets. After a series of processing, the output of the algorithm contains four 32-bit data packets, which will successfully create a 128-bit hash value.

SSF33, SSF28, SCB2 (SM1): It is not allowed to use hidden and undisclosed commercial algorithms of national cryptography used in domestic civil and commercial applications.

3 SVM-Based Multi-protocol Abnormal Traffic Detection Experiment

3.1 Experimental Background

At present, there is no public power EMS network communication data set available for testing and evaluation at home and abroad. Therefore, in order to verify the feasibility analysis and algorithm verification of the multi-protocol abnormal traffic detection proposed in this paper, a simulation experiment environment designed and built based on this paper.

3.2 Experimental Data Collection

This experiment uses TMW Protocol Test Harness simulation software to simulate power EMS dispatching data network communication. This software can not only simulate DNP3.0 protocol communication, but also simulate IEC60870-5-104 protocol communication. Set up the DNP3.0 Master and IEC60870-5-104 Master on the SCADA server, install the DNP3.0 Slave and IEC60870-5-104 Slave on the RTU simulation device and the simulated attack source, and the RTU simulation device simulates the power EMS to dispatch normal communication Process, simulate attack source to simulate deceptive attacks such as stealing and tampering on SCADA server and RTU equipment.

4 Experimental Analysis of Multi-protocol Abnormal Traffic Detection Based on SVM

4.1 SVM-Based Multi-protocol Abnormal Traffic Detection Results

The SVM-based multi-protocol abnormal traffic detection module in this experiment is implemented using the Libsvm library, which runs fast, is easy to use, supports multiple languages and environments, and is currently the most commonly used SVM library. By selecting different kernel functions, the above 1,260 training samples are used as the input of the abnormal traffic detection model to train the anomaly detection model.

Table 1. Multi-protocol abnormal traffic detection results based on SVM

Kernel function	Accurate classification	Classification accuracy	False negatives	False negative rate	Number of false positives	False alarm rate	Total testing
Linear kernel function	867	85.91%	45	4.56%	96	9.52%	1008
Polynomial kernel function	882	87.60%	39	3.77%	87	8.63%	1008
Radial basis kernel function	924	91.77%	31	2.98%	53	5.26%	1008

Using the simulation environment to simulate communication, input 1008 groups of data to be detected, including 504 abnormal data and 504 normal data, and complete the test of abnormal flow detection modules with different kernel functions. The results are shown in Table 1. The accuracy of the results is above 80%, indicating that the test is very effective.

4.2 Analysis of Simulation Test Results

The security protection module, IEC60870-5-104 deep packet filtering module, DNP3.0 deep packet filtering module and SVM-based multi-protocol abnormal traffic detection module for plant and station equipment network attacks were tested separately. According to the test results, it can be seen For ARP spoofing attacks against plant-side equipment, SYN flood attacks, and spoofing attacks against the power EMS control layer network, the power EMS network security protection strategy has a good protection effect and can ensure the security of the power EMS network.

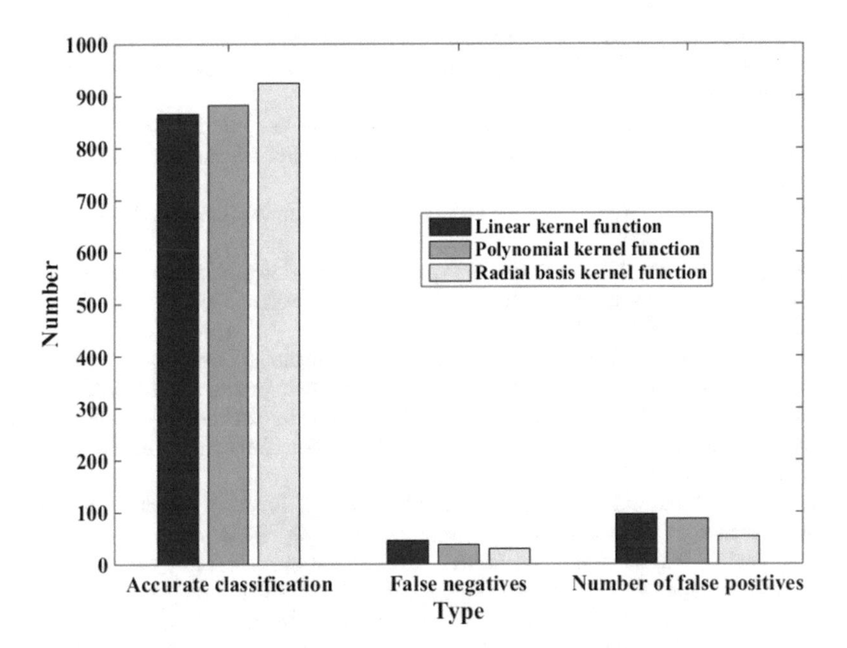

Fig. 1. Accurate classification, false negatives and false positives

It can be seen from the above simulation test results that the SVM-based multi-protocol abnormal traffic detection can detect abnormal traffic in the power EMS regulation layer network. Among them, the multi-protocol abnormal traffic detection model using the radial basis kernel function has a higher classification. The accuracy rate and low false alarm rate and false alarm rate show a good classification effect. Using this kernel function, a classification accuracy rate of 91.77% can be achieved, which effectively detects abnormal traffic in the power EMS control layer network.

5 Conclusions

Web applications face the risks of attacks from external and internal networks, as well as misuse and unauthorized operations by legitimate users. Therefore, in order to ensure the security of Web applications, a set of in-depth security protection systems must be established to conduct multi-level and multi-method Detection and protection. This paper studies the principles related to web applications and understands the origin of its vulnerabilities. Through the research of firewall technology, vulnerability detection technology and encryption algorithm technology, a simulation experiment of multi-protocol abnormal traffic detection based on SVM is designed. Through the experiment, it is learned that the simulation experiment can be very good. Detect web application vulnerabilities to achieve the ability to protect security.

References

1. Periyasamy, K., Arirangan, S.: Prediction of future vulnerability discovery in software applications using vulnerability syntax tree (PFVD-VST). Int. Arab J. Inf. Technol. **16**(2), 288–294 (2019)
2. Zonouz, S.A., Berthier, R., Khurana, H., et al.: Seclius: an information flow-based, consequence-centric security metric. IEEE Trans. Parallel Distrib. Syst. **26**(2), 562–573 (2015)
3. Joh, H.C., Malaiya, Y.K.: Modeling skewness in vulnerability discovery. Qual. Reliab. Eng. Int. **30**(8), 1445–1459 (2015)
4. Narang, S., Kapurt, P.K., Damodaran, D., et al.: Bi-criterion problem to determine optimal vulnerability discovery and patching time. Int. J. Reliab. Qual. Safety Eng. **25**(1), 1850002:1–1850002:16 (2018)
5. Young, J.H., Michael, P., Hyun, S.K., et al.: Computational discovery of pathway-level genetic vulnerabilities in non-small-cell lung cancer. Bioinformatics **9**, 1373–1379 (2016)
6. Wood, A., He, Y., Maglaras, L.A., et al.: A security architectural pattern for risk management of industry control systems within critical national infrastructure. Int. J. Crit. Infrastruct. **13**(2–3), 113 (2017)
7. Sharma, R., Sibal, R., Shrivastava, A.K.: Vulnerability discovery modeling for open and closed source software. Int. J. Secure Software Eng. **7**(4), 19–38 (2016)
8. Gonzales, D., Kaplan, J.M., Saltzman, E., et al.: Cloud-trust - a security assessment model for infrastructure as a service (IaaS) clouds. IEEE Trans. Cloud Comput. **3**, 1–1 (2017)
9. Mooney, A., Quille, K., Bergin, S.: PreSS#, A web-based educational system to predict programming performance. Int. J. Software Eng. Knowl. Eng. **4**(7), 178–189 (2015)
10. Kapur, P.K., Singh, O., Khatri, S.K.: Preface. Int. J. Reliability Qual. Saf. Eng. **23**(6), 1602002:1 (2016)

Comprehensive Evaluation of Rural Development Level Based on Data Mining

Mengqi Xu(✉)

School of Management, Shanghai University, Shanghai, China
leemengxx@163.com

Abstract. The rural economy is restricted by many factors. It is necessary to integrate data analysis technology and dig into comprehensive evaluation indicators that meet the level of rural development. In this way, we can make accurate judgments on rural development and put forward constructive opinions on development. Based on the statistical data of the Statistical Yearbook from 2009 to 2018, this paper takes the new urbanization pilot city Ningbo as the research object and uses MATLAB software for data analysis. Through the analytic hierarchy process and entropy method to calculate the comprehensive weight value, this paper constructs a three-level index comprehensive evaluation system, and calculates the comprehensive evaluation score. The study found that as the years change, the level of rural development in Ningbo has shown an upward trend. At the same time, this paper calculates and compares the scores of the primary and secondary indicators, and puts forward suggestions for rural development after analyzing the reasons.

Keywords: Rural development · Analytic hierarchy process · Entropy method · Comprehensive evaluation

1 Introduction

With the continuous development of information technology, the connection between rural and urban areas has become closer. In the era of the Internet of Everything, the level of rural economic development has become more and more difficult to assess. The influencing factors are complex. Finding valuable elements from redundant information is the key to comprehensive evaluation. The aspects of rural issues are quite complex. The reform of rural areas must be based on the needs of farmers. Targeted investment construction will produce half the effort. The material and spiritual needs of farmers must be met and farmers' sense of happiness must be met. This is the starting point for this article to study rural development.

At present, many scholars have conducted research on topics such as rural construction, agricultural development and rural revitalization through data analysis. The huge development that China's rural areas has achieved is at the cost of destroying the ecological environment [1]. Reform and opening policies, continuous accumulation of human and material capital, and technological progress have promoted the continuous growth of agriculture and rural economy [2]. More and more villages in China have realized industrialization, urbanization and peasant citizenization [3]. Jiang Yuansheng

J. MacIntyre et al. (Eds.): SPIoT 2020, AISC 1282, pp. 649–656, 2021.
https://doi.org/10.1007/978-3-030-62743-0_93

[4] conducted a comprehensive evaluation of the construction of an all-round well-off society in China's rural areas through the Rural Overall Well-off Index (ROXI) and Rural Human Development Index (RHDI). Zhang Ping [5] used the principal component analysis method to evaluate the development level of China's rural service industry and found that the development level of each province is quite different. It is necessary to increase financial support, increase farmers' income, and improve the level of education. situation. Yan Zhoufu [6] established an evaluation index system for rural revitalization through principal component analysis and expert scoring method for comprehensive weighting. Han Lei [7, 8] measured and compared the China Rural Development Index from the five aspects of economic development, social development, living standards, ecological environment, and urban-rural integration on the country's and regional rural development processes. The level of rural development at the provincial and provincial levels has been continuously improved. Xu Fang [9] conducted a comprehensive evaluation on the level of rural revitalization in Anhui, and believed that the level of social governance and welfare is worse than other aspects, and the effect of rural revitalization is not obvious.

The existing literature evaluates certain aspects of rural development from a specific perspective, and there are relatively few related studies on the comprehensive evaluation of rural development level. This paper takes Ningbo, a pilot city for new urbanization, as the research object, constructs a comprehensive evaluation index system through analytic hierarchy process and entropy method for comprehensive evaluation. Finally, through data analysis, this article evaluates the level of rural development in Ningbo, with a view to providing guidance for correct adjustments to its further development.

2 Comprehensive Evaluation System

2.1 Selection of Indicators and Data Sources

The evaluation index system of rural development level should reflect the whole picture of rural development. This paper selects five first-level indicators: urban-rural integration level, villagers' living standards, agricultural development level, people's livelihood security level and ecological environment level. As shown in Table 1, based on the comprehensive consideration of the inherent logic of the rural development level and the availability of data, 12 secondary indicators such as the urban-rural gap are selected to conduct a comprehensive analysis of the level of rural development in Ningbo. The data comes from the statistical yearbook, and the relevant data of Ningbo City from 2009 to 2018 are selected.

2.2 Weight Assignment Method

This paper investigates the indicators of rural development in Ningbo through literature surveys and field visits, and constructs a comprehensive evaluation system using a comprehensive weighting method combining analytic hierarchy process and entropy method.

Table 1. Index classification table

I	II	III
Urban-rural integration level	Urban-rural gap	Household registration urbanization rate
		Urban-rural disposable income ratio
		Urban-rural consumption expenditure ratio
Villagers' living standards	Financial ability	Per capita disposable income increase rate
		Per capita consumption expenditure increase rate
		Engel coefficient
	Quality of life	Durable electrical appliances per 100 households
		Housing area per capita
		Electricity consumption per capita
		Car ownership per 100 households
		Per capita expenditure on education, culture and entertainment
	Social insurance	Health care expenditure per capita
Agricultural development level	Agricultural scale	Agricultural investment
		Crop planting area
		Aquaculture area
		Forest nursery area
		Year-round live pigs
	Industry growth	Agricultural appreciation rate
		Forestry value-added rate
		Fishery appreciation rate
		Value-added rate of animal husbandry
	Farming modernization	Per thousand hectares of agricultural machinery
		Number of agricultural industrialization companies
Livelihood security level	Infrastructure	Number of villages benefiting from running water
		Number of broadband villages
	Grassroots organizations	Number of village committees
		Number of village clinics
	Financial construction	Number of rural cooperative financial institutions
		Number of new rural financial institutions
	Social assistance	Number of low-income families
Ecological environment level	Environmental governance	Rural environment improvement Funds
		Number of sewage treatment villages
		Soil and water loss treatment area

When using the Analytic Hierarchy Process (AHP), the factors at the same level are compared with each other, and the relative importance is expressed by the 1–9 scale method proposed by the American operations researcher T.L. Saaty to establish a judgment matrix. Calculate the maximum eigenvalue and its corresponding eigenvector for each paired comparison matrix, and use the consistency index and the random consistency index ratio to check the consistency. When using the entropy method, it is first necessary to construct a data matrix, standardize the data, obtain the standardized matrix, and calculate the information entropy value and the difference coefficient, and finally calculate the index weight after normalizing the difference coefficient. Subjective and objective weighting is a method that combines subjective weight with objective weight. Referring to the research method of Shen Yuting [10], this paper determines the target equation according to the principle of minimum relative entropy, and solves the optimal comprehensive weighting through the Lagrangian multiplier method.

3 Empirical Analysis

3.1 Comprehensive Empowerment Results

The comprehensive weighting results are calculated through the analytic hierarchy process and the entropy method. From the summary analysis in the sixth column of

Table 2. Comprehensive index weight assignment results

I	II	III	AHP	EM	COM
URIL (0.0502)	URG (1.0000)	HRUR (0.6301)	0.0316	0.0477	0.0461
		URDIR (0.2184)	0.0110	0.0547	0.0291
		URCER (0.1515)	0.0076	0.0470	0.0224
VLS (0.4452)	FA (0.5695)	PCDIIR (0.1265)	0.0321	0.0091	0.0203
		PCCEIR (0.1865)	0.0473	0.0277	0.0430
		EC (0.6870)	0.1742	0.0358	0.0939
	QL (0.3331)	DEA (0.1530)	0.0227	0.0435	0.0373
		HA (0.4283)	0.0635	0.0165	0.0384
		EC (0.2252)	0.0334	0.0529	0.0499
		CO (0.0836)	0.0124	0.0501	0.0296
		ECE (0.1099)	0.0163	0.0521	0.0346
	SI (0.0974)	HCE (1.0000)	0.0434	0.0455	0.0528
ADL (0.1682)	AS (0.2098)	AI (0.3259)	0.0115	0.0373	0.0246
		CPA (0.2463)	0.0087	0.0113	0.0118
		AR (0.1861)	0.0066	0.0091	0.0092
		FNA (0.1209)	0.0043	0.0489	0.0172
		YLP (0.1209)	0.0043	0.0062	0.0061
	IG (0.2402)	AAR (0.4350)	0.0176	0.0121	0.0173
		FVR (0.1465)	0.0059	0.0263	0.0148
		FAR (0.3092)	0.0125	0.0182	0.0179
		VRAH (0.1094)	0.0044	0.0476	0.0172
	FM (0.5499)	AM (0.6667)	0.0617	0.0335	0.0540
		AIC (0.3333)	0.0308	0.0277	0.0347

(continued)

Table 2. (*continued*)

I	II	III	AHP	EM	COM
LSL (0.1781)	I (0.2816)	NVBRW (0.7500)	0.0376	0.0065	0.0186
		NBV (0.2500)	0.0125	0.0057	0.0100
	GO (0.2553)	NVCOM (0.2500)	0.0114	0.0087	0.0119
		NVCLI (0.7500)	0.0341	0.0566	0.0522
	FC (0.1074)	NRCFI (0.8333)	0.0159	0.0075	0.0130
		NNRFI (0.1667)	0.0032	0.0503	0.0151
	SA (0.3557)	NLF (1.0000)	0.0634	0.0095	0.0291
EEL (0.1583)	FG (1.0000)	REIF (0.5396)	0.0854	0.0121	0.0382
		NSTV (0.2970)	0.0470	0.0348	0.0480
		SWLTA (0.1634)	0.0259	0.0476	0.0417

Table 2, the villagers' living standard accounts for 40%, the agricultural development level accounts for 22%, the livelihood security level accounts for 15%, and the ecological environmental level accounts for 13%, and the urban-rural integration level accounts for 10%. In the comparison of comprehensive weight values, the Engel coefficient, the amount of agricultural machinery used per thousand hectares, the number of village clinics, the per capita health care expenditure, and the per capita electricity consumption account for a relatively large proportion. These indicators have a high contribution to the evaluation of rural development levels.

3.2 Analysis of Evaluation Results

As shown in Fig. 1, the overall evaluation score shows a fluctuating upward trend, indicating that the level of rural development in Ningbo has gradually improved. It can be seen from Fig. 2 that the continuous improvement of the comprehensive score comes from the continuous improvement of the villagers' living standards. The level of ecological environment and the level of urban-rural integration only slightly improved

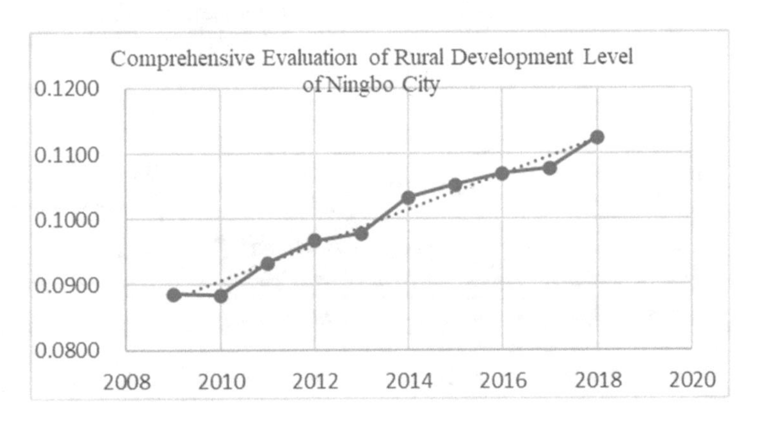

Fig. 1. Comprehensive score trend graph

after 2013. The level of people's livelihood security and the level of agricultural development will gradually decline after 2013.

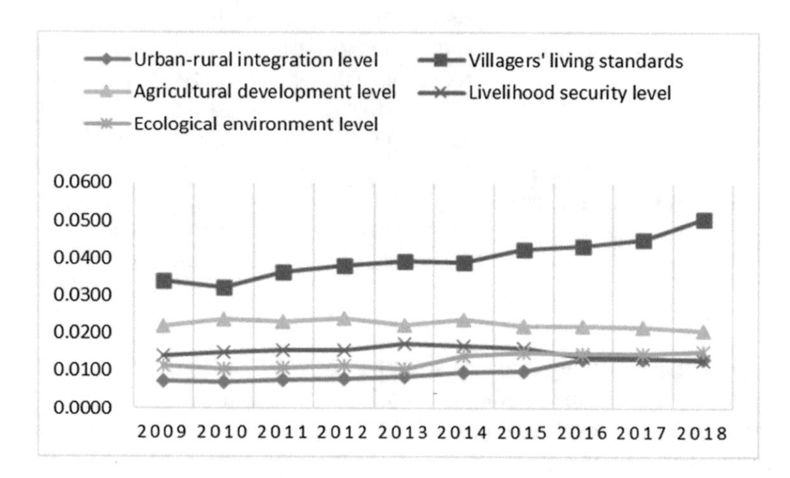

Fig. 2. Change trend chart of first-level indicators of rural development level

Figure 3 shows that the urban-rural gap score is getting higher and higher. The urbanization rate of household registration has gradually increased, the level of urbanization has increased, and the inverse indicators of the urban-rural disposable income ratio and the urban-rural consumption expenditure ratio have become smaller and smaller, indicating that the urban-rural gap has gradually narrowed. The economic

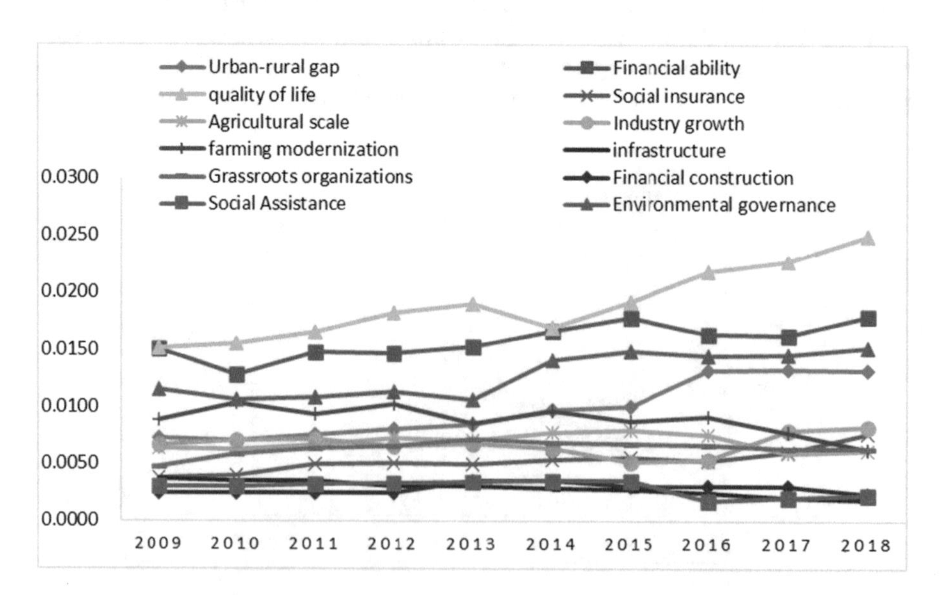

Fig. 3. Change trend chart of secondary indicators of rural development level

ability and quality of life scores increase year by year. The scores of grassroots organizations and financial construction have changed little, and the scores of environmental governance have increased year by year, indicating that the importance of the ecological environment has been recognized.

4 Conclusions and Recommendations

According to the comprehensive evaluation results of the comprehensive weighting of the analytic hierarchy process and the entropy method, during the period from 2009 to 2018, only 2010 was slightly lower than 2009, and the scores of other years maintained growth. It shows that under the support of national policies, the level of rural development in Ningbo has been continuously improved in recent years, the gap between urban and rural development has been narrowed, and an all-round well-off society has been gradually realized, and a new socialist countryside has been built. The reason for the decline in the comprehensive evaluation score in 2010 was mainly because the economic capacity of villagers fell by 15% compared with 2009, environmental governance fell by 8%, and the urban-rural gap fell by 3%, indicating that the development of rural areas should be mainly people-oriented and focus on improving farmers' living standards. Both urban and rural development should be taken into account and the gap between urban and rural areas should be narrowed. At the same time, the protection of the ecological environment should be emphasized. The decline in the quality of the ecological environment restricts rural development. The research on the change trend of the score structure found that although the increase in the number of subsistence households led to a decrease in the score of social assistance, it was more reflected in the government's vigorous poverty alleviation policy in the process of realizing new urbanization. Rural development is all-round and multi-level, and it is necessary to consider how to improve the level of rural development from many aspects. Focus on increasing the income of villagers, improving the quality of life, and narrowing the gap between urban and rural development. It is necessary to make full use of geographical advantage resources to fully develop fishery and tourism. At the same time, we will develop rural areas and agriculture in a balanced manner, pay attention to environmental protection, and continue to promote rural development.

References

1. Xiwen, C.: Environmental issues and rural development in China. Manage. World **01**, 5–8 (2002). (in Chinese)
2. Wang, S.: Overcoming poverty in development—summary and evaluation of China's 30 years of mass poverty reduction experience. Manage. World **11**, 78–88 (2008). (in Chinese)
3. Wang, J.: New stage of rural development in China: urbanization of villages. Chinese Rural Econ. **10**, 4–14 (2015). (in Chinese)
4. Jiang, Y., Heping, J., Delin, H.: Comprehensive evaluation of the construction of a comprehensive well-off society in rural China. Agric. Econ. Probl. **S1**, 61–69 (2005). (in Chinese)

5. Zhang, P., Sun, W.: Comprehensive evaluation and empirical study on the development level of interprovincial rural service industry in my country. Rural Econ. **11**, 81–85 (2015). (in Chinese)
6. Yan, Z., Wu, F.: From dual division to integrated development—research on evaluation index system of rural revitalization. The Economist **06**, 90–103 (2019). (in Chinese)
7. Han, L., Liu, C.: China rural development process evaluation and regional comparison under the background of rural revitalization. Rural Econ. **12**, 44–48 (2018). (in Chinese)
8. Han, L., Wang, S., Liu, C.: China's rural development process and regional comparison—based on the 2011-2017 China rural development index. China Rural Econ. **07**, 2–20 (2019). (in Chinese)
9. Fang, X., Shen, Y.: Study on the level of rural revitalization in Anhui province based on fuzzy comprehensive evaluation. Int. J. Educ. Econ. **2**(1) (2019)
10. Shen, Y., Jin, H.: Research on China's local government debt risk early warning system—based on the analytic hierarchy process and entropy analysis. Contemp. Finan. **06**, 34–46 (2019). (in Chinese)

Intelligent Monitoring System of High-Voltage Cabling

Yingmin Huang(✉), Chaoqiang Hu, and Cuishan Xu

Guangzhou Panyu Cable Group Co., Ltd., Guangzhou, Guangdong Province 511442, China
standfist@163.com

Abstract. As the society and information technology has developed, the intelligent monitoring system of high-voltage cabling has been comprehensively applied. Most high-voltage cables have adopted measures for safe protection to protect the conductive core multi-layer protective layer. And high-voltage cables are buried deep underground, the enclosed space is not susceptible to be disturbed from external factors, and the line failure may not occur. But it is precisely because it is buried underground and requires a more elaborate laying process, it can eventually cause the failure of the entire circuit system once there are small problems. China is overpowered and the power grid covers a large area. Once there is a problem with the high-voltage cable, it is difficult to find the location and type of cable fault. The traditional cable troubleshooting method requires a lot of manpower, material and financial resources, and the efficiency is not high, which seriously affects the lives of residents and the production of enterprises. To transform cable circuits is an effective way to solve circuit faults. Therefore, it is of great practical and significant to use high-tech to implement real-time monitoring of cable circuits, find the faults at the first time, and perform fault handling in a timely manner.

Keywords: High-voltage circuit · Intelligent monitoring · Circuit

China is about to enter a well-off society, people's living standards have improved accordingly, and the requirements for power supply are getting higher and higher. On the one hand, the power is required to be safer in operation and to protect people's lives and property; on the other hand, it is expected that the quality of the power will be higher to reduce the occurrence of power failures and speed up the resolution of power failures. The development of information technology provides a new way for monitoring system of electricity. Information technology can provide an effective and safe monitoring system for high-voltage cables and implement comprehensive monitoring of high-voltage cables. It is the main direction of high-voltage cable monitoring in the future to detect faults in a timely manner, and quickly locate the fault location and implement effective solutions.

J. MacIntyre et al. (Eds.): SPIoT 2020, AISC 1282, pp. 657–662, 2021.
https://doi.org/10.1007/978-3-030-62743-0_94

1 Importance of Installing Intelligent Monitoring System for High-Voltage Cabling

(1) To improve the monitoring of power operation state

The change of life style makes people attach great importance to the safety and economy of electric power. The change of transmission market puts forward new requirements for the monitoring equipment of distribution network switch cabinet and distribution box [1]. Both of them are easy to be affected by external factors, such as changes of humidity and temperature. These two kinds of equipment are equipped with precise measuring instruments. Usually, these two kinds of equipment are installed in the open air. The change of environment and climate will easily damage the precision instruments in the box, and then cause failure. In the hot summer, the temperature in the box is too high, and the circuit in the long-term load state is easy to burn due to cable fault, resulting in the paralysis of the whole circuit. Traditional monitoring is mainly done by running patrols, which is extremely dependent on the professional judgment of the inspectors [2]. However, since the switchgear cannot be regularly shut down for maintenance, inspectors cannot see the internal components of the switchgear and cannot make correct judgments. And once a fault occurs, it is difficult to find the fault location by patrol inspection, which makes the work cumbersome and seriously affects the time of circuit rescue. To install reliable monitoring instruments for these two devices with the use of high-tech information technology, such as humidity sensor and temperature sensor, can grasp the temperature and humidity around the cable in real time, and make better preventive measures to avoid cable burning out.

(2) To improve the accuracy of circuit fault location

With the emphasis on cable protection, the probability of circuit failure is relatively low, but it still exists. There are three common circuit faults. The first one is low-resistance fault. When a low-resistance fault occurs, it indicates that the insulation layer of the cable has been seriously damaged. At this time, the fault-resistance is extremely small. It is usually 10 times less than the wave impedance of the line itself. And low-voltage pulse method can be used to accurately locate low-resistance faults [3]. The second one is high-resistance and flashover fault. In case of high-resistance fault, it indicates that the insulation protection layer of the cable has failed and can not meet the requirements of insulation and withstand voltage. However, under normal circumstances, the cable core is intact, and its insulation resistance is larger than 10 times of the wave impedance of the line itself. High-resistance and flashover fault are the most common circuit faults, accounting for up to 90%. The last one is the disconnection fault. It refers to the voltage transmitted at the beginning. During the transmission, the end fails to receive normally, but it is not because of a problem with the cable insulation protection layer. However, due to more and more complex lines and large coverage of power grid, it is difficult to locate the fault location for any kind of fault. The implementation of intelligent monitoring system for high-voltage cabling can quickly locate the fault, reduce the investment of manpower and material resources, carry out rescue quickly, and ensure the normal life of people and the production of enterprises.

2 Overall Plan of Monitoring System for High-Voltage Cabling

(1) Analysis of functional requirements and overall structure

The intelligent monitoring system for high-voltage cabling can provide power supply protection, so that residents' lives are not affected. The distribution network of China covers a wide area and the assembly structure is complex. Once the circuit fails, the fault must be located quickly and accurately. The intelligent monitoring system has effectively completed this goal, and can carry out comprehensive monitoring of the entire distribution network to ensure the stability of the circuit operation. The application of intelligent monitoring system is a distributed structure, which is composed of multiple links, including ring network cabinets, integrated monitoring terminals and substation systems. In the intelligent monitoring system, components for monitoring temperature, line load, and humidity are installed, which can be monitored in real time [4]. The GPS is used for data collection and transmission by the wireless communication module. The signals are classified and transmitted to each master station communication terminal, which is displayed and fed back by each master station communication terminal and stored in the background database. The intelligent monitoring system can automatically compare the collected temperature, humidity and current signals. When the signal is abnormal, the annunciator is triggered to automatically alarm. When it is determined that there is a circuit failure, the intelligent monitoring system starts to calculate automatically in different regions and display it in a graphical form, and then the corresponding staff will determine the location of the failure. The main station of the intelligent monitoring system mainly transmits signals through two forms, that is wired communication and independent wireless networking. They connect different monitoring terminals accordingly to output the signals effectively. The specific monitoring system structure is shown in Fig. 1.

(2) Functional structure classification of main station of monitoring system

The function of main station is to provide better power supply for the circuit and meet the operation requirements of the urban distribution network. The function modules of main station of the monitoring system should be clearly divided, so that the monitoring system can play a maximum role. The functional structure of intelligent monitoring system for high-voltage cabling is mainly divided into four categories, each of which has its specific role [5]. In the communication module, it is mainly responsible for collecting monitoring terminal data and protecting switch value acquisition and GPS time synchronization; in the logic identification module, it is mainly responsible for operation status monitoring, status alarm, fault identification, fault location and fault branch and section discrimination; in the line information database, it is mainly responsible for static database, real-time database and historical database; in the transient waveform data processing module, it is mainly responsible for calculating the transient content system and identifying the arrival time of wave head.

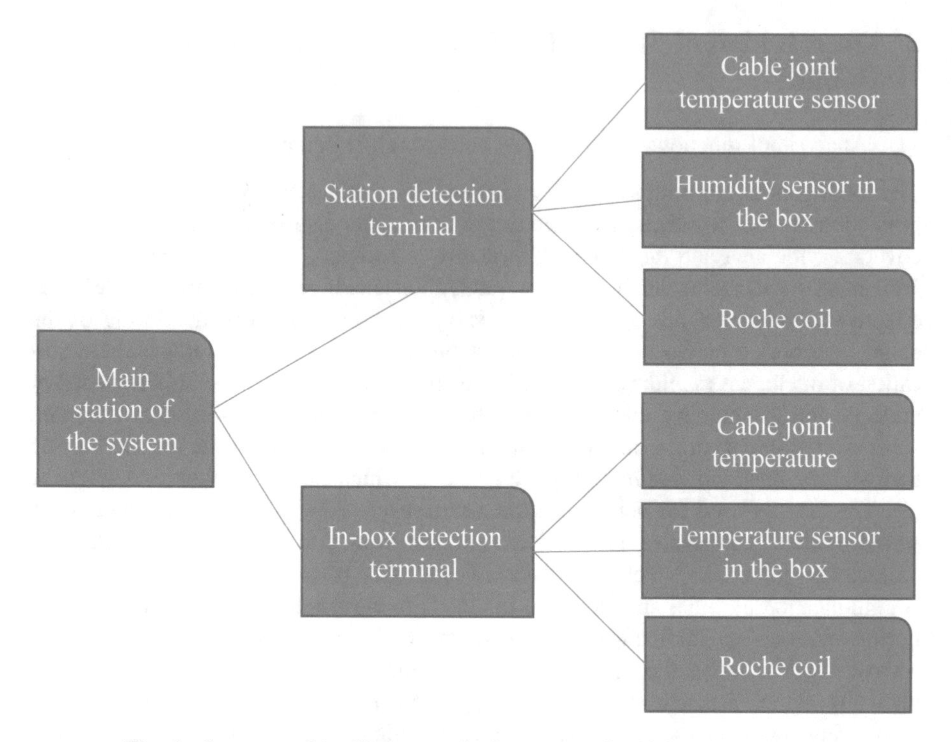

Fig. 1. Structure of intelligent monitoring system for high-voltage cabling

(3) Acquisition of intelligent monitoring information and fault location information

The acquisition of monitoring information and fault location information of intelligent monitoring system for high-voltage cabling is mainly based on the traveling wave transmission characteristics of three core cable. In order to ensure the safety of current transmission, the three core cable is used in the high-voltage cable, and the phase separation structure is adopted. Due to the structure and responsibility of the three core cable, the material made of has electrical properties, so it is much more complicated to carry out intelligent monitoring. In the three core cable, it has the characteristics of transient traveling wave signal transmission. In general, the circuit system fault will cause large traveling wave frequency, and has a wide distribution frequency band, so it must be analyzed in the transient field. In order to ensure the safety of power supply, the high-voltage cable usually operates in an open-loop mode. On the one hand, the main part of the circuit is connected with the ring network cabinet, and on the other hand, it needs branch lines. Once the circuit fails, the circuit can be closed for the use of the fault area section, and the incoming line of the connected branch cable is powered on, so as not to affect the power consumption of users temporarily. The intelligent monitoring of high-voltage cabling with the principle of traveling wave location can quickly locate the circuit fault. When the intelligent monitoring system is used to locate the fault point, it is necessary to make full use of the good enclosed space inside the ring main unit, box type transformer and cable distribution box, in which the extraction elements are placed to collect the transient signal and measure the fault point effectively.

3 Fault Location Algorithm for Distributed Measurement of Circuits

When the cabling is faulty, a traveling wave will form. When it starts to form, it is a transient signal with a relatively wide frequency. The signal spreads outward with the circuit. In the process of diffusion, the signal will gradually decay, and finally there will be none. The wave head of the signal is steep at the beginning, and gradually decreases with the degree of attenuation [6]. Therefore, HHT is used as a mathematical tool to calculate the fault location to mainly measuring the distance between the fault point and the monitoring point, and be characterized by the transient content coefficient. The intelligent monitoring system will number each monitoring point and rank the numbers according to the order. When the circuit fails, the three terminals are distanced, and the number sequence is used to check, and then the fault location is found.

(1) HHT

HHT is a mathematical method for analyzing time series signals. Its basic principle is derived from the Fourier principle, which is different from the Fourier linear principle. It is nonlinear. The HHT signal is taken from itself and has good adaptability to signal analysis. And it can also analyze local signals. When the circuit fails, the generated traveling wave signal is non-linear and has a good fit with HHT. It is a good choice to use HHT to analyze the circuit fault.

(2) Basis for the sequencing of monitoring points

After the circuit breaks down, the traveling waves formed begin to spread to the surroundings. During the diffusion process, the emission will continue to occur until the traveling waves reach each monitoring point. During the transmission of the traveling wave, the refracted or reflected traveling wave will be affected by the dispersion effect, and the signal attenuation is particularly weak when it reaches the monitoring point [7]. In addition, when the fault point in the circuit is farther away from the monitoring point in the circuit, the wave impedance of the traveling wave is greater, the attenuation of the signal formed is more, and the signal received by the monitoring point is weaker. The numbers of intelligent monitoring points are sorted using this feature, which ensures the correctness of the numbers [8].

(3) Distance measurement of circuit failure

When measuring the distance of a circuit fault, there are two main influencing factors, namely the length of the line and the waveform attenuation. The accuracy of the intelligent monitoring system is related to the distance of the fault. When the fault point is located exactly in the middle of the measurement points at both ends, the accuracy of the monitoring result is the highest. Therefore, once the circuit fails, the monitoring system first selects the three closest monitoring points for measurement, and then takes the two endpoints where the signal attenuation is smaller. And the distance and wave speed of the fault point are recalculated to effectively improve the accuracy of the intelligent monitoring system. When selecting a monitoring point to monitor a fault route, several important principles need to be considered: to select the fault point at both ends of the monitoring point; to meet the symmetrical distribution of two signals at the fault point; to ensure that the cable and the fault point cable are the same model [9, 10].

4 Conclusion

In a word, the safety of high voltage cable line is the most important. How to find and solve the fault point in time is an important step to protect the line safety. The installation of intelligent monitoring system for high-voltage cabling can effectively improve the monitoring of power operation state and improve the accuracy of circuit fault location. According to the city's demand for intelligent monitoring system, this paper puts forward the corresponding functional requirements and overall structure, and introduces the functional structure of main station of the monitoring system in detail. This paper also analyzes the fault location algorithm of circuit distributed measurement, and understands HHT, the basis for the sequencing of monitoring points and the distance measurement of circuit failure.

Acknowledgements. The project of 2017 Guangzhou Panyu District Innovation Leading Team–The R&D and Industrialization of The Intelligent Power Grid Transmission and Distribition Line Connection Products (2017-R01-7).

References

1. Shi, Q.: Research on Intelligent Monitoring System of High-voltage Cabling. North China Electric Power University (2014)
2. Tan, D., Wang, W., Bai, H.: Application and research of intelligent control platform for operation management of high-voltage cable. In: National Power Cable Installation and Operation Experience Exchange Meeting (2011)
3. Yuan, Y., Gao, Z., Chen, X., et al.: On-line monitoring of high-voltage cable line grounding system. Power Grid Clean Energy **32**(199(02)), 82–87 (2016)
4. Chen, H., Wang, D., Liu, Z.: Application research of real-time online monitoring master station system for high-voltage power cable. Inf. Recording Mater. **019**(006), 96–97 (2018)
5. Xu, X., Mai, X.: Research on intelligent expert system for high-voltage cable operation state. In: The Ninth National Power Cable Operation Experience Exchange Meeting, Institute of High-Voltage Research, China Electric Power Research Institute, Wuhan, Hubei Province (2012)
6. Xu, P., Wei, G., Liu, Z., et al.: Research on external breaking system for high-voltage cable based on cloud intelligent well cover system. Power Big Data (9) (2017)
7. Cao, W., Xu, W.: Monitoring System for Intelligent High-voltage Cable
8. Peng, L., Jiang, M., Zheng, X., et al.: Research on cable trench dehumidification technology based on wireless sensors. Electron. Test. (2020)
9. Liu, Z., Su, F., Wang, X., et al.: Cable terminal partial discharge monitoring system based on distributed wireless TEV sensor. Electr. Measur. Instrument. (17) (2019)
10. Wang, T.: Research on Cluster Routing Algorithm and Credibility Enhancement Technology of Wireless Sensor Network (2018)

Strategic Thinking on High-Quality Development of Digital Economy in Yangtze River Delta Region of China in the Era of Big Data

Ruitong Zhang(✉)

Department of Economics, Shanghai University, Shanghai, China
tt541438186@163.com

Abstract. As the world has entered the era of big data, the digital economy, as an important part of the development of the real economy and innovation-driven development, has become a key factor in China's implementation of major national strategies. In this paper, the Yangtze River Delta region, which ranks the top among Chinese three major urban agglomerations, is taken as the research object. At the same time, this paper combines with the experience of high-quality development of digital economy of other countries, the key elements of the development of big data industry are analyzed and summarized, and a package of enlightening suggestions for the development of big data industry are given.

Keywords: Big data · Yangtze River Delta region · Digital economy

1 Introduction

With the advent of the era of big data and the upgrading of electronic computer technology, countries have successively focused their future development strategies on information technology. At present, digital economy has become an important economic model. Digital economy refers to the use of the rising of network infrastructure and information tools such as smart phones, the Internet - cloud computing - block chain - the Internet of things such as information technology, human beings deal with large data quantity, quality and speed, so as to promote the human economic formation from industrial economy to information economy and wisdom economy form transformation, improve the efficiency of resource optimize configuration, eventually to promote rapid development of social productivity [1–4].

Chinese digital economy has been growing rapidly since its inception, but it lacks a clear digital economy development index system, and has not yet formed a reasonable planning and development scale. This paper aims to sort out the development index system of digital economy, analyze the development situation of digital economy in China quantitatively, and take the urban agglomeration of Yangtze River Delta as an example, so as to understand the development status and existing problems of digital economy in China. At the same time, it combines some advanced ideas and successful

J. MacIntyre et al. (Eds.): SPIoT 2020, AISC 1282, pp. 663–669, 2021.
https://doi.org/10.1007/978-3-030-62743-0_95

experience of other countries to provide development experience for the high-quality development of China's digital economy [5, 6].

2 The Development Status of Digital Economy in the Yangtze River Delta Region

The Yangtze River Delta region is rich in big data application scenarios, and the application benefits initially appear. Since 2015, the tertiary industry in the Yangtze River Delta has accounted for more than 50% for the first time, and the industrial structure is stable in the state of "three, two and one". In the tertiary industry, a large amount of data and a large number of scientific research backbone are gathered. Therefore, big data is the first to be applied in the tertiary industry, with the most prominent effect [7] (Fig. 1).

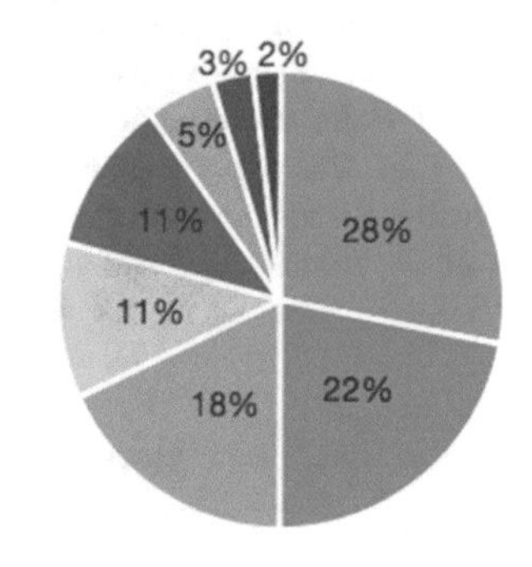

Fig. 1. Big data application distribution in the Yangtze River Delta Region

According to data released by the China Academy of Information and Communications Technology in 2019, the general development status of the digital economy in China's three major urban agglomerations can be seen. Yangtze river delta, the pearl river delta, Beijing -Tianjin -Hebei digital economy grew 18%, 18% and 14% respectively, and the digital economy of the country's digital economy proportion is 28%, 14%, 11%, digital economy accounted for the proportion of the local economy is 41%, 44%, 41%, Yangtze river delta urban agglomeration significantly high level of development of urban agglomeration area, total scale is far more than the pearl river delta and the Beijing -Tianjin -Hebei region [8].

The urban agglomeration of The Yangtze River Delta includes three provinces and one city: Shanghai, Jiangsu, Zhejiang and Anhui, and 41 prefecture-level cities. The comparison range of the Statistics and Economic index of CITIC Research Institute covers 27 central cities of the Yangtze River Delta, such as Shanghai, Hangzhou, Nanjing, Wuxi, Changzhou, Hefei, Wuhu, Anqing and Xuancheng. The following comparison method is used to compare the development of digital economy in these central cities based on four dimensional indicators: digital economy index, basic index, industrial index and smart people's livelihood index [9].

2.1 Digital Economy Index

Digital economy index is the product of the combination of economic theory and big data, and its source includes three parts: the growth of information technology-related industries themselves; Technology industries enter other industries as inputs and help them to accelerate; The ability of the whole society to make use of all output driven by information technology. In terms of the overall digital economy index, Hangzhou and Shanghai constitute the first echelon of digital economy development, with index scores of 83 and 74 respectively, which have a good foundation for digital economy development, followed by Nanjing, Ningbo and Jinhua. Anqing city, Yancheng City and Yangzhou city are affected by their weak economic foundation. In the trough area of digital economy index, the score is under 40 points, and the development difference of each city is significant. Table 1 shows the top 10 cities in the digital Economy index [10].

Table 1. The top 10 cities in the digital Economy index

City	Digital economy index	Rank
Hangzhou	83	1
Shanghai	74	2
Suzhou	63	3
Nanjing	59	4
Ningbo	58	5
Jinhua	58	6
Hefei	54	7
Jiaxing	51	8
Wenzhou	50	9
Changzhou	48	10

2.2 Basic Component Index

The basic index of digital economy includes logistics infrastructure index, information infrastructure index and digital office index. The top five basic indexes are Hangzhou, Nanjing, Suzhou, Shanghai and Changzhou. At the top of Shanghai in logistics infrastructure index, its electricity business state reversed transmission of mature logistics construction, leading industrial city of Suzhou in the information infrastructure, shows the wisdom city future development, digital office index was used to measure wisdom government communication foundation, Hangzhou has a lot of advantages, software nailing help provinces county rural realize five administrative online.

2.3 Industry Index

The digital industry Index measures the development status of the digital economy industry in the region through the enterprise data level, including regional trade index, enterprise sales index, agricultural digital economy development index and industrial digital economy development index. The top three cities are Jinhua, Jiaxing and Hangzhou. The regional trade index reflects the trade activity of a city's e-commerce and other digital economy. Jinhua and Wenzhou rank top two, with the world's largest small commodity distribution center and e-commerce ecosystem. According to statistics, Jinhua in Yiwu has a total of 334 Taobao villages.

2.4 Intelligence and People's Livelihood

The livelihood index reflects the online use of the four major life services, namely education, medical care, transportation and living expenses, and to some extent shows the convenience of the digitalization with Nanjing, Hangzhou, Suzhou and Shanghai ranking first respectively.

3 Other Countries' Digital Economy Strategies in the Era of Big Data

3.1 OECD Digital Economy Strategy

OECD governments have formulated policies in specific areas and sectors, and the relevant strategies are based on information industry policies, intersecting with existing broadband network, digital government and security governance policies. The Austrian Digital Agenda is based on several information industry policies such as the Austrian Broadband Plan, digital health and digital education, in order to provide better digital products and services and create a more secure and robust digital network environment for individuals and businesses.

3.2 Digital Economy Strategy of the UK

In 2009, the British government launched the Digital Great Britain Action Plan. The aim is for the UK to lead the world digital economy and promote the development of broadband infrastructure and ICT technologies and industries in the UK. To implement this plan, the UK government promulgated in April 2010 the Digital Economy Act 2010, which aims to protect the development of creative industries for digital content. In 2015, the UK government launched the Digital Economy Strategy, which aims to adopt digital technology for innovation and set five goals for the strategic plan. Building a user-centered digital society; Helping digital innovators; Promoting the development of infrastructure, platforms and ecosystems; Ensuring the sustainability of the innovative development of the digital economy.

3.3 US Digital Economy Development Strategy

The strategic layout of the Internet industry in the United States has two characteristics: first, it attaches great importance to intellectual property rights, and attaches great importance to the protection of technology research and development and patents of the Internet industry; second, it attaches great importance to the application of the Internet to all fields of production and life. As for the implementation of big data, in 2009, the big data strategy was proposed to publicize a large amount of data, and enterprises were encouraged to use public data to set up an innovation center for digital services and continuously implement strategic deployment in cloud computing.

The previous development strategies of digital economy in various countries mainly focus on the narrow sense of digital economy, mostly in the field of information and communication technology. But we can see with the continuous development of digital economy, digital economy has not only the level of information communication, it has been deeply integrated into all walks of life, including industry, agriculture, entertainment, and other areas of the wider all digital, so countries about the development strategy of the digital economy also give full attention to some traditional industries, and strengthen the relevant supervision and digital economy. In addition, it can be seen that America's strategic deployment of cloud technology and property rights is worthy of China's great learning. The development of digital economy not only stops at infrastructure construction, but also needs technological innovation. At this time, paying attention to patent protection policy is a guarantee for innovators. The application of cloud technology is the focus of the future development of digital economy, which can realize the information-based management of the whole society, industry and individuals, and finally realize the digital China.

4 Thinking of Development Strategy

It can be seen from the above analysis that, first of all, the position of a city in the digital economy system is not only caused by the economic level, but also determined by the digitalization and informatization degree of the city. In addition, the development status of various regions in China's digital economy is similar to that of the Yangtze River Delta region, with great spatial differences. Each region will have its own innate advantages and main development points. However, the improvement of digital economy is more driven by high-tech industries, but the threshold of innovation is often compared high. Regions with poor economic foundation can first rely on upgrading information infrastructure to promote industrial upgrading and then develop the digital economy. Therefore, the three provinces and one city should strengthen the top-level design, overall planning layout, deepen comprehensive cooperation, orderly promote all aspects of the work, highlight the following key points and difficulties.

4.1 Collaborative Promotion of Digital Technology Innovation and Application

Digital technology innovation is the first driving force of the digital Yangtze River Delta. Yangtze river delta should make full use of hierarchical coordination mechanism, to further improve the cooperative innovation network of digital technology, digital innovation resources, the full integration of the three provinces and one city at the same time make full use of the world conference on the Internet platform, such as strengthening and Beijing- Tianjin- Hebei, the pearl river delta region complementary advantages, collaborative innovation, deepen and the Yangtze river economic belt, Numbers of the silk road, precision "One Belt And One Road" and other regional cooperation, build the perfect digital technology innovation ecosystem, we long triangle digital technology.

High level innovation and integrated application, control the digital economy of the new standards to set the leading power for the digital Yangtze River Delta inexhaustible power.

4.2 Jointly Build the Yangtze River Delta Data Port

Data is the treasure mine of digital triangle. Yangtze river delta should be based on a "Shanghai international data center" platform, coordination promote the construction of basic database and regional information hub port, unified interface, data format, directory, diameter, etc., to speed up across the industry data, government affairs, social data warehouse according to unified standards and integrated transportation, medical, tourism, environmental protection and other key industries in the field of data resources, build dynamic, vast amounts of data port of Yangtze river delta, promote across provinces and cities, cross-sectoral data resource sharing and application development, urban development and support the digital economy and wisdom.

4.3 Jointly Create an Open and Standardized Market Environment

Digital Yangtze River Delta needs a good market environment. The three provinces and one city should jointly promote the construction and opening up of the digital market, formulate supportive policies to support the development of the digital economy, encourage the real economy to "go online through the cloud", guide digital enterprises to carry out global digital cooperation, constantly explore the domestic and foreign digital economy market, and form an open and cooperative market environment. We will further improve the management mechanism of the digital market, explore the establishment of laws and regulations on data rights confirmation, openness, circulation, traceability, control and privacy protection, strengthen the protection of intellectual property rights in the digital economy, balance the contradiction between data open circulation, privacy protection and data security, and create a standardized and secure market environment.

4.4 Focus on the Cultivation of Digital Talents

Talent is the core element of the digital Yangtze River Delta. The three provinces and one city should optimize the distribution of productive forces in the digital economy, build core digital economy areas, characteristic small towns and national Internet Innovation and Development pilot zones at a high level, and build a multi-layer development platform for the digital economy to attract digital talents. We will jointly implement the project to attract talents for the digital economy, actively introduce leading talents for the digital economy, and jointly cultivate a number of leading digital masters and practical digital craftsmen. We will promote the mutual recognition of standards for digital talents among the three provinces and one city, and promote the cross-regional mobility and optimal allocation of digital talents, so as to strengthen the support for the development of digital economy. We should strengthen the cultivation of digital leading enterprises, vigorously develop platform enterprises, promote the rapid growth of unicorn enterprises, and promote the gathering of digital talents and the development of digital economy with dynamic market players.

References

1. McGlosson, C., Enriquez, M.: Financial industry compliance with Big Data and analytics. J. Financ. Compliance **3**(2), 103–117 (2020)
2. Jin, G.Z., Wagman, L.: Big data at the crossroads of antitrust and consumer protection. Inf. Econ. Pol. (2020)
3. Netcracker Technology: Netcracker 2020 Puts Service Providers at the Center of the Digital Economy. Medical Letter on the CDC & FDA (2020)
4. Science-Social Science: Reports Summarize Social Science Study Results from Stockholm School of Economics (Digital Disruption beyond Uber and Airbnb-Tracking the long tail of the sharing economy). Sci. Lett. (2020)
5. Athique, A.: Integrated commodities in the digital economy. Media Cult. Soc. **42**(4), 554–570 (2020)
6. Bertani, F., Ponta, L., Raberto, M., Teglio, A., Cincotti, S.: The complexity of the intangible digital economy: an agent-based model. J. Bus. Res. (2020)
7. Science - Management Science: Study Findings from Toulouse Business School Provide New Insights into Management Science (Turning information quality into firm performance in the big data economy). Sci. Lett. (2019)
8. Engels, B.: Detours on the path to a European big data economy. Intereconomics **52**(4), 213–216 (2017). https://doi.org/10.1007/s10272-017-0677-4
9. Ji, Y., Shao, B., Chang, J., Bian, G.: Privacy-preserving certificateless provable data possession scheme for big data storage on cloud, revisited. Appl. Math. Comput. **386** (2020)
10. Senousy, Y., Hanna, W.K., Shehab, A., et al.: Egyptian social insurance big data mining using supervised learning algorithms. Rev. Intell. Artif. **33**(5), 349–357 (2020)

Data Distributed Storage Scheme in Internet of Things Based on Blockchain

Yin Zhang and Jun Ye(✉)

School of Computer Science and Cyberspace Security, Hainan University, Haikou, China
yejun@hainanu.edu.cn

Abstract. With the explosive growth of Internet of Things data, the storage of large amounts of data has become a problem that plagues people. These IoT devices generally do not have powerful computing and storage capabilities. Most of them assume the role of data collection and information transfer, which also brings challenges to data storage. And a large proportion of the large amount of data is the user's privacy data, so under the premise of storing a large amount of data, we must also ensure the security of the data. In view of the above problems, this article proposes an effective IoT data storage model based on secret sharing, which effectively guarantees the security of IoT data, and proposes a secure distributed storage of IoT data based on blockchain. This solution has the advantages of a complete blockchain. Under the supervision of the edge computing nodes of the entire network, it can ensure that each transaction data is not tampered with, and it can also obtain complete original data.

Keywords: Internet of Things · Secret sharing · Blockchain · Edge computing · Distributed storage

1 Introduction

The Internet of Things, the Internet of Everything, is to allow all human-related devices to access the Internet, so as to classify information, classify sharing, and information aggregation, so that people-centric devices have more functions, give them life, and create people-centric Ecosphere. According to Gartner's prediction, the number of IoT device will grow to around 30 billion by 2020, and will more than double by 2025 [1]. The explosive growth of Internet of Things data needs to be stored in a secure manner, which will undoubtedly be a huge challenge in the future.

Since its establishment, blockchain technology has broad application prospects [2]. This article combines blockchain with IoT to provide privacy protection for IoT data to achieve secure storage of data. Blockchain has many advantages to satisfy the above functions:

- Distributed layout: It replacing the previous centralized service center, transactions in this blockchain network are initiated, confirmed and maintained by full blockchain network nodes, without worrying about the problem of low fault tolerance brought by the centralized server;

J. MacIntyre et al. (Eds.): SPIoT 2020, AISC 1282, pp. 670–675, 2021.
https://doi.org/10.1007/978-3-030-62743-0_96

- Unique block structure: The newly generated blockchain will store the hash value of the previous block. This hash value is generated by the transaction data stored in the block. Once the data in a block is changed, it will discovered by nodes across the network, so the blockchain has a unique block structure and has the function of preventing malicious tampering;
- Proof of work: The process for generating new blocks is implemented in accordance with the mechanism of proof of work. The miner node gain the right to write new blocks into the blockchain only when a specific required hash value is generated and the proof of work is completed.

The initial design of IoT devices is to collect data and transfer information, but at the same time is limited by power consumption, it is impossible to have strong computing power. Edge computing is a good solution. The edge device can be any computing resource between the data source and the cloud. They have powerful computing capabilities and at the same time shorten the distance from the Internet of Things devices and respond quickly. The edge computing node can take over the data storage request from the IoT device, enter the blockchain network for the IoT device, and complete a series of work such as transaction propagation/transaction verification/workload proof/block writing, etc.

There are many problems with the traditional centralized data storage method. When a database is compromised, it means that all users' information is leaked. Distributed cloud storage servers can solve this problem well. This article will combine distributed storage and secret sharing plans. The core idea of both is decentralization, which is where they fit best. The server of the distributed storage cluster is a participant in the secret sharing plan. The feature of secret sharing is that even if any participant gets a sub-secret, he cannot guess the complete secret, this form guarantees the security of the data. The distributed storage also provides a guarantee for the secret sharing scheme. Even if some servers (as long as not all) are attacked and some sub-secrets are lost, it will not affect the recovery of the final secret.

2 Related Work

At present, there are some researches on data security storage of Internet of things. This article analyzes these schemes.

Based on the existing research, Mo [3] proposed a distributed file system architecture. This solution is very effective in optimizing task scheduling, but it does not consider the issues of data access control in the Internet of Things, which is a job we have to do. For the integrity of data, Tian et al. [4] presents a tailor-made public auditing scheme for data storage in fog-to-cloud based IoT scenarios. This solution can meet the performance and security requirements, but there are still some significant problems for ensuring data integrity and usability in the fog-to-cloud scenario. In order to store and protect a large amount of Internet of Things data, Li et al. [5] introduced the concept of distributed data storage, combined with blockchain to achieve data protection, but the process of blockchain transactions was not carefully described. Fu et al. [6] studied data processing and secure data storage. Based on fog computing and cloud computing, they designed a secure, flexible, and efficient data storage and retrieval system, but only supported two data retrieval methods. In view of the new security challenges faced by the Internet of Things between the two parties, Wang et al. [7] proposed a method that management secure cloud-assisted Internet of Things data that uses the cloud to collect, store, and access data, but the solution uses the encryption method is too complicated. Xiong et al. [8] constructed a new storage model based on CP-ABE for data storage and secure access in IoT applications, but the solution is too complicated. Yang et al. [9] proposed a medical big data storage system with adaptive access control based on the intelligent Internet of Things, which can save storage space under the premise of ensuring data security and access control. Xia et al. [10] proposed a secure, trust-oriented edge storage model that can ensure data security, but this scheme increases the energy consumption of IoT devices.

3 IoT Data Storage Model Based on Secret Sharing

In this article, Shamir's secret sharing scheme is used in the IoT data storage model based on secret sharing [11]. The process of constructing (t, n) Shamir secret sharing scheme will be introduced below.

3.1 Initialization

Firstly, selecting n elements $x_i(i = 1, 2, \ldots, n)$ in the finite field q, and distribute x_i to n different participants $Z_i(i = 1, 2, \ldots, n)$, the value of x_i is open, not secret.

3.2 Sub-secret Generation

Constructing n polynomials of degree $t - 1$, the following is the general formula

$$f(x_i) = s + a_1x_i + a_2x_i^2 + \cdots + a_{t-1}x_i^{t-1},\ a_i \in q,\ i = 1, 2, \ldots, t-1 \tag{1}$$

The sub-secret $f(x_n)$ obtained by the nth participant is constructed as

$$f(x_n) = s + a_1x_n + a_2x_n^2 + \cdots + a_{t-1}x_n^{t-1} \tag{2}$$

Among them, s is the secret that we will share, P is a large prime number, and $s < P$, the n unequal x obtained in the previous step of initialization are taken into $f(x)$ to get n groups $(x_i, f(x_i))$, Z_i will be assigned to n different participants, and open P, destroy the polynomial, each participant is responsible for keeping his own $(x_i, f(x_i))$.

3.3 Secret Recovery

Let $Z_1, Z_2, \ldots, Z_t$ among the participants participate in the secret recovery, collect the sub-secrets $f(x_i)$ of these participants involved in the secret recovery, and get

$(x_1,f(x_1)),(x_2,f(x_2)),\ldots,(x_t,f(x_t))$ such t points. Finally, the Lagrange interpolation method can be used to restore the polynomial

$$f(x) = \sum\nolimits_{i=1}^{t} f(x_i) \prod\nolimits_{j=1, j \neq i}^{t} \frac{x - x_j}{x_i - x_j} \tag{3}$$

Then restore the secret s

$$s = f(0) = \sum\nolimits_{i=1}^{t} f(x_i) \prod\nolimits_{j=1, j \neq i}^{t} \frac{-x_j}{x_i - x_j} mod(P) \tag{4}$$

When $x = 0, f(0) = s$, we can recover s. It is worth noting that when there are less than t participants in the recovery of the secret, the polynomial cannot be recovered, but only a single sub-secret cannot infer s. So any information about the secret is not available.

4 IoT Data Storage Scheme Based on Blockchain

Combining the blockchain with the Internet of Things to provide privacy protection of the Internet of Things data, so as to realize the safe storage of data. However, there will be some changes in the operation mode of the blockchain. This article simplifies the blockchain transaction process to ensure security while reducing resource consumption. The following will introduce the blockchain-based IoT data storage solution.

4.1 Transaction Generation and Broadcasting

Taking the fast-growing body area network as an example, Internet of Things devices, such as smart watch W, are ready to store a certain amount of data in distributed cloud storage when they are idle to relieve local storage pressure. Firstly, connecting the paired mobile phone M (here is a Bluetooth connection, in other IoT types may also be other connection methods), after M receives the original data S, perform a simple string segmentation of S to generate $s_1', s_2', \ldots, s_i'$, $i = 1, 2, \ldots, m$, the size of m depends on the data size, and then let s_i' decimalized to get $s_i, i = 1, 2, \ldots, m$.

After making the above preparations, M will send data storage requests to the distributed cloud storage server separately. The cloud server participating in the storage is equivalent to the participant $Z_i (i = 1, 2, \ldots, n)$, after the storage is successful, Z_i returns the unique address *Addr* stored in the data block, M is ready to generate a new transaction to broadcast to the blockchain network, M writes the address *Addr* into this new transaction, stores the data and generates.

The input parameter *ID* represents the unique identifier of W, where the data comes from, and *Addr* where the data goes. The *version* specifies the version rule referenced by the transaction, and *lockTime* represents the transaction lock time. There is also an important step, M will hash the transaction data of *Tran* to get H_1, and then use $W's$ private key SK_W to encrypt H_1, get an encrypted document D, D is added to the transaction as a digital signature of W, the resulting $Tran_1$ will be broadcast by M. The above process is visible in process 1.

4.2 Transaction Verification and Writing

In the previous step, after M broadcasts the new transaction $Tran_1$ to the full blockchain network, the nodes in the network will verify $Tran_1$, and after successful verification, it will be written to a candidate block B. Then, the node that wins the proof of work will connect the candidate block to the longest chain in the blockchain. The following steps need to be performed during the transaction verification process: (1) Create the Hash value H_2 of the transaction $Tran_1's$ data to be verified, except for the digital signature itself; (2) Use the public key of the account agreed to store PK_W to decrypt the encrypted document D (digital signature) in the transaction $Tran_1's$ data, and get H_1; (3) Contrast H_1 and H_2, if the two values are the same, the transaction is authorized by the owner of the private key SK_W corresponding to the account W that agreed to store, the verification is successful, otherwise it is not, the verification fails. The verification and writing of the transaction are shown in process 1.

```
Process 1 Transaction Verification And Writing
Input Tran1,PKW,D,H1 Output null
1.  start process Hash(Tran1) -> Obtain H2
2.     Decrypt(PKW,D) -> Obtain H1
3.     Judgment(H1,H2)
4.     if H1 = H2
5.       Write(Tran1)
6.     else Abort
7.     return -> end process
```

4.3 Data Acquisition and Recovery

Regarding data acquisition, it is necessary to add that an access control list needs to be added to the distributed cloud storage server to limit external access to data. Due to space limitations, this article will not make too many explanations. Then there is data recovery. According to the secret recovery process in Sect. 3, the block data of the Internet of Things can be recovered, that is, the secret s_i', and then M is spliced to form the final data S. Data acquisition and recovery are as in the process 2 shows.

```
Process 2 Data Acquisition And Recovery
Input ID Output S
1.  start process Request(ID) -> Obtain Addr
2.     Request(ID,Addr)
3.     if Allow access -> Obtain si
4.       Restore(si) -> Obtain S
5.     else Abort
6.     return S -> end process
```

5 Conclusion

Aiming at the problem of massive data security storage caused by the surge of IoT data, this article proposes a new distributed storage solution for IoT data based on secret sharing and blockchain, which has theoretical reliability, efficiency and feasibility. And data access is under regulatory control, and only authorized users can obtain target data.

Acknowledgments. This work was partially supported by the Science Project of Hainan University (KYQD(ZR)20021).

References

1. Internet of Things (IoT) connected devices installed base worldwide from 2015 to 2025 (2016). https://www.statista.com/statistics/471264/iot-number-of-connected-devices-worldwide/
2. Li, X., Jiang, P., Chen, T., et al.: A survey on the security of blockchain systems. Future Gener. Comput. Syst. **107**, 841–853 (2020)
3. Mo, Y.: A data security storage method for IoT under Hadoop cloud computing platform. Int. J. Wirel. Inf. Netw. **26**(3), 152–157 (2019)
4. Tian, H., Nan, F., Chang, C.C., et al.: Privacy-preserving public auditing for secure data storage in fog-to-cloud computing. J. Netw. Comput. Appl. **127**, 59–69 (2019)
5. Li, R., Song, T., Mei, B., et al.: Blockchain for large-scale Internet of Things data storage and protection. IEEE Trans. Serv. Comput. **12**(5), 762–771 (2018)
6. Fu, J.S., Liu, Y., Chao, H.C., et al.: Secure data storage and searching for industrial IoT by integrating fog computing and cloud computing. IEEE Trans. Industr. Inform. **14**(10), 4519–4528 (2018)
7. Wang, W., Xu, P., Yang, L.T.: Secure data collection, storage and access in cloud-assisted IoT. IEEE Cloud Comput. **5**(4), 77–88 (2018)
8. Xiong, S., Ni, Q., Wang, L., et al.: SEM-ACSIT: secure and efficient multiauthority access control for IoT cloud storage. IEEE Internet Things J. **7**(4), 2914–2927 (2020)
9. Yang, Y., Zheng, X., Guo, W., et al.: Privacy-preserving smart IoT-based healthcare big data storage and self-adaptive access control system. Inf. Sci. **479**, 567–592 (2019)
10. Xia, J., Cheng, G., Gu, S., et al.: Secure and trust-oriented edge storage for Internet of Things. IEEE Internet Things J. **7**(5), 4049–4060 (2019)
11. Yannuzzi, M., Milito, R., Serral-Gracià, R., et al.: Key ingredients in an IoT recipe: Fog Computing, Cloud computing, and more Fog Computing. In: 2014 IEEE 19th International Workshop on Computer Aided Modeling and Design of Communication Links and Networks (CAMAD), pp. 325–329. IEEE (2014)
12. Shamir, A.: How to share a secret. Commun. ACM **22**(11), 612–613 (1979)

Design of Unified Management Platform for Power Operation and Maintenance Based on Cloud Computing Technology

Mingming Li(✉), Ruiqi Wang, Zhigang Yan, Tao Wan, Xin Tang, and Hengshuo Huang

State Grid Zhumadian Power Company, Zhumadian 463000, China
kjxxbl104@163.com, 13839917959@163.com,
13513991868@163.com, 1360342210@139.com,
13783961150@139.com, dushenhuang@163.com

Abstract. On March 25, 2015, several opinions of the Central Committee of the Communist Party of China on further deepening the reform of the electric power system (Zhongfa [2015] No. 9) was issued, abbreviated as "No. 9 document of power reform". The focus of the new power reform is "three deregulation, one independence and three strengthening": the liberalization of operating electricity price, electricity sales business, distribution business, power generation and power supply plans other than public welfare and regulatory nature, and the transaction is stable. This electricity reform will be conducive to restoring the commodity attribute of electric power, realizing the marketization of electric power trading, thus stimulating the market vitality of electric power enterprises, reshaping the value chain of electric power industry, and promoting the development of the following businesses.

Keywords: Cloud computing · Power operation and maintenance · Management

1 Introduction

With the rapid development of power operation and maintenance technology, the growing scale of the network, 4G formal commercial, network operation and maintenance management and application support issues are becoming increasingly important. Network quality has become the lifeline of the major operators' attention. The stable operation of network equipment is directly related to the normal use of customer mobile services and user perception. How to effectively manage and ensure the safe operation of the core equipment to the greatest extent is the direction that network maintenance personnel have been striving for. Relying on cloud computing technology to carry out the planning and design of operation and maintenance management and application system of power operation and maintenance network, support the actual production application and management practice, and quickly respond to the needs of network operation and maintenance, is one of the important means to effectively improve the management level and operation and maintenance capacity of power operation and maintenance network [1].

J. MacIntyre et al. (Eds.): SPIoT 2020, AISC 1282, pp. 676–681, 2021.
https://doi.org/10.1007/978-3-030-62743-0_97

2 Cloud Computing Technology

With the popularization of Internet technology, the continuous improvement of information technology, and the maturity of virtual technology, cloud computing is increasingly mature on the basis of grid. Cloud computing platform forms a resource pool of mutual sharing and cooperation with huge infrastructure, data storage, various platforms and software, and abstracts hierarchical services on this basis, providing users with services such as infrastructure, platform, software and so on in the way of on-demand use, so that users do not have to care about the specific implementation of the underlying. In a narrow sense, cloud computing refers to the delivery and use mode of IT infrastructure. It refers to obtaining the required resources (hardware, platform, software) through the network in an on-demand and easy to expand way. The network that provides resources is called the cloud. In the view of users, the resources in "cloud" can be infinitely expanded, and can be obtained at any time, used on demand and expanded at any time. Guangyi cloud computing refers to the delivery and use mode of services, which means to obtain the required services through the network in an on-demand and easy to expand way. This service can be it and software, Internet related, or any other service. "Cloud" is some virtual computing resources that can be self maintained and managed, including computing servers, storage servers, etc. Cloud computing centralizes all computing resources and is managed automatically by software. Cloud computing includes three levels of services: infrastructure as a service (IAAs), platform as a service (PAAS) and software as a service (SaaS). Cloud products at relevant levels are as follows: (1) application layer corresponds to SaaS, such as Google Apps, software + services; (2) platform layer corresponds to PAAS, such as IBM it factory, Google APPEngine, Force.com (3) the infrastructure layer corresponds to IAAs, such as Amazon EC2, IBM blue cloud and sun grid;

As a new network service mode, cloud computing realizes the automatic and on-demand supply of IT infrastructure, greatly improves work efficiency and saves human resources. At the same time, virtualization technology also greatly improves the utilization of resources and reduces maintenance costs [2].

2.1 Basic Operation and Maintenance Platform

For network management and support, based on the network management virtualization platform, build a scene oriented, service-oriented, open sharing, personalized customization, scalable and scalable cloud management and support service, and realize the whole process, full cycle, integrated network management application service and hierarchical support.

2.2 Virtualized IT Infrastructure Platform

Through the construction of distributed and modularized cloud computing resources, the network management virtualization platform has built a virtualized IT infrastructure platform and realized the transformation of Cloud Computing Oriented Architecture. According to the SOA system, services at all levels are planned and managed uniformly, supporting smooth migration of applications, efficient management and utilization of resources. Relying on the resource pool to allocate on demand and obtain services on demand, it not only simplifies management and maintenance costs, but also improves scalability, reduces development time, optimizes resources and reduces risks [3].

2.3 Virtualized Cloud Application Services

Relying on the network management cloud computing platform, the network operation and maintenance basic platform should become the integrated support portal of the whole network, and build the integrated network management and support ability. It adopts modular and component-based design methods and relies on virtualization technology to form a complete network operation and maintenance basic application service directory, so as to realize the personalized network management ability of on-demand service, flexible customization and elastic expansion.

3 GSA Algorithm for Bad Data Identification in Power System

The quality of a large number of real-time data in the power system determines the safety and stability of the power system operation. By monitoring and identifying the two processes, eliminating a small amount of bad data in the measurement data is an important part of ensuring the power system operation in a safe and stable state. The purpose of bad data identification of power system is to find and eliminate a small amount of bad data in the measured values based on the obtained state estimates and the redundant information of the system, so as to improve the reliability of state estimation of power system. Power system bad data identification mainly includes state estimation and data mining. There are two kinds of data mining based on neural network and clustering analysis. The state estimation method will produce the phenomenon of residual inundation and residual pollution, and the amount of calculation is very large. The research of its application in power system is gradually declining. At present, data mining technology is widely used in power system research, with a large number of academic achievements. The bad data identification algorithm based on BP neural network has simple structure and fast identification speed, but the accuracy of identification depends on the selection of error threshold, which is subjective and empirical. The weight matrix needs training and training set. Once the network model changes, the training set needs to be used to retrain the weight matrix. Clustering analysis is an observational learning method, which classifies the data by comparing the similarity between data, and searches out the potential relationship between data objects. It is an important method applied to large database search in data mining theory.

3.1 Identification of Bad Data in Power System Based on GSA

K-means clustering method is mainly used in the identification of bad data in power system. The algorithm based on K-means clustering analysis is sensitive to the initial clustering center and often ends at the local optimum. It is not suitable for finding non convex clustering, sensitive to noise and abnormal data, and only suitable for the case where the mean value of clustering is significant (when the data set contains symbolic attributes), the cluster center cannot be calculated). The K value of clustering number needs to be determined artificially by experience judgment, which affects the identification accuracy and makes the identification miss detection or false detection. A gap statistical algorithm (GSA) is proposed to estimate the optimal number of clusters. The gap statistical algorithm and cluster analysis method are combined to identify the bad data in power system, and good identification results are achieved. GSA algorithm solves the problem of determining the optimal number of clusters in the process of bad data identification, which is essentially K-means clustering algorithm.

3.2 Bad Data Identification Based on K-Means Clustering

Principle of K-means algorithm: k-means algorithm is a clustering algorithm based on hierarchical division, which minimizes the variance sum of N data objects.

The standard is divided into k class (k is the input), so that the clustering results meet the following requirements: the similarity of objects in the same cluster is higher, while the similarity of objects in different clusters is lower. Clustering similarity is calculated by using the mean value of each cluster object - a "central object" (cluster center). The specific steps are as follows:

(1) Select the initial cluster center. Select k objects from N data objects as cluster centers.
(2) The Euclidean distance of N data objects to K clustering centers is calculated, and the data objects are divided into the class with the smallest distance.
(3) The cluster mean of K classes is selected as the clustering center.
(4) Calculate the standard measure function, and repeat two or three steps until the clustering result (measure function) no longer changes, then the clustering ends. The calculation method of Euclidean distance is as follows:

$$\mathrm{d}(i,j) = \left[\left(x_{i1} - x_{j1}\right)^2 + \left(x_{i2} - x_{j2}\right)^2 + \cdots + \left(x_{ip} - x_{jp}\right)^2 \right]^{1/2}$$

$i = \left[x_{i1}, x_{i2}, \cdots, x_{ip}\right]$ *and* $j = \left[x_{j1}, x_{j2}, \cdots, x_{jp}\right]$. In the clustering algorithm, I represents the data object and j represents the clustering center.

Mean square deviation is used as the standard measure function, which is defined as follows:

$$E = \sum_{i=1}^{k} \sum_{peC_1} |p - m_i|^2$$

P represents a point in the space, represents the cluster center of class I (P and can be multidimensional), and E is the sum of mean square deviation of all data objects and corresponding cluster centers.

4 Power Operation and Maintenance Platform

According to the characteristics of power operation and maintenance, the basic platform of network operation and maintenance is planned and designed to achieve the goal of three-tier hierarchical architecture: unified collection layer, shared data layer, operation and maintenance management application layer.

Unified acquisition layer: it provides data acquisition and instruction issuing for the upper platform, so as to shield the data interface or instruction interface between the underlying network equipment and the external system.

Shared data layer: provide unified storage and sharing management of shared data (especially OSS big data) for the upper application platform.

Operation and maintenance management application layer: the application service layer of network operation and maintenance management. According to the different life cycle of support business and network operation and maintenance, it is divided into resource opening management, fault discovery and processing, performance analysis and management, network optimization management, and unified management of information, human, task and support system related to network operation and maintenance.

5 Platform Application Analysis

Compared with the traditional network management applications, the power operation and maintenance network operation and maintenance management platform based on cloud computing has the following advantages: (1) hardware virtualization. Virtual management, scheduling and application of hardware resources are realized based on network management cloud computing platform. Through the virtual platform users use computing resources, database resources, hardware resources, storage resources, etc., greatly reduce maintenance costs and improve resource utilization; (2) three tier architecture division. A unified, clear and clear three-tier architecture Division will be established to form a unified planning, complete interface and full-featured operation and maintenance management support capacity, and improve the production process and operation and maintenance management efficiency; (3) customization and expansion as required. According to different requirements of network operation and maintenance support, users can customize corresponding services, functions and resources according to actual needs, and deploy corresponding applications and application scenarios according to users' needs. Users do not need to care about where and how to deploy the resources, only need to analyze the network operation and maintenance requirements clearly, then they can carry out personalized and targeted support work; (4) high reliability and security. It provides reliable and secure data storage management at the device level and application level, and adopts data redundancy configuration, disaster recovery strategy and strict authority management strategy to ensure data security.

6 Conclusions

With the rapid development of power operation and maintenance technology, the growing scale of network, 4G formal commercial, network operation and maintenance management and support issues become increasingly important. Based on cloud computing virtualization technology, the planning and design of the power operation and maintenance network operation and maintenance management platform is carried out to support the actual production application and management practice, quickly respond to the network operation and maintenance requirements, and build the network wide [first paper network (www.dylw.Net) to provide professional paper writing and publishing services, Welcome to] the unified, service-oriented, open and sharing, personalized customization, and scalable expansion of the power operation and maintenance network management platform can realize the whole process, full cycle and integrated network management application support, which can effectively improve the network management level and operation and maintenance ability.

References

1. Wikipedia. Cloud computing [EB/OL] (2011)
2. Xu, Y.: TD - Analysis of development status and trend of power operation and maintenance industry. Mob. Commun. (05) (2010)
3. Zhao, X., Lin, H., Zhang, M.: 3GPP Long Term Evolution (Power Operation and Maintenance) System Architecture and Technical Specifications. People's Posts and Telecommunications Press, Beijing (2010)
4. Xu, H., Chen, X., Liu, N.: Coping with challenges and establishing a sound network and information security system. Telecommun. Sci. (02) (2007)
5. Chen, K., Zheng, W.: Cloud computing: system examples and research status. J. Softw. (05) (2009)

Research on Economic Growth Based on MCMC Principal Component Regression Analysis

Long Guoqi(✉)

Shanghai Urban Construction Vocational College, Shanghai 201415, China
kyxsbjbl@126.com

Abstract. The MCMC algorithm is integrated into the principal component regression analysis model, and a new principal component regression analysis method is proposed, which not only has the advantages of effectively avoiding multicollinearity between explanatory variables and simplifying the structure of regression equation, but also can make full use of the fusion of the algorithm. According to the advantages of prior information, model information and sample likelihood function, the method is applied to the data modeling analysis of economic development indicators of Jiaxing City from to. The results show that the method can effectively overcome the shortcomings of existing analysis methods and build a model with higher prediction accuracy.

Keywords: Principal component regression · Algorithm · Economic index

1 Introduction

In the regression model of dependent variable with respect to multiple independent variables, when there is a strong correlation between independent variables, i.e. there is multicollinearity between commonly referred variables, the regression coefficient obtained by using least square estimation is generally poor, mainly because some parameters cannot be considered by significance test The principal components of variables are uncorrelated. If we can use the obtained principal components to build regression model, we can avoid the disadvantage of the above situation. Specifically, we can use the observation data of independent variables to calculate the scores of the first several principal components. When we use the observation value of the principal components, we can establish the dependent variable relationship. The meaning of principal component regression analysis is to obtain the regression of score value from principal component and then convert it to the regression equation of original independent variable.

2 Analysis Model

Let $X = (X_{1,} \cdots, X_p)'$ be a p-dimensional variable, assuming that it has second-order moment, the mean vector and covariance matrix are recorded as $\mu = E(X)$,

J. MacIntyre et al. (Eds.): SPIoT 2020, AISC 1282, pp. 682–688, 2021.
https://doi.org/10.1007/978-3-030-62743-0_98

$\sum = D(X)$, remember $Z = (Z_1, \cdots, Z_p)'$, $T = (T_1, \cdots, T_p)'$, $T_i = (t_{1i}, \cdots, t_{pi})'$. Orthogonal transformation:

$$\begin{cases} Z_1 = t_{11}X_1 + t_{21}X_2 + \cdots + t_{p1}X_p = T_1'X \\ Z_2 = t_{12}X_1 + t_{22}X_2 + \cdots + t_{p2}X_p = T_2'X \\ \cdots \\ Z_p = t_{1p}X_1 + t_{2p}X_2 + \cdots + t_{pp}X_p = T_p'X \end{cases} \tag{1}$$

The abbreviation is Z = T′X. Considering a higher cumulative contribution rate (e.g. over 85%), select the value m ($\leq p$), find $Z_1, \cdots, Z_m$, which can fully reflect the information of the original variable $(X_1, \cdots, X_p)$ and is not related to each other. Considering the dimensional difference of the actual data, the data is usually standardized first [1]. The specific steps are as follows: standardization of the original data; solving the correlation coefficient matrix of variables; finding out the characteristic roots of the correlation coefficient matrix and sorting them from large to small and their corresponding eigenvectors; determining m principal components according to the cumulative contribution rate, and obtaining $Z_i = T_i'X (i = 1, \cdots, m)$.

2.1 Principal Component Regression Analysis Process

According to the above principal component analysis method, m(<p) principal components $Z_1, Z_2, \cdots, Z_m$. Transform the standardized data matrix into the principal component score matrix of the sample $Z_{n\times m}$.

$$Z_{n\times m} = \begin{pmatrix} Z_{11} & \cdots & Z_{1m} \\ \vdots & \vdots & \vdots \\ Z_{n1} & \cdots & Z_{nm} \end{pmatrix} = (Z_1, \cdots Z_m) = \begin{pmatrix} Z'_{(1)} \\ \vdots \\ Z'_{(n)} \end{pmatrix} \tag{2}$$

3 Bayesian MCMC Analysis of Linear Regression Model

The meaning of mark y_i, $Z_{(i)}$, β, σ^2 is as before:

$$P(y_1, \cdots, y_n \mid z_1, \cdots z_n, \beta, \sigma^2) = (2\pi\sigma^2)^{-\frac{\pi}{2}} \exp\left\{-\frac{1}{2\sigma^2}\sum (y_i - \beta' z_{(i)})^2\right\} \tag{3}$$

If the prior distribution of β is multivariate normal distribution, and is recorded as multN(β_0 Σ_0), then:

$$P(\beta|Z,Y,\sigma^2) \propto P(Y|Z,\beta,\sigma^2).P(\beta) \propto \exp\left\{\frac{1}{2}\left(-2\beta' Z' Y/\sigma^2 + \beta' Z' ZY\beta/\sigma^2\right) - \frac{1}{2}\left(-2\beta' \sum\nolimits_0^{-1} \beta_0 + \beta' \sum\nolimits_0^{-1} \beta\right)\right\}$$
$$= \exp\left\{\beta'\left(\sum\nolimits_0^{-1} \beta_0 + \frac{Z/Y}{\sigma^2}\right) - \frac{1}{2}\beta'\left(\sum\nolimits_0^{-1} \beta_0 + \frac{Z'Z}{\sigma^2}\right)\beta\right\} \tag{4}$$

Therefore, the variance and mean of the posterior distribution of β are:

$$\upsilon ar(\beta|Y,Z,\sigma^2) = \left(\sum\nolimits_0^{-1} + \frac{Z'Z}{\sigma^2}\right)^{-1} \tag{5}$$

$$E(\beta|Y,Z,\sigma^2) = \left(\sum\nolimits_0^{-1} + \frac{Z'Z}{\sigma^2}\right)^{-1}\left(\sum\nolimits_0^{-1} \beta_0 + \frac{Z'Y}{\sigma^2}\right) \tag{6}$$

If the prior $\sum_0^{-1}$ value is very small, then E $(\beta|Y,Z,\sigma^2)$ is approximately equal to the least square estimator $(Z'Z)^{-1}(Z'Y)$, If σ^2 value is very large, then the expectation of β is approximately equal to β_0 in fact, it is equivalent to its prior value. Let $\Gamma = 1/\sigma^2$, let its prior distribution $\Gamma \sim$ gamma $\upsilon_0/2$, then the subsequent distribution is $P(\Gamma Z, Y, \beta) \propto P(Y|Z,\beta,\Gamma) \cdot P(\Gamma) \propto \left[\Gamma^{\upsilon_0/2-1}\exp\left(-\Gamma\upsilon_0\sigma_0^2/2\right)\right]$, which is the density function of gamma distribution.

$$(\sigma^2|Z,Y,\beta) \sim \text{Inv - gamma}(\left[n + \frac{\upsilon_0}{2}, \upsilon_0\sigma_0^2 + SSR(\beta)\right]/2) \tag{7}$$

3.1 Basis of Algebra Parameter Estimation of Linear Regression Model Based on Gibbs Sampling Algorithm

Given the initial value of the parameter to be estimated $\left\{\beta^{(0)}, \sigma^{2(0)}\right\}$, t = 1, ···, M, M + 1, · · · , M + N, where M is the sample size of pre burning, and the total number of M + N sampling Step 1 update β.

Calculate V = $\upsilon ar(\beta|Y,Z,\sigma^2)$, m = $E(\beta|Y,Z,\sigma^2)$;

Sample multN(m,V) from multivariate normal distribution $\beta^{(t)}$;

Step 2 update σ^2:

a) Calculate SSR($\beta^{(t)}$), SSR($\beta^{(t)}$) = $Y'Y - 2\beta' X' Y + \beta' X' X\beta$;

b) Sample $\sigma^{2(t)}$ from Inυ-gamma($\left[n + \frac{\upsilon_0}{2}, \upsilon_0\sigma_0^2 + SSR\left(\beta^{(t)}\right)\right]/2$);

Step 3 make: $\hat{\beta} = \frac{1}{N}\sum_{t=M+1}^{M+N} \beta^{(t)}$ $\hat{\sigma} = \frac{1}{N}\sum\limits_{t=M+1}^{M+N} \sigma^{2(t)}$

In this way, the parameters (β, σ^2) of the regression model can be estimated. According to the principal component analysis method, principal component regression model and sampling principle in the above algorithm, the sampling algorithm is nested in the parameter estimation of the principal component regression model. The algorithm steps are as follows:

The first step is to calculate the sample mean and standard deviation of each variable according to the data, and standardize the data matrix; The second step is to find out the standardized data covariance matrix of explanatory variables, that is, the correlation coefficient matrix of explanatory variables in the original data; the third step is the principal component analysis based on the correlation coefficient matrix of the second step [2]. The fourth step is to get the principal component with a higher cumulative contribution rate; the fifth step is to calculate the principal component of each standardized sample data according to the functions of each principal component and standardized data The sixth step is to use the posterior distribution of algorithm and parameters to establish the regression model of standardized posterior dependent variable to the scores of samples in the fifth step without the prior distribution of information; the seventh step is to use the principal component regression equation to calculate the function of standardized dependent variable to standardized explanatory variable; the eighth step is to use the inverse function of standardized function, that is, to add the average value, Multiply the standard deviation to calculate the regression function of the dependent variable relative to the original variable;

4 Empirical Analysis of Economic Development Trend Data

Considering the total import amount Y of Jiaxing and three explanatory variables: GDP X_1, The relationship between total household savings X_2 and total retail sales of social consumer goods X has been collected. The data of the above indicators for 14 years from 1999 to 2000 are shown in Table 1 for details. The data comes from Zhejiang statistical yearbook. Based on the data in Table 1, the total imports are related to GDP, total household savings and retail sales of social consumer goods General regression analysis modeling, principal component regression analysis modeling and principal component regression analysis modeling based on MCMC method are used to establish the quantitative relationship expression of total imports with respect to regional GDP, total household savings and total retail sales of social consumer goods [3].

According to the algorithm steps described above, first standardize the observation data, use R software to program and process the principal component analysis based on the covariance matrix after the standardization of three explanatory variables, calculate the principal component score data of each sample, and then use the principal component score data to carry out the principal component regression. The three characteristic roots of the correlation coefficient matrix of independent variables are 2.9837, 0.01438, 0.001843. According to the calculation method of the cumulative contribution rate, the cumulative contribution rate of the first two principal components is more than 99.8%. Take two principal components and use $x_t^*(t = 1, 2, 3)$ to represent the independent variables after standardization, $z_i (i = 1, 2)$ to represent the i the principal component and get $z_1 = 0.5766\ x_1^* + 0.5769\ x_2^* + 0.5786\ x_3^*$; $z_2 = 0.7322\ x_1^* - 0.6791\ x_2^* - 0.05257\ x_3^*$. From the calculation results, we can see that the three explanatory variables are in the first principal component. The contribution to total imports is basically the same. One of the major factors affecting total imports is the second principal component, such as GDP, total household savings, and total retail sales of consumer goods. Of course, we should consider the impact of excluding total household savings based on The main component expression calculates the score of each standardized sample on each main component, and then uses the score data of each sample. The regression estimation equation of the standardized data obtained by the classical binary linear regression model is (y^* represents the variable after y Standardization). In the $y^* = 0.5712\ z_1 - 1.118\ z_2 = 1.1442\ x_1^* - 0.4262\ x_2^* + 0.27196\ x_3^*$ regression equation, the coefficient test of z_1 and z_2 has significant p-value below 0.05, using The regression estimation equation of the dependent variable represented by the explanatory variable based on the original data is: $\hat{y} = -132640.3964 + 291.6634\ x_1 + 40.023\ x_2 + 180.2991\ x_3$, the statistics ($R^2$, F, P) are significant, and the RMSE is 22896. From the calculation results, x_1, x_3 If the coefficient is positive, it means that the total import is positively related to the gross national product and the total retail sales of social consumer goods, and the gross national product has a greater impact on the import than the total retail sales of social consumer goods; because the saving, consumption and purchase of imported goods belong to the payment category, in general, the total import and the total savings should be negatively related, while x_2 The coefficient of is a positive number, which shows that the positive correlation is not in line with the common sense of economics, indicating that the established model needs to be further improved.

Table 1. Data of total imports and its influencing factors in Jiaxing City from to

particular year	GDP x_1 (100 million yuan)	Total savings of residents x_2 (100 million yuan)	Retail of social consumer goodsx_3(100 million yuan)	Total imports y(US $10000)
1997	408.83	103.45	159.4817	44915
1998	433.39	127.21	168.3971	45083
1999	460.31	146.43	180.9869	52978
2000	524.03	172.31	200.3275	83832
2001	586.73	200.56	223.7435	105168
2002	677.51	250.34	249.2095	114800
2003	823.54	345.43	279.3875	209555
2004	1002.41	394.63	325.3423	281902
2005	1158.38	459.48	375.5923	287861
2006	1345.78	546.47	432.3735	347696
2007	1586	620.4	505.8651	443610
2008	1819.78	764.63	607.0139	572852
2009	1918.03	1003.98	694.296	525357
2010	2300.2	1196.94	799.3625	673421

This model shows that the total import is positively correlated with the gross national product and the total social consumer goods, but negatively correlated with the total household savings. The impact of the gross national product index on the total import is stronger than that of the total social consumer goods. The economic significance is very obvious. Using the data of 2009 and 2010, the general principal component regression method and MCMC are used The predicted value of principal component regression modeling is compared with root mean square error, as shown in Table 2. The result of data calculation shows that the predicted value based on MCMC principal component regression model is closer to the actual value than the predicted value based on general principal component regression model. From the perspective of RMSE, which reflects the prediction accuracy of the model, the method is also better than the general principal component regression method.

In order to compare the results of various modeling methods, the multiple linear regression model based on the same data based on the classical method is: $\hat{y} = -11069942 + 456.5x_1 - 119x_2 - 158x_3$, the coefficient of x_2 and x_3 is not significant, P value is more than 0.6; the coefficient of x_3 is negative, indicating that the total import and the total retail sales of social consumer goods are negatively correlated, which is not consistent with the general economic knowledge The key to the problem is that there is a serious multicollinearity between these explanatory variables, so the regression equation is actually pseudo regression, so the model is meaningless. If we model and estimate other possible linear models, such as stepwise regression of explanatory variables, we find the root mean square error of the model (RMSE) is greater than the root mean square error of the principal component regression model. Therefore, the principal component regression model based on MCMC algorithm is more important. Therefore, the most accurate model is established based on the principal component regression of MCMC algorithm.

Table 2. Comparison of prediction results of general principal component regression and MCMC principal component regression

particular year	Total value of import and export(US $10000)	Prediction value of general principal component regression model (US $10000)	Prediction value of MCMC principal component regression model (US $10000)
2009	526357	560458.6797	554660.956
2010	678421	684032.2372	680448.2065
RMSE	/	24437.78367	20065.18741

5 Conclusions

In order not to omit the information of each index in the study of index relations of economic and financial problems, people begin to choose more initial explanatory variables. In the set of these explanatory variables, the information components are often very complex, including the parts that have strong explanatory effect on the explained variables and whether they have explanatory effect. There is also a part of repeated and redundant information. If they are directly modeled by classical multivariable linear regression, the resulting model often has problems, such as violating the economic common sense, the significance test of coefficient can not be passed, etc., and the stability and prediction ability of the model are generally poor. The author applies the method to principal component regression. In modeling, the method-based principal component regression analysis method is proposed, which can not only reduce the dimension of explanatory variables, simplify the structure of regression equation, but also eliminate the adverse effects of the correlation between variables, fully integrate all aspects of important information, and construct a good model that is more in line with the economic laws Compared with the classical multiple linear regression model, principal component regression model and principal component regression model, it is found that the principal component regression method can build a model which is more consistent with the actual economy and has higher prediction accuracy under the model complexity equivalent to the principal component regression method.

References

1. Huixuan, G.: Applied Multivariate Statistical Analysis, pp. 265–276. Peking University Press, Beijing (2005)
2. Wang, H., Wu, Z., Meng, J.: Linear and nonlinear methods of partial least squares regression, pp. 117–127. National Defense Industry Press, Beijing (2006)
3. Wang, H., Wang, J., Huang, H.: Research on modeling strategy of principal component regression. J. Beijing Univ. Aeronaut. Astronaut. **34**(6), 662–668 (2008)
4. Cheng Maolin's analysis of economic growth factors based on principal component model. Oper. Res. Manag. **21**(1):175–179 (2012)
5. Hongyan, W., Shibing, L.: An empirical study on the income difference of urban residents in China based on the principal component regression analysis. Prediction **28**(1), 77–80 (2009)
6. Zhu, H., Han, Y., Wu, Z.: Bayesian prediction analysis of multiple linear regression model. Oper. Res. Manag. **14**(3), 44–48 (2005)

Data Confidentiality and Privacy in IoT

The Construction and Implementation of the Security Defense System of University Campus Network

Haiwei Wu and Hanling Wu(✉)

Hainan Vocational University of Science and Technology, Haikou, Hainan Province, China
haicn163@eiwhy.com

Abstract. Today, campus networks with different scales have been built in colleges and universities. As an infrastructure platform for information construction, the campus network has penetrated into all aspects of college teaching, scientific research, management, etc., offering convenience to teachers' and students' work, study and life. At the same time, the security problem of campus network is becoming more and more prominent; the safety of campus network becomes particularly important. To develop campus informatization, the construction of college campus networks should be strengthened. At the same time, to guarantee the safety and stability of campus networks, schools need to strengthen the construction and application of security defense systems. The article focuses on analyzing the common security problems in college campus networks, and gives relevant security protection measures.

Keywords: Universities · Campus networks · Network security · Defense systems

1 Introduction

At present, colleges and universities should strengthen the construction of campus network and campus network application to promote the realization of informatization development of current college campuses. However, in the process of Internet access, relevant staff need to pay attention to Internet security issues. There are various risk factors in the Internet, including viruses, hacker attacks, Trojan horses, malware and other negative factors. In order to avoid serious and effectively ensure the safety and stability of campus network, the construction of a security defense system should be strengthened, to prevent the campus network from being negatively affected, affecting the normal work and life of campus students and teachers [1].

2 Analysis of Main Security Problems in Campus Network

The campus network is in an open network environment, with many users, complex network structure (as shown in Fig. 1), and relatively difficult maintenance and management. These characteristics determine that it must have security problems.

J. MacIntyre et al. (Eds.): SPIoT 2020, AISC 1282, pp. 691–696, 2021.
https://doi.org/10.1007/978-3-030-62743-0_99

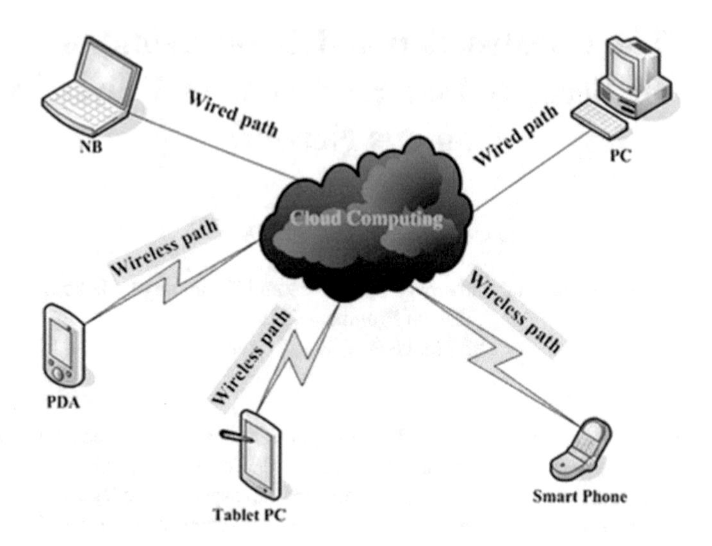

Fig. 1. Campus network topology

2.1 Physical Destruction

There are many hardware devices and network cables and other facilities in the campus network. They may suffer from natural disasters like fires, earthquakes, and some human factors during their use. These physical damages may affect the operation, and even cause serious consequences of the campus network paralysis [2].

2.2 Computer Virus Threat

Computer viruses are a major hidden danger that threatens the security of campus networks. They can spread within campus networks through storage media, system vulnerabilities, and e-mails. Common examples include Trojan horses, worms, and backdoors. Computer viruses spread quickly, are destructive, and difficult to remove, which poses a serious threat to the security of campus networks [3].

2.3 Cyber Attacks

The university campus network has the characteristics of openness, which makes it vulnerable to some malicious attacks from external networks, such as DDos attacks, ARP attacks, and SQL injection attacks. In addition, there are also aggressive behaviors within campus networks. For example, some students, out of curiosity or desire for practice, will use the knowledge learned in class to try to attack the campus network. What's more, driven by some interests, illegally attacked servers and stole internal data [3]. These internal and external attacks will interfere campus network and even cause more serious consequences.

2.4 Terminal Vulnerabilities

The terminal vulnerabilities in the campus network mainly include operating system vulnerabilities, hardware device vulnerabilities, websites and information systems based on Web applications. These vulnerabilities are potential security risks in the campus network. Hackers and other criminals can use these vulnerabilities to launch malicious attacks on the campus network. Computer viruses can also use these loopholes to spread, threatening the security of campus networks [4].

2.5 Dissemination of Bad Information

The Internet is full of unhealthy information such as spam, violent pornography, and even reactionary information. This information can be disseminated in college campus networks through the Internet, which has an unfavorable impact on the physical and mental development of students [4]. In addition, this information often carries Trojan horses and viruses, which also poses a serious threat to the security of campus networks.

2.6 Weak Awareness of Cybersecurity

Most campus network users believe that network security is specifically responsible for the school network center and other departments and has nothing to do with individuals. Therefore, they generally lack network security awareness. For example, some students often browse some unsafe web pages, open spam emails of unknown origin, download files that hide viruses, and install virus-bearing software. This negligence in security awareness has also laid a hidden danger to the security of the campus network [5].

2.7 Defects in Campus Network Security Management

The security threat of campus network also comes from management problems. Higher education institutions generally emphasize operation and neglect management. The upper-level leaders have insufficient awareness and attention to network security work, and their investment in security work is relatively low. The campus network management team is short of personnel, especially professional network security personnel. Most network administrators have low professional skills and lack strong skills in preventing and handling network security risks [5]. In addition, the safety management system of the campus network is not perfect, which cannot effectively restrict and regulate the online behavior of students, faculty and staff.

3 Safety Protection Measures for College Campus Network

Based on these security issues faced by college campus networks, this research will propose relevant protective measures from the physical environment, technology and management.

3.1 Do Physical Security Protection

Physical security protection is the basic prerequisite for ensuring the safe operation of college campus networks. It mainly refers to the protection of network servers, computer systems, network switching routes and other network equipment and Hardware entities such as network cables protect them from natural disasters, physical damage, electromagnetic leakage, operational errors, human interference and wire-line attacks [6], as shown in Fig. 2. Schools should centralize core network equipment such as servers, switches, and firewalls in the network center computer room to provide a good computer room environment, including UPS power supply, normal temperature and humidity, fire prevention, waterproof, anti-static, anti-theft, etc. For network communication lines such as optical fibers, methods such as overhead, threading or deep burial are adopted to prevent the line from being physically damaged.

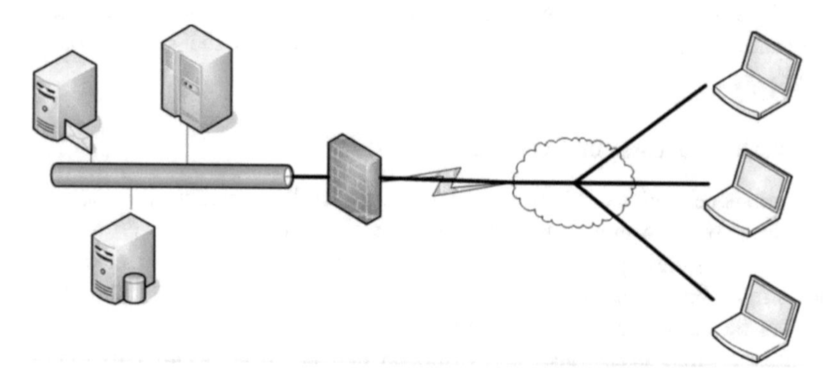

Fig. 2. Computer network protection system

3.2 Implementation of Campus Network Real-Name Authentication

Colleges and universities should do a good job of real-name authentication for campus network users. Through the deployment of the campus network user authentication and accounting system, all terminals and devices connected to the campus network are uniformly authenticated and authorized [7]. All campus network users must pass real-name authentication to open campus network accounts. Faculty and staff can use personal work numbers as online accounts, and students use student numbers as online accounts.

3.3 Purchasing Professional Network Security Services

University network information centers usually face the dilemma of insufficient personnel and insufficient security expertise. For this reason, universities can expand channels and outsource part of the network security work to professional security companies by purchasing services. Especially in some major events or sensitive periods, these professional, credible, and excellent network security companies can be used to jointly ensure the security of college campus networks [5].

3.4 Establish a Safety Management Organization

The development of network security work requires an authoritative, top-down management organization as an organizational guarantee. Schools should establish a dedicated network security work leadership group, with the school's top leader as the group leader [6]. There are offices and network security working groups and other institutions, which clearly divide labor and perform their duties to jointly to keep the safety of campus network.

3.5 Develop a Safety Management System

To promote the orderly progress of campus network security work, schools should formulate and strictly implement effective security management systems in accordance with relevant national laws and regulations and the actual situation of the school, and standardize and clarify the scope and scope of each unit and position in network security work. Responsibilities and other content [7].

3.6 Strengthen Cyber Security Education

An important aspect of network security management is to strengthen the network security education and improve their security awareness and prevention capabilities. Different measures are adopted for different campus network user groups [8]. For school leaders, emphatically introduce them to the importance and urgency of network security, and continuously improve their awareness and importance of network security work. For teachers and students, through the promotion of cybersecurity knowledge education and training activities, on the one hand, they will understand the laws and regulations of cybersecurity and improve their awareness of cybersecurity [9]. On the other hand, they will be encouraged to learn and master some basic cybersecurity skills. Improve their cyber security defense capabilities.

3.7 Establish a Safety Inspection Mechanism to Investigate and Promote Changes

Safety inspection is a very important part of safety management and cannot be ignored. Universities should comprehensively consider the actual conditions of different departments and establish a set of reasonable safety inspection mechanisms. For example, during important activities and a fixed period of time each month, all school websites and application information systems are inspected for safety. In addition, security inspections such as penetration and testing can also be carried out from time to time [9]. Through these normalized work, problems can be found in a timely and effective manner, and rectification can be supervised and implemented.

3.8 Establish an Emergency Response Mechanism

Many network security threats cannot be prevented. For this reason, schools should establish a network security emergency response mechanism. The school should base on the characteristics of its own network and formulate an emergency response plan that suits the school's actual situation. Carry out emergency drills regularly, and continuously improve and perfect the emergency response mechanism in the specific practical process [10].

4 Conclusion

The security problem of college campus network is a dynamic and complex system, which requires comprehensive and multi-level prevention. Schools must attach great importance to network security, and improve users' security awareness and prevention capabilities through publicity, education, and training. From both technical and management aspects, security measures supported by advanced network security technology and scientific security management methods are established. At the same time, we should keep pace with the times, constantly update and improve the security protection system, continuously improve the security defense capabilities of the campus network, ensure that the campus network operates more safely, and better serve the school.

Acknowledgement. Project Fund: This paper is supported by the Foundation for the Second Batch of Education Projects on University-industry Collaboration in 2019. The project number is 201902015008.

References

1. Hongli, W.: The current situation and prevention of computer network security. Inf. Secur. Technol. **3**(4), 13–14 (2018). (in Chinese)
2. Meiyu, X.: Research on the construction of campus network security. China Electron. Commer. **7**(3), 52–54 (2018). (in Chinese)
3. Xiushi, H., Xiaomei, L.: Discussion on network security issues. Sci. Technol. Get. Rich Guide **11**(30), 48 (2017). (in Chinese)
4. Mengxiao, L.: Analysis of the development status of campus network and the construction of security defense system. Mod. Trade Ind. **12**(26), 185–186 (2017). (in Chinese)
5. Du, Yu.: The construction and implementation of a security defense system for college campus networks. Commun. World **4**(05), 38–39 (2015). (in Chinese)
6. Chun, H.: The construction and implementation of a security defense system for college campus networks. Comput. Telecommun. **12**(11), 27–28 (2015). (in Chinese)
7. Kaixue, Y.: Construction of a security defense system for college campus networks. China Sci. Technol. Inf. **8**(02), 61–62 (2015). (in Chinese)
8. Wei, Z.: Research on the security defense system of college campus network. Electron. Technol. Softw. Eng. **5**(06), 252–254 (2017). (in Chinese)
9. Zewei, X.: Analysis of campus network management and security protection schemes. Comput. Knowl. Technol. **2**(16), 72–74 (2018). (in Chinese)
10. Zhanqi, Y.: Research on the security risk response strategy of campus network. Fujian Comput. **9**(3), 102–103 (2018). (in Chinese)

Study on Outlet Height Optimization of Behind-Levee Drainage Pumping Stations

Bo Dong(✉)

Nanjing Automation Institute of Water Conservancy and Hydrology, Ministry of Water Resources Jiangsu, NAIWCH Co., Ltd., Nanjing, China
dongbo@nsy.com.cn

Abstract. In order to save operation cost of the behind-levee drainage pumping station, aiming at large amplitude of variation of the outer river water level, we put forward optimization method of outlet height of the pumping station. The construction cost of outlet projects of the pumping station, such as outlet sump, culvert through levee and sluice gate et al., were calculated. The calculation method of operation cost of the pumping station was proposed according to the energy performances of pump system, the combination of inlet and outlet water levels and annual operation hours. The relational model between the construction cost of outlet projects, the operation cost of the pumping station and the position height of the culvert through levee were established. Taking a typical project as an example, under the premise of guaranteeing the engineering safety and drainage requirements, the optimal position height of the culvert through levee was determined with the target of minimum sum value of construction cost of outlet project and operation cost of pumping station in the project lifetime. The results indicate that when the outer river water level and its variation and annual operation hours are certain, the position height of outlet sump and culvert is the main factor which affect construction cost of outlet project and operation cost, and there must be an optimal position height of outlet sump and culvert that can make the total cost minimum. In the example, assuming that lifetime of the pumping station is 25 years and the annual operation hours are 360 h, when the depth from the levee crest to the top of culvert is optimal height 1.75 m, the total cost is minimum; the annual operation cost can be reduced by 8.13% and the total cost can be saved by about 72.6 thousand yuan compared with using outlet open channel.

Keywords: Behind-levee drainage pumping station · Minimum cost · Culvert through levee · Position height · Optimization

1 Introduction

The construction cost of outlet projects and operation cost of behind-levee drainage pumping station are influenced by position height of the outlet sump and culvert. When outlet projects are in high positions, constructions cost is low because of shallow excavation depth in the levee, but the pump device head, the operation power and the

J. MacIntyre et al. (Eds.): SPIoT 2020, AISC 1282, pp. 697–708, 2021.
https://doi.org/10.1007/978-3-030-62743-0_100

operation cost are high. On the contrary, when outlet projects are in low positions, the pump device head, the operation power and the operation cost are low, but the construction cost is high [1–3]. In order to achieve the optimal excavation depth in the levee in construction of outlet projects and make the total cost of construction and operation maximum, a variety of different layout schemes should be made according to the hydrologic characteristics, the construction difficulty and cost, the performances of the pump device, operation time and other factors, the optimal outlet layout scheme can be chosen by technical and economic comparison. But, according to the survey, almost all this kind of pumping stations adopted outlet open channel, the outlet project with the highest position, which raised the pump head and caused energy waste. At present, researches on outlet form of this kind of pumping stations mainly focus on optimal design of section size and shape of water transfer canal or culvert [2, 4, 5, 7], calculation and check of conveyance capability [6], anti-seepage, reinforcement and leak-proof measures of water transfer canal or culvert [1, 8]. However, researches on economic analysis and comparison of different kinds of outlet project forms are relatively few. The relational model between construction cost of outlet projects, operation cost of the pumping station and the position height of culvert through levee will be established in this paper, with the target of the minimum total cost, through optimizing and solving the optimal position height of the outlet sump and culvert, which will thus improve the design of the pumping station [4–7].

2 Construction Cost of Culvert Through Levee

The outlet projects of behind-levee drainage pumping station mainly include outlet sump, outlet culvert through levee, sluice gate, energy dissipator, bottom protection et al. The profile of outlet sump and outlet culvert is shown in Fig. 1, where H_s is the total height of outlet sump, H_h is the total height of outlet culvert, h is the height from the levee crest to the top of culvert, L_0 is the horizontal distance from the levee crest inside to the outlet sump, L_1 is the length of outlet sump, L_2 is the length of outlet culvert, L_3 is the length of energy dissipation slope, L_4 is the length of bottom protection, b is the width of the levee crest, m_1 is the upstream dam slope rate, m_2 is the downstream dam slope rate, m_3 is the energy dissipation slope rate, δ_0 is the side wall thickness of outlet sump, δ_1 is the bottom wall thickness of culvert, δ_2 is the top wall thickness of culvert, δ_3 is the bottom wall thickness of energy dissipation slope, δ_4 is the thickness of bottom protection, ∇_{sb} is the bottom elevation of outlet sump, ∇_{cb} is the bottom elevation of outlet culvert, ∇_{lt} is the crest elevation of levee, ∇_{bt} is the road surface elevation of buried culvert, ∇_{rt} is the bottom elevation of outer river, ∇_{max} is the maximum operation water level of outer river, ∇_{min} is the minimum operation water level.

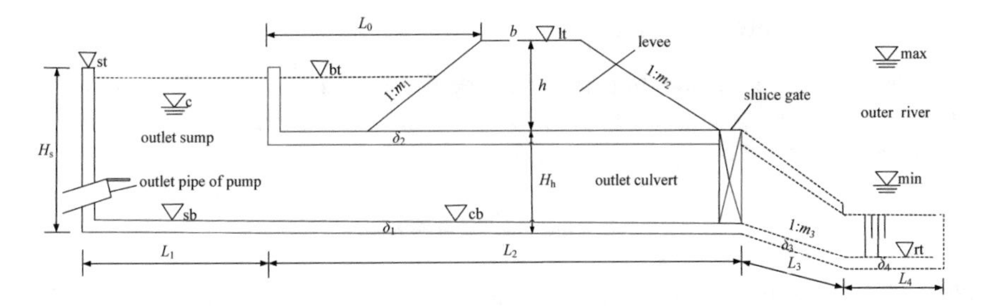

Fig. 1. Profile of outlet sump and outlet culvert

2.1 Components of Construction Cost of Outlet Project

The construction cost of outlet project of the pumping station is mainly composed of four parts including the construction costs, costs of metal structure equipments and project installation, temporary project costs and independent costs, in which the construction costs mainly include the costs of building materials and labors, and the costs of excavation, backfill and tamping earthwork. The costs of metal structure equipments and project installation mainly include the costs of the purchase and installation of sluice gates and headstock gear. The temporary project costs mainly include the costs of construction diversion, closure works, land earthwork et al. The independent costs mainly include the costs of construction management, production preparation, research, survey and design et al. Because the costs of purchase and installation of metal structure equipments, temporary works and independent costs of different layout schemes of outlet constructions are almost the same, when analyzing and comparing the costs of outlet project construction for varies of layout schemes, we just need to calculate and analyze the construction costs.

2.2 Costs of Excavation and Backfill Earthwork

The total cost of earthwork can be calculated by the following mathematical model:

$$C_{ew} = \sum V_i f_i = \sum (\bar{A}_{wi} f_{wi} h_{wi}(h) + \bar{A}_{ti} f_{ti} h_{ti}(h)) \tag{1}$$

Where, C_{ew} is the cost of excavation and backfill earthwork, yuan; V_i is the volume of outlet construction for part i, m^3; f_i is the budget unit price of outlet construction for part i, yuan/m^3; f_{wi}, f_{ti} is the budget unit price of excavation, backfill and tamping earthwork respectively, yuan/m^3; $\overline{A}_{wi}$, $\overline{A}_{ti}$ is the average area of excavation, backfill and tamping earthwork for part i respectively, m^2; $h_{wi}(h)$, $h_{ti}(h)$ is the function relation between the depth of excavation and backfill earthwork and the excavation depth of culvert which was the height from the levee crest to the top of culvert h for part i respectively, m.

2.3 Building Material Costs

The costs of outlet construction material engineering mainly include costs of stone masonry engineering, reinforced concrete and plain concrete engineering, which can be calculated by the following mathematical model:

$$C_{\mathrm{mt}} = \sum V_j \times f_j \tag{2}$$

Where C_{mt} is the total cost of outlet construction material engineering, yuan; V_j is the quota amount of material for part j, m^3; f_j is the budget price of the corresponding material which includes costs of labors and constructions, yuan/m^3.

2.4 Total Cost of Outlet Project Construction

Total cost of outlet project construction $C(h)$ is the function of the position height from the levee crest to the top of culvert h, which consists of the costs of outlet sump, outlet culvert, energy dissipator facilities and bottom protections, all of which are C_{s}, C_{c}, C_{d} and C_{p} respectively, among which the costs of sluice gate and headstock gear were not taken into account because the sluice gate and headstock gear are the same for different position height of outlet construction, and the costs of bottom protection can also be treated as a constant irrelevant to h, thus it can be described by the following mathematical model:

$$C(h) = C_{\mathrm{s}} + C_{\mathrm{c}} + C_{\mathrm{d}} + C_{\mathrm{p}} = \sum \left(\bar{A}_{wi} f_{wi} h_{wi}(h) + \bar{A}_{ti} f_{ti} h_{ti}(h)\right) + \sum V_j \times f_j \tag{3}$$

Where, C_{s}, C_{c} and C_{d} are relevant to the total height of outlet sump H_{s}, the length of outlet culvert is L_2, the length of energy dissipation slope is L_3, all of which are the functions of the depth from the levee crest to the top of culvert represented. According to Fig. 1, the following mathematical model can be obtained:

$$\begin{cases} H_{\mathrm{s}} = H_{\mathrm{h}} + h - (\nabla_{\mathrm{lt}} - \nabla_{\mathrm{st}}) \\ L_2 = L_0 + b + m_2 h \\ \nabla_{\mathrm{cb}} = \nabla_{\mathrm{lt}} - h - H_{\mathrm{h}} + \delta_1 \\ L_3 = \sqrt{1 + m_3^2} \cdot (\nabla_{\mathrm{cb}} - \nabla_{\mathrm{rt}}) \end{cases} \tag{4}$$

The width of excavation, backfill and tamping earthwork for each part of outlet construction of the pumping station is irrelevant to h. The horizontal surface areas of excavation, backfill and tamping earthwork of outlet sump and bottom protection are also irrelevant to h, and the corresponding costs can be described as a linear function of h and a constant respectively; the horizontal surface areas of excavation, backfill and tamping earthwork of the outlet culvert and the energy dissipation slope are the linear function of h, and the corresponding costs can be described as a quadratic function and a linear function of h respectively. Thus total cost $C(h)$ of outlet construction can be described as the following quadratic function of h:

$$C(h) = a_0h^2 + b_0h + c_0 \tag{5}$$

Where, a_0, b_0, c_0 are the coefficients,

$$a_0 = \bar{B}_{w2}m_2f_{w2} + \frac{\bar{B}_{t2}m_2f_{t2}}{2},\ b_0 = \bar{B}_{w1}L_1f_{w1} + \bar{B}_{w2}m_2(H_h - \nabla_{lt} + \nabla_{st})f_{w2} + \bar{B}_{w2}(L_1 + b)f_{w2}$$
$$+ \frac{L_0 + L_1 + 2b}{2}\bar{B}_{t2}f_{t2} - \frac{m_3\bar{B}_{w3}f_{w3}H_h}{2},\ c_0 = \bar{B}_{w1}L_1f_{w1}[H_h - m_1\nabla_{lt} + \nabla_{bt} + (m_1 - 1)\nabla_{st}]$$
$$+ \bar{B}_{w2}(L_1 + b)f_{w2}(H_h - \nabla_{lt} + \nabla_{st}) - \bar{B}_{t2}f_{t2}[L_0(\nabla_{lt} - \nabla_{bt}) - \frac{m_1(\nabla_{lt} - \nabla_{bt})^2}{2}]$$
$$+ \frac{m_3\bar{B}_{w3}f_{w3}H_h}{2}(\nabla_{lt} + \delta_1 - h - H_h - \nabla_{st}) + C_{t1} + C_{t3} + C_{w4} + C_{t4} + C_m,$$

Where, $\bar{B}_{w1}$ is the average width of excavation earthwork of the outlet sump, f_{w1} is the average budget unit price of excavation earthwork of the outlet sump, $\bar{B}_{w2}$ is the average width of excavation earthwork of the culvert, f_{w2} is the average budget unit price of excavation earthwork of the culvert, $\bar{B}_{t2}$ is the average width of backfill and tamping earthwork of the culvert, f_{t2} is the average budget unit price of backfill and tamping earthwork of the culvert, $\bar{B}_{w3}$ is the average width of excavation earthwork of the energy dissipation slope, f_{w3} is the average budget unit price of excavation earthwork of the energy dissipation slope, C_{t1}, C_{t3}, C_{t4} are the costs of the backfill and tamping earthwork of the outlet sump, the energy dissipation slope and the bottom protections respectively which are irrelevant to h, C_{w4} are the costs of the excavation earthwork of bottom protection which is irrelevant to h.

3 Operation Cost of Pumping Station

3.1 Time Characteristics of Pump Device Operation Head

Once the water level of inland river rises to a certain height, the behind-levee drainage pumping station needs to operate and pump water immediately. The resistant coefficients of pump inlet and outlet pipe of different layout schemes can be regarded as the same value. The pump head H can be calculated by the following mathematical model:

$$H = \nabla_c - \nabla_{js} + SQ_z^2/n^2 \tag{6}$$

Where, ∇_c is the actual water level of outlet sump, m; ∇_{js} is the design water level of inlet sump, m; S is the pipe resistant coefficient of pump, s^2/m^5; Q_z is the entire flow rate pumped by the pumping station, m^3/s; n is the number of operating pump units. When the sum of the actual water level of outer river ∇_{rt} and the water head loss of outlet culvert Δh is larger than the design water level of outlet sump ∇_{cs}, than $\nabla_c = \nabla_{rt} + \Delta h$; Otherwise $\nabla_c = \nabla_{cs}$.

According to the characteristics that drainage pumping station operates a longer time near the average head and a shorter time near the maximum or the minimum head, the functional relation which is about actual pump device operation head H_z and the time density distribution of annual operation head of a pumping station (the operation time of unit head in a certain head span) was established [3], as shown in Fig. 2.

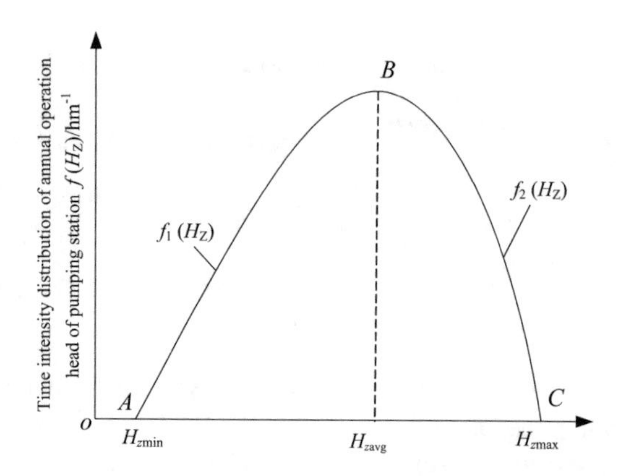

Fig. 2. Time intensity distribution of annual operation head of pumping station

Point A, B and C (Fig. 2) correspond to the minimum operation head H_{zmin}, the average operation head H_{zavg} and the maximum operation head H_{zmax} respectively. Assuming that the function $f(H_Z)$,which is about the density distribution of annual operation head and operation time of pumping station, consists of piecewise functions $f_1(H_Z)$ and $f_2(H_Z)$, thus it can be described by the following quadratic polynomial approximately:

$$f(H_z) = \begin{cases} f_1(H_z) = a_1 H_z^2 + b_1 H_z + c_1 \\ \qquad (H_{zmin} \le H_z < H_{zavg}) \\ f_2(H_z) = a_2 H_z^2 + b_2 H_z + c_2 \\ \qquad (H_{zavg} \le H_z \le H_{zmax}) \end{cases} \tag{7}$$

Where a_1, b_1, c_1, a_2, b_2 and c_2 are the coefficients of the quadratic polynomial.

If the annual operation time of pumping station is T, the functions $f_1(H_Z)$ and $f_2(H_Z)$ should satisfy the mathematical model bellow:

$$\begin{cases} f_1(H_{zmin}) = f_2(H_{zmax}) = 0 \\ f_1'(H_{zavg}) = f_2'(H_{zavg}) = 0 \\ f_1(H_{zavg}) = f_2(H_{zavg}) \\ \int_{H_{zmin}}^{H_{zavg}} f_1(H_z)\mathrm{d}H_z + \int_{H_{zavg}}^{H_{zmax}} f_2(H_z)\mathrm{d}H_z = T \end{cases} \tag{8}$$

The coefficients a_1, b_1, c_1, a_2, b_2 and c_2 of the functions $f_1(H_Z)$ and $f_2(H_Z)$ and $f\,(H_z)$,which are about the density distribution of annual operation head and time of pumping station, can be calculated by solving the equations set (8).

3.2 Selection of Pump Units

The selection of pump units should ensure that the pumps can supply the design flow and design head of the pumping station, and pump type should be chosen identically in one pumping station as far as possible. The pumps should be selected according to average head, operate in high-efficient areas in most time and operate safely and stably at the maximum and the minimum heads. When the amplitude of variation of outer river level is great, the pumps with wide high-efficient areas should be chosen for behind-levee drainage pumping station, in the premise of safe and stable operation of the pump units and enough drainage flow, construction cost and operation cost of the pumping station should be decreased as far as possible.

3.3 Determination of Operation Point of Pump Unit

The operation point of the pump unit can be determined according to pump head curve and required head curve of the pump device, which contains pump flow Q, pump device head H_z and pump head H, pump shaft power P_p, pump efficiency η_p and pumping station flow Q_z. Through calculating the energy consumption of the transmission gear, the motor, power transmission lines and transformer, the total input power from power grid P and the efficiency of pumping station system η_{xt} can be calculated respectively by the following formula:

$$p = \Delta p_t + \Delta p_w + n \cdot \Delta p_m + n \cdot \Delta p_{dr} + n \cdot \Delta p_p \tag{9}$$

$$\eta_{xt} = \eta_t \cdot \eta_{xt} \cdot \eta_w \cdot \eta_m \cdot \eta_{dr} \cdot \eta_z \tag{10}$$

Where, ΔP_t, ΔP_w, ΔP_m and ΔP_{dr} are the power losses of transformer, transmission lines, motors and transmission gears respectively, kW; n is the number of operating pump units; η_{xt}, η_t, η_w, η_m, η_{dr} and η_z represent the efficiencies of pumping station system, transformer, transmission lines, motors, transmission gears and pump devices respectively, %.

3.4 Calculation of Operation Cost of Pump Units

The annual operation cost of pump units in the pumping station F_a can be obtained with the following mathematical model:

$$F_a = p\int_{t_1}^{t_2} \frac{\rho g Q H_z}{1000\eta_{xt}} n\mathrm{d}t = p\int_{t_1}^{t_2} \frac{\rho g Q H_z}{1000\eta_{xt}} \cdot \frac{Q_Z}{Q}\mathrm{d}t \tag{11}$$

Where, p is the electricity unit price, yuan/(kW·h); t_1 and t_2 are the lower and upper limits of time integration, h. Because d$t = f(H_z)$ dH_z, then formula (11) can be turned into formula (12):

$$F_a = p\int_{H_{zmin}}^{H_{zavg}} \frac{\rho g Q(H_z)H_z}{1000\eta_{xt}(H_z)} \cdot \frac{Q_z}{Q(H_z)} \cdot f_1(H_z)\mathrm{d}H_z + p\int_{H_{zavg}}^{H_{zmax}} \frac{\rho g Q(H_z)H_z}{1000\eta_{xt}(H_z)} \cdot \frac{Q_z}{Q(H_z)} \cdot f_2(H_z)\mathrm{d}H_z \tag{12}$$

4 Cost of Outlet Projects and Operation Cost of Pumping Station in Lifetime

4.1 Expressions

Combining formula (5) and formula (12), the total cost of pumping station in lifetime W, which is the sum of outlet project cost $C(h)$ and operation cost F_a, can be calculated using dynamic analysis method shown as the following formula:

$$W = C(h) + F_a = C(h) + \sum_{1}^{k} \frac{F_j}{(1+\gamma)^j} \tag{13}$$

Where, total operation cost in lifetime F can be calculated according to accumulation of annual cost after completion of the pumping station, k is lifetime of pumping station, it is 25 years here; F_j is operation cost of pumping station in the year of № j, yuan; γ is the annual interest rate, %.

4.2 Analysis of Influence Factors of Construction Cost of Outlet Project

The construction cost of outlet project mainly includes the costs of excavation, backfill and tamping earthwork, stone masonry engineering and concrete engineering, labors et al. The influence factors include geological and geomorphologic conditions, soil type of the pumping station site, section sizes and construction requirements of outlet sump and culvert, the price of construction materials and labors, the position height of culvert through levee, the quality requirements of the project et al.

4.3 Analysis of Influence Factors of Operation Cost

Formula (12) shows that operation cost of pumping station is affected by operation points of pump units, energy consumption for each part of pumping station system, total operation time et al. among which the operation points are relevant to the type selection and match of pump units, the hydraulic characteristics of pumping system, the position height of outlet culvert, the water levels and their fluctuation of inland and outer rivers et al. The relation was investigated in this paper between operation cost of pumping station considering water level fluctuation of outer river and position height of outlet culvert together with annual operation hours.

5 Optimization of Position Height of Outlet Culvert of Example Pumping Station

5.1 General Situation of Pumping Station

The drainage pumping station locates in Hangji town of Yangzhou city, whose main task is to drainage the inner waterlogging from Jiuwei River (inland river) to Jiajiang River (outer river). The pumping station is composed of head pond, inlet sump, pump house, outlet sump, culvert, backwater gate, energy dissipation buildings et al. The equipments of the pumping station contain three sets of vertical and axial-flow pump units of 32ZLB-125 pump matched three-phase squirrel-cage asynchronous motor of JSL-12-10-95 and an amorphous alloy electricity transformer of S11-315. Power cables of VV-3 × 95 and V-3 × 240 + 1 × 120 are chosen as the transmission lines of the pumping station and the lengths are 60 m and 50 m respectively. The design water level and minimum water level of inland river are 2.8 m and 2.0 m respectively. The average water level, the maximum water level and the minimum water level of outer river are 5.4 m, 6.5 m and 3.6 m respectively. The design drainage flow of the pumping station is 4.0 m^3/s, the average annual operation hours are 360 h.

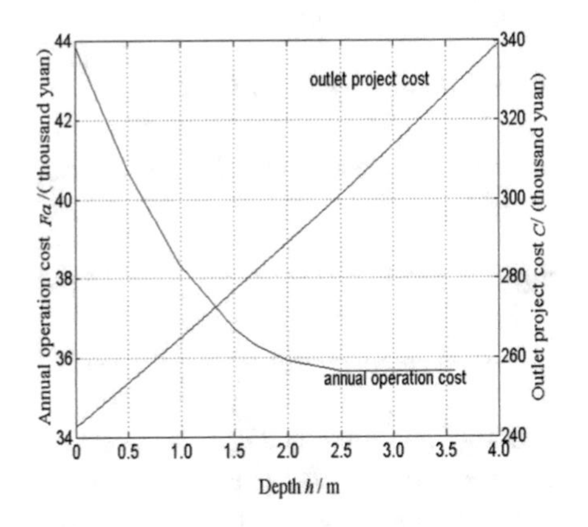

Fig. 3. The construction costs of outlet project and annual operation cost of pumping station

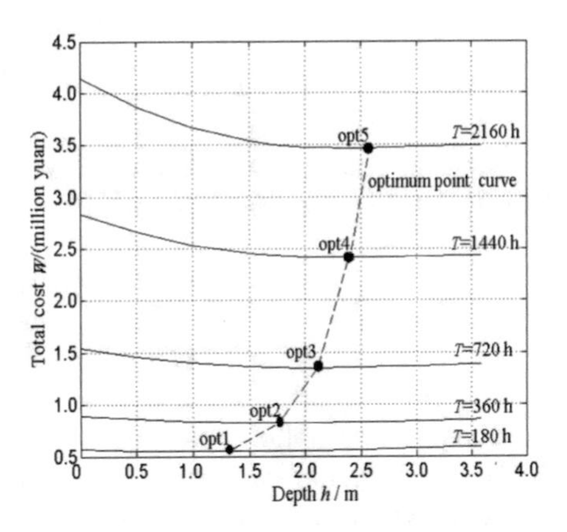

Fig. 4. Total construction costs of outlet project and operation cost in different annual operation time of pumping station

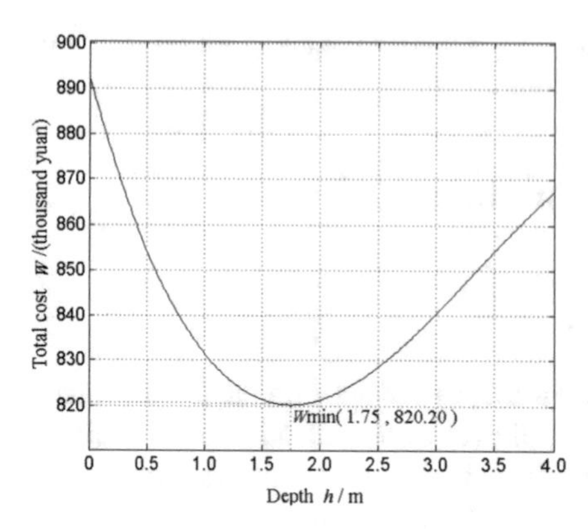

Fig. 5. Total construction costs of outlet project and operation cost of pumping station, while T = 360 h

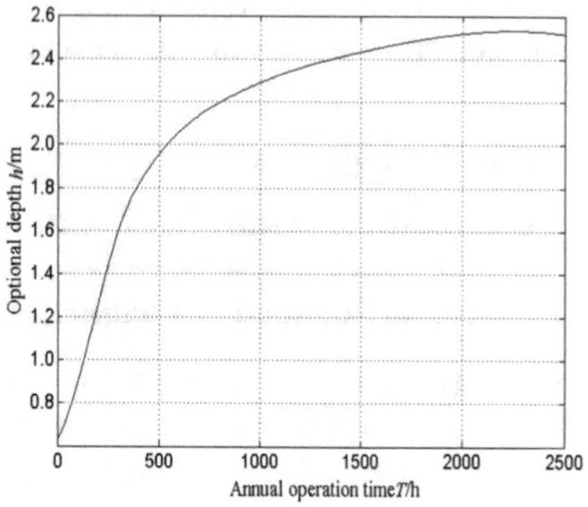

Fig. 6. Relation between the optimal depth from levee crown to culvert top and annual operation time

5.2 Calculation of Outlet Project Cost and Operation Cost

Combining the running records, the distribution function of time intensity of annual operation head for different water levels of outer river of the pumping station was established and solved, as shown in the following formula:

$$f(H_z) = \begin{cases} -57.47H_z^2 + 620.69H_z - 1489.66, (3.6 \le \nabla_{rt} < 5.4) \\ -153.89H_z^2 + 1662.01H_z - 4301.26, (5.4 \le \nabla_{rt} \le 6.5) \end{cases} \tag{14}$$

Thus the annual operation cost Fa and the construction cost of outlet project C can be calculated by the following mathematical model:

$$C = 478.04h^2 + 22290h + 242398 \tag{15}$$

$$F_a = -244.11h^3 + 2425.4h^2 - 7861.9h + 43948 \tag{16}$$

The pumping station was built in 2010, assuming its lifetime is 25 years, selecting 2010 as calculation base point, the annual interest rate is 5%. Thus the relations between investment of the outlet project, the annual operation cost and the depth h from levee crest to top of culvert is shown in Fig. 3. The relation between the total cost of outlet projects and operation in lifetime of the pumping station and h is shown in Fig. 4. When the pumping station operates 360 h yearly, the total cost of outlet projects and operation in lifetime is shown in Fig. 5. The relation between the optimal depth h from levee crown to culvert top, which makes the total cost of outlet projects and operation in lifetime minimum, and the annual operation hours T is shown in Fig. 6.

5.3 Analysis of Calculation Results

As shown in Fig. 3, construction cost of outlet projects presents an increase tendency with the increase of h. With the increase of h, the costs of building materials, earthwork engineering and labors are increased, and thus total construction cost of outlet projects increases.

As shown in Fig. 3, with the increase of h, the operation cost decreases sharply first and then approaches to and reaches a constant. When h is small, the culvert is above the water level of outer river all the time or in most time, the pump head is raised up artificially, and consequently, the operation cost increases. When h increases to a certain value in which the culvert is below the water level of outer river all the time, the pump head will be determined by the water level of outer river completely, a further increase of h will have no influence on the pump head and the annual operation cost. In the example, when h exceeds 2.785 m, the annual operation cost is almost constant, if we continue to increase h, there will be a significant increase of construction cost of outlet projects, which is uneconomical obviously [8].

As shown in Fig. 4–5, total cost of the outlet projects and operation tends to decrease first and increase later with the increase of h overall. When h is small, increasing h can reduce the operation cost remarkably while construction cost of outlet projects will have a small increase, thus the total cost tends to decrease; when h exceeds a certain value, with the increase of h, the operation cost will have a small decrease and reach to a constant while the construction cost of outlet projects will have a big increase, the total cost tends to increase. Thus there's a critical value for h which makes the total cost of outlet projects and operation minimum. In the example, when the depth from the levee crest to the top of culvert h is 1.75 m, the total cost gets the minimum value about 820.20 thousand yuan, comparing with the total cost of outlet open channel, 892.77 thousand yuan will be saved and total cost will be saved by 8.13%. For annual operation hours T = 180, 360, 720, 1440, and 2160 h, compared with adopting open channel, adopting outlet projects with optimal position height could save the total cost of outlet projects and operation by 2.55%, 8.13%, 12.08%, 15.24% and 16.39% respectively.

As shown in Fig. 6, the optimal depth from levee crown to culvert top was calculated when the annual operation time are 180 h, 360 h, 720 h, 1440 h and 2160 h respectively. The optimal depth from levee crown to culvert top tends to increase with the increase of annual operation hours and the amplitude of increase tends to decrease. The reason is that, with the increase of annual operation hours, the proportion of operation cost among the total cost of outlet projects and operation increases, and the great the h is, the fewer and reaching to a constant the operation cost is, and the more the outlet project cost is.

6 Conclusions

The optimal height from levee crest to top of culvert for the behind-levee drainage pumping station, which makes the total cost of construction and operation minimum, can be calculated by establishing mathematical model of construction cost of outlet projects and operation cost according to forms of outlet projects, annual operation time, time density distribution of operation head of pump devices and lifetime of the pumping station. For the example pumping station, for annual operation hours T = 180, 360, 720, 1440, and 2160 h, compared with adopting open channel, adopting outlet projects with optimal position height could save the total cost of outlet projects and operation by 2.55%, 8.13%, 12.08%, 15.24% and 16.39% respectively.

The optimal height from levee crest to top of culvert of the behind-levee drainage pumping station tends to increase with the increase of annual operation time of the pumping station, but the reliability of sluice gate at culvert outlet must be guaranteed, in case of the flood of outer river pours into inland river when the pumping station is not running, meanwhile, the pump units should operate stably and safely in the amplitude of fluctuation of water level of outer river.

The construction cost of outlet projects occupies a leading position in total cost of outlet projects and operation of the pumping station with few operation hours. The optimal height from levee crest to top of culvert is small, so as to cost of outlet projects goes down and safety and reliability of the projects could be improved, it is advisable to choose the small optimal height of h, even outlet open channel.

References

1. Mei, C., Shang-gui, Q.: Application of anti-seepage reinforcement techniques in dike-crossing structures. Jilin Water Resour. (12), 30–33 (2003) (in Chinese with English abstract)
2. Kai, F., Xiao-hong, J., Jian-hua, J.: Application of acceleration genetic algorithm in design of lined canal transversal section. China Rural Water Hydropower **10**, 109–110 (2008) (In Chinese with English abstract)
3. Xiao-li, F., Bao-yun, Q., Xing-li, Y., et al.: Quantitative selection of pump units and adjusting modes for water diversion and drainage tubular pumping station. J. Hydroelectric Eng. **27**(4), 12–17 (2012)
4. Hwang, J.H., Kikumoto, M., Kishida, K., et al.: Dynamic stability of multi-arch culvert tunnel using 3-D FEM. Tunn. Undergr. Space Technol. **21**, 3–4 (2006)
5. Shi-chao, N.: Research on canal cross-section optimization design based on extended differential evolution algorithms. Yangtze River **40**(13), 82–86 (2009)
6. Xiao-ping, S., Shu-wen, L., Zi-feng, H.: Flow capacity calculation of embankment culvert for self and strong discharge pumping stations. Heilongjiang Sci. Technol. Water Conservancy **33**(6), 37–38 (2005) (In Chinese with English abstract)
7. Yun-fei, T., Ke-qin, W., Wei-feng, S., et al.: Research on Optimal Design and Method of Irrigation Channel. China Rural Water Hydropower **4**, 58–60 (2012) (In Chinese with English abstract)
8. Si-fa, X., Hao, Z.: Experiment and evaluation on temperature sensitivities and tensile forces of anti-seepage geomembrane of farmland irrigation canal. Trans. CSAE, **27**(4), 7–11 (2011) (In Chinese with English abstract)

Marketing Mode of Wechat Public Account for College Students' Job Searching-Based on Computer Informationization

Li Li, Wenlin Yi(✉), Ziyi Hu, Shanshan Hu, Lingyun Fu, and Xiaojie Wang

School of Management, Wuhan Donghu University, Wuhan, China
2849996146@qq.com

Abstract. In the Internet age, WeChat recruitment is widely used in in college students' job searching by virtue of its efficient and convenient characteristics. This article conducted a questionnaire survey on the use of job searching WeChat public account by college students. Taking three types of job searching WeChat public accounts—"51Job", "Huawei recruitment" and "UniCareer" as examples, this paper analyzed the WeChat public account marketing model and existing problems. Some measures was proposed to improve the marketing effect by using information technology.

Keywords: Informatization · Wechat public account · Job searching · Marketing model

1 Introduction

According to the latest news released by the Ministry of Education of the People's Republic of China, in 2019, the number of college graduates in China is estimated to be 8.34 million, more than 140000 in 2018, of which 40.8% of college students have difficulties in finding jobs. Especially in the special period of COVID - 19 outbreak, the application scope and user group of online recruitment are further expanded,WeChat public accounts for job searching, has also become a common tool for job hunting. According to the "Economic Impact report on the Code" released by the Institute for Global Industry of Tsinghua University and Tencent Center for Internet&Society in January 9, 2020, WeChat has more than 1 million 500 thousand developers, more than 8200 third party platforms, and WeChat official account has broken 20 million. In 2019, Wechat generated 26.01 million jobs.

2 Analysis of the Status Quo of University Graduates' Use of WeChat Official Accounts for Job Searching

In order to understand the usage and marketing effect of the WeChat public account for college students to search job, this article conducted a questionnaire survey on college graduates through the questionnaire star. The survey content includes several aspects of

J. MacIntyre et al. (Eds.): SPIoT 2020, AISC 1282, pp. 709–716, 2021.
https://doi.org/10.1007/978-3-030-62743-0_101

the use frequency and satisfaction of the WeChat recruitment public account. A total of 196 valid questionnaires were received, including 30 junior college students, 102 undergraduates, and 64 postgraduates.

2.1 Survey of Online Job Search Channels for College Students

As shown in Fig. 1, among the 196 college students surveyed, 136 have used WeChat public account for job searching, 165 have used recruitment website for job searching, 130 have used recruitment app, and 27 have used microblog for job searching.

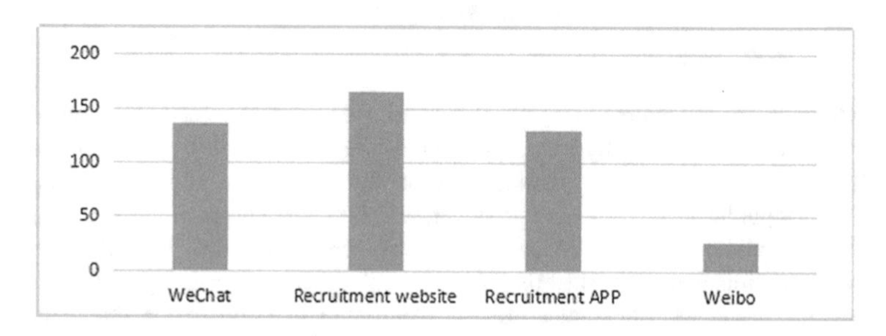

Fig. 1. The number of users of online recruitment channels

2.2 Survey on the Use Frequency of Wechat Public Account for Job Searching

The survey results show that 8% of college graduates are very concerned about the job search WeChat public account, and they will browse for attention every day; 23% of college students generally follow and read 4 to 5 times a week; 52% of college graduates do not pay much attention within a week Read 2 to 3 times; 17% of college graduates do not pay attention to the WeChat public account, they have almost forgotten these job search WeChat public accounts (Fig 2).

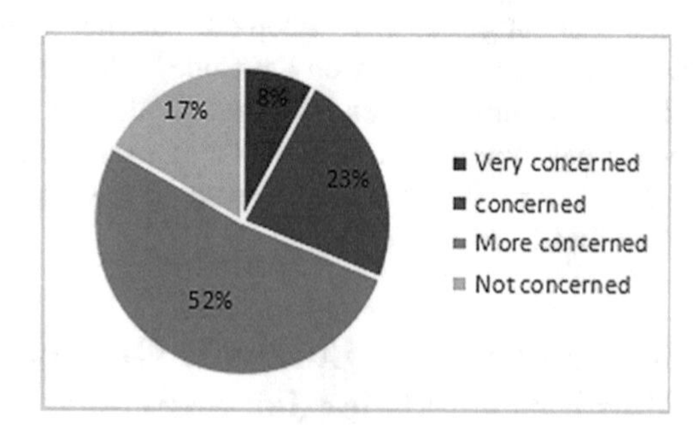

Fig. 2. Frequency chart of using job search WeChat official account

2.3 Survey on the Attractiveness of Marketing Tweets on Wechat Public Account

As shown in Fig. 3, 56% of college graduates will open and read WeChat public account for job searching because of tweets, 12% of them read shorter in WeChat public account (Less than three minutes); and 44% of college graduates ignore the tweets because it's title is not attractive enough. Thus it can be seen, the attraction of tweets needs to be improved.

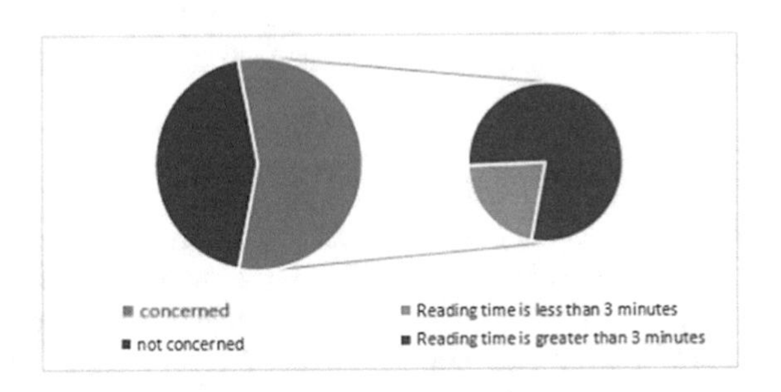

Fig. 3. Diagram of attractive analysis of marketing tweets

3 Analysis of the Marketing Mode of Wechat Public Account for Job Searching

3.1 Characteristics of WeChat Public Account for Job Searching

Under the background of the era of mobile Internet,Wechat recruitment inherits the advantages of low-cost, wide spread and convenient use of social network recruitment, but there are also many problems, such as the large gap between the number of active fans and the number of readers, and little interaction with users.

Table 1. Operating data of the WeChat public account of job searching

WeChat public account name	Active fans	Headline average reading	Headline average likes	Headline average message
51Job	105723	10032	10	8
Huawei recruitment	411723	18486	102	12
UniCareer	163788	13045	18	5

Table 2. Characteristics of WeChat official account for job hunting

WeChat public account name	Type	Features
51Job	Built by the large recruitment website	Various job types, rich recruitment information, complicated information and weak pertinence
Huawei recruitment	Built by a well-known enterprises	Recruitment information is time-sensitive, Outstanding brand advantages and clear targeted customer groups
UniCareer	Established by the media	Low visibility, rich training courses, clear for user groups, and more interaction with users

This article divides WeChat public accounts for job searching into three types: the first type WeChat official account is built by the large recruitment website, the second type WeChat public number is built by well-known enterprises, and the third type of WeChat public number is established by the media. A representative WeChat official account is selected from each type: "51job", "HUAWEI recruitment" and "UniCareer". The number of fans (Table 1) and their characteristics (Table 2) were analyzed. Table 1 data comes from the statistical data of "watermelon data" 2020.03.28-2020.04.02.

3.2 Product Strategy Analysis Based on Information Technology

Although each public account has its own characteristics, the marketing mode shows great similarity. The following will be analyzed from two aspects: product strategy and marketing strategy.

3.2.1 Using Information Technology to Classify and Integrate Information

The ability of job information integration is an important aspect of improving marketing effect. The information provided by "HUAWEI recruitment" is timely, "51job" and "UniCareer" pay more attention to the universality of information. All kinds of WeChat public accounts for job searching have set up a small program of quick job search, which effectively reduces the search time of users. For example, "Huawei Recruitment" has set up quick search options such as position, city and industry in the campus recruitment column.

3.2.2 Provide Important Knowledge and Information Related to the Workplace

Most of the WeChat public accounts for job searching will provide job searching skills, resume writing, workplace survival skills, career development skills and other relevant workplace knowledge. For example, "HUAWEI recruitment" will provide information related to the industry and enterprises; "51job" provides more job-hunting skills and an introduction to the current situation of various industries. "UniCareer" promotes the job-hunting course by pushing articles.

3.2.3 Analyze Users' Needs and Offer Training Courses

Most job search WeChat official account will offer various kinds of paid video courses, covering job hunting skills, occupation planning courses, office software operation courses and so on. Such as "51job" offers more courses on interview skills and communication skills; "UniCareer" provides training courses according to the characteristics of college students; The HUAWEI recruitment official account provides less training courses, because its training courses are offered on other platforms of HUAWEI.

3.3 Marketing Strategy Analysis Based on Information Technology

3.3.1 Using the Internet, Marketing in the Network Community

Virtual community is an organic system composed of users, common goals and network media. [1]. WeChat public accounts for job searching adopts the online community marketing mode, which uses the strong ties between consumers to enhance the popularity and promote the sales of goods through sharing and word of mouth [2]. Common marketing methods include: users can forward product information to their WeChat Moments and send screenshots to auditors after more than 30 min, so as to obtain the qualification of joining the group or get job information or training courses free of charge.

3.3.2 Using Big Data for Accurate User Portrait

Big data can accurately match recruitment candidates with positions. Big data makes a person's portrait more and more clear, which can provide recruitment enterprises with "nine dimensional radar chart" of candidates, and display the potential characteristics and comprehensive information of candidates. We can also analyzes the needs and preferences of job seekers through big data, so as to conduct precise marketing. For example, "51job" and "Huawei Recruitment" classify the positions provided by the enterprise in a detailed way, and provide job introduction and interactive Q&A; "UniCareer" evaluates the competitiveness of job seekers through private customized services, formulates exclusive job-hunting programs, conducts one-on-one personalized counseling, and improves their soft power and skills in job-hunting.

3.3.3 Combine Online and Offline to Expand Advertising

WeChat public accounts for job searching are promoted and marketed through various means such as advertising in the WeChat Moments, tweet publicity, air publicity, offline job fairs and other means. For example, "UniCareer" will provide two-dimensional codes of corresponding job-hunting products according to the theme of daily headlines; "Huawei Recruitment" held propaganda meetings in Colleges and universities, and set up "Huawei open day" to strengthen corporate culture publicity; "51job" improves the number of College Students' job seekers by combining online propaganda and offline campus recruitment, and also increases the exposure by setting up a question activity to the stars in the small program.

4 Survey and Analysis of the Satisfaction of Recruitment Wechat Public Account for College Students

According to the survey, 9% of college students are very satisfied with WeChat official account, 64% of them are generally satisfied, 27% of them are not satisfied. Thus, the official account of WeChat is still improving. As shown in Fig. 4, there are the following problems in WeChat official account: 40% of college students think that information and content are homogenized; 21% of college students feel that information quality is not high; 22% of college students feel that information update is too slow; 17% of college students feel that communication effect is not good.

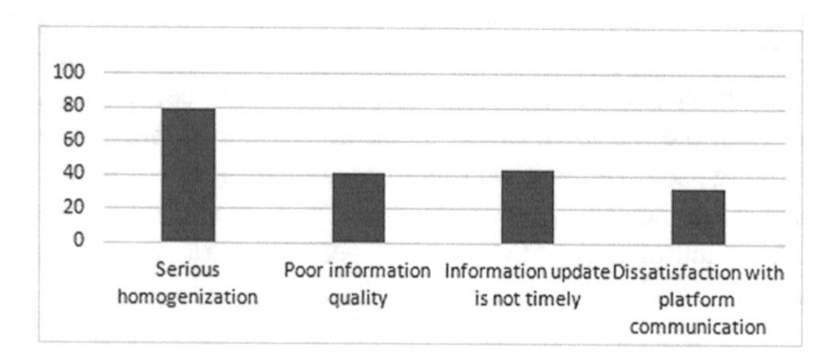

Fig. 4. Problems in WeChat public account for job searching

5 Measures to Improve the Marketing Strategy of Wechat Public Accounts for Job Searching Based on Computer Informatization

5.1 Using Information Technology to Provide Quality and Accurate Services

The operators of WeChat public accounts for job searching can use mature cloud computing, cloud storage, data mining, big data technology to collect and analysis college students' job-searching needs, reading habits [3]. According to the results of the analysis and user's reading needs and preferences, We can precisely adjust the types of articles and content, and increase the viscosity of users.

We should pay attention to the user's reading experience and improve the overall visual beauty of the page. For example, use H5 micro scene and current affairs hot spot to enrich the content and form of tweets and improve the click rate of users.

5.2 Use Big Data to Accurately Market and Reduce Costs

By means of using big data technology, we can quickly classify, recommend and match the massive job search information and recruitment information, which greatly reduces the operating cost and time cost of the platform [4]. For example: filter low matching resume to save recruitment time, accurately published advertisements and job information to reduce publicity costs.

5.3 Enhance the Efficiency of Information Search and Enhance User's Experience

The continuous progress of information technology makes the efficiency and effect of information retrieval greatly improved. The operators of WeChat public accounts for job searching should invest and update information technology, improve search engines, add interface quick search, information classification and other functions so that different customers can obtain information accurately, purchase courses and so on.

5.4 Enhanced Information Feedback and Communication by Using New Technologies

Based on the rapid development of information technology, operators can use "Intelligent job assistant" to accurately match the recruitment needs through human-computer interaction, provide users with one-stop intelligent job-hunting services, including job search, progress check and Q & A, and strengthen information feedback and communication between users; "Intelligent job assistant" can also urge recruitment companies to update recruitment status in time, set up reminder function, improve the transparency of enterprise and job seekers, establish credit evaluation mechanism and record system, and improve the trust between WeChat official account and users [5–8].

6 Conclusions

At present, computer information technology has been widely used in recruitment WeChat public account marketing activities, such as community network marketing, big data for user portrait, information classification and screening, but the depth and breadth of applications need to be improved. Various recruitment WeChat public account platforms also have problems such as serious homogeneity of information and content, low quality of information, untimely update of information, and unsatisfactory communication effects between users and the platform.

We need to make full use of computer information technology to promote the marketing effect of WeChat and other technologies to provide accurate and high-quality information services; Use big data to accurately market and reduce costs; and improving user satisfaction through improved search functions; using the latest information technology to improve communication efficiency.

Acknowledgment. This work was supported by the grants from Ministry of Education of the People's Republic of China's first batch of Industry-University Cooperative Education project in 2019, Project No. 201901266016.

This work was also supported by the grants from Hubei Provincial Collaborative Innovation Centre of Agricultural E-Commerce (Wuhan Donghu university research [2019] No. 17 Document).

References

1. Plant. R: Online communities. Technol. Soc. **26**(1), 51–65(2004)
2. Leung, L.: Impacts of Net-generation attributes, seductive properties of the Internet, and gratifications-obtained on Internet use. Telematics Inform. **20**(2), 107–129 (2003)
3. Huijun, C.: The evolution from information-based human resource management to big data human resource management takes tencent as an example. China Manage. Inf. **22**(1), 89–91 (2019). (in Chinese)
4. Qiang, F., Zhiping, Z.: Thinking on the innovation of recruitment mode in the era of big data. China Manage. Inf. **22**(16), 81–82 (2019). (in Chinese)
5. Yongbo, D., Tong, L., Shiyun, T.: Research on network recruitment information ecosystem model construction and its operating mechanism——based on mechanism design theory. Inf. Sci. **37**(9), 85–89 (2019). (in Chinese)
6. Michaelidou, N., Siamagka, N.T., Christodoulides, G.: Usage, barriers and measurement of social media marketing: An exploratory investigation of small and medium B2B brands. Ind. Mark. Manage. **40**(7), 1153–1159 (2011)
7. Chetna Priyadarshini, S. Sreejesh, M.R.: Anusree: Effect of information quality of employment website on attitude toward the website. Int. J. Manpower **38**(5), 729–745 (2017)
8. Berthon, P.R., Pitt, L.F., Plangger, K., Shapiro, D.: Marketing meets Web 2.0, social media, and creative consumers: Implications for international marketing strategy. Bus. Horiz. **55**(3), 261–271 (2012)

Public Credit Information Data Quality Evaluation and Improvement

Yan Zhao[1], Li Zhou[1], Zhou Jiang[1], Fang Wu[1], Bisong Liu[1], and Zhihong Li[2(✉)]

[1] Social Credit Branch, China National Institute of Standardization, Beijing, China
[2] Department of Human Resources Management, Beijing Institute of Petrochemical Technology, Beijing, China
lizhihong@bipt.edu.cn

Abstract. Public credit is an important part of social credit, which plays an important role in accelerating the transformation of government functions, innovating supervision methods, improving the market credit environment and promoting social governance. In the process of collecting, sharing and exchanging public credit information, the "garbage information" and the incorrect, incomplete and inconsistent data have seriously affected the quality of the public credit data, which is not conducive to giving full play to the value and advantages of the public credit information. Therefore, it is urgent to improve the quality of public credit data, establish an effective data quality management mechanism, and improve the availability and use value of data. In this paper, from the perspective of data standardization, completeness, accuracy, consistency, accessibility and timeliness, we put forward the overall data quality evaluation indexes of public credit. Then the paper puts forward strategies for improving the quality of public credit data from the perspective of establishing the public credit information data standard system and improving the data quality continuously with PDCA Cycle.

Keywords: Public credit information · Data quality · Public credit information collection · Public credit information sharing

1 Introduction

The social credit system is the cornerstone of the market economy. And the collection, sharing and application of credit information are the important foundation of the construction of the social credit system, which play important roles in accelerating the transformation of government functions and innovation of regulatory methods, helping the economic transformation and upgrading, improving the market credit environment, and improving social governance and public service level [1–4].

Public credit information and market credit information are important parts of the social credit information, and the effective integration of them is conducive to the formation of a strong joint force in the construction of social credit system. Among them, public credit information refers to the data or information produced or obtained

J. MacIntyre et al. (Eds.): SPIoT 2020, AISC 1282, pp. 717–723, 2021.
https://doi.org/10.1007/978-3-030-62743-0_102

by administrative organs, judicial organs, other organizations that perform public management (service) functions according to law, and public enterprises and institutions in the process of performing their duties to reflect the credit behavior of natural persons, legal persons and other organizations [5].

As early as June 2014, the outline of social credit system construction planning (2014–2020) issued by the State Council put forward clear requirements for public credit information management, namely: "Each region should record, improve and integrate the credit information generated in the process of performing the public management function of each department and unit in this region, and form a unified credit information sharing platform. It is necessary to establish the open catalogue of government credit information, vigorously promote the exchange and sharing of government credit information, strengthen the application of credit information in public management and improve the efficiency of performance of duties [6, 7].

The public credit information has the typical characteristics of rich data types, complex structure and diverse sources. As the key content and important carrier of public credit system management, public credit information platform is the infrastructure for collecting credit information, sharing credit information, conducting credit supervision, providing credit information services, and creating an honest and trustworthy environment [8].

However, at present, the collection of credit information of each department is not perfect, and the collection and publicity of credit information are mainly based on administrative permission and administrative penalty information, and the lack of unified standards and management norms for data affects the application of data.

In terms of cross departmental data sharing, public credit information is mainly scattered in local administrative organs, judicial organs, other organizations performing public management (service) functions in accordance with the law, and public enterprises and institutions. Information among departments, industries and regions is still shielded from each other. There are a large number of valuable information resources that are idle and wasted, unable to meet the exchange requirements for resource sharing. At the same time, it is difficult to connect platform information, which is not conducive to expanding the application scope. In the process of collecting, sharing and exchanging credit information, there are some worthless and meaningless "junk information". Sometimes, data collection and sharing are not timely, cannot be accessed, and cannot effectively meet the needs of governments, enterprises, consumers and other interested parties. Incorrect data, incomplete data and inconsistent data have seriously affected data quality, which is not conducive to giving full play to the value and advantages of credit information. Therefore, it is urgent to improve the data quality, establish an effective data quality management mechanism, and improve the data availability and use value.

2 Public Credit Data Quality Evaluation Index

To improve the quality of public credit data, first of all, we need to evaluate the quality of public credit data. Based on the investigation of the current situation of public credit data quality and customer demand, this paper puts forward the evaluation indexes of public credit overall data quality from the perspective of data standardization, completeness, accuracy, consistency, accessibility and timeliness. The above six indexes have been defined as C_i (where i = 1, …, 6) and the weights have been defined as w_i (where i = 1, ···, 6); total data quality of public credit has been defined as C.

$$C = \sum w_i \times C_i \times 100 \quad (1)$$

See Table 1 for public credit data quality evaluation indexes.

Table 1. Public credit data quality evaluation indexes

Serial number	Evaluation index	Meaning	Evaluation index measurement	Calculation formula	Weight
1	Standardization	The extent to which the public credit data format (including data type, value range, data length, precision, etc.) meets the requirements of data standards, data models, business rules, etc.	C_1 (standardization of public credit data)	$C_1 = X/Y \times 100\%$ X = the amount of data that meets the requirements of data format in the statistical period Y = total data submitted in the statistical cycle	w_1
2	Completeness	The extent to which public credit data records are assigned values as required by data rules	C_2 (completeness of public credit data records)	$C_2 = X/Y \times 100\%$ X = total number of data records actually submitted in the statistical cycle Y = total number of data records to be submitted in the statistical cycle	w_2
3	Accuracy	The degree to which the public credit data can accurately represent the real value of the public credit information subject	C_3 (accuracy of public credit data content)	$C_3 = X/Y \times 100\%$ X = the amount of data with correct content in the statistical cycle Y = total data submitted in the statistical cycle	w_3

(*continued*)

Table 1. (*continued*)

Serial number	Evaluation index	Meaning	Evaluation index measurement	Calculation formula	Weight
4	Consistency	The extent to which public credit data does not contradict data used in other specific contexts	C_4 (consistency ratio of public credit data)	$C_4 = X/Y \times 100\%$ X = the amount of data meeting the consistency requirements in the statistical cycle Y = total data submitted in the statistical cycle	w_4
5	Accessibility	According to the requirements of business rules, public credit data meets the user's ability to query and use	C_5 (access rate of public credit data)	$C_5 = X/Y \times 100\%$ X = the number of successful data acquisition by users in the cycle Y = the number of times users query data in the cycle	w_5
6	Timeliness	According to the requirements of business rules, the extent to which public credit information data is submitted and updated within the specified time	C_6 (the ratio of public credit data submitted in time)	$C_6 = X/Y \times 100\%$ X = actual number of records submitted on time Y = number of records to be submitted	w_6
Total data quality of public credit C				$C = \sum w_i \times C_i \times 100$	

3 Strategies for Improving the Quality of Public Credit Data

The improvement of public credit data quality is not only a technical problem, but also a complex management problem. This study will propose solutions from the following two aspects:

3.1 Establishing Standard System of Public Credit Information and Formulating the Key Technology and Management Standards

From the perspective of the whole process of information flow, establishing a standard system covering the whole process of public credit information collection, sharing and use, creating a "common language" of data, improving the quality of public credit data will lay a solid foundation for data exchange, information sharing and business collaboration. The standard system of public credit information data is shown in Fig. 1.

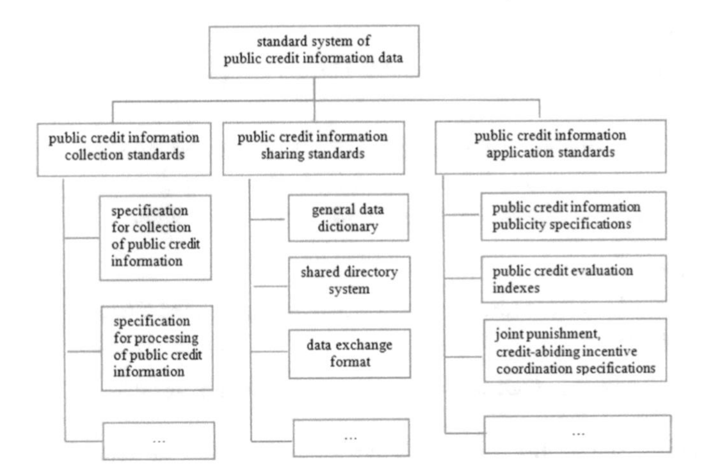

Fig. 1. The standard system of public credit information data

Among them, public credit information collection standards mainly regulate the collection, processing of public credit information, and provide standardized technical guidance for the subsequent storage and use of public credit information. The public credit information sharing standards mainly provide business standards related to information sharing for the establishment of public credit information sharing mechanism, including general data dictionary, shared directory system, data exchange format, exchange method and interface specification. The application standards of public credit information are mainly based on the needs of information application, and relevant business standards are formulated to support the effective use of credit information, such as public credit information publicity specifications, public credit evaluation indexes and joint punishment, credit-abiding incentive coordination specifications.

3.2 Continuous Improvement Process of Public Credit Data Quality Based on PDCA Cycle

According to the process management method of Deming Cycle (PDCA Cycle), this paper evaluates and continuously optimizes the quality of public credit data from the Plan stage, the Do stage, the Check stage and the Action stage.

(1) Plan stage

The main tasks of public credit data quality management in the Plan stage include but are not limited to:

a) Defining the goal of public credit data quality management;
b) Establishing data quality management organization;
c) Evaluating the current situation of data quality management resources, environment and personnel capacity;
d) Defining data quality evaluation index and data standard of public credit information system.

(2) Do stage

The main tasks of public credit data quality management in the Do stage include but are not limited to:

a) Evaluating the quality of public credit data from various dimensions according to the data standards and indexes defined in the Plan stage;
b) Listing all problem data according to the evaluation results.

(3) Check stage

The main tasks of public credit data quality management in the Check stage include but are not limited to:

a) Listing all problem data according to the evaluation results, and analyzing the reasons one by one;
b) Developing data quality improvement plan according to the data quality evaluation indexes, problems lists and cause analysis results, develop the data quality improvement plan.

(4) Action stage

The main tasks of public credit data quality management in the Action stage include but are not limited to:

a) Implementing the data quality improvement scheme proposed in the Check stage;
b) Solving the existing problems through improvement measures;
c) Improving the effectiveness and solving the incremental problems through control measures;
d) Verifying whether the improvement measures are effective and checking whether the improvement objectives have been achieved.

4 Conclusions

Public credit is an important part of social credit, which plays an important role in accelerating the transformation of government functions, innovating supervision methods, improving the market credit environment and promoting social governance. In the process of collecting, sharing and exchanging public credit information, the "garbage information" and the incorrect, incomplete and inconsistent data have seriously affected the quality of the public credit data, which is not conducive to giving full play to the value and advantages of the public credit information. Therefore, it is urgent to improve the quality of public credit data, establish an effective data quality management mechanism, and improve the availability and use value of data. In this paper, from the perspective of data standardization, completeness, accuracy, consistency, accessibility and timeliness, we put forward the overall data quality evaluation indexes of public credit. Then the paper puts forward strategies for improving the quality of public credit data from the perspective of establishing the public credit information data standard system and improving the data quality continuously with PDCA Cycle.

Acknowledgements. This research was supported by National Key R&D Program of China (Grant No. 2017YFF0207600, 2017YFF0207601, 2017YFF0207602) and Dean fund project of China National Institute of Standardization under grant No. 552019Y-6660, No.552020Y-7461.

References

1. Outline of Social Credit System Construction Plan (2014–2020), Guofa, vol. 21 (2014)
2. Li, Z., Bi-song, L.: Research on construction and system framework of china's social credit standardization. Stand. Sci. **1** (2014)
3. Zhouk, L., Jiang, Z., Zhao, Y.: Research on the public credit information development and its standards system. Stand. Sci. **11** (2018)
4. Information technology service-Governance-Part 5: Specification of data governance (GB/T 34960.5-2018)
5. Credit-General vocabulary (GB/T 22117-2018)
6. Information technology-Evaluation indicators for data quality (GB/T 36344-2018)
7. Zheng, Z., Lili, Z.: The construction of China's public credit system: characteristics, problems and Countermeasures. Expand. Horiz. **2** (2017)
8. Lili, Z., Zheng, Z.: On the connotation of government credit and the path selections for the construction of governmental affairs integrity in China. Credit Ref. **3** (2020)

Reform of Database and Data Warehouse Course Based on OBE-CDIO Model

Yao Li(✉), Jingjing Jiang, Sheng Guan, and Weijia Zeng

Department of Information Management and Information System, Dalian University of Science and Technology, Dalian 116052, China
liyao@dlust.edu.cn

Abstract. In the era of big data, many traditional courses had to keep pace with the times. The Database and Data Warehouse course is a new course derived from the course of Database Principles. Aiming at the problems of emphasis on theory, light practice and light process in traditional teaching of Database course, this paper proposes a teaching design model based on the combination of OBE and CDIO. It aims at the teaching goals, teaching content, teaching links, teaching of Database and Data Warehouse courses Methods and course evaluations were designed in all directions. At the same time, the course design is practiced through a semester of real course teaching. According to the evaluation system, analyze the teaching effect and draw teaching reflection.

Keywords: OBE · CDIO · Database and Data Warehouse

1 Introduction

In the era of big data, the teaching of many traditional courses has failed to meet the graduation requirements of college students. Traditional Database courses emphasize theory, practice, and process. The result is often that students' theoretical knowledge assessment results are very good, but their practical skills are weak. Students cannot adapt to corporate work well after graduation.

In this context, our school replaced the previous course of Database Principles and Applications with the course of Database and Data Warehouse. On the basis of the traditional course content, add data warehouse related content, at the same time, the content of theoretical knowledge has been streamlined and practical content has been increased [1].

Database and Data Warehouse course is the core basic course for information management and information system majors, and provides important support for subsequent professional courses such as information system analysis and design, and comprehensive information management training. In order to enable students to flexibly apply theoretical knowledge to complex engineering practice, to achieve the training goal of our university to cultivate innovative applied talents, the introduction and integration of the educational concepts of OBE and CDIO are of great significance to the reform of Database and Data warehouse course [2].

J. MacIntyre et al. (Eds.): SPIoT 2020, AISC 1282, pp. 724–730, 2021.
https://doi.org/10.1007/978-3-030-62743-0_103

2 OBE-CDIO Teaching Mode

OBE (Outcome-Based Education) [3] The educational concept originated in the United States, and its essence is to take the student's main performance in the learning process as an important indicator of education quality assessment, emphasize the student's learning results in the process of talent training, and As an important factor to feedback the quality of teaching activities [4].

The concept of CDIO (Conceive-Design-Implement-Operate) was put forward by educators such as MIT, that is, the "conception-design-implementation-operation" teaching model. Its four processes are derived from products and systems The life cycle process covers the necessary professional abilities of the vast majority of engineers [5], and uses this as the basis for assessing the achievement of student abilities.

OBE is a teaching model oriented on "Expected Learning Results", while CDIO is a model that drives the design of curriculum content, teaching methods, education culture, etc., based on the "Expected Learning Results" set represented by the "CDIO Competency Outline" [6]. The two complement each other. This article combines the concepts of OBE and CDIO, and explores the course teaching design of Database and Data Warehouse.

3 Model of Course Teaching Design Based on OBE-CDIO

Under the educational concept of OBE-CDIO, it is necessary to design the expected results of curriculum learning, curriculum teaching methods, and evaluation methods in combination with the teaching indicators in CDIO. First, we should investigate the application capabilities of Databases and Data Warehouses in the form of questionnaires, seminars, etc. for students' employment enterprises, school-enterprise cooperative enterprises, and graduates. Collect, compare and analyze the curriculum outlines of standard universities in different levels. Collect the survey results and get a survey analysis report. According to the research and analysis report, combined with the training objectives of the information management and information system professional training program for graduates, the teaching goals of the Database and Data Warehouse course is formulated. According to the teaching objectives, optimize on the basis of the original curriculum content, reduce theoretical class hours, increase experimental and practical hours. At the same time, the teaching methods should be as diversified as possible. It can take the form of homework, personal experiments, team experiments, team projects, etc. to optimize the content of different courses. Promote students' understanding of the principle by applying the reverse. According to the course teaching objectives, course teaching content, course teaching methods, establish an evaluation system for the course teaching effect [6]. Carry out specific teaching links according to the course content and teaching methods. At the same time, quantification methods are used for the different parts of the lectures to perform score statistics. For assignments, experiments, projects, etc., specific scoring criteria need to be given according to the course goals. In setting the questions for the final exam, you also need to correspond to the course objectives, and set questions with corresponding scores for different knowledge points. After the semester's actual teaching process is over, the results obtained by different assessment methods are statistically analyzed based on the evaluation system to obtain the achievement rate of each course goal. According to the course goal achievement rate, analyze the teaching effect, form a course analysis report, and give guiding suggestions for the next round of course content optimization. The course teaching design model based on OBE-CDIO is shown in Fig. 1.

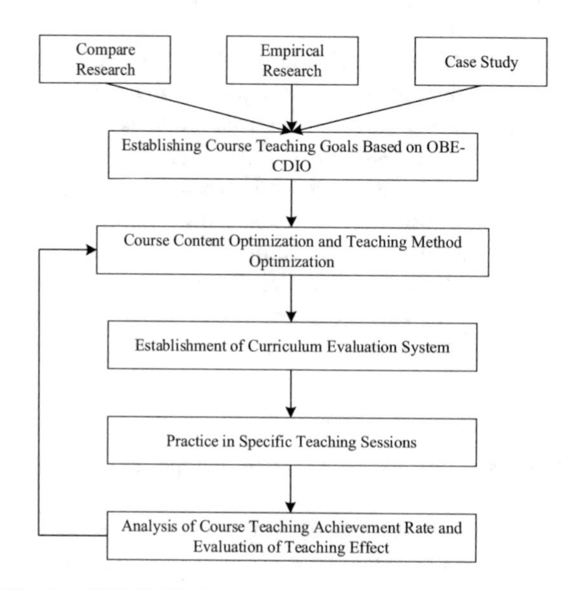

Fig. 1. OBE-CDIO-based curriculum teaching design model

4 Design and Practice of Database and Data Warehouse Course

4.1 Design of Teaching Goals

Under the guidance of the OBE education concept, based on information management and information system professional talent training goals and industry and enterprise research, clarify the talent training specifications for Database and Data Warehouse, and establish teaching goals of the course as follows.

The course develops students' understanding of Databases and Data warehouses, analysis and design skills, and development and application skills of Databases and Data warehouses. Through the teaching of the course, students can understand and master the basic principles and basic concepts of Databases and Data warehouses, can perform Database requirements analysis, ER modeling, relational Database modeling, can standardize the design of Database systems, and can manipulate relationships using SQL language Database. At the same time, can create and apply Data warehouses, can write simple and complex SQL statements according to a given Database system, can master the DBMS environment using SQL Server as an example, and can implement operations on Databases and Data warehouses in it. Able to perform basic maintenance of the Database system. Be able to interact well with the teachers of the class, strict requirements on themselves, and form good study habits.

According to the professional talent training ability system and social research, establish the curriculum quality, knowledge and ability training requirements, derive the corresponding index points in the CDIO, and design the expected learning effect. The correspondence between graduation requirements and expected learning results is shown in Table 1.

Table 1. Correspondence between graduation requirements and expected learning outcomes

Graduation requirements		CDIO indicator points	Expected learning outcomes
Knowledge	Basic professional knowledge	1.3.1 basic professional knowledge	ILO-1 Understand and master the basic knowledge of Database and Data Warehouse, can apply basic knowledge into practical problems
Ability	Basic professional ability	2.2.1 analysis problems	ILO-2 Understand the significance and function of analysis problems, master the basic methods of analysis problems, and use analysis methods to solve the practical problems of Database and Data Warehouse
	Core professional ability	2.1.1 overall thinking	ILO-3 Identify and define a Database or Data warehouse system, understand the behavior of Database and Data Warehouse, identify the interaction between Database and the outside world and the impact on Database behavior, and ensure a comprehensive understanding of Database and Data Warehouse
		2.3.1 ability to conceptualize and abstract	ILO-4 Grasp the conceptualization and abstraction methods of Database and Data Warehouse, and solve the practical problems in Database construction
		3.1.2 modeling	ILO-5 Master the method of Database and Data Warehouse modeling, and have the ability to establish scientific and reasonable Database and Data Warehouse model
		3.3.1 ability to process information	ILO-6 With basic DBMS software skills, can quickly and accurately collect, process, retrieve and organize Database and Data Warehouse information
		8.8.1 design and implement	ILO-7 According to the design of Database and Data Warehouse structure, the implementation of Database and Data Warehouse is carried out to ensure the smooth implementation process
Quality	Professional quality	5.1.2 learning attitudes and habits	ILO-8 Develop positive learning attitude and correct learning habits, and have good autonomous learning ability

4.2 Design of Teaching Content

In order to achieve the expected learning results, the teaching content of the Database and Data Warehouse course relies on the teaching materials but is not limited to the inherent contents of the teaching materials. In the course of designing the course content, organize the teaching content around the expected learning results and teaching goals of the course. The specific design of the teaching unit of Database and Data Warehouse course is shown in Table 2. Table 2 gives the main content of the course teaching units and teaching units. More specific details are given in the curriculum standards. The difficulties, priorities and ability development requirements of each course unit are clearly given.

Table 2. Teaching unit design

No.	Unit name	No.	Unit name
1	Introduction	6	Implementation and use of databases
2	Database requirements and ER modeling	7	Data warehouse concept
3	Relational database modeling	8	Data warehouse and data mart modeling
4	Normalization	9	Implementation and use of data warehouse
5	SQL	10	Overview of DBMS functions and database management

4.3 Design of Teaching Process

Database and Data Warehouse is implemented through a variety of teaching links. The specific teaching links are as follows.

(1) Lecture in class: In the basic knowledge of the Database, the basic knowledge of the Data warehouse is explained in the classroom.
(2) Case: In the Database modeling, standardization, Data warehouse modeling, and SQL statement programming, the design is based on the real case background of ZAGI retail company throughout the teaching content.
(3) Exercises: After students have mastered the basic ability to solve database problems, they will arrange and explain exercises in class, and count them into formative assessment results in the form of classroom performance.
(4) Homework: In order to make full use of students' extracurricular time, an online question assignment system was introduced. Answer questions online at the same time point at the beginning, middle, and end of the semester. The results of the three assignments are included in the formative assessment results.
(5) Experiment: In the SQL statement and Data warehouse practice part, design the experimental link. Students carry out the experimental process according to the project instruction book, the teacher guides them, and the results are submitted in the form of experimental reports. The teacher gives 2 experimental results and counts them into formative assessment results.

(6) Project: In the final stage of the course, students work on the Database implementation and Data warehouse analysis projects of the enterprise procurement system in small groups. A comprehensive project is used throughout the course content, and the experimental report is used as the result. The results are included in the formative assessment results.

Results-oriented, strengthened process monitoring as a whole guideline for the design of the teaching cycle, the results of exercises, homeworks, experiments and projects will be included in the course's usual grades, and students will be assessed from multiple perspectives to see if they have achieved the expected learning effect.

4.4 Design of Teaching Methods

According to the needs of the teaching process, the teaching methods adopted are as follows.

(1) Teaching method: Teachers mainly use teaching methods when explaining the basic principles and knowledge of Databases and Data warehouses.
(2) Discuss teaching method: In the basic concepts of Database and Data Warehouse, Database requirements and ER modeling, use the discussion teaching method to teach.
(3) Practice teaching method: After the theoretical knowledge is taught, the content of classroom teaching is consolidated with practice teaching methods.
(4) Experimental teaching method: The teacher instructs the students to use SQLSERVER2012 and carry out relevant experiments in the SQL language practice part according to the project instruction book.
(5) Case teaching method: Through the real case of the sales department of ZAGI retail company, run through the various chapters of the course. In each chapter, small cases are used to supplement the teaching.

4.5 Design of Course Evaluation

The total score of the course of Database and Data Warehouse is composed of formative assessment and final assessment in a proportion of 40% and 60%. In order to strengthen the effect of practical teaching, the assessment of this round of teaching has increased the proportion of scores in formative assessment. In the formative assessment, classroom performance, homework, experiments, and projects are scored separately. The final assessment is performed in the form of a closed-book exam. According to the expected learning effect, the final exam questions are set accordingly. Finally, the comprehensive formative assessment and final assessment are deduced to obtain the degree of achievement of the learning effect of the course. For the 10 classes of information management and information system majors of our school in 2018, a total of 323 students in Database and Data Warehouse courses learned the goal achievement degree analysis. The specific analysis results are shown in Table 3.

According to the Data in Table 3, the degree of course achievement is 0.65, and the evaluation of the achievement of the course goal is "basically achieved." In general, through the study of this course, the vast majority of students have mastered the basic theoretical knowledge of the course, and possessed a certain ability to combine theory with practice, cultivated students' professional literacy, and basically achieved various teaching goals. However, there is still room for further improvement in the achievement of expected learning results and graduation requirements.

5 Summary

Based on the teaching concept of combining OBE and CDIO, this paper proposes a course teaching design model, and designs and practices the course for Database and Data Warehouse courses. Judging from the actual teaching results of a semester, the course design of the results-oriented engineering education concept can more accurately and effectively support the professional graduation requirements, and can find the weaknesses in the teaching process according to the evaluation system, and in the subsequent teaching continuous improvement.

Table 3. Teaching unit design

ILO	ILO-1	ILO-2	ILO-3	ILO-4	ILO-5	ILO-6	ILO-7	ILO-8
Weights	0.1	0.2	0.1	0.1	0.2	0.1	0.1	0.1
Assessment method	Achieving degree of single assessment in ILO							
Usual performance								0.72
Project		0.76	0.76	0.76	0.76	0.76	0.76	0.76
Exercise	0.74		0.75	0.74	0.74			0.75
Experiment			0.74			0.74	0.74	0.74
Final exam	0.54	0.58	0.7	0.7	0.52	0.28	0.54	0.55
Expected achievement	0.67	0.62	0.64	0.73	0.64	0.73	0.64	0.6
Total achievement	0.65							

References

1. Yang, Y., Zhou, F.: Reform and practice of data warehouse and data mining curriculum based on CDIO. Pioneering Sci. Technol. Mon. **9**(9), 51–59 (2015). (in Chinese)
2. Shan, D., Zhang, Y., Yang, Y., Sun, X.: Exploration of teaching reform of digital circuit course under OBE-CDIO mode. Comput. Educ. **282**(6), 86–89 (2018). (in Chinese)
3. Junpeng, W.: Reform and practice of teaching and principles of single chip microcomputer based on OBE teaching mode. Sci. Technol. Inf. **423**(30), 93–95 (2015). (in Chinese)
4. Wang, C., Ie, J.: Discussion on the teaching mode of database technology course based on computational thinking and CDIO. Educ. Watch **7**(3), 89–90 (2018). (in Chinese)
5. Zhao, H., Li, X., Kesheng, X.: Research on database principles and application course teaching reform methods based on cdio model. High. Educ. J. **23**, 132–133 (2016). (in Chinese)
6. Zhao, H., Ke, H., Meng, X.: Exploration and practice of teaching reform of "principle and application of database system". Heilongjiang Educ. **23**(4), 58–59 (2015). (in Chinese)

Enhancement of Risk Prevention Ability of Students Funding Subjects in the Age of All Media

Yun Lu(✉)

West Yunnan University, Lincang 677000, China
511315926@qq.com

Abstract. The all-media is an important manifestation of technological development and conceptual progress in the new era, and plays an important role in the daily learning and life of college students. Various subjects of student funding work have made full use of modern multimedia technologies and methods, and have made many new attempts at improving funding effectiveness and reducing funding risks. Education authorities use data sharing to avoid fair risks, border colleges and universities use real-time multimedia tools to avoid procedural risks, recipients obtain funding information through information platforms to avoid beneficiary risks, and local student credit loan banks strictly abide by loan application procedures to avoid review risks. Through the establishment of a joint risk prevention and control mechanism for funding entities, the establishment of a sound scientific and standardized risk management system, and the strengthening of guidance and education for funding recipients, various measures have been taken to further enhance funding risk prevention capabilities.

Keywords: All media · Student funding · Risk prevention

1 The Connotation of Student Funding in the Era of All Media

1.1 The Meaning of All Media

All-media is a comprehensive, all-time, all-group integration of traditional media and new media under the circumstances of more advanced communication tools, comprehensive coverage of information dissemination, highly developed network technologies, and the emergence of new media platforms. It is "Under the development of digital network technology, the traditional and emerging media methods are comprehensively used to achieve the integration and interoperability of media methods, content and functions". Based on the concept of human-oriented design, the service is refined according to the different needs of the audience. "The wide-channel, three-dimensional form provides a more diverse, open, and dynamic form of communication of information and services, and the resulting media field" [1]. The advent of the all-media era has changed the relationship between people and information. The limitations of information manufacturing and communication have narrowed or even

J. MacIntyre et al. (Eds.): SPIoT 2020, AISC 1282, pp. 731–736, 2021.
https://doi.org/10.1007/978-3-030-62743-0_104

disappeared. College students have begun to use information to create new lifestyles and gradually control the right to speak in the media world. The era of total media has completely changed the "How college students live, learn and obtain information" [2].

1.2 The Impact of the All Media on Funding

The characteristics of communication in the all-media era have profoundly changed the communication mechanism and communication path of student funding policies, and put forward new requirements for traditional education. On the one hand, all-media provides a new paradigm for student discourse practice: convenient funding information dissemination tools and a new field of funding discourse. On the other hand, the "full process", "holographic", "full effect" and "full staff" of information dissemination and access in the all-media era have also exposed every link and process of funding work under the magnifying glass, further increasing funding risks. Aiming at risk management issues such as alienation of funding, lack of integrity, and increased moral hazard in the evaluation of scholarships, green channels, and the development of student loans, how to establish an effective risk management constraint and prevention mechanism for funded projects has put forward more difficult requirements for funding work in the new era.

2 Risk Prevention Content of Student Funding Subjects in the All-Media Era

2.1 Educational Authorities Use Data Sharing to Avoid Fair Risks

Affected by the quality of the students, the geographical environment, and the orientation of running the school, border colleges and universities often have many minority students, many students with financial difficulties, and many students from registered families. The funding work involves a wide range of tasks and arduous tasks. In order to save procedures, simplify administration and delegate power, the practice of allowing students to issue poverty certificates at all levels and various types is now abolished. In the process of identifying poor students and awarding scholarships, the data that can be referred to by universities has been reduced, which adds the difficulty of the job. Under such circumstances, through sharing real-time data provided by educational departments or civil affairs departments, realizing data communication and docking, and even implementing platform exchanges across provinces and departments, colleges and universities can greatly save manpower costs, improve funding efficiency, and avoid unfair funding work. For example, the education department should set up a data sharing platform with the civil affairs department, and provide data on urban subsistence family students, family members who set up cards, and family members who are targeted for care to colleges and universities to ensure those poor students can enjoy funding policies in schools and to ensure the accuracy and fairness of funding. If there is no platform data, or it is difficult to build a data platform across provinces and departments, the link of providing original supporting materials can be appropriately added during the audit to ensure that all the students from families with financial difficulties can be funded.

2.2 Frontier Colleges and Universities Use Real-Time Multimedia Tools to Avoid Funding Process Risks

Frontier colleges and universities should standardize procedures and improve processes in the green channel, the identification of students from families with financial difficulties, and the selection and distribution of scholarships. The practices and results of each process will be made public in due course. To the extent permitted by national policies, and under the premise of protecting the privacy of students with financial difficulties in the family, timely publicize the progress of the funded projects, the list of aided students, and the release of funds through a variety of media methods, and offline the combination of publicity and platform publicity can promptly resolve student doubts, alleviate negative emotions of students, and avoid risks caused by imperfect procedures and irregularities. The school funding management department shall review the funding data in new media such as the campus website, WeChat public account, Weibo, and Tik Tok, strengthen the training of relevant management personnel, and give full play to the role of data media in supporting funding.

2.3 Grantees Obtain Funding Information Through Information Platforms to Avoid Grantee Risks

Students from families with financial difficulties should get rid of the negative thoughts of "waiting, relying on, and asking", pay attention to the data platforms and notices of funding departments at all levels, understand the national funding policy, be familiar with the application time of each funding project, the application process and the availability of provided materials. Students should take the initiative to inquire and understand relative policies, and make corresponding preparations to ensure the smooth flow of information about the funding project's publicity and acceptance, and ensure that students who meet the conditions of the needy can get the relevant information in time. For example, for the projects of the rain and dew program carried out in various parts of Yunnan Province, the requirements of the counties and districts are not consistent, and the publicity and coverage are also different, students who meet the application requirements should timely understand the relevant information and go to the education funding departments of each county to complete the application work themselves; Students from financially disadvantaged families who have received financial aid projects such as student origin loans must have integrity and gratitude after graduation, log on to the loan system to repay the principal and interest of the loan, and often do good deeds to help students when they are able to carry forward [4–6].

2.4 Strict Loan Application Procedures of Credit-Student Loan Banks in Student Places to Avoid Review Risks

The student origin credit loan is mainly borne by the China Development Bank. In 2005, the China Development Bank started issuing student credit loans to students 'hometowns. In 2007, the China Development Bank started the student credit loans to students' schools in Yunnan. In the promotion of the student loan policy, the student loan management department should further simplify the loan application procedures and strict loan eligibility review, so as to truly implement "the policy of giving full loans to eligible people" and avoid the situation that truly poor students have no money to lend. And some students from non-poor families divert their student loans to other sources. Banks should make full use of the advantages of the all-media era to break the information asymmetry pattern. The situation in some backward areas should be changed that, because of the undeveloped logistics and information, the admission notice was received late, and the application time for the student loan was missed. The loan handling bank should actively contact the students studying at colleges and universities, contact students' education departments and other administrative units to conduct a detailed review of the qualifications of the loaned students, instead of "accepting the receipts in full" as a result of the preliminary review by the related educational department; What's more, the loan repayment risk assessment should be conducted to further reduce the funding risk of the student credit loan.

3 The Risk Prevention Measures of the Funding Body's Funding Work

3.1 Establish a Joint Risk Prevention and Control Mechanism Between Funding Entities

The most distinctive feature of the whole media is that it can realize information sharing and instant interaction. Through information sharing, a series of processes such as "fast search, efficient screening, accurate positioning, convenient downloading, information processing, and re-dissemination" are realized, and "the flow, sharing and memory of information resources in the time and space dimensions" is achieved. Finally, "breaking the geographical restrictions on information exchange and communication" [3]. The key to preventing and controlling the risks of university funding work from the perspective of the whole media is to strengthen the linkage and improve the mechanism of information sharing and real-time interaction between the funding bodies such as the Education Bureau, Civil Affairs Bureau, banks, universities and poor students. On the one hand, the education and civil affairs departments in the province and the border colleges and universities should further strengthen the information linkage, especially the family information of poor students, the difficult situation, the family's enjoyment of precision poverty alleviation treatment, and civil affairs subsidies to avoid students' repeated submission of their difficult information due to poor information and lead to funding risks. On the other hand, with the improvement of the level of school running, more and more students from other provinces are studying in frontier universities. Due to the differences in the implementation of funding policies between provinces, it is difficult for students from other provinces to implement some of their funding projects in universities in the province. For example, Yunnan Province now includes all filed families of registered cardholders as part of the state's first-class grants. However, the implementation of this policy has been difficult due to the fact that filed family card information has not yet been effectively shared and transmitted in the civil affairs department. It is prone to omission or over reporting, which leads to the occurrence of "pseudo-poor students", which affects the effectiveness of precision funding.

3.2 Establish and Improve a Scientific and Standardized Risk Management System

Nothing can be accomplished without norms or standards. All frontier colleges and universities must adapt to local conditions, formulate and improve effective methods for the evaluation and management of scholarships, management methods for work-study programs, green channels, methods for the identification of students with financial difficulties in families, measures for supervision and review of funding work, and funding risks prevention and control management measures, etc., gradually establish and improve the funding work system so that all funding work is standardized and supported by evidence. And as far as possible, the various management systems are promoted through the campus network, WeChat public account, Tik Tok and other media, and strive to make every student understand the state and school's funding policies and application methods. In addition, it is necessary to increase the intensity of risk prevention and control of funding work, and incorporate the funding risk prevention and control work into the important indicators of the anti-corruption work of all levels and departments. Teacher and student representatives and aided student representatives should be convened regularly to have a discussion to understand their voices and resolve risks in the bud. Finally, it is possible to establish a sound, complete, standardized, open, transparent, preventable and controllable funding risk prevention and control system.

3.3 Strengthen the Guidance and Education of the Recipients

In the implementation body of the funding work, the funding target is the key role, that is, students from families with financial difficulties. Schools should strengthen the guidance and education of the recipients, strengthen the funding effectiveness to the greatest extent, and avoid funding risks. First, we must strengthen the supervision of funding work. Let recipients know that they are receiving funding, not donations. The aided funds should first be used to meet normal life and study. Colleges and universities should set up a special network platform by professionals for funding work, build a three-level management mechanism for schools, departments, and classes, and manage all-media accounts, such as mailboxes, forums, QQ groups, blogs, WeChat, online consulting and survey platforms, to further open up a dedicated network channel for students to feedback funding issues. Second, we must improve the psychological assistance measures for poor students. We should speed up the transformation from economic aid to educational aid through the combination of economic assistance and psychological counseling, so that the poor students with financial difficulties in the family can get out of the shadow of poverty as soon as possible, and build a spirit of consciousness, self-confidence, self-reliance, and self-improvement to study hard and keep improving. Third, we must strengthen gratitude education and honesty education for aided students. Through various media means, we should guide the aided students to establish correct moral cognition, firm moral will, and good moral habits, and inspire them to learn to be trustworthy, learn to be grateful, repay in time, and work hard to feed back.

The report of the Nineteenth National Congress of the Communist Party of China clearly pointed out that it is necessary to improve the student funding system so that the vast majority of new rural and urban labor force will receive high school education and more higher education. Improving the quality and effectiveness of student funding is an important means and main measure to ensure fair education. In the era of all media, we should mobilize the implementation subjects of student funding, make full use of modern multimedia technology and methods to avoid risks in the work of student funding in colleges and universities, and further improve the effectiveness of student funding, ensuring the smooth implementation of student funding as an important project of "protecting people's lives and warming people's hearts"; it can also further advance the work of eliminating poverty and promoting social equity and justice. All media is a new manifestation of human intelligence and technological progress. In the new period, each student funding subject should find the risk prevention and control points of student funding, give full play to the role of technology support in all media, improve the efficiency of border college students' funding, and reduce the risk of funding projects.

Acknowledgements. This work was supported by Scientific Research Fund Project of Yunnan Provincial Department of Education (2020J0753).

References

1. Wang, L.: On the education of university students' values in the new media age. J. Party School Jinan Municipal Committee CPC **5**, 97 (2019). (in Chinese)
2. Zhao, M., Yang, Z.: A study on the model of ideological and political education on the internet of border colleges and universities in the perspective of "great ideology". Western China Qual. Educ. **26**, 11 (2019). (in Chinese)
3. Liang, Q., Bao, N.: Reflection on dilemma of ideological and political discourse in the age of all media. J. China Univ. Min. Technol. (Soc. Sci.) **6**, 57 (2019). (in Chinese)
4. Meng, L., Ma, A., Guo, L., Ma, Y., Li, J.: Risk analysis and resolution on the financial aid to the poor college students. J. Hebei Normal Univ. Sci. Technol. (Soc. Sci.) **11**(1), 87–91 (2012). (in Chinese)
5. He, H.: Integrity risks and prevention and control of college student funding. J. Tongling Voc. Tech. College **2**, 96–98 (2017). (in Chinese)
6. Song, X.: Targeted financial aid: the path choice of financial risk management in universities from the perspective of educational equity. J. Jiujiang Voc. Tech. College **1**, 56–59 (2018). (in Chinese)

Privacy Protection Under the Construction of Smart Cities in China—From the Perspectives of Legislation

Ruiyi Liu(✉)

Institute for Disaster Management and Reconstruction,
Sichuan University, Chengdu, China
ruiyi_liu@163.com

Abstract. Taking a normative analyzing approach, this study summarized the framework of laws and regulations concerning privacy protection and analyzed the legislative problems in the privacy protection in a smart city scenario. Since the laws and regulations for privacy protection are scattered in the Constitution and other existing laws, the privacy protection in China is inevitably scattered in laws of particular areas. Major problems exist in almost all sections in collecting, spreading, storing and using personal information. Basic principles, including collection limitation, data quality, purpose specification, security safeguards, openness, should be followed for the improvement of the legislation of Chinese concerning privacy issues in smart cities.

Keywords: Privacy protection · Smart city · Legislation

1 Introduction

Smart services have been widely used in constructing smart cities to create more convenient and efficient livelihood due to rapid growth of urban population and increasing demand of better management of smart cities. However, the passive use of information technology and hardware, such as internet of things, surveillance equipment, cloud technology and so on, also brings great concerns of privacy protection due to the leakage and unlawful use of privacy. Thus, the security of privacy protection is one of the most important issues in the construction of smart cities. Researchers have achieved great proceedings in their attempts to the understanding of smart cities and have applied various approaches to examining smart city practices in the protection of privacy [1]. In China, due to the lacking of comprehensive legal principles for privacy protection and working definition of privacy, the privacy protection in smart cities is more difficult and more complicated [2].

Based on the practices in China and the general principles for privacy throughout the world, two prerequisites should be considered in improving the privacy protection in constructing smart cities in China. First, it is necessary to establish a general notion of the right, which benefits both public and individuals, to protect privacy; [3] second, China should establish a privacy protection regime which is compatible to the rest of

J. MacIntyre et al. (Eds.): SPIoT 2020, AISC 1282, pp. 737–743, 2021.
https://doi.org/10.1007/978-3-030-62743-0_105

the world. Therefore, this paper will focus on a further study from a perspective of privacy protection in a smart city scenario in China from a legislative perspective.

In this article, I identify existing stipulations scattered in China's laws and regulations and their problems in fulling the needs of privacy protection. More importantly, on the analysis of existing laws and regulations concerning privacy protections, I'll try to promote some advices for the improvement of China's legislation in a Smart City Scenario. In particular, this article addresses three questions:

- What is the framework of China's legislation concerning privacy?
- What are the problems in protecting privacy in China's laws and regulations in a smart city scenario?
- How could we make better legislation in construction smart cities?

2 The Framework of China's Legislation Concerning Privacy Protection

In developed countries, the regulatory for privacy protection tends to be more centralized and coordinated of government authority [4]. Having a different political, political and legal context with developed countries, in China, the protection of privacy is mainly relied on current laws, instead of a specially enacted law. Thus, the articles protecting privacy in China are inevitably scattered in different existing laws, such as the Constitutions and laws of particular areas, such as the civil law and finical laws.

2.1 Privacy Protection by the Chinese Constitution

Although there is no clear notion of privacy right in China's constitution, article 37 to 40 of the Constitution, which protect the basic citizen's rights of personal freedom, dignity, residence and the freedom and privacy of correspondence, provide a constitutional framework for the legal protection of privacy. Article 37 and article 40 protect the freedom of citizens and the freedom and privacy of correspondence. According to Article 37, citizens are free from unlawful detention, deprivation or restriction by any means. Besides the general freedom protected in article 37, the freedom and privacy of correspondence is also protected by the constitution, unlawful examination of personal correspondence is prohibited. Compared with article 37, article 40 provides a special protection for the freedom of thoughts. Article 38 and article 39 extends the protection of citizen's privacy from body to territory [5]. By prohibiting the intrusion of personal dignity and residence, the constitution provides a broader protection for the fundamental rights of a citizen and his/her privacy.

However, considered as the constitutional foundation, the four articles are inadequate for privacy protection and limited for privacy protection for following reasons: first, these articles are not explicit enough and need to be further defined by other laws and regulations; second, they are limited to private citizens, the privacy of public bodies are not clearly defined [6].

To sum up, the constitution of China has just provided a limited constitutional framework for privacy protection, moreover, the protection by the constitution is indirect and need to be clearly defined by other laws and regulations. Thus, the constitutional protection for privacy is inadequate in China.

2.2 Privacy Protection by Civil Law and Other Laws and Regulations

2.2.1 Privacy Protection by Civil Law

The Civil law of China provides three forms of protection of privacy: personal name, portrait and reputation. According to article 99, Chinese citizens shall enjoy the right of personal name, any interference and unlawful use of personal name shall be prohibited; article 100 requires that the use of citizen's portrait for commercial purposes shall be under the permission of the owner; article 101 requires that the damage of citizen's reputation shall be prohibited and relieved. All the three articles of the Civil Law are the detailed application of the fundamental rights of the Chinese constitution and provide legal protection of citizen's name, portrait and reputation by defining them as personal privacy. More importantly, both the right to name and portrait are limited to the unauthorized commercial use, which means that if a citizen's name or portrait is used unlawful or improperly by another citizen for non-commercial, he/her can only be relieved by suing the latter for the violation of article 101 as invading the right of reputation. This is a loophole of the Civil Law, because the unauthorized use of a citizen's name or portrait for noncommercial purposes without damage to the citizen's reputation, there will be no relief means for the owner [7].

2.2.2 Privacy Protection by Other Laws and Regulations

Some other laws and regulation also provide protection of privacy in special industries or economic sectors. NPC and its standing committee have enacted a great amount of laws to regulate financial transactions. For example, article 29 of the Commercial Bank Law requires that commercial banks should ensure confidentiality for depositors in handling citizen's savings and protect their personal privacy. At the same time, it also requires that commercial banks should also prevent entity or individual to inquire, freeze or deduct individual saving accounts without the permission by law [8]. Another example is the Announcement of the Securities Association of China on Issuing the Interim Measures for the Management of Bona Fide Information of the Members of the Securities Association of China (2003), which strictly prohibit the disclosure of personal privacy without the permission of law [9].

3 Problems in Protecting Privacy in China's Laws and Regulations in a Smart City Scenario

3.1 Legislative Problems Concerning Privacy Security

In smart services, sensitive information such as health condition, income, individual activities and so on, can be easily traced. By correlation and data processing, the privacy of an individual is in great risk to be exposed and used illegally by others. Thus, the rights of the owner, including his/her rights to be informed of the use of personal information, to control the use of his/her personal information, as well as the rights of relief, should be fully protected by law. However, due to the over generalization of regulations, a lot of practices are not well regulated by law.

3.2 Legislative Problems Concerning Government Services

As the administrator and service provider, government plays an important role in protecting privacy in the construction of smart cities. On the one hand, government is the major promotor of the application of smart technologies, followed by all walks of life, communities, enterprises, families and so on. It has to take into consideration not only the information security of individual, enterprises, as well as the national level, but the prevention of the outflow of personal information across the border in a globalized era [10]. On the other hand, policies and laws concerning privacy are essential for the construction of a smart city. However, compared with those in advanced countries, the laws and regulations in China protecting privacy are incomplete and not well designed.

3.3 Legislative Problems Concerning Service Providers

Generally, while a user is using smart services, his/her sensitive information such as location, inquiry, content, identity, etc. are collected at the same time., which may cause potential hazards of being unlawfully used. These problems are mainly caused by four reasons: (a) insufficient investment in privacy protection technologies and hardware. Pursuing maximum profit, enterprises are unwilling to invest on privacy protection technologies and hardware which seldom bring direct benefits. This may result in great loopholes in smart services both in their hardware and in their management. For example, in many industries concerning smart services, users' passwords are not encrypted and stored in servers in plaintext, leaving users sensitive information in great danger of be leaked and illegally used. (b) insufficient management level. The internal management loopholes of employees' personal information are one of the major reasons for privacy leakage. Although most employers are required to sign an agreement for privacy protection, it usually cannot function properly due to the absence of liability. In a Smart City Scenario, everyone could be a smart service provider, putting even more challenges to the protection of privacy.

3.4 Legislative Problems Concerning the Third Parties

A third party concerning smart services is an individual or organization besides service provider, government. A third party might have no relation to privacy; however, it might collect, steal, or purchase personal information for business benefits, personal revenge and so on [11]. In China, underground buying and selling personal information has grown into an industry, which causes great loss for both individuals and the society. At present, only Amend (VII) to the Criminal Law stipulates the crime of illegally selling and buying personal information. However, the punishment is too light compared to the benefits it brings by violating the law, [12] thus, it creates a grey zone for information collector. With the growth of smart industry, these benefit-driven black market tend to boost even faster with more accurate, detailed, uncontrolled exposure of personal information, if it is not constrained by sufficiently and carefully designed laws and regulations.

4 Improving the Legislation Concerning Privacy Protection in a Smart City Scenario in China

4.1 Improving the Legislation for Privacy Security

For the protection of privacy, the principles below should be followed in the collection, spreading, storage and usage of personal information: (a) notification/transparency. While collecting personal information, the owner should be notified of the time, content, method, identity of the collector, and purpose of usage. (b) lawfulness/appropriateness. Personal information must be collected by law and should be strictly within the appointed scope. It is illegal to collect personal information through implanted RFID. Buying or selling personal information without the consent of the owner should be deemed to be unlawful. (c) controllability/self-determination. Service provider must ensure that the owner can be involved in the whole process of collecting, processing and analyzing personal information. The owner should be able to retrieve, update, revise or delete incorrect, outdate information as well information collected through illegal means. (d) security/confidentiality. Service provider should ensure the security and confidentiality of personal information collected. Technology and hardware for protection privacy should be applied and the owner should have the rights to require his/her personal information should be protected in accordance with certain level of standards or restrictions. Besides, some other principles concerning the owners' right such as actionability, inalienability, inheritance, and so on, should also be also be protected by laws and regulations.

4.2 Improving the Legislation for Regulating Government Management

As administrator and service provider of smart city, government plays a different role in the protection of privacy, thus, the strategy for legislation should differ from that for individual, service provider and the third-party. In essence, the power of government is public power, which should not do anything not authorized by law. For the legislation of government concerning privacy protection in smart cities, the principles below should be followed: (a) the authorization of law. Compared with service provider, government is in charge of the infostructure which contains greater volume, more accurate and sensitive information of citizens, which need to be more strictly restricted by so that the collection, processing, storage and usage are effectively ruled by law. (b) responsibility for oversight. In order to prevent personal information outflow across boarder or to those who do not have sufficient means to provide protection of personal information, more strict rules should be applied to prevent intrusion, leakage, or other illegal usage by internal personnel. (c) public interest. In legislation, public power and private right should be balanced which is especially important in the scenario of smart city. (d) special protection for sensitive information. Special protection should be applied to sensitive information in certain areas such medicine, finance, biology, judiciary and so on.

4.3 Improving the Legislation for Regulating Service Provider

Compared with government, the greatest difference of service provider is that it pursuits profits through smart service. Thus, the qualification and liability of service in protecting privacy is the most things in regulating service provider. (a) minimum entry standard. The minimum standard on privacy concerning hardware and technologies of service provider should be clearly defined. Service provider could not meet the minimum standard should not allowed to collect, distribute, store and analyze users' information. (b) purpose specification. The purpose for privacy collection should be clearly declared and the use of collected privacy should be strictly restricted to the purpose declared. (c) collection limitation. According to OECD, privacy should be collected with fair and appropriate means and with the knowledge of the subject.

4.4 Improving the Legislation for Restricting the Exchange of Personal Information Among Third-Parties

As is described above, third-parties might also collect, distribute or use personal information illegally for benefits or other purposes. Thus, like that of government and service providers, the rules for the collection, usage, storage of personal information by the third-parties should also be strictly restricted. (a) punishment. The third-parties are direct cause for personal information leakage, thus, the punishment for the violation of law and invasion of privacy should be strengthened, so that the intuition for their pursing of unlawful benefits could be restrained. (b) discriminatory treatment. The punishment for different type of leakage, purpose, methods should be varied, and should appropriate to their seriousness and harmfulness. (c) presumption of fault. Due to the elusiveness, timeliness of the collection and distribution of personal information in smart cities, the leakage of personal information is difficult to be traced. Thus, for the protection of privacy, it should be applied the presumption of fault, as long as an individual can provide clear evidence for the invasion of privacy by a service provider or a third-party, then the service provider or the third-party should be presumed to be fault and should take legal responsibility.

References

1. Mani, Z., Chouk, I.: Impact of privacy concerns on resistance to smart services: does the 'Big Brother effect' matter? J. Mark. Manage. **35**, 15–16 (2019)
2. Singh, A.: Advances in smart cities: smarter people, governance, and solutions. J. Urban Technol. **26**(4), 85–88 (2019)
3. Caird, S.P., Hallett, S.H.: Towards evaluation design for smart city development. J. Urban Des. **24**(2), 188–209 (2019)
4. Feng, Y.: The future of China's personal data protection law: challenges and prospects. Asia Pac. Law Rev. **27**(1), 62–82 (2019)
5. Yang, C., Kluver, R.: Information society and privacy in the People's Republic of China. J. E-Govern. **2**(4), 85–105 (2006)
6. Huang, J.: Chinese private international law and online data protection. J. Private Int. Law **15** (1), 186–209 (2019)

7. Martínez-Ballesté, A., Pérez-Martínez, P.A., Solanas, A.: The pursuit of citizens' privacy: a privacy-aware smart city is possible. IEEE Commun. Mag. **51**(6), 45–51 (2013)
8. Huang, J.: Chinese private international law and online data protection. J. Private Int. Law **15**(1), 186–209 (2019)
9. Caruana, M.M., Cannataci, J.A.: European Union privacy and data protection principles: compatibility with culture and legal frameworks in Islamic states. Inf. Commun. Technol. Law **16**(2), 99–124 (2007)
10. Han, D.: Search boundaries: human flesh search, privacy law, and internet regulation in China. Asian J. Commun. **28**(4), 434–447 (2018)
11. Qin, C., Qian, Z., Wang, J., Zhang, X.: New advances of privacy protection and multimedia content security for big data and cloud computing. IETE Tech. Rev. **35**(1), 1–3 (2018)
12. Joss, S., Cook, M., Dayot, Y.: Smart cities: towards a new citizenship regime? A discourse analysis of the British smart city standard. J. Urban Technol. **24**(4), 29–49 (2017)

The Application of Virtual Reality Technology in Environmental Art Design

Kunming Luo(✉) and Xiaoyu Wang

Jiangxi Tourism and Commerce Vocational College, Nanchang, Jiangxi, China
Yinxing_tx3@163.com

Abstract. The continuous growth of national economy and the continuous improvement of virtual reality technology (VR technology for short), the development of the environmental art design industry has made a qualitative leap. As a new technology, VR technology has been widely used in various industries and created Numerous realistic values. By applying virtual reality technology reasonably in the work of environmental art design, the design content can be presented visually and vividly. Moreover, it can fully guarantee the accuracy of budget control, effectively facilitate the communication between designers and clients, and promote the design of each link. This paper will further analyze and discuss the application of virtual reality technology in environmental art design. This article will further analyze and explore the use Application of VR technology in environmental art design.

Keywords: Environmental art design · Virtual reality technology · Practical application

1 Introduction

The development of environmental design industry in China must keep pace with the times in the current era of technological innovation. Virtual reality technology is a vital means of modern environmental art design work process, is an indispensable key content. It can directly affect the scientific accuracy and efficiency of environmental art design, and can effectively promote the innovative development of environmental art design, to ensure that it can create the greatest social and economic benefits at the lowest cost. Therefore, modern environmental art designers must establish advanced design concepts, correctly recognize the need for the reasonable use of VR technology, and effectively combine this technology with environmental art design content, so as to create and design better works, and satisfy the design requirements of market customers at the greatest extent.

2 Virtual Reality Technology and Characteristics

Virtual reality technology is a new technology that surpasses network technology, media information technology and architectural animation. It combines the advantages of modern artificial intelligence, graphics, and computers. With the production of

J. MacIntyre et al. (Eds.): SPIoT 2020, AISC 1282, pp. 744–751, 2021.
https://doi.org/10.1007/978-3-030-62743-0_106

advanced computer equipment as the basis of work, specific helmets, glasses, and gloves are used to motivate people and experience the physical activities of the objects in the virtual environment. Precisely, based on the assistance of VR technology applications, it can bring human consciousness into a brand-new virtual world that is close to real life. In this virtual world of three-dimensional space, it can provide people with sensory simulations of vision, hearing and touch [1].

The characteristics of VR technology mainly cover the following aspects: 1) Immersion. Relevant personnel can effectively build a completely virtual real environment by using VR technology. People in the virtual environment can have various sensory simulation functions, and it brings an immersive feeling. This is the immersion of VR technology. It combines the characteristics of human senses and psychology, and uses advanced computer technology and equipment to effectively generate a three-dimensional effect image. As long as you wear matching sensory facilities, you can fully immerse yourself in the virtual world environment without feeling any interference from the outside world; 2) Interactivity. As a new technology, VR technology is a modern computing technology that interacts with humans. Users can effectively obtain relevant data information by using VR facilities and equipment. In addition, users can directly use the keyboard and mouse to achieve interaction with the computer. In environmental art design work, designers use VR technology to interact with the computer. The computer will actively follow the instructions of the user to adjust to the actual graphics of environmental art design. Designers can also respond to external data and information corresponding to language and actions, effectively completing the investigation and practical operation of external objects; 3) Multi-perception. Multi-perception is one of the salient features of VR technology also. The practical application of this technology can provide users with different human sensory effects such as vision, hearing and touch, and realize different sensory experiences of users in virtual environments. It is precisely because of these characteristics, VR users are enable to have an immersive feeling in the virtual environment. The study of the environmental art design is about the relationship between human and the environmental spaces to achieve the harmonious unity of man and nature. The users of the space environment pay great attention to the sensory feelings of the public. The sensory experience and the application of VR technology can meet the needs of people's sensory experience and fully guarantee the quality of environmental art design works of designers.

3 The Necessity of Virtual Reality Technology in the Application of Environmental Art Design

3.1 Visual Image Display the Environmental Art Design

With the continuous acceleration of the pace of urbanization in China, a large number of public environmental spaces in cities and towns need environmental art design to satisfy the people's needs of green living experience. To ensure the maximum use of urban environmental space, design high-quality art Landscape works must make full use of modern VR technology. Compared with traditional environmental art design methods, the use of VR technology can effectively create a completely virtual environment by improving the construction of software and hardware platforms. The realistic space environment provides a variety of sensory experiences for market customers, prompting customers to have an immersive feeling in the virtual space environment [2]. Through the stimulation of different senses, the client's brain is required to complete the memory of the environmental art design display content in a highly concentrated state, which fully guarantees the persistence and depth of the memory content. As shown in Fig. 1, it enables customers to see different objects in the VR environment vividly and enjoy the stimulation of different sensory information clearly.

Fig. 1. Application of VR technology in garden landscape design

3.2 Enhancing the Interaction in Environmental Art Design

In the process of environmental art design, the use of VR technology can reflect its interactive advantages. Compared with traditional expressions, VR technology is more interactive in terms of customer communication. Customers can be invited to the virtual room, experiencing the different environment art design parts and the close relationship between different design elements in the virtual environment, so that customers can not only fully and truly understand the design solutions provided by the designer but also timely feedback the improvement opinions to designer. With the assistance of the application of VR technology, customers and designers can conduct in-depth discussions and exchanges, timely discover various problems in the design scheme, and take effective solutions. As shown in Fig. 2 below, customers enter the virtual room for experience operations. In a mobile virtual environment, customers can observe and feel every detail of the design features and then communicate with the designer closely [3].

Fig. 2. Demonstration of virtual reality technology in a virtual room

3.3 Realize the Display of Character Image and Scenery

When designers use VR technology for environmental art design work, they should not only pay close attention to the optimal design of the main scene but also lay emphasis on the scientific setting of the scenery plan. Integrating a variety of morphological elements into the overall design plan to ensure a good environmental atmosphere. For example, common green plants, character scenes, buildings, etc. In traditional environmental art design, architectural animation and design renderings can not directly operate autonomously [4], but can scientifically virtualize various character images and scenes in different scenes by using advanced VR technology. Based on the assistance of VR technology systems, designers can virtualize various historical characters, animals, and plants in the design scene. Thereby it constructs the humanized environment effectively, realizes the goal of displaying different characters and scenes, and meets the relevant design needs of different customers. As shown in Fig. 3 below, it shapes different characters and animals effectively.

Fig. 3. Application of VR technology in historical scene

4 Application of Virtual Reality Technology in Environmental Art Design

4.1 Application of Virtual Reality Technology in Landscape Design

4.1.1 Preparation Phase

In garden landscape design, the designer must clarify the actual geological situation on the site firstly, and determine the design requirements and conditions based on the relevant on-site geological information. The designers must thoroughly understand the various information combined with the environmental conditions, climate, construction height and customer requirements, etc. They should design goals in a targeted manner and use it as the core benchmark for the entire garden landscape design creation. If the designer wants to fully understand the relevant conditions of the garden landscape design, he must analyze the project separately, such as different needs, goals, facts, etc. [5] and scientifically summarize the problems and limitations in the design. The designer should make good use of the VR system for on-site simulation and preliminary exploration, and related the relationship between the form relationship and the landscape design condition closely, which is conducive to the subsequent design and communication with customers.

Designers can use the VR system to carry out a full-scale simulation of the nature of the base, scientifically analyze the rivers, roads, buildings, and trees around the base, and understand the actual slope and topographic trend of the base, which can help the designer determine whether the base is used as a limiting condition for different uses. In addition, designers can simulate the wind direction and sunlight of the surrounding environment of the base, providing basic requirements for indoor and outdoor space creation effectively. As shown in Fig. 4 below, the VR system is used to analyze the base.

Fig. 4. Analysis of the base by using the virtual reality system

4.1.2 Design Phase

At the stage of garden landscape design, the designer must clearly know that the elements that make up the garden landscape mainly include greenery, water paving, and sculpture landscapes. Although different landscapes are in relatively fixed positions, people's perception of them will vary with the angle, distance and light. For example, a stationary tree in a garden landscape, people watching it from nearby and distance will produce different effects. Designers apply VR technology in the design of garden landscapes. It is possible to obtain more comprehensive information in the VR environment, and to simulate and evolve various environmental space things, which is conducive to improving the level of designer's modeling concept [6]. What's more, designer can observe and analyze the garden landscape design plan in the VR space at any time during the design phase. It is more convenient to explore the virtual space and the perspective can be automatically changed. The objects in the virtual scene can be used efficiently at real size, such as generate, copy, move, zoom, etc. In practice, the designer can optimize and modify at any time and save effectively. As shown in Fig. 5 below, it is a VR design drawing of a garden landscape space. The designer can switch between different design schemes at any time in the VR garden landscape space, which is conducive to sensory experience of different scene images at different observation points and ensures that the most artistic garden landscape can be designed.

Fig. 5. VR design scheme of garden landscape

4.1.3 Verification Phase

The specific process of the designer's work in practice is as follows: Firstly, carry out the overall concept. Secondly, consider the details, and carry out the optimization and revision of the overall concept. Only after the overall concept of the garden landscape is finally cleared, can the in-depth consideration be made from the whole to the details. Designers also need to strengthen communication and contact with customers, which fully ensure the design scheme scientifically and normatively.

In the landscape design verification stage, through the use of VR systems to carry out technical evaluations with customers, it is convenient to comprehensively and accurately find contradictions between different construction service systems and structural designs in the communication process, and jointly check the design scheme from different details. In order to ensure that the functions and aesthetic perspectives of the design scheme can meet the requirements of customers. In traditional garden landscape design, designers usually use computer-aided design and computer-based drawing, which can reduce the work of design and drawing. But it is difficult to connect computer equipment and garden landscape design structure closely. Based on the application of VR technology, it breaks the disadvantages of traditional garden landscape design procedures effectively, and combines the economic and feasibility of judging theory scientifically. This can avoid subsequent on-site modification in the initial work [7]. As shown in Fig. 6 below.

Fig. 6. Application of virtual reality technology in landscape restoration

4.2 Application Prospects of Virtual Reality Technology in Environmental Art Design

With the continuous innovation and improvement of VR technology in the future, VR technology will play a greater role in environmental art design. Compared with traditional environmental art design methods, the use of VR technology can break through the limitations of traditional dimensions effectively and construct a complete environmental art design expression system scientifically. Not only designers can obtain rich data information, but also perform a variety of sensory experiences. What's more, it can design the best work in accordance with the user needs. In the process of multi-dimensional virtual environment art design, it integrate different types of environmental spatial data effectively. Besides, designers can process various data information more conveniently, and observe the structural changes of design graphics visually [8]. The functional characteristics of VR technology itself promote the reform and innovation development of the essence of environmental art design, further improve the methods and content of environmental art design. What's more, it cultivate designers' good creative ability, and give full play to their imaginative thinking capabilities. In the field of environmental art design, thė use of this technology will be developed accordingly, but the majority of designers must correctly recognize the use of VR technology. The fundamental purpose is to serve the public and satisfy the interaction with customers [9]. Therefore, designers should fully consider the customers' perceptual tendencies and carry out targeted environmental art design in the process of applying VR technology.

5 Conclusion

In summary, for the sake of ensure the stable and sustainable environmental art design industry development in China, designers must study the application of VR technology thoroughly. VR technology can allow designers to effectively engage in a realistic virtual three-dimensional space environment, which can break through the limitations of two-dimensional graphic design of traditional environmental art scientifically and distribute designers' creative thinking [10]. It can encourage designers to design solutions from multiple aspects and angles and taking full account of details to design higher-quality works. The auxiliary application of virtual reality technology can not only strengthen the interaction between customers and designers, but also deepen the understanding of the relationship between the elements of environmental art design and space through sensory experience in a virtual environment.

References

1. Minjie, Z.: On the application of virtual reality technology in environmental art design. Archit. Eng. Technol. Des. **7**(20), 647 (2016)
2. Manzhong, L.: Virtual world · perfect expression—a brief discussion on the application of virtual reality technology in the performance of environmental art design. J. Huangshi Inst. Technol. (Humanit. Soc. Sci.) **09**, 35–37 (2016)

3. Meng, M.: On the integrated application of virtual reality technology in environmental art design. Mod. Decor. Theory **10**, 279 (2016)
4. Bozhi, H.: Research on the application of virtual reality technology in environmental art design. Pop. Lit. **7**, 95–96 (2015)
5. Yihan, L.: Analysis on several problems in the application of virtual reality technology to environmental art design. Mod. Decor. Theory **08**, 90–91 (2013)
6. Zeng, W.: Application of virtual reality technology in environmental art design. Shandong Ind. Technol. **02**, 275–277 (2016)
7. Yingjie, G.: Application research of virtual reality technology in art practice teaching platform. J. Hebei Norm. Univ. (Educ. Sci. Ed.) **06**, 95–96 (2013)
8. Zhe, C.: Application analysis of virtual reality technology in environmental art design. Art Sci. Technol. **30**(8), 313 (2017)
9. Zhanjun, W.: Application research of virtual reality technology in environmental art design. J. Chifeng Univ. Nat. Sci. Ed. **31**(24), 49–51 (2015)
10. Jiajun, L.: Application research of virtual reality technology in environmental art design. Art Sci. Technol. **32**(7), 220 (2019)

Application of VR Technology in Teaching Archery

Kai Zhang(✉)

Physical Education Department of Tianjin University of Technology and Education, Tianjin 300222, China
zhangkai787878@126.com

Abstract. Objective: This paper aims to explore the effectiveness of using VR in archery teaching, and to accumulate experience in order to promote the integration of physical education and information technology in tertiary education, improve the teaching effect, and further promote the reform of physical education. Methods: In this study, literature review, questionnaire and experiments were used to establish an experimental group and a control group of 20 subjects each. The experimental group used VR technology in archery teaching and relaxation exercises, whereas the control group used traditional teaching methods. Results: the results show that compared with the control group, the relevant data of the experimental group is significantly stronger. A significant difference exists in the students' learning interest and control of heart rate. The P value is less than 0.05. There is a very significant difference in the acceptability of teaching methods. The P value is less than 0.01. conclusion: the use of VR technology in archery teaching plays an important role in enhancing students' learning interest and acceptability of teachers' teaching methods, inspiring students' potential and enhancing their ability to regulate their own heart rate. VR technology can very effectively assist archery teaching.

Keywords: VR technology · Archery · Physical education at tertiary level

1 Introduction

The "13th Five-Year Plan" for Education Informatization issued by the Ministry of Education of China in 2016 clearly stated that in order to promote the widespread and in-depth application of information technology in education, it is of great necessity to gradually deeply integrate information technology and education to innovate teaching and reform the education system. The emergence of VR technology as a new information technology brings users an efficient and high-quality information experience and application. After searching domestic literature using the key words of VR technology and physical education, it is found that most are theoretical discussions of the important role of VR technology on physical education and physical training. Few are touched upon from the perspective of practical experiments and methods. Archery in sports is a niche sport, which is why research on it is relatively scarce, especially teaching methods [1–3]. Applying VR technology to archery teaching is bound to enrich classroom teaching methods, and train more teachers in archery, and fix

J. MacIntyre et al. (Eds.): SPIoT 2020, AISC 1282, pp. 752–757, 2021.
https://doi.org/10.1007/978-3-030-62743-0_107

problems such as lack of reference materials, monotonous teaching methods, and poor teaching results. The combination of VR technology and physical education will provide a strong impetus for the reform of physical education, provide portable and effective auxiliary teaching methods for educators, and create better learning conditions for students.

2 Definition of VR Technology

According to the research results of scholars at home and abroad, VR technology is a virtual simulation system based on computer technology. Users in this system can obtain real-world simulations and control of virtual environments. The sensory experience of sight, touch hearing etc. Make users feel immersive.

3 Value of VR Technology in Physical Education

3.1 Optimizing PE Teaching Methods

To master sports skills, students must first establish the correct concept and visual appearance of movements. In the traditional physical education process, teaching is mainly conducted through pictures, sounds and teachers' demonstrations. Traditional teaching methods are incompetent in demonstrating difficult movements. By applying VR technology to physical education to display in a three-dimensional or even multi-dimensional way, students will find it easier to understand and memorize certain skills [4–6].

3.2 Innovating Sports Teaching Concepts

In a VR system, students can use the virtual environment for autonomous learning. Students can repeatedly watch and strengthen certain movements if they are inaccurate, non-standard, and unclear through the VR system. Previews and reviews by pre-class and post-class express system is conducive to further study and consolidation to sports skills, improve classroom learning efficiency and master the knowledge.

3.3 Making up for the Impact of Weather and Sports Field

In the course of physical education in universities, outdoor teaching courses are often affected by severe weather such as wind, rain, smog and low temperature. Teaching and competitions can be affected due to the irregular teaching venues. However, students can get a relatively real experience in the virtual environment, and carry out reinforcement learning and exercises to make up for the impact of weather and sports field.

3.4 Overcoming Fear and Avoid Sports Injuries

With the continuous improvement of students' sports skills and performance, the desire for learning of difficult and tactical moves gradually increases, and the experience of failure is bound to increase the fear of learning. Factors such as the students' physical fitness level, teachers' inadequate ability, and improper protection and assistance may all lead to sports injuries. With the use of VR system, students can view the difficult movements clearly and multi-dimensionally and perform effective simulation exercises, improve the self-confidence of conquering difficult movements, improve the effective experience of correct technical movement exercises, and reduce the probability of sports injuries during the learning process [7–9].

4 Research Results

4.1 Impact on Learning Interest

Table 1. Comparison of student learning interest data

	Experimental group		Control group			
	$\bar{X}$	S	$\bar{X}$	S	t	P
Statistics after the experiment	4.25	0.85	3.55	0.90	2.597	<0.05*

Note: A p-value less than 0.05 is statistically significant.

Interest is the driving force of student learning. Teachers can fully mobilize students' interest in learning in order to establish a good learning effect. Table 1 shows a comparison of the study interest data of the two groups of students. The experimental group students' interest in learning the archery course was significantly higher than that of the control group, and the data comparison has a significant difference. The questionnaires also show that students in the experimental group liked sports very much, and the proportion was 75%. In comparison, the number of students who liked the sports in the control group increased slightly, but some students disliked it very much. The data show that the use of VR technology can generally increase learning interest, meet the learning needs of students, and motivate students in their learning.

4.2 Impact of Teaching Mode Acceptability

Table 2. Comparison of the degree of teaching mode acceptability

	Experimental group		Control group			
	$\bar{X}$	S	$\bar{X}$	S	t	P
Statistics after the experiment	3.56	0.65	2.25	0.40	7.945	<0.01**

Note: A p-value less than 0.01 is statistically very significant.

Interest is the driving force of learning, and the teaching mode plays an important role in fully mobilizing students' motivation. Table 2 shows the comparison and analysis of data concerning teaching mode acceptability of the two groups of students. The experiment shows that the experimental group students' acceptability of the teaching mode of archery course was significantly higher than that of the control group. The data comparison has a very significant difference. The questionnaire also shows that 85% students of the experimental group were very satisfied with the teaching mode, whereas only 35% students of the control group were very satisfied with the teaching mode. The proportions of the students who were not satisfied with in the experimental and the control group were 10% and 25%. It can be clearly seen that VR technology is more preferable [10]. In the interview, the students in the experimental group said by using VR technology in teaching they could better focus their attention, and VR technology gave clearer instructions, displayed movements more clearly, and made learning easier.

4.3 Impact on Heart Rate During Relaxation

Table 3. Comparison of heart rate data during relaxation

	Experimental group		Control group			
	$\bar{X}$	S	$\bar{X}$	S	t	P
Statistics after the experiment	66.4	7.35	70.6	7.64	3.275	<0.05*

Note: A p-value less than 0.05 is statistically significant.

Good archery exercisers requires not only stable skills, but also a stable mental state. Every archer must have the ability to adjust the breathing and heart rate. Table 3 shows the comparison of heart rate data between two groups of students after the experiment. The average heart rate of the students in the experimental group was significantly lower than that in the control group, and the data comparison had significant differences. Through interviews with the experimental group, VR technology showed that the relaxation instructions were clear, the pictures were refreshing and comforting, and the relaxation effect was very good. VR technology plays a positive role and has significant effects.

5 Conclusion

Through the combination of VR technology and physical education to in archery teaching, experiments results show that VR technology can effectively stimulate students' learning interest and enthusiasm to practice. It can make them more satisfied with the teaching modes and promote participation in teaching activities, and it significantly strengthens the learning efficiency of students and the teaching effect of teachers. Students' motivation for autonomous learning and desire for knowledge are stronger. VR technology effectively facilitates students' preview and review and consolidation of the technical movements taught by teachers after class, playing an important role in teaching difficulty movements. VR technology has enriched teachers' teaching methods and modes, and promoted a more active classroom atmosphere and improving the effectiveness of archery teaching.

6 Suggestions

6.1 If VR technology is to be better applied to physical education, teachers need to learn the state-of-the-art VR technology, the working principles and main functions of VR equipment, how to achieve the teaching goals through it and be prepared for solving teaching problems.

6.2 Teachers should strengthen their sports skills, improve their own athletic ability, master the laws of movement and technical requirements, learn more professional knowledge, and make early preparations for producing more effective VR teaching materials.

6.3 Teachers should plan the teaching content and plan in a detailed way, carefully summarize difficult knowledge points that are difficult for students to experience and understand. They must know where to VR equipment to assist teaching so as to use VR technology more effectively in their teaching activities.

6.4 Teachers should stablish teaching, research and technical teams, resort the help of computer professionals in the team, and strengthen communication and exchanges. In this way, they are able to make the VR teaching content more in line with the actual conditions and requirements of physical education, better solve problems in physical education, and improve the effectiveness of teaching.

6.5 Teachers should strengthen the shared use of VR equipment in the teaching of different sports and effectively improve the utilization of VR equipment in physical education. Teachers are advised to save expenses to purchase equipment as well as the time cost to produce VR teaching materials.

Acknowledgements. This paper is the research result of the *"13th Five-Year Plan" of Tianjin Education Sciences Planning*. Project Number: VE1052.

References

1. Nissim, Y., Weissblueth, E.: Virtual reality (VR) as a source for self-efficacy in teacher training. Int. Educ. Stud. **10**(8), 52 (2017)
2. Mel, S., Sanchez-Vives, M.V.: Enhancing our lives with immersive virtual reality. Front. Robot. AI **3**, 74 (2016)
3. Allison, C., Redhead, E.S., et al.: Factors influencing orientation within a nested virtual environment. External cues, active exploration and familiarity. J. Environ. Psychol. **51**, 158–167 (2017)
4. Giblin, S., Collins, D.J., Button, C., et al.: Practical precursors to reconsidering objective for physical activity in physical education. Br. J. Sports Med. **51**(21), 1572 (2017)
5. Neumann, D.L., Moffitt, R.L., Thomas, P.R., Loveday, K., Watling, D.P., Lombard, C.L., Antonova, S., Tremeer, M.A.: A systematic review of the application of interactive virtual reality to sport. Virtual Reality **22**(3), 183–198 (2017). https://doi.org/10.1007/s10055-017-0320-5

6. Rynarzewska, A.I.: Virtual reality a new channel in sport consumption. J. Res. Interact. Mark. **12**(4), 472–488 (2018)
7. Marusak, H.A., Peters, C.A., Hehr, A., et al.: A novel paradigm to study interpersonal threat-related learning and extinction in children using virtual reality. Sci. Rep. **7**(1), 110 (2017)
8. Kavanagh, S., Luxtonreilly, A., Wuensche, B., et al.: A systematic review of virtual reality in education. Themes Sci. Technol. Educ. **7**, 10 (2017)
9. Real, F., Deblasio, D., Ollberding, N., et al.: Resident perspectives on communication training that utilizes immersive virtual reality. Educ. Health **30**(3), 228–231 (2017)
10. Janeh, O., Langbehn, E., Steinicke, F.: Walking in virtual reality: effects of manipulated visual self-motion on walking biomechanics. ACM Trans. Appl. Percept. **14**(2), 1–15 (2017)

Public Opinion Guidance with New Government Media in the Smart City

Yunqing Ji(✉)

Shandong University Finance and Economics, Jinan, People's Republic of China
94760405@qq.com

Abstract. Internet changed people's life, the emergence of new media is changing the people's habit of reading the news and information, the government has also gradually increased the construction of new government media, they hope to take use of the new media platform to guide the network public opinion. This paper studies the new characteristics of public opinion in new media age, analyzes the problems of the new government media in the public opinion guiding, and puts forward some optimization strategy.

Keywords: New government media · Public opinion · Public opinion guidance

1 Introduction

The development of the Internet has promoted the emergence of new media, which refers to a new type of media based on technology that can greatly expand the dissemination of information, accelerate the transmission speed and enrich the transmission mode. It is quite different from traditional media [1]. People use new media platform can access to information and message faster, it increases the frequency of people using it. According to the new media trends report in 2018 from the penguin, among Chinese Internet information consumers, the average time spent on information consumption (all channels) per day reached 76.8 min. And about 73.7% of the users can take out more than 30 min a day to look at information, 47.1% of the users read information more than 1 h a day [2]. The new media has become an important part in people daily life. Therefore many local and central governments have seen the space for the development of new media, and have set up official microblog or WeChat to release relevant news information, so as to shorten the distance with the public.

The rapid development of new media provides convenience for people to gain information, not only that, also changes the spread of public opinion. Such as chengdu school canteen messy problem a while ago, the recent Hong Kong waste young events, and the college sky-high accommodation costs events, these events aroused heated discussion online as soon as they were reported by the new media platform. The new media platform provide a environmental where everyone can be involved in the spread of public opinion, it prompted the hot public opinion emerge in endlessly. On the other hand, it is also becoming more and more important also has brought certain challenges for the government to control public opinion. Therefore it is becoming more and more

J. MacIntyre et al. (Eds.): SPIoT 2020, AISC 1282, pp. 758–764, 2021.
https://doi.org/10.1007/978-3-030-62743-0_108

important so for the government to consider about how to use new media tools to guide public opinion correctly.

Public composed of individuals and social groups always comment on the incidents are closely related to their own concerns or interests. These views constitute public opinion. Public opinion is the interlaced sum of various emotions, intentions, attitudes and opinions held by the public in a certain historical stage and social space [3]. New media provides a more free communication platform. It is a new way for government to guide public opinion, and many scholars are researching about it. Li Zongjian etc. [4] think we should change the traditional way of public opinion guidance and take advantage of the Internet technology and the fast spread of new media to create the conditions for public opinion guidance. Xie Jin-lin [5] think public opinion has the characteristics of diversification and intensification in transition network, the public attitude and advice should be considered in public opinion guidance. Zheng Lei etc. [6] research the current Chinese government official webio's role, problems and challenges in crisis management through the empirical study. Li Enlin etc. [7] think government official WeChat can play a good role in government public opinion work. They all see the chance of new media in public opinion guidance, and put forward some practical solutions. Therefore, it has certain feasibility that taking advantage of new media to guide public opinion.

2 Characteristics of Public Opinion in the New Media Era

New media has changed the communication mode of traditional public opinion and also endowed public opinion with some new features. Cao Xiaoyang et al. [8] believed that opinions expressed online are anonymous and direct, sometimes even emotional or irrational. Pan Xu et al. [9] believed that the discussion of online public opinion is generally negative and emotional. Wang Yang et al. [10] believed that public opinion on the Internet has the characteristics of polyphyly multiple and multanimity. It is not difficult to see from the summary and analysis that the current network public opinion has the following characteristics:

2.1 The Transmission Speed Is Fast and the Transmission Mode Is Diverse

With the help of the new media's sensitivity, the rapidity and extensity of information transmission, public opinion was processed soon when it was exposed to public view. Unlike previous event passed through layer upon layer process to reach people, public opinion events can do real-time reports even now. At the same time, people can collect lots of information related to event from the new media's search engine more conveniently, public can understand event deeply and share comment with others.

2.2 One-Sided, Emotional and Irrational

For some public opinion information, because the new media workers may focus on different aspect or intentionally processing for some details, one can only understand partial information and do not know the whole story. If some of the new media use emotional color or introductory statement, or touch the netizen on your idea, it's easy to arouse the netizens' emotions. In addition, because the comment published online is anonymous and no constraint, some people through the keyboard's comments may be a radical thought, instead of rational thinking.

2.3 Easy to Participate and Easy to Take Advantage of

The availability of the Internet makes people participate in the discussion of public opinion events easier, as long as there is network, everyone can learn a lot of information and publish their views. For some professional knowledge people don't know or understand, people are likely to be puzzled by some media publishers word and be used.

2.4 Group Is Strong

People tend to seek people with similar views when expressing their own views on an event, and some people will follow the thoughts and opinions of most people to establish their own views, showing a strong group.

3 Use Government Affairs New Media to Guide Public Opinion Problem

The new features of public opinion brought by new media make people see the difference between online public opinion and traditional public opinion, and make the government pay more and more attention to use the new media to guide the public opinion. However, there are still some problems in the guidance of public opinion with new government media.

3.1 The Timeliness

Although the speed of information transmission is very fast, it is still a minority that can deal with public opinion in time. The first one to two hours after the public opinion was reported is the best time to explain the truth and guide public opinion. However, due to conservative ideas and complex processes of many governments, the official media accounts are often unable to deal with public opinion formally and make responses within this period, which is easy to make the public question. Moreover, after several hours of fermentation, it maybe appear some word with strong incendiary content, it is easy to arouse the public mood and make the public opinion deviate from the direction. Then, the effect of public opinion guiding will be poor [7]. Some people will think that the long-term processing is to discuss countermeasures to hide the public's eyes and ears.

3.2 Authenticity

The official new media account as a representative of the government should send the appropriate response after confirming the authenticity in the face of the occurrence. But the first information received may be one-sided, not entirely accurate. However some authorities convey some information in advance before the fact fully investigate clearly, sometimes it is not real. After fact appears, it maybe cause that the public's trust in the official news has declined. For example, an official account published a piece of news with directivity when the incident of "chongqing bus falling off a bridge" occured, which caused netizens to attack the female driver. After the investigation, the accident had nothing to do with the female driver, but the official account did not further explain or apologize, so it caused the public to question the account.

3.3 The Way to Deal with It

Many official media still adopt traditional ways to deal with public opinion, such as cold treatment and non-response, which will arouse people's opposition and make people think that the government is evading and weaken their trust in the authority of the government. For recent events on the network, some media accounts take some way to forbid people to discuss events, such as remove video, shielding subject. Although using this approach can reduce the massive discussions for a short period of time to some extent, it's easy to cause the negative feelings of people, then the official notice may not be believed by public. In addition, some government media will delete the wrong news and deal with it coldly after making mistakes, which will reduce the popularity of the public.

3.4 Professional

For management of new government media, they need to have some professional knowledge related to public opinion. But in fact, most of the new government media don't meet this condition. Daily operation of many government media accounts lacks the popularization of some professional knowledge, or they can't solve some basic professional problem from user comments [6]. In addition, some departments' media account's operation is even outsourced to other enterprises, which improves the efficiency of account operation but may reduce the professionalism of the account. The account generally adopts shift system, and everyone's professional knowledge is also different. Therefore, the operation of the account cannot fully meet the needs of the public.

3.5 Workflow

Recently some new government media is still the continuation of the traditional way in the approval process [6]. It process feedback in a hierarchical manner. Daily operator collect public opinion from comments or messages and send these to high-level people. Then they handled information according to instruction and feedback from high-level. This can guarantee the seriousness of the official account, but this will waste a lot of time and manpower. Some new government media account not only audit the information that is published every day, but also audit every reply, which is inefficient. The same is true for the process of collecting and reporting public opinion. Complicated procedures will increase workload and reduce work efficiency.

4 Optimization Strategy of Guiding Public Opinion with New Media

4.1 Find and Deal with Public Opinion in Time

When finding public opinion, we should respond positively and deal with them in a timely manner. For public opinion events that can be concluded quickly, we can release information when the result of the event is confirmed. For public opinion events that last for a long time, we should release announcements to appease the public. Don't let the public mistake the government for being perfunctory, it will provide a readily exploitable loophole for people with ulterior motives. It is also possible to build a public opinion early warning system by big data technology and user analysis. This helps to find and deal with public opinion timely. And this makes it possible to block divergent sources.

4.2 Verify Information and Guide Public Opinion Correctly

We should pay attention to the authenticity of information when we pursuit the information timeliness. The new government media should pay attention to investigate the real situation of the incident when it reports news. For events that is not final, the description of the news should be objective. Editors do not use ambiguous language or make overly directional assumptions as far as possible when they report news. It is more likely to guide people's thought correctly, and calm their mood. The media should use positive words to reduce the living space of negative words when things caused discussion, so as to achieve the effect of correct guidance.

4.3 Treat Public Opinion with a Rational and Objective Attitude

For the occurrence of public opinion events, we should not react negatively or simply block them. IN addition, we should not be afraid of the discussion of the public, but take the right way to face the discussion of the public. The blind blocking may lead to the adverse psychology of the public, which may have a negative impact. When there are some mistakes in the operation of the official account, we should not be afraid to face the mistakes. We should sort things out and learn from experience, and admit mistakes frankly.

4.4 Train Operators to Improve Their Professional Level

Conducting training and testing for operators of new media regularly. It can strengthen their professional knowledge and skills on public opinion, so that they can identify public opinion information. Then the operation team can deal with some small fluctuations of public opinion in the initial stage, so as to reduce the occurrence of common sense errors. If the operation team takes the form of outsourcing, it is necessary to contact with companies which have some professional knowledge and experience.

4.5 Optimize Workflow and Improve Efficiency

The operation team should establish better management mechanism and optimize workflow. For some daily news releasing or simple replying to the public, some complicated process can be omitted, it can reduce the waste of time, manpower and other resources generated by the audit. The government can generate good interaction and establish a more harmonious relationship with the public through replying them timely. In addition, the importance of public opinion events should be distincted [6]. For low-level events, they are handled by the common operator rather than reported to the high-level manager. When major public opinion appear, they are reviewed through a hierarchical review process. This way can greatly save time and improve efficiency.

5 Conclusion

Network changed people's life in many ways, the new government media has brought more opportunities and conditions for public opinion guiding. Attaching great importance to the construction and maintenance of the new government media can better adjust the relationship between the government and the people. And it is beneficial for setting up the image of the government. This paper simply analyzes the strategy of guiding public opinion with the new government media in theory. In the following study, we will study how to guide public opinion with big data and other technical.

References

1. Jianghong, X.: Introduction to New Media. Shanghai Jiao Tong University Press, Shanghai (2006)
2. Penguin Intelligenie: 2018 New Media Trend Report [DB/OL] (2018). http://www.199it.com/archives/804544.html
3. Yi, L.: An Introduction to Internet Public Opinion Research. Tianjin People's Publishing House, Tianjin
4. Zongjian, L., Zhuru, C.: Challenges and countermeasures of public opinion guidance in the new media era. J. Shanghai Adm. Inst. **17**(5), 76–85 (2016)
5. Jinlin, X.: Government governance of network public opinion: concept, strategy and action. Theor. Invest. **2**, 8–12 (2010)
6. Lei, Z., Yinghao, W.: Crisis management of government microblog: roles, challenges and problems. E-Government **6**, 2–7 (2012)
7. Enlin, L., Congo, W.: Research on the strategies of government WeChat to deal with public opinion crisis in the practice of social governance innovation. Study Explor. **9** (2017)

8. Xiaoyang, C., Shujin, C., Guihong, C.: Research on the characteristic structure and model of network opinion. Inf. Sci. **2**, 231–234 (2010)
9. Pan, X., Qingyun, W.: Analysis of the characteristics of internet public opinion based on new media events. Southeast Commun. **2**, 32–34 (2010)
10. Yang, W., Cui, S.: Characteristics and guidance of internet public opinion in mobile internet era. Youth Journal. **14**, 30–31 (2015)

Needs Analysis of the AR Application in the Packaging Design of Ya'an Tibetan Tea

Jie He(✉), Lijun Zhang, and Xia Jiang

Sichuan Agricultural University, No. 46 Xinkang Road, Ya'an, Sichuan, China
985998259@qq.com

Abstract. Purpose Based on the current situation of packaging design of Ya'an Tibetan tea, this paper explores its application needs of AR technology. **Method.** According to the statistics of the existing packaging design data of Ya'an Tibetan tea, this paper analyzes the production protection strategy of intangible cultural heritage, the image building of regional public brands and the urgent need for consumption upgrading. Then it discusses the necessity of AR in the packaging design of Ya'an Tibetan tea. **Result.** The intervention of AR provides a new design direction for its packaging design. **Conclusion.** AR packaging design can expand the layout space of traditional packaging, enhance the interactive experience, strengthen consumer brand memory, etc., with broad application prospects.

Keywords: AR technology · Ya'an Tibetan tea · Packaging design · Interactive experience

1 Introduction

At present, with the in-depth implementation of consumption upgrading, consumers' health awareness is constantly improving. Ya'an Tibetan tea attracts great attention because of its good health-care effect in weight reduction, lipid lowering and digestion. However, its packaging has problems such as lack of interactive experience and weak brand recognition.

In the context of the Internet of Things, the function of packaging is no longer limited to product protection and simple information transmission. Instead, it should focus more on the individualized needs of the new generation of consumer groups, enhance interactive experience, and promote packaging upgrading through technology. The prototype of Augmented Reality (AR), originated in the 20th century [1], is a human-computer interaction that places virtual elements into the direct or indirect real environment to enhance users' perception [2]. There have been some successful cases of the AR application in packaging, such as the AR expression bottles jointly launched by Pepsi and QQ, the AR musical bottles of Coca-Cola, and the AR packaging of Heinz tomato ketchup. At present, AR has not been introduced into the packaging of Ya'an Tibetan tea.

J. MacIntyre et al. (Eds.): SPIoT 2020, AISC 1282, pp. 765–771, 2021.
https://doi.org/10.1007/978-3-030-62743-0_109

2 Analysis of Product Packaging Design of Ya'an Tibetan Tea

Tibetan tea should have two levels of meanings, and one is broad and another is narrow. In a broad sense, Tibetan tea refers to the tea that Tibetans have drunk in their history. In a narrow sense, it refers to the brick tea that Tibetans have been drinking since the Tubo era, with Ya'an as its production center and Ya'an Benshan (small leaf tea) in its raw material [1]. This paper discusses the Tibetan tea in the narrow sense.

With the successful registration of "Ya'an Tibetan tea" geographical indication certification trademark, the brand awareness of business operators has been gradually improved. At present, there are 9 enterprises that are qualified to use this trademark. The paper selects Ya'an Youyi Tea Co., Ltd, Sichuan Auspicious Tea Co., Ltd., Ya'an Yixing Tibetan Tea Co., Ltd., Ya'an Tea Factory Co., Ltd., Ya'an Cailong Tea Factory, Ya'an Helong Tea Factory Co., Ltd., Ya'an Zhougongshan Tea Industry Co., Ltd., Tibetan Langsai Tea Factory, and Ya'an Xikang Tibetan Tea Group Co., Ltd. Then it analyzes packaging samples of 244 products of these enterprises. In the charts below, A, B, C, D, E, F, G, H and I refer to these nine enterprises in turn.

2.1 The Form of Tea

Ya'an Tibetan tea mainly has the shape of brick, loose tea, strip and cake. According to the data (Table 1), this kind of tea is mainly traditional bricks. Among the 244 products, 111 are brick tea, accounting for 45.5%; 90 are loose tea for 37%; 25 are strip tea for 10.2%; Other forms like small block, cake, and zongzi are relatively few for 7.3%.

Table 1. Ya'an Tibetan tea packaging design-form

Form	Brick	Loose tea	Strip	Tea cake	Small block	Zongzi shape
Number	111	90	25	11	5	2
Percentage	45.5%	37%	10.2%	4.5%	2%	0.8%

Brick tea belongs to the traditional form of Tibetan tea. Though it reflects the traditional tea making process and Tibetan tea cultural characteristics to some degree, it has a poor experience in the actual drinking, mainly in the segmentation, boiling (brewing) of tea, storage and other links.

2.2 Packaging Structure

Ya'an Tibetan tea is now mainly packaged in boxes, bags, cans, barrels, and bamboo. According to the data (Table 2), among the 244 packages, 178 are packed in boxes, accounting for 73%; 30 in bags for 12.3%; 19 in bamboo for 7.8%; Other structures, like cans, barrels and scrolls, account for a relatively small proportion of 6.9%.

Table 2. Ya'an Tibetan tea packaging design-structure

Structure	Box-packed	Bag-packed	Bamboo-packed	Can-packed	Barrel-packed	Scroll-packed
Number	178	30	19	13	3	1
Percentage	73%	12.3%	7.8%	5.3%	1.2%	0.4%

2.3 Packaging Material

Table 3. Ya'an Tibetan tea packaging design-material

Material	Number	Percentage
Paper	123	50.4%
Paper + Others	68	27.9%
Aluminum foil + Others	22	9%
Cloth + bamboo strips	13	5.3%
Others	18	7.4%

According to the data (Table 3), among the 244 products, 123 are paper packaging, accounting for 50.4%, mainly involving the external packaging printing paper and yellow paper, corn paper, cotton paper, etc. There are 68 types of paper plus other materials for 27.9%, mainly including paper + aluminum foil, paper + bamboo strip, etc.; 22 types of aluminum foil plus other materials for 9%; 13 types of cloth + bamboo strip for 5.3%; 18 types of mixed materials for 7.4%.

2.4 Packaging Vision

2.4.1 Graphic

The graphic elements in the packaging design of Ya'an Tibetan tea are mainly 8 auspicious patterns and flower and grass patterns. According to the data (Table 4), among the 244 product packaging designs, 8 auspicious patterns are used 130 times, accounting for 28.5%; flower and grass patterns 76 times for 20.4%.

Table 4. Ya'an Tibetan tea packaging design-graphic

	A	B	C	D	E	F	G	H	I	All	%
8 auspicious patterns	8	3	7	27	36	1	32	16		130	34.9%
Tibetan architecture	1			2	2	1			1	7	1.9%
Prayer wheel		1		1	2					4	1.1%
Thangka		1		8				14	1	24	6.4%
Geometric graphic	11	4	4	6	9	11	1	11	5	62	16.6%
Animal pattern	4		1	2		4				11	2.9%
Moire	1			2	1	2		3		9	2.4%
Flower / grass	14	8	4	14	6	4	5	17	4	76	20.4%
Related to tea	3	4	7	5	4	4	4	5	14	50	13.4%

2.4.2 Color

Table 5. Ya'an Tibetan tea packaging design-color

	A	B	C	D	E	F	G	H	I	All	%
Yellow	25	14	11	27	11	8	9	13	9	127	52%
Red	12	11	4	6	4	8	7	3	1	56	23%
Blue		1		1	1	2			1	6	2.5%
Green		1	3		2	5	2	2		15	6.1%
White	3	1		6			2			12	4.9%
Black		4	4	1		4	2	11	2	12	11.5%

The packaging color of Ya'an Tibetan tea covers 5 original symbolic colors of white, blue, red, yellow and green in the Tibetan primitive religion. According to the data (Table 5), yellow is used 127 times, accounting for 52%; red is 56 times for 23%. The application proportion of yellow and red is as high as 75%. This will inevitably lead to the lack of brand differences, and competitiveness.

2.4.3 Script

Ya'an Tibetan tea packaging is mainly in black, Tibetan, and Song fonts. According to the data (Table 6), black font is used 208 times, accounting for 27%; Tibetan font is 159 times for 20.7%; Regular script font is 110 times for 14.3%; Song font is 100 times for 13%; official script font is 69 times for 9%; Other fonts, like calligraphy, Cuqian and Yao, are 123 times for 16%.

Table 6. Ya'an Tibetan tea packaging design- font

	A	B	C	D	E	F	G	H	I	All	%
Black system	32	25	22	38	14	25	13	26	13	208	27%
Song system	21	7	6	19	3	4	7	13	3	100	13%
Tibetan type	22	23	9	10	13	23	20	26	13	159	20.7%
Official script	4	12	8	18	12	3	3	9		69	9%
Regular script type	12	18	10	12	9	7	16	20	6	110	14.3%
Others	12	4	3	37	6	15	3	30	13	123	16%

3 Needs Analysis of AR Application in Ya'an Tibetan Tea Packaging

"With the change of consumption structure and the influence of 'Internet +', the function and significance of modern packaging design have changed dramatically" [2]. The design methods need to be continuously upgraded to meet new needs [3].

3.1 Need for Productive Protection Strategy

Productive protection is an important way to protect intangible cultural heritage. In the Guidance on Strengthening the Productive Protection of Intangible Cultural Heritage, the Ministry of Culture points out that: “The productive protection of intangible cultural heritage refers to keeping the authenticity, integrity and inheritance of intangible cultural heritage as the core in the process of productive practice. It is based on the premise of effective transmission of intangible cultural heritage skills, and with the help of production, circulation, and sales. It is a protection method that transforms intangible cultural heritage and its resources into cultural products.” In 2008, the traditional production techniques of Ya’an Tibetan tea (Southern Route Tea) was included in the national intangible cultural heritage list. Packaging is not only an important link in the commodity circulation and sales, an important carrier for consumers to contact with brands, but also an important starting point for productive protection strategies. Thus, AR’s introduction in the product packaging design of Ya’an Tibetan tea is significant to the implementation of productive protection strategies.

3.2 Need for Building Regional Public Brand Image

In recent years, the regional public brand value of Ya’an Tibetan tea has increased significantly, with frequent brand image building and promotion. According to the “2018 Evaluation Research Report of China Tea Regional Public Brand Value” jointly issued by Zhejiang University CARD and China Tea Brand Value Evaluation Research Group, the brand value of “Ya’an Tibetan tea” reaches 1.845 billion yuan, an increase of 364 million yuan or 25% over last year. Meanwhile, Ya’an Tibetan tea ranks first in the “most powerful brand development”, and its brand building has achieved remarkable results.

Apart from the basic functions of product protection, storage and transportation, promotion and extension of shelf life, packaging also has the function of transmitting information, which is the carrier of brand and product information [4]. Packaging is one of the contact points for brand advertising. The intervention of AR plays a vital role in building and promoting the regional public brand image of Ya’an Tibetan tea. Compared with traditional product packaging, AR packaging can enhance consumers’ memory to brands [5].

3.3 Need for Consumption Upgrading

With the continuous development of China’s economy and society, the level of national culture is gradually improved, and the consumption demand is also constantly upgraded. The improvement of contemporary people’s quality of life includes, but not limited to, the material level. More importantly, it is reflected in the spiritual level of satisfaction [6]. According to McKinsey 2017 China Consumer Survey: “Double-click” Chinese Consumers, “post-90s” is becoming a new generation of consumption engines, who accounts for 16% of Chinese population. From now until 2030, they will contribute more than 20% of China’s total consumption growth, higher than any other groups. This is a remarkably diverse group. These consumers have a significantly strong self-awareness, younger, emphasis on health, a more mature brand cognition. Besides, they have higher requirements for products and services purchased [7].

At present, Ya'an Tibetan tea packaging lacks interactive experience, which limits the consumption desire of the new generation of consumers. The most striking feature of AR is the interactive experience with multiple senses [8]. Thus, the rise of "AR + brand" marketing corresponds to consumers' needs for novel experience in this era.

4 Conclusion

Ya'an Tibetan tea has experienced from the "route tea" with Tibetans as the main consumers to the domestic tea consumed by the public, from the intangible cultural heritage to the regional public brand, from the minority to the masses. The change of its sales regions, consumers and product positioning set higher requirements for product packaging design.

Currently, the packaging design of Ya'an Tibetan tea should not be limited to the basic functions of product protection, transportation, storage and information transmission. Instead, it should be an object that can interact with consumers deeply. In the context of the Internet of Things, "packaging + technology" has received increasing attention. The intervention of AR can effectively expand the layout space, enhance the interactive experience, and strengthen brand memory. Thus, it is necessary to apply AR to the packaging design of Ya'an Tibetan tea.

Acknowledgements. 1. Study on Packaging Design of Special Agricultural Products Based on Regional Cultural Features, Social science research project of Sichuan Agricultural University; No. 2015YB08.

2. Study on the Visual Symbols of Regional Cultural Features in the Packaging Design of Special Agricultural Products in Rural Tourism Destinations, Project funded by Sichuan research center of agricultural special brand development and communication, No. CAB1611.

3. Study on the Public Brand Image Design of "Ya'an Tibetan Tea" under the Background of Consumption Upgrading, Special art research project of double branch plan of Sichuan Agricultural University.

References

1. Zhaogui, L., Gengdong, L.: Tibetan Tea, p. 33. Sichuan Ethnic Publishing House, Chengdu (2007)
2. Xu, L.: From packaging design to branding design: the strategies for branding design of Chinese tea during the period of consumption updating. Art Des. **2**, 30–36 (2018)
3. Kolko, J.: Thoughts on Interaction Design. Fang Zhou(trans). China Machine Press, Beijing (2012)
4. Gaimei, Z., Yue, C., Lu, W., Mingzhi, C., Xiaoli, S., Jiandong, L.: Research and development of intelligent packaging for the organic tomato based on augmented reality technology. Packag. Eng. **21**, 2 (2018)
5. Qiong, X.: Modern brand design packaging innovation based on AR technology. Packag. Eng. **2**, 61 (2017)

6. Li, J., Tang, Y.: Vision, function and culture: the emotional value of modern packaging design. Art Des. **2** (2018)
7. McKinsey 2017 China Consumer Survey: “Double-click” Chinese Consumers (2017). https://www.mckinsey.com.cn/McKinsey
8. Li, G., Xue, X.: Modal analysis of advertising touchpoint virtualization in the view of AR media. A Vast View on Publishing, p. 12 (2019)

Construct Sharing Mechanism of College Students' Physical Health Monitoring Service Based on Geographic Information System

Mingang Guo and Yizhao Wu(✉)

Department of PE, Wuhan University of Technology, Wuhan, China
wyz@whut.edu.cn

Abstract. Constructing the sharing mechanism of College Students' physical health monitoring service is not only an important resort to do a good job in college sports work, but also an important means to optimize the network system of College Students' physique health monitoring. Starting from the importance and practical significance of the sharing mechanism in China, this paper puts forward the idea of introducing geographic information system (GIS) technology into the construction of this sharing mechanism. By means of literature, expert interview, mathematical statistics and other research methods, this paper taking 7 subordinate universities in Wuhan City as an example, makes a deep inquiry into the sharing mechanism, and considers that it is very necessary and feasible to build a sharing mechanism for physical health monitoring service of university students. And the best location of the platform is Wuhan University of Technology. The purpose is to provide feasible reference for the humanize, informatization, intellectualized and efficient construction service of College Students' physique health monitoring service in China.

Keywords: Physical health · Geographic information system technology · Sharing mechanism

1 Introduction

Reasonable monitoring the physical health of college students is an effective measure to comprehensively grasp the physical condition of college students, to ensure that colleges successfully train and transport sport talents for society, and also to promote the implementation of the "Healthy China Strategy".

As the concept of sharing deepens, education resource sharing has also become a hot area of academic research. Scholars have focused on how to deal with problems such as waste of resources, repeated construction, and inefficient utilization during the development of physical education [1–4].

Under the dual context of educational resource sharing and "Healthy China Strategy", constructing an efficient sharing mechanism of physical health monitoring services for college students will not only help to promote the monitoring become humane, informative, efficient and intelligent, but also help to promote the overall improvement of college students' physical fitness.

J. MacIntyre et al. (Eds.): SPIoT 2020, AISC 1282, pp. 772–778, 2021.
https://doi.org/10.1007/978-3-030-62743-0_110

2 Logical Starting Point

2.1 Scientifically Evaluate the Effectiveness of Physical Education Work

The establishment of the sharing mechanism has imposed stricter requirements on university sports workers from an objective level. No matter from the technical level or the theoretical level, it will urge college sports workers to continuously improve their physical fitness Capability. In addition, according to the data monitored by the service sharing mechanism, it can not only feedback the college student's physical health status, but also reflect the advantages and disadvantages of physical education in different colleges, and provide data support for scientific evaluation of the quality and effectiveness of physical education.

2.2 Important Source of Data for Sports Research

The data monitored by the monitoring service sharing mechanism provides data support for the research of college students' physical health field. Physical fitness of college students can be used as a test to check the results of sports scientific research, which can objectively reflect the effectiveness of sports scientific research and indicate the direction of development. The theoretical result of physical education research and the monitoring data is an important guarantee for college students' physical fitness. Which can also effectively guarantee the authenticity of research data and the reliability of sports scientific research conclusions.

2.3 Powerful Measure for College Students to Carry Out Scientific Fitness

The establishment of the mechanism provide college students with a set of targeted, scientific, and time-effective fitness prescriptions. Also, it eliminates the tediousness of physical testing procedures and the lagging of data, making the fitness monitoring become more humane, efficient, and convenient, and then provides timely targeted fitness prescriptions [5–7].

3 Feasibility

3.1 Policy Document Provides a Legal Basis

In 2016, the State Council successively issued the "Opinions of the General Office of the State Council on Strengthening Student Sports to Promote the Comprehensive Development of Students' Physical and Mental Health", the "National Fitness Plan (2016–2020)", and the "Healthy China 2030 Plan", etc. Strengthen evaluation and monitoring and other means to improve the physical fitness of adolescents in our country, integrate the physical fitness of students into the school assessment system and administrative accountability. The promulgation of the above rules and regulations will undoubtedly provide a basic legal system guarantee for the establishment of the monitoring service sharing mechanism.

3.2 Resource Sharing Provides a Strong Guarantee

Establishing the monitoring service sharing mechanism can not only effectively reduce the cost of repetitive construction of physical health testing centers in universities, but also save the labor and equipment cost of colleges that participate in fitness testing. Through the professional intervention of third-party institutions, it can avoid potential testing risks in physical fitness monitoring, and provide more professional services and guidance for improving the quality of monitoring work.

3.3 Application of GIS Provides a Scientific Basis

GIS technology can not only effectively manage various information resource that with spatial attributes, but also facilitate the decision-making and effectively monitor and analyze resource information, and even compare data. Apart from that, the data collection, spatial analysis, and decision-making processes can be integrated into a common information flow by GIS, so as to significantly improve work efficiency [8–10].

4 Paths

4.1 Establishing a Service Sharing Platform

1) Theoretical model for selecting the sharing platform based on GIS:
   The P-median model refers to the selection of the best location under a given number and location of demand sets and multiple candidate facility locations, assigning other demand points to the current facility, and achieve the lowest combined costs and the most benefits. In this study, it is used to solve the problem of location selection for students' physical health sharing center.

$$Result_i = w1 * DTotal_i + w2 * SCenter_i + w3 * STeache_i \tag{1}$$

$$DTotal_i = 6\sum \mathrm{f} = \mathrm{o}\ \mathrm{D_y} \tag{2}$$

$$SerPop = AllPop/day \tag{3}$$

$$SerTeacher = AllPop/PopEverday\ /\ day \tag{4}$$

$$SCenter_i = SerPop - Center_i \tag{5}$$

$$STeacher_i = SerTeacher - teacher_i \tag{6}$$

$$x = x - \mu/\sigma \tag{7}$$

i = 0, 1, 2, 3, 4, 5, 6, that's mean seven universities.

Based on research needs, seven universities under the Ministry of Education were selected as the research objects. Which including Wuhan University, Huazhong University of Science and Technology, Wuhan University of Technology, Central

China Normal University, China University of Geosciences, Huazhong Agricultural University, and Zhongnan University of Economics and Law. Take the distance between universities, the total umber of students served per day, and the total number of test teachers in the fitness test center as the weight. Resulti is the final result, w1, w2, w3 are given weights, Dtotali is the sum of the distance from the i-th school to the other 6 universities, D is the distance from university i to j, AllPop is the total number of 7 universities, Day is the days required for the first fitness test, Pop Every Day is the maximum number of students that a fitness test teacher can serve per day, SerPop is the number of students that the target fitness test center needs to serve per day, SerTeacher is the number of teachers that required. Scenteri is the number of students that the test center need to serve when one day difference between the i-th university and the target physical testing center. STeacheri is the number of fitness test teachers between the i-th university and the target physical testing center. Equation 7 is a normalized formula, which needs to be used after the Dtotal, Scenter, and Steacher were calculated.

2) Visualization of site selection for the sharing platform based on GIS:
   Using the GIS visualization technology, to study and plot the spatial location of the monitoring service sharing platform and its spatial relationship with other universities. As shown in the Fig. 1 below, pink indicates that Wuhan University of Technology, black indicates Central China Normal University, blue indicates Wuhan University, purple indicates China University of Geosciences, green indicates Huazhong University of Science and Technology, dark green indicates Zhongnan University of Economics and Law, and red indicates Huazhong Agricultural University. It can be seen from the figure that among the 7 subordinate universities in Wuhan, the university student physical health monitoring service sharing platform should take Wuhan University of Technology's physical fitness test center as the best location based on GIS technology.

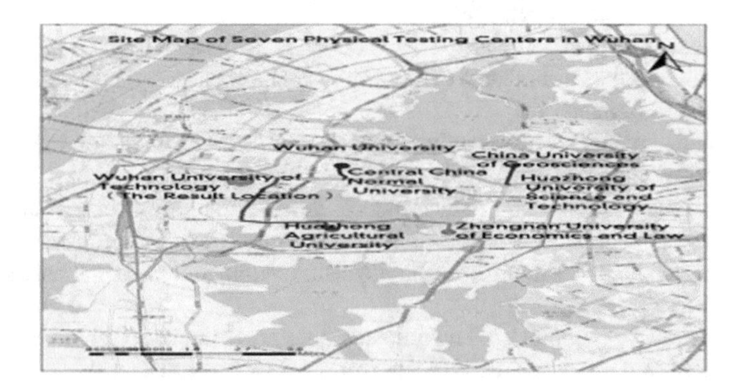

Fig. 1. Site map of seven physical testing centers in Wuhan

4.2 Optimize the Shared Operating Mechanism

1) Establish the shared decision-making mechanism:
   First, establish a decision-making procedure system and organizational structure. College students' physical health monitoring decision-making process consists of investigation, research, preparation of plan, analysis demonstration, selection of plan, organization and implementation. From the perspective of the establishment of the organization, Wuhan Hongshan District Sports Department will lead the physical fitness testing authorities of seven universities to jointly establish a shared service decision-making center.
   Second, establish a survey and follow-up feedback system for decision-making. Specific methods: timely and scientific methods should be adopted to conduct in-depth investigations of the actual situation of college student physical health monitoring, to obtain reliable first-hand information, and to provide support for the formulation of decision-making programs.
   Third, establish an expert consultation and demonstration system. Specific approach: Establish an expert advisory group or advisory committee, and organize authoritative experts to conduct in-depth investigations and research demonstrations on the sharing mechanism, which can absorb opinions and reduce the blindness of decision-making.
2) Establish the shared information management mechanism:
   The construction of the management mechanism should be improved from the following modules:
   Selection: Registration module: Use the information of the student ID card to register through the intelligent physical testing platform. Physical examination students can make appointments online or offline according to their actual needs. After confirmation, the management software will generate a physical examination ID number (or student ID), and the tester can enter the corresponding ID number to execute the physical examination.
   Physical test project template: This module is to assign teachers according to physical test projects. Students can start physical tests according to the prompts of physical test projects.
   Physical examination report module: After the physical examination information is reviewed by the main examination teacher, it will be printed out in the form of a physical examination report, which reflects the detailed information of all physical examination items and exercise prescription suggestions of the physical examination students, and submit it to the physical examination students themselves or their units.
   Statistical query module: This module can query the number of physical tests, test items, etc. And can also calculate the workload of test teachers.
   Intervention service module: Establish student-specific health files based on statistics information, and provide intervention services for individual students.
   System maintenance module: This module is to manage and maintain the large amount of proposed data. Where should arrange specific person to perform maintenance through password authority.

3) Establish the shared evaluation mechanism:
   First, establish the performance evaluation mechanism. It is necessary to generate a complete evaluation index system, and incorporate the monitoring service sharing indicators into the annual evaluation index system, and conduct regular performance appraisals to related universities. Second, establish the supervision and evaluation mechanism. Actively set up a supervision group for the sharing monitoring services, and regularly conduct comprehensive evaluations and inspections. Third, establish the evaluation theory research mechanism. Through systematic theoretical and empirical research, improve the institutional system of the scientific development evaluation mechanism. Fourth, establish an education publicity evaluation mechanism. Strengthen the education and publicity so that the theories, systems, and various assessment mechanisms of the physical fitness monitoring services won the support and understanding of students, at same time create a campus public opinion environment. Fifth, establish an organization evaluation mechanism. Strengthen the construction of the organization team and management of the evaluation, and clarify the powers or responsibilities of the personnel and related departments. Finally, establish leadership assessment, job performance assessment and student satisfaction assessment to promote assessment diversity, gradually improve the shared evaluation system.
4) Establish the shared reward mechanism:
   First, establish a performance-oriented incentive mechanism, so that universities can reward students with outstanding performance based on the physical condition of the students between the 7 universities. Second, establish a human security incentive mechanism. A special reward fund for physical test personnel can be established to reward advanced individuals and teams that have made outstanding contributions to promoting the sharing of physical fitness monitoring services. Third, establish an internal regulatory incentive mechanism. Strengthen the supervision and form a mechanism that integrates supervision, management and incentives. Fourth, establish a whole-process incentive mechanism. Such as salary and promotion incentives, outstanding talents incentives, transformation of scientific research achievements incentives, evaluation incentives.

References

1. Jarrad, T., Dry, M., Semmler, C., Turnbull, D., Chur-Hansen, A.: The psychological distress and physical health of Australian psychology honours students. Aust. Psychol. **54**(4), 302–310 (2019)
2. Pang, B., Varea, V., Cavallin, S., Cupac, A.: Experiencing risk, surveillance, and prosumption: health and physical education students' perceptions of digitised health and physical activity data. Sport Educ. Soc. **24**(8), 801–813 (2019)
3. MahmoodyVanolya, N., Jelokhani-Niaraki, M., Toomanian, A.: Validation of spatial multicriteria decision analysis results using public participation GIS. Appl. Geogr. **112**(11), 23–31 (2019)

4. Marie, B.T., McFarland, A., Miller, J.W., Lowe, V., Hatcher, S.S.: Physical and mental health experiences among African American college students. Soc. Work Public Heal. **34** (02), 145–157 (2019)
5. Xue, B., Liu, T.: Physical health data mining of college students based on DRF algorithm. Wirel. Pers. Commun. **102**(4), 3791–3801 (2018)
6. Paraskevis, N., Roumpos, C., Stathopoulos, N., Adam, A.: Spatial analysis and evaluation of a coal deposit by coupling GIS techniques. Int. J. Min. Sci. Technol. **29**(6), 943–953 (2019)
7. Liu, F.-F.: Computer multimedia technology applied in physical education teaching with intuitionistic fuzzy information. Int. J. Digit. Content Technol. Appl. **6**(23), 119–125 (2017)
8. Fouladgar, M.M., Yazdani, A., Zavadskas, E.K.: An integrated model for prioritizing strategies of the Iranian mining sector. Technol. Econ. Dev. Econ. **17**(3), 459–483 (2011)
9. Qizhen, W.: Automatic control system regulation scheme of textile air conditioning. Sci. Technol. Inf. **22**, 395–396 (2010)
10. Xiaoyun, M.: Research on the construction of urban stadium construction GIS platform. Comput. Knowl. Technol. **15**(28), 80–82 (2019)

Probability Programming and Control of Moving Agent Based on MC-POMDP

Yongyong Zhao(✉) and Jinghua Wang

College of Mechanical and Electric Engineering, Changchun University of Science and Technology, Changchun 130022, China
znkz610zyy@163.com

Abstract. A design scheme for probabilistic planning and decision-making of mobile agents is proposed, which realizes the functions of probabilistic planning and grab control of mobile agents. How to use MC-POMDP algorithm to perform sensing and control in an unknown environment is described in detail. Combined with the particle filter algorithm to approximate the confidence state space, the existing POMDP technology was improved, and the probability planning was optimized. The actual operation results show the feasibility and effectiveness of the scheme. The system uses its own sensors to sense the environmental information, performs dynamic probabilistic path planning, successfully approaches the target object, and implements the mobile grabbing function of the mobile agent. Based on the example application of reinforcement learning principles, a new direction for the probability planner to deal with the generalized uncertainty of mobile agents is prospected.

Keywords: Mobile agent · Markov process · Probabilistic programming

1 Introduction

One of the main functional requirements of a mobile agent is to collect, sense and analyze the external environment through its own sensors, so as to achieve the purpose of autonomously moving towards the target, and complete some specific grabbing tasks under obstacle conditions. In some areas, intelligent mobile robots based on deep reinforcement learning [1] have been promoted and applied, such as autonomous driving decisions, star and moon detection, and military mobile robots in the field. These have important academic value and application significance in the fields of production, scientific research and exploration. In view of these situations, this paper focuses on the probabilistic planning and control decision of mobile agents based on Monte Carlo's approximate partial Markov process (MC-POMDP) [2], and introduces our research work and results in this area.

J. MacIntyre et al. (Eds.): SPIoT 2020, AISC 1282, pp. 779–784, 2021.
https://doi.org/10.1007/978-3-030-62743-0_111

2 System Structure Framework

In order to ensure the stability and scalability of the system, the modular design is adopted. The probability planning and decision control system of mobile agents based on MC-POMDP designed in this paper is divided into three modules: information perception module, behavior selection module and motion control module, as shown in Fig. 1.

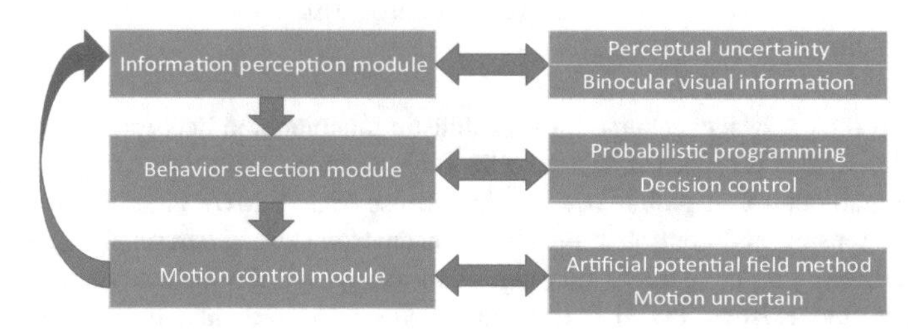

Fig. 1. Block diagram of probabilistic planning and control system for mobile agents based on MC-POMDP

For the perceived uncertainty and behavioral uncertainty of mobile agents, this solution performs probabilistic planning and control of agent motion capture based on the probabilistic algorithm's estimation of sensor data. After the motion control module detects the probability change of the environment, it sends all state information such as pose, translation, rotation, and shape to the behavior information perception layer. The path probability planning module uses the transmitted information to construct a two-dimensional map, and generates obstacle avoidance paths for obstacle avoidance control. When approaching the target, it can adjust the posture of the agent's arm to control its successful grasping of the target.

3 Partial Observable Probability Planning for General POMDP

In terms of the ability to design optimal control strategies, the main advantage of probabilistic planning and control over traditional deterministic omniscient methods lies in the practical application of perceptual uncertainty and behavioral uncertainty. Two types of indeterminacy of the value iteration algorithm [3] for used to believe said, its underlying framework is part can be observed markov decision process (Partially observables MDP, POMDP). POMDP algorithm can predict uncertainty, actively collect information, and pursue the best measurement and control of any performance index.

For continuous cases, the POMDP approximation is used to deal with the dimensional disaster problem of iterative calculation of MDP values. POMDP computes the general core equation of the value iteration over the confidence space:

$$V_T(b) = \gamma \max_u [r(b, u) + \int V_{T-1}(b')p(b'|u, b)\mathrm{d}\,b'] \tag{1}$$

Considering that the confidence space is more complicated to calculate than in the state space, it is not clear whether the integration can be performed correctly. Therefore, we discuss the key factors of POMDP, the update and conversion of conditional probability:

$$p(b'|u, b) = \int p(b'|u, b, z)p(z|u, b)\mathrm{d}z \tag{2}$$

Then, the internal integral of formula (1):

$$\int V_{T-1}(b')p(b'|u, b, z)\mathrm{d}b' \tag{3}$$

Equation (3) is derived by normalizing through Bayesian filtering [4]. Equation (1) can be transformed into the following two equations:

$$\begin{aligned} V_T(b, u) &= \gamma[r(b, u) + \int V_{T-1}(B(b, u, z))p(z|u, b)\mathrm{d}z] \\ V_T(b) &= \max_u \ V_T(b, u) \end{aligned} \tag{4}$$

The above formula is more convenient than the formula (1). It is not integrated over all possible confidence distributions b', reducing a large number of integration operations. In the formula, as the value function of the period T on the confidence b, it is assumed that the next action is u. POMDP is characterized by a variety of uncertainties: uncertainty of control effects, perceived uncertainty, and environmental uncertainty. The POMDP value iterative algorithm defines the basic update mechanism and becomes the basis of the efficient decision algorithm MC-POMDP.

4 POMDP Improved Observable Probability Planning and Control

Although POMDP is highly adaptable in terms of sensor limitations, its calculations are overly complex and impractical when used for accurate planning. The improved approximate POMDP algorithm using particle approximation confidence can represent value functions in any state space. By dynamically constructing the confidence point set, MC-POMDP can maintain a relatively small confidence set without the limitation of a limited state space.

The MC-POMDP algorithm relies on some approximations: the use of a set of particles constitutes an approximation; the local learning algorithm is used to approximate V; the Monte Carlo approximation function is a backup step. These approximate reasons make MC-POMDP greatly improve the simplicity of the algorithm, save computing costs, helping solve the problem of actual mobile agent grabbing.

The easy-to-implement particle filter sample set represents a discrete approximation of continuous confidence. Due to the random nature of particles, the probability of any particle set being observed twice is zero. Therefore, we need an update that can represent V with a certain particle set. Here, the Neighborhood component analysis (NCA) learning algorithm [5] is used to perform the NCA distance measure learning on the original data set to complete the dimensionality reduction and reduce the data calculation and storage burden. Then, when interpolation is performed between two different confidences, a k-nearest neighbor algorithm (KNN, k-NearestNeighbor) is used to perform local weighted interpolation.

A backup of the Monte Carlo approximation function. Since each nested integral corresponds directly to a Monte Carlo sampling step [6], Monte Carlo sampling is used to approximate the following integral equation. The measurement probability is broken down as follows:

$$p(z|u,b) = \iint p(z|x')p(x'|u,x)b(x)\mathrm{d}x\mathrm{d}x' \tag{5}$$

Therefore, Eq. (5) can be rewritten as:

$$\begin{aligned} V_T(b) &= \gamma \max_u \left[\int r(x,u)b(x)dx + \int V_{T-1}(B(b,u,z))[\iint p(z|x')p(x'|u,x)b(x)\mathrm{d}x\mathrm{d}x']\mathrm{d}z \right] \\ &= \gamma \max_u \iiint [r(x,u) + V_{T-1}(B(b,u,z))p(z|x')p(x'|u,x)]b(x)\mathrm{d}x\mathrm{d}x'\mathrm{d}z \end{aligned} \tag{6}$$

The Monte Carlo approximation of the integration is a multivariate sampling algorithm that samples $x \sim b(x), x' \sim p(x'|u,x)$ and $z \sim p(z|x')$. We calculate $B(b,u,z)$ by Bayesian filtering. Then, use the local learning algorithm to calculate $V_{T-1}(B(b,u,z))$ and look up the table to get $r(x,u)$.

5 Experimental Results and Analysis

MC-POMDP algorithm implementation results are shown in Fig. 2. The robot is placed somewhere outside its sensing range, but near the object that can be grasped. Here a three-phase strategy is proposed: the search phase, the robot rotates until it perceives the object; the motion phase, the robot walks near the center of the object to grab it; the final step is the final grasping action. The experimental results of mobile agent search, motion and grabbing are as follows:

In the simulation diagram, the success rate is a function of the number of planning steps. The results show that the mobile agent has completed the initial requirements of the system. The MC-POMDP algorithm is used to plan the optimal or suboptimal path

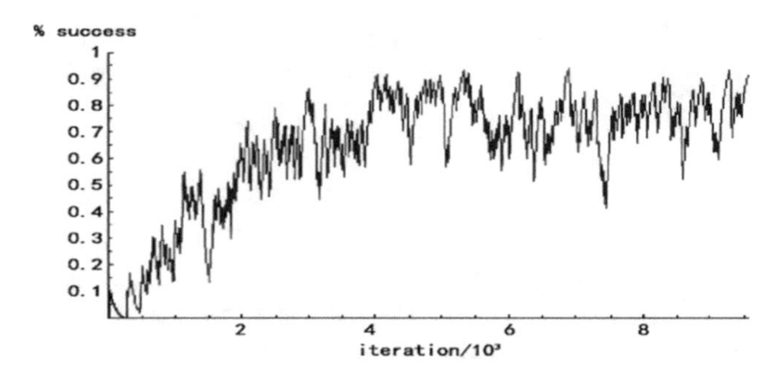

Fig. 2. Experimental success rate of mobile agent search and capture tasks

from the starting point to the target point, and to locate itself during the movement process, and finally successfully capture the target.

6 Outlook

Inspired by the ideas from active perception to deep reinforcement learning [7] and reinforcement learning in artificial biological systems [8], deep reinforcement learning is used for probability measurement and control research, so that agents have the same capabilities as human autonomous detection and decision-making schemes.

First, the mobile robot can infer the probability distribution information of the entire implementation effect space through various assumptions based on the current conditions, and then obtain the ambiguity and confidence of the goal. Then, the probabilistic robot retrieves relevant data and historical experience, and feedback corrects the ambiguity and confidence of the effect of the hypothetical pathway through deep reinforcement transfer learning. Finally, in the process of achieving the goal, the mobile robot can verify before and after, and iterate the real-time approach to quickly reach the expected goal.

7 Conclusion

Using zoom measurement technology to measure depth information can effectively expand the system's measurement range. The improved Monte Carlo sampling algorithm is used to replace the unknown integral solution, which improves the system speed. The neighborhood component analysis (NCA) algorithm is used to effectively realize the confidence state space learning of particle swarm. By using the improved artificial potential function, the local minimum point phenomenon in the artificial potential field method is eliminated. Experiments show that the designed and implemented path probability planning and grabbing control system has good performance.

Acknowledgments. Project supported by the Funds for the "13th Five-Year Plan" for scientific and technological research projects of the Education Department of Jilin Province, China (Grant No. JJKH20181139KJ).

References

1. Zhao, L., Wang, J., Liu, J., et al.: Routing for crowd management in smart cities: a deep reinforcement learning perspective. IEEE Commun. Mag. **57**(4), 88–93 (2019)
2. Wang, C., Ju, P., Lei, S., et al.: Markov decision process-based resilience enhancement for distribution systems: an approximate dynamic programming approach. IEEE Trans. Smart Grid **PP**(99), 1 (2019)
3. Heydari, A.: Stability analysis of optimal adaptive control under value iteration using a stabilizing initial policy. IEEE Trans. Neural Netw. Learn. Syst. **29**(9), 4522–4527 (2018)
4. López-Araquistain, J., Jarama, Á.J., Besada, J.A., et al.: A new approach to map-assisted Bayesian tracking filtering. Inf. Fusion **45**, 79–95 (2018)
5. Wang, D., Tan, X.: Bayesian neighborhood component analysis. IEEE Trans. Neural Netw. Learn. Syst. **29**(7), 3140–3151 (2017)
6. Chen, H.N., Mao, Z.L.: Study on the failure probability of occupant evacuation with the method of Monte Carlo sampling. Procedia Eng. **211**, 55–62 (2018)
7. Kragic, D.: From active perception to deep learning. Sci. Robot. **3**(23), eaav1778 (2018)
8. Neftci, E.O., Averbeck, B.B.: Reinforcement learning in artificial and biological systems. Nat. Mach. Intell. **1**, 133–143 (2019)

Investigation on Power Marketing Model Based on Hierarchical Management System in a Smart City

Liangfeng Jin[1(✉)], Ruoyun Hu[1], Weihao Qiu[2], Shining Lv[1], and Haina Chen[3]

[1] Power Science Research Institute of State Grid Zhejiang Electric Power Co., Ltd., Hangzhou 31200, China
nnn88nnn@126.com
[2] State Grid Zhejiang Electric Power Co., Ltd., Hangzhou 31200, China
[3] Hangzhou Power Supply Co., Ltd., Hangzhou 31200, China

Abstract. In recent years, with the rapid development of market economy and global economy, the development of China's power marketization process is accelerating, power marketing in the process of power marketization has undergone great changes. This article first to the layered management system related to the concept of narrative, then combining clustering analysis algorithm and layered management is introduced into the electric power marketing, and then on the basis of electricity marketing theory of marketization of electric power under the background of our country electric power marketing model, carried on the thorough analysis of the current situation of the layered management system theory as the guiding ideology, together with the present situation of the development of Chinese electric power marketing model to build the corresponding marketing mode of electric power. The experiment of this paper shows that the power marketing mode under the layered management system has reversed the rigid management situation caused by the traditional "one-size-fits-all" management and promoted the further development of the power industry.

Keywords: Hierarchical management · Electricity marketization · Marketing model · Clustering analysis algorithm

1 Introduction

With the continuous acceleration of the global economic marketization trend, China's electric power industry has developed rapidly. At present, China's electric power industry shows the trend of marketization and stratification. Power marketization promotes the complexity of power marketing level and the expansion of marketing scale. Therefore, corresponding marketing mode must be established to help realize the further development of power industry. As far as it goes. The marketing mode of electric power in China has not got rid of the traditional "one-size-fits-all" mode, and cannot adapt to the development trend of electric power marketization, which to some

J. MacIntyre et al. (Eds.): SPIoT 2020, AISC 1282, pp. 785–790, 2021.
https://doi.org/10.1007/978-3-030-62743-0_112

extent causes obstacles to the effective management of electric power industry. Layered management provides a new method and idea for the marketing mode of power marketization, which can realize the effective and scientific management of power industry.

The concept of hierarchical management has been applied in various industries. With the continuous expansion of its application scope, people pay more and more attention to this management style. At present, domestic and foreign scholars have conducted in-depth research on hierarchical management. In [1], the author systematically discusses layered management from the aspects of management mode, management function, task and management effect, which promotes people's comprehensive understanding of layered management. In [2], the author lists three commonly used analysis algorithms in hierarchical management system, including principal component analysis, factor analysis and cluster analysis, and describes the specific concepts and application scenarios of these three methods in detail. In [3], the author constructs the theoretical framework and logical model of hierarchical management, providing a more scientific computing method for hierarchical management.

Because electric power marketing occupies an important position in economic and social construction, Chinese scholars have also paid more attention to it and conducted a series of researches on electric power marketing. In [4], the author explains the necessity of studying the power marketing model, and explains the current situation and main problems of China's power marketing. On this basis, the author makes an in-depth discussion on the improvement of the power marketing model. In [5], the author establishes the power marketing model from the perspective of power supply enterprises and promotes the management innovation of power enterprises. In [6], the author applies the clustering analysis algorithm to power marketing and realizes the partition and analysis of the subsystem of power marketing customers.

To achieve effective management of electric power industry, promote the transformation and innovation of marketing mode of electric power, this article first to the layered management system related to the concept of narrative, then combining clustering analysis algorithm and layered management is introduced into the electric power marketing, and then on the basis of electricity marketing theory to the marketization of electric power under the background of our country electric power marketing model, carried on the thorough analysis of the current situation of the layered management system theory as the guiding ideology, together with the present situation of the development of Chinese electric power marketing model to build the corresponding electric power marketing model [7, 8]. The research in this paper not only promotes the innovation of China's electric power marketing model and the rapid development of the electric power industry, but also provides certain reference significance for future relevant researches [9, 10].

2 Method

To explore the power marketing model under the layered management system, it is necessary to master the power marketing data at different levels, so as to realize the effective mining of data information. Due to the current market-oriented development of power, power marketing data is extremely large and involves various types of data. Clustering analysis algorithm can solve this problem well. This algorithm is usually used to study the classification problem of various factors. According to different attributes of data, it can establish the relationship between different data according to relevant indicators, such as similarity and difference indicators, with the help of mathematical algorithm, so as to carry out relevant clustering analysis. It can be seen that there is a consistency between the basic principle of cluster analysis and hierarchical management, so this method can be applied to the study of power marketing mode under hierarchical management system. The clustering analysis algorithm mainly realizes the effective recognition of the hierarchy of management objects by means of data mining. With the help of this algorithm, it can also observe the characteristics of management objects effectively, and then analyze the corresponding hierarchy of management objects more intensively. The common formula of clustering analysis algorithm is as follows:

$$Si = SP(P) - SP - Pi\,(P) \tag{1}$$

Where, SP(P) represents the importance determination function of cluster analysis, and Pi represents the set of attribute sets. When SP(P) = 1, the specific calculation formula of sp-pi (P) is as follows:

$$SP - Pi(P) = \frac{cardPOSSP(P)}{cardU^{Si}} \tag{2}$$

In the above formula, the Si value is proportional to the importance of the Pi attribute set.

3 Necessity Analysis of Studying Power Marketing Mode

At present, in order to improve the international competitiveness of our country must strive to promote the development of the national economy, in the process of power marketization, the development of the power industry and the development of the national economy are closely related, to promote the overall development of the power industry must continue to improve the power marketing model. Since the power marketing mode involves a wide range of aspects, such as the development direction, methods and objectives of the power industry, and related to the final development effect of the power industry, it requires that the power marketing mode has a certain degree of professionalism and standardization. Under the background of the new era, the traditional "one-size-fits-all" electric power marketing mode carries out unified measurement on all aspects of electric power development, and fails to consider the hierarchy and particularity of industrial development, thus failing to meet the needs of the rapid development of electric power marketization. Therefore, continuously promoting the transformation of power mode and realizing the layered management of power marketing mode can improve work efficiency to the maximum extent, reduce the marketing mistakes caused by unclear hierarchical management, improve the utilization rate of power resources, and finally realize the all-round rapid development of the power industry.

4 Discuss

4.1 Construction Strategy of Power Marketing Model Under Hierarchical Management System

(1) Hierarchical analysis and management of power marketing
With the rapid development of power marketization, power marketing involves different subjects. Therefore, in order to construct the power marketing model of hierarchical management, hierarchical analysis and management must be realized first. Through consulting relevant materials, it is found that the stratification analysis of power marketing includes the following aspects. The specific management stratification is shown in Table 1 below. The data in the table are the results of the author's investigation and arrangement.

Table 1. Analysis and management of power marketing

Serial number	Analysis management module	Proportion
1	Market analysis	10.1%
2	Customer analysis	11.8%
3	Electricity price analysis	14.1%
4	Analysis of sales quantity	13.2%
5	Electricity bill recovery analysis	11.2%
6	Marketing strategy analysis	15.6%
7	Demand side management analysis	13.1%

*Data were collected from the survey results

Data from Table 1 can see, electricity management involves wider marketing analysis, mainly related to the market analysis, customer analysis, analysis, marketing power, electricity, recycling, marketing strategy analysis and the analysis of the demand side management in seven aspects, including the analysis of the marketing strategy, including the formation of marketing management, marketing activities, the evaluation of part of the marketing effect. The analysis and management of each module is targeted at different levels of power marketing, from power itself to power market and its objects, and the seven levels are organically linked through analysis and management, which is conducive to the establishment of a hierarchical management system in the power industry and the orderly operation of the system.

4.2 Application Analysis of Power Marketing Model Under Hierarchical Management System

In order to test the scientificity and feasibility of the electric power marketing model proposed in this paper, it must be tested before it is put into use. In this paper, the matable data analysis system is used to carry out the simulation and test experiment of the power marketing mode under the layered management system. The final test results are shown in Fig. 1. The data in the figure is the result of the author's experimental arrangement.

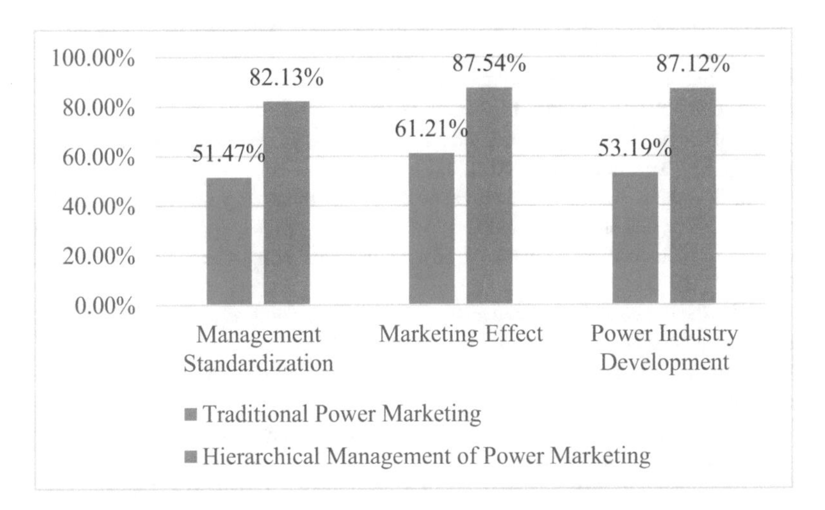

Fig. 1. Hierarchical management of power marketing compared with traditional marketing

From the data in Fig. 1, it can be seen that compared with the traditional power marketing model, the hierarchical management of power marketing far exceeds the traditional power marketing model in terms of management standards and marketing effects, and contributes 34% more than the traditional power marketing model in the development of the power industry, effectively promoting the overall development of the power industry. Therefore, the layered management power marketing model proposed in this paper is scientific and feasible.

5 Conclusion

Through the research of this paper, the following conclusions are drawn: first, under the background of the development of power marketization, the traditional power marketing model has been unable to adapt to the development of the power industry; Second, the implementation of hierarchical management of power marketing model, a comprehensive consideration of the customer, market, demand and marketing strategies, such as different aspects, can promote the comprehensive development of the power industry. The power marketing mode is closely related to the development of the power industry, and the implementation of layered management of the power marketing mode ADAPTS to the development needs of power marketization, which is conducive to the future sound development of the power industry.

References

1. Zhang, J., Zhang, W., Guo, W.: Main problems and countermeasures of power marketing in developing power market. Power Demand Forecast Manage. **13**(5), 70–71 (2017)
2. Yang, R.: Thinking on fine management of power marketing. Inner Mongolia Sci. Technol. Econ. **13**(7), 70–71 (2018)
3. Pan, S., Liang, J.: Current situation and prospective analysis of power marketing management. Mod. Trade Ind. **11**(36), 47 (2018)
4. Li, C.: Strategic innovation and technical analysis of power marketing management. Shandong Ind. Technol. **11**(5), 216 (2017)
5. Yan, L., Song, X., Xiong, W.: Innovation of power marketing management based on big data. Sci. Manage. **13**(2), 186 (2017)
6. Chen, P.: Implementation and innovation of power marketing management. Technol. Mark. **25**(1), 187 (2018)
7. Cheng, J.: Status quo and management measures of marketing strategy management of Tunli enterprises. China Sci. Technol. **13**(22), 56–66 (2017)
8. Liu, H.: Problems in power marketing management and countermeasure. Manage. Sci. **14**(4), 91–92 (2017)
9. Chen, X.: Research on the construction of digital integrated management system for tunli marketin. Sci. Technol. Entrepreneur **11**(5), 87 (2017)
10. Jo, P.: Thinking on marketing management of tunli market. Private Sci. Technol. **11**(12), 249 (2017)

Application Research of Virtual Reality Technology

Lei Li, Yuanxi Wang, and Wei Jiang(✉)

Harbin Finance University, Harbin, China
lilei_112009@163.com, 13980894@qq.com, jw9@qq.com

Abstract. Virtual reality technology makes users immerse in it because of its realistic simulation effect and vivid interactive experience. If this state is applied to the education industry, it will certainly lead to a profound change in the education field. The current application level of virtual reality is relatively narrow. With the continuous development of science and technological progress, we believe that VR technology will be widely used in the future.

Keywords: Virtual Reality Technology · Financial field · Higher education

Virtual Reality Technology is be loved by gamer, especially young people, because of its vivid simulation effect and real experience. Meanwhile, the classroom in university is occupied by many dynamic games in smart phones. Are students don't like the class, or because the teachers' class is so bad? Actually, it is the game too attractive to move their eyes that teachers can't compare with. Can this vivid VR technology be used in our education? Can we pull students' attention back to our classroom by using this kind of new cognitive tools? And can we use it to do what other things or only games? This is what this paper wants to discuss.

1 Introduction of Virtual Reality Technology

1.1 Development History

In 1965, Ivan Sutherland first proposed the basic idea of VR System in a paper titled "The Ultimate Display". From then on, human beings began to explore the research of Virtual Reality System. Lincoln Laboratory of MIT in the United States began the development of helmet-mounted displays officially in 1966. The first fully functional system emerged, in 1970. In the 1980s, Jaron Lnaier put forward the term "Virtual Reality" (VR) officially. At the same time, the NASA and the US Department of Defense organized a series research on this kind of technology, and achieved a remarkable result. The computer hardware technology had developed rapidly in the 1990s, and it made the computer software system continuous improvement too. It makes real-time animation which based on large data and images possible. The design of human-computer interaction system has innovated continuously. Meanwhile, input and output equipment which is novel and practical constantly enter the market. All these make the development of Virtual Reality System have a good technical foundation.

J. MacIntyre et al. (Eds.): SPIoT 2020, AISC 1282, pp. 791–798, 2021.
https://doi.org/10.1007/978-3-030-62743-0_113

1.2 The Meaning of Virtual Reality Technology

Virtual Reality Technology is a new computer simulation systems, based on computer technology, which fuses with traditional display, servo, simulation, database technology and modern sensor technology, interaction, human-computer interface, three-dimensional graphics technology, generating a realistic three-dimensional world, which makes users browse the objects or scenes in a interactive way in the unreal world from their own perspective.

1.3 Application of Virtual Reality Technology Researching

Since the 21st century, Virtual Reality Technology has been supported by both hardware and software very well, due to the rapid development of computing technology. Therefore, the application area of Virtual Reality Technology is more extensive, and also makes the technology become more mature. Although Virtual Reality Technology has only been produced for more than fifty years, its development speed is quite astonishing. It has been widely used in people's daily life, for instance, military, modeling and simulation, visualization of scientific computing, education and training, medicine, art and entertainment, etc.

The research and application of Virtual Reality Technology start earlier in abroad than in China. Research on technology, theory and application has reached a certain height, especially in the military, medical and educational fields, in which the application has begun to mature. NASA and the U.S. Department of Defense are important departments in the virtual reality technology which famous in its research and application. It is mainly used for space simulation and virtual operation exercises, as well as manipulating space robots in outer space to perform tasks that humans can not do. In addition, the Royal Air Force uses Virtual Reality Simulator to simulate the training of British parachutists to achieve the whole process of parachuting. The most famous institution which applied VR technology to medical surgery is the University of North Carolina in the United States who uses simulation technology to simulate the key links of surgery. VR technology has made great contributions to human medicine.

Domestic research on Virtual Reality Technology started in the early 1990s which is late than abroad a lot. Because VR is a research field with large investment scale and high difficulty in technology, it is difficult to get the return in short term. However, due to its tremendous impact on scientific and technological research and application, especially in military, medical, educational and other fields, involving human well-being, our government and relevant departments attach great importance to it. In the Tenth Five-Year Plan, the Eleventh Five-Year Plan, the Twelfth Five-Year Plan, the National 863 Plan, the National Natural Science Foundation and the National Defense Science and Technology Commission, the research has been listed as the key funding areas. And the national "973" plan places VR technology as the top priority. After nearly 30 years of efforts, China's VR technology has made some achievements. For example, in military industry, industrial simulation, medical treatment, emergency rescue, education and training, real estate and home, VR technology highlights its unpredictable technical advantages. Beijing Palace Museum launched the "Forbidden City beyond Time and Space" - Virtual Palace Museum Project in 2008. Using VR technology, the buildings and cultural relics in the Palace Museum are made into three-dimensional images through image acquisition, digital processing and compression technology, integrating high-definition, ultra-wide screen and surround stereo digital audio, so that people can view and appreciate the buildings and cultural relics of the Palace Museum at will from various angles. Moreover, Virtual Reality Technology has been used in digital libraries, large-scale performance lighting and scene customization widely. In addition, a large number of enterprises and institutions have invested in technology development, not only actively carry out applied research, but also make full use of their websites to provide learning and communication platform for the much number of domestic fans and college students (Table 1).

Table 1. Research progress of key universities about VR in China

Research Subject	Big Event	Remarks
National University of Defense Technology	In 1994, the first image-based virtual information space generation platform was successfully developed in China.	One of the earliest research institutes of image-based virtual reality in China.
	In 1999, a collaborative virtual reality system, virtual space conference system, was successfully developed.	
Zhejiang University	The project of "Basic Theory, Algorithms and Implementation of Virtual Reality" in 2003 studies the establishment of virtual environment, natural human-computer interaction, enhanced VR, distributed VR and the application of VR in product innovation.	Approved National Key Basic Research Development Project, 973 Project.

2 Current Application of Virtual Reality Technology

2.1 Virtual Reality Technology in Educational Application

The application of Virtual Reality Technology in education has played a main role in promoting the education level and the sharing of educational resources. Through using Virtual Reality Technology, students can visit the seabed, travel in space, visit historical castles, and even go deep into the atom to observe the trajectory of electrons and experience Einstein's relativistic world, acquire knowledge vividly, stimulate thinking, and greatly shorten the distance from theory in book to practical application [1].

Overseas: For example, cyber Math, a research project in the field of mathematics education, is a shared virtual environment based on avatars. It is suitable for teachers and students to teach and explore when they are separated on the spot or in space. Dr. David Warner of Loma Linda University Medical Center and his team used computer graphics and virtual reality devices to explore neurological problems successfully and pioneered virtual reality pediatric therapies.

Virtual Reality Technology is applied in Higher Education in China mainly concentrates on the following three aspects: the application of VR in the scope of scientific research; in virtual simulation campus; virtual teaching and experiment.

2.2 The Application of VR Technology in Scientific Research

Universities in China have carried out research on VR Technology in many fields, which has a great promoting function in scientific and technological research. Some key institutions of higher learning and universities in China are also actively involved in this research field. At present, many systems have been implemented and are being developed [2] (Table 2).

Table 2. Research progress of key universities about VR in China (Continued)

Research Subject	Big Event	Remarks
Beihang University	In 2007, the State Key Laboratory for New Technologies was approved.	In cooperation with the Armored Forces Engineering Institute of the PLA, the overall design objective is to provide a multi-weapon cooperative or confrontational tactical drilling system for military simulation training and exercises in China.
Tsinghua University	Research VR, presence and face objects in images, including face detection, registration and labeling methods.	It is hoped that computers can process and understand human visual images like human vision.
Wuhan University	In 1989, the State Planning Commission officially approved the establishment of the National Key Laboratory of Surveying, Mapping and Remote Sensing Information Engineering, which is the only National Key Laboratory of Surveying and mapping discipline in China.	It presided over the formulation of the Ministry of Housing and Urban-Rural Construction Standard of the People's Republic of China, Technical Specification for Three-dimensional Modeling of Cities, which was officially released, and developed a number of three-dimensional simulation systems.
Harbin Institute of Technology	With "Face Recognition Theory, Technology, System and Application" won the second prize of the National Science and Technology Progress Award in 2005.	They will develop a further study on the head posture, hand movements, speech and intonation synchronization in human speech.
University of Science and Technology Beijing	Pure Interactive Vehicle Simulated Driving Training System, which was patented by the State in 2004.	The three-dimensional graphics developed by the system are very realistic, and there is little difference between the virtual environment and the real driving environment.

2.3 Virtual Reality Technology in Campus's Application Focus on Virtual Simulation

Virtual campus is the earliest application of Virtual Reality Technology and network education. At present, this kind of usage mainly in virtual campus which is realized browse function. This is its actual function. With the deepening use of network in education, three-dimensional visualization virtual campus based on teaching, administration and campus life functions is coming out. Real, interactive and situational technologies in VR will inevitably lead to a revolution in the way of education.

2.4 Virtual Teaching and Experiments

In practical application, virtual teaching and experimental application are early in science and engineering, especially in the fields of architecture, machinery making, physics, chemistry, computer and other disciplines, has substantial breakthroughs. Using the virtual experiment system, students make all kinds of difficult experiment can be true, feel the same experience like real experiments, enhance intuitive knowledge, make the content of teaching understand deeply, avoid all kinds of dangers brought by real experiments or operations, and break the limitation of space and time. This is the most widely used of Virtual Reality Technology in education, And teaching will become more vividly under this current technological conditions.

In fact, VR technology can do more than this in education. It can help a lot of children who are not health in ability to enjoy in a virtual world and learn much knowledge which they can't in real world. And importantly, they can don't mind other people's attitudes.

2.5 VR Technology in Travel

We can use VR Technology to enjoy different places and cultures without going to there and many complicated travels. For example, we don't need to go to Niagara Falls, but we can feel the spray from the waterfall and look at the majestic scenery.

This maybe make aircraft company closed, but we can help a lot of people who can't travel easily. And make some new industries like travel promotion services.

In fact, we can do a lot of things in using VR. No matter in which fields, technology still can make working efficiency. And more importantly, technology should make our lives more easily and happier.

2.6 VR Technology in Medical Care

In this world, our medical resources are not very equilibrium. Especially, in some less-developed regions, many people can't get effective treatment when they have ills. But VR technology can help them. For instance, doctors can treat patients in a remote hospital using VR. And we also can use medical equipment to transfer our information to doctors or hospital. They can get this import messages to help us.

3 Expansion of Virtual Reality Technology Application

3.1 Theoretical Basis

The concept of modern higher education advocates student-centered, while the new cognitive tools represented by VR help learners to achieve a profound understanding of knowledge, so that students become the main body of information processing, active learners of knowledge, and teachers become helpers, promoters and guiders of students' active learning [3]. From the three perspectives of participation, generation and controllability, the new cognitive tools should show positive, creative and active learning ideas [4]. Virtual reality technology meets the basic needs of modern higher education for new cognitive tools. The learning environment created by virtual reality technology is free and autonomous, and the learning process is explored by learners [5, 6]. The learning content needs to be created by using various tools and information resources.

3.2 Practical Application Value

First, it has an intuitive teaching form. Using this kind of technology to make a learning environment that cannot be easily created under normal teaching conditions, learners can observe and learn in a more intuitive way, breaking the limitations of time and space in traditional teaching. Virtual reality technology creates a realistic virtual learning environment, makes know-how learning intuitive and vivid, and realizes visualized teaching under "real situation".

Second, strengthen the entertainment of learning. Game teaching theory holds that games can arouse learners' motivation, guide learners to actively solve difficult problems, and try to think from different perspectives by playing different roles and identities in games. Virtual reality technology is the best learning tool to achieve this goal at present. The learning environment constructed by virtual reality technology often contains game elements, so as to enjoy the pleasure of learning.

Third, taste the real environment. Virtual reality technology can give people a real feeling. In the process of learning, learners can explore freely and learn independently, providing learners with a sense of real environment experience and various tools and information resources [7].

Fourth, save educational resources and make up for the shortage of teaching conditions. The virtual learning environment constructed by virtual reality technology can be repeatedly used in the same field, which will not lead to the loss of education resources and save social energy. At the same time, it makes up for the inadequacy of teaching conditions in some economic backward areas.

In addition, virtual reality technology can highlight its huge technological advantages in training students' thinking and exploring abilities, teaching students in how to accordance with their aptitude and so on.

3.3 Financial Field Application

Most of the existing applications are concentrated in engineering courses, but less in liberal arts courses.

First of all, virtual reality technology also has some applications in financial courses. Students enter the virtual financial environment in the role of Central Bank, President of Commercial Bank, Bank Clerks, enterprises, consumers, Third Party Institutions, CA Certification Center, etc., deeply understand the process of currency circulation in the financial system. Students use the virtual experiment system to gain the same role experience as the real life, avoiding some problems that the closeness, confidentiality and high cost of in financial system.

Secondly, virtual reality technology is also very useful in simulating stock exchange. Because of the financial sensitivity of the securities industry, students can not personally practice the process of stock speculation, nor can they know the operating mechanism behind the stock market. Virtual reality technology can make this come true.

Lastly, in some law-related courses, students play the roles of judges, prosecutors, lawyers and so on. They act themselves as the real parties, participants, judges, prosecutors, and strive for the best results of the case judgment. Virtual reality technology can promote the authenticity of case trial in this process. And physical courses in university, like skiing, diving, mountaineering, equestrian, parachute jumping etc. can also use virtual reality technology to make students experiment different sports.

In addition, under the level of existing virtual reality technology, further exploit the available space, such as in the teaching and experiment process of various courses in universities, which deep-seated content can also get more teaching output through the upgrading of technology. And through virtual reality technology help university to solve the problem of high cost, security and other issues in practice process (Fig. 1).

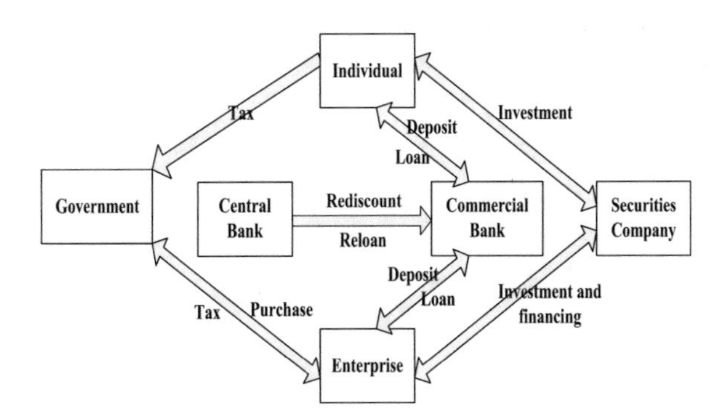

Fig. 1. Through virtual reality technology help university to solve the problem of high cost, security and other issues in practice process

4 Summary

In the ecological environment of higher education, a large number of scientific research forces have been gathered and have most of the young people in society who are easy to accept new things. Universities should be the leader in the application of new cognitive tools. Therefore, the usage of cognitive tools which is new in the field of education should start with universities and be widely applied. The application of virtual reality technology in the field of Liberal Arts in universities will broaden the scope of application, and open up a larger market where virtual reality technology takes root. Therefore, in-depth exploration in this field will bring considerable social benefits to society, as well as new ideas for teaching reform in universities. Through the use of new cognitive tools, universities have made progress in teaching methods, updated teaching content, more flexible and convenient assessment methods, ultimately helping universities to improve the existing teaching efficiency and provide more senior talents to adapt to social development.

References

1. Austin, Texas. 2017 NMC Technology Outlook for Chinese Higher Education: A Horizon Project Regional Report, The New Media Consortium (2017)
2. Ministry of Education of the People's Republic of China. http://www.moe.gov.cn/s78/A08/
3. Csikszentmihalyi, M., Csikszentmihalyi, M.: Flow: the psychology of optimal experience. Des. Issues **8**(1), 75–77 (1991)
4. Shaoping, Z.: Research on Influencing Factors of community user acceptance behavior of cloud logistics information platform based on immersion theory. Anhui University (2017)
5. Brilliant, Z.H.: Application of immersion theory in online learning. Softw. Guide **12**(09), 185–187 (2013)
6. Yu, Z.: Design and development of research-based learning based on virtual reality technology. Central China Normal University (2009)
7. Li, J., Zhao, Y., Huang, L., Yang, Q.: Construction and exploration of innovation and entrepreneurship platform for urban underground space engineering students based on DCLOUD technology. Int. J. Front. Eng. Technol. **1**(1), 48–55 (2019)

Online Booking Laboratory Management System

Jian Huang(✉)

Xijing University, Xi'an 710123, China
565200245@qq.com

Abstract. With the progress of science and technology and the times, the teaching level is constantly innovating. Colleges and universities are also further improving the practical application ability of students. Laboratory construction is an essential link to improve the practical ability of students. This design is mainly based on JSP technology for business logic development, using MySQL database for data storage and call, using hypertext markup language (HTML) to build a web page interface, achieving five management functions, namely system management, appointment record management, teacher and student management, laboratory management and appointment management. The object is teachers and students in Colleges and universities. The traditional laboratory managed by hand is replaced by online reservation laboratory. The test shows that the design of the reservation laboratory system has basically reached the expected goal, and will further expand the system function and improve the browser compatibility in the later stage.

Keywords: Demand analysis · Online booking · Open experiment

1 Requirement Analysis

1.1 Background Analysis

At present, in order to improve the utilization rate of laboratories in Colleges and universities at all levels, enhance the practical ability of students, make the theoretical knowledge fully practiced, improve the teaching quality, and cultivate students' ability of scientific and technological innovation. It is necessary to establish a more convenient, open and multi-functional laboratory appointment system.

The goal of the system design is to improve the utilization rate of the college laboratory, enhance the students' practical ability, deepen the understanding of the theoretical knowledge of the classroom, at the same time, let our school and even more colleges and universities go further in the process of information construction, so as to make appointment move towards information, standardization and humanization [1–3].

Before the opening of this project, we have consulted the current situation of laboratory utilization in China, and combined with the actual situation of the school, we need to meet the following requirements for the design and implementation of this project, the laboratory appointment management system:

J. MacIntyre et al. (Eds.): SPIoT 2020, AISC 1282, pp. 799–804, 2021.
https://doi.org/10.1007/978-3-030-62743-0_114

The designed system should be more orderly, more reasonable, human-oriented and information-based in the laboratory management, make full use of the laboratory, protect the equipment in the laboratory more safely, and reduce the excessive damage to the laboratory equipment.

Through network appointment, teachers and students can communicate more easily, improve teachers' work efficiency and quality, and reduce teachers' work pressure.

Let students use the laboratory reasonably, improve students' practical ability and theoretical understanding ability. During the period of school, students can improve their knowledge reserve and basic practical ability. At the same time, let students use the laboratory more humanized. Through the functions of the reservation laboratory, we can realize the transformation from manual management to information technology, improve the speed and efficiency of reservation processing, increase the interest of students, and thus change passive learning into active learning.

1.2 Feasibility Analysis

1.2.1 Technical Feasibility

Feasibility analysis requires a comprehensive and systematic analysis as the main method, with economic benefits as the core, around the various factors affecting the project, using the data obtained from demand analysis, the feasibility of the proposed project. At the same time, it puts forward the comprehensive analysis and evaluation of the whole feasibility, points out the advantages and disadvantages and suggestions for improvement.

MVC (model view controller) is the design mode of this project. MVC (model view controller) is mainly to separate the user interface and business logic of the software, so that the code can achieve scalability, reusability and maintainability, and enhance the flexibility of the system. Its design mode is relatively clear, and it is more convenient in application, which improves the feasibility of the system [4, 5].

Business logic design uses JSP (Java Server Pages) language. It is a dynamic web page development technology based on the servlet specification. In JSP files, HTML code and Java code exist together. HTML code is used to display static content in web pages, and Java code is used to display dynamic content in web pages. The main purpose is to improve the relationship between data and database, and at the same time, using browser to display the preset effect perfectly, which has high feasibility.

Finally, MySQL is a database management system, which is a data warehouse organized, stored and managed according to the database structure. Using database system to store and manage a large number of data. MySQL is a database based on the relational model. It processes the data in the database with the help of mathematical concepts and methods. The operation mode is relatively simple and easy to develop, and the code used is relatively simple and easy to understand when calling the database, which is one of the completely feasible technologies.

1.2.2 Operational Feasibility

The lab reservation management system is developed based on Web technology, database technology and related languages. In use, as long as you log in the corresponding account and password, you can have the corresponding functions. For example: after the teacher logs in, he/she can make an appointment for the laboratory. In this process, he/she can use and master the functions of the system without having professional computer knowledge. The operation process is relatively simple and will not damage the data. Different roles have different permissions to ensure the security of the system. The interface is in the form of a web page, which is the same as the way of accessing information on the Internet in peacetime. It does not need to install additional software, so it is very convenient to operate.

1.2.3 Economic Feasibility

If the system is put into colleges and universities, it will greatly improve the mode of laboratory appointment and the waste of laboratory resources, improve the utilization rate of the laboratory, reduce the management of personnel, and only need 1–2 people to complete. In addition, in the process of research and development of the system, considering the funds, combined with the relevant knowledge of software engineering and the learning of online materials, the labor input cost in the system is about 5000–10000 yuan. At the same time, the server required by the system does not need to be newly built, but can be deployed to the server of the University, saving the cost investment. In later maintenance, only regular maintenance is needed, so it is also economically feasible.

1.3 System Functional Requirements

1.3.1 System Management

System management mainly realizes the functions of user information management, system administrator management and data backup operation management.

The information management of users mainly refers to the management of teachers and students on their own users and passwords, the management of reservation laboratories, the management of administrators on part of teachers and students' information, the management of students on their own users and passwords, and the processing operations on the reservation of laboratories (which can be reserved or cancelled).

The system administrator mainly manages the registration operation of user identity, including the login name and initial password of user name, information entry and modification of user role, and deletion of students and teachers who are not in school. At the same time, it will make statistics on the quality of the equipment [6–8].

1.3.2 Appointment Record Management

The main function of appointment record management is to query and modify your own appointment, mainly including data query, modification and deletion.

1.3.3 Management of Teachers and Students

The management of teachers and students is mainly the management of students and laboratories, as well as the maintenance of personal information of teachers and students.

Through the two-way management of administrators and teachers and students, it can make communication more convenient, at the same time reduce the work pressure, and improve the security and maintainability of the system.

1.3.4 Laboratory Management

According to the existing distribution of laboratories, we mainly complete the effective management of each laboratory in Colleges and universities. Management includes: entry, modification, query and deletion of laboratory information.

1.3.5 System Design

System design usually includes structure design, module design, data design and user interface design. The main task is to use the appropriate technology and method to design the structure of the system. System design is the technology and method of system structure design, which defines the structure realization process of new system.

The system design is based on the analysis of software requirements, and the results of structural analysis provide the most basic input information for it. The tasks and contents of the overall design mainly include the following:

1) The overall structure of the software and the external design of the modules.
2) Software processing flow design, determine the relationship between subsystems, between modules to transfer data and call.
3) Determine the functions of the software and assign them.
4) Data structure design, determine logical structure design, physical structure design, data structure design and program Relationship.
5) Interface design, interface design among user interface, software and hardware interface and module.
6) Operation design mainly refers to the combination of operation modules and operation time.
7) Error handling design, including error output information, error handling countermeasures.

In the system design stage, the top-down gradual refinement method is adopted. The principle and process of system design: the principle of gradual refinement from macro to micro, the combination of qualitative and quantitative analysis, the combination of decomposition and coordination, and the modeling method, and the generality, relevance, integrity and hierarchy of the system should be taken into account. Decompose the system according to the requirements of the overall structure, function, task and objective of the system, make each subsystem coordinate with each other, and realize the overall optimization of the system [9, 10].

2 Test

The test is mainly carried out in black box test mode, mainly for the evaluation of various functions of the system, including performance management, appointment management and reliability of the system. During the test, in the operating system environment, the performance test and compatibility test are carried out through different browsers, and at the same time, the laboratory appointment is made for the system in different browsers. At this time, the load of the system is large, and the corresponding data is relatively large, which can test the real performance status and appointment management of the system, and the system can be determined during the test whether the reliability meets the requirements.

Software testing is a key component of software quality assurance. The understanding of software testing can be divided into the following stages: testing is debugging stage, testing is to prove the software is correct stage, testing is to find the error stage in software, testing is to reduce the risk of software not working, is to measure the process stage of software quality elements, and testing can produce low risk software understanding stage.

3 Summary

The design and implementation of online reservation open laboratory management system mainly uses JSP development technology, MySQL as the background database and Tomcat server as the running platform. The system has a good interface and running environment.

The system can complete system management, appointment record management, teachers and students management, laboratory management and appointment management functions. System management is the management of all users and data. Reservation record is a record and storage of all teachers and students' reservation laboratories, which is convenient for teachers and students and administrators to query. Teacher and student management is the management and maintenance of teachers and students' information by administrators. Laboratory management is the maintenance of laboratory types and other information. Reservation management is a function that allows teachers and students to reasonably reserve laboratories.

The system basically meets the needs of teachers and students to make an appointment for the laboratory, so that the learning time of teachers and students can be reasonably arranged. And the interface designed by the system is simple and easy to operate, which is convenient for teachers and students to use. The principles of system design are as follows:

1) Modularization, consisting of modules and sub modules.
2) Abstraction and gradual refinement. Abstraction focuses on process and data and ignores low-level details. Refinement helps to design.
   Timing reveals low-level details, and the combination of the two is conducive to the integrity of the design.
3) Cohesion and coupling, in the actual design, it is difficult to achieve the ideal module relationship, that is, the internal modules are
   Function cohesion (all processing elements in the module belong to a whole, and complete a single function), and the modules are all data coupling, but the design of the module should maximize the pursuit of high cohesion and low coupling.

References

1. Zhu, X.: Design and implementation of online booking platform for open laboratories. J. Zhoukou Normal Univ. **35**(2), 109–113 (2018)
2. Hu, H., Hu, Y.: Development and design of laboratory appointment and management system. Communication World **1**, 220–221 (2016)
3. Li, G.: University open laboratory appointment management system. J. Changchun Univ. Technol. **37**(4), 411–416 (2016)
4. He, Z.: Design and implementation of laboratory appointment management system of Guangdong Maoming Health Vocational College. Jilin University (2017)
5. Geng, X., Zhang, Y.: JSP Practical Course (3rd Edition). Tsinghua University Press, Beijing (2015)
6. Zakas, N.C.: JavaScript Advanced Programming (3rd Edition). People's Post and Telecommunications Press, Beijing (2012)
7. Tomorrow Technology. HTML5 from Introduction to Mastery (2nd Edition). Tsinghua University Press, Beijing (2017)
8. Tomorrow Technology. MySQL from Introduction to Mastery. Tsinghua University Press, Beijing (2017)
9. Chen, Z.: Database access optimization based on hirburnate. Comput. Appl. Softw. **29**(7), 145 (2012)
10. Hu, Y., Zhang, Q.: Pagination display of web database based on ASP.NET. Comput. Syst. Appl. (1), 30–31 (2004)

University Education Management Based on Internet of Things Technology

Xiong Qi(✉)

Department of Economics and Management,
Yunnan Technology and Business University, Kunming, Yunnan, China
xiongchumoqi@163.com

Abstract. With the continuous promotion of Internet of Things technology, resources and tools, the informatization of education management in colleges and universities Construction is an irreversible trend. This experiment is aimed at the continuous development of the current Internet of Things technology and the research trend of education management informatization in my country's universities. This article takes the Internet of Things technology as the center, combines the Internet of Things and the educational management of colleges and universities, links the Internet of Things, schools, students, and teachers together, innovates in education management under the background of the Internet of Things, and builds things. Network education management model and build platform. Incorporate WSN (Wireless Sensor Network Technology) into the campus LAN. Use radio frequency identification technology (RFID) to read the identity information of teachers and students (campus card electronic tags), aiming at the current situation of college education management, and learning from the successful cases of other colleges and universities, and how to realize automatic and remote fault detection and equipment maintenance Perform basic analysis and improve various functions. The experimental research results show that. As a new technology and field, the Internet of Things has greatly strengthened the communication between people and things. In the cultivation of talents in universities, it is necessary to strengthen the construction of connotation and strengthen the construction of laboratory technical teams, with the goal of cultivating high-quality talents.

Keywords: Internet of Things technology · RFID technology · Education management · Colleges and universities

1 Introduction

As a national strategic emerging industry, the Internet of Things plays a very crucial role in the current social and economic development [1]. Under the requirements of the continuous development of the industry, a large number of skilled talents who combine theory and practice are required. In order to meet the development needs and teaching needs of emerging industries, many universities in China have proposed a hierarchical experimental teaching platform construction plan, which includes the technology foundation of the Internet of Things The experimental platform, the Internet of Things

J. MacIntyre et al. (Eds.): SPIoT 2020, AISC 1282, pp. 805–810, 2021.
https://doi.org/10.1007/978-3-030-62743-0_115

technology research experimental platform, the Internet of Things technology comprehensive experimental platform, and the Internet of Things technology application demonstration platform have brought China's education and teaching field into a brand new stage [2, 3]. As an interdisciplinary subject, the Internet of Things is related to multiple disciplines such as computer, communication, and electronics [4].

With the rapid development of science and technology, the application of Internet of Things technology in universities has become more and more extensive [5]. The use of campus cards is the epitome of the application of Internet of Things technology in campuses. All teachers and students have a campus electronic card with RFID tags [6]. They can buy rice in the cafeteria, daily necessities in the supermarket in the school, bottled drinking water in the dormitory, and water in the school bathhouse. The core principle is to upload the read information to the background database through the RFID reader, and then deduct the corresponding amount from the campus card. This can save a lot of manpower and material resources and improve work efficiency. Student attendance is one of the key tasks of school education management. It needs to be counted every day, and the workload is huge [7].

Education management is an important part of talent education in contemporary colleges and universities. Education reform under the Internet of Things technology creates favorable conditions for the long-term advancement of college education management [8]. With the widespread popularization of Internet of Things technology, seizing the opportunities of the times, and highlighting the characteristics of the era of education management, it will be the core focus of the future development of education management in universities, and it is important for solving many contradictory problems in university education management, education practice and student management at this stage. It is of great significance to promote the training level of modern advanced talents to a new height [6]. Accurate positioning is an important prerequisite for the development of IoT technology. Educational management in colleges and universities must use the precise characteristics of cloud computing to scientifically carry out innovative planning of educational management practices. The use of big data resources in education management information is essentially an impact on the traditional education management system of colleges and universities, by changing the traditional form of education management structure [9].

2 Method

2.1 Application of AK-Means Algorithm

A reasonable determination of the number of K-means clusters k has a great impact on the effect of clustering. Generally, the main method to determine the value of k is to use an evaluation function. The selection range of k value until the best clustering effect is found is [2, $\mathrm{int}\sqrt{n}$], so to find a suitable k value, multiple iterations are required, and the method to find the best k value will also be run multiple times The K-means algorithm increases the time complexity, and the clustering time becomes longer, which is not conducive to processing large-scale data. Therefore, this paper proposes a k-value adaptive method (AK-means for short). The purpose of AK-means is to obtain the value of k through the K-means algorithm once, and achieve a better clustering effect. Several parameters of the AK-means algorithm are defined as follows.

(1) Dispersion between clusters:

$$\mathrm{Disp} = \frac{\sum_{i=1}^{k} \mathrm{Disp_i}}{k}$$

In the above algorithm formula (1), Disp represents the degree of dispersion between clusters, $\mathrm{Disp_i}$ represents the position of the i-th cluster center point, and k represents the number of clusters.

(2) The degree of aggregation within the cluster:

$$\mathrm{Aggr} = \frac{\sum_{i=1}^{k} \mathrm{Aggr_i}}{k}$$

In the algorithm formula of (2) above, Aggr represents the degree of aggregation within all clusters, $\mathrm{Aggr_i}$ represents the average distance from all elements in the i-th cluster to the center point, and k represents the number of clusters.

(3) The evaluation value of clustering:

$$E = \frac{\mathrm{Aggr_k} - \mathrm{Aggr_{k=1}}}{\mathrm{Disp_k} - \mathrm{Disp_{k=1}}}$$

In the formula for the clustering evaluation value of (3) above, $\mathrm{Disp_k}$ represents the degree of dispersion between clusters at the kth time, $\mathrm{Disp_{k=1}}$ represents the degree of dispersion between clusters at the k = 1th time,$\mathrm{Aggr_k}$ represents the k-th intra-cluster aggregation degree, and $\mathrm{Aggr_{k=1}}$ represents the k = 1 intra-cluster aggregation degree. E is the cluster evaluation value

2.2 Apply Iot-Related Technologies to Realize Intelligent Education Management in Colleges and Universities

Teaching equipment is a necessary condition for teaching, scientific research and service of college students, and it is the material basis for cultivating talents. With the increasing application of teaching equipment in teaching, the number of teaching equipment in colleges and universities has greatly increased. Under the conditions of informatization, how to manage and use teaching equipment well is a process of continuous exploration and progress. Using the Internet of Things technology, an automatic management system has been established to track and manage important teaching equipment throughout the entire process to improve their service life. The label is embedded in the device and has an identification code set. The system can track and record in real time [6, 8, 9].

3 Experimental Design Ideas

First of all, briefly explain the related concepts of the Internet of Things, education management model and management innovation, and on this basis explain the connection between the Internet of Things technology and education management and its role and significance in current college education. Second, consult the relevant literature and clarify the current research status of the problem and the common problems existing in major domestic universities and the reasons for the problems. Taking the University of Political Science and Law as a case, it combines theoretical research with actual operation to find the characteristics and shortcomings of the current practical education management of colleges and universities under the Internet of Things technology. Third, a questionnaire survey was conducted on the selected 70 students, and the data obtained by the questionnaire survey was analyzed by comparison. A total of 70 questionnaires were distributed and 68 valid questionnaires were recovered, with a recovery rate of 97.14%.

4 Results

4.1 AK-Means Algorithm Validity Experiment

In order to verify the validity of the AK = Means data, this experiment downloads the experimental data from the database building and selects the University of Political Science and Law as the experimental data set. The number of them is 130 and 136 respectively, and the attributes of the data are 5 and 4 respectively. The types of data are all 3. Record the clustering process Aggr, Disp and E of AK-Means on the two data sets respectively. See Table 1 and Table 2 for details.

Table 1. List of data sets Aggr, Disp and E

	8	7	6	5	4	3	2	1
Disp	5.66	6.22	8.36	11.11	12.13	13.45	23.22	65.21
Aggr	0.91	0.975	1.08	1.15	1.32	1.47	1.67	1.98
E		13.22	21.33	25.67	13.22	14.55	16.77	18.33

Table 2. List of Aggr, Disp and E of the ZFSC data set

	9	8	7	6	5	4	3	2
Disp	433.22	451.56	557.44	632.11	998.58	1132.43	1653.22	2108.54
Aggr	68.22	75.78	125.21	137.88	155.25	161.33	167.89	210.23
E		2.446	2.978	4.763	4.990	5.325	6.331	7.213

In order to clearly show the changes in the evaluation value E and the value of $\log_2 E - \log_2 E_0$ in the clustering process, the drawn line chart is shown in Fig. 1:

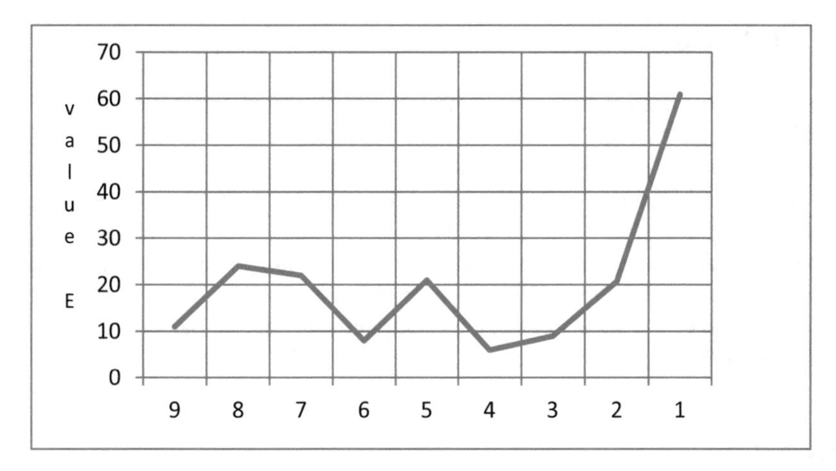

Fig. 1. The change trend of the evaluation value E during the clustering process

As shown in Fig. 1, when the value of k is a number greater than 2, the value of E has a small change. The value of $\log_2 E - \log_2 E_0$ is always less than 1, indicating that the combination is reasonable and the algorithm can continue to iterate. When the value changes from 3 to 1, the E value suddenly increases, and the $\log_2 E - \log_2 E_0$ value is greater than 1, which does not meet the merging conditions, indicating that the merging is unreasonable. Therefore, the number of clusters in both data sets is 3. The experimental results are consistent with the real number 4 of the two data sets, which proves the effectiveness of AK-means.

4.2 Constructing an Experimental Education Management Platform for the Internet of Things

Colleges and universities need to start from the training goals of practical technical talents and cultivate College graduates with independent analysis and problem-solving abilities, entrepreneurial and innovative abilities, have established a comprehensive base for experimental teaching reform, scientific research and social consultation, student innovation ability training, and skills training, so that experimental teaching, scientific research, and school-enterprise Cooperation provides a strong guarantee. It can also assess students' practical ability, which is a major achievement of experimental teaching innovation and reform. Practical teaching, as a link in cultivating students' subject literacy, scientific spirit, and practical ability, can achieve a huge improvement in students' comprehensive ability through an effective talent training model, and then deliver a large number of high-quality talents to the society. The Internet of Things can realize the comprehensive connection between people and people, people and things, and things and things. It is also called the third wave of world information industry after computers and the Internet in the field of science and technology.

5 Conclusion

Although the application prospect of the Internet of Things technology is broad, we should also realize that the Internet of Things from understanding to development is not achieved overnight, and the Internet of Things technology is constantly updated and breakthroughs. As the commercialization of the Internet of Things matures, there will be more and more applications of the Internet of Things in the education field. The Internet of Things is an important direction and driving force for a new round of education industry revolution. Use the new technology of the Internet of Things to transform some of the drawbacks of education into advantages, thereby promoting the reform and upgrading of education.

References

1. Feng, X., Hu, J.: Research on the identification and management of vehicle behaviour based on Internet of things technology. Comput. Commun. **156**, 68–76 (2020)
2. Gnotthivongsa, N., Huangdongjun, Alinsavath, K.N.: Real-time corresponding and safety system to monitor home appliances based on the Internet of Things technology. Int. J. Mod. Educ. Comput. Sci. **12**(2), 1–9 (2020)
3. Putri, A.O., Ali, M., Saad, M., et al.: Wearable sensor and internet of things technology for better medical science: a review. Int. J. Emerg. Technol. Learn. (iJET) **7**(4), 1–4 (2019)
4. Liu, J.W., Chen, S.H., Huang, Y.C., et al.: An intelligent identification model for the selection of elite rowers by incorporating Internet-of-Things technology. IEEE Access **PP**(99), 1 (2020)
5. Ezma, M.E., Ugwu, C.I.: A framework for monitoring blood pressure of patients in the rural area using Internet of Things technology. J. Theoret. Appl. Inf. Technol. **96**(11), 3153–3159 (2018)
6. Elkaleh, E.: Leadership curricula in UAE business and education management programmes: a Habermasian analysis within an Islamic context. Int. J. Educ. Manage. **33**(6), 1118–1147 (2019)
7. Charungkaittikul, S., Wang, V.X.: Guidelines for lifelong education management to mobilize learning community. Int. J. Adult Vocat. Educ. Technol. **9**(1), 31–41 (2018)
8. Pisonova, M., Brecka, P., Papcunova, V., et al.: Historical-philosophical aspects of professional communication in education management. XLinguae **13**(3), 171–184 (2020)
9. Lyu, H., Faoro, N., Mcdonald, M., et al.: Evaluation of RFID technology to capture surgeon arrival time to meet american college of surgeons committee on trauma verification guidelines. ACI Open **03**(01), e13–e17 (2019)

The Influencing Mechanism of Richness of UGC on Purchase Intention – An Empirical Analysis Based on "Content + Transaction" APP Users

Yao Li and Zhenzhen Song(✉)

Business School of Henan University, Kaifeng 475000, Henan, China
2812510425@qq.com

Abstract. "Content + transaction" apps play a pivotal role in the mobile application market. How to develop potential buyers based on user-generated contents has also become the key for enterprise operation. This paper proves by research that the variety of UGC in forms and quality reliability will affect the digital self-presentation effect of users, which in turn will promote users' willingness to reuse, including willingness to visit and share, and ultimately affect their willingness to buy. In the final stage, user thinking mode plays a regulating role.

Keywords: User-generated contents · Digital self-presentation · Purchase intention

1 Introduction

With the development of information technology and the change of life style, entertainment and business are gradually integrated. Websites with user-generated contents and various themes have emerged in China, such as Mafengwo APP, Little Red Book APP. Companies accumulate users through generation of contents and develop toward to online trading, and "content + transaction" has become their main mode of operation. As Zhang Depeng put forward that user-generated content has become an important way and method for enterprises to improve marketing effect [1]. With the rise of show culture, people seek to get social attention through active self-presentation. These "content + transaction" based APP communities are the best display platforms. This paper attempts to explore the relationship between user-generated content behavior and subsequent purchase intention to reveal the internal mechanism by integrating these two factors.

2 Hypothesis Presentation and Model Construction

2.1 UGC Richness and Digital Self-presentation

Based on media richness theory, Tang Xiaobo et al. divided information richness into three dimensions: information content richness, information expression richness and information quality richness [2]. Based on the research results of this scholar, this paper

J. MacIntyre et al. (Eds.): SPIoT 2020, AISC 1282, pp. 811–816, 2021.
https://doi.org/10.1007/978-3-030-62743-0_116

also classifies the user-generated content from these aspects. In studying the self-construction of consumers on WeChat platform, Dai Fu points out that various symbolic expressions such as text, pictures, music, emojies and short videos are convenient for users to display their images [3]. With the "content + transaction" APP, users can set labels for themselves according to the type of content they share and establish their identity. What's more, high-quality content can get more adoption and feedback, and help users build their image in the process of interaction. Therefore, the hypotheses are put forward as:

Assuming 1a/1b/1c: Variety of UGC in forms/variety in category/quality reliability have a positive impact on digital self-presentation.

2.2 Digital Self-presentation and Reuse Intention

Huang Lili et al. believe that the sharing behavior of virtual communities is the behavior of sharers taking the initiative to build their self-image, and the self-presentation motivation promotes the continuous sharing behavior of individuals [4]. The judgment of sharing behavior on the outcome of an individual, such as image, reputation, identity, etc., will also affect the willingness to re-share. Therefore, this paper believes that the positive digital self-presentation will affect the willingness of sharer to reuse, and the willingness to use in the "content + transaction" APP platform includes the willingness to access, share and purchase. As a deeper willingness to buy, the study is proposed to be divided into two parts for verification.

Hypothesis 2a/2B: Digital self-presentation has a positive effect on users' willingness to access/share.

Willingness to reuse is a process of forming of users' habits and a kind of trust for the platform, and trust will have a positive impact on purchase intention [5].

Hypothesis 3A/3B: The user's willingness to access/share has a positive impact on the purchase intention.

2.3 The Mediating Effect of Digital Self-presentation

From the perspective of expectation confirmation theory, when users create content out of the motivation of self-presentation, digital self-presentation results are regarded as a degree of expectation confirmation, and the degree of expectation confirmation will affect the subsequent use intention. Throughout the process, the digital self-presentation perception results will connect pre- and post-behaviors.

Hypothesis 4a/4b4c/4d/4e/4f: Digital self-presentation plays a mediating role in form diversity of UGC /variety richness/quality reliability and willingness to access/share.

2.4 The Moderating Effect of Thinking Mode

Mode of thinking includes the integrated way of thinking and analytical thinking, some scholars pointed out that the perception of similarity on the brand is affected by the different thinking style, consumers with analytical thinking mode tend to focus on the difference between the brands, and consumers with integrated way of thinking tend to focus on the similarities between brand [6]. Therefore, this paper believes that when consumers are of integrated thinking based personality and when they have a high willingness to access and share APP, they will pay attention to the connection between platforms and have a higher willingness to buy in the APP mall.

Hypothesis 5a/5B: Thinking mode has a moderating effect on users' willingness to visit/share and purchase: compared with analytical thinking, users' willingness to visit/share has a greater influence on the purchase intention of individuals with overall thinking.

According to the research hypothesis, the model is constructed as shown in Fig. 1:

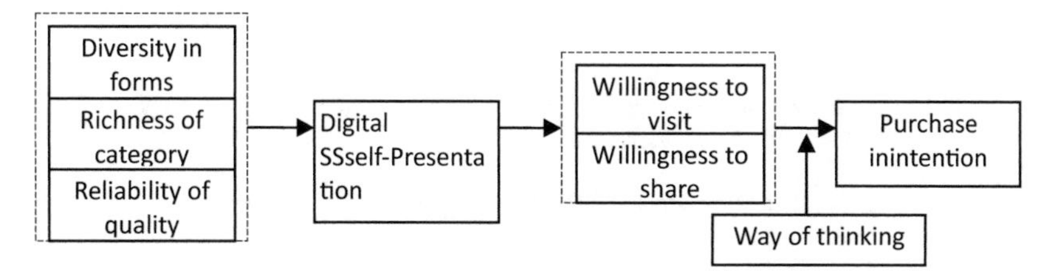

Fig. 1. Conceptual model

3 Research on Design and Data Collection

In this paper, online questionnaire survey was adopted to collect data. The UGC richness was measured based on the research results of Chinese scholar Tang Xiaobo on information richness. The measurement of digital self-presentation was made with reference to Sheer scale [7]; The research scales of Lin, Bhattacherjee and Kim [8–10] were used for visiting intention, sharing intention and purchase intention. Holistic thinking was measured by Choi (2007)'s scale [11]. 238 valid questionnaires were collected, and the ratio of samples from male and female was close to 1:1. Most people visited the APP at 3 times with each visiting time lasted less than 45 min.

4 Data Analysis and Testing for Hypothesis

4.1 Confirmatory Factor Analysis

AMOS 21.0 was used in this study to conduct confirmative factor analysis for the model. The goodness of fit index of the model showed that the chi-square degree of freedom ratio was 1.679, close to 2. RMSEA value is 0.054, which is less than 0.08. GFI value is 0.854, greater than 0.8. TLI, IFI and CFI were respectively 0.924, 0.936 and 0.935, which were all greater than 0.9. The indexes of goodness of fit of the model reach ideal values, which indicates that the goodness of fit of the model is in good state.

4.2 Analysis on Path

The path analysis of the model is shown in Fig. 2.

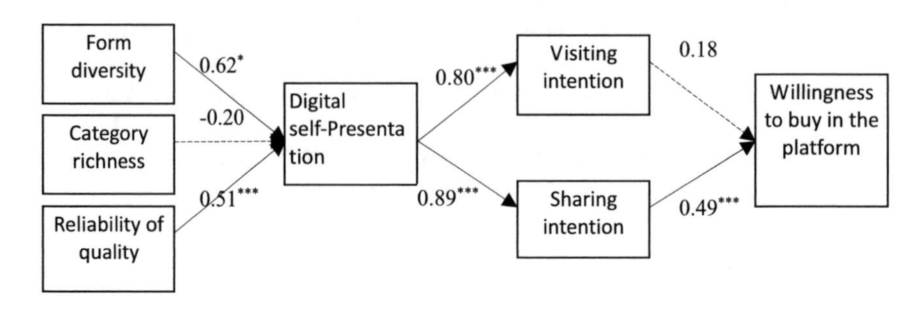

Note: The dotted line indicates that the test is not passed;** means P<0.05, ** means P<0.01;* * * indicates that P < 0.001

Fig. 2. Path diagram for standardized path coefficient estimation

4.3 Testing Results of Mediating Effects

The mediating effect of digital self-presentation was examined using the SPSS BootStrap method, and the results were shown in Table 1.

Table 1. Testing results of mediating effect

Mediation path		Direct effect			Indirect effect			Results
		Effect size	95% confidence interval		Effect size	95% confidence interval		
			Lower limit	Upper limit		Lower limit	Upper limit	
H4a	FD → DSP → VI	0.1403	0.03	0.2506	0.2589	0.1624	0.3655	Part of the intermediary
H4b	FD → DSP → SI	0.1483	0.0402	0.2565	0.3516	0.2525	0.4677	Part of the intermediary
H4e	QR → DSP → VI	0.1815	0.0435	0.3196	0.321	0.203	0.4423	Part of the intermediary
H4f	QR → DSP → SI	0.1892	0.0538	0.3246	0.439	0.2992	0.5772	Part of the intermediary

4.4 Testing Results of Moderating Effect

SPSS 21.0 was used for multi-layer linear regression to test the moderating effect, and the results are shown in Table 2.

Table 2. Test for moderating mechanism

Variable	Purchase intention		Purchase intention	
	Model 1		Model 2	
	Standardized coefficient β	Significance	Standardized coefficient β	Significance
Willingness to share	0.378	0.000	0.401	0.000
Holistic thinking	0.243	0.000	0.264	0.000
Interaction			0.124	0.030
Rsquare	0.299		0.313	
Fvalue	50.203**		35.581**	

5 Conclusion and Suggestions

5.1 Research Conclusions

The results show that richness of UGC type has no significant impact on digital self-presentation, and the positive impact of visit intention on purchase intention is not significant. Some studies have pointed out that the richness of information types can help users shape diversified and vertical personality, but it is not suitable to shape professional and flat personality, so there will be results deviation in the measurement of different samples. Compared with sharing intention, access intention lacks user participation and user integration.

5.2 Guidance

The research results of this paper can provide guidance for marketing activities of "content + transaction" APP platform. In order to promote the purchase willingness of platform users, enterprises need to:

A. Provide technologies to support user-generated content in various forms. For example, media technology is developed to support uploading and production for short video of users, and video, picture and text can be uploaded.
B. Innovating platform incentives. The platform is required to implement incentive measures according to characteristics in content sharing for users, users are encouraged to share content and display themselves, so as to efficiently transform potential buyers.

References

1. Zhang, D., Lin, M., Chen, C., Liu, S.: Do emotions and relationships in brand communities inspire recommendation?—Research on the influence of customers' psychological dependence on recommendation by mouth. Manag. Rev. **31**(02), 155–168 (2009)

2. Tang, X., Wen, P., Cai, R.: Empirical analysis on influencing factors of social media user behavior. J. Tongji Univ. (Nat. Sci.) **43**(03), 475–482 (2015)
3. Dai ,F., Jiang, S.: Study on self-construction of college students in WeChat moment—From the perspective of symbolic interaction theory. J. Beijing Univ. Posts Telecommun. (Soc. Sci. Ed.) **20**; No. 103(04), 11–17 (2018)
4. Huang, L., Feng, W., Qu, X.: Factors influencing information sharing in virtual communities: from perspective of multi-layer analysis. Chin. J. Journal. Commun. CJJC **36**(09), 20–34 (2014)
5. Guan, Y., Tao, H., Wang, Z., Song, Y.: Trust in online shopping. Adv. Psychol. Sci. **19**(08), 1205–1213 (2011)
6. Monga, A.B., John, D.R.: What makes brand elastic? Influence of brand concept and styles of thinking on brand extension evaluation. J. Mark. **74**(3), 80–92 (2010)
7. Sheer, V.C.: Teenagers' use of MSN features, discussion topics, and online friendship development: the impact of media richness and communication control. Commun. Quart. **59** (1), 82–103 (2011)
8. Lin, J.C.-C.: Online stickiness: is antecedents and effect on purchasing and effect on purchasing intention. Beh. Inf. Technol. **26**(6), 507–516 (2007)
9. Bhattacherjee, A.: Understanding information systems continuance: an expectation-confirmation model. MIS Quart. **25**(3), 351–370 (2001)
10. Kim, Y.J., Han, J.: Why smartphone advertising attracts customers: a model of web advertising, flow, and personalization. Comput. Hum. Beh. **33**(9), 256–269 (2014)
11. Choi, I., Koo, M., Choi, J.A.: Individual differences in analytic versus holistic thinking. Pers. Soc. Psychol. Bull. **33**(5), 691–705 (2007)

Path Planning Algorithm of Mobile Robot Based on IAFSA

Yi Guan(✉) and Lei Tan

Department of Information Management, Dalian Neusoft University of Information, Dalian, Liaoning, China
guanyi@neusoft.edu.cn

Abstract. In order to improve the search speed and shorten the search time of robot path planning, the algorithm and its characteristics of mobile robot in path planning are summarized. Firstly, the path planning of mobile robots is classified and summarized. From the perspective of mobile robots' mastery of the environment, the path planning of mobile robots is replanned through IAFSA. Meanwhile, the development status, advantages and disadvantages of relevant algorithms are summarized. Then it points out the future development trend of robot path planning technology in improved algorithm, hybrid algorithm, multi-robot collaboration, complex environment and multi-dimensional environment.

Keywords: Path planning · IAFSA algorithm · Hierarchy of obstacles · Mobile robot

1 Introduction

With the continuous development of science and technology, mobile robot technology arises at the historic moment [1]. When studying this technology, the research focus is mainly on the optimization of path planning technology. At present, the commonly used methods for global path planning mainly include genetic algorithm, fast random search tree algorithm and bee colony algorithm, etc.

In China, the research time of mobile robot path is relatively short, but China's research speed in this aspect is relatively fast, and more research experience is summarized [2]. Through the investigation of mobile robot market sales, actual service effect, and control sensitivity, Chinese researchers provide Suggestions for path planning to ensure that the path planning scheme can better meet the application practice [3].

2 Improved Method

2.1 Mobile Robot Path Planning

Path planning technology is one of the key core technologies in mobile robot research and development. In fact, the path planning technology of the robot refers to the indicator of some arguments (such as the lowest value of the working generation, the shortest path selection, the shortest computing time, etc.), and selects an optimal or sub-optimal

J. MacIntyre et al. (Eds.): SPIoT 2020, AISC 1282, pp. 817–822, 2021.
https://doi.org/10.1007/978-3-030-62743-0_117

obstacle avoidance path that connects from the starting line to the end line in the task area [4]. According to the understanding level of mobile robots to their working area information, the author divides robot path planning into two parts: path planning based on understanding of partial area information (also known as local path planning) and path planning based on complete area information understanding (also known as global path planning). Local path planning is a real-time dynamic path planning based on the local environmental information collected by the sensors carried by the robot during the execution of the task. Global path planning first requires the establishment of an abstract global environment map model based on the known global environment information, and then the use of an optimization search algorithm to obtain the global optimal or better path on the all-region map model, and finally guides the mobile robot to move safely to the target point in real conditions. It mainly involves two parts:

First, environmental information understanding and map model construction [5].

The second is global path search and robot guidance. In the path planning of mobile robots, both global and local path planning should be combined. The former aims at finding the global optimal path, while the latter aims at avoiding obstacles in real time.

2.2 Problems Faced by Path Planning of Mobile Robots

(1) Limitations

The path planning method of mobile robot has some limitations, so the path planning technology of mobile robot should be explored on the basis of limitation compensation [6].

(2) Regional activities

Current mobile robot research activities mainly focus on the land, for example, harvesting robot, robot soccer, etc., aiming at bad environment robot path planning, the limited test activities, which will increase the difficulty that path planning, is not suitable for degrees of freedom robot in complex environment planning, therefore, need to delve into professional and technical personnel.

(3) There are problems in the path planning of mobile robots, which is applied to nonholonomic constraint planning [7].
(4) Noise, energy consumption and detection accuracy [8].

3 Analysis Method

3.1 IAFSA Algorithm Introduction

The IAFSA algorithm is quite flexible and adaptable, and is a very popular path planning algorithm. IAFSA is defined as follows: g(n) represents the actual path cost from the starting point Start to a certain node m, h(n) is a heuristic function, which represents the estimated cost from node m to the end point End, f(n) = g(n) + h(n) is the comprehensive evaluation value of the algorithm. Commonly used heuristic functions for grid maps are Manhattan distance, Euclidean distance and diagonal distance. A* search is to continuously compare the f(n) value of the expansion node in the moving process, and save the minimum f(n) value point each time into the list as a candidate node.

There are two extreme cases in the IAFSA algorithm: 1. When the heuristic function is 0, f(n) = g(n), then IAFSA evolves into the Dijkstra algorithm, the planning path is optimal but the algorithm efficiency is reduced; 2. When heuristic When the formula function is much larger than g(n), f(n) is approximately equal to h(n). At this time, IAFSA evolves into a BFS algorithm, and the optimal path may not be obtained. If the heuristic function value is less than the actual cost value, IAFSA can find the optimal path, but there are more expansion nodes and lower operating efficiency. If the heuristic function value is higher than the actual cost value, the IAFSA extension node will be less and the operation will be faster, but the path planning result may not be the best.

In maps with many obstacles, the theoretical cost calculated by the heuristic function h(n) is often much smaller than the actual cost, which leads to the need for IAFSA to expand more redundant nodes in different directions, making the IAFSA algorithm The operating efficiency is greatly reduced. Therefore, it is necessary to appropriately increase auxiliary conditions in the path planning process to restrict and guide the expansion direction, thereby reducing the number of redundant nodes and improving the efficiency of the algorithm.

3.2 Necessary Point Path Planning

To effectively restrict the search process under the premise of satisfying the shortest path and reduce the amount of visits to redundant nodes, a shortest path must be found for direction guidance in the path planning process. Obviously, there is a point where the shortest path must pass in the edge area of an obstacle. As shown in Fig. 1, from the start point S to the end point G, five different paths can be formed through five points A, B, C, D, and E. Obviously, S-A-G is longer than S-B-G, and S-C-G and S-D-G are two. All paths need to detour point B or point E, and the length must be greater than S-B-G or S-E-G, so it can be seen that at least one point of the two points B and E is the shortest path.

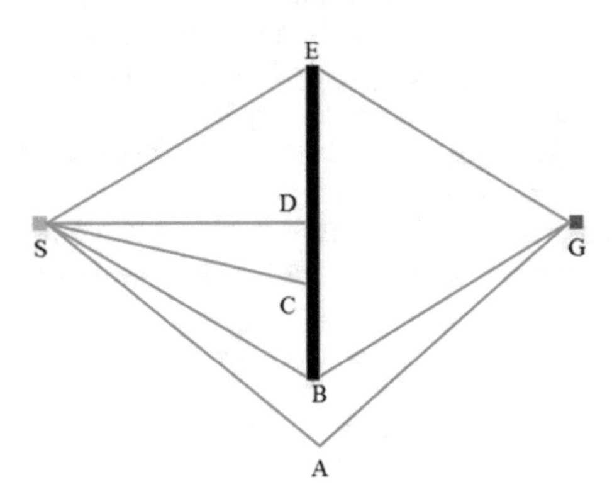

Fig. 1. Path example diagram

According to the distribution characteristics of obstacles in the grid map, they are divided into different groups, namely obstacle blocks, and then according to the relative position relationship between each obstacle block and the starting point, a hierarchy of obstacle blocks is formed from near to far. Using the "start-end point" vector as the reference system, calculate the distance from each grid point in each obstacle block to the reference system, select the largest point on both sides of the obstacle block that has an intersection with the reference system as the landing point, and then connect the adjacent All landing points between obstacle blocks form all simulated paths with landing points at various levels as intermediate nodes. Select the top k simulation paths with the least cost as the sample, calculate the comprehensive score of all landing points in the sample, and the highest score is the shortest path.

3.3 Establish a Hierarchy of Obstacles

The IAFSA algorithm automatically avoids obstacles when it encounters obstacles when searching for paths, and continues searching in other directions, regardless of the distribution characteristics and hierarchical relationships of obstacles, and has no effective control over the direction of node expansion, so it needs to perform grid map Pre-processing, associating all obstacle grid points, and establishing a hierarchical system. First traverse each grid point in the map, find all obstacle points, and classify the obstacles that appear in the neighborhood of each obstacle point into the set of the current obstacle point. Execute in a loop until all obstacle points are classified, so as to obtain all obstacle sets, namely obstacle blocks.

Specific steps are as follows:

Step1: Establish the two-dimensional array G rid corresponding to the grid map, store the list of nodes to be detected ListO bstacles, the current level obstacle list Obstacle B lock and the obstacle block hierarchy list ListO bstacle Block;

Step2: According to the map grid sequence, save the first undetected obstacle grid point into the list ListO bstacles, and write true to the corresponding bit of the array G rid;

Step3: Check the list ListO bstacles cyclically. If ListO bstacles is empty, enter Step4; otherwise, let Current ListO bstacle. First and delete the first node of ListO bstacles. Detect the 8 neighboring grid points of Current. If an obstacle is detected and the corresponding bit of the obstacle in the array G rid is false, it is added to the list ListO bstacles, and the corresponding bit of the array G rid is set to true. Add the Current node to Obstacle B lock and enter Step3;

Step4: Insert Obstacle B lock into ListO bstacle Block, if all values of G rid are true, the program ends, otherwise enter Step2.

The grid map is preprocessed through the above steps to form a hierarchical system of obstacles that can be identified by the system, as shown in Fig. 2.

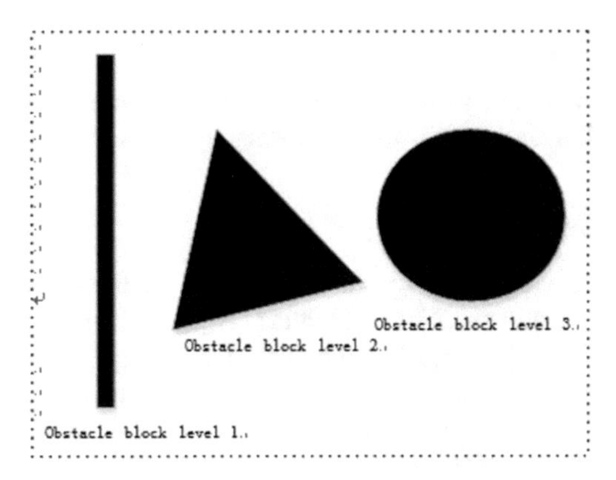

Fig. 2. Obstacle block hierarchy

3.4 Technical Level and Competitive Advantage

(1) Product technical level

In China, the path planning algorithms for mobile robots are generally artificial fish swarm algorithm (AFSA) and genetic algorithm (GA), but there are many problems, such as its easy to fall into the optimal, the result accuracy is not high, and the convergence speed of genetic algorithm is slow.

(2) Competitive advantage

In specific space (where the obstacles are relatively clear), the path planning algorithm of mobile robot based on IAFSA can shorten the operation time, and the static path planning has fast aging time. It does not need the mechanism model and the real problem description, so it can be applied in more fields.

4 Conclusion

(1) Project innovation points

By improving the original algorithm, dynamic path planning can be realized, and master the application skills of global and local planning methods, which can not only improve the application scope of path planning technology, but also expand the thinking of relevant researchers in path planning.

(2) Novelty of the product

Artificial fish swarm algorithm (AFSA) and genetic algorithm (GA) are improved to reduce the problems such as easy to get into the optimal situation, low result accuracy, and slow convergence speed of genetic algorithm. The applicability of the selected method should be considered when the path planning is carried out, so that the dynamic path planning can be realized.

(3) Advancement and uniqueness

Local exchange between parts has a good ability to overcome the local extremum and obtain the global extremum, thus speeding up the optimization speed and the success rate [9].

(4) Competitive advantage

Compared with the original algorithm, it does not require high initial value, which can be randomly generated or set as fixed value [10]. The requirement of parameters is not high and the allowable range is large. The disadvantages of artificial fish aggregation in non-global extremum point are overcome. The ability of global search and local mining is effectively taken into account. The algorithm of clustering analysis can obtain the number of categories and category centers of clustering problems, and adopt a more convenient way of individual description, from region to region center.

References

1. He, S., Belacel, N., Chan, A., Hamam, H., Bouslimani Y.: A hybrid artificial fish swarm simulated annealing optimization algorithm for automatic identification of clusters. World Sci. Pub. Comp. **15**(5), 67–69 (2016)
2. Chen, L., Zhang, H.: Orderly discharging strategy for electric vehicles at workplace based on time-of-use price. Inventi J. Pvt. Ltd. 222–224 (2017)
3. Chen, L., Zhang, H., Schuster, T.: Orderly discharging strategy for electric vehicles at workplace based on time-of-use price. Hindawi Pub. Corp. 376–377 (2016)
4. Jiaan, W., Ancheng, X., Jintao, J., Linyang, G.: Optimization lighting layout of indoor visible light communication system based on improved artificial fish swarm algorithm. J. Optics **22**(3), 88–90 (2020)
5. Xingming, Z., Peng, W., Yanping, B., Yuchao, H.: DOA estimation of array signal based on iafsa-music algorithm. Math. Pract. Understand. **49**(22), 163–170 (2019) (in Chinese)
6. Lin, C., Yang, B.: Simulation of flow line scheduling of production enterprises based on improved artificial fish swarm algorithm. Int. J. Simulat. Model. **17**(3), 112–114 (2018)
7. Wawrzyniak, J., Drozdowski, M., Sanlaville, É.: Selecting algorithms for large berth allocation problems. Euro. J. Operat. Res. **283**(3), 844–862 (2020)
8. Sheikholeslami, A., Mardani, M., Ayazi, E., Arefkhani, H.: A dynamic and discrete berth allocation problem in container terminals considering tide effects. Iran. J. Sci. Technol. Trans. Civil Eng. **44**(1), 369–376 (2020)
9. Tufano, A., Accorsi, R., Manzini, R.: A simulated annealing algorithm for the allocation of production resources in the food catering industry. Brit. Food J. **122**(7), 2139–2158 (2020)
10. Yang, L., Xinyu, L., Jianhui, M.: Large-scale permutation flow shop scheduling based on improved simulated annealing algorithm. Comput.-Integra. Manufact. **26**(02), 366–375 (2020) (in Chinese)

Virtual Machine Software in Computer Network Security Teaching

HaiJun Huang(✉)

School of Intelligent Systems Science and Engineering, Yunnan Technology and Business University, Kunming 657100, Yunnan, China
zstxujing@163.com

Abstract. The development of network makes its security more and more important, but there are many problems in the teaching of network security. Network security has become an important guarantee of network construction. Mastering the technical ability of network security has become the necessary quality of computer and network professionals. Based on this, this paper studies the application of virtual machine software in the teaching of computer network security. Through questionnaire survey, this paper understands the current situation of computer network security course in Colleges and universities, and summarizes the problems existing in the current college computer network security course. Among them, 22.1% of the problems are caused by the fear of system security risks in the teaching process, and the students' operation practice content is relatively high The reason of less is 19.3%, and the problem caused by the limitation of university equipment resources accounts for 18.6%, and so on. Therefore, the use of virtual machine technology in computer network security in Colleges and universities can effectively solve the problems.

Keywords: Virtual machine · Computer · Network security · Network teaching

1 Introduction

Network security course is an important professional course for computer related majors, with high practical requirements [1, 2]. The understanding and mastery of it directly affects students' professional level and whether they can engage in network management and security maintenance and other related work [3, 4]. With the continuous promotion of the new curriculum reform, the school began to pay more attention to the computer network security technology course as an important course, and considered that this course has a key role in training information security professionals [5, 6].

The use of network technology in teaching classroom has been everywhere, but there are still many problems in network technology teaching. On the one hand, the complexity of network teaching environment leads to network teaching security problems; on the other hand, due to the limitations of school teaching conditions, there are great problems for students' practical operation [7, 8]. The application of virtual machine technology provides a new idea for network security management. Virtual

J. MacIntyre et al. (Eds.): SPIoT 2020, AISC 1282, pp. 823–829, 2021.
https://doi.org/10.1007/978-3-030-62743-0_118

machine technology not only solves the problem of network security in the teaching process, but also effectively relieves the tension of teaching equipment, thus reducing the cost of equipment procurement in Colleges and universities. At the same time, virtual machine technology restores a real practical teaching environment and further improves the teaching quality [9, 10].

This paper first introduces the virtual machine, outlines the nature of virtual machine and introduces the relevant knowledge of virtual machine software, analyzes the importance of computer network security technology course, and then through the questionnaire survey method to interview college teachers and students, to understand the current situation of college computer network security teaching, summarize the existing problems, and then through the virtual machine This paper analyzes the application of virtual machine technology in college computer network security teaching, and summarizes the advantages of virtual machine software in computer network security teaching.

2 Introduction to Relevant Knowledge

2.1 Introduction to Virtual Machine

The essence of virtual machine technology is the simulation of computer hardware. The application of virtual machine technology in computer network teaching is an important use of virtual machine technology. It is to connect multiple computers through the way of networking, and the teacher's host computer controls the students' application of the slave computer. In fact, virtual machine is a complete computer system, including all the computer systems, including CPU, memory, hard disk, optical drive, graphics card and network adapter. In the same computer can run multiple virtual machine systems at the same time to ensure their normal operation.

2.2 The Importance of Computer Network Security Technology Course

1. The security of social information needs to be improved, Since the Internet appeared in people's lives, the problem of information security has always existed. The frequency of information leakage and data loss is very high. China's information security problems are becoming more and more serious, even the data within the government will be invaded and deleted, resulting in very serious consequences. This makes the country gradually realize the importance of improving information security.
2. Teaching objectives of computer network security technology course, The main purpose of the school to carry out this course is to teach students the basic network security technology, so that the new generation of people realize that information security is urgently needed to be improved. Taking this course as the main way to cultivate professional talents, it can effectively improve the quality and efficiency of information security management. When the school carries out the course, it is based on the social needs. The main goal of teaching is to help students form the awareness of protecting information security, improve their ability to protect

information security, and enhance their competitiveness, so as to lay a solid foundation for the realization of information security in China.

3 Research Method

(1) Questionnaire survey, In this paper, through the interview of a number of domestic colleges and universities, 450 teachers and students in the process of interview questionnaire survey, to understand the university computer network security course teaching status, and then sort out the analysis, summed up the current university computer network security course problems, the interview process learned that some colleges and universities have used virtual machine technology for computer network security teaching This paper also carries out a questionnaire survey on the teachers and students of these colleges and universities to understand the current situation of using virtual machine teaching, comparative analysis of the teaching status of these colleges and universities, summed up the advantages of virtual machine technology in the computer network security course. (2) Literature review method, In this paper, by consulting the relevant literature, we understand the relevant content of virtual machine technology, and compared with the domestic and foreign universities which are in the forefront of virtual machine technology teaching, and further optimize and summarize the specific application of virtual machine technology in computer network security.

4 The Performance of Virtual Machine in Computer Network Security Teaching

4.1 Analysis of Problems in Computer Network Security Teaching

In this paper, through the questionnaire survey, to understand the current situation of college computer network security course teaching, and then sort out the analysis, summed up the current university computer network security course problems, the specific analysis is shown in Fig. 1.

As can be seen from Fig. 1, the problems existing in the computer network security course of domestic colleges and universities mainly include the following aspects: (1) Hidden dangers in teaching system, accounting for 22.1%, In order to let students truly feel the security risks in the computer network, usually, in the teaching of computer network security, the computer hardware system is used to demonstrate the situation of the terminal PC being invaded by the network. But this kind of demonstration itself, also can have certain security hidden danger, once to the network Trojan horse, the virus control or the later stage elimination is not good, also can cause the threat teaching network system risk. (2) Less content of operation practice, accounting for 19.3%, Hacker intrusion is a compulsory content in the teaching of computer network security. Usually, there are many ways of hacking into network terminals. Demonstrating the intrusion method in the teaching process is also the teaching focus to guide students to understand the network background rules. In the teaching process, teachers can only demonstrate the similar background intrusion mode for students, and

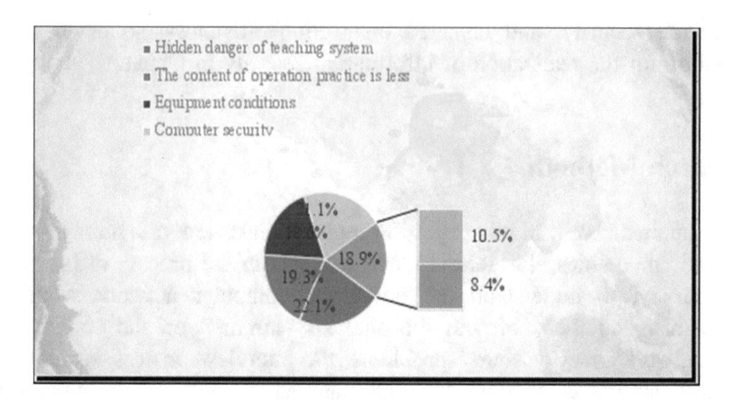

Fig. 1. Problems in computer network security teaching

it is not possible to require students to invade a running real website or publish its network background login link port. Therefore, the content of operation practice in the teaching of computer network security is less, and the PPT or video micro class produced by students for less observation, the content that can really carry out operational training does not have the corresponding teaching conditions. (3) Limited equipment conditions, accounting for 18.6%, In the teaching of network security, the attack and defense experiment requires each student to have two computers. At the same time, the computer must also be able to realize real-time communication. This requirement will greatly increase the financial burden of the school, so it is difficult to achieve, which will affect the efficiency and effect of teaching to a certain extent. (4) Computer security problems, accounting for 21.1%, In the attack and defense experiment of network security teaching, students not only need to simulate and deal with hacker attacks, but also need to deal with virus attacks. It is difficult to effectively guarantee the security of computer system, and it is easy to cause system damage in the process of experiment. This is also the reason why many schools are not willing to carry out attack and defense experiments. Students lack of practical operation and can not effectively apply the knowledge they have mastered.

4.2 Application of Virtual Machine Software in Computer Network Security Teaching

By comparing domestic and foreign universities which are in the forefront of using virtual machine technology teaching, this paper summarizes the specific application of virtual machine technology in computer network security. The analysis is shown in Table 1.

From Table 1, we can see that the application of virtual machine in the teaching of computer network security is mainly manifested in five aspects: first, sorting out the application of virtual machine file name to ensure the normal teaching. The second is to construct the attack and defense operation platform to improve the teaching effect. The third is to carry out the Trojan horse experiment activity, let the students master the network security guard skill. Fourth, establish a network simulation environment to

Table 1. Application of virtual machine software in computer network security teaching

Application measures	Objective
Organize virtual machine file name application	Ensure normal teaching
Construction of attack and defense operation platform	Improve teaching effect
Carry out Trojan horse experiment	Master the skills of network security
Establish a network simulation environment	Protect computer security
Building multiple operating systems	Address resource constraints

protect computer security. Fifth, build multiple operating systems to solve the problem of resource constraints. (1) Organize virtual machine file name application, In order to carry out network security teaching on virtual machine, it is necessary to reorganize the file names stored in the virtual machine. When the virtual machine is used in the network, the memory of the virtual machine terminal of the student is relatively high, and it is easy to slow down the networking speed here. Or after the teachers' terminals are connected to the Internet, the picture sequence of non real name authentication is confused. Therefore, it is necessary to rearrange the subordinate file names of all terminal PCs when starting the virtual machine for teaching networking. If the teaching time is not enough, the teacher can encode the virtual image file uniformly, then upload it through the internal LAN, and download it on the terminal PC. (2) Construction of attack and defense operation platform, The ultimate goal of using virtual machine in network security teaching is to present the key teaching contents such as network security risks, hidden dangers and intrusion means in front of students. Therefore, in the process of using virtual machine, it is necessary to further build the attack and defense operation platform to strengthen the students' autonomous learning ability. After the course is close to the real network operation, the students' ability to handle the real network security can be improved. Make the virtual machine play the technical function of guiding students to operate the drill, and then explain the key knowledge points by reviewing the attack and defense training environment. (3) Carry out Trojan horse experiment, In the network security teaching, teachers can make full use of virtual machine to carry out Trojan horse experiment, so that students can learn how to deal with the Trojan horse in the network and master the network security prevention skills.

4.3 Advantages Analysis of Virtual Machine Software in Computer Network Security Teaching

By comparing the colleges and universities that use virtual machine technology and do not use virtual machine technology for computer network security teaching, the advantages of virtual machine software in computer network security teaching are summarized, as shown in Fig. 2.

It can be seen from Fig. 2 that the advantages of virtual machine technology in the teaching of computer network security are mainly shown in the following aspects:

(1) Strengthen students' practical operation ability, 22.5%, Due to the use of virtual machine technology in network teaching, every student can have the opportunity to

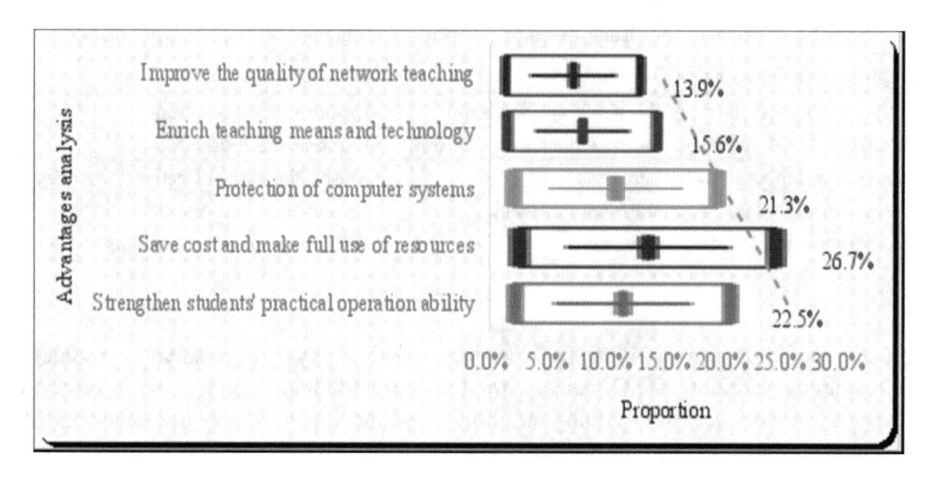

Fig. 2. Advantages of virtual machine software in computer network security teaching

practice. It can create an environment for students to learn theoretical knowledge and practice at the same time, which increases students' practice opportunities and improves their operation ability. In the study of some problems, it is necessary to combine theory with practice. In the previous classroom, students only understand knowledge through the teacher's narration, because there is no practical operation, students' understanding of some knowledge may be biased. However, through the use of virtual machine technology, students can immediately carry out practical operation after learning theoretical knowledge. Through practical operation, they can better understand some problems, at the same time, deepen the impression of some problems, and have a more comprehensive understanding and mastery of these problems. Through virtual machine technology, students can get a more real experience. (2) Saving cost and making full use of resources, accounting for 26.7%, The school's teaching praise is limited, which leads to the school's teaching conditions are not perfect. However, virtual machine technology can simulate real software system and various hardware devices. And these software systems and hardware devices in the virtual machine is a file, use and stop is a file open and close. In this way, the time of installing and unloading software is reduced, and the purchase of software is also saved. After adopting virtual machine technology, the management and maintenance of computer are more convenient. (3) Protection of computer system, accounting for 21.3%, In the network security training teaching, there are many operations are destructive, if carried out on the physical computer, it is likely to cause computer damage. Because the students are new to the computer, it is easy to have wrong operation, if it is in the computer, it is easy to cause large-scale damage. If students use virtual machine technology to carry out various experiments, even if there are important errors, they can also use the snapshot technology of virtual machine to quickly restore the system. In this case, the application of virtual machine software, the corresponding operation on the virtual machine, can effectively protect the computer security.

5 Conclusions

Based on the analysis of the application of virtual machine software in the teaching of computer network security, this paper summarizes that the application of virtual machine technology in teaching can do more with one stroke, and also solves the conditions and environment needed for teaching. While explaining the course principle, students' learning interest, motivation and efficiency have been greatly improved through demonstration and students' hands-on practice, Students can understand the principles more thoroughly, and at the same time, they can gradually improve themselves with knowledge in the teaching of computer network security, and become a comprehensive professional.

References

1. Lee, Y.S., Jeong, J., Son, Y.: Design and implementation of the secure compiler and virtual machine for developing secure IoT services. Future Gen. Comput. Syst. **76**(nov.), 350–357 (2017)
2. Ashraf, Z., Sohail, A., Yousaf, M.: Performance analysis of network applications on IPv6 cloud connected virtual machine. Int. J. Comput. Netw. Inf. Secur. **11**(12), 1–9 (2019)
3. Han, J., Zang, W., Liu, L., et al.: Risk-aware multi-objective optimized virtual machine placement in cloud. J. Comput. Secur. **26**(5), 1–24 (2018)
4. Shi, J., Xu, H., Lu, L.: Research on the migration queue of data center's virtual machine in software defined networks. Dianzi Yu Xinxi Xuebao/J. Electron. Inf. Technol. **39**(5), 1193–1199 (2017)
5. Gao, J.: 77. A support vector machine model for computer network security technology. Boletin Tecnico/Tech. Bull. **55**(12), 564–568 (2017)
6. Gao, H., Wu, J., Wei, C., et al.: MADM method with interval-valued bipolar uncertain linguistic information for evaluating the computer network security. IEEE Access **7**(99), 151506–151524 (2019)
7. Sadeanit, M.: Guide to computer network security (4th ed.). Comput. Rev. **59**(2), 70 (2018)
8. Stojanov, Z., Dobrilovic, D., Zoric, T.: Exploring students' experiences in using a physical laboratory for computer networks and data security. Comput. Appl. Eng. Educ. **25**(2), 290–303 (2017)
9. Tobin, P., Le-Khac, N.A., Kechadi, M.T.: A lightweight software write-blocker for virtual machine forensics. J. Inf. Secur. Res. **8**(1), 1–9 (2017)
10. Shi, H., Mirkovic, J., Alwabel, A.: Handling anti-virtual machine techniques in malicious software. ACM Trans. Inf. Syst. Secur. **21**(1), 2.1–2.31 (2018)

Key Technologies of Virtual Reality Fusion and Interactive Operation for Intelligent Learning in Network Environment

Shuo Wang[1(✉)], Yujuan Yan[1], Jialin Li[2], and Yujia Zhai[1]

[1] Jilin Engineering Normal University, Changchun, Jilin, China
15876031@QQ.com
[2] Changchun Institute of Land Surveying and Mapping, Changchun, Jilin, China

Abstract. With the globalization development of educational informatization, information technologies such as Internet of Things, cloud computing and big data are constantly emerging. As the core and fundamental cornerstone of wisdom education, intelligent learning has come into people's vision and received great attention from many foreign countries. However, the study of wisdom learning in China has just started. Based on this, this paper studies the further development of wisdom learning by combining the key technologies of virtual and real fusion and interactive operation. This paper firstly analyzes the research status of intelligent learning, discusses various interaction technologies, and proposes virtual and real interaction methods based on Kinect sensor instruction type and multi-channel type. These interaction methods can effectively control virtual objects and improve the stability, real-time and authenticity of the augmented reality interaction system. Then, this paper combined with the network learning platform to carry out an experiment on the scheme, the results show that the scheme effectively improves the learning efficiency of students in the network intelligent learning platform, the attendance rate increased by 11.27%.

Keywords: Intelligent learning · Virtual and real integration · Virtual and real interaction · Network learning platform

1 Introduction

With the rapid development of cloud computing, Internet of Things, big data and other new generation of intelligent information technologies, the "Internet+" era has quietly emerged, exerting a huge and far-reaching impact on our study and life [1, 2]. The cross-border integration of "Internet + education" gives rise to a brand new education space combining online and offline, virtual and real, which provides an engine driving force for promoting smart learning and cultivating smart talents [3, 4].

Deihimi Ali proposed a 24-h/day solution for MG management and flexibly chose the time solution. HOD and HSOC charts were used to balance the cost of ESSs in one day, and a multi-objective uniform water cycle optimization algorithm for MG short-term OM was proposed. Extreme distance iteration, reasonable comparison of multi-objective optimization algorithm [5]. Wang Hesheng proposed an interactive control algorithm for the operation of the auxiliary robot operator on various constraints.

J. MacIntyre et al. (Eds.): SPIoT 2020, AISC 1282, pp. 830–837, 2021.
https://doi.org/10.1007/978-3-030-62743-0_119

The method decouples motion and force control in a constrained coordinate system and modifies the motion velocity online. Firstly, the constraint frame is determined online according to the previous motion direction. Then the selection matrix is dynamically adjusted and the direction of constrained motion is selected as the driving axis. Therefore, the driving shaft and the non-driving shaft are decoupled; Finally, speed control and impedance control are respectively carried out on the above shaft [6]. In order to ensure the stable operation of the power grid in case of failure, auxiliary services such as LVRT should be added to the power grid. Xin Zhao proposed a power grid interactive micro motor LVRT control strategy based on positive/negative sequence droop control. By adopting the proposed control strategy, the power grid voltage can be supported throughout the fault period, profits can be made, and voltage can be dropped.

As an effective way to cultivate intelligent talents, intelligent learning plays a vital role in the implementation of intelligent education. Based on this, this paper, in the context of the network environment, combines virtual and real fusion and interactive operation technology to conduct in-depth research on intelligent learning, and makes technological improvements on the network learning platform, so as to improve the efficiency of intelligent learning.

2 Intelligent Learning Interacts with Reality

2.1 Intelligent Learning

Intelligent learning environment is the effective support to promote the change of learning and teaching methods and the new demand of students for the development of learning environment in the context of social informatization. As a high-end form of digital learning environment, intelligent learning environment can perceive learning situation, identify learners' characteristics, provide appropriate learning resources, automatically record learning process and evaluate learning results. Internet of Things technology, cloud technology and intelligent recommendation technology provide technical support for building smart learning environment. Compared with the traditional campus learning environment, learning environment from wisdom, blended learning theory of constructivism learning theory, modern teaching theory, using the appropriate equipment, tools, technology, media, and integration of open learning resources, curriculum resources in micro subject, mobile learning resources, resource sharing project, and establish a kind of intelligent, open and integrated the new service model.

2.2 Virtual and Real Fusion and Interactive Operation

Determine the position relationship between virtual objects and real scenes, register virtual objects to the target location calculated by the target recognition and tracking module, and realize the perfect fusion of virtual objects and real scene images; The virtual and real interaction module obtains the user's operation instructions and realizes the control of the virtual object. The unit length of coordinates is expressed in millimeters. The intersection of the optical axis of the camera and the image is defined as the origin of the image coordinate system. Let the physical size of unit pixel be x and y in the X-axis direction and Y-axis direction respectively, and the conversion formula of pixel coordinate system to image coordinate system is as follows:

$$u = \frac{x}{\Delta x} + u_0, v = \frac{y}{\Delta y} + v_0 \tag{1}$$

The translation matrix and rotation matrix of the depth camera and RGB camera are shown in the formula:

$$T = T_{rgb} - R_{rgb} R_{iF}^{-1} T_{ir} = T_{rgb} - RTir \tag{2}$$

3 Design of the Experiment

3.1 Experimental Background

The cultivation of intelligent talents is an unavoidable practical problem in the process of educational reform and informationization. It aims to cultivate talents with good value orientation, strong action ability, good thinking quality and deep creative potential. It involves the innovation of educational and teaching concepts, the optimization of learning environment and the innovation of teaching methods. Intelligent education cultivates man-machine collaborative data wisdom, teaching wisdom and cultural wisdom, so that teachers can display effective teaching methods and students can obtain appropriate and personalized learning services and good development experience. With the continuous development of "Internet + education", various kinds of teaching software and platforms are increasing day by day. However, the diversity, heterogeneity and variability of platforms and software also lead to the emergence of data "gap" and information "island", which is not conducive to the mastery of subject knowledge and cultivation of students' innovation ability.

3.2 Experimental Design

Table 1. Experimental results

The test number	The average score	Satisfaction with the network platform (%)	Attendance rate of specialized courses
Group 1	72.1	86.2	92.8
Group 2	76.1	79.6	91.3
Group 3	782	88.4	86.4
Group 4	74.9	87.9	85.1
Group 5	88.4	87.8	86.9

In this paper, a network learning platform as a sample, in-depth study. Based on virtual interaction technology, RGB image needs to be mapped into depth image. For this purpose, Ge Body Frame component is designed, which contains 7 Data members and 6 member functions. Kinect Manager members and Sensor Data members are the aggregation relationship of the Body Frame class. In view of the use of the smart learning platform, the questionnaire survey method was adopted to understand the Suggestions of students and teachers on the construction objectives and functional requirements of the smart academic platform, so as to further clarify the design objectives and core functions of the smart academic platform. A total of 300 questionnaires were sent out to investigate the design objectives and functional requirements of the platform. In this study, SPSS was mainly used for analysis, and Cronbach's Alpha reliability coefficient method was used to determine whether there was a high internal consistency. The results are shown in Table 1.

4 Virtual and Real Interaction Technology for Intelligent Learning

4.1 Analysis of Key Technologies of Virtual and Real Fusion and Interactive Operation for Intelligent Learning

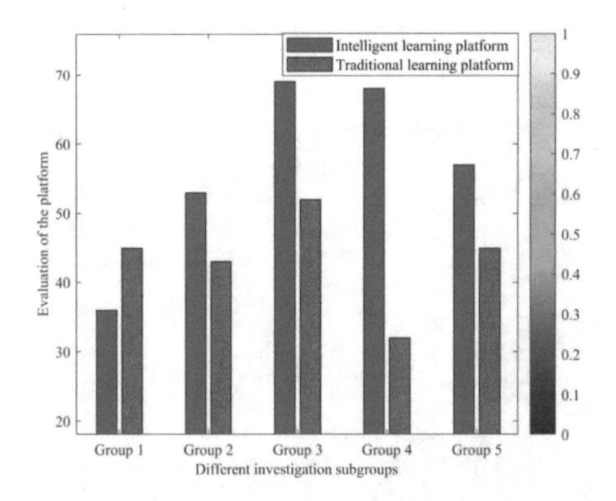

Fig. 1. Students' evaluation of traditional learning platform and smart learning network platform

As shown in Fig. 1, in the survey on the frequency of academic platforms in the traditional academic platforms, this paper finds that more than half of the users only use the academic platforms occasionally, with 13.18%, 22.48% and 12.02% of the users using the platforms once a week, two or three times a day, respectively. This also shows that the current situation of academic platforms construction is unsatisfactory. Option statistical results from high to low in turn for resources are of variable quality (58.91%), the function is not fully (41.47%), the lack of academic exchange platform (40.31%), information retrieval, difficult (33.33%), the lack of subject gateway (32.95%), single (29.46%), the lack of resources simulation practice (27.91%) and other (fees, too much advertising, the lack of experts, not 5.04% accurate retrieval) with single format support. The survey results show that users have low satisfaction with the existing academic platform, and the existing academic platform has deficiencies in resources, social contact, practice and other aspects, which need to be improved. Students' highest recognition for mentality, learning and research ability, and learning style is to expand academic vision, cultivate academic innovation and mobile ubiquitous learning, accounting for 67.05%, 59.69% and 51.16%, respectively. The data show that the average score of each item is above 3, indicating that users have a high degree of recognition of the design item, pay great attention to and attach great importance to the construction of intelligent academic platform, and have high expectations for the construction target.

In the traditional classroom, teachers and students meet face to face to impart knowledge and answer questions. However, once outside the classroom, this simplest and most direct form of communication no longer exists. Pc-based network learning platform has improved the situation that teachers and students cannot communicate directly when they are separated from each other geographically. However, the interaction between teachers and students in this way is still limited by time and space. Teachers and students still need to use desktops or laptops to enter the learning platform at a specific place. The intelligent learning network platform not only gets rid of the limitation of traditional classroom, but also breaks through the limitation brought by fixed PC.

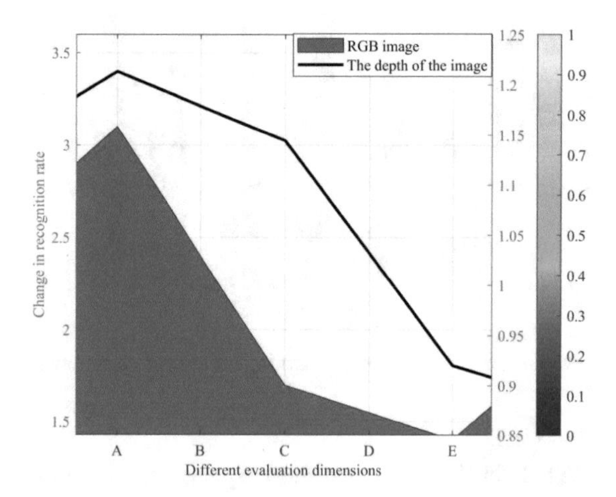

Fig. 2. Changes in RGB image recognition rate

As shown in Fig. 2, virtual and real interaction technology has been well applied in the intelligent learning network platform. Taking professional courses as an example, we first acquire the depth image data of a classroom scene and use RGB image data to calculate the size of RGB image and depth image, and establish the cache area with the same size as the depth image. The depth image and RGB image coordinate transformation matrix were calculated, the pixels in RGB image were traversed, the pixel points in the corresponding depth image of RGB image were found by using the transformation matrix, and whether they belonged to human body was judged, and mapped to the same coordinate position point and saved. Finally, the RGB image information of human body was obtained by using the interpolation algorithm. The depth image can be used to obtain human joints, calculate the position information of human hands in the depth image, segment the human hands image, and finally use the neural network for gesture recognition. In addition, by mapping the coordinates of the depth image of the hand to the world coordinate system, the activities of the human hand model in the scene can be one-to-one corresponded to the activities of the human hand in the real scene image. Through the recognition of the clenching and releasing movements of the human hand, the function of grasping virtual objects can be realized. At the interactive operator side, the depth image information and RGB image information are first obtained. The depth image information can be segmented into the depth image of the body's position, and the RGB image information of the body is obtained according to the coordinate mapping of the depth image and RGB image. This step is completed in the GetBodyFrame component. After image segmentation to get the body depth, first the pixels in the image is mapped to the world coordinate system, so as to get point cloud information, and then using the Mesh component in the Unity3D, will the body part of the depth of the image and the RGB image into the Mesh, for triangular Mesh, three-dimensional human body model building work is completed, then set the attribute information on the human body model to imitate the physical properties of the real object.

4.2 Suggestions on Key Technologies of Virtual and Real Fusion and Interactive Operation for Intelligent Learning

As far as the design of the platform itself is concerned, the transmission and processing of system data involves the information exchange between two processes (namely instruction recognition and interactive operation process), which is realized by using socket long connection mode. At the interactive operation end, the Texture2D type of Unity3D needs to be converted into the byte stream type, the byte stream and data length of the image are sent to the instruction recognition end, and the image data information is obtained through decoding of the data, and the result is sent back to the interactive operation end after calculation. The instruction recognition end is implemented by Python language, and the interactive operation end is implemented by C# language.

Also note that there are three main components in the system layer that control the basic functions of the system and are responsible for docking the interfaces provided by KinectSDK. KinectInterop components provide basic tools for system operation, such as coordinate system mapping between RGB images and deep images, file and external resource reading and writing, and multi-sensor configuration. The Interface between hand control and mouse control is provided in the InteractionManager component. Its functions include the location of left and right hands, left-right grab, touch events, and mouse click and drag information. The KinectGestures component provides interface functions for dynamic gesture controllers.

In addition to individual learning, mobile virtual learning community can be built in intelligent learning network platform. The learning community encourages learners to carry out interactive and collaborative learning with the quick and immediate interactive functions of mobile devices. Learners can carry out various interactive activities with teachers and other learners, and can organize students who share the same interests and learning needs with themselves to form a learning community. Members of the community can urge each other, communicate at any time, collaborate in learning and make progress together. Learners in the Shared community generally share the same knowledge and emotional needs, similar learning goals and similar interests. They are willing to abide by the community rules established through the joint consultation of learners and to do their part to build a better learning platform. While actively and autonomously researching and sharing learning resources, learners also have the opportunity to discuss and share their confusion in the learning process with like-minded people in the community, timely release information and share learning experience, and jointly construct social knowledge. This intangible resource-sharing learning mechanism greatly deepens the information exchange and mutual understanding between learners and community members, and creates a positive atmosphere for common learning.

5 Conclusions

Intelligent learning applies network technology to higher education, which can realize learners' free learning, personalized learning, interactive and cooperative learning, and knowledge sharing of social construction. Based on wisdom study expounds related concepts, on the basis of summarize its characteristics and put forward the principle of the construction of a learning platform, and detailed design the route of system environment and architecture, in order to improve the authenticity of augmented reality, the actual interaction system and immersive, using depth image data and RGB image data to improve the utilization rate of interaction design. Finally, the intelligent learning network platform is developed according to the construction and design of the system.

Acknowledgements. This work was supported by the project of Jilin Engineering Normal University: Key Technologies of Virtual Reality Fusion and Interactive Operation for Intelligent Learning in Network Environment.

References

1. Esserman, L.J.: The WISDOM study: breaking the deadlock in the breast cancer screening debate. NPJ Breast Cancer **3**(1), 34 (2017)
2. Owens, J.E., Menard, M., Plews-Ogan, M., et al.: Stories of growth and wisdom: a mixed-methods study of people living well with pain. Glob. Adv. Health Med. **5**(1), 16–28 (2016)
3. Kumar, C., Mishra, M.K.: Operation and control of an improved performance interactive DSTATCOM. IEEE Trans. Industr. Electron. **62**(10), 6024–6034 (2015)
4. Gunduz, H., Jayaweera, D.: Reliability assessment of a power system with cyber-physical interactive operation of photovoltaic systems. Int. J. Electr. Power Energy Syst. **101**, 371–384 (2018)

5. Deihimi, A., Zahed, B.K., Iravani, R.: An interactive operation management of a micro-grid with multiple distributed generations using multi-objective uniform water cycle algorithm. Energy **106**, 482–509 (2016)
6. Hesheng, W., Bohan, Y., Weidong, C.: Unknown constrained mechanisms operation based on dynamic interactive control. Caai Trans. Intell. Technol. **1**(3), 259–271 (2016)

An Approach to Observing China's Higher Educational Equity: The Application of Multimedia and Database in Liberal Arts Education After Lockdown

Zongye Jiao[1(✉)], Shali Shao[2], and Yixuan Wang[3]

[1] College of Arts, The Chinese University of Hong Kong, Hong Kong, China
jzy286@outlook.com

[2] Student Services Centre, Xi'an Peihua University, Xi'an, China

[3] The Dresden College of Music, Orchester Conductor, Dresden, Germany

Abstract. By studying the use of Multimedia and Database in Chinese liberal arts education, this paper argues that science and technology have a significant impact on the fairness of liberal arts education in China apart from those universities in developed regions. Through analyzing the use of two academic tools provided by universities in different regions, it can be concluded that in the field of liberal arts, the quantity and quality of resources available to liberal arts students in various universities are basically equal. This suggests that the use of these two technological tools could help universities narrow a range of serious inequalities caused by uneven educational classes and geographical disparities. However, for the three regions, all universities in the developed regions, at any level, are significantly better than those in the other regions in the number of traditional paper archives and technological tools. That is to say, with the help of science and technology, the differences for universities in developed regions can be largely ignored, but the gap between universities in developed regions and other regions, west and middle of China, has not been narrowed by the use of technological tools. Therefore, this paper argues that the government should invest more in big data and multimedia in universities in the middle and western regions, which will help realize the relative fairness of education in all regions and at all levels with the minimum cost.

Keywords: Double First Class · West of China · Covid-19 · Paper Materials · Library · Government investment

1 Introduction

Due to the impact of the epidemic, the Chinese government first took strict measures to shut down the city to prevent and control the epidemic. Various universities of China use distance learning service for students, which also provides an excellent way to observe the fairness of university education in China. For nearly forty years, because of the rapid development of economy, China's higher education has a rapid development. But relying on the levels of development in different regions, there is more and more big gap between the brand-name Universities, such as "Double First Class

J. MacIntyre et al. (Eds.): SPIoT 2020, AISC 1282, pp. 838–843, 2021.
https://doi.org/10.1007/978-3-030-62743-0_120

Universities", and Other Class Institutions, especially that schools offer students the resources quantity and quality of the gap is very big [1], for example the teachers holding doctoral degree, percentage of teachers having studied abroad, the number of state-level scientific research rewards etc. [2]

After the epidemic, because of the lockdown, universities adopt the model of distance education, so that technology becomes the core force dominating university education. It also provides a way to think the relationship between technology and the fairness of university education in China. In order to explore this issue, this paper focuses on the two main roles of science and technology in higher education: Multimedia and Database. Furthermore, compared with the education of Science and Engineering in universities, laboratory fails to play a big role in liberal arts education, which means most of its tasks can be resolved by distance learning. [3] Thus, through exploring the application of Multimedia and Database in liberal arts education, the influence of science and technology on the equity of higher education in China is able to be investigated.

2 Methodology

In the writing of this paper, a series of steps and research methods are taken in order to obtain more accurate data and then to complete the tasks of the thesis.

2.1 Data Collection

The Information Retrieval Method [4] is used to collect information from universities; several factors are therefore utilized and then divided into several groups (see Table 1).

The data collection has fully taken into account the impact of the regional gap, so in the specific operation, the surveyed universities are respectively from the developed eastern region, the central region and the northeast region, and the less developed western region.

At the school level, both elite and non-elite universities are treated as open source data, and internal gaps are fully noted. For example, for elite universities, one is the best university in the region, the other is the average university, and the third is the worst. For infamous universities, choosing the best university, a mediocre university, and a private university.

According to the special situation of Chinese universities, this paper also collects relevant information of private universities, so as to fully understand the current situation of Chinese universities in terms of Multimedia, digital Database, and printed materials which can provide a good standard to estimate the influences of two scientific factors when employing distance-learning to teach students.

Because this paper focuses on Liberal Arts Education, the universities selected in this paper are all comprehensive or Liberal Arts (Arts) universities. Additionally, Professional schools, such as conservatory of music, are also listed separately to facilitate comprehensive analysis and reach a more comprehensive conclusion.

For digital resources, in order to better understand the role of Database in education equity, this paper divides digital resources into two parts and lists them in Chinese Resources and English Resources to understand specific differences.

2.2 Data Processing

In the first step, this paper will collect relevant data according to the Method indicated by the data Collection, and then use the Comparative Analysis Method [5] to analyze the collected data. Specifically, data processing was separated by two categories:

Compare data from the same region to understand the specific characteristics of each region; Compare the specific situation of schools of the same nature in different regions to understand the situation of universities of the same level in different regions.

In the second step, this article uses Inductive Reasoning Method [6] to analyze the data obtained from the first step to understand the use of Multimedia and Database for liberal arts education throughout the Chinese mainland, thus to understand the role of the two tools in higher education equality in China (see Table 1).

Table 1. Data processing

University class	Regions	Universities	Digital database (Number)		Multimedia (Radio, Video, Ppt) (Number)	Paper Materials (10000, Updated Data)
			Chinese	English		
Double	Advanced Region (East of China)	1. Peking University	187	814	21	800 (2017)
		2. Central Conservatory of Music	19	13	2890	27(2015)
		3. Jinan University	68	63	15	440.2 (2019)
First	Others (Middle and Northeast of China)	4. Wuhan University	83	147	8	687 (2019)
		5. Hunan University	80	64	5	730.54 (2018)
		6. Henan University	78	68	2	370 (2015)
Class	Developing region (West of China)	7. Lanzhou University	47	51	2	360 (2014)
		8. Xinjiang University	70	43	20	184 (2016)
		9. Qinghai University	78	26	16	90 (2019)
	Advanced region (East of China)	10. Yangzhou University	35	53	24	382
		11. Nanjing University of the Arts	78	13	11	80 (2015)
		12. Nanfang College of Sun Yat-Sen University	15	4	4	188.57 (2019)

(*continued*)

Table 1. (*continued*)

University class	Regions	Universities	Digital database (Number)		Multimedia (Radio, Video, Ppt) (Number)	Paper Materials (10000, Updated Data)
			Chinese	English		
Other	Others regions	13. Bohai University	17	5	3	282 (2019)
		14. Hubei University	31	17	13	241.5 (2018)
		15. Xi'an Peihua University	4	1	2	229.1 (2019)
Classes	Developing region (West of China)	16. Guangxi Normal University	57	15	10	323.672 (2019)
		17. Longdong University	18	1	1	170.2 (2019)
		18. Xingjian College of Science and Liberal Arts	8	2	3	106 (2019)
Notes	1. All data are from the official web of related universities. 2. Because the "Double First Class" Art Universities are all located in developed regions, this paper only selects the two Universities of Arts in developed regions as the research objects, and other Universities of Arts located in other regions are not chosen. 3. Three of these universities are Private universities in Other Classes, including No. 12, 15, and 18.					

3 Data Analysis

From the statistics belonging to Double First Class, It is clear that the universities from Developed Areas and others do not have significant differences in three factors (Digital Database, Multimedia, and Paper Materials) except the English Digital Database from the Peking University and the Multimedia from the Central Conservatory of Music compared with the universities from Developing areas whose numbers from three factors are clearly lower.

For the data from Other Classes, although the number of Paper Materials are nearly the same in three places, the number of two technological factors from institutions in developed regions are very higher than universities from other two regions, which means students from those common universities situated in east of China have high level access to touching more research results from the world after Lockdown.

In terms of the overall data, the gap between private colleges and second-rate public universities in the region is gradually narrowing. This data is helpful to fully understand the status and level of private colleges, at least the gap between private colleges and public universities is not as big as people imagine.

Furthermore, to analyze kinds of factors from the Table 1, the gap among Double First-Class Universities is really little than that among Other Classes Universities, although for the element of Printed Materials, east and other regions of China's colleges from First-Classes have more books than any other colleges from the First Class and Other Classes. And, for two tech-factors, their situations are nearly the same as the situation from the Paper Materials, while the Colleges of Arts in advanced regions have particular advantages.

4 Conclusion with Advises

Through analyzing collected Data, although students in universities in developed regions study by Internet, their learning resources are still sufficient to support their study and research, but universities in other regions are not able to compete with those in developed regions, whether or not they are first-class universities or not. Moreover, compared with the number of traditional Paper Materials, it can be seen that the universities with a dominant number of printed Materials also have an advantage in the number of Database and Multimedia.

In other words, the full use of technology has narrowed the gap between universities in the middle and western regions, but widened the gap with universities in all developed regions, and has not changed the resource advantage of universities in the traditional developed regions.

The government should increase investment in higher education in the middle and western regions, especially in the construction and sharing of Databases and Multimedia. Because for the liberal arts education, the investment cost of two kinds of resources is small, and both can be rapid construction in the short term. This way allows students to access the same data resources as top universities in developed regions and the latest scientific research and knowledge information through the website of the school library in distance education. Not only is it very helpful to improve the competitiveness of universities in the middle and western regions, but also in a short time, to some extent, to narrow the resources' gap among universities across the country.

Universities in the Middle and west of China should pay more attention to the ability of liberal arts students to learn by themselves. Because middle and west of China's universities, whether famous or not famous, have the resources of digital Database and Multimedia nearly at the same level, students in these institutions are equally competitive. Of course, premise is that the students can use the data provided by the network actively explore the library resources, and then to improve their learning ability. If these schools can encourage their students to do so, even under the blockade, students who are good at self-learning can improve themselves and compete with students in developed areas by making full use of the power of two tec-tools.

References

1. Wang, Z., Liu, H., Wang, S.: "Birds of a feather flock together" University scientific research cooperation in double first-class construction—based on the perspective of literature analysis. Adm. Rev. (10), 99–109 (2019). (in Chinese)
2. Zhang, N.: "–From the perspective of incentive theory, the causes and countermeasures of brain drain under the background of" double first-class "construction in colleges and universities." J. Liuzhou Vocat. Tech. Coll. **18**(5), 46–52 (2018). (in Chinese)

3. Yang, X., et al.: Cross-domain feature learning in multimedia. IEEE Trans. Multimedia **17**(1), 64–78 (2014)
4. Bunawan, A., Nordin, S., Rahman, S.A.: Demonstrating the electronic resources information retrieval: analysis on the most preferable method. https://www.researchgate.net/publication/320225183_Demonstrating_the_Electronic_Resources_Information_Retrieval_Analysis_on_the_Most_Preferable_Method (2017)
5. Sadeghi, S.H.: E-learning practice in higher education: a mixed-method comparative analysis. Studies in Systems Decision & Control, vol. 122 (2018)
6. Ming, L., et al.: The relationship theory and model of inductive reasoning and deductive reasoning. Psychol. Sci. **2018**(4), 1017–1023 (2018)

Key Technologies of Automobile Assembly Intelligent Transfer Platform Based on Visual Perception

Yuefeng Lei(✉) and Xiufen Li

School of Hyundai Auto, Rizhao Polytechnic, Rizhao 276826, Shandong, China
leiyuefeng280@163.com

Abstract. This paper designs a hanging carrier platform based on visual perception technology. The main purpose of the platform is to automatically carry the load to a place like a ceiling and hang it. In this paper, binocular stereo vision technology is used to detect the position of the load's mounting point. Based on the principle of the Zhang type calibration method, MATLAB software programming was used to complete the calibration of the binocular camera, and then in-depth study of the Surf stereo matching algorithm, which is an accelerated and improved version of the scale-invariant feature transformation algorithm (Sift). This article obtains positioning errors through visual perception, and realizes closed-loop control of the movement of the transfer platform based on visual servos; the intelligent transfer platform control unit developed based on the Internet of Things technology can not only achieve precise positioning and positioning between the suspended conveyor line and the transfer platform Material transfer can also have the information interaction function between the transfer platform and the AGV, which can eliminate the positioning error with the AGV and automatically transfer the material. Studies have shown that after readjusting the installation position of the CD camera, the sudden change of the center of the circle is reduced after the lifting table is raised to the secondary positioning point. The transfer cycle of less than 45 s can effectively meet the cycle requirements of the annual output of 240,000 vehicles on the automobile assembly line.

Keywords: Visual perception · Car assembly · Intelligent transfer platform · Stereo matching

1 Introduction

With the development trend of intelligence and automation in all walks of life in the society, the demand for robots with high precision and fast response in the intelligent robot industry is increasing. The binocular vision ranging technology can make the robot have visual ability, which can greatly simplify the complexity of various operations of the robot. As a result, the degree of automation of various equipment is greatly improved, and the demand for human resources of the equipment is reduced. At the same time, the labor intensity is reduced, the working environment is improved, and the labor efficiency is improved.

J. MacIntyre et al. (Eds.): SPIoT 2020, AISC 1282, pp. 844–849, 2021.
https://doi.org/10.1007/978-3-030-62743-0_121

With the development of computer technology and the maturity of information transmission and processing technology, machine vision is gradually widely used in all aspects of industrial production, especially in image recognition and size measurement such as fingerprint recognition, face recognition, part size measurement, assembly Machine vision technology is widely used in the parts position perception in the process. Pratama used the Internet of Things technology to obtain heat transfer data inside the can, and established a heat conduction model of canned food in thermal sterilization. And apply the Internet of Things technology to MES to realize real-time information collection during the manufacturing process [1].

This paper studies and designs a set of mounted carrying platform with automatic ranging function based on binocular vision. By simulating the principle of human visual imaging, the system uses two cameras to obtain images of the same scene from two different viewpoints, and then uses the principle of binocular vision ranging to detect the parallax of the target point in the left and right images taken by the camera. The mathematical model of the visual principle calculates the depth between the target point and the camera, i.e. the carrying platform, and the three-dimensional space position of the target point, and transmits the calculation result to the controller through the serial port. The controller controls the motors of the carrying platform to make corresponding movements and finally transports the objects to be hung to the corresponding positions.

2 Key Technologies of Intelligent Assembly Platform for Automobile Assembly Based on Visual Perception

2.1 The Principle of Binocular Vision

(1) Camera imaging model

The nonlinear model is a model that considers the objective factors that exist in reality, such as lens distortion and assembly errors, which cause the projection transformation to not conform to the linear model. Therefore, the actual imaging model is mostly a nonlinear model. Lens imaging f is the focal length of the lens, v is the image distance, and u is the object distance [2, 3]. From the similarity principle of triangle, we can get:

$$\frac{1}{f} = \frac{1}{u} + \frac{1}{v} \tag{1}$$

It can be seen from Eq. (1) that when u is much larger than f, so that v $\approx$ f, the lens imaging model is similar to the small hole imaging model. Light rays obey the law of refraction when passing through a lens:

$$n_1 sin\theta_1 = n_2 sin\theta_2 \tag{2}$$

Where n_1 and n_2 are the refractive indices of the two media, and θ_1 and θ_2 are the angle of incidence and the angle of refraction, respectively. The lens linear imaging model is based on the assumption of lens paraxial refraction. Paraxial refraction means that the angle between the incident light and the optical axis is small [4]. According to the Taylor series, $\sin\theta \approx \theta$ in paraxial refraction, at this time the law of refraction can be expressed as:

$$n_1\theta_1 = n_2\theta_2 \tag{3}$$

In actual imaging, the greater the angle of incidence, the greater the difference between sin θ and θ, resulting in the light rays closer to the edge collecting earlier, causing spherical distortion of the lens, also known as radial distortion.

(2) Mathematical model of binocular distance measurement

The binocular horizontal convergence mode needs to use geometric principles, use at least six feature points, and then use a complex equation calculation to calibrate the camera matrix. At the same time, its mathematical model is also more complicated. Considering that this paper has certain requirements for the rapid response of the system, the binocular horizontal parallel mode is selected [5, 6].

2.2 Camera Calibration

(1) Corner detection algorithm

The corner detection algorithm based on the gray image does not need to extract the edge contour, but only needs to process the gray level of the image, so the corner extraction error caused by the error of the edge contour extraction can be avoided. Therefore, this method has a small amount of calculation, high accuracy, and a wide range of applications. Harris corner detection algorithm is a corner detection algorithm based on gray image [7]. The algorithm is simple, stable, uniform, and highly accurate, and it is a widely used algorithm. Therefore, Harris corner detection algorithm is used in the process of camera calibration in this paper [8].

(2) Production of calibration board

In order to realize the Zhang's calibration method, we must first make a black and white checkerboard for calibration. The checkerboard grid made in this paper is a square with a size of 2 26 mm, with a total of 7 × 9 squares. The dimensional accuracy and flatness of the checkerboard have a great influence on the calibration accuracy. In order to improve the calibration accuracy, the calibration board should be made as much as possible so that the checkerboard grid is in the same plane, so the middle part 5 × 7 grid is used for calibration during the calibration to avoid the area with higher unevenness around the chessboard [9, 10].

3 Experimental Research on Key Technologies of Intelligent Assembly Platform for Automobile Assembly Based on Visual Perception

3.1 System Hardware Design

The power module of the intelligent transfer platform uses a switching power supply with a power of 10 W, an output voltage of 48 V, and a rated current of 20 A. A DC power conversion module is used to step down the 48 V to 24 V, 12 V, 5 V for CD cameras, DSP image processing boards, ARM motion control board, ZigBee wireless communication module, RF card reader, photoelectric sensor, touch screen, motor driver and stepper motor power supply.

The vision module is mainly composed of a CD camera and a DSP image processing board. The CD camera adopts the front-end CD, which is widely used in the field of automobile safety, and has the characteristics of small size, high resolution and low power consumption. The ICETK-DM642-BR evaluation board is used as the DSP development platform, and the processor chip adopts the C6416 high-performance fixed-point DSP chip of TMS320DM642 produced by TI.

3.2 Hardware Circuit Design

The lower computer STM32 control chip has three main modules: data acquisition module, stepper motor control module, and data transmission module. The data acquisition module consists of a linear displacement sensor on the ball screw cloud transmission platform. The sensor can collect the position parameters of the slider. The motor control module is responsible for transmitting pulse control signals and angle information to the four motor drivers. The data transmission module is responsible for transmitting the real-time position of the platform collected by the lower computer, the motor speed and other information to the upper computer.

4 Experimental Research and Analysis of Key Technologies Based on Visual Perception of Intelligent Assembly Platform for Automobile Assembly

4.1 Analysis of Calibration Experiment of CD Camera

When the secondary positioning point is reached, the center of the circle has moved a distance, and the deviation data has changed. After the second precise positioning, the center of the circle returns to within ±1 pixel of the center point of the image to complete the final positioning of the circle mark. According to the calibration results of the CD camera, the actual distance represented by one pixel during coarse positioning is about 0.54 m, and the actual distance represented by one pixel during fine positioning is about 0.15 m. Identify and position the dot mark 50 times at 4 different starting positions, all of which can meet the requirement of positioning accuracy less than 1 m.

Table 1. Tact time of transfer platform

Coarse positioning	Rise to the second anchor point	Fine positioning	Jack up material tray	Descend to initial position	Total time
<6	18.4	<3	6	24 s	<52
<4	16.4	<1	4	18 s	<43

After readjusting the installation position of the CD camera, the center of abrupt distance decreases after raising the lifting table to the secondary positioning point. Table 1 shows the time taken by each stage of the transfer platform positioning during the experiment. Experiments show that the transfer cycle of less than 43 s can effectively meet the cycle requirements of an annual output of 240,000 vehicles on the automobile assembly line.

4.2 Positioning Experiment of Hanging Conveyor Line

A positioning error detection experiment was carried out on the hanging conveyor line. In the experiment, the hanging conveyor line was run forward at a speed of 0.1 m/s. The CD camera was used to detect the stopping error after the hanging conveyor line reached the station. The stopping error of the hanging conveyor line is shown in Fig. 1.

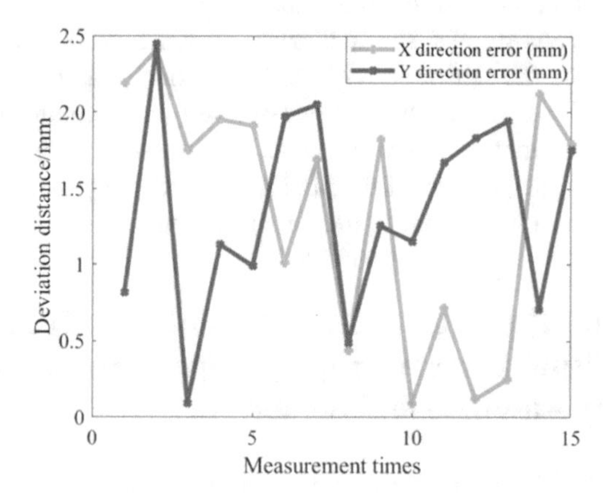

Fig. 1. Stop error of suspended conveyor line

It can be seen from Fig. 1 that the error of the running direction of the suspended conveyor line (X direction) is about 3 m, and the error of the running direction of the suspended conveyor line (Y direction) is about 2 m due to the limit of the running track of the hanging conveyor line.

5 Conclusions

This paper proposes a vehicle target detection algorithm based on deep hierarchical saliency network. This method is based on the VGG model and uses the convolutional neural network to extract the saliency features of the image to obtain a rough saliency map; the hierarchical circular convolution network is used for further layer-by-layer refinement to obtain the final fine saliency map. For the detection of vehicle target areas, this method detects the accuracy rate (0.91), recall rate (0.88) and F value (0.89) on the KITTI test library, this method is more accurate than the traditional detection method and hierarchical segmentation method, and can provide an effective method for the intelligent car to perceive the target in front of it.

References

1. Pratama, P.S., Nguyen, T.H., Kim, H.K., et al.: Positioning and obstacle avoidance of automatic guided vehicle in partially known environment. Int. J. Control Autom. Syst. **14**(6), 1–10 (2016)
2. Song, C., Chen, Z., Qi, X., et al.: Human trajectory prediction for automatic guided vehicle with recurrent neural network. J. Eng. **16**, 1574–1578 (2018)
3. Huang, Z., Yang, L., Zhang, Y., et al.: Photoelectric scanning-based method for positioning omnidirectional automatic guided vehicle. Opt. Eng. **55**(3), 034105 (2016)
4. Zhimin, C., Xinyi, H., Zhenxin, C., et al.: Position estimation of automatic-guided vehicle based on MIMO antenna array. Electronics **7**(9), 193 (2018)
5. Wei, D., Wenjie, C., Sandhya, C., et al.: Controlled gene and drug release from a liposomal delivery platform triggered by X-ray radiation. Nat. Commun. **9**(1), 2713 (2018)
6. Michaelides, A., Raby, C., Wood, M., et al.: Weight loss efficacy of a novel mobile diabetes prevention program delivery platform with human coaching. BMJ Open Diabetes Res. Care **4**(1), e000264 (2016)
7. Jinju, L., Er, S.P., Vipul, G., et al.: Mono-arginine cholesterol-based small lipid nanoparticles as a systemic siRNA delivery platform for effective cancer therapy. Theranostics **6**(2), 192–203 (2016)
8. Greer, R.D., Han, S.A.: Establishment of conditioned reinforcement for visual observing and the emergence of generalized visual-identity matching. Behav. Dev. Bull. **20**(2), 227–252 (2015)
9. Ahveninen, J., Huang, S., Ahlfors, S.P., et al.: Interacting parallel pathways associate sounds with visual identity in auditory cortices. Neuroimage **124**(Pt A), 858–868 (2016)
10. Su, C., Sicheng, Z., Zhijun, D., et al.: Visual identity-based earthquake ground displacement testing method. Shock Vibr. **2019**(PT.1), 2585423.1–2585423.10 (2019)

Personnel Management Optimization of University Library Based on Complex Network Theory

Xinyu Wu[1,2](✉)

[1] International College of NIDA, National Institute of Development Administration NIDA, 118 Moo3, Serithai Road, Klong-Chan, Bangkapi, Bankok 10240, Thailand
[2] Library of SWU, Southwest University, No. 2 Tiansheng Road, Beibei District, Chongqing 400700, China
xygirl@swu.edu.cn

Abstract. University library is the auxiliary institution of the university, which mainly provides various services for teachers, students and readers. However, with the advent of the information age, the establishment of modern digital libraries makes library management more difficult. Human resource management of university library is an important measure to improve the comprehensive quality of librarians, an important step to promote the development of libraries, and also of great significance to the development of universities. Therefore, it is an issue that needs to be paid attention to in the current university management process. The purpose of this paper is to study the optimization of university library personnel management based on the complex network theory. In this paper, the definition of complex network is briefly described, and then the degree of complex network is analyzed. The experiment part takes the library graduate student reader's borrowing information construction network as the research object, carries on the library borrowing network research, emphatically studies the library borrowing network complex network characteristic, carries on the thorough analysis to the library borrowing network community division. The experimental results show that the module degree Q value of the first community partition is the largest, which is 0.3029.

Keywords: Complex network · University library · Personnel management · Management optimization

1 Introduction

With the continuous development of modern multimedia technology, the application of computer information system in university libraries has become more and more popular. The emergence of the Internet has changed people's cognition of the world and brought some new changes to the work of university libraries [1, 2]. The emergence of computer networks has greatly changed the working mode and work content of university libraries [3, 4]. The current university library is developing towards the direction of combining the traditional mode and the digital mode, so for the university

J. MacIntyre et al. (Eds.): SPIoT 2020, AISC 1282, pp. 850–856, 2021.
https://doi.org/10.1007/978-3-030-62743-0_122

library, both in the overall personnel structure setting and in the internal management process have undergone great changes [5]. Limited by the types of resources, traditional libraries show a single collection of resources. Paper documents are the basic composition of traditional libraries. Meanwhile, limited by the speed and scope of searching materials, traditional libraries have a low utilization rate [6].

In the university library human resources management still exist many defects in the process, although there are many libraries have the human resources management as an important content of the library management, but most of the library only in the thoughts and ideas with this kind of consciousness, and in the process of actual management did not fully carry out to the actual, this whether from personnel allocation, recruitment, system settings, etc., can be found that the influence of the old system, these problems have become the important obstacle of the current library human resource management [7, 8]. Although some scholars have carried out corresponding researches on the human resource management of university libraries and made some achievements, most of the researches are based on the problems. In combination with specific research objects and by combining theory with practice, corresponding measures should be taken on the basis of investigation to further improve library human resource management, which plays a positive role in promoting actual management [9, 10].

In this paper, the definition of complex network is briefly described, and then the degree of complex network is analyzed. The experiment part takes the library graduate student reader's borrowing information construction network as the research object, carries on the library borrowing network research, emphatically studies the library borrowing network complex network characteristic, carries on the thorough analysis to the library borrowing network community division. The experimental results show that the module degree Q value of the first community partition is the largest, which is 0.3029.

2 Complex Network Theory

2.1 Complex Network

Complex network, as its name implies, should be a network composed of a large number of real and complex individuals according to their mutual relations; In terms of scale, network is much more complex than random network and regular network, and the network scale is also relatively large. From the current research, there is no simple and effective method that can directly generate the complex network conforming to the real system, so that we can more accurately study the statistical characteristics of the network.

Community structure of complex network: all vertices in the network are divided into different groups according to rules such as whether there are edge connections and the strength of the connections. The connections between the vertices belonging to the same group are relatively close, while there are almost no connections between the vertices belonging to different groups.

2.2 Complex Network Analysis

In this paper, degree is used to analyze the network, because node degree is the most direct index to determine and quantify the connection between nodes. The degree or connectedness of a node is the number of other nodes in a network connected to that node. Using the adjacency matrix, the degree of a node in the network can be defined as:

$$k_i = \sum\nolimits_{j \in V} a_{ij} \tag{1}$$

The degree distribution of nodes in the network is described by the degree distribution function P(k). P(k) represents the probability that a node of degree K is randomly selected in the network, or the proportion of nodes with degree K in the network. In the real world, the degree distribution characteristics of complex networks with different properties mainly include scale-free distribution, single scale distribution and wide scale distribution, which correspond to different distribution function characteristics.

Because sometimes in the actual situation, the sample is limited, the data will appear quite strong noise, that is, the data in the degree distribution of the tail oscillation amplitude is large, showing the phenomenon of fat tail. For this phenomenon, the cumulative degree distribution function of the network can be measured, namely

$$P_{Cum}(k) = \sum\nolimits_{k'}^{\infty} p(k') \tag{2}$$

It can be seen from the expression of the cumulative degree distribution function that the cumulative degree distribution can be obtained by summation of the original degree distribution P(k) in terms of degrees, and the statistical shocks in the tail of the original degree distribution can generally be eliminated.

3 Experimental Design of Library Lending Network

3.1 Data Sources

The research object of this paper is the borrowing records of graduate students in the library of Our Province Normal University, and the statistical time is from March 2019 to December 2019. The G (Culture, Science, Education, Sports) and Q (biological Sciences) books were selected for graduate readers who had borrowed books during this period. 246 graduate readers had borrowed G or Q books at least once.

3.2 Construction of Library Lending Network

Based on the borrowing information of graduate readers, a complex network model of the library borrowing network is established, and the following basic assumptions are made: (1) Each reader is regarded as a node in the complex network; (2) In the construction of library lending network, the heavy edge formed by multiple borrowing records between readers is not considered; (3) Without considering the weight of edges, the whole network is undirected and has no right to heavy network. Pajek is used to make relevant statistics on the characteristic parameters of library lending network. The statistical results are shown in Table 1.

Table 1. Statistical results

Readership	Average path length	Clustering coefficient	<k>	Degree distribution index
246	8	0.6338	4.6260	1.4999

4 Discussion of Experimental Results of Library Lending Network

4.1 Experimental Results and Discussion

According to the borrowing records of graduate student readers of the library of Normal University of Our province from March 2019 to December 2019, the borrowing records of Q books are selected, and 62 graduate students have borrowed Q books at least once. According to the method of building borrowing network in this paper, the complex network model of Q books is established. The complex network of Q books is divided into communities by spectrum bisection. The value of module degree Q after each community partition is shown in Table 2 and Fig. 1.

Table 2. Q value of module degree after community division

Spectral plane community division	Module degree Q value
The 1st community division	0.3029
The 2nd community division	0.2862
The third community division	0.1688
The 4th community division	0.1740
The 5th community division	0.1089

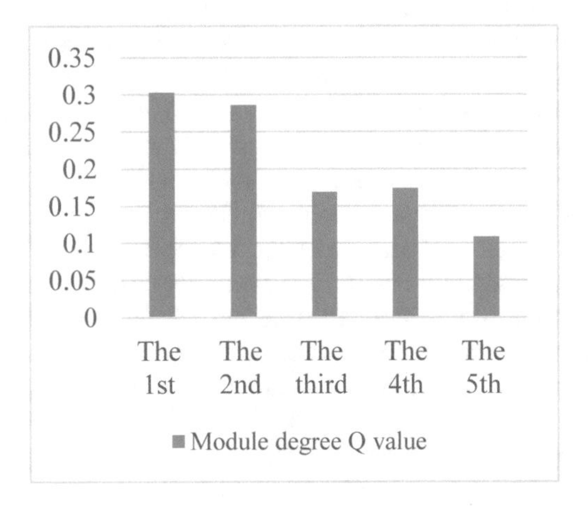

Fig. 1. Q value of module degree after community division

The closer the module degree Q is to the upper limit 1, the more obvious the community structure is. In Table II, the module degree Q value of the first community partition is the largest. The result of community division corresponding to the maximum value of Q is the most ideal network community structure, so the result of the first community division is the result of complex network community division constructed by using spectrum bisection method for Q books. In the divided communities, some “key readers” will spread the borrowing information to another community, which not only achieves the purpose of information dissemination, but also continuously expands the scope of communication between readers.

4.2 Suggestions to Promote the Optimization of University Library Personnel Management

Strengthen the analysis of the nature of librarians' work. In order to ensure that the library in the process of recruitment to recruit talents to accomplish the library's development, so the first thing you need to comprehensive analysis of the position of the nature of the work, the work is the work of library human resource management department, mainly analyzes the working skills requirements, knowledge content such as demand, through a content analysis can recruitment work lay the foundation for the library, at the same time can also provide theoretical basis for candidates for future work.

Rationally plan the structure of human resources. University library is a knowledge-intensive unit, so the future development direction of library should be planned first, and then reasonable personnel organization should be worked out to provide help for human resource management. Before the recruitment, the library managers need to research of existing staff, analysis of the existing talent structure, and then based on the library development plan to improve the talent introduction plan, including short term talent introduction plan and long-term talent introduction plan two aspects, which exists in view of the current university library human resources age structure is not reasonable, the problem such as the irrational proportion of men and women to improve.

Optimize the mechanism of talent selection and recruitment. For university library, the traditional personnel management system is the important reasons of unreasonable human resource management, so first have to be improved on this point, so as to talented person's power to grasp in their own hands, so as to prevent individual staff go through relations will not be returned to the qualified personnel to library management job. Secondly, it is necessary to develop a sound talent recruitment system. Each procedure, such as written test and interview, should be set up in detail during the recruitment process, so as to observe whether the applicant is competent for this job through objective investigation. Finally, a perfect talent competition system should be established. Only in the competition can excellent talents be found.

Strengthen the performance appraisal of human resources. Performance appraisal refers to the process of using scientific and reasonable methods to investigate the actual performance and attitude of employees, and reflecting these contents into specific data, so as to help managers better judge the effectiveness of each employee's work. Performance appraisal is an important basis to ensure the development of the organization. It can effectively stimulate the work potential of employees and is an important tool for human resource management in university libraries. Performance appraisal mainly investigates the professional ethics, working ability, actual performance and attendance of the staff. Therefore, when setting up the performance appraisal system of university libraries, the system can also be formulated by referring to these four aspects.

5 Conclusion

By constructing the library lending network model of graduate readers in a university, this paper considers each graduate reader as a node in the complex network, and the edge between nodes represents two readers who have borrowed the same book. University library borrowing was validated by empirical study network with small-world and scale-free properties, which has the characteristics of complex networks, and then using the method of complex network community partition to divide the community library network, to divide the good community, according to the size of the readers relevant similarity, similar to scale to provide targeted personalized recommendation service. Complex networks theory is applied to the analysis of university library's circulation message, studying the characteristics of a complex network of library network and community, is a complex network theory in the field of book intelligence application and extension, and provides a new research direction in the field of intelligence for books, and to better carry out the library's business has important theoretical significance and practical application value.

Acknowledgements. This work was supported by 203771.

In 2020, chongqing higher education teaching reform project (steering committee) project (203771) under the "major public emergency mechanism of emergency management in university library and services research".

References

1. Lin, Z., Wen, F., Xue, Y.: A restorative self-healing algorithm for transmission systems based on complex network theory. IEEE Trans. Smart Grid **7**(4), 2154–2162 (2017)
2. Yushu, S., Xisheng, T., Guowei, Z., et al.: Dynamic power flow cascading failure analysis of wind power integration with complex network theory. Energies **11**(1), 63 (2017)
3. Zhang, L., Lu, J., Zhou, J., et al.: Complexities' day-to-day dynamic evolution analysis and prediction for a Didi taxi trip network based on complex network theory. Mod. Phys. Lett. B **32**(9), 1850062 (2018)
4. Wang, Y.H., Shen, X.R., Yang, S.Q., et al.: Three-dimensional dynamic analysis of observed mesoscale eddy in the South China Sea based on complex network theory. EPL (Europhys. Lett.) **128**(6), 60005 (2020)
5. Strozzi, F., Garagiola, E., Pozzi, R., et al.: Length of stay reduction in the emergency department and its quantification using complex network theory. Int. J. Oper. Res. **36**(1), 1 (2019)
6. Zhou, C., Ding, L., Zhou, Y., et al.: Visibility graph analysis on time series of shield tunneling parameters based on complex network theory. Tunn. Undergr. Space Technol. **89**(Jul), 10–24 (2019)

7. Zekun, W., Xiangxi, W., Minggong, W.U.: Identification of key nodes in aircraft state network based on complex network theory. IEEE Access **7**, 60957–60967 (2019)
8. Li, G.J., Hu, J., Song, Y., et al.: Analysis of the Terrorist Organization Alliance Network based on complex network theory. IEEE Access **7**(99), 103854–103862 (2019)
9. Chong, P., Yin, H.: Analysis on optimization strategies of hazardous materials road transportation network using complex network theory. Complex Syst. Complex. **15**(3), 56–65 (2018)
10. Gao, L., Cao, J.H., Song, T.L., et al.: Evolution model of equipment support system of systems based on complex network theory. Binggong Xuebao/Acta Armamentarii **38**(10), 2019–2030 (2017)

Load Optimization Scheme Based on Edge Computing in Internet of Things

Fanglin An and Jun Ye(✉)

School of Computer Science and Cyberspace Security, Hainan University, Haikou, China
yejun@hainanu.edu.cn

Abstract. Ubiquitous Internet of Things devices are increasingly being used by people. However, cloud computing alone cannot solve all the usage scenarios. Edge computing is created to enable real-time computing of small-scale data or Internet of Things devices in underdeveloped regions to function and use normally. Edge computing is essentially the use of its own or surrounding computing centers with powerful computing power to serve itself, It's not like cloud computing, unified control of resources and achieve the goal of energy saving, it also makes some calculations carrier got a lot of waste, not very friendly to the global push for green computing, load optimization research for this problem, this paper calculated according to the characteristics of the edge, the edge of the predictable cognitive load optimization scheme is put forward, thus reduce the energy consumption, promote green computing.

Keywords: The Internet of Things · Edge calculation · Load optimization · Green computing

1 Introduction

A context resource manager for a virtual machine deployed in an Internet of Things network is a group of applications that allocate resources such as computing processes, networks, nodes, and storage to a cloud computing environment on demand. It can migrate some computations that can be migrated within the limited range to compute nodes that have spare time. However, computation requests include computation time limit. For example, unmanned computation requests must be completed in a very short time, otherwise millisecond delay will directly lead to accidents.

Performing tasks on network edge devices is a new computing model. Edge computing refers to any resource between the data source generated by the network edge terminal device and the data path of the cloud computing center. The basic idea is to transfer the task to a resource center near the data source [1, 2]. The relationship between edge computing and cloud computing is not one or the other, but complementary. They complement each other and jointly promote the development of the Internet of Things.

Edge computing has been proposed for a long time. However, with the rapid development of Internet of Things technology, cloud server receives hundreds of thousands of computing requests all the time. As for the hot topic of energy waste, energy conservation and environmental protection has become a new research hotspot.

J. MacIntyre et al. (Eds.): SPIoT 2020, AISC 1282, pp. 857–861, 2021.
https://doi.org/10.1007/978-3-030-62743-0_123

A recent study shows [3] that data centers alone will account for 2% of global energy consumption by 2020. W. Shi et al. [4] proposed that with the development of mobile network to 5G stage, MEC, as its key technology, will bring unlimited possibilities for network services and other services. MEC technology transforms the traditional cloud computing model of centralized data storage into mobile edge data storage. This technology can reduce the network operation and business delivery delay. Providing user experience is the key technology for 5G networks in the future. Qadri Y A et al. [5] proposed that the use of artificial intelligence (AI) has changed almost every level of medical IoT system. The fog/edge paradigm brings computing power closer to deployed networks, alleviating many of the challenges in the process. Zhang Y et al. [6] proposed that both edge computing and cloud computing provide computing services for mobile devices to enhance their performance. The edge computing can reduce transmission latency by providing local computing services, while the cloud can support huge computing requirements.

Since the terminal of the Internet of Things does not need to work all the time in some specific environments, its demand for edge computing can be divided into continuous and intermittent. The energy-saving optimization of the Internet of Things for continuous work can be achieved by optimizing the calculation algorithm to reduce the operating frequency of CPU. This aspect is not studied in this paper. In this paper, the feasibility of hibernating energy saving of computing nodes in the intermittent Internet of Things network is optimized, and the node deployment of edge computing and the regular prediction of calculated amount with the duration of time series are mainly studied.

The method in this paper aims at the perception of edge computing environment and enables cloud servers to conduct machine learning through historical computing records, Thus deploying the entire Internet of Things server computing network, which can meet the computing requirements at any time and reduce the power consumption of a part of computing nodes as far as possible. Finally, by balancing machine learning and power consumption, the optimal energy saving strategy is made.

2 Related Work

Various research teams around the world are increasingly interested in the power consumption in edge computing and cloud computing environments. In view of the research on data migration from edge computing to cloud computing nodes, new energy-saving computing networks are designed to reduce power consumption.

In the past, virtual machine resource management was treated as a packing problem in cloud server environment, and the methods of best fitting, first fitting, best fitting decreasing, genetic algorithm and simulated annealing [7] were proposed. But the previous methods only give the optimal solution of the problem, not the global optimal solution. X. Meng et al. [8] suggested using traffic patterns between virtual machines and optimizing the location of virtual machines to improve network scalability. However, this method is only applicable to the part of high bandwidth demand, and cannot solve the problem of high edge computing power. W. Wang et al. [9] solved the problem of energy saving through virtual machine allocation. The problem of centralized allocation and migration of virtual machines is solved by adopting the method based on multiple neural intelligence and considering the cost. For each computing host, an agent needs to be deployed, which requires constant communication between these agents and increases the cost of additional operations and computing time. A. Wolke et al. [10] discussed workload perception migration, but it needs to be carried out in A deterministic environment where the workload is known A priori. In Wang Y et al.'s scheme [11], an improved genetic algorithm is adopted to carry out fine-grained task migration, which achieves a good effect of reducing energy consumption. However, his hypothesis is aimed at the situation of high bandwidth and low task migration cost in the 5G era. Mohiuddin I et al. [12] designed a technology that considers server load and migration cost. This method migrates computing with lower migration cost to computing virtual machine of other cloud servers in the same class by classification, so as to achieve the purpose of closing several servers and saving energy. The premise of this method is to ignore the conversion time difference between the cloud server sleep and standby and the power consumption on startup, and not to comprehensively consider multiple factors in the complex environment. Although it is more effective than genetic algorithm and annealing algorithm, the factors considered are too simple, and the actual implementation does not save energy very well.

3 Our Scheme

Here, we analyze that the energy consumption of the current edge computing equipment is mainly caused by transmission power and redundant standby energy waste caused by several server nodes processing a small amount of computation. On the one hand, it can reduce the interference to neighbor nodes by reducing transmitting power, improve network throughput, reduce energy consumption of nodes, and also extend the life of nodes and networks. On the other hand, through machine learning analysis of computing records, cloud servers can be deployed in advance in the next state, enabling them to maintain normal usage while reducing energy consumption. As shown in Fig. 1, computing nodes are deployed to optimize energy saving through machine learning prediction of the next state.

Here we discuss in detail the machine learning algorithm used in the above figure: For ease of understanding, the function $N = f(t) = 1100 + t + t^2 + t^3$ sin t is used here to represent the total number N of calculation times for all edge computing nodes corresponding to each time t in a day, where the function image is as shown in the Fig. 2 shows.

Of course, this is not the case in the real environment. Here is just to describe the scheme of this article more vividly, but in the real environment, similar data information $\{t_1, N(t_1)\}, \{t_2, N(t_2)\}, \cdots, \{t_n, N(t_n)\}$ can be obtained. What we need is a record of all the computations that have gone on.

Setting: Divide the time of day into k different Spaces according to computing capacity A of server node, In any space, there will be an upper limit of $N_{i-\max}$ and a lower limit of $N_{i-\min}$, so that it meets: $N_{i-\max} - N_{i-\min} = r \times A$, of which $r = m - s$ and m are the minimum number of sleepable nodes set by the system administrator in advance, s is the number of standby server nodes set by the system administrator.

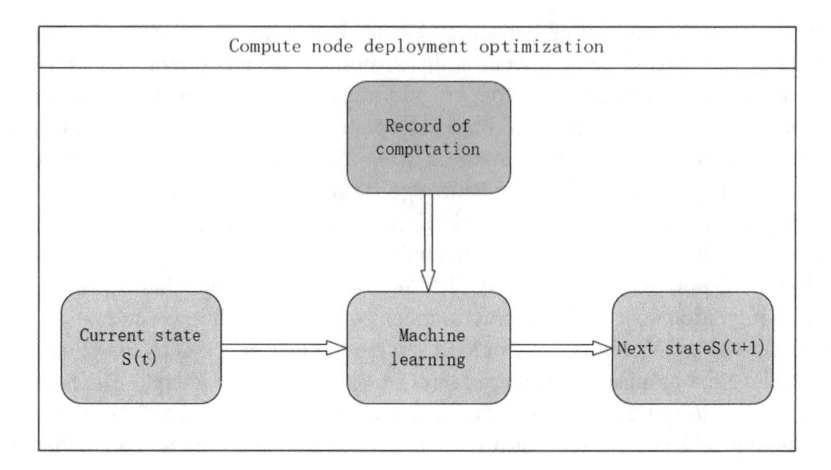

Fig. 1. Optimized deployment of compute nodes

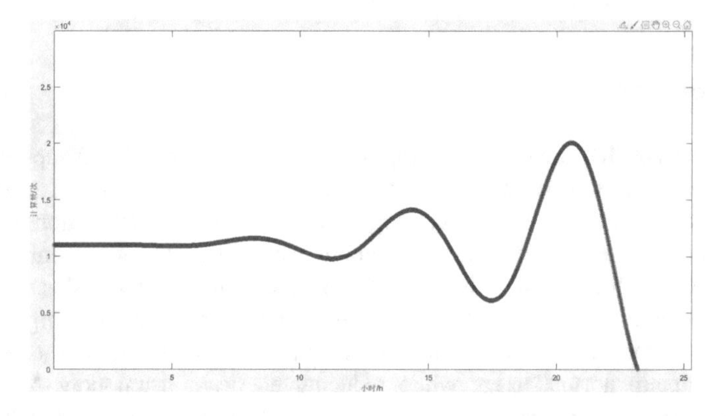

Fig. 2. Diagram of the variation of simulated calculated amount over time series

Machine learning: Build a machine learning network prediction model. By fitting the daily data, the machine learning model output $F_{i-\max}$ will continue to tend to the actual output $N_{i-\max}$.

Prediction: For the space i where the current time is t, the node calculation amount $F_{(i+1)\max}$ at the next time will be predicted by the machine learning model.

Deployment: According to the prediction result $F_{(i+1)\max}$ in the previous step, considering the prediction error, in order to ensure that the computing nodes can still meet the daily computing needs, at the next moment, $F_{(i+1)\max} + s$ edge computing server computing nodes in the IoT network are reserved, and the remaining nodes are shut down to achieve the goal of saving energy.

Every day in a real environment, the data do not necessarily have similarity, so not only to study the data in a day, and what's more, machine learning should be carried out on the calculation records of each quarter and month. Through the special holiday, the weather, the temperature parameters such as the overall study, to make the training of the model can more accurately predict the next moment of node calculation, further reduce the waste of energy consumption of redundant servers.

4 Conclusion

In recent years, sensor networks have developed rapidly. However, the current research and discussion focus is not on sensors themselves. Instead, various sensors form an IoT edge computing network through various low-power information transmission computing technologies, thus providing various applications. Aiming at the problem of energy consumption waste of computing nodes in the edge computing network of the Internet of Things, this paper uses machine learning model to predict, deploy node network in the next stage in advance, close or sleep the unnecessary computing nodes, and reduce energy consumption waste.

Acknowledgments. This work was partially supported by the Science Project of Hainan University (KYQD(ZR)20021).

References

1. Yong, C., Jian, S., Cong-Cong, M., et al.: Mobile cloud computing research progress and trends. Chin. J. Comput. **40**(2), 273–295 (2017)
2. Mao, Y., You, C., Zhang, J., et al.: A survey on mobile edge computing: the communication perspective. IEEE Commun. Surv. Tutorials **19**(99), 1 (2017)
3. Vasu, R., Nehru, E.I., Ramakrishnan, G.: Load forecasting for optimal resource allocation in cloud computing using neural method. Middle-East J. Sci. Res. **24**(6), 1995–2002 (2016)
4. Shi, W., Cao, J., Zhang, Q., et al.: Edge computing: vision and challenges. IEEE Internet Things J. **3**(5), 637–646 (2016)
5. Qadri, Y.A., Nauman, A., Zikria, Y.B., et al.: The future of healthcare internet of things: a survey of emerging technologies. IEEE Commun. Surv. Tutor. **22**(2), 1121–1167 (2020)
6. Zhang, Y., Lan, X., Ren, J., et al.: Efficient computing resource sharing for mobile edge-cloud computing networks. IEEE/ACM Trans. Netw. (2020)
7. Jangiti, S., Shankar Sriram, V.S. : Scalable and direct vector bin-packing heuristic based on residual resource ratios for virtual machine placement in cloud data centers. Comput. Electr. Eng. **68**, 44–61 (2018)
8. Meng, X., Pappas, V., Zhang, L.: Improving the scalability of data center networks with traffic-aware virtual machine placement. In: Proceedings IEEE INFOCOM (2010)
9. Wang, W., Jiang, Y., Wu, W.: Multiagent-Based resource allocation for energy minimization in cloud computing systems. IEEE Trans. Syst. Man Cybern. **47**(2), 205–220 (2016)
10. Wolke, A., Ziegler, L.: Evaluating dynamic resource allocation strategies in virtualized data centers. In: 2014 IEEE 7th International Conference on Cloud Computing (CLOUD) (2014)
11. Wang, Y., Zhu, H., Hei, X., et al.: An energy saving based on task migration for mobile edge computing. EURASIP J. Wirel. Commun. Netw. **2019**(1), 133 (2019)
12. Mohiuddin, I., Almogren, A.: Workload aware VM consolidation method in edge/cloud computing for IoT applications. J. Parallel Distrib. Comput. **123**, 204–214 (2019)

Fire Detection Method Based on Improved Glowworm Swarm Optimization-Based TWSVM

Ke Fu, Zhen Guo, and Jun Ye(✉)

School of Computer Science and Cyberspace Security, Hainan University, Haikou, China
yejun@hainanu.edu.cn

Abstract. In recent years, fires occurred frequently, and the solutions of fire detection and recognition are required automatically. Although there are various method of fire detection such as temperature and smoke sensor, the efficient real-time detection is not guaranteed. In order to solve this problem, a quick and effective fire detection method of twin support vector machine (TWSVM) based on improved glowworm swarm optimization algorithm (GSO) is proposed. Compared with SVM, twin support vector machine not only extends the range of the kernel function selection, but also obviously enhance the generalization ability of support vector machine. However, there are also some problems of twin support vector machine including the disability of dealing with the parameter selection problem properly which could reduce the classification capability of the algorithm. In this paper, we use improved glowworm swarm optimization algorithm to search the optimal penalty parameter and kernel parameter of TWSVM. Reasonable parameters will improve the performance of twin support vector machine and increase the accuracy. Experimental results show that the proposed approach is efficient and has high classification accuracy compared with traditional support vector machine and so on. Finally, the proposed method of twin support vector machine based on improved glowworm swarm optimization algorithm effectively improves the accuracy and real-time performance of flame identification, and settles the problems of TWSVM such as the difficulty of parameter selection in flame recognition and long optimization time of common parameter optimization algorithms.

Keywords: Fire detection · Twin support vector machine · Glowworm swarm optimization

1 Introduction

At present stage, fire is one of the most common and universal threats to public safety and social development. With the continuous development of society, the risk of fire is also increased due to the implement of various flammable materials and so on. Traditional automatic fire alarm system collect relevant data and analyze them through temperature and smoke sensors to predict fire probability, but the real-time detection

J. MacIntyre et al. (Eds.): SPIoT 2020, AISC 1282, pp. 862–867, 2021.
https://doi.org/10.1007/978-3-030-62743-0_124

cannot be guaranteed. With more intuitive and faster response, the fire image recognition is widely studied and increasingly applied to diverse circumstances.

As for the problem of how to detect fire image accurately and timely under various interferences, many researchers contribute many solutions to this problem. Mao, W et al. (2018), used multi-channel convolutional neural network technology to recognize flame [1]. Image processing, which is based on support vector machine (SVM) was applied to forest fire detection and identification by Mahmoud et al. (2019) [2]. Teng Z et al. (2010), proposed a hidden Markov models for fire detection [3]. Compared with other machine learning methods, such as artificial neural network and convolutional neural network, support vector machine (SVM), a classical method based on statistical learning theory which can handle classification and regression, can greatly improve the machine learning's performance of high dimension and local minima. Although support vector machine has excellent performances in some aspects, it requires a lot of quadratic programming calculation in dealing with large-scale data which may lead to large amount of classification calculation and slow classification speed. Twin support vector machines, which was proposed by Jayadeva et al. (2007), has faster training and learning speed and lower computational complexity and time costing [4]. In this paper, we apply TWSVM theory to fire detection which not only greatly improves the performance of flame recognition but also largely reduces the time complexity of the algorithm.

Similar with SVM, TWSVM has a poor performance on kernel function selection. One of the common kernel functions of TWSVM is Gaussian radial basis kernel function but a lot experimental data show that its generalization ability still needs perfection. For this issue, many scholars contribute to the study of TWSVM algorithm, there are still problems we can solve to improve the performance of TWSVM algorithm [5–7]. Wang Z et al. (2013), designed and implemented the smooth twin parametric-margin support vector machine based on genetic algorithm (GA) on model selection [8]. Bian, Yongming et al. (2018), put forward an improved fire detection algorithm based on tchebichef moment invariants and PSO-SVM [9]. Bi, Fangming et al. [10], improved the fire detection method through improved fruit fly optimization-based SVM. Gao, Yikai et al. (2020), proposed to put twin support vector machine based on improved artificial fish swarm algorithm (IAFSA) into flame recognition and analyzed different algorithm's performance on fire detection [11]. In this paper, we use an improved glowworm swarm optimization (GSO) algorithm to search the optimal penalty parameter and kernel parameter of TWSVM which Ding, Shifei et al. (2017) have proposed and apply it to flame image recognition [12]. As for glowworm swarm optimization, there are many contributions to its improvement such as RF-GSO and GSOS-ELM [13, 14]. In this paper, a adaptive Glowworm Swarm Optimization algorithm with Changing Step (CSGSO) is presented to solve the problem that the Glowworm Swarm Optimization (GSO) algorithm to optimize the multi-modal function existing slow convergence and low precision defects.

Through the above improvements, we have developed a qualified method of TWSVM based on GSO which could overcome the basic shortcomings of low accuracy and low optimization speed, realize the rapid and accurate selection of TWSVM parameter thus help TWSVM achieve higher classification accuracy. In this paper, we apply this feasible and effective method to the flame recognition field.

2 Materials and Methods

2.1 Twin Support Vector Machine

Twin support vector machine, proposed by Jayadeva et al. in 2007, is developed based on traditional SVM which can handle binary classification through solving two quadratic programming problems (QPPs) [4]. TWSVM generates a hyperplane for each class of samples to make each hyperplane is as close as possible to the samples belonging to one of classes and as far as possible from the samples of the other class. Since the size of the quadratic programming problem solved by TWSVM is a quarter of the original SVM, theoretically, the computational efficiency is 4 times that of the traditional SVM, which greatly improves the classification efficiency. A simple geometric interpretation is given in Fig. 1. In Fig. 1, two lines represent two hyperplanes of TWSVM, red points represent the samples of Class +1 and blue points represent the samples of Class −1 [12].

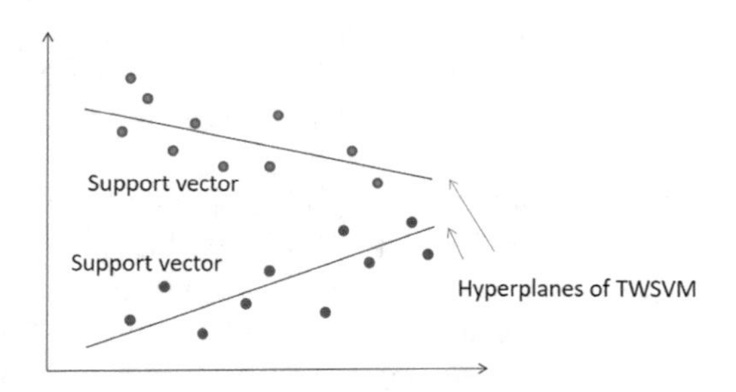

Fig. 1. The basic thought of TWSVM.

2.2 Glowworm Swarm Optimization

In the GSO algorithm, each artificial glowworm walks in the solution space with fluorescence and has its own line of sight, which is called local-decision range. Their brightness depends on the target value in their position. The brighter the glowworm is, the better its position, the better its target function value. To put it simply, GSO algorithm mainly consists of four stages: glowworm deployment (initialization), luciferin update, position update and decision radius update. Among these four stages, the luciferin update and position update stand for a core part of the algorithm and following formula are presented as below:

$$l_i(t) = (1 - \rho)l_i(t - 1) + \gamma J(x_i(t)) \tag{1}$$

$$X_i(t+1) = X_i(t) + s\left(\frac{X_j(t) - X_i(t)}{\left\|X_j(t) - X_i(t)\right\|}\right) \tag{2}$$

where $l_i(t)$ represents the luciferin level associated with glowworm i, ρ $(0 < \rho < 1)$ is the luciferin decay constant, and γ is the luciferin enhancement constant.

Glowworm Swarm Optimization Algorithm with Changing Step. Anew adaptive Glowworm Swarm Optimization algorithm with Changing Step (CSGSO) is presented to solve the problem that the Glowworm Swarm Optimization (GSO) algorithm to optimize the multi-modal function existing slow convergence and low precision defects. Glowworm i is attracted to the brighter glowworm j to update its position. The improved position update formula is as follows:

$$x_i(t+1) = x_i(t) + \beta * (l_{ij})(x_j(t) - x_i(t)) + \alpha * (rand - 1/2) \tag{3}$$

where α is the step size factor which takes the constant above [0, 1] and rand is a random factor uniformly distributed on [0, 1].

2.3 CSGSO-TWSVM

In this section, CSGSO will be applied to the parameter selection of TWSVM. The position of each glowworm is a vector of dimension corresponding to a set of parameters of TWSVM. The objective function of CSGSO-TWSVM is to calculate the average classification accuracy of TWSVM with current position of each glowworm. CSGSO will generate an extreme value as the optimal parameters of TWSVM.

The algorithm of CSGSO-TWSVM is as follows:

(S1) Initialize the considering dataset and divide them into two parts, one for training of 80% and another for testing of 20%.
(S2) Initialize the parameters of GSO and the positions of each glowworm with random values.
(S3) Calculate the objective function with position of each glowworm, output the optimal value in this iteration and record the position of each optimal glowworm.
(S4) Update the local optimal value and judge whether it is the best one in recorded global optimal value. If so, update the global optimal value, if not continue.
(S5) Calculate the luciferin of each glowworm by (1).
(S6) Compare each glowworm's luciferin in the decision domain and pick out the larger one to consist of its set of neighbors. The local-decision range is as follows:

$$N_i(t) = \{j : \|x_j(t) - x_i(t)\| < r_d^i(t); l_i(t) < l_j(t)\} \tag{4}$$

In this formula, $N_i(t)$ represents the set of neighbors of glowworm i at time t and $r_d^i(t)$ represents the dynamic decision domain of glowworm i at time t.

(S7) Calculate the transition probability of each glowworm in the set of neighbor and glowworms are attracted by the brighter one in the neighbors. Then glowworms update their position by (3).
(S8) Update the radiuses of dynamic local decision ranges of each glowworm and judge whether the iteration number has reached the maximum or not. If so, continue; otherwise, go back to Step 3.
(S9) Update the parameters of TWSVM as the final global optimal value and build TWSVM model.
(S10) End the algorithm.

3 Results and Analysis

In order to demonstrate the effectiveness of GSO-TWSVM, we collected training samples by acquiring fire images on the network and intercepting public fire video data sets as following. More than 10,000 images were collected, and 4,730 images were finally selected as the actual training test data set.

The experiments in this paper were trained and tested under the following working conditions: MATLAB 2016 on a computer with 8G memory, Intel (R) 2.2 GHz Intel Core i7 8750H CPU and 1T hard disk. The performances of fire flame recognition test are compared among the SVM, TWSVM and CSGSO-TWSVM. We use grid method with cross validation to determine the parameters of SVM and TWSVM as well as the kernel parameter. Because the results obtained by the heuristic algorithm may have randomness for each running cycle, each algorithm is run ten times, and the maximum number of iterations of each algorithm is one hundred. The corresponding optimal fitness value (c_1, c_2, σ) in the 10 times is taken as the optimal parameter.

The fitness function of all intelligent optimization algorithms in the experiment is the average classification accuracy obtained by using the ten-fold cross validation method. The optimal parameters found by different algorithms (c_1, c_2, σ), the optimal fitness value (p) corresponding to the optimal parameters, and the average parameter optimization time (t) of the algorithm running ten times are shown in Table 1.

Table 1. The comparison of different algorithm.

Algorithm	c_1/c_2	σ	p(%)	t(s)
SVM	1.757	0.138	79.547	2959.547
TWSVM	1.783	0.089	80.815	1586.763
CSGSO-TWSVM	1.341	0.355	82.643	987.351

It can be shown from Table 1 that CSGSO-TWSVM has obviously advantages in the optimal fitness value and average parameter optimization time compared with SVM and TWSVM.

4 Conclusion

CSGSO-TWSVM uses glowworm swarm optimization with changing step to find the optimal solutions of TWSVM and sets the accurate objective function to the function of the classification algorithm which can determine extreme value domain with a high speed and has good generalization ability. However, it may trap into local optimum and low convergence speed in the later period of the optimization but it still can provide an effective new method for applying TWSVM to flame recognition.

Acknowledgments. This work was partially supported by the Science Project of Hainan Province (No. 619QN193), the Science Project of Hainan University (KYQD(ZR)20021).

References

1. Mao, W., Wang, W., Dou, Z., et al.: Fire recognition based on multi-channel convolutional neural network. Fire Technol. **54**(2), 531–554 (2018)
2. Mahmoud, M.A.I., Ren, H.: Forest fire detection and identification using image processing and SVM. J. Inf. Process. Syst. **15**(1), 159–168 (2019)
3. Teng, Z., Kim, J., Kang, D.: Fire detection based on hidden Markov models. Int. J. Control Autom. Syst. **8**(4), 822–830 (2010)
4. Jayadeva, Khemchandani, R., Chandra, S.: Twin support vector machines for pattern classification. IEEE Trans. Pattern Anal. Mach. Intell. **29**(5), 905–910 (2007)
5. Huang, H., Ding, S., Shi, Z.: Primal least squares twin support vector regression. J. Zhejiang Univ. Sci. C Comput. Electron. **14**(9), 722–732 (2013)
6. Chen, W.J., Shao, Y.H., Li, C.N., et al.: MLTSVM: a novel twin support vector machine to multi-label learning. Pattern Recognit. **52**, 61–74 (2016)
7. Zhang, X., Ding, S., Sun, T.: Multi-class LSTMSVM based on optimal directed acyclic graph and shuffled frog leaping algorithm. Int. J. Mach. Learn. Cybern. **7**(2), 241–251 (2016)
8. Wang, Z., Shao, Y., Wu, T.: A GA-based model selection for smooth twin parametric-margin support vector machine. Pattern Recognit. **46**(8), 2267–2277 (2013)
9. Bian, Y., Yang, M., Fan, X., et al.: A fire detection algorithm based on Tchebichef moment invariants and PSO-SVM. Algorithms **11**(6), 79 (2018)
10. Bi, F., Fu, X., Chen, W., et al.: Fire detection method based on improved fruit fly optimization-based SVM. Comput. Mater. Continua **62**(1), 199–216(2020)
11. Gao, Y., Xie, L., Zhang, Z., Fan, Q.: Twin support vector machine based on improved artificial fish swarm algorithm with application to flame recognition. Appl. Intell. **50**(8), 2312–2327 (2020)
12. Ding, S., An, Y., Zhang, X., et al.: Wavelet twin support vector machines based on glowworm swarm optimization. Neurocomputing **225**, 157–163 (2017)
13. Su, Z., Li, L., Li, W., Meng, F., Sigrimis, N.A.: Design and experiment on adaptive dimming system for greenhouse tomato based on RF-GSO. Nongye Jixie Xuebao/Trans. Chin. Soc. Agric. Mach. **50**, 339–346 (2019)
14. Liu, F., Zhong, D.: GSOS-ELM: an RFID-based indoor localization system using GSO method and semi-supervised online sequential ELM. Sensors **18**(7), 1995 (2018)

A Data Privacy Protection Scheme for Underwater Network

Dandan Chen[1], Shijie Sun[2], Xiangdang Huang[1], and Qiuling Yang[1](✉)

[1] School of Computer Science and Cyberspace Security, Hainan University, Haikou 570228, China
18736895993@163.com, 990709@hainanu.edu.cn

[2] School of Information and Communication Engineering, Hainan University, Haikou 570228, China
sun@hainanu.edu.cn

Abstract. Underwater sensor network is deployed in unattended underwater environment, which can be used to collect sensitive information. It has problems such as low bandwidth, high energy consumption and difficulty in ensuring data security. In order to reduce energy consumption and ensure data security, this paper proposes a data privacy protection scheme for underwater network. This scheme makes an upload decision when the node collects data, and compresses the uploaded data in the ordinary node to reduce the data transmission volume and energy consumption; furthermore, the cluster head node aggregates the data after receiving the data packets sent by the node to reduce the data redundancy; finally, the data packets are segmented to ensure the data security.

Keywords: Data security · Data aggregation · Packet segmentation

1 Introduction

In underwater sensor network, the ultimate goal is to protect data security and save energy consumption. In the early research, due to the bad environment of nodes, the privacy and security of their data has not been paid enough attention [1]. Data aggregation is a basic process of underwater sensor network, which aims to effectively collect and aggregate data to improve the network life [2]. Anes Yessembayev et al. Proposed a scheme to detect the presence of malicious attacks on good or bad sensor nodes and its application to data aggregation. This scheme can effectively detect malicious nodes [3], but it can not prevent data from being attacked. Neng - Chung Wang et al. proposed an efficient data aggregation in wireless sensor network scheme based on the grid, the scheme can effectively reduce the total energy consumption of network [4], but the limitation is that all ordinary sensor nodes within the data transmission is only allowed in the local community, only the data aggregation and data encryption can no longer meet the needs of data security. Wang Jun et al. [5] proposed a method of security data fusion based on dynamic segmentation technology. This method is to segment data packets to improve the security of data, but the redundancy

J. MacIntyre et al. (Eds.): SPIoT 2020, AISC 1282, pp. 868–873, 2021.
https://doi.org/10.1007/978-3-030-62743-0_125

of data packets is uncertain, and the uncertainty of redundancy will not determine the repeatability of data. In this case, nodes may consume redundant energy.

To sum up, the current research is deficient in data security and network energy consumption, which cannot guarantee data security while reducing network energy consumption. In this paper, a data privacy protection scheme for underwater networks is proposed. In this paper, a data privacy protection scheme for underwater network is proposed. The idea of the scheme is to optimize and upload the collected data, and then further compress the data after successful uploading. The compressed data packets are transmitted to the cluster head, where the data packets are further aggregated and segmented. Packet is divided into small packet slices and sent to different nodes, and a data slice is kept at the sharding node. This scheme can not only reduce the total energy consumption of the network, but also ensure the security of data.

2 Related Work

While protecting the privacy and integrity of data, we should also reduce the total energy consumption of the network and extend the life of the network. Data aggregation is usually used to reduce transmission energy consumption, and data packets are segmented to ensure data security.

Literature [6] proposed a data aggregation scheme for wireless sensor networks that can identify the attacked nodes in a timely manner. In order to ensure the security of the whole network, it is necessary to determine the nodes that are attacked in wireless sensor network. Danyang Qin et al. [7] proposed a DCFDAS, which is mainly aimed at the delay problem and consists of DAT and FCFS. DAT is the control of the number of child nodes that provides the data aggregation topology for FCFS. FCFS is to reduce the energy consumption required to complete the data aggregation of the whole network. The whole mechanism uses the way that nodes change their working state regularly to reduce the energy consumption of nodes. M. Shobana et al. [8] proposed a CSDAM for data processing and aggregation of large-scale wireless sensor networks. The scheme adopts multi-level data aggregation method, which can effectively reduce energy consumption and increase network life. Hongzhi Lin et al. [9] proposed an energy-saving UASN two-layer compressed data aggregation framework. The lower layer is the compressed sampling layer based on compression sensing while the upper layer is the data aggregation layer based on traditional sampling methods. The scheme can effectively reduce the number of sampling nodes and reduce the energy consumption of the whole network. The data aggregation framework CPSN [10] also introduces a two-layer algorithm, which can optimize the data transmission.

This paper presents a data privacy protection scheme for underwater network, which can effectively protect the data security and reduce the total energy consumption. The content of this paper is mainly divided into three parts: data upload and compression, data aggregation, data segmentation and transmission.

3 Data Privacy Protection Scheme

3.1 The System Model

In this paper, a hierarchical underwater network structure model is adopted. According to different functional types of nodes, there are mainly base station nodes, cluster-head nodes and ordinary nodes. Ordinary nodes are nodes in the cluster, whose main work is data processing and transmission, the main job of the cluster head is to receive the data sent by the nodes in the cluster and aggregate it, base station is to receive all the hair sent data recovery restore. In the selection of cluster head, when the node receives the "HELLO" message sent by the base station, it decides whether it is a cluster-head node according to the probability PC, which is a parameter with selection sensitivity of all nodes [11]. The network model formed after node deployment and cluster head selection is shown in Fig. 1. Small white circles are common nodes, purple circles represent cluster heads, and red circles represent a cluster.

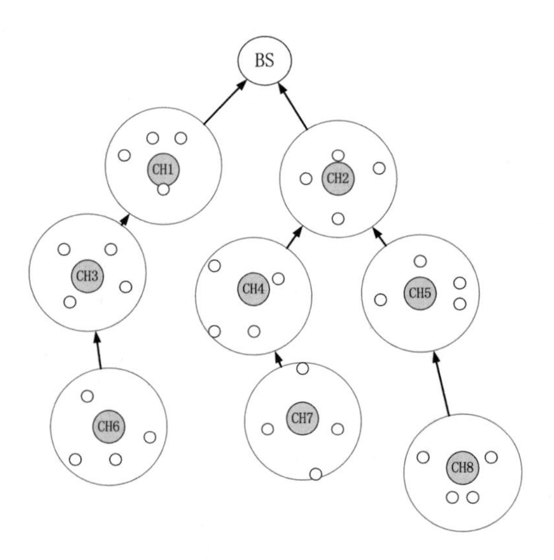

Fig. 1. Structure model of hierarchical underwater network

3.2 Data Upload and Compression

Due to the long-term monitoring of nodes, the data collected in each cluster after partition has a high similarity [12], and a large number of data with high similarity will greatly increase network energy consumption in the transmission and encryption process. To solve this problem, this paper adopts the data uploading decision mechanism [13] and the sensor network data compression algorithm [14] based on space-time correlation before data uploading. The data upload and compression process is as follows:

Step 1: Collect the object data;
Step 2: Establish a two-dimensional table to store the collected information, compare the data before uploading with the data already uploaded, if the similarity is as high as 95%, consider the high similarity, and give up the upload; if the similarity is less than 95%, decide to upload;
Step 3: Use space transformation technology to analyze the correlation of data in space and reduce noise;
Step 4: According to the time correlation of sensor network data, compressed sensing algorithm is introduced to compress the spatial coefficient and reduce the redundancy of sensor network data.

3.3 Data Aggregation

The cluster head aggregates the data sent by nodes in the cluster, and the aggregation process is as follows:

Step 1: Aggregation request: Whenever the base station is ready to collect node-aware data, it will first send the information request of data aggregation to the nearest cluster hair, and the request information will be passed down from the cluster head successively. In this network model, the base station sends information request of data aggregation to CH1, CH2, and then CH1 and CH2 send this request to other cluster heads;
Step 2: Data encryption: In order to prevent attackers from attacking the data, IAES [15] encryption algorithm is adopted to encrypt the data packets of nodes in the cluster and transmit them to the cluster head;
Step 3: Data aggregation: The improved CPDA [16] algorithm is used to reduce the amount of data transmission and avoid additional network overhead.

3.4 Data Segmentation and Transmission

After receiving the acquisition command from the base station, the ordinary node begins to preprocess and collect the target data, and then encrypts and sends the data to the cluster head. After receiving the information, the cluster head validates the data, which fails to discard the data. Validation passes, the data is aggregated, and the aggregated data packets are segmented. In order to reduce the workload of cluster head and ensure data privacy through data segmentation, each packet is divided into two sections. Using the Euclidean decomposition method as the basis for slicing, the original data of each node is decomposed into two parts, one of which is kept by the node itself, and the other is sent to the adjacent node [17]. And so on until all the data is sent to the base station. Take the network CH3, CH4, CH6, CH7 for demonstration, as shown in Fig. 2.

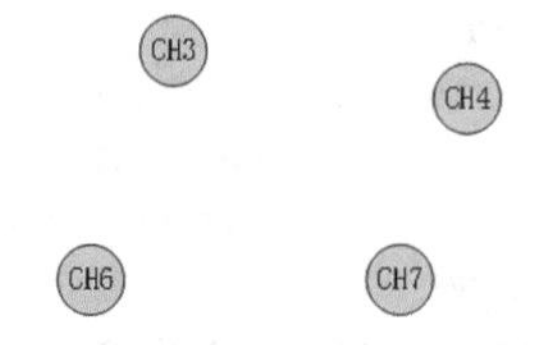

Fig. 2. Illustrates the section

As shown in Fig. 3 after data processing, CH6 and CH7 are divided into two segments, a6, b6, a7 and b7 respectively. One segment is reserved for the cluster head of each segment, and the other segment is sent to the nearest upper cluster head. Here CH6 keeps a6 and sends b6 to CH3, which is nearest to the upper layer; CH7 retains a7 and sends b7 to CH4, which is nearest to the upper layer. Similarly, CH3 divides data packets into a3, b3, and CH3 sends b3 to the nearest upper cluster head, retaining a3 and receiving b6. At this time, the data packets in CH3 are c3 = a3 + b6; similarly, CH4 contains c4 = a4 + b7. Finally, the data slices are sent to the base station along the shortest path for data recovery.

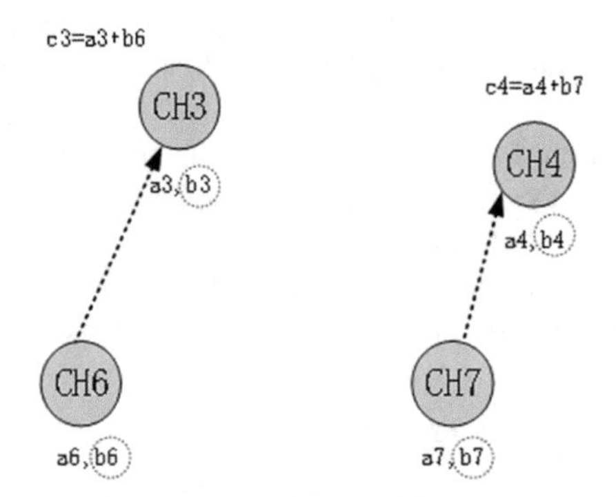

Fig. 3. Data segmentation process

4 Conclusion

Data aggregation is a very important operation in sensors to eliminate redundant data, but at the same time, the security of data is also an issue that can not be ignored, which is also an issue that researchers pay more and more attention to. This paper makes an upload decision when the node collects data, and compresses the uploaded data on the common node, which greatly reduces the data transmission. After receiving the packet sent by the node, the cluster head aggregates the data to reduce the repeatability of the data again. The aggregated packets are then sliced up to ensure the security of the data. The above three steps can effectively reduce the total network energy consumption and prolong the network life.

Acknowledgment. This work was supported by the following projects: the National Natural Science Foundation of China (61862020); the key research and development project of Hainan Province (ZDYF2018006); Hainan University-Tianjin University Collaborative Innovation Foundation Project (HDTDU202005).

References

1. Fu, S., Jiang, Q., Ma, J.F.: A privacy protection data aggregation scheme for wireless sensor networks. J. Comput. Res. **053**(009), 2030–2038 (2016)
2. Sran, S.S., Kaur, L., Kaur, G.: Energy aware chain based data aggregation scheme for wireless sensor network. In: Icesa. IEEE (2015)
3. Yessembayev, A., Sarkar, D., Sikder, F.: Detection of good and bad sensor nodes in the presence of malicious attacks and its application to data aggregation. IEEE Trans. Signal Inf. Process. Netw. **4**(3), 549–563 (2018)
4. Wang, N., Chiang, Y., Hsieh, C., et al.: An efficient grid-based data aggregation scheme for wireless sensor networks. J. Internet Technol. **19**(7), 2197–2205 (2018)
5. Shenyang University of Chemical Technology. A security data fusion method based on dynamic segmentation technology: CN201710584414.7, March 11 2017
6. Zou, Y.F., Zhao, J.: Discussion on data aggregation scheme of wireless sensor network that can determine the attacked node in time. China New Commun. **21**(4), 80–81 (2019). https://doi.org/10.3969/j.issn.1673-4866.2019.04.064
7. Qin, D., Zhang, Y., Ma, J., et al.: A distributed collision-free data aggregation scheme for wireless sensor network. Int. J. Distrib. Sens. Netw. **14**(8) (2018)
8. Shobana, M., Sabitha, R., Karthik, S., et al.: Cluster-based systematic data aggregation model (CSDAM) for real-time data processing in large-scale WSN. Wirel. Pers. Commun., 1–19 (2020)
9. Lin, H., Wei, W., Zhao, P., et al.: Energy-efficient compressed data aggregation in underwater acoustic sensor networks. Wirel. Netw. **22**(6), 1985–1997 (2016)
10. Harb, H., Makhoul, A., Tawbi, S., et al.: Comparison of different data aggregation techniques in distributed sensor networks. IEEE Access **5**, 4250–4263 (2017)
11. Fang, W., Wen, X., Xu, J., et al.: CSDA: a novel cluster-based secure data aggregation scheme for WSNs. Clust. Comput. **22**(3), 5233–5244 (2019)
12. Shen, L.; Yi, X.M.; Chen, X.Y.: Abnormal data driven WSNs data acquisition and routing algorithm. Sens. Microsyst. **39**(4), 140–143+147 (2020). https://doi.org/10.13873/J.1000-9787(2020)04-0140-04
13. Huang, X., Sun, S., Yang, Q., et al.: Data uploading strategy for underwater wireless sensor networks. Sensors **19**(23), 5265 (2019)
14. Wang, L.J., Gao, Z.Y., Yao, P.S.: Data compression algorithm based on spatiotemporal correlation in sensor network. J. Jilin Univ. (Sci. Ed.) **58**(2), 337–342 (2020). https://doi.org/10.13413/j.cnki.jdxblxb.2019115
15. Jin, L.: Analysis of data transmission encryption algorithm based on WSN environment. Digit. World (010), 96–97 (2017)
16. Ye, M.L.: Privacy protection scheme research based on data aggregation technology of wireless sensor network. Electron. Des. Eng. **26**(18), 127–131 (2018)
17. Zhang, X., Liu, X., Yu, J., et al.: Energy-efficient privacy preserving data aggregation protocols based on slicing. In: Green Computing and Communications, pp. 546–551 (2019)

Optimization of Production and Operation Plan of Su CuO Oil Pipeline

Xue-tong Lu(✉)

China Petroleum Engineering Co., Ltd. Beijing Company,
Beijing 100085, China
13501376032@163.com

Abstract. Su CuO oil pipeline is the first long-distance oil pipeline in Hulunbuir oilfield, which undertakes the main task of crude oil transportation. In order to reduce the cost of crude oil transportation, a mathematical model for the optimization of the production and operation scheme of the Su CuO oil pipeline is established with the objective of minimizing the production and operation cost. The model is a nonlinear optimization problem with mixed variables. According to the structural characteristics of the model, two-level hierarchical optimization method is used to solve the problem. The results show that the comprehensive cost can be saved by 18% in winter and 28% in summer.

Keywords: Su CuO oil pipeline · Operation scheme · Optimization mixed · Variable nonlinear optimization

Chinese Library Classification Number: 'Ie832.2 · Document code a

1 Introduction

The total length of the sulao oil pipeline is 87.4 km, with a diameter of 9 × 5 mm and a buried depth of 1.6 M. There are three oil transportation stations in the whole line, namely the first Suyi heat pump station, the second heating station in China and the zuogangzhuan oil transportation terminal. The designed capacity is 50q50000 tons/year, and the heating mode is adopted for operation. The energy consumption equipment in the first station of Jiangsu Province includes two export pumps and one split phase change heating furnace, and the energy consumption equipment in the second station includes two split phase change heating furnaces. The pipeline was officially put into operation in September 2009, with short production time and insufficient operation experience, resulting in high energy consumption of the system. Therefore, it is of great significance to optimize the production and operation plan of sucuo oil pipeline for reducing the energy consumption of system operation and improving the economic benefits of pipeline transportation.

In the 1950s, chernikin first put forward the concept of optimal working condition of hot oil pipeline. Chinese scholars Wu Changchun and Yan Dafan [21 put forward the two-stage model for the optimal operation of hot oil pipeline in the 1980s, and achieved good results in practical application.

J. MacIntyre et al. (Eds.): SPIoT 2020, AISC 1282, pp. 874–880, 2021.
https://doi.org/10.1007/978-3-030-62743-0_126

In this pa̍per, combined with the technological process of sucuo oil pipeline, the mathematical model of pipeline operation scheme optimization is established, and the solution method is given.

2 Establishment of Mathematical Model

The purpose of optimizing the operation plan of Su CuO oil pipeline is to ensure Received on April 21, 2011, and received on May 23, 2011 the support of Heilongjiang University Science and technology innovation team construction plan (2 ∞ 9tdd8). Brief introduction of the author: Yang Lin (19 pulp 1). From Heze, Shandong Province, engineer, hard man. Research direction: Oilfield Surface Engineering and energy saving and consumption reduction technology. On the premise of completing the transportation task, the optimal opening scheme of energy consumption equipment and the optimal outbound parameters of each station are determined, so as to minimize the operating energy consumption cost of the system. The cost of energy consumption includes the cost of thermal energy consumption and the cost of power energy consumption. Therefore, the optimization mathematical model can be described as follows:

$$\min F = e_1 B + e_2 W \tag{1}$$

$$\text{s.t} \quad \sum_{j=1}^{2} x_{1j} \cdot q_{1j} = q \tag{2}$$

$$q_{1j}^{\min} \leq q_{1j} \leq q_{1j}^{\max}, j = 1, 2 \tag{3}$$

$$\mathrm{H}_{1j}^{\min} \leq \mathrm{H}_{1j} \leq \mathrm{H}_{1j}^{\max}, j = 1, 2 \tag{4}$$

$$\mathrm{Q}_1 \leq \mathrm{Q}_1^{\max} \tag{5}$$

$$\mathrm{Q}_{2\mathrm{k}} \leq \mathrm{Q}_{2\mathrm{k}}^{\max}, \mathrm{k} = 1, 2 \tag{6}$$

$$\mathrm{P}_{\mathrm{i}}^{\mathrm{out}} \leq \mathrm{P}_{i}^{\max}, i = 1, 2 \tag{7}$$

$$\mathrm{T}_{\mathrm{i}}^{\mathrm{out}} \leq \mathrm{T}_{i}^{\max}, i = 1, 2 \tag{8}$$

$$\mathrm{P}_{\mathrm{i}+1}^{in} \geq \mathrm{P}_{in,\mathrm{i}+1}^{min}, i = 1, 2 \tag{9}$$

$$\mathrm{T}_{\mathrm{i}+1}^{in} \geq \mathrm{T}_{in,\mathrm{i}+1}^{min}, i = 1, 2 \tag{10}$$

$$\mathrm{f}(\mathrm{T}, X, Q_{\mathrm{x}}) = 0 \tag{11}$$

Equation (1) in (11), e_1 It refers to the unit price of fuel cost; e_2 refers to the unit price of electricity cost; F refers to the total energy consumption cost; daily refers to the fuel consumption; W refers to the power consumption; x_{1j} refers to the operation state of the first pump in Jiangsu first station, which is deaf. When $X_{ij} = 1$, the pump is on. When $X_{ij} = 0$, q_{ij} is shut down; q is the planned capacity; q_{1j}^{min} and q_{1j}^{max} are the lower and upper limits of the first pump capacity constraint; H_{1j} is the lift of the first pump; H_{1j}^{min} and H_{1j}^{max} are the lower and upper limits of the first pump capacity constraint; Q_1 is the heating capacity of the heating furnace of the first station; Q_1^{max} refers to the maximum heating capacity of the heating furnace in the first station of Jiangsu Province; milk refers to the heating capacity of the K heating furnace in the second station of Jiangsu Province; Q_{2k} side refers to the maximum heating capacity of the K heating furnace in the second station of Jiangsu Province; p_i^{out} refers to the outbound pressure of the I station; p_1^{max} refers to the upper limit of the outbound pressure of the I station; T_i^{out} mountain refers to the outbound temperature of the I station; T_i^{max} refers to the upper limit of the outbound temperature of the I station; p_{i+1}^{min} refers to the inbound pressure of the $I+1$ station; $p_{in,i+1}^{min}$ Is the lower bound of the inbound pressure of station $I+1$; T_{i+1}^{min} Is the entry temperature of station $I+1$; T_{i+1}^{min} It is the lower limit of the inbound temperature of the $I+L$ station; R is the outbound temperature vector, $r = \{I//II = L, 2\}$; X is the pump opening scheme vector, $x = \{Geli// = L, 2\}$; K is the displacement vector of each pump under the pump opening scheme $x, v = \{g, towel = L, 2\}$; Eq. (2) is the planned capacity constraint; Eq. (3) is the displacement constraint of the pump; Eq. (4) is the head constraint of the pump; Eq. (5) is the first Jiangsu pump The heating capacity of the station heating furnace is limited; Eq. (6) is the heating capacity of the second station heating furnace; Eq. (7) and Eq. (8) are the pressure and temperature constraints of the outgoing station; Eq. (9) and Eq. (10) are the pressure and temperature constraints of the incoming station; Eq. (11) is the hydraulic and thermal balance constraints.

In the objective function, the calculation formulas of power and thermal energy consumption are as follows:

$$W = \sum\nolimits_{j=1}^{2} X_i \frac{\rho g q L_{1j} H_1}{1000 \eta_{1j}^{p} \eta_{1j}^{e}} \tag{12}$$

$$B = \frac{G_1 c(t_1^{out} - t_1^{in})}{Q_r \eta_1^f} + \sum\nolimits_{i=1}^{2} \frac{G_{2i} c(t_{2i}^{out} - t_{2i}^{in})}{Q_r \eta_{2i}^f} \tag{13}$$

In formula (12) and formula (13), η_{1j}^{p} and η_{1j}^{e} are the pump efficiency and motor power of the first pump in Jiangsu first station; G_1 I t refers to the mass flow of the heating medium of the first station of Jiangsu Province; G_{2i} refers to the mass flow of the heating medium of the second station of China; C refers to the specific heat under the average temperature; t_1^{out} and t_1^{in} refers to the temperature at the inlet and outlet of the first station of Jiangsu Province; t_{2i}^{out} and t_{2i}^{in} taste refer to the temperature at the inlet and outlet of the first station of Jiangsu Province; η_1^f refers to the temperature at the inlet and outlet of the first station of Jiangsu Province Thermal efficiency: η_{2i}^f of the I heating furnace in the second station; Q_r the low calorific value of the fuel.

In the process of optimization, the temperature constraint of oil products has a great influence on the final optimization results. Whether it can not only ensure the safe operation of the system, but also minimize the energy consumption cost of the system. The temperature constraint plays a decisive role. According to the average values of several groups of test results of the crude oil transported by sucuo oil pipeline, the oil condensation point is 26 °C, and the peak of wax precipitation is 5 °C–28 °C. At the peak of wax precipitation, a large amount of wax crystallization will seriously affect the fluidity of oil products. Therefore, considering the range of oil condensation point and wax precipitation peak, the limit value of oil inlet temperature is determined as 31 °C.

3 Mathematical Model Solving Method

In the optimization mathematical model of Su CuO oil pipeline, the oil outlet temperature belongs to continuous variable, and the pump combination belongs to discrete variable [3]. Therefore, the optimization problem belongs to mixed variable nonlinear optimization problem [4, 5]. According to the structural characteristics of the model, the original optimization model is divided into two sub models: the pipeline optimal pump pipe matching model and the outbound temperature optimization model, which are solved by iteration.

Pipeline Optimal Pump Pipe Matching Sub Model
The model is based on the given temperature of crude oil at each heat station of the pipeline to determine the optimal pump on scheme X′. The model is described as:

$$\min W(X, Q_x) \tag{14}$$

$$p_i^{\mathrm{out}} \leq p_i^{\max}, i = 1, 2 \tag{15}$$

$$p_{1j}^{\mathrm{in}} \geq p_{in,i+1}^{\min}, i = 1, 2 \tag{16}$$

$$q_{1j}^{\min} \leq q_{1j} \leq q_{ij}^{\max}, j = 1, 2 \tag{17}$$

Where, Q_x is the displacement of pump under pump combination X.

This optimization problem belongs to the mixed variable optimization problem, which is solved by implicit enumeration method and dynamic programming method [2, 3].

Outbound Temperature Optimization Sub Model
This model is to determine the optimal outlet temperature of crude oil in each heat station on the basis of certain pump start-up scheme. The model is described as:

$$\min S = B(\mathrm{T}) \tag{18}$$

$$T_i^{\mathrm{out}} \leq T_i^{\max} \leq q_i^{\max}, i = 1, 2 \tag{19}$$

$$T_{i+1}^{\mathrm{in}} \geq T_{in.i+1}^{\min}, i = 1, 2 \tag{20}$$

This optimization problem belongs to the nonlinear programming problem. The penalty function method is used to transform the model into unconstrained optimization problem, and then the direction acceleration method '6' proposed by p0wei and improved by Zangwill is used to solve it.

The algorithm block diagram is shown in Fig. 1.

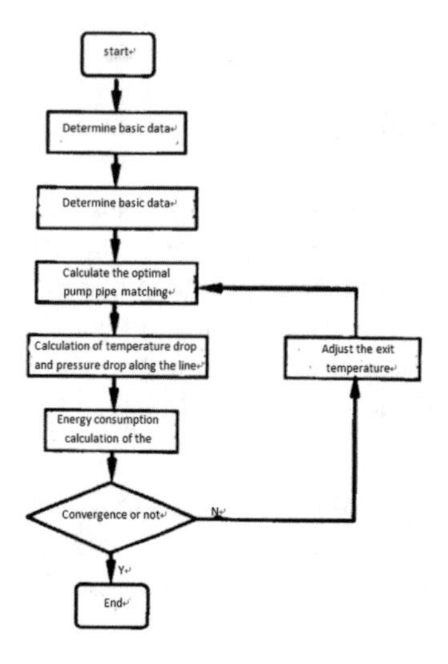

Fig. 1. Algorithm block diagram

4 Example Calculation

Because the energy consumption of the pipeline is also affected by the environmental change, the energy consumption in summer is the least, while the energy consumption in winter is the largest. Therefore, take the transportation scheme of 69.23 m^3/h on February 15, 2010 and 73.19 m^3/h on August 15, 2010 as an example for optimization calculation. See Table 1, 2 and 3 for optimization results. It can be seen from Table 1 and Table 2 that under the condition of meeting the minimum allowable operation parameters, the parameters have been optimized, the oil transportation temperature has been significantly reduced, and the situation of saving thermal energy consumption and power energy consumption is shown in Table 3.

Table 1. Comparison of production parameters before and after optimization of operation scheme when the daily output of February 1 s in winter is 23 m^3/h

	Incoming temperature/°C		Outgoing temperature/°C		Inlet pressure/MPa		Outgoing pressure/MPa	
	Before optimization	After optimization	Before optimization	After optimization	Before optimization	After optimization	Before optimization	After optimization
Su first stop	44	–	74.87	65.55	0.3	–	2.97	2.98
Two stops	43.73	36.642	75.41	66.94	1.55	1.52	1.45	1.41
Gang Gang turne								
Oil terminal	40.7	36.96	–	–	0.11	0.1	–	–

Table 2. Comparison of production parameters before and after optimization of operation scheme when the output on August 15 is 73.19 $m^{3/}h$ in summer

	Incoming temperature/°C		Outgoing temperature/°C		Inlet pressure/MPa		Outgoing pressure/MPa	
	Before optimization	After optimization	Before optimization	After optimization	Before optimization	After optimization	Before optimization	After optimization
Sufirst stop	44	–	66.42	55.88	0.1	–	2.18	2.18
Two stops	43.69	37.315	71.2	59.49	1.12	1.2	1.18	1.18
Oil terminal	43	37.83	–	–	0.34	0.31	–	–

Table 3. cost comparison before and after optimization

Season month	Throughput		Power consumption (yuan/m^3)		Fuel quantity (yuan/m^3)		Total cost (yuan/m^3)	
	$m^{3/h}$	Before optimization	After optimization	Before optimization	After optimization	Before optimization	After optimization	%
	69.23	0.466	0.466	15.323	12.312	15.789	12.799	18
	3.19	0.438	0.438	12.933	9.631	13.371	10.069	28

5 Conclusion

(1) Taking the minimum production and operation cost as the objective, the temperature and pressure of the outgoing station as the design variables, and the equipment working load as the constraints, the optimization mathematical model of the production and operation scheme of sucuo oil pipeline is established. According to the structural characteristics of the model, two-level hierarchical optimization method is used to solve the model.

(2) The production and operation schemes of Su CuO oil pipeline in winter and summer are optimized. Compared with those before optimization, the operation energy consumption cost is reduced by 18% and 28%, respectively.

References

1. Chernikin: Transportation of condensable crude oil. Trans. Yan Dafan. Petroleum Industry Press Club, Beijing (1959)
2. Wu, C., Yan, D.: Two level hierarchical optimization model for steady-state operation of hot oil pipeline. J. Pet. **3**(10), 109–117 (1989)
3. Editorial board: Handbook of Modern Applied Mathematics. Operation research and optimization theory volume. Qing Dynasty Hua University Press, Beijing (1998)
4. Wei, L.: Research on Optimization Technology of oilfield surface pipe network based on Intelligent Computing. Large Qing Petroleum Institute, Daqing (2005)

5. Wei, L., Liu, Y., Sun, H.: Optimization of operation and operation of water injection system in multi-source oilfield. Pet. Drill. Prod. Technol. **29**(3), 59–626 (2007)
6. Averil: Nonlinear programming (Volume II). Shanghai Science and Technology Press, Shanghai (1980)
7. Wei, L., Chen, M., et al.: Optimization of production and operation scheme of Qingha oil pipeline. Sci. Technol. Eng. **29**(10), 7263–7266 (2010)

Author Index

J. MacIntyre et al. (Eds.): SPIoT 2020, AISC 1282, pp. 881–884, 2021.
https://doi.org/10.1007/978-3-030-62743-0

Printed in the United States
By Bookmasters